Hans-Georg Elias
Macromolecules

Macromolecules. Hans-Georg Elias.
Copyright © 2005 WILEY-VCH Verlag GmbH & Co. KGaA, Weinheim
ISBN: 3-527-31172-6

Related Titles

Matyjaszewski, K., Gnanou, Y., Leibler, L. (eds.)

Macromolecular Engineering

Precise Synthesis, Materials Properties, Applications

4 Volumes
2007
ISBN 3-527-31446-6

Kemmere, M. F., Meyer, Th. (eds.)

Supercritical Carbon Dioxide

in Polymer Reaction Engineering

2005
ISBN 3-527-31092-4

Meyer, Th., Keurentjes, J. (eds.)

Handbook of Polymer Reaction Engineering

2 Volumes
2005
ISBN 3-527-31014-2

Xanthos, M. (ed.)

Functional Fillers for Plastics

2005
ISBN 3-527-31054-1

Elias, H.-G.

An Introduction to Plastics

2003
SBN 3-527-29602-6

Wilks, E. S. (ed.)

Industrial Polymers Handbook

Products, Processes, Applications

4 Volumes
2000
ISBN 3-527-30260-3

Elias, H.-G.

An Introduction to Polymer Science

1997
ISBN 3-527-28790-6

Hans-Georg Elias

Macromolecules

Volume 1: Chemical Structures and Syntheses

WILEY-VCH

WILEY-VCH Verlag GmbH & Co. KGaA

The Author

Prof. Dr. Hans-Georg Elias
Michigan Molecular Institute
1910 West St. Andrews Road
Midland, Michigan 48540
USA

Volume 1   Chemical Structures and Syntheses
Volume 2   Industrial Polymers and Syntheses
Volume 3   Physical Structures and Properties
Volume 4   Processing and Applications of
Polymers

All books published by Wiley-VCH are carefully produced. Nevertheless, authors, editors, and publisher do not warrant the information contained in these books, including this book, to be free of errors. Readers are advised to keep in mind that statements, data, illustrations, procedural details or other items may inadvertently be inaccurate.

Library of Congress Card No.:
applied for

British Library Cataloguing-in-Publication Data
A catalogue record for this book is available from the British Library.

Bibliographic information published by Die Deutsche Bibliothek
Die Deutsche Bibliothek lists this publication in the Deutsche Nationalbibliografie; detailed bibliographic data is available in the Internet at <http://dnb.ddb.de>.

© 2005 WILEY-VCH Verlag GmbH & Co. KGaA, Weinheim

Printing   Betz-Druck GmbH, Darmstadt
Binding   J. Schäffer GmbH, Grünstadt
Cover Design   Gunther Schulz, Fußgönheim

Printed in the Federal Republic of Germany
Printed on acid-free paper

ISBN-13:   978-3-527-31172-9
ISBN-10:   3-527-31172-6

Didici in mathematicis ingenio,
in natura experimentis,
in legibus divinis humanisque auctoritate,
in historia testimoniis nitendum esse.

Gottfried Wilhelm Leibniz
(1646-1716)

(I learned that
    in mathematics one depends on inspiration,
    in science on experimental evidence,
    in the study of divine and human law on authority, and
    in historical research on authenticated sources.)

Preface

Macromolecular science has grown substantially since the first German edition of this book was written (1962-1970). The German editions ballooned from the single volume of the first edition (1971) with 856 pages to the four volumes of the sixth edition (1999-2002) with a total of 2564 pages. This English edition is based on the latest (sixth) German edition. It is not, however, a cover-to-cover translation, but has been completely checked, modified, corrected, and updated where necessary. More than 25 % of the figures of volume I are new and so are many tables and derivations.

Like its German predecessor, the English-language edition comprises four volumes. Volume I is concerned with the fundamentals of chemical structure and synthesis of polymers, Volume II with individual polymers and their industrial syntheses, Volume III with fundamentals of physical structures and properties, and Volume IV with processing and applications of polymers as plastics, fibers, elastomers, thickeners, etc.

Chapter I of Volume I introduces the field by discussing basic terms and the arguments and counter-arguments from which the science of macromolecules arose. It is followed by four chapters that outline the constitution (Ch. 2), configuration (Ch. 4), and conformation (Ch. 5) of macromolecules, and the most common methods for the characterization of macromolecules (Chapter 3) that are used most frequently in the synthesis of polymers.

The remaining chapters are concerned with general aspects of polymerization, including statistics and polymerization of chiral monomers (Ch. 6), polymerization equilibria (Ch. 7), ionic polymerizations (Ch. 8), coordination polymerizations (Ch. 9), free-radical polymerizations (Ch. 10), polymerizations by radiation or in ordered states (Ch. 11), copolymerizations (Ch. 12), polycondensations and polyadditions (Ch. 13), biological polymerizations (Ch. 14), reactions of macromolecules (Ch. 15), and molecular engineering for the preparation of new polymer architectures (Ch. 16). The Appendix lists SI units and their prefixes, IUPAC symbols for physical quantities, multiplicative prefixes for numerals used in systematic names, definitions of concentrations, and names of ratios of physical quantities. Most chapters contain historical notes which list the groundbreaking early papers in the field. All chapters are followed by extensive lists of books and reviews, including some older ones that contain information which is rarely found in newer books on the same subject.

I am indebted to my good friends and former colleagues at Michigan Molecular Institute, Professors Petar R. Dvornic and Steven E. Keinath, who read and checked the final draft of all chapters and made many helpful suggestions. The remaining factual and judgmental errors are, of course, all mine.

Midland, Michigan   Hans-Georg Elias
Summer 2005

List of Symbols

Symbols for physical quantities follow the recommendations of the International Union of Pure and Applied Chemistry (IUPAC), symbols for physical units those of the International Standardization Organization (ISO). Exceptions are indicated.

I.Mills, T.Cvitas, K.Homann, N.Kallay, K.Kuchitsu, Eds., (International Union of Pure and Applied Chemistry, Division of Physical Chemistry), "Quantities, Units and Symbols in Physical Chemistry", Blackwell Scientific Publications, Oxford 1988.

Symbols for Languages

D = German (deutsch), F = French, G = (classic) Greek, L = (classic) Latin.
The Greek letter υ (upsilon) was transliterated as "y" (instead of the customary "u") in order to make an easier connection to written English (example: $\pi o \lambda \upsilon \varsigma$ = polys (many)).
For the same reason, χ was transliterated as "ch" and not as "kh."

Symbols for Chemical Structures

A: symbol for a monomer or a leaving group (polycondensations)
B: symbol for a monomer or a leaving group (polycondensations)
L: symbol for a leaving molecule, for example, H_2O from the reaction of $-COOH + HO-$
R: symbol for a monovalent ligand, for example, CH_3- or C_6H_5-
Z: symbol for a divalent unit, for example, $-CH_2-$ or $-p$-C_6H_4-
Y: symbol for a trivalent unit, for example, $-C(R)<$ or $-N<$
X: symbol for a tetravalent unit, for example, $>C<$ or $>Si<$
*: symbol for an active site: radical ($^\bullet$), anion ($^\ominus$), cation ($^\oplus$)
*p*Ph   *para*-phenylene (in text)
p-C_6H_4 *para*-phenylene (in line formulas)

Mathematical Symbols (IUPAC)

=	equal to		>	greater than
≠	not equal to		≥	greater than or equal to
≡	identically equal to		>>	much greater than
≈	approximately equal to		<	less than
~	proportional to (IUPAC: ~ or ∝)		≤	less than or equal to
≙	corresponds to		<<	much less than
→	approaches, tends to		±	plus or minus
Δ	difference		sin	sine of
δ	differential		cos	cosine of
f	function of (IUPAC: *f*)		tan	tangent of
Σ	sum		cot	cotangent of
∫	integral		sinh	hyperbolic sine of
Π	product		arcsin	inverse sine of

lg   logarithm to the base 10   (IUPAC: lg or $\log_{10}$)
ln   logarithm to the base e (natural logarithm) (IUPAC: ln or $\log_e$)

Averages and Other Markings

—   line above letter indicates common average, for example, $\overline{M}_\mathrm{n}$ = number-average of molar mass (note: subscript is not italicized since it does *not* represent a physical quantity that is kept constant)

~   tilde indicates a partial quantity, for example, $\tilde{v}_\mathrm{A}$ = partial specific volume of component A

[ ]   square brackets surrounding the symbol of the substance indicates the amount concentration ("mole concentration"), usually in mol/L

⟨ ⟩   angled brackets surrounding a letter indicate spatial averages, for example, $\langle s^2 \rangle$ = mean-square average of radius of gyration

Exponents and Superscripts

°   degree of plane angle [= $(\pi/180)$ rad]

′   minute of plane angle [= $(\pi/10\ 800)$ rad]

″   second of plane angle [= $(\pi/648\ 000)$ rad]

°   pure substance

∞   infinite, for example, dilution or molecular weight

$^\mathrm{m}$ amount-of-substance related quantity if a subscript is inexpedient. According to IUPAC, $^\mathrm{m}$ can be used as either superscript or subscript

$^\mathrm{(q)}$ qth order of a moment (always in parentheses since it does not represent a power)

‡   activated quantity, for example, $E^\ddagger$ = activation energy

$^\alpha$ exponent in the intrinsic viscosity–molecular weight relationship, $[\eta] = K_\mathrm{v} M^\alpha$

Indices and Subscripts

o   standard or original state, for example, T_o = reference temperature

0   state at time zero

1   solvent

2   solute (usually polymer)

3   additional component (for example: salt, precipitant, etc.)

∞   final state

A   substance A, e.g., M_A = molar mass of substance A

a   group, monomeric unit from substance A, e.g., mass m_a

a   number of functional groups A in a molecule $A_\mathrm{a}B$

am   amorphous

B   substance B

b   group, e.g., monomeric unit from substance B

b   number of functional groups B in a molecule AB_b

bd   bond, especially chain bond

be   effective bond length (= length of monomer unit projected onto the chain direction)

bp   boiling temperature (boiling point)

br   branch, branched

c   ceiling, for example, T_c = ceiling temperature

ch   chain (L: *catena*)

comb   combination

cr   crystalline
crit   critical
cryst crystallization
cycl   cyclic

D   related to diffusion

eff   effective
end   end-group
eq   equilibrium

f   free or functional

G   glass transformation
g   any statistical weight, e.g., n, m, z or x, w, Z

H   hydrodynamically effective property or hydration
h   hydrodynamic average

I   initiator, $I^{\bullet}$ = initiator radical
i   *i*th component (variable)
i   initiation
i   isotactic diad (IUPAC recommends m = *meso*; see Chapter 4)
ii   isotactic triad (IUPAC: mm)
is   heterotactic triad (IUPAC: mr)
<u>is</u>   sum of heterotactic triads, <u>is</u> = is + si
it   isotactic

j   variable

k   variable

L   liquid, melt (L: *liquidus*)
l   liquid

M   melting
M   monomer molecule
Mt   metal
m   molar (also as superscript)
md   median
mol   molecule
mon monomer
mu   monomeric unit

n   number average (note: not in italics since it does not refer to a physical quantity that is held constant)

P   polymer
p   quantity at constant pressure
p   propagation
pm   polymerization, especially propagation
pol   polymer

q	index, defined differently for each section or chapter
R	reactive entity
R	retention
r	relative (only in M_r = relative molecular mass = molecular weight)
r	based on end-to-end distance, e.g., α_r = linear expansion coefficient of a coil (with respect to the end-to-end distance)
red	reduced
rel	relative
rep	repeating unit
S	solvating solvent
s	syndiotactic diad (IUPAC recommends r = _racemo_)
s	related to radius of gyration
seg	segment
si	heterotactic triad (IUPAC: rm)
soln	solution
ss	syndiotactic triad (IUPAC: rr)
_T_	quantity at constant temperature
t	termination
tr	transfer
u	monomeric unit in polymer
u	conversion
_V_	quantity at constant volume
v	viscosity average (solutions)
w	mass average ("weight average"); note: not in italics since it does not refer to a physical quantity that is held constant
x	crosslink(ed)
z	z average

Prefixes of Words (in systematic polymer names in _italics_)

alt	alternating
at	atactic
blend	polymer blend
block	block (large constitutionally uniform segment)
br	branched. IUPAC recommends sh-branch = short chain branch, l-branch = long chain branch, f-branch = branched with a branching point of functionality _f_
co	joint (unspecified)
comb	comb
compl	polymer–polymer complex
cyclo	cyclic
ct	cis-tactic

eit	erythrodiisotactic
g	graft
ht	heterotactic
ipn	interpenetrating network
it	isotactic
net	network; μ-net = micro network
per	periodic
r	random (Bernoulli distribution)
sipn	semi-interpenetrating network
star	star-like. f-star, if the functionality f is known; f is then a number
st	syndiotactic
stat	statistical (unspecified distribution)
tit	threodiisotactic
tt	trans-tactic

Other Abbreviations

AIBN	N,N'-azobisisobutyronitrile
BPO	dibenzoylperoxide
Bu	butyl group (iBu = isobutyl group; nBu = normal butyl group (according to IUPAC, the normal butyl group is not characterized by n, which rules out Bu as an unspecified butyl group); sBu = secondary butyl group; tBu = tertiary butyl group)
Bz	benzene
C	catalyst; C* = active catalyst or active catalytic center
cell	cellulose residue
Cp	cyclopentadienyl group
DMF	N,N-dimethylformamide
DMSO	dimethylsulfoxide
Et	ethyl group
G	gauche conformation
glc	glucose
GPC	gel permeation chromatography
I	initiator
IR	infrared
L	solvent (liquid)
LC	liquid-crystalline

Me   methyl group
Mt   metal atom

Np   naphthalene
NMR   nuclear magnetic resonance

P   polymer
Ph   phenyl group
Pr   propyl group

SEC   size exclusion chromatography

THF   tetrahydrofuran

UV   ultraviolet

Quantity Symbols (unit symbols, see Chapter 17, Appendix)

Quantity symbols follow in general the recommendations of IUPAC.

A absorption ($A = \lg (I_0/I) = \lg (1/\tau_i)$); formerly: extinction
A area; A_c = cross-sectional area of a chain
A Helmholtz energy ($A = U - TS$); formerly: free energy
$A^{\ddagger}$ pre-exponential constant (in $k = A^{\ddagger} \exp(- E^{\ddagger}/RT)$)
a thermodynamic activity
a linear absorption coefficient ($a = (1/L) \lg (I_0/I)$)

b bond length; b_{eff} = effective bond length

C number concentration (number of entities per total volume); see also c
$[C]$ amount-of-substance concentration of substance C = amount of substance C per total volume = "molar concentration of C"
C transfer constant (always with index, e.g., C_r of a regulator, C_s of a solvent)
C heat capacity (usually in J/K); C_p = isobaric heat capacity (heat capacity at constant pressure p); C_V = isochoric heat capacity (heat capacity at constant volume V); C_m = molar heat capacity (heat capacity per amount-of-substance n)
C_{tr} chain transfer constant of polymerizations ($C_{tr} = k_{tr}/k_p$)
c specific heat capacity (usually in J/(g K)); c_p = isobaric specific heat capacity; c_V = isochoric specific heat capacity. Formerly: specific heat
c concentration = mass concentration (= mass-of-substance per total volume) = "weight concentration." IUPAC calls this quantity "mass density" (quantity symbol ρ). The quantity symbol c has, however, traditionally been used for a special case of mass concentration, i.e., mass-of-substance per volume of solution and the quantity symbol ρ for another special case, the mass density ("density") = mass-of-substance per volume of substance.

d diameter; diameter of blob (d_{blob}), of sphere (d_{sph}), etc.

E energy
e parameter in the Q,e copolymerization equation
e cohesion energy density

F force

f fraction (unspecified), see also x, w, ϕ, etc.

f functionality

G Gibbs energy ($G = H - TS$); formerly: free enthalpy

G statistical weight fraction ($G_i = g_i/\Sigma_i\, g_i$)

g statistical weight (for example: n, x, w). IUPAC recommends k for this quantity which is problematic because of the many other uses of k. Similarly, K cannot be used for the statistical weight fraction because of the many other meanings of K.

g parameter for the ratio of dimensions of branched macrmolecules to those of unbranched macromolecules of equal molecular weight (branching index); g_h = branching index from hydrodynamic measurements

H height

H enthalpy; ΔH_{mix} = enthalpy of mixing, $\Delta H_{mix,m}$ = molar enthalpy of mixing

h Planck constant ($h = 6.626\ 075\ 5 \cdot 10^{-34}$ J s)

I light intensity

i radiation intensity of a molecule

i variable (ith component, etc.)

J flux (of mass, volume, energy, etc.)

K general constant; K_n = equilibrium constant

k rate constant (always with index); k_i = rate constant of initiation; k_p = rate constant of propagation, k_t = rate constant of termination, k_{tr} = rate constant of transfer

k_B Boltzmann constant ($k_B = R/N_A = 1.380\ 658 \cdot 10^{-23}$ J K^{-1})

L length (always geometric); L_{chain} = true (historic) contour length of a chain (= number of chain bonds times length of valence bonds); L_{cont} = conventional contour length of a chain (= length of chain in the all-trans macroconformation); L_K = length of a Kuhn segment; L_{ps} = persistence length; L_{seg} = segment length

l length

M molar mass of a molecule (= physical unit mass/amount of molecule, e.g., g/mol). $\overline{M}_n$ = number-average molar mass; $\overline{M}_w$ = mass-average molar mass; M_{crit} = critical molar mass; $\overline{M}_{R,n}$ = number-average molar mass of reactants (= polymer plus monomer)

M_r relative molar mass = relative molecule mass = molecular weight (physical unit: 1 = "dimensionless"); $\overline{M}_{r,n}$ = number-average molecular weight

m mass; m_{mol} = mass of molecule

N number of entities

N_A Avogadro constant ($N_A = 6.022\ 136\ 7 \cdot 10^{23}$ mol^{-1})

n amount of substance (in mol); formerly: mole number

n refractive index

p conditional probability

p pressure

p extent of reaction (fractional conversion); p_A = extent of reaction of A groups

p number of conformational repeating units per completed helical turn

Q electric charge = quantity of electricity
Q heat
Q parameter in the Q,e copolymerization equation
Q polymolecularity index (= "polydispersity index"), e.g., $Q = \overline{M}_\mathrm{w}/\overline{M}_\mathrm{n}$
Q intermediate variable or constant, usually a ratio; varies with section
q charge of an ion

R molar gas constant ($R \approx 8.314\ 510$ J K^{-1} mol^{-1})
R radius; R_d = Stokes radius, R_v = Einstein radius
R rate of reaction; R_p = rate of propagation, etc.
R dichroic ratio
r spatial end-to-end distance of a chain, usually as $\langle r^2 \rangle^{1/2}$ with various indices
r copolymerization parameter
r_o initial ratio of amounts of substances in copolymerizations

S entropy; ΔS_mix = entropy of mixing, $\Delta S_\mathrm{mix,m}$ = molar entropy of mixing
S solubility coefficient

T temperature (always with units). In physical equations always as thermodynamic temperature with unit kelvin; in descriptions, either as thermodynamic temperature (unit: kelvin) or as Celsius temperature (unit: degree Celsius). Mix-ups can be ruled out because the physical unit is always given. IUPAC recommends for the Celsius temperature either t as a quantity symbol (which can be confused with t for time) or θ (which can be confused with Θ for the theta temperature). T_c = ceiling temperature, T_G = glass temperature, T_M = melting temperature
T transparency
t time
t rotational angle around helix axis

U internal energy
u fractional conversion of monomer molecules (p = fractional conversion of groups; y = yield of substance)
u excluded volume

V volume; V_h = hydrodynamic volume, V_m = molar volume; $\tilde{V}_\mathrm{m}$ = partial molar volume
v specific volume; $\tilde{v}$ = partial specific volume
v linear velocity ($v = \mathrm{d}L/\mathrm{d}t$)

W work
w mass fraction = weight fraction

X degree of polymerization of a molecule with respect to monomeric units (not to repeating units); $\overline{X}_\mathrm{n}$ = number-average degree of polymerization of a substance; $\overline{X}_\mathrm{w}$ = mass-average degree of polymerization of a substance
x mole fraction (amount-of-substance fraction); x_u = mole fractions of units, x_i = mole fraction of isotactic diads, x_ii = mole fraction of isotactic triads, etc.
x_br degree of branching

Y refractive index increment (= $\mathrm{d}n/\mathrm{d}c$)

Y	degree of polymerization with respect to repeating unit
y	yield of substance
Z	z fraction ($Z_i = z_i/\sum_i z_i$)
z	z-statistical weight
z	coordination number, number of neighbors
α	angle, especially rotational angle of optical activity
α	linear thermal expansion coefficient of materials ($\alpha = (1/L)(dL/dT)$). Note: in literature often as β
α	linear expansion of coils (α_s if radius of gyration; α_r if end-to-end distance; α_h if hydrodynamic dimensions; α_D if diffusion; α_v if viscosity (dilute solutions))
α	degree of crystallinity (with index for method: X = X-ray, d = density, etc.)
$[\alpha]$	"specific" optical rotation
β	angle
β	cubic thermal expansion coefficient [$\beta = (1/V)(dV/dT)$]; in literature usually as α
γ	angle
γ	crosslinking index
δ	solubility parameter
δ	chemical shift
ε	cohesive energy
ε_r	relative permittivity (formerly: dielectric constant)

η dynamic viscosity, e.g., η_0 = viscosity at rest (Newtonian viscosity),

η_1 = viscosity of solvent, η_e = extensional viscosity,

η_r $= \eta/\eta_1$ = relative viscosity,

η_i $= (\eta - \eta_1)/\eta_1$ = relative visc. increment   (= specific viscosity η_{sp}),

η_{inh} $= (\ln \eta_r)/c$ = inherent viscosity   (= logarithmic visc. number),

η_{red} $= (\eta - \eta_1)/(\eta_1 c)$ = reduced viscosity   (= viscosity number η_{sp}/c),

$[\eta]$ $= \lim \eta_{red,c \to 0}$ = limiting visc. number   (= intrinsic viscosity)

Θ	theta temperature
θ	torsional angle (conformational angle in macromolecular science)
ϑ	angle, especially scattering angle or torsional angle (organic chemistry)
κ	isothermal (cubic) compressibility
Λ	aspect ratio
λ	wavelength (λ_o = wavelength of incident light)
μ	moment of a distribution
μ	chemical potential
ν	moment of a distribution, related to a reference value
ν	kinetic chain length
ν	frequency
ν	effective amount concentration of network chains
ν	velocity

$\mathit{\Xi}$ zip length

ξ frictional coefficient

$\mathit{\Pi}$ osmotic pressure

π mathematical constant pi

ρ density (= mass/volume of the same matter)

σ_n (number) standard deviation

σ cooperativity

σ wave number

ς degree of coupling of chains

τ bond angle, valence angle

τ relaxation time

τ light transmission; τ_{it} = internal transmission; τ_{et} = external transmission

$[\mathit{\Phi}]$ molar optical rotation

ϕ volume fraction; ϕ_f = free volume fraction

ϕ angle

χ Flory–Huggins interaction parameter

$\mathit{\Omega}$ angle

$\mathit{\Omega}$ thermodynamic probability

$\mathit{\Omega}$ skewness of a distribution

ω angular velocity

Table of Contents

1 Survey

1.1 Introduction

The first volume of this 4-volume work "Macromolecules" is concerned with the principles of chemical structures and syntheses of **macromolecules (polymer molecules)** and **macromolecular substances (polymers)**. Macromolecules are natural or synthetic molecules composed of a great number of atoms, thus having high molecular weights of several thousands or millions. They are the basic elements of life forms; they also serve man as materials for food, shelter, clothing, energy, and many amenities (Table 1-1).

Life would not be possible without **natural macromolecules**: deoxyribonucleic acids contain the code for hereditary factors, celluloses and lignins form the skeletal materials of plants, scleroproteins constitute the muscles of animals, enzymes catalyze chemical reactions, starch stores energy for plants, the protein hemoglobin transports oxygen in the body, etc.

Naturally occuring macromolecular substances ($\equiv$ **biopolymers**) have been used by humankind since times immemorial. Proteins of meat and polysaccharides of grain are important foodstuffs. Wool and silk, both proteins, serve for clothing. Amber, a high-molecular weight resin, was used as jewelry by the old Greeks. Asphalt, also polymeric, is mentioned as an adhesive in the Bible.

Nature recycles biopolymers via low-molecular weight substances. Since biopolymers degrade fairly easily under the action of water, oxygen, light, and microorganisms, man has tried to prolong the useful life of biopolymeric materials by chemical transformations, for example, of skins and furs by tanning, a crosslinking reaction of collagen proteins resulting in leather. Unacceptable use properties such as the stickiness of raw natural rubber, a 1,4-*cis*-poly(isoprene), led in 1839 to rubber vulcanization, a chemical crosslinking of long rubber molecules by sulfur that converts the rubber into a nonsticking elastomer. Use of a greater proportion of sulfur resulted in Ebonite, a hard black material, now called a thermoset.

Table 1-1 Annual world consumption of organic polymeric materials (exclusive of food).

Type	Annual consumption in million tons						
	1940	1950	1960	1970	1980	1990	2000
Fuel (fire wood, charcoal)				400	500	600	
Building materials (lumber, etc.)				1100	1300	1300	
Paper and cardboard				300	350	400	
Plastics (incl. additives)	0.36	1.6	6.7	31	59	100	172
Fibers, synthetic	0.005	0.069	0.70	5.0	11.4	15.7	31.7
regenerated natural	1.1	1.6	2.6	3.4	3.3	3.2	2.8
cotton, wool, silk	8.1	8.0	11.8	14.0	16.4	21.0	21.1
Rubbers, synthetic	0.043	0.54	1.94	5.9	8.7	9.9	10.8
natural	1.44	1.89	2.02	3.1	3.9	5.2	7.3
Thickeners, natural						6.5	
synthetic						4.6	
Adhesives, sealants						5.5	
Resins for graphic purposes						1.0	
Population (in billions)	*2.25*	*2.53*	*3.04*	*3.70*	*4.43*	*5.29*	*6.08*

Necessity is the mother of invention: scarcity of rags used as raw material for the manufacture of paper induced the search for a substitute, resulting in paper-making from wood cellulose. The accidentally discovered nitrocellulose, when mixed with camphor, led to the first semisynthetic plastic as substitute for ivory (1869). When silkworms were killed by a disease, nitrocellulose was spun into the first semisynthetic fiber (1884).

The first fully synthetic plastic was invented in 1906. Without the excellent electrical insulting properties of this phenolic resin, the newly established electrical industry would not have grown so fast. Since the central powers lacked natural rubber during World War I, a synthetic substitute based on dimethylbutadiene was developed.

All these semisynthetic and synthetic polymers were prepared without knowledge of their macromolecular structure, which started to be explored in earnest only after ca. 1920, especially by Hermann Staudinger (1884-1965; Nobel prize 1953). The resulting scientific foundation allowed the production of the first economically and technically successful synthetic rubber in 1929 (poly(chloroprene)), the first thermoplastic in 1930 (poly(styrene)), the first fully synthetic fiber in 1931 (poly(vinyl chloride)), and the first synthetic textile fiber in 1939 (poly(hexamethylene adipamide) = nylon 6.6).

1.2   Basic Terms

With time, many scientific and technical terms changed their meanings (see Section 1.3). It is therefore advisable to introduce modern terms first and then mention older or different terms, usages, or definitions in the appropriate later sections. The following definitions are based, in general, on the recommendations of the polymer nomenclature committee of the International Union of Pure and Applied Chemistry (IUPAC). Different terms are sometimes used by the Chemical Abstracts Service (CAS).

The word **macromolecule** is not only a Greek-Latin hybrid but also an oxymoron because it refers to a small mass (G: *molecula*, diminutive of *moles* = mass) that is large (L: *makros*). The word "molecule" no longer denotes a small mass but a "stable" entity composed of two or more atoms that are held together by **chemical bonds** (Section 2.1).

According to present usage, macromolecules are simply large molecules. Soluble synthetic macromolecules usually possess **molar masses** (physical unit: g/mol) or **molecular weights** (= relative molecular masses; physical unit: 1) between several hundreds and several millions, some biological macromolecules even up to several billions. Insoluble macromolecules have even greater molecular weights. For example, diamonds consist of covalently bound carbon atoms. A one-carat diamond weighs $m = 0.2$ g; its molar mass M is thus $M = mN_A \approx (0.2 \text{ g})(6.02 \cdot 10^{23} \text{ mol}^{-1}) = 1.204 \cdot 10^{23}$ g/mol (N_A = Avogadro constant). Such *insoluble* entities are not considered "molecules" by many chemists.

According to IUPAC, a macromolecule is defined as

"a molecule of high relative molecular mass, the structure of which essentially comprises the multiple repetition of units derived, actually or conceptually, from molecules of low relative molecular mass."

Because of this definition, "macromolecule" and "polymer molecule" are considered synonyms by IUPAC.

In general usage, "macromolecule" and "polymer molecule" are not synonyms, however. A **macromolecule** is simply a very large molecule; the word implies nothing about its chemical structure, especially *not a repetition* of units. The term **polymer molecule**, on the other hand, indicates a molecule that is composed of several or many parts (G: *polys* = several, many; *meros* = part). The nature of the "parts" is left open (see Section 1.3) but the parts are usually thought to mean "equal" or "similar" ones.

For example, a poly(ethylene) *molecule* with the constitution $H[CH_2CH_2]_NH$ is composed of N *equal* ethylene units $-CH_2CH_2-$ and 2 endgroups H–; it is a polymer molecule. The *substance* "poly(ethylene)", $H[CH_2CH_2]_nH$, contains many poly(ethylene) molecules of usually widely varying numbers of ethylene units per molecule; the subscript n here symbolizes the *average* amount-of-substance ("mole number") of ethylene units.

A simple enzyme molecule, on the other hand, possesses the generalized constitutional formula $H[NH–CHR–CO–....–NH–CHR'–CO]OH$. It combines many *different* amino acid residues $-NH–CHR–CO-$ (R = H (glycine), CH_3 (alanine), etc.) in an irregular, biologically important sequence. For the same type of enzyme, the number and sequence of these amino acid residues does not vary from molecule to molecule. Such enzyme molecules are macromolecules but not polymer molecules in the common (non-IUPAC) meaning of the words. In other words: a polymer molecule is always a macromolecule but a macromolecule is not necessarily a polymer molecule.

"Polymer molecule" is also not synonymous with "polymer". A polymer molecule is a molecule but a **polymer** is a *substance* consisting of many polymer molecules (the counterpart is "macromolecular substance"). Since N is the IUPAC symbol for a number and n is the IUPAC symbol for the amount-of-substance (physical unit: mole with the symbol "mol"), the constitutional structure of poly(styrene), for example, will be depicted as $+CH_2–CH(C_6H_5)+_N$ for the molecule and as $+CH_2–CH(C_6H_5)+_n$ for the substance.

N is always a positive whole number whereas n may be a whole or fractional number since it is an average over many molecules. The type of average will not be specially indicated in structural formulas.

It is important to distinguish between *molecules* and *substances*. Solid-state properties are, for example, very often not those of molecules but those of substances, i.e., of molecules *and* their arrangements and interactions. In chemical writing, this distinction is often not made because it is frequently assumed that the context indicates whether a chemical name denotes a molecule or a substance.

Polymers are generated from **monomers** (G: *monos* = single, sole, alone) by a process called **polymerization** (see Section 1.1.3). Styrene $CH_2=CH(C_6H_5)$ is, for example, the monomer for poly(styrene), $+CH_2–CH(C_6H_5)+_n$, with the **monomer(ic) unit** or **mer** $-CH_2–CH(C_6H_5)-$. A **macromonomer** is a "large" monomer with higher molecular weight. An example is $CH_2=CH\{COO[CH_2CH(C_6H_5)]_{25}H\}$.

The number N of monomer units per macromolecule is called the **degree of polymerization**, X, of the macromolecule (sometimes abbreviated as **DP**). The degree of polymerization of a polymer (= a polymeric substance), usually also symbolized by X, equals *numerically* (but not with respect to the physical unit) the average amount of monomeric units per polymer molecule. In contrast to N, the degree of polymerization, X, of a polymeric substance may be a fractional number since it represents an average if the number of monomer units varies from polymer molecule to polymer molecule.

The simplest macromolecules are composed of so-called **linear chains** which are, by definition, macromolecules with no extrinsically (i.e., by polymerization) generated branching points. By definition, intrinsic branching of monomer molecules in the sense of organic chemistry does not lead to branched macromolecules (see p. 51).

Examples of linear chains are

Polymeric sulfur	$\bullet$S–S–S–S–....–S–S-S$\bullet$ (where $\bullet$ is a free radical)
Poly(methylene)	R–CH$_2$–CH$_2$–CH$_2$–....–CH$_2$–R'
Poly(oxymethylene)	R–O–CH$_2$–O–CH$_2$–....–O–CH$_2$–O–CH$_2$–R'
Poly(styrene)	R–CH(C$_6$H$_5$)–CH$_2$–CH(C$_6$H$_5$)–CH$_2$–....–CH(C$_6$H$_5$)–CH$_2$–R'

The chains of these molecules consist of connected **chain atoms** which may all be identical (S in polymeric sulfur, C in poly(methylene) and poly(styrene)), or not identical (C and O in poly(oxymethylene)). Chain atoms may be organic or inorganic. They may also be unsubstituted in the inorganic sense (polymeric sulfur), unsubstituted in the organic sense (poly(methylene), poly(oxymethylene)), or substituted in the organic sense (poly(styrene)). Both ends of linear chains carry endgroups (radicals $\bullet$, R, R').

Chain atoms (CAS: **skeletal atoms**) with their substituents are called **chain units**. In the above examples they are –S–, –CH$_2$–, –O–, and –CH(C$_6$H$_5$)–. Other common chain units of polymers are –CO–, –NH–, and –C$_6$H$_4$–.

The term "linear chain" was originally coined by Hermann Staudinger, the father figure of macromolecular chemistry (see Section 1.3). He assumed that such chains are always completely stretched, i.e., that chains assume the macroconformation of zig-zag chains with all-trans conformations of the chain atoms (see Chapter 5 and the two examples shown below). Such zig-zag chains of *whole* molecules are extremely rare in the crystalline state and they are completely absent in solution (volume III). The term "linear" now refers to the constitutionally one-dimensional assembly of chain units.

| poly(methylene) | poly(oxymethylene) |

The bonds between chain atoms need not be necessarily covalent. Chain units may also be interconnected by coordinative bonds or electron deficient bonds (examples are palladium chloride and polyesters with ferrocene as chain unit (see also Section 2.3.3). By definition, ionic bonds and metallic bonds do not lead to macromolecular chains (Section 2.1.1).

Linear chains possess two **endgroups**. The designation of the chemical structure of the endgroups is a matter of definition. For example, poly(hexamethylene adipamide) = nylon 66, H$\leftarrow$NH(CH$_2$)$_6$NHCO(CH$_2$)$_4$CO$\rightarrow_n$OH, possesses the endgroups H– and –OH if one looks at the monomeric units –NH(CH$_2$)NH– and –CO(CH$_2$)$_6$CO– but the endgroups H$_2$N– and –COOH if one considers its chemical reactivity. In the literature, chemical structures of endgroups of polymer molecules are often not given because they are not easy to determine in high-molecular weight polymers. The structure of poly-(hexamethylene adipamide) is then simply written as $\leftarrow$NH(CH$_2$)$_6$NHCO(CH$_2$)$_4$CO$\rightarrow_n$.

Chains contain **constitutional units** consisting of one or more chain units. This term relates to the various chemical groups in polymer molecules and *not* to the chemical structure of the monomers used for polymer syntheses. Poly(ε-caprolactam) = polyamide 6 (PA 6) = nylon 6 with the chemical structure $\leftarrow$NH(CH$_2$)$_5$CO$\rightarrow_n$ thus contains not only NH(CH$_2$)$_5$CO as constitutional unit but also the constitutional units NH, CH$_2$, CO, NHCH$_2$, CH$_2$CH$_2$, CH$_2$CO, CONH, NH(CH$_2$)$_2$, (CH$_2$)$_3$, etc. (Table 1-2).

Table 1-2 Monomers, monomeric units, constitutional units, und constitutional repeating units of poly(ethylene) (PE), poly(methylene) (PM), and the polyamides PA 6 and PA 66. The monomers are examples: PA 6 is usually synthesized by polymerization of ε-caprolactam and not of ε-aminocaproic acid as shown here. PA 66 was assumed to be synthesized from hexamethylenediamine and adipic acid.

Term	PE	PM	PA 6	PA 66
Process-based terms				
Monomers	$CH_2=CH_2$	CH_2N_2	$H_2N(CH_2)_5COOH$	$H_2N(CH_2)_6NH_2$ and $HOOC(CH_2)_4COOH$
Monomeric units (Mers)	CH_2CH_2	CH_2	$HN(CH_2)_5CO$	$HN(CH_2)_6NH$ and $OC(CH_2)_4CO$
Structure-based terms				
Chain atoms	C	C	N, C	N, C
Chain units	CH_2	CH_2	NH, CH_2, CO	NH, CH_2, CO
Constitutional units	CH_2	CH_2	$NH, CH_2, CO, NHCO,$ $NHCH_2, CH_2CO,$ etc.	$NH, CH_2, CO, NHCO,$ $NHCH_2, CH_2CO,$ etc.
Constitutional repeating units	CH_2	CH_2	$HN(CH_2)_5CO$	$HN(CH_2)_6NHCO(CH_2)_4CO$

The **constitutional repeating unit (CRU)** is the smallest constitutional unit whose repetition in a single sequential arrangement describes a **regular polymer molecule**. In polyamide 6, it is the constitutional unit $-NH(CH_2)_5CO-$; here, the constitutional repeating unit is identical with the monomeric unit. In PA 66, on the other hand, the constitutional repeating unit $-HN(CH_2)_6NHCO(CH_2)_4CO-$ is also identical with the monomeric unit if the polymer was synthesized from $H_2N(CH_2)_6NHCO(CH_2)_4COOH$. PA 66 has two types of monomeric units, $-HN(CH_2)_6NH-$ and $-CO(CH_2)_4CO-$ if it was obtained from hexamethylenediamine $H_2N(CH_2)_6NH_2$ and adipic acid $HOOC(CH_2)_4COOH$.

CRUs are called **structural repeating units (SRUs)** by CAS. This term is ill-defined since "structure" also includes configuration, conformation, and various other physical arrangements.

Note that "constitutional unit" and "constitutional repeating unit" are **structure-based terms** whereas "monomer(ic) unit" is a **process-based term**. This distinction is important for the definition of the **degree of polymerization of a molecule**, X, which, as a process-based term, refers to the number of *monomeric units* per molecule and *not* to the number of CRUs. The molecule $H-[NH(CH_2)_6NHCO(CH_2)_4CO-]_N OH$ thus has a degree of polymerization $X = N$ (if obtained from N $H_2N(CH_2)_6NHCO(CH_2)_4COOH$) or $X = 2N$ (if synthesized from N $H_2N(CH_2)_6NH_2 + N$ $HOOC(CH_2)_4COOH$). The same reasoning applies to the **degree of polymerization of a polymer** (as a substance).

Polymer molecules which differ only with respect to the number N of otherwise constitutionally identical repeating units are considered members of a **polymer homologous series**. This terminology differs from the one of low-molecular weight organic chemistry where "homology" (G: *homo* = same; *logos* = word, from *legein* = to speak) refers to the presence of the same functional group in a series of similar chemical compounds, for example, the hydroxy group in unbranched aliphatic alcohols: CH_3OH, CH_3CH_2OH, $CH_3CH_2CH_2OH$ ... $CH_3(CH_2)_NOH$. The homology of this series of compounds is caused by the presence of the OH group in the terminology of organic chemistry, whereas in macromolecular chemistry, it is due to methylene groups as chain units.

No definite demarcation line exists between low and high molecular weight representatives of a polymer homologous series. In studies of polymerization kinetics and physical polymer properties, it is useful to define polymer molecules as those with a degree of polymerization of 2 and higher. In preparative and industrial chemistry, it is customary, however, to distinguish polymers with a "small" number of monomeric units per molecule as **oligomers** (G: *oligos* = few) from "genuine" polymers. "Small" is relative: less than about 20 for synthetic polymers but several hundred for nucleic acids.

Oligomers with "functional" (= reactive) endgroups that are introduced by so-called chain transfer reactions (see pages 221 and 344) are called **telomers** in preparative macromolecular chemistry (G: *telos* = end). In biochemistry, **telomeres** refers to the peptide and deoxynucleic acid containing end sections of chromosomes. **Telopeptides** are not oligomers consisting of peptide units –NH–CHR–CO– in general but end sections of collagen molecules that are made up of peptide units that are not arranged in triplet structures (Section 14.3).

The theoretically and practically important **degree of polymerization**, X, cannot be measured directly. For a linear macromolecule, it can be calculated, however, from the experimental **molecular weight**, M_r (= relative molecular mass; see p. 84) and the molecular weights $M_{r,u}$ of monomeric units and $M_{r,end}$ of endgroups. The contributions of endgroups can be neglected for higher molecular weights:

(1-1) $X = (M_r - M_{r,end})/M_{r,u} \approx M_r/M_{r,u}$

Most polymers do not consist of identical polymer molecules. They are rather **inhomogeneous** ("disperse"; see p. 44) with respect to the constitution and/or the degree of polymerization of their molecules. The degrees of polymerization and molecular weights of such polymers are thus always **averages** whose numerical values depend on the statistical weight imposed on the degrees of polymerization, $X_{r,i}$, and molecular weights, $M_{r,i}$, of the molecules of class i.

There are many different statistical weights and very many different types of averages (Chapter 3 and volume III). However, for polymer science, most important are the number and the weight average. The number average (index n) counts the number N_i of molecules with the degree of polymerization, X_i, or the molecular weight $M_{r,i}$, respectively. The weight average (index w) employs the mass m as statistical weight:

(1-2) $\overline{X}_n = \Sigma_i N_i X_i / \Sigma_i N_i$; $\overline{M}_{r,n} = \Sigma_i N_i M_{r,i} / \Sigma_i N_i$ (if $M_{end} << M_{r,i}$)

(1-3) $\overline{X}_w = \Sigma_i m_i X_i / \Sigma_i m_i$; $\overline{M}_{r,w} = \Sigma_i m_i M_{r,i} / \Sigma_i m_i$ (if $M_{end} << M_{r,i}$)

Polymer properties often vary systematically with the degree of polymerization or the molecular weight. Some properties, such as the boiling temperature (Fig. 1-1) or the melt viscosity, increase continuously with the degree of polymerization. Other properties, such as the melting temperature (Fig. 1-1) and the tensile strength first increase but then become constant and independent of the degree of polymerization and the molecular weight. The manner in which the properties vary with X or M depends not only on the molecule mass but also on the interaction and the geometric arrangement of chains and chain sections (volume III).

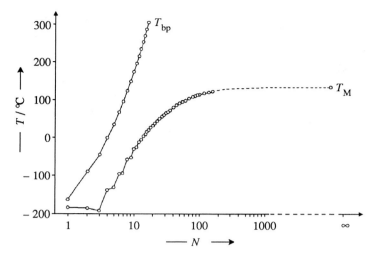

Fig. 1-1 Dependence of melting temperatures, T_M, and boiling temperatures, T_{bp}, of alkanes and poly(methylene)s, $H(CH_2)_NH$, on the number N of methylene groups per molecule [1, 2].

Polymer molecules may not only be linear but also branched or crosslinked, either on purpose by using suitable monomer molecules or unintended by side reactions during the synthesis. **Branching** can be single or multiple, orderly or disorderly, and solitary or consecutive so that a great number of different branching types exist (Chapter 2.5).

The **crosslinking** of chains during or after polymer synthesis leads to polymer **networks** where **network chains** are interconnected by **crosslinks (junctions)**. Low crosslink densities and high segment mobilities lead to **elastomers**. Conversely, high crosslink densities and low segment mobilities are typical for **thermosets**. Once prepared, neither elastomers nor thermosets can be brought into a fluid state without chemical degradation. They differ in this respect (and many others) from so-called **thermoplastics** (G: *plassein* = to form) which are linear or branched polymers that can be remolten and reshaped without change of chemical structure. **Fibers** are often, but not always, thermoplastics.

1.3   Development of the Macromolecular Hypothesis

The term "polymer" has a long history. The Swede Jøns Jacob Berzelius discovered in 1831 [3] that the three tartaric acids exhibited non-identical physical properties despite their identical chemical compositions. In modern chemical (Fischer) representation:

<pre>
     COOH   COOH   COOH
      | | |
  H —C—OH   HO —C—H   H —C—OH
      | | |
  HO—C—H   H —C—OH   H —C—OH
      | | |
     COOH   COOH   COOH

 D-tartaric acid   L-tartaric acid   meso-tartaric acid
</pre>

(the racemic mixture of D and L is known as racemic acid = uvic acid = D,L-tartaric acid)

Since the same chemical composition has *several* (here: three) different "parts", Berzelius called such chemical compounds "polymers" (G: *polys* = many, several; *meros* = part). Note that he used *polys* in the sense of "several" and not as "many" (see the quotation below). "Polymer" thus originally meant that *all* these compounds were polymeric to *each other* but *not* that *each* of them was composed of *many* (identical) parts.

One year later, Berzelius changed his terminology and introduced the term "isomer" in addition to "polymer" [4]. The original quotation in (antiquated) German says (footnotes [1-5] are added; they are not in the original):

"Isomere sind "... Körper..., die aus einer gleichen absoluten und relativen Atomen-Anzahl derselben Elemente zusammengesetzt sind und gleiches Atomgewicht [1] haben, ..., womit nicht der Fall zu verwechseln ist, wo die relative Anzahl der Atome gleich ist, die absolute aber ungleich. So ist z.B. die relative Anzahl von Kohlenstoff- und Wasserstoff-Atomen im ölbildenden Gas [2] und im Weinöl [3] absolut gleich, nämlich die Zahl der Wasserstoff-Atome ist doppelt so groß, als die der Kohlenstoff-Atome, allein in dem einem Atom [4] vom Gase sind bloss 1 Atom Kohlenstoff und 2 At. Wasserstoff enthalten, CH^2, während dagegen im Weinöl 4 Atome Kohlenstoff und 8 Atome Wasserstoff enthalten sind, = C^4H^8. Um diese Art von Gleichheit in der Zusammensetzung [5], bei Ungleichheit in den Eigenschaften, bezeichnen zu können, möchte ich für diese Körper die Benennung *polymerische* (von πολνς mehrere) vorschlagen...".

The translation into English tries to use a style similar to the German original:

"Isomers are ... bodies [1] ..., which are composed of an equal absolute and relative atom-number of the same elements and have an equal atomic weight [2], ..., which is not to be confused with the case where the relative number of atoms is equal but the absolute number is unequal. For example, the relative number of carbon atoms and hydrogen atoms is absolutely equal in the oil-forming gas [3] and in wine oil [4], namely, the number of hydrogen atoms is twice as large as those of the carbon atoms but in the one atom [5] of the gas are only 1 atom of carbon and 2 atoms of hydrogen, CH^2, whereas, however, wine oil contains 4 atoms of carbon and 8 atoms of hydrogen = C^4H^8. In order to characterize this type of equality in composition [6], with unequality in the properties, I would like to propose for these bodies the term *polymerics* (from πολνς several) ...".

The word "polymer" was thus restricted to those compounds that had the same relative composition (in today's meaning) but different molecular weights. Note that "composition" at that time referred to what we now call "relative composition" and that this "relative composition" is a function of the atomic weights assigned to the various types of atoms (see below).

---

[1] "Körper" (body) is now called "substance". [2] Now: molecular weight. [3] Now: ethene. [4] "Weinöl" is literally translated as "wine oil" but I do not know whether that was at that time the English term for what is now called "1-butene". [5] "Atom of gas": according to Avogadro the smallest "composition" of gases. [6] At the time of Berzelius, "composition" meant the *relative proportion* of the numbers of the various types of atoms per molecule; today, it refers to the *absolute numbers* of atoms per molecule.

Note that "molecule" (L: *molecula* = small mass, diminutive of *moles* = mass) was first used by Daniel Sennert (1572-1637) [5a]. This term originally comprised both today's "molecules" and "atoms". In 1805, John Dalton (1766-1844) restricted the term "molecule" to the smallest parts of compounds [5b, 6] but "compound" was not used with the same meaning as today's "chemical compound". In 1814, for example, André Marie Ampère called a present-day atom a "molecule" and a present-day molecule a "particle". In 1858, Carl von Nägeli used "molecule" for the smallest morphological parts of starch in plants. The word "molecule" adopted its present meaning after the Karlsruhe conference (1860). Von Nägeli thus replaced "molecule" by "micelle" (which later changed its meaning, too) [7].

Isomers

$$CH_2\!=\!CH_2 \qquad CH_2\!=\!CH-CH_2-CH_3 \qquad CH_2\!=\!C(CH_3)_2$$

Ethene C_2H_4 1-Butene C_4H_8 Isobutene C_4H_8

Polymers

Ethene and 1-butene, however, are not polymers in the modern sense.

Seven years after Berzelius' proposal, an interesting phenomenon was observed by the German pharmacist Eduard Simon [8]. The trees *Liquidambar orientalis* and *L. styraciflua* from Asia Minor deliver storax, an aromatic smelling resin, which was already used in Egypt 3000 years ago for the embalming of corpses. Storax was traded in many grades, for example as liquid storax (*styrax liquidus*), a dark grey, highly viscous mass from which threads can be pulled [9]. On heating storax with water, a clear organic liquid distilled off [10] which Simon called "Styrol" (E: styrene). However, additional heating converted "Styrol" into a gelatinous material (which is today known to be a concentrated solution of poly(styrene) in styrene). Since Simon suspected a chemical reaction and since oxygen was the only possible other reactant, he postulated an oxidation and called the resulting gel "Styroloxid" (E: styrene oxide) [8]. Six years later, however, an analysis of "styrene oxide" by Blyth and Hofmann [9] showed that it did not contain oxygen. Since "styrene oxide" evolved "after" styrene, it was now renamed "metastyrene" (G: *meta* = between, with, besides, after). Note that "meta" is presently used in the sense of "between" and that "metastyrene" is not a meta compound in the modern sense.

$$CH_2\!=\!CH \qquad \{CH_2-CH\}_n \qquad \overset{O}{\overset{/\backslash}{H_2C-CH}}$$
$$C_6H_5 \qquad \phantom{\{CH_2-}C_6H_5 \qquad C_6H_5$$

Simon's name	Styrol (styrene)	Styroloxid (styrene oxide)	-
Blyth-Hofmann name	Styrol (styrene)	Metastyrol (metastyrene)	-
Modern trivial English name	styrene	poly(styrene)	styrene oxide

Simon's observations were remarkable in two respects. First, a liquid (styrene) was converted upon heating into a "solid" and not into a gas as is usually the case. Second, metastyrene had the same composition (in the Berzelius sense, i.e., relative composition) as styrene: 8 carbon atoms and 8 hydrogen atoms, but different properties. Styrene and metastyrene were thus considered "polymeric" to each other.

Later, the word "polymer" changed its meaning again. Researching isomers (in the Berzelius sense of the word) of acetylene CH≡CH, Berthelot [14] reported that styrene converts on heating into a resinous polymer ("polymère resineux") (resinous because it contained less styrene than Simon's gelatinous material). He ascribed both styrene and metastyrene the "composition" C/H = 16/8 (instead of 8/8) since he assigned carbon the atomic weight 6 instead of 12 as was usual in his time. Since acetylene correspondingly had the "composition" C/H = 4/2 (instead of 2/2), both styrene and metastyrene appeared to Berthelot as polymeric to acetylene. All these compounds were either low-molecular weight materials (acetylene, styrene) or oligomeric ("metastyrene"). When truly high-molecular weight materials were discovered later, they were called "**high polymers.**"

Since "metastyrene" and styrene were polymers according to the Berzelius terminology, Berthelot called the conversion of styrene into metastyrene a "**polymerization**" [14]. "Polymerization" thus meant originally that molecules were interconverted without change of their (relative) composition; it did *not* mean the joining of small molecules.

The next fifty years saw the term "polymerization" used for so many different processes that no agreement upon the meaning existed. The situation changed after Hermann Staudinger in 1920 [15] proposed a new definition (translation from the German):

> Polymerization processes, in the broader sense, are all processes where two or more molecules are joined to a product with equal (relative) composition but higher molecular weight.

This definition paved the way to the discovery of true macromolecular structures. It also caused a rift in the use of the word "polymerization". In Germany, it was (and still is) used exclusively in the sense of Staudinger's definition (no leaving molecules; equal chemical composition of monomers and monomeric units of polymers). In Anglo-Saxon scientific papers, however, especially in the United States, "polymerization" was (and is) used for *all* processes leading to macromolecules, i.e., including both "addition polymerization" (no low-molecular weight leaving molecules) *and* "condensation polymerization" (process with leaving molecules) (see below).

Berthelot also observed that the conversion of liquid styrene to resinous "metastyrene" could be reversed by applying higher temperatures [14]. Later, this simple, temperature-induced interconversion of styrene and poly(styrene) became a main support for the so-called micellar theory of such compounds (see below).

Several years before Berthelot, Charles Adolphe Wurtz [16] converted ethylene oxide (= oxirane) into low molecular weight poly(ethylene oxide)s = poly(oxyethylene)s, in today's notation

$$(1\text{-}4) \qquad N \; H_2C\!-\!CH_2 \;\longrightarrow\; +O\!-\!CH_2\!-\!CH_2 +_N$$
$$\phantom{(1\text{-}4) \qquad N \; } \backslash\,/$$
$$\phantom{(1\text{-}4) \qquad N \; H_2C\!-\!}O$$

At about the same time, A.-V. Lourenço [17] reacted ethylene oxide with ethylene halides. From the resulting mixture, he isolated substances with different numbers N of oxyethylene units per molecule. He also noted that, with increasing N, the composition of these substances approaches that of ethylene oxide. This finding was indeed remarkable since oligo(oxyethylene)s $R(O\!-\!CH_2\!-\!CH_2)_{1\text{-}6}R'$ are liquids at room temperature but ethylene oxide is a gas. Lourenço also observed that the viscosity of the oligomers increased with increasing degree of polymerization, N. He furthermore suggested a chain structure for his products.

Such chain structures were also postulated later for a series of substances that were polymers of acetyl salicylic acid [18], *p*-hydroxybenzoic acid [19], or hydroxypivalic acid [20]. These polymers with the general structure $+O\!-\!Z\!-\!CO+_n$ are now called polyesters. None of the suggested chain structures was proven, however.

Shortly thereafter, Thomas Graham [21] discovered that bone glue (= depolymerization product of the protein collagen) and some other chemical compounds diffused only slowly in water and permeated with difficulty through membranes. Since this behavior was characteristic for all noncrystallizing, glue-like materials known at that time, and in sharp contrast to the known crystallizing substances that diffused and permeated fast, Graham distinguished between crystalloids and colloids (G: *kolla* = glue). He thus ascribed the colloidal character to *substances* and not to physical *states*.

Colloids differed in appearance and behavior, however, which led to many proposals for their subclassifications. In 1903, for example, A. Müller [22] distinguished between "suspensions" with physical flocculations and "high-molecular molecules" with chemical precipitations. Examples given for "high-molecular weight materials" were proteinaceous bodies and colloidal silicic acid, both indeed truly macromolecular substances.

The oxide hydrates of iron or aluminum are other examples of inorganic colloids. The reactivities of these colloids do not differ very much from that of their crystalline forms, however. It followed that all substances can be transformed into their colloidal states and back into their non-colloidal states if the conditions are right. Colloids must be therefore general *states* of matter and not substances (Wolfgang Ostwald [23], P.P. von Weimarn [24]). The correct conclusion that all low-molecular weight substances can be converted to their colloidal states also led to the incorrect reversal of this statement that all colloids are associations or aggregates of small molecules, i.e., "physical molecules." Colloids should thus have the same formula weight as their non-colloidal forms.

For non-colloidal substances, formula weights and molecular weights are identical. However, the molecular weight of colloids could not be obtained by the *chemical methods* available at that time without additional structural information. *Physical methods* based on Raoult's law [25] and van't Hoff's law [26], on the other hand, resulted in very high molecular weights of 10 000–20 000 for rubber, starch, and cellulolose nitrate [27, 28]. These high molecular weights seemed implausible to most researchers since the same methods gave molecular weights of covalently built crystalloids which agreed very well with their chemical formula weights. Hence, since the chemical formula weight of colloids could not be determined unambiguously, molecular weights by physical methods were also suspect.

Furthermore, proportionality is required for vapor pressure and concentration by Raoult's law and osmotic pressure and concentration by van't Hoff's law. Both demands were well fulfilled within limits of error for the then known covalently built crystalloids but not for colloids. This "violation" of the laws of Raoult and van't Hoff made the observed high molecular weights of colloids even more suspect.

We know today that both laws are limiting laws for small concentrations and that a concentration dependence of apparent molecular weights, i.e., those calculated from data obtained at finite concentrations by limiting laws, is the rule, even for low-molecular weight substances. In 1900, Nastukoff [29] recognized this effect which is caused by interactions between solute and solvent; he also suggested extrapolating to zero concentration the values obtained by ebullioscopy at finite concentrations. A similar extrapolation of osmotic pressure data [30] yielded a molecular weight of 100 000 for natural rubber.

However, the formulation of Raoult's and van't Hoff's laws as limiting laws for infinite dilution was not acceptable at that time. Several observations seemed to contradict it. Firstly, these laws always applied regardless of concentration to covalent crystalloids and to low-molecular weight substances that could be converted into colloids (reason: the concentration dependence of vapor pressure and osmotic pressure was within limits of error). Secondly, colloids were not the only class of substances that showed strong deviations from Raoult's law. Similar discrepancies were also found for electrolytes. Since all known electrolytes were inorganic compounds and since all of them could be converted into colloids, at least in principle, it was concluded that there must exist some special forces which were keeping the chemical molecules together in their colloids.

Thirdly, all experiences of organic chemistry seemed to contradict the assumption that high-molecular weight compounds consist of covalently bonded atoms. Classical (low-molecular weight) organic chemistry is based on the principle of the smallest change in constitution in chemical reactions, elementary analysis as the base for establishing the chemical constitution of a compound, and crystallizability as the measure of the purity of chemical compounds. True, there were also other compounds like alcohols and sugars that could not be crystallized or only with difficulty but these were considered unexplained exceptions. Fourthly, organic colloids failed another purity criterion of organic chemistry: a substance is pure if it can be characterized by a single chemical structure with a single molecular weight. Several organic colloids exhibited different molecular weights, however, despite the same chemical structure.

The question thus was: which special forces are keeping the organic colloids together? One knew about the existence of intermolecular forces from the study of gases [31]. Similar forces could surely act in solution. For organic colloids, the idea of partial valences seemed attractive. Such valences supposedly kept together molecules with conjugated double bonds [32]. Indeed, the discovery [33] of the molecular complex quinhydrone (from benzoquinone and hydroquinone) seemed to be proof of the existence of partial valences.

Partial valences also offered an easy explanation of the curious behavior of natural rubber. In 1826, Michael Faraday [34] showed that rubber has the composition C_5H_8 which implied one double bond per molecule. In 1904-1909, Carl Harries [35] confirmed this composition by ozonization and subsequent hydrolysis of the ozonide. Since he found the same composition for this product and the original rubber, endgroups were ruled out. In addition, the low molecular weight of the product suggested a ring structure which was first assumed to consist of two isoprene units and later of five to seven units (Fig. 1-2).

The assumption of low-molecular weight ring molecules that are held together by partial valences also seemed to explain the fact that natural rubber cannot be distilled: it was known that associated molecules showed much higher boiling temperatures than the non-associated ones.

In 1910, however, Samuel Shrowder Pickles suggested a chain structure for natural rubber [36]. He studied the addition of bromine to the double bonds of rubber and found that the brominated rubber had the same degree of polymerization as natural rubber. He concluded, therefore, that natural rubber must be a chemical compound and not a physical molecule (an associate). His suggestion was ignored, however.

Ring structures, on the other hand, were suggested for many other organic colloids. For example, the proposed ring structure of cellulose (Fig. 1-2), with three hydroxyl groups and a half-acetal group, represented the chemical behavior well. The colloidal character of cellulose and its derivatives was explained by a physical association of many rings to greater entities. Ring structures also explained the apparent absence of endgroups. We know today that analytical methods known at that time were simply too insensitive to detect the small proportion of endgroups in high-molecular weight polymers.

In the years between 1910 and 1920, "proof" of organic colloids as physical molecules was overwhelming. Organic colloids had the same composition and the same reactivity as their non-colloidal counterparts. They were not crystallizable and could be

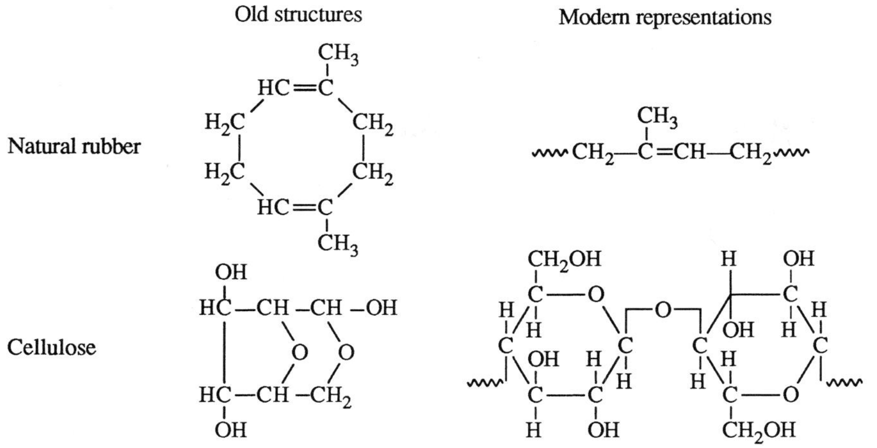

Old structures   Modern representations

Fig. 1-2 Old and modern structures of natural rubber and cellulose. ⌇⌇ indicates "chain".

easily converted into their non-colloidal base materials. The absence of endgroups was proof of small rings. These colloids showed anomalies in molecular weight determinations but such phenomena were also known from inorganic colloids. The colloidal character was easily explained by the presence of special forces, the Thiele partial valences.

The one lonely caller in the desert opposing this view was Hermann Staudinger (Nobel prize 1953). While investigating ketenes in the years before 1912 [37], he found "polymeric products" (in the sense of Berzelius) which he ascribed to cyclobutane derivatives. Since in 1916 another author [38] considered these dimers to be molecular complexes, Staudinger countered in 1920 by pulling together in a survey article [15] all the reasons for covalent bonds in general and chain structures of colloidal organic polymerization products in particular, including Pickles' rubber chains [36]. He called such polymerization products "**macromolecules**" [39].

However, chains of atoms held together by valence bonds require the presence of endgroups. Nevertheless, their apparent absence was thought not to contradict this requirement since it was assumed that reactivities decrease with increasing molecular weight and that therefore "... high-molecular weight compounds are by far less reactive than their monomolecular starting materials ..." [15]. Here, the reactivity of a molecule was confused with the reactivity of a reactive group.

In later papers, Staudinger focussed on proving experimentally his idea that organic colloids are true macromolecules. He first attacked the so-called "first micelle theory" which claimed that organic colloids consist of small rings that are held together by partial valences. "Micelle" was originally the name for submicroscopic, long, crystalline entities prevalent in starch particles (L: *mica* = crumb) [7]; the term was later used for all associated particles.

In 1922, Staudinger and Fritschi [40] hydrogenated natural rubber. Since the hydrogenated product, $+CH_2-CH(CH_3)-CH_2-CH_2+_n$ no longer contained the carbon–carbon double bonds of natural rubber, $+CH_2-C(CH_3)=CH-CH_2+_n$ it should not be able to display "partial valences" and thus should not lead to colloidal properties. However, the colloidal character still existed as was also found for brominated rubber by Pickles long before [36]. Similarly, the removal of double bonds by hydrogenation of poly(styrene), $+CH_2-CH(C_6H_5)+_n$, to poly(vinyl cyclohexane), $+CH_2-CH(C_6H_{11})+_n$, also removed the supposed carriers of "partial valences" but still left intact the colloidal character of these materials.

Therefore, Staudinger concluded that organic colloids consist of many covalently bonded atoms and that they are true "macromolecules" [39]. Since the bond strength of primary (covalent) bonds is much greater than that of the secondary ones (van der Waals, etc.), such "molecule colloids" should preserve their colloidal character in all solvents, in contrast to "association colloids" [41, 42].

Nevertheless, these experimental findings were not accepted as proof by most other researchers. For example, the molecular weight of natural rubber in camphor was found to be 1400-2000 [43] whereas Staudinger obtained values of 3000-5000 for hydrogenated rubber. This difference was not acceptable for an organic compound but it was ignored that the two samples were not the same *substances* but different *materials*.

Other experimental findings and interpretations also seemed to speak against Staudinger's idea of true macromolecules. The X-ray pictures of many organic colloids more resembled those of liquids than those of low-molecular weight crystalloids. Some did have crystallite-like X-ray features but these indicated relatively small crystallographic unit cells. It was known, however, that the size of unit cells of crystallized *low-molecular weight* compounds is proportional to the molecular weight.

X-ray measurements did speak against small rings [44], however, since rings were incompatible with the observed structures of the elementary cells. The widths of X-ray reflections corresponded to crystallite lengths of 30-60 nm, and since Kurt H. Meyer and Herman(n) F. Mark [44] assumed that the length of a crystallite is identical with the length of a molecule, they calculated the molecular weight of cellulose as ca. 5 000 and that of natural rubber as 5 000–10 000. The far higher molecular weights of 150 000–380 000 obtained for natural rubber by membrane osmometry were first erroneously explained as molecular weights of solvated chains and later as those of physical molecules, i.e., as micelles (or associates). This "second micelle theory" thus assumed chains instead of rings and higher molecular weights (but still not very high ones) than the "first micelle theory" but still insisted on an association of chains.

Staudinger countered that crystallite lengths have nothing to do with molecule lengths [45]. It was indeed found 30 years later that polymer chains can backfold on crystallization so that the crystallite length becomes independent of the molecule length (vol. III).

Another problem was the unknown mechanism of polymerization. Staudinger's hypothesis was based on the behavior of either natural colloids such as natural rubber, cellulose, and amylose and on synthetic polymers such as poly(styrene) and poly(oxymethylene). Staudinger and Urech [46] suspected that during polymerization styrene molecules are added one after the other at an active center of unknown structure of the growing poly(styrene) chain; this center was identified later as a free radical [47, 48]. However, the nature of the initiation step remained unclear. Hypotheses included the addition of radicals to monomer molecules [49] or the formation of activated complexes from styrene and dibenzoyl peroxide [50]. The problem was finally solved in 1941-1943 by labeling initiators [51-53] which proved that initiator fragments become endgroups of polymer molecules.

To bypass this obstacle, Wallace Hume Carothers [54] decided to prepare macromolecular substances stepwise by known condensation reactions of low-molecular weight chemistry, for example, by reacting glycols HO–R–OH with dicarboxylic acids HOOC–R'–COOH (Eq.(1-5)). Note that these *condensation polymerizations* differ from the *addition polymerizations* of Eq.(1-4) in that small molecules (here: water) are liberated and that monomeric units are no longer identical with monomer molecules.

(1-5)   HOZOH + HOOCZ'COOH $\rightleftharpoons$ HOZO–OCZ'COOH + H_2O
HOZO–OCZ'COOH + HOZOH $\rightleftharpoons$ HOZO–OCZ'CO–OZOH + H_2O
HOZO–OCZ'COOH + HOOCZ'COOH $\rightleftharpoons$ HOOCZ'CO–OZO–OCZ'COOH + H_2O
2 HOZO–OCZ'COOH $\rightleftharpoons$ HOZO–OCZ'CO–OZO–OCZ'COOH + H_2O etc.

- -

n HO–Z–OH + n HOOC–Z'–COOH $\rightleftharpoons$ H[O–Z–O–OC–Z'–CO]$_n$OH + $(2n-1)$ H_2O

Carothers thus showed that macromolecules can be prepared by the known methods of organic chemistry and that, at least for condensation polymerizations, there is nothing mysterious about their formation. His work delivered further proof for covalent bonds between monomeric units and also led to the first commercially successful synthetic textile fiber, poly(hexamethylene adipamide) H[HN(CH$_2$)$_6$NHCO(CH$_2$)$_4$CO]$_n$OH (polyamide 6.6, nylon 66), from hexamethylene diamine and adipic acid.

Further arguments against the micelle theory were provided by biochemistry but this was not recognized by organic and physical chemists since they hardly read biochemical literature. For example, in 1926, James B. Sumner [55] crystallized the enzyme urease and in 1930, John H. Northrop [56] the enzyme pepsin (both were awarded Nobel prizes in 1946, together with Wendell M. Stanley). The crystallization of these colloids without loss of their colloidal character after redissolving the crystals was further proof that "colloid" was a state and not a substance. Furthermore, The(odor) Svedberg [57] showed in 1926-1940 with his new ultracentrifuge that proteins in their colloidal solutions were uniform with respect to their molecular weight regardless of temperature and/or salt concentration (Nobel prize 1926 for disperse systems). Electrophoresis, developed in 1938 by Arne Tiselius [58] (Nobel prize 1948), proved that a protein always has the same charge per mass, which is quite different from the behavior of inorganic association colloids. These findings were contrary to the assumption of association colloids.

It is interesting that Staudinger never utilized these results to support his assertion that organic colloids are true macro*molecules*. He undoubtly knew about Svedberg's work since he asked for a reseach grant in order to buy a Svedberg ultracentrifuge for the determination of molecular weights of synthetic macromolecules. His proposal was turned down, however. Staudinger was thus forced to search for a less expensive method for the determination of molecular weights.

That method was found in dilute solution viscometry (Section 3.6 and vol. III). Staudinger assumed linear polymer molecules to be rods, consisting of N_u monomeric units with molecular weights $M_{r,u}$. He further postulated that each monomeric unit provides the same contribution to the specific viscosity, $\eta_{sp} = (\eta - \eta_1)/\eta_1$, of the solution. The specific viscosity indicates the excess, $\eta - \eta_1$, of the viscosity η of the solution over that of the solvent, η_1, normalized by the viscosity of the solvent. Since the solution viscosity increases with increasing polymer concentration, the relative viscosity increment must be further reduced by the polymer concentration to give the reduced viscosity, $\eta_{red} = \eta_{sp}/c$ (formerly called viscosity number). Since each monomeric unit u delivers the same contribution, it followed that $\eta_{red} = KN_uM_{r,u} = KM_r$ where K is a constant that depends on the type of polymer and solvent as well as the temperature.

However, many researchers, especially Werner Kuhn [59], pointed out in the early 1930s that linear polymer molecules are not rigid rods. Segments can rather rotate around their chain bonds which leads to a coiling of molecules. Hence, the Staudinger "viscosity law", $\eta_{red} = KM_r$, should be replaced by $\eta_{red} = K'M_r^{\alpha}$ where $\alpha < 1$ (p. 98, 104).

The statistical reasoning underlying this assertion was, however, alien to Staudinger, an organic chemist. For many years, he stuck to his nice and simple "law", so that even in 1941, at the party in honor of his 60th birthday, his students rhymed "Die Kuhnschen Knäuel sind uns hier ein Gräuel" (Kuhn's coils are detested here).

The importance of statistics in polymer science was first recognized by Werner Kuhn in 1930 as he studied the chemical degradation of cellulose chains [61]. It is known today that statistical considerations play major roles in the constitution of macromolecules (Chapter 2), molecular weight distributions (Section 3.4), microconformations (p. 137), polymer syntheses and reactions (Chapters 6-16), macroconformations in solutions, melts, crystalline states, and liquid-crystalline states and all physical properties of polymers (volumes III and IV).

1.4 Discoveries and Inventions

Natural, semisynthetic, and synthetic macromolecular compounds were used industrially many years before their macromolecular structure became known (Table 1-3). These industrial developments were mostly empirical, simultaneous in the various industrial branches (rubber, fibers, working materials, adhesives, thickeners, etc.), and basically independent of the insights of science. Who could suspect at about 1900 that rigid wood, elastic rubber, tough leather, soft wool, stiff sisal fibers, glittery silk, powdery starch, sticky bone glue, and hard Bakelite® were all composed of macromolecules?

The developments accelerated in 1925-1930 after the macromolecular structure of many of these materials became known (Table 1-3). A second period of development began about 1950 after petrochemicals such as ethene and propene became available in large amounts. These petrochemicals pushed older raw materials to the side, such as acetylene obtained from coal (vol. II).

At about the same time, new polymerization processes were discovered. In 1939, ethene was polymerized industrially by ICI using traces of oxygen as initiator. This high-pressure, low-density poly(ethylene) was important to the development of radar in World War II. The process required high pressures and high temperatures, however. In 1954, Phillips Petroleum and Standard Oil disclosed that ethene can be polymerized at far lower temperatures and pressures to less branched, high-density poly(ethylene) if supported metal oxides such as CrO_3 or reduced MoO_3 were used as catalysts.

In the same year, Karl Ziegler (Nobel prize 1963) obtained a patent for the synthesis of high-density poly(ethylene) at room temperature and low pressure by polymerization of ethene with "mixed catalysts" consisting of combinations of metal compounds of elements of transition elements (groups 4-8) of the Periodic System of elements, for example, $TiCl_4$, with compounds of elements of groups 1-3, for example, $Al(C_2H_5)_3$. The same catalysts (which he obtained from Ziegler) allowed Guilio Natta (Nobel prize 1963) to prepare stereoregular polymers from 1-olefins or dienes (Chapter 9).

At the same time and during the following years, new theories of physical structures and properties were developed. These theories were based on modern physical concepts and mathematical procedures. The leaders in this development were awarded Nobel prizes: Paul J. Flory in 1974 and Pierre-Gilles de Gennes in 1991.

Table 1-3   Early industrial polymers. * No longer produced; NR = natural rubber.

Polymer	Discovery	Production	Present main applications
Vulcanized rubber (elastomer)	1839	1851	tires
Cellulose nitrate	1846	1889	gun cotton
Vulcanized rubber (thermoset)	1851		casings
Cellulose nitrate (+ camphor)	1865	1869	celluloid
Crosslinked casein (Galalith)	1897	1904	haberdashery
Phenolic resins (crosslinked)	1906	1909	thermosets (electrical insulators)
Dimethylbutadiene rubbers	1909	1915	elastomers *. Blends with NR: 1912
Alkyd resins	1847	1926	thermosets (coatings)
Amino resins	1904	1928	thermosetting working materials
Poly(butadiene)	1911	1929	elastomers (number Buna rubber)
Poly(chloroprene)	1925	1929	elastomers (Chloroprene)
Poly(vinyl acetate)	1912	1930	adhesives, poly(vinyl alcohol)
Poly(styrene)	1839	1930	thermoplastics, expanded plastics
Poly(vinyl chloride)	1838	1931	synthetic fibers*; thermoplastics: 1938
Poly(ethylene oxide)	1859	1931	thickeners, sizes
Poly(methyl methacrylate)	1880	1928	thermoplastics (organic glass)
Poly(vinyl ether)s	1928	1936	adhesives, plasticizers
Unsaturated polyesters	1930	1936	thermosets
Poly(isobutylene)	1934	1937	elastomers
Styrene-butadiene rubbers	1926	1937	elastomers (letter Buna rubber)
Poly(hexamethylene adipamide)	1934	1938	fibers, thermoplastics (1941)
Poly(vinylidene chloride) copolymers	1838	1939	thermoplastics (packaging films)
Poly(N-vinyl pyrrolidone)		1939	hair setting lotion, flocculant, thickener
Polysulfide rubbers (Thiokol)	1926	1939	elastomers
Poly(ethylene), low-density	1932	1939	thermoplastics
Poly(ε-caprolactam)	1938	1939	fibers (Perlon®), thermoplastics
Polyurethanes	1935	1940	plastics, elastomers, plastic foams
Poly(acrylonitrile)	1894	1942	fibers
Silicones	1901	1942	liquids, resins, elastomers
Epoxy resins	1939	1946	adhesives
Poly(ethylene terephthalate)	1941	1953	fibers, bottles
Poly(tetrafluoroethylene)	1939	1950	plastics, fibers
ABS polymers		1954	plastics
Poly(ethylene), high-density	1953	1955	thermoplastics, plastic foams
Poly(butadiene), *cis*-1,4-		1956	elastomers
Poly(isoprene), *cis*-1,4-	1909	1958	elastomers *
Poly(propylene), isotactic	1954	1959	thermoplastics, fibers
Poly(oxymethylene)	1859	1959	thermoplastics
Bisphenol A polycarbonate	1953	1959	thermoplastics
Aromatic polyamides		1961	high modulus fibers
SBS triblock copolymers		1965	thermoplastic elastomers
Poly(acetylene)	1976		semiconductors

Literature to Chapter 1

DEFINITIONS

International Union of Pure and Applied Chemistry, Macromolecular Division, Commission on
Macromolecular Nomenclature, Compendium of Macromolecular Nomenclature, Blackwell
Scientific, Oxford 1991; Glossary of Basic Terms in Polymer Science, Pure Appl. Chem.
68 (1996) 2287-2311

HANDBOOKS

Houben-Weyl, Methoden der organischen Chemie (E. Müller, Ed.), 4th ed., vol. XIV (2 parts),
    Thieme, Stuttgart 1961-63, 5th ed., vol. E 20, Makromolekulare Stoffe (3 parts), Thieme 1987
G.W.Becker, D.Braun, Eds., Kunststoff-Handbuch, Hanser, Munich, 2nd ed., 11 vols. in 17 parts
    (1983 ff.)
G.Allen, J.C.Bevington, Eds., Comprehensive Polymer Science, Pergamon, Oxford, 7 vols. (1989),
    First Supplement (1992), Second Supplement (1996)
J.E.Mark, Ed., Physical Properties of Polymers Handbook, AIP Press, Williston (VT), 1996
J.C.Salamone, Ed., Polymeric Materials Encyclopedia, CRC Press, Boca Raton (FL) 1996, 12 vols.;
    short version: Concise Polymeric Materials Encyclopedia, CRC Press, Boca Raton (FL) 1999
H.F.Mark, Encyclopedia of Polymer Science and Engineering, Wiley, Hoboken (NJ), 3rd ed. in
    3 parts of 4 volumes each (2003-2004)
E.S.Wilks, Ed., Industrial Polymers Handbook (relevant chapters of Ullmann's Encyclopedia of
    Industrial Chemistry, 5th ed.), Wiley-VCH, Weinheim 2001 (5 vols.)
J.L.Atwood, J.W.Steed, Eds., Encyclopedia of Supramolecular Chemistry, Dekker, New York 2004

DATA COLLECTIONS

O.Griffin Lewis, Physical Constants of Linear Homopolymers, Springer, Berlin 1968
W.J.Roff, J.R.Scott, Handbook of Common Polymers, Butterworths, London 1971
J.Brandrup, E.H.Immergut, E.A.Grulke, Eds., Polymer Handbook, Wiley, New York, 4th ed. 1999
–, CAMPUS®, diskettes with properties of plastics that are produced by the ca. 30 companies
    belonging to the CAMPUS system
–, PLASPEC, Fachinformationszentrum Chemie, Berlin (data bank, ca. 7000 plastics)
–, Plastics Databases, ASM International, Materials Park, Ohio, USA
B.Ellias, S.H.J.Thompson, Eds., Polymers - The Database of Fundamental Properties on CD-ROM,
    Blackie, Glasgow 1998
J.E.Mark, Ed., Polymer Data Handbook, Oxford University Press, New York 1999

DICTIONARIES

PARAT (Fachinformationszentrum Chemie, Berlin), Index of Polymer Trade Names, VCH,
    Weinheim 1987
M.Ash, I.Ash, Encyclopedia of Plastics, Polymers and Resins, Chem.Publ.Co., New York 1982-
    1988 (4 vols.)
A.M.Wittfoht, Plastics Technical Dictionary, Hanser, Munich 1992
M.B.Ash, I.A.Ash, Handbook of Plastics Compounds, Elastomers, and Resins, An International
    Guide by Category, Tradename, Composition and Supplier, VCH, Weinheim 1992
R.J.Heath, A.W.Birley, Dictionary of Plastics Technology, Blackie and Son, Glasgow 1992
D.V.Rosato, Rosato's Plastics Encyclopedia and Dictionary, Hanser, Munich, 2nd ed. 1993
T.Whelan, Polymer Technology Dictionary, Chapman and Hall, London 1994
M.S.M.Alger, Polymer Science Dictionary, Chapman & Hall, London, 2nd ed. 1997

BIBLIOGRAPHIES

J.Schrade, Kunststoffe (Hochpolymere), bibliography of German language books on plastics (high
    polymers), J.Schrade, Schweiz.Aluminium AG, Zurich, First issue 1911-1969 (1976), Second
    issue (1980)
O.A.Battista, The Polymer Index, McGraw Hill, New York 1976
E.R.Yescombe, Plastics and Rubbers, World Sources of Information, Appl. Sci. Publ., Barking
    (Essex), England, 2nd ed. 1976
S.M.Kaback, Literature of Polymers, Encycl. Polym. Sci. Technol. **8** (1968) 273
J.T.Lee, Literature of Polymers, Encycl. Polym. Sci. Eng., 2nd ed., **9** (1987) 62
R.T.Adkins, Ed., Information Sources in Polymers and Plastics, K.G.Saur, New York 1989

HISTORICAL DEVELOPMENT

R.Olby, The Macromolecular Concept and the Origin of Molecular Biology, J.Chem.Educ.
    4 (1970) 168
J.H.DuBois, Plastics History-U.S.A., Cahners Books, Boston (MA) 1972
J.K.Craver, R.W.Tess, Eds., Applied Polymer Science, Am.Chem.Soc., Washington (DC) 1975
C.Priesner, "H.Staudinger, H.Mark und K.H.Meyer - Thesen zur Grösse und Struktur der Makromole-
    küle", Verlag Chemie, Weinheim 1980

F.M.McMillan, The Chain Straighteners: Fruitful Innovation. The Discovery of Linear and Stereo-regular Polymers, MacMillan, London 1981

R.B.Seymour, Ed., History of Polymer Science and Technology, Dekker, New York 1982 (reprint of papers in J.Macromol.Sci.-Chem. **A 15** (1981) 1065-1460)

R.Friedel, Pioneer Plastic. The Making and Selling of Celluloid, University of Wisconsin Press, Madison (WI) 1983

H.Morawetz, Polymers: The Origins and Growth of a Science, Wiley-Interscience, New York 1985

R.B.Seymour, G.A.Stahl, Eds., Genesis of Polymer Science, Am.Chem.Soc., Washington (DC) 1985

R.B.Seymour, G.S.Kirshenbaum, Eds., High Performance Polymers: Their Origin and Development, Elsevier, New York 1986

E.Jostkleigrewe, Ed., Makromolekulare Chemie - Das Werk Hermann Staudingers in seiner heutigen Bedeutung, Schnell und Steiner, Munich 1987

D.A.Hounshell, J.K.Smith, Jr., The Nylon Drama, Amer.Heritage of Invention and Technology 4/2 (1988) 40

R.B.Seymour, Ed., Pioneers in Polymer Science, Kluwer, Boston (MA) 1989

J.Alper, G.L.Nelson, Eds., Polymeric Materials: Chemistry for the Future, Amer.Chem.Soc., Washington (DC) 1989

S.T.I.Mossman, P.J.T.Morris, Ed., The Development of Plastics, Royal Society of Chemistry, Cambridge 1994

J.L.Meikle, American Plastic: A Cultural History, Rutgers Univ. Press, New Brunswick (NJ) 1995

S.Fenichell, Plastic-The Making of a Synthetic Century, Harper Business, New York 1996

Y.Furukawa, Inventing Polymer Science. Staudinger, Carothers, and the Emergence of Macromolecular Chemistry, University of Pennsylvania Press, Philadelphia (PA) 1998

T.M.Brown, A.T.Dronsfield, P.J.T.Morris, In the Beginning ... was Ebonite, Educ.Chem. **39**/1 (2002) 18

BIOGRAPHIES and AUTOBIOGRAPHIES

H.Staudinger, Arbeitserinnerungen, Hüthig, Heidelberg 1961; -, From Organic Chemistry to Macromolecules: A Scientific Autobiography Based on My Original Papers, Wiley-Interscience, New York 1970

J.D.Watson, The Double Helix. A Personal Account of the Discovery of the Structure of DNA, Athenaeum, New York 1968

E.Trommsdorf: Dr. Otto Röhm, Chemiker und Unternehmer, Econ, Düsseldorf 1976

E.Guth, Birth and Rise of Polymer Science - Myth and Truth, J.Appl.Polym.Sci.-Appl.Polym. Symp. **35** (1979) 1

S.Carra, F.Parisi, I.Pasquon, P.Pino, Eds., Giulio Natta: Present Significance of His Scientific Contribution, Ed. Chim., Milan, Italy 1982

H.Kuhn, Leben und Werk von Werner Kuhn, 1899-1963, Chimia **38** (1984) 191

H.-F.Eicke, Peter J.W.Debye's Beiträge zur Makromolekularen Wissenschaft - ein Beispiel zukunftsweisender Forschung, Chimia **38** (1984) 347

H.F.Mark, From Small Organic Molecules to Large: A Century of Progress, American Chemical Society, Washington (DC) 1993 (mainly autobiographic)

M.Hermes, Enough for One Lifetime. Wallace Carothers, Inventor of Nylon, Chemical Heritage Foundation, Philadelphia 1996

-, Japanese History of Polymer Science and Technology, Society of Polymer Science, Japan, Tokyo 1998 (personal histories of 24 leading Japanese polymer scientists)

A.Gaines, Wallace Carothers and the Story of Dupont Nylon, Mitchell Lane Publ., Hockessin (DE), 2001

COLLECTED PAPERS

H.Mark, G.S.Whitby, Collected Papers of Wallace Hume Carothers, Interscience, New York 1940

G.Natta, F.Danusso, Stereoregular Polymers and Stereospecific Polymerization, Pergamon Press, New York, 2 vols. 1967

H.Staudinger, Das wissenschaftliche Werk, Hüthig und Wepf, Heidelberg, 7 vols. (in several parts) (1969-1976) (scientific papers of H.Staudinger)

L.Mandelkern, J.E.Mark, U.W.Suter, D.Y.Yoon, Eds., Selected Works of Paul J.Flory, Stanford University Press, Stanford (CA) 1985

References to Chapter 1

[1]   R.C.Weast, Ed., Handbook of Chemistry and Physics, The Chemical Rubber Co., Cleveland
         (OH), 46th ed. 1964
[2]   B.Wunderlich, Macromolecular Physics, vol. 3, Crystal Melting, Academic Press, New York
         1980, Table VIII.4
[3]   J.J.Berzelius, Jahres-Bericht über die Fortschritte der physischen Wissenschaften (Tübingen) **11**
         (1831) 44
[4]   J.J.Berzelius, Jahres-Bericht über die Fortschritte der physischen Wissenschaften (Tübingen) **12**
         (1832) 63
[5]   O.-A.Neumüller, Römpps Chemie-Lexikon, Franckh'sche Verlagshandlung, Stuttgart, 8th ed.
         1979-1988; [5a] p. 3806; [5b] p. 2642; [5c] p. 415
[6]   A.Thackray, "John Dalton", Critical Assessments of His Life and Science, Harvard University
         Press, Cambridge (MA) 1972
[7]   C(arl Wilhelm von) Naegeli, Die Micellartheorie, Akad.Verlagsgesellschaft, Leipzig 1928
         (Ostwald's Klassiker, Neue Ausgabe, Vol. 227). Naegeli lived 1817-1891.
[8]   E.Simon, Ann.Chem.Pharm. **31** (1839) 265
[9]   J.Blyth, A.W.Hofmann, Liebigs Ann.Chem. **53** (1845) 292
[10]   This clear liquid was apparently discovered by a chemist called Neuman (cited in [11]); it was
         described by other authors [12, 13] before Simon's discovery [8].
[11]   W.Nicholson, A Dictionary of Practical and Theoretical Chemistry (1786), 2nd ed. 1808;
         see A.J.Warner, in R.H.Boundy, R.F.Boyer, Eds., Styrene - Its Polymers, Copolymers and
         Derivatives, Reinhold Publ. Corporation, New York 1952, p. 1 ff.
[12]   M.Bonastre, J.pharm.chim. **17** (1831) 338
[13]   F.d'Arcet, Ann.Chim.Phys. **66** (1837) 110
[14]   P.E.M.Berthelot, Bull.Soc.Chim.France [2] **6** (1866) 289. The masthead of this paper refers to
         the author as M.Berthelot, not P.E.M.Berthelot. The "M" is, however, not the abbreviation
         of the Berthelot's first name but stands for *Monsieur* as it was custom in France (and some-
         times still is). The true Christian names of Berthelot are Pierre Eugéne Marcelin. Berthelot
         (1827-1907) was the author of ca. 1800 scientific papers and ca. 20 books [5b].
[15]   H.Staudinger, Ber.Dtsch.Chem.Ges. **53** (1920) 107
[16]   A.Wurtz, C.R.Hebd.Séances Acad.Sci. **49** (1859) 813, **50** (1860) 1195
[17]   A.-V.Lourenço, C.R.Hebd.Séances Acad.Sci. **49** (1859) 619, **51** (1860) 365;
         Ann.Chim.Phys. [3] **67** (1863) 273
[18]   A.Kraut, Ann.Chem.Pharm. **150** (1869) 1
[19]   A.Kiepl, J.Prakt.Chem. **28** (1983) 193
[20]   E.E.Blaise, L.Mareilly, Bull.Soc.Chim.France **3** (1904) 33, 308
[21]   Th.Graham, Phil.Trans.Royal Soc. [London] **151** (1861) 183, J.Chem.Soc. [London]
         1864, 318
[22]   A.Müller, Z.Anorg.Chem. **36** (1903) 340
[23]   Wo.Ostwald, Kolloid-Z. **1** (1907) 291, 331
[24]   P.P.von Weimarn, Kolloid-Z. **2** (1907/1908) 76
[25]   F.-M.Raoult, C.R.Hebd.Séances Acad.Sci. **95** (1882) 1030, Ann.chim.phys. [6] **2** (1884) 66,
         C.R.Hebd.Séances Acad.Sci. **101** (1885) 1056
[26]   J.H.van't Hoff, Z.Phys.Chem. **1** (1887) 481, Phil.Mag. [5] **26** (1988) 81
[27]   J.H.Gladstone, W.Hibbert, J.Chem.Soc. [London] **53** (1888) 679, Phil.Mag. [5] **2** (1889) 38
[28]   H.T.Brown, G.H.Morris, J.Chem.Soc. [London] **55** (1889) 462
[29]   A.Nastukoff, Ber.Dtsch.Chem.Ges. **33** (1900) 2237
[30]   W.A.Caspari, J.Chem.Soc. [London] **105** (1914) 2139
[31]   J.D. van der Waals, Die Kontinuität des gasförmigen und flüssigen Zustands, Diss. U. Leiden
         1873; dito, J.A.Barth, Leipzig, 2nd. ed. 1895 (vol. I) and 1900 (vol. II)
[32]   J.Thiele, Justus Liebigs Ann.Chem. **306** (1899) 87
[33]   P.Pfeiffer, Liebigs Ann.Chem. **404** (1914) 1, **412** (1917) 253
[34]   M.Faraday, Quart.J.Sci. **21** (1826) 19
[35]   C.D.Harries, Ber.Dtsch.Chem.Ges. **37** (1904) 2708, **38** (1909) 1195, 3985
[36]   S.S.Pickles, J.Chem.Soc. [London] **97** (1910) 1085
[37]   H.Staudinger, Die Ketene, F.Enke, Stuttgart 1912, p.46
[38]   G.Schroeter, Ber.Dtsch.Chem.Ges. **49** (1916) 2697
[39]   H.Staudinger, Ber.Dtsch.Chem.Ges. **57** (1924) 1203

[40]   C(arl Wilhelm von) Naegeli, Die Micellartheorie, Akad.Verlagsgesellschaft, Leipzig 1928
        (Ostwalds Klassiker, Neue Ausgabe, vol. 227). Naegeli lived 1817-1891.
[41]   H.Staudinger, H.Fritschi, Helv.Chim.Acta **5** (1922) 785
[42]   H.Staudinger, Ber.Dtsch.Chem.Ges. **59** (1926) 3019
[43]   H.Staudinger, K.Frey, W.Starck, Ber.Dtsch.Chem.Ges. **60** (1927) 1782
[44]   R.Pummerer, H.Nielsen, W.Gündel, Ber.Dtsch.Chem.Ges. **60** (1927) 2167
[45]   K.H.Meyer, H.Mark, Ber.Dtsch.Chem.Ges. **61** (1928) 593, 1939
[46]   H.Staudinger, Ber.Dtsch.Chem.Ges. **61** (1928) 2427
[47]   H.Staudinger, E.Urech, Helv.Chim.Acta **12** (1929) 1107
[48]   W.Chalmers, J.Am.Chem.Soc. **56** (1934) 912
[49]   H.Staudinger, W.Frost, Ber.Dtsch.Chem.Ges. **68** (1935) 2351
[50]   H.W.Melville, Proc.Royal Soc. [London] **A 163** (1937) 511
[51]   G.V.Schulz, E.Husemann, Z.Phys.Chem. **B 39** (1938) 246
[52]   C.C.Price, R.W.Kell, E.Kred, J.Am.Chem.Soc. **63** (1941) 2708, **64** (1942) 1103
[53]   W.Kern, H.Kämmerer, J.Prakt.Chem. **161** (1942) 81, 289
[54]   P.D.Bartlett, S.C.Cohen, J.Am.Chem.Soc. **65** (1943) 543
[55]   W.H.Carothers, Chem.Rev. **8** (1931) 353
[56]   J.B.Sumner, J.Biol.Chem. **69** (1926) 435
[57]   J.H.Northrop, J.Gen.Physiol. **13** (1930) 739
[58]   T.Svedberg, R.Fahraeus, J.Am.Chem.Soc. **48** (1926) 430 (hemoglobin);
        T.Svedberg, J.B.Nichols, J.Am.Chem.Soc. **48** (1926) 3081 (egg albumin)
[59]   A.Tiselius, Kolloid-Z. **85** (1938) 129
[60]   W.Kuhn, Kolloid-Z. **68** (1934) 2
[61]   W.Kuhn, Ber.Dtsch.Chem.Ges. **63** (1930) 1503

2 Constitution

2.1 Atoms and Molecules

2.1.1 Definition of a Molecule

Macromolecules are large molecules (Section 1.2). But what are molecules? Are diamonds macromolecules? What then is a supramolecule? A physical molecule? The answers obviously depend on the *definition* of the term "molecule".

The word "molecule" (L: *molecula* = small mass, diminutive of *moles* = mass) was introduced into the chemical literature by Daniel Sennert (1572-1637) [1a]. The term comprised originally both our present "molecule" and our present "atom". In 1805, John Dalton (1766-1844) restricted the meaning of "molecule" to the smallest parts of "compounds" [2], but "compound" was not used in the modern sense as "chemical compound". For example, André Marie Ampère in 1814 called our present-day "atom" a "molecule" and our present-day "molecule" a "particle". Carl von Naegeli in 1858 named the smallest morphological units of starch "molecules". The modern meaning of "molecule" became established at the Karlsruhe conference in 1860 which prompted Naegeli to change the name of his units to "micelles" from "molecules" [3].

The modern meaning of "molecule" can be traced to the 1805 proposal by Amadeo Avogadro, Conte di Quaregna (1776-1856). **Molecules** are now defined as particles composed of two or more atoms that are held together by chemical bonds; these particles are the smallest entities that exhibit the characteristic chemical properties of the corresponding stoichiometrically composed chemical substance [1b]. A **chemical substance** (L: *substantia* = substance, essence), as defined by chemistry, is matter that consists of either the many atoms of a **chemical element** or the many molecules of a **chemical compound** of two or more atoms. The physical quantity of a substance is the **amount of substance**, measured in the physical unit "mol", which is independent of the physical state, shape, etc., of the chemical compound.

Chemical substances are a subclass of **matter** (L: *materia* = lumber) which is defined as a space-occupying object that has a mass; it differs from antimatter, vacuum, and energy. A **material** is matter consisting of one or more than one substance. The properties of a material depend on the properties of the substance(s) *and* on their physical states.

Chemical bonds are subdivided into primary bonds with bond energies of ca. 50-1000 kJ/(mol bond) and secondary bonds with bond energies of less than ca. 50 kJ/(mol bond) [4]. **Primary bonds** are further subdivided into three limiting types: ionic bonds, covalent bonds, and metallic bonds. Coordinative bonds can be defined as intermediates between ionic and covalent bonds and electron-deficient bonds as intermediates between covalent and metallic bonds.

Secondary bonds comprise hydrogen bonds, B, dispersion forces. The chemical literature describes **van der Waals bonds** either as those caused by dispersion forces (vdW bonds in the narrower sense) or as bonds that are not primary bonds, i.e., secondary bonds (vdW bonds in the broader sense). Hydrogen bonds are sometimes considered weak primary bonds. **Hydrophobic bonds** are not chemical bonds but entropy effects ("**entropy bonds**").

Macromolecular science usually follows Staudinger (Section 1.3) who restricted the term "macromolecule" to those large molecules where the atoms of the actual or hypothetical main chains are held together by covalent bonds [4] and not by any other type of bonds. Linus Pauling, on the other hand, defined "chemical bonds as bonds between two atoms or groups of atoms in the case where the forces acting between them are such as to lead to the formation of an aggregate with sufficient stability to make it convenient for the chemist to consider it as an independent molecular species" [5]. Pauling points out that "chemical bond" is a general term that comprises electrostatic, covalent, and metallic bonds as well as hydrogen bonds but not dipole-dipole bonds and van der Waals bonds (in the narrow sense). According to Pauling, "valence bonds" are those bonds that correspond to the classic ideas about the valence of atoms. They comprise, for example, the C–H bonds in methane CH_4 and the bonds between H and Cl in an HCl molecule in the gaseous state but not the bonds between $Na^\oplus$ and $Cl^\ominus$ in NaCl crystals. Ionic bonds in solid substances as well as metallic bonds are therefore not secondary *valences*.

Pauling's definition is far reaching. It considers as molecules a number of gaseous and solid inorganic compounds that organic chemists would not regard as such. The bonds Be–O in gaseous $(BeO)_3$ are covalent; $(BeO)_3$ is therefore a molecule. Sodium chloride exists in the gaseous states not only in monomeric and dimeric species but also in trimeric and tetrameric ones that are held together by ionic bonds. If the type of chemical bonding does not matter for molecules in the *gaseous* state, then the *solid* $(NaCl)_n$ must be considered an inorganic polymer [6] although it has no valence bonds.

Similar reasoning must be applied to metallic bonds. Elemental boron is usually considered to be an inorganic polymer since the boron atoms are held together by electron-deficient bonds in some of its allotropes. Similar delocalized bonds are also present in some metalloids that therefore should also be classified as inorganic polymers.

Closely connected to the problem of bonding is the size of molecules. Already in 1938, diamond and some other one-aggregate compounds were called "giant molecules" [7]. Staudinger [8] opposed this view since these "molecules" are (a) too big, (b) impossible to break down into smaller "molecules," and (c) completely undefined with respect to size. It is easy to see that none of his counter arguments are valid since there is no size restriction for macromolecules, and diamonds can easily split into defined smaller ones.

2.1.2 Chemical and Physical Molecules

The definitions of a "molecule" presented above are not very useful for the topics of this series of books. They all consider a molecule as a corporeal entity with a time-averaged stable arrangement of atoms for a certain state of aggregation. A molecule should however preserve its identity in *various* states, for example, not only in the gaseous state or the crystalline state but also in the liquid state or in solution. Such behavior is obviously possible only if *directed* bonds are present.

Molecules are therefore defined as stable entities composed of two or more atoms that are held together by directed bonds in which the binding electrons are shared by the bound atoms. Such a definition limits the type of bonds between two atoms to the covalent bond and its intermediates to the ionic and metallic bonds, i.e., the coordinative bonds and the electron-deficient bonds but not the ionic and metallic bonds themselves. Entities held together by metallic bonds are not considered macromolecules since the bonds are not directed although the binding electrons are shared by all atoms. The same reasoning applies to ionic crystals since the bonds are not directed *and* the binding electrons are not shared by the bonded atoms.

The existence of a *molecule* is, however, not only determined by the stability of the bond between two individual atoms but also by the number and strength of the bonds that cause the stability of the molecule itself. One therefore has to distinguish between chemical and physical molecules.

Chemical molecules conserve their identities in various solvents (if chemical reactions are absent) and in the melt (if the thermal energy is insufficient for a degradation).

Physical molecules are "stable complexes" (associates, aggregates, micelles, etc.) of chemical molecules that behave like chemical molecules in the solid state, the melt, and in concentrated solutions but disband into constituent chemical molecules at infinite dilution. Poly(ethylene glycol) $HO[CH_2CH_2O]_{n-1}CH_2CH_2OH$ is an example. Its molecules are present as single chemical molecules in dilute aqeous solutions. However, they associate via their OH endgroups to larger physical molecules in benzene solutions where the degree of association increases with increasing polymer concentration:

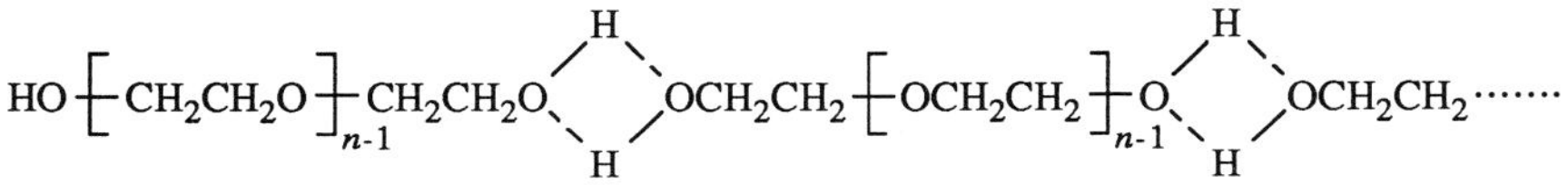

Melts of these chemical compounds contain hydrogen bonds between endgroups (as shown above) as well as hydrogen bonds between hydroxyl endgroups $-OH$ and in-chain ether groups $-CH_2-O-CH_2-$ (not shown).

Physical molecules with considerably higher stability are obtained if the ends of molecules are connected by several hydrogen bonds. Such **supramolecular polymers** (= *linear physical polymers*) are relatively stiff and therefore exhibit liquid-crystalline behavior in melts and concentrated solutions (see volume III). An example is

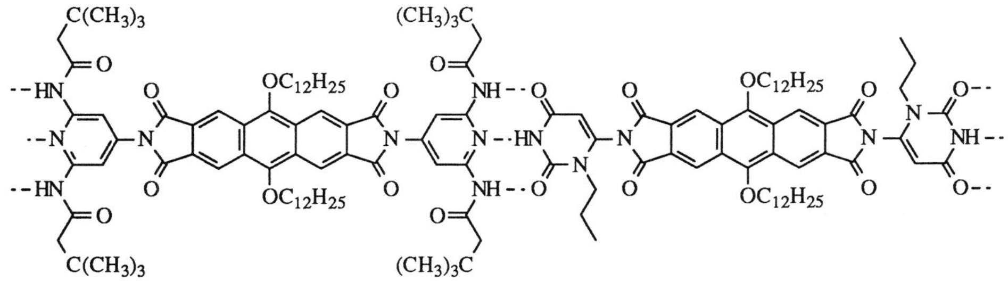

Nonlinear supramolecular polymers have been known in the biopolymer kingdom for many years although they were and are not called that way. Examples are the quarternary structures of certain corpuscular enzymes (Section 14.3), the triple helices of collagen (Section 14.3), and the double helices of deoxyribonucleic acids (Section 14.2).

Supramolecular polymers are distinct from superpolymers and hyperpolymers. **Superpolymers** are chemical polymer molecules with exceptionally high molecular weights. **Hyperpolymer** is a rarely used term for polymer molecules that have been coupled to form greater entities. Examples are hemoglobin molecules that have been connected to each other by bifunctional chemical bonds.

2.2   Nomenclature of Chemical Structures

2.2.1   Overview

Macromolecules are the smallest chemical units of a polymeric substance. Their chemical structure is determined by their constitution (Chapter 2), configuration (Chapter 4), and conformation (Chapter 5). The molecular weight is not an independent structural feature since it is dictated by the constitution of the macromolecule. It is discussed in a separate chapter (Chapter 3) and, in greater detail, in volume III because one is normally concerned with the molecular weight of polymeric *substances* and not with that of polymeric molecules.

The **constitution** of a molecule is determined by the type, number, and sequence of its atoms without regard to their spatial arrangements. "Constitution" is not synonymous with "chemical structure" but rather a subclass since chemical structure comprises constitution, configuration, and conformation.

The constitution of macromolecules must be described by many more parameters than that of low-molecular weight compounds: type and sequence of chain atoms, monomeric units, and constitutional repeating units; type and number of endgroups; type, number, arrangement, size, and distribution of chain bonds; and, for substances, the molecular weight and molecular-weight distribution.

2.2.2   Types of Polymer Names

Macromolecules and macromolecular substances are described by different types of names: structure-based names (= systematic names), source-based names (= generic names = poly(monomer) names), group names, trivial names, and trade names.

Trade names are names by which commodities are known to the trade. They are artificial names of *materials* and not necessarily those of *substances*. For example, Vestamides® are plastics based on several polyamides that may or may not contain inorganic fillers. The ® indicates that the trade name is officially registered with the appropriate government office (in the United States: United States Patent and Trademark Office); this **trademark** is legally restricted to the use of the owner or licensee. Trade names carry the symbol ™ if legal rights are claimed without or before registration. Several trade names have become free names because the owner of the trade mark either gave up or let lapse the legal rights to that name worldwide or in certain countries. An example is "nylon," formerly a trademark and now a free trade name for aliphatic polyamides.

The word "nylon" is a **trivial name**. Such names do not necessarily indicate molecules or substances but may also refer to materials. "Catalase" is, for example, the trivial name for a certain catalyzing protein with a defined chemical structure. "Cellulose," on the other hand, is not necessarily the name of a certain, well-defined glucose polymer but often refers to a *group* of substances that contain various proportions of other sugar units besides glucose units (Section 14.4). Trivial names often indicate the origin (e.g., cellulose), the occurrence or property (e.g., nucleic acid), or the function (e.g., catalase) of the molecule, substance, or material. Other trivial names are artificial; they may, for example, be whimsical (for example, Questra®, a PE of Dow Chemical) or indicate the manufacturer (for example, Lupolen® = a *pol*y(eth*ylen*e) from BASF, *Lu*dwigshafen).

Synthetic polymers usually carry **source-based names** that indicate the monomer(s) from which the polymers were generated. The monomer name is put in parentheses (but see below) and prefixed with "poly". Polymers of ethene $CH_2=CH_2$ (formerly: ethylene) are thus called "poly(ethylene)s". Note that such names, when used industrially, do *not* implicate ideal chemical structures of polymer *molecules* such as $+CH_2-CH_2+_n$ but rather *groups* of *substances* that have the same *major* monomeric unit $-CH_2-CH_2-$.

The spelling of poly(monomer) names is not systematic in common English. Simple polymer names are usually written as one word (examples: polyethylene, polystyrene). more complex names as two words (examples: polyvinyl chloride, polyethylene terephthalate), and highly complex names with the name of the monomer(s) in parentheses (example: poly(2,6-dimethyl-1,4-phenylene ether) (PPE); this polymer is commonly known as polyphenylene oxide (PPO) but PPO® is a registered trademark). Many exceptions exist to this "rule": polyethyleneimine, for example, is written as one word.

This book will always put monomer names in source-based polymer names in parentheses, in part, because it is systematic, and, in part, to avoid confusion with the "poly" compounds of low-molecular weight organic chemistry. The prefix "poly" describes in organic chemistry chemical compounds with *two or more identical substituents* whereas in macromolecular chemistry it denotes chemical compounds with many *polymerized monomeric units*. Examples:

| an isocyanate | a polyisocyanate (organic chemistry) | a poly(isocyanate) (macromolecular chemistry) |

Historically, source-based names were short forms of names; they did not indicate *many* monomeric *units*. "Polymeric" was originally the label for all substances that exhibited the same relative composition (see Section 1.3). Monomeric vinyl chloride (VCM) and poly(vinyl chloride) (PVC) were considered "polymeric" to each other. Accordingly, PVC was first called "polymeric vinyl chloride" and later "polyvinyl chloride" (original German: polymerisches Vinylchlorid → polymeres Vinylchlorid → Poly(vinylchlorid); "polymerisches" and "polymeres" are both translated as "polymeric").

The monomer names in source-based polymer names do *not* indicate the constitution of the monomeric unit. For example, the monomeric unit of poly(vinyl chloride) does not contain the vinyl group $CH_2=CH-$ but rather a chlorine-substituted alkane group, $-CH_2-CHCl-$. While "chemical experience" often lets one guess the constitution of the monomeric units from the name of the monomer(s), in some cases this is difficult and often impossible. An example is acrolein that can polymerize not only via the vinyl group and via the aldehyde group but may also form six-membered rings if the vinyl and aldehyde group of the same monomer molecule react successively:

| acrolein | various monomeric units of poly(acrolein) |

Structure-based names are **systematic names** that ought to be unambiguous. Unfortunately, they are not. Not only have there been different approaches to naming by the International Union of Pure and Applied Chemistry (IUPAC; a non-profit organization) and the Chemical Abstracts Service (CAS; a for-profit organization owned by the non-profit American Chemical Society) but these two organizations have considerably changed their own "rules" and "recommendations" over time (see below).

Systematic names of linear macromolecules consist of the name(s) of the repeating unit(s) (in parentheses), a prefix, and the names of the two endgroups. Different rules exist for inorganic and organic macromolecules.

Organic macromolecules are named as follows. *Constitutional repeating units* (p. 5), e.g., $-CH_2-CHCl-$, are regarded as biradicals that are named according to the substitution principles of the nomenclature rules of organic chemistry. The names of these units are subsequently added following the additivity rules of inorganic chemistry.

The sequence of chain units in repeating units follows certain IUPAC and/or CAS seniority rules. The implied chain direction is thus mostly not identical with the direction of chain growth. An example is the IUPAC structural formula and the IUPAC systematic name for poly(hexamethylene adipamide) molecules, $-[NH(CH_2)_6NH-OC(CH_2)_4CO]_{\overline{n}}$, (as substance: PA 66, PA 6.6, nylon 6.6) according to the 1975 rules (for the 1996 rules, see p. 29):

$-[NH(CH_2)_6NH-OC(CH_2)_4CO]_{\overline{n}}$ poly(imino(1,6-dioxohexamethylene)iminohexamethylene)
                                                   poly(iminoadipoyliminohexamethylene)

Inorganic macomolecules with predominantly covalently bonded chain atoms may be named according to either organic or inorganic nomenclature rules. An example is the name of the poly(diphenylsiloxane) molecule, $-[Si(C_6H_5)_2-O]_{\overline{n}}$:

$-[OSi(C_6H_5)_2]_{\overline{n}}$ poly[oxy(diphenylysilylene)]   :   organic nomenclature (1975)

$-[Si(C_6H_5)_2O]_{\overline{n}}$ *catena*-poly[(diphenylsilicon)-μ-oxo]   :   inorganic nomenclature

The *prefix* consists of the Greek word (in Latin spelling) for the number of the **constitutional repeating units** (**CRUs**) per molecule. An unknown large number is represented by "poly". In structural formulas, an unknown *number* of CRUs *per molecule* is usually symbolized in the literature by the subscript n, the IUPAC symbol for the amount of *substance* instead of N, the IUPAC symbol for numbers.

Endgroups are usually not specified in polymer names because they are generally not known. Names of known endgroups, however, are written to the left of the name of the poly(constitutional repeating unit). The name of the endgroup to the left of the CRU is prefixed by α, the name of the endgroup to the right of the CRU by ω. Examples:

$CCl_3(CH_2)_NCl$ α-(trichloromethyl)-ω-chloro-poly(methylene)

$Cl(S)_8Cl$ α-chloro-ω-chloro-*catena*-octa(sulfur)

$CH_3CO[OCH_2CH_2OCO(p\text{-}C_6H_4)CO]_NOCH_3$ α-acetyl-ω-methoxy-poly(oxyethyleneoxyterephthaloyl);
                                                   poly(monomer) name: poly(ethylene terephthalate)

However, the polymer literature does not only contain these 1975 IUPAC names but also other names, both older and newer ones. An older nomenclature constructed polymer names from the prefix poly, the name of the monomer, and the suffix amer. These names are still used by industry for those polymers that were developed during the years in which this nomenclature was promoted. An example is the polymer from the ring-opening polymerization of cyclopentene:

$-\!\!\left[\text{CH=CHCH}_2\text{CH}_2\text{CH}_2\right]_{\overline{n}}$ poly(pentenamer)   old IUPAC nomenclature
                                  poly(1-pentenylene)   1975 IUPAC nomenclature
                                  poly(1-pentene-1,5-diyl)   1996 IUPAC + CAS nomenclature

The newer nomenclature rules of CAS (1993) were also adopted by IUPAC. These rules consider all CRUs as biradicals, characterized by the syllable -diyl.

Thus, the literature contains the following names for polyamide 6.6 (PA 6.6):

poly[imino(1,4-dioxo-1,4-butanediyl)imino-1,6-hexanediyl]   (CAS, outdated)
poly[imino(1,6-dioxohexamethylene)iminohexamethylene]   (IUPAC, old (1975))
poly(iminoadipoyliminohexamethylene)   (IUPAC, old, alternative (1975))
poly(iminoadipoyliminohexane-1,6-diyl]   (CAS + IUPAC, new (1996))
poly[imino(1,6-dioxohexamethylene)iminohexane-1,6-diyl]   (CAS + IUPAC, alternative (1996))

The bewildering number of names for the same chemical compound caused CAS to introduce so-called **CAS Registry Numbers**. These numbers are apparently arbitrary; an example is 32131-17-2 for the polyamide 6.6. Unfortunately, this system also seems not to be unambiguous since this author experienced no less than ten CAS registry numbers for poly(propylene oxide).

This book uses mainly modern source-based names. In some cases, older source-based names have been retained, for example, poly(ethylene) for the name of the polymer from the polymerization of ethene instead of poly(ethene). It happens that poly(ethylene) is the modern structure-based name since the biradical $-\text{CH}_2-\text{CH}_2-$ is called "ethylene" and not "ethene-1,2-diyl."

All source-based and CAS structure-based names refer to *idealized structures* of molecules. Like syntheses of low-molecular weight substances, polymer syntheses are also accompanied by side reactions. In low-molecular chemistry, such side reactions lead to byproducts that can by removed during the work-up. In macromolecular chemistry, side reactions often do not generate *removable* byproducts but "wrong" structural units that become part of the polymer chain. For example, poly(ethylene) from the free-radical polymerization of ethene contains not only the constitutional repeating unit $-\text{CH}_2-\text{CH}_2-$ but also constitutional units $-\text{CH}_2-\text{CH}\{(\text{CH}_2\text{CH}_2)_i\text{H}\}-$ with $i = 1, 3, 4, 5$, and $i \gg 5$ (see Volume II).

Industry goes a step further. Here, a polymer containing comonomeric units usually carries only the name of the majority component. An example is the so-called linear low-density poly(ethylene) from 92 % ethene and 4, 6, or 8 % 1-octene or 1-butene.

Plastics often carry the names of polymers although they are not substances but rather materials that contain adjuvants such as fillers, plasticizers, antioxidants, and the like within the polymeric base material. These adjuvants vary in their proportions between a few fractions of a percent (certain plastic films) and 90 % (barium ferrite-filled ethene-vinyl acetate copolymers for magnetic adhesive tapes). Industry also sometimes uses the same trade name for very different polymers that are then distinguished by letters, numbers, or combinations thereof. An example is Bexloy® (DuPont) that may be an amorphous polyamide (Bexloy C), a high-impact polyester (Bexloy J), an ionomer (Bexloy W), or a thermoplastic elastomer (Bexloy V).

2.2.3   Systematic Names of Inorganic Macromolecules

The name of the constitutional repeating unit (CRU) of an inorganic macromolecule consists of the name of the central atom and the names of sidegroups and bridge groups. The central atom is the atom with the highest seniority (Fig. 2-1); radon has the highest seniority and fluorine the lowest. The seniorities of inorganic chemistry thus do not agree with those of organic chemistry. For example, the butyl lithium C_4H_9Li of organic chemistry is called lithium butyl LiC_4H_9 in inorganic chemistry.

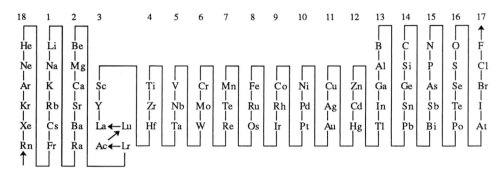

Fig. 2-1 Seniorities of central atoms in inorganic compounds. Numbers indicate the numbers of groups in the periodic system of elements.

The atoms and groups connected to central atoms are called **ligands**. They are always arranged in alphabetical order, independent of their number and their type (sidegroup or bridge group). Names of bridge groups are prefixed by μ. They are also separated by a dash from the names of sidegroups.

If the same group is present as a sidegroup and a bridge group, it is listed first as a bridge group. A bridge group with more than one bond to the central atom is also a chelating group. The italicized symbols of coordinating groups are arranged after the name of the bridge group.

In names of inorganic macromolecules, an italicized term for the dimensionality of the molecule is inserted between the names of the endgroups and the expression for the number of constitutional repeating groups. These terms are

> *cyclo* for ring-shaped molecules;
> *catena* for single-strand molecules;
>   (*ino* in silicate chemistry; G: *ino* = genitive of *is* = fiber);
> *phyllo* for two-dimensional molecules (G: *phyllon* = leaf);
> *tecto* for three-dimensional molecules (G: *tecton* = carpenter, builder).

For example, the ring-opening polymerization of *cyclo*-octa(sulfur) (S_8) leads to *catena*-poly(sulfur) = $\overline{+S_8+}_n$ (numbers 1 and 2 in Table 2-1).

Inorganic and semi-inorganic polymers with predominantly covalent bonds may be named according to the rules of organic chemistry (numbers 4 and 5 in Table 2-1). The seniorities for central atoms differ, however (compare Figure 2-1 with Section 2.2.4). Note also that nomenclature rules change from time to time.

Table 2-1 Constitutional repeating units (CRUs), trivial names, and structure-based names of inorganic macromolecules. * Structure-based name according to organic-chemical nomenclature.

No.	CRU	Trivial name	Structure-based name
1	–S– in S_8	sulfur	*cyclo*-octa[sulfur]
2	–S– in–S_N–	polymeric sulfur	*catena*-poly[sulfur]
3	–SiF$_2$–	siliconedifluoride	*catena*-poly[difluorosilicone]
4	–Si(C$_6$H$_5$)$_2$–O– –O–Si(C$_6$H$_5$)$_2$–	poly(diphenylsiloxane)	*catena*-poly[(diphenylsilicone)-μ-oxo] * poly[oxy(diphenylsilanediyl)]
5	⋯PCl$_2$⋯N⋯ –N=PCl$_2$–	phosphonitrilic chloride; poly(dichlorophosphazene)	*catena*-poly[(dichlorophosphorus)-μ-nitrido]; * poly[nitrilo(dichlorophosphoranylidyne)]
6	–Ag–NC–	silver cyanide	*catena*-poly[silver-μ-(cyano-*N:C*)]
7	NH$_3$ | –Zn–Cl–		*catena*-poly[(amminechlorozinc)–μ-chloro]
8	–C$_6$H$_5$ | –O–Si– | O | –O–Si– | C$_6$H$_5$	poly(phenylsesquisiloxane)	* poly(2,4-diphenyl-1,3,5-trioxa-2,4-disila-penta-1,5:4,2-tetrayl)
9	OC$_2$H$_5$ | ⋯P⋯N⋯ | OC$_2$H$_5$	poly(diethoxyphosphazene)	*catena*-poly[(diethoxyphosphorus)-μ-nitrido] * poly[nitrilo(diethoxyphosphoranylidyne)]

2.2.4 Systematic Names of Organic Macromolecules

Names of organic macromolecules consist of the name of the constitutional repeating unit (CRU) and the prefix "poly". The CRU of a regular single chain is a bifunctional unit that is sometimes composed of two or more subunits. The name of the subunit is the name of the corresponding low-molecular weight biradical. For example, the repeating unit –NH(CH$_2$)$_6$NH–CO(CH$_2$)$_4$CO– of polyamide 6.6 contains two imino groups –NH–, one hexamethylene group –(CH$_2$)$_6$–, and one 1,6-dioxohexamethylene group –CO(CH$_2$)$_4$CO– (but not two carbonyl groups –CO– and one tetramethylene group –(CH$_2$)$_4$–, see below).

The names of the subunits are added to give the name of the CRU (additivity method); no dashes between the names are allowed. The names of biradicals of subunits or CRUs of heterochains with two or more heteroatoms per unit may also be formed by the replacement method. The sequence of biradicals within a CRU is determined by seniority rules that apply regardless of the type of substituents. The chain directions implied by the structural names thus do not conform to the direction of chain growth.

The following seniority rule applies:

$$\text{heterocycles} > \text{heterochains} > \text{carbocycles} > \text{carbon chains}$$

Carbon Chains

Names of subunits with four or more saturated alicyclic carbon chain segments consist of either the corresponding systematic name or the accepted trivial name of the subunit, followed by the name of the position of the radical ends and the suffix diyl (if biradical), triyl (if triradical), tetrayl (if tetraradical), etc. Table 2-2 shows the systematic names of some common aliphatic CRUs and Table 2-3 the structures, structural names, and trivial names of some carbon polymers.

Note that some old trivial names have become new systematic names. Other old trivial names have also been maintained, for example, "oxalyl" for $-COCO-$, "malonyl" for $-COCH_2CO-$, "succinyl" for $-COCH_2CH_2CO-$, "adipoyl" for $-CO(CH_2)_4CO-$, and "terephthaloyl" for $-CO(1,4-C_6H_4)CO-$.

Table 2-2 Old trivial names and new structural names of some common aliphatic CRUs.

Structure	Old trivial name	New systematic name
$-CH_2-$	methylene	methylene
$-CH_2-CH_2-$	ethylene	ethylene
$-CH(CH_3)-CH_2-$	propylene	propylene (not: 1-methylethylene)
$-CH_2-CH_2-CH_2-$	trimethylene	propane-1,3-diyl
$-CH=CH-$	vinylene	ethene-1,2-diyl
$=CH-CH=$		ethanediylidene
$CH_3CH<$	ethylidene	ethane-1,1-diyl

The old source-based name poly(ethylene) is replaced by the new source-based name poly(ethene) whereas poly(ethylene) is now the structure-based name of the polymer of ethene, i.e., $+CH_2-CH_2+_n$. The same applies to the polymer of propene: poly(propene) is source-based and poly(propylene) is structure-based.

Substituents have the same names as in organic chemistry. They are characterized by position numbers and precede the name of the aliphatic subunit. Examples of substituents are "methyl" ($-CH_3$), "hydroxy" ($-OH$), "acetoxy" ($-OCOCH_3$), "methoxycarbonyl" ($-COOCH_3$), "sodium carboxylato" ($-COONa$), "chloro" ($-Cl$), "cyano" ($-CN$), "amino" ($-NH_2$), and "imino" ($-NH-$ or $=NH$).

The direction of the alicyclic subunits is determined by the seniority rule (i) > (ii) > (iii) if the CRU possesses two or more acyclic subunits that are separated by carbon rings, hetero atoms, and/or heterocyclics:

(i)   acyclic subunits with the greatest number of substituents;

(ii) subunits whose substituents have the lowest position number;

(iii) substituents with names that appear earlier in the alphabet.

Table 2-3 Some structure-based names and trivial names.

Source-based structure (1-4) Systematic structure (5-6)	Structure-based name	Trivial name
1 $\quad$ +CH$_2$+$_n$	poly(methylene)	poly(methylene)
2 $\quad$ +CH$_2$CH$_2$+$_n$	poly(methylene)	poly(ethylene)
3 $\quad$ +CH$_2$CH$_2$CH$_2$+$_n$	poly(propane-1,3-diyl)	poly(trimethylene) *not* poly(propylene)
4	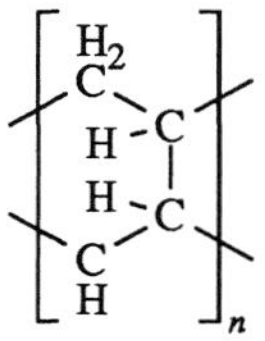 poly(butane-1,4:3,2-tetrayl)	cyclized 1,2-poly(butadiene)
5 $\quad$ +CH=CHCH$_2$CH$_2$+$_n$	poly(1-butene-1,4-diyl)	1,4-poly(butadiene)
6	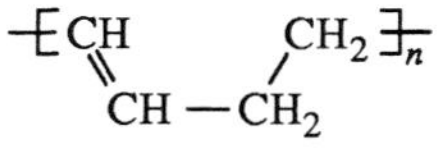 poly(*trans*-1-butene-1,4-diyl)	*trans*-1,4-poly(butadiene)

Carbocyclic Rings

CRUs may also contain single carbon rings or ring systems consisting of several rings that are joined at two or more ring positions. Ring systems also include those where two rings are connected by only one atom (spiro compounds) or by only one bond (e.g., biphenyl) provided that trivial names are preserved in the latter case. These ring systems and rings are cited in the following order of decreasing seniority:

ring (system) with the highest complexity > another of the same carbocycle
> ring (system) of lower seniority > acyclic group appearing earliest in the alphabet.

Complexities decrease in descending order I > II > III > IV > V. Examples:

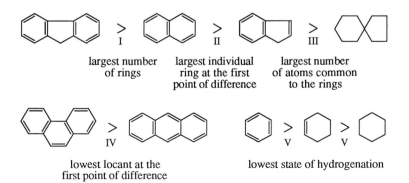

largest number $\quad$ largest individual $\quad$ largest number
of rings $\qquad$ ring at the first $\qquad$ of atoms common
$\qquad$ point of difference $\quad$ to the rings

lowest locant at the $\qquad$ lowest state of hydrogenation
first point of difference

CRUs with ring systems carry the trivial names of the system. Heterocyclic biradicals have the suffix "diyl" attached.

An example is (IUPAC 1975)

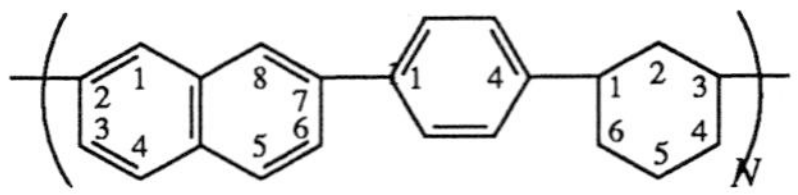

poly(2,7-naphthylene-1,4-phenylene-1,3-cyclohexylene)

Heterochains

The subunit with the highest seniority is farthest left, that with the lowest seniority farthest right. Seniorities in organic compounds are the reverse of those in inorganic macromolecules (Table 2-1). Oxygen has the highest seniority in heterochains (but not in heterocycles!), followed by sulfur, selenium ..., and, finally, boron and mercury:

$$O > S > Se > Te > N > P > As > Sb > Bi > Si > Ge > Sn > Pb > B > Hg$$

The subunit with the heteroatom of the highest seniority is followed by (a) the next following other heteroatom of the same seniority, (b) the heteroatom with the next highest seniority, (c) the carbocycle with the highest seniority, and (d) the alicyclic unit with the highest seniority. The seniority and the length of the path between two heteroatoms as well as the number, position, and name of substituents is also considered, similarly to the seniority rules for alicyclic units.

The shortest path has to be chosen if an aliphatic unit is embedded between two rings or a heteroatom and a ring. An example is the polymer molecule with the structure

$$-\!\!\left[O-CH_2-NH-CHCl-CH_2-SO_2-(CH_2)_6\right]_{\overline{n}} \text{ and not } -\!\!\left[O-(CH_2)_6-SO_2-CH_2-CHCl-NH-CH_2\right]_{\overline{n}}$$

The CRU is written in this way because the shortest path between the two heteroatoms O (highest seniority) and S (second highest seniority) is via $-CH_2-NH-CHCl-CH_2-$ with 4 chain atoms. The reverse order, $-O-(CH_2)_6-SO_2-CH_2-CHCl-NH-CH_2-$ does have the correct seniority $O \rightarrow S \rightarrow N$ but a longer path with 6 chain atoms between O and S.

Heterochains can be named by the addition method or by the substitution method. The **addition method** lines up the names of the subunits without dashes. The **substitution method** treats the CRU as an alicylic chain whose chain atoms are, in part, replaced by heteroatoms or heterogroups. Both methods use different designations:

	–O–	–S–	–SO–	–SO$_2$–	=N–	–NH–	–N=N–	–N=N–NH–
Addition method (old)	oxy	thio	sulfinyl	sulfonyl	nitrilo	imino	azo	azoimino
Addition method (new)	oxy	sulfandiyl	sulfindiyl	sulfonyl	nitrilo	imino	diazendiyl	
Substitution method	oxa	thia				aza		diazoamino

An example is the polymer molecule with the constitution

$$-\!\!\left[OCH_2OCH_2NHCH_2CH_2SCH_2NHCH_2CH_2\right]_{\overline{n}}$$

Addition method: poly(oxymethyleneoxymethyleneiminoethylenethiomethyleneiminoethylene)
Substitution method: poly(1,3-dioxa-8-thia-5,10-diazadodecane-1,12-diyl)

Heterocycles

The name of heterocyclic subunits consists of the name of the heterocycle and the suffix "diyl". The connecting points of an unsubstituted heterocycle with other subunits of a CRU are charcterized by position numbers *in front* of the heterocycle. Substituted heterocycles or internal connections however carry the position number *following* the name of the heterocycle and before the suffix "diyl". The connection point at the left side of the hetereocycle should have the lowest positiion number that is compatible with the numbering of the heterocycle.

Deviating from the rule for heterochains, nitrogen has the highest seniority in heterocycles. The seniorities of the remaining heteroatoms remain the same:

$$N > O > S > Se > Te > P > As > Sb > Bi > Si > Ge > Sn > Pb > B > Hg$$

Ring systems follow each other with the following seniorities (i) → (vii) if the CRU contains more than one heterocycle:

(i)   the ring system with nitrogen in the ring;
(ii)   the ring system with other hetero atoms in the ring;
(iii) the ring system with the greatest number of rings;
(iv)   the ring system with the greatest individual rings;
(v)   the ring system with the greatest number of hetero atoms;
(vi)   the ring system with the greatest variety of hetero atoms;
(vii) the ring system with the greatest number of hetero atoms which possesses the highest seniority.

Structure-based names can be very complicated (Table 2-4). They are thus mainly used for archival purposes (CAS) and for complex new polymer molecules. Even CAS and IUPAC do not intend to replace source-based names (poly(monomer) names) by structure-based names.

2.3   Atomic Structure and Chain Formation

2.3.1   Overview

Chains composed of chain atoms of the same chemical element are called **homochains,** those of chain atoms from two or more elements, **heterochains**. Homochains are easily formed from carbon, rarely however from other elements. Chains are called **inorganic** or **semi-inorganic** if 50 % or more of the chain atoms are not carbon.

Homochains and heterochains may be substituted. Note that "substitution" has different meanings in inorganic and organic chemistry. A truly unsubstituted homochain is *catena*-poly(sulfur), $+S+_n$, Poly(silane)s (silanes), $H+SiH_2+_n H$, are called "substituted" in inorganic chemistry since silicon Si_n is the basic body. In organic chemistry, on the other hand, poly(methylene) $H+CH_2+_n H$ is "unsubstituted" since alkanes C_nH_{2n+2} are considered to be the basic body here and not the diamond, C_n.

Table 2-4 Idealized structures, poly(monomer) names, and corresponding structure-based names.

Common structural formula of CRU	Source-based name	Structure-based name
$-CH_2-$	poly(methylene)	poly(methylene)
$-CH_2-CH_2-$	poly(ethene)	poly(methylene)
$-CH_2-CH(CH_3)-$	poly(propene)	poly(propylene)
$-CH_2-CH(C_2H_5)-$	poly(1-butene)	poly(1-ethylethylene)
$-CH(CH_3)-CH(CH_3)-$	poly(2-butene)	poly(1,2-dimethylethylene)
$-CH_2-CH(C_6H_5)-$	poly(styrene)	poly(1-phenylethylene)
$-CF_2-CF_2-$	poly(tetrafluoroethylene)	poly(1,2-difluoroethylene)
$-CH_2-CF_2-$	poly(vinylidene fluoride)	poly(1,1-difluoroethylene)
$-CH_2-CHCl-$	poly(vinyl chloride)	poly(1-chloroethylene)
$-CH_2-CH(CN)-$	poly(acrylonitrile)	poly(1-cyanoethylene)
$-CH_2-CH(OCOCH_3)-$	poly(vinyl acetate)	poly[1-acetoxyethylene]
$-CH_2-CHOH-$	poly(vinyl alcohol)	poly(1-hydroxyethylene)
$-CH_2-CH(COOCH_3)-$	poly(methyl acrylate)	poly[1-(methoxycarbonyl)ethylene]
$-CH_2-CH(CH_3)(COOCH_3)-$	poly(methyl methacrylate)	poly[1-(methoxycarbonyl)-1-methylethylene]
$-CH_2-O-$	poly(formaldehyde)	poly(oxymethylene)
$-CH_2-CH_2-O-$	poly(ethylene oxide)	poly(oxyethylene)
$-NH(CH_2)_5CO-$	poly(ε-caprolactam)	poly[imino(1-oxyhexamethylene)]
$-NH(CH_2)_6NH-OC(CH_2)_4CO-$	poly(hexamethylene-adipamide)	poly[iminoadipoyl-iminohexane-1,6-diyl]
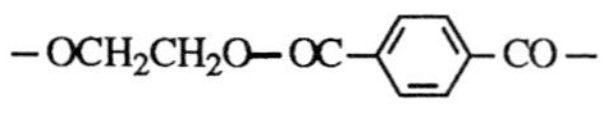 $-OCH_2CH_2O-OC-\langle\bigcirc\rangle-CO-$	poly(ethylene terephthalate)	poly(oxyethyleneoxyterephthaloyl)
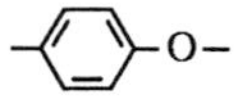 $-\langle\bigcirc\rangle-O-$	poly(phenylene oxide)	poly(oxy-1,4-phenylene)
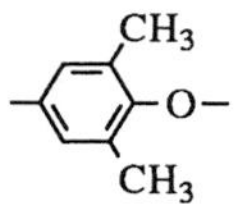 $\begin{matrix} CH_3 \\ -\langle\bigcirc\rangle-O- \\ CH_3 \end{matrix}$	"poly(phenylene oxide)"	poly(oxy-(2,6-dimethyl-1,4-phenylene)]
$\begin{matrix} -CH_2-\!\!\!\frown\!\!\!- \\ O \quad O \\ | \\ CH_3CH_2CH_2 \end{matrix}$	poly(vinyl butyral)	poly[(2-propyl-1,3-dioxane-4,6-diyl)methylene]
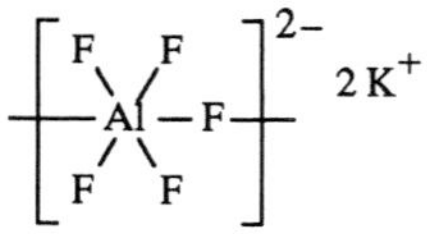 $\begin{bmatrix} F \quad F \\ \backslash / \\ -Al-F- \\ / \backslash \\ F \quad F \end{bmatrix}^{2-} 2\,K^+$	*trans-catena*-poly[dipotassium[[(tetrafluoroaluminate)-μ-fluoro](2–)]] *trans-catena*-poly[dipotassium[[tetrafluoroaluminate(III)]-μ-fluoro]]	

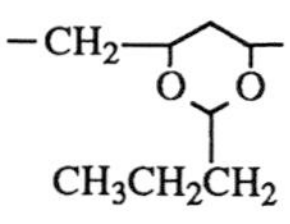

2.3.2   Homochains

Unsubstituted Homochains

Very few elements form unsubstituted homochains in the inorganic meaning of the word. The best known example is sulfur. Most other unsubstituted **elemental polymers** are tectopolymers (three-dimensional) and the rare phyllopolymers (two-dimensional).

Groups 1-12. All elements of these groups are metals at normal conditions (Fig. 2-1). Hydrogen is the exception (it becomes metallic at very high pressures).

Group 13. The element **boron** is a polymer in the solid state. The other elements of this group are metals (Al, Ga, In, Tl).

Group 14. Elemental **carbon** exists in several polymeric forms: as a three-dimensional polymer in diamond (two modifications); as a two-dimensional planar polymer in graphite and as a tube polymer in buckytubes; "one-dimensional" as spheres or rotational ellipsoids in the series of fullerenes, C_{60} and higher (Fig. 2-2, see also volume II).

Fig. 2-2 Polymeric modifications of carbon. From left to right: three-dimensional diamond, two-dimensional graphite, and "one-dimensional" C_{60} fullerene.

Silicon forms a tectopolymer in the solid state. No graphite and fullerene analogs are known for this element. **Germanium** is also polymeric in the solid state.

Tin has an amorphous polymeric modification below 13°C, a tetragonal-crystalline one between 13°C and 161°C, and an orthorhombic-crystalline one above 161°C. The conversion crystalline-amorphous is very slow. It is known as "tin pest" since the lower density (greater volume) of the amorphous polymer destroys tin objects such as beakers.

Lead is a metal but a hexaaryl-cyclotriplumbane, *cyclo*-Pb_3Ar_6, has been synthesized.

Group 15. **Nitrogen** forms a diamond-like polymer structure (cubic lattice) with single bonds between atoms if N_2 is compressed above 115 GPa and 2500 K. Poly(nitrogen) is stable at $p > 42$ GPa.

Phosphorus, arsenic, and **antimony** are present as polymers in some of their so-called allotropic modifications. The best known example is black phosphorus.

Group 16. **Oxygen** exists at very low temperatures in an ozone modification that probably consists of homochains.

At room temperature, elemental **sulfur** forms S_8 rings whose orthorhombic α-modification converts at 98°C into the monoclinic β-modification. At the melting temperature of this modification (120°C), 95 wt% of sulfur exists as S_8 rings and 5 wt% as S_i rings with $6 \leq i \leq 26$ ($i \neq 8$). At $T_f = 159$°C, S_i rings cleave homolytically; the resulting biradicals initiate the polymerization of the S_8 rings to *catena*-poly(sulfur) biradicals, $^\bullet[S_8]_\infty{}^\bullet$, which manifests itself by the first appearance of radicals and a strong increase of the melt viscosity (Fig. 2-3). At this "floor temperature" (see Section 7.3.2) of 159°C, the melt consists of 95 % S_8, 5 % S_i, and only 3 % S_∞.

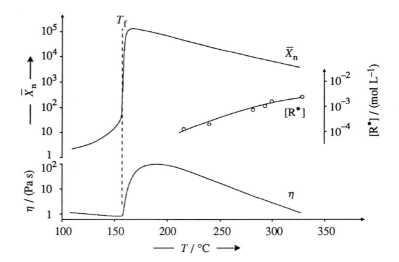

Fig. 2-3 Temperature dependence of the melt viscosity η of sulfur [9] and the concentration $[R^\bullet]$ of free sulfur radicals ([10], by electron spin resonance). The number-average degree of polymerization, $\overline{X}_n$, was calculated by assuming $n\,S_8 \rightleftarrows [S_8]_n$ [11] (see text, however).

The melt viscosity (and thus the degree of polymerization) increases strongly with increasing temperature, passes through a maximum at ca. 190°C, and then decreases. At 250°C, the melt consists of 37 % S_8, 9 % S_i, and 54 % S_∞. On pouring the melt into cold water, a so-called **plastic sulfur** is obtained, a *catena*-poly(sulfur) that is plasticized by low-molecular weight sulfur rings.

In the solid state, **selenium** and **tellurium** are polymers with probably the same structure as polymeric sulfur.

Group 17. The elements of this group apparently do not exist as elemental polymers in the solid, liquid, or dissolved state. **Iodine**, however, forms iodine chains as clathrates in the interior of helical amylose chains.

Substituted Homochains

Elements of the first period have homochains with higher numbers of chain atoms than elements of the second, third, etc. periods. Within each group of elements, the maximum observed number of chain atoms decreases with increasing period number.

In the simplest case, homochains are substituted by hydrogen. **Boranes** are oligomeric boron hydrides; high-molecular weight boranes are unknown. The "trivalent" boron does not form boranes $H(BH)_N H = B_N H_{N+2}$ but a series of compounds of different compositions, for example, decaboranes $B_{10}H_8$, $B_{10}H_{12}$, $B_{10}H_{14}$, and $B_{10}H_{16}$. In these compounds, boron is bound to hydrogen not only by regular valences but also by so-called 3-center bonds (1 electron pair per 3 boron atoms or per 2 boron atoms and one hydrogen atom via electron-deficient bonds, e.g. $>B\cdots H\cdots B<$).

The ability of **carbon** to exist in long "hydrogenated" carbon chains forms the basis of organic chemistry. Alkanes $H\text{-}[CH_2]_n$ exist with n up to several millions. These high-molecular weight alkanes are known as poly(methylene)s (structure-based name), poly(methylene)s (source-based; if diazomethane, CH_2N, was polymerized), and poly-(ethylene)s (source-based; if ethene, $CH_2=CH_2$, was polymerized). Commercial poly-(ethylene)s are usually not linear but branched (Sections 9.2.2, 10.4.4; volume II).

Many substituted saturated carbon chain polymers $+CH_2-CHR+_n$ are known. Like poly(ethylene)s, they are produced in large amounts and serve as plastics, fibers, coatings, and the like. Examples are poly(propylene) ($R = CH_3$), poly(styrene) ($R = C_6H_5$), poly(vinyl acetate) ($R = OCOCH_3$), and poly(vinyl chloride) ($R = Cl$). Disubstituted polymers of the type $+CH_2-CRR'+_n$ are not so common; examples are poly(vinylidene chloride) ($R = R' = Cl$) and poly(methyl methacrylate) ($R = CH_3$, $R' = COOCH_3$).

Unsaturated carbon chains may by unsubstituted. An example is 1,4-poly(butadiene) $+CH_2-CH=CH-CH_2+_n$. Substituted unsaturated carbon chains include 1,4-poly(isoprene) $+CH_2-C(CH_3)=CH-CH_2+_n$, poly(chloroprene) $+CH_2-CCl=CH-CH_2+_n$, and poly(2,3-dimethylbutadiene) $+CH_2-C(CH_3)=C(CH_3)-CH_2+_n$.

Silicon forms self-igniting silanes, $H(SiH_2)_nH$, with $n \leq 45$. Silanes $+SiR_2+_n$ carrying organic substituents R are stable, however (Volume II). Dimethyl substituted poly(silane) $+Si(CH_3)_2+_n$ is known by many names: poly(dimethylsilane) (source-based), *catena*-poly(dimethylsilicon) (inorganic, structure-based), poly(dimethylsilylene), and poly(dimethylsilanediyl) (both organic structure-based).

There are also high-molecular weight silanes with the composition $(SiH)_n$ that are probably *phyllo*-poly(silane)s with six-membered, annellated rings (L: *anellus* = small ring). The structure of these compounds resembles graphite.

The tendency to form long chains is further reduced in germanes, $H+GeH_2+_nH$ ($1 \leq n \leq 6$), and stannanes, $H+SnH_2+_nH$ ($1 \leq n \leq 6$).

Nitrogen group. A blue mass of composition $(NH)_n$ is obtained by decomposition of hydrazoic acid (hydrogen azide), HN_3, at 1000°C and quenching the reaction product by liquid nitrogen at temperatures below −196°C. The resulting mass probably consists of homochains $+NH+_n$. It converts to ammonium azide, NH_4N_3, at −125°C.

The corresponding phosphanes, arsanes, and stilbanes are known only as short chains or small rings with N = 1-10 (P), 1-5 (As), or 1-3 (Sb) member atoms.

Sulfur group. The elements of the 16th group form even smaller hydrogen-substituted oligomers (sulfanes, selanes, telluranes) than the elements of the 15th group. Sulfur compounds include chain-like poly(sulfane)s $H+S+_nH$ with n = 2-8 and halogen-poly-(sulfane)s $R+S+_nR$ with R = Cl, Br, I as well as sulfur cations $+S+_n^{2\oplus}$ with N = 4, 8, and 16, and sulfur anions $+S+_n^{2\ominus}$ with n = 4 and 9.

Effect of Atomic Structure

As pointed out in the preceeding sections, homochains are formed only by a cluster of elements of the 13th to 17th groups of the periodic system. This experimental observation can be explained as follows.

Elements of the first period do not have available d orbitals. They can thus form only four or less σ bonds per atom. Only carbon and the elements to the right of carbon possess sufficient electrons to contribute at least one electron to each complete bond (σ, π_x, π_y). Elements to the right of carbon have smaller bond energies than carbon (Table 2-5). They are electron donors that easily form hetero chains with electron acceptors. Nitrogen, oxygen, and fluorine have relatively low bond energies that are caused by the strong mutual repulsion of free electron pairs (Pitzer strain).

Table 2-5 Bond energies in kJ/(mol bond) of bonds between atoms of like elements.

						H–H	435
C–C	348	N–N	161	O–O	139	F–F	153
Si–Si	177	P–P	215	S–S	213	Cl–Cl	243
Ge–Ge	157	As–As	134	Se–Se	184	Br–Br	193
Sn–Sn	143	Sb–Sb	126	Te–Te	138	I–I	151
		Bi–Bi	105				

Elements of the first period to the left of carbon possess fewer electrons than occupiable orbitals. Since atoms try to fill their energetically accessible outer orbitals, boron and beryllium form bonds with hydrogen, methyl groups, etc., resulting in overlapping orbitals.

Elements of the second period have d orbitals with sufficiently low energies that can be used for the formation of bonds between like atoms. d orbitals are, however, usually used for π bonds in corresponding hybrids, and not for σ bonds because hybridization increases the stability of molecules. Such hybridizations are strongest for silicon, phosphorus, and sulfur. These elements are therefore present as polymers in the solid state, and, in part, also in their melts.

Elements of the fourth period use orbitals more for σ bonds than for π bonds. The bonding character of elements of the third period is intermediate between those of the second and fourth period.

It is thus expected that bond energies decrease within each group and each period with increasing atomic numbers (Table 2-5). The ability to form macromolecules thus decreases with increasing atomic numbers.

The bond energies of Table 2-5 correspond approximately to expectations. They are, however, not unproblematic because they were not measured for true chain structures but for low-molecular weight compounds with only one chain bond, i.e., dimers. The strength of such bonds is greatly influenced by substituents. The energies of Table 2-5 are thus bond energies that are averaged over the bond energies of many differently substituted compounds and not true dissociation energies. Effects are even expected for bond energies of pairs of two chain atoms since one-, two-, and three-dimensional structures are influenced differently by polarization effects. They thus exhibit different bond stabilities.

2.3.3 Heterochains

Heterochains are known in great numbers, especially inorganic and semi-inorganic ones. Inorganic heterochains occur in many more types than organic heterochains, for example, as (I) linear chains without rings in the chain, (II) linear chains with rings in the chain, and (III) "spiro" compounds with smaller and/or larger ring structures (Table 2-6). Most of the inorganic polymer compounds are however not linear chains but phyllo polymers and especially tecto polymers.

Carbon forms with oxygen, sulfur, and nitrogen many different heterochains in which the heteroatoms either alternate with the carbon chain atoms or separate larger segments composed of carbon atoms as chain units.

Table 2-6 Some types of constitutional repeating units of inorganic heterochains.

$$
\begin{array}{ccc}
\overset{\displaystyle Q}{\underset{\displaystyle Q'}{-\mathrm{X}-\mathrm{Z}-}} & \overset{\displaystyle R}{\underset{\displaystyle R'}{-\mathrm{Ar}-\mathrm{Z}-\mathrm{Z}'-}} & \mathrm{X} \\
\quad\quad \mathrm{I} & \quad\quad \mathrm{II} & \quad\quad \mathrm{III}
\end{array}
$$

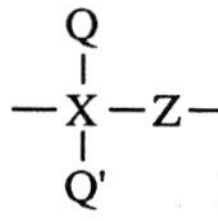

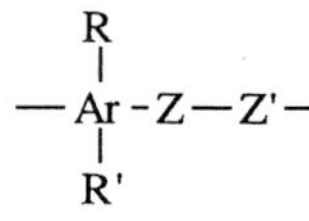

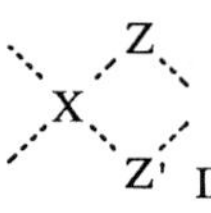

X =	B, Al, Si, Ge, Sn, Pb, P, Sb, Ti, V, Cr, Fe, Co	Be, Si, Pd, N
Z =	O, S, Se Si, P, B	if X = Be: H, F, Cl, CH$_3$ if X = Si: O, S if X = Pd: Cl if X = Nb: I
Z' =	O, N	
Q, Q' =	O, F, Cl, Br, organic substituents	
R, R'	CH$_3$, C$_6$H$_5$	
Ar =	C$_6$H$_2$, borazine, phosphazine	

Important examples of CRUs of organic heterochains comprise the following repeating units where R, R' = monofunctional substituents; Z, Z' = bifunctional non-aromatic or aromatic groups; and Ar = aromatic group:

–O–CRR'–	–O–(CRR')$_i$– –O–Ar–	–O–OC–Z– –O–Z–O–OC–Z'–CO–	–SO$_2$–Z–
polyacetals	polyethers	polyesters	polysulfones
–NH–OC–Z– –NH–Z–NH–OC–Z'–CO–	–NR–Z–	–NH–CO–NH–Z–	–CO–NH–CO–Z–
polyamides	polyimines	polyureas	polyimides

The formation and stability of heterochains depends mainly on the electronegativity of the chain atoms that is a measure of the ability of an element to compete with the other bonded element for the greater proportion of the electron charge. Electronegativities E cannot be measured directly. They are rather estimated from ionization potentials, atomic radii, force constants, or bond energies. Best known is the series of electronegativities that was established by Linus Pauling.

Fluorine, the most electronegative element, is given $E = 4.0$, carbon 2.5, and hydrogen 1.0. Combination of an element with $E > 2.5$ with elements of smaller electronegativity should lead to heterochains, provided certain selectivity rules are obeyed. Oxygen ($E = 3.5$), nitrogen (3.0), and sulfur (2.5) thus form heterochains with phosphorus (2.1), boron (2.0), arsenic (2.0), antimony (1.9), bismuth (1.9), silicon (1.8), germanium (1.8), tin (1.8), lead (1.8), vanadium (1.6), titanium (1.5), aluminum (1.5), and zirconium (1.4).

The tendency to form heterochains competes with the tendency to establish multiple bonds which comprise π bonds in addition to σ bonds. π bonds have degrees of bonding of 2. Degrees of bonding can be calculated from force constants which in turn are obtained from vibrational spectra.

π bonds exist if the following empirical conditions are fulfilled:

1. Both atoms of a bond must be electron-deficient.
2. The sum of Pauling's electronegativities must be at least 5.
3. The difference of electronegativities should be small, preferably below 1.5.

For example, nitrogen has an electronegativity of 3.0. For two nitrogen atoms, the sum of their electronegativities is 6, the difference 0. The two nitrogens therefore form a very stable triple bond. Polymeric nitrogen, $+N=N+_n$ on the other hand, is not stable under normal conditions (see page 39).

The carbon-nitrogen bond in hydrogen cyanide $H-C\equiv N$ has a sum of 5.5 and a difference of 0.5. This triple bond should be stable. It is, however, weaker than the nitrogen-nitrogen triple bond and, indeed, derivatives of polymeric hydrogen cyanide do exist (volume II).

The sum of electronegativities decreases to 5.0 and the difference increases to 1.0 in boron-nitrogen bonds. HBNH is not monomeric but exclusively trimeric.

Very large differences of electronegativities lead to ionic bonds and thus not to macromolecules. Correspondingly, the bond energy of bonds between elements of the first and second period with fully occupied orbitals increases with increasing difference of electronegativities (Table 2-7). Bond energies of bonds between boron and carbon (440 kJ/(mol bond)) and between boron and nitrogen (830 kJ/(mol bond)) are significantly higher because of the electron structure of boron. Comparisons of various types of bonds always need to consider the position of the element in the periodic system, i.e., the "ionic percentage" of the bond.

Bond energies, i.e., dissociation energies, relate primarily to the average thermal cleavability of bonds and thus to the thermal stability of molecules. The stability of a bond against other reagents depends, on the other hand, on the ionic percentage of the bond and the number of unoccupied orbitals and free electron pairs, respectively. The higher these quantities, the lower is the activation energy for chemical transformation. The resistance against reduction, oxidation, hydrolysis, etc., decreases with increasing atomic number of the element within each group of the periodic system. For example, hydrocarbons C_nH_{2n+2} are not hydrolyzed but silanes Si_nH_{2n+2} are because Si has a coordination number of 6 and only 4 positions are occupied.

In compounds with phosphorus, oxygen, sulfur, selenium, and halogens, carbon atoms are slightly positive and are thus easily attacked by nucleophilic reagents. A carbon atom bound to a metal atom, on the other hand, is slightly negative and assailable by electrophilic reagents.

Table 2-7   Bond energies and differences of electronegativities.

Bond	Difference of electronegativities	Bond energies in kJ/(mol bond)
C–S	0.0	260
C–N	0.5	290
C-Si	0.7	290
C–O	1.0	350
Si-O	1.7	370
B–C	0.5	440
B–N	1.0	830

All macromolecules with heteroatoms as chain atoms are therefore more labile than carbon chains. They are more easily attacked; they also enter exchange equilibria. Substituents act either as electron donors (e.g., methyl) or as electron acceptors (e.g., halogens). Main chain bonds are thus either weakened or strengthened by substituents, depending on their nature.

Similar considerations apply to carbon substituents at heteroatoms. Si–CH$_3$ bonds are only slightly polarized and therefore sufficiently stable in the industrially important poly(dimethylsiloxane)s, $+Si(CH_3)_2-O+_n$. The bonds Ti–C and Al–C, on the other hand, are vulnerable to attacks of oxygen or hydrogen because of the electronegativity of the metal atoms.

Elements of groups 2-12 (see Fig. 2-1) often possess dynamic equilibria between the various coordination numbers. Elements such as fluorine, chlorine, oxygen, etc., can add an electron pair or even two electron pairs to the free low-energy orbitals of group 2-12 elements. Oxygen may be mono-, bi-, tri-, or even tetrafunctional, depending on the partner. In all cases, coordination numbers increase. Fluorine bridges exist, for example, in the complex anion $+(TlF_4)^{2\ominus}-F+_n$ (from thallium fluoride and aluminum fluoride), chlorine bridges in palladium(II) chloride, and iodine bridges in niobium(II) iodide. Even methyl groups form bridges with beryllium.

All these compounds have many unoccupied orbitals. They are easily attacked and dissociate into smaller units in all common solvents. Because they only exist in the solid state, they were not considered to be macromolecules in the past.

2.4 Classification of Polymers

2.4.1 Structure-Based Terms

Polymers are either classified according to their structure or according to the constitution of the original monomers (process-based terms). According to their structures, polymer chains are subdivided into homochains with identical chain atoms (G: *hómoios* = equal) and heterochains with two or more types of chain atoms (G: *heteros* = the other (of two)) (Sections 2.3.2 ff.). Examples of homochain polymers are poly(acrylonitrile) $+CH_2-CH(CN)+_n$, $+S+_n$, and $+SiR_2+_n$. Examples of heterochain polymers are polyethers $+O-Z+_n$, polyamides $+NH-Z-CO+_n$ or $+NH-Z-NH-OC-Z'-CO+_n$, polyurethanes $+NH-CO-O-Z+_n$, polyphosphazenes $+PR_2=N+_n$, polysiloxanes $+SiR_2+_n$, and many others.

Polymers are **uniform** ("**monodisperse**") if their macromolecules are identical with respect to their constitution, molecular weight, and configuration. Enzymes are an example. Each *molecule* of an enzyme consists of the same number of different α-amino acid residues –NH–Z-CO– (and sometimes also imino acid residues) in the same sequence. The *substance* "enzyme" is thus uniform with respect to its molecules.

The uniformity of enzymes is caused by matrix-controlled syntheses. Practically all polymerizations leading to synthetic polymers are controlled statistically, however. Such polymers are **non-uniform** ("polydisperse") with respect to the number, arrangement, relative configuration, and maybe also the constitution of monomeric units (see below) even if they originated from the same monomer. Polymers whose molecules differ only in the number of monomeric units per molecule are called **non-uniform with respect to molar mass** (IUPAC) ("molecularly non-uniform", "polymolecular").

IUPAC recommends the adjective "uniform" if all molecules of a polymer are identical regarding the constitution, configuration, and molar mass, and "non-uniform" if they are not. The use of "monodisperse" and "polydisperse" is discouraged since "disperse" (G: *dispergare* = to distribute) is used in natural science exclusively for the distribution of one matter in another matter, i.e., *multiphase* systems. "Monodisperse" is furthermore an oxymoron since "disperse" relates to "more than one" and it cannot be "mono" what is already multiple ("disperse").

Molecules of many polymers have only one type of constitutional repeating unit in the same arrangement. Such polymers are called **regular**. An example is poly(oxycarbonylethylidene) $+OCOCH(CH_3)+_n$ which is either obtained as poly(lactide) from the ring-opening (chain) polymerization ("addition polymerization") of lactide or as poly(lactic acid) from the polycondensation ("condensation polymerization") of lactic acid:

(2-1)

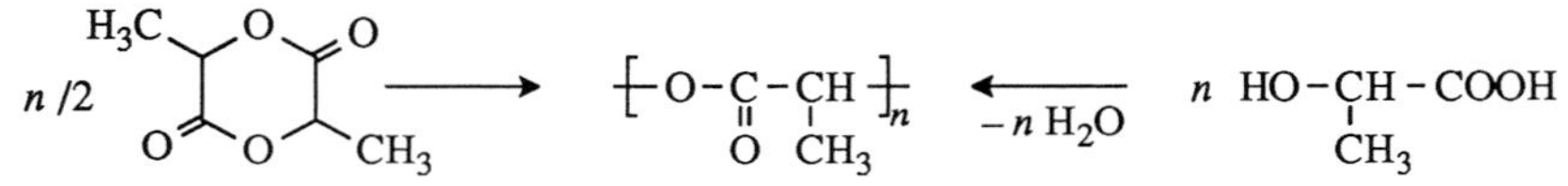

$$n/2 \quad \text{(lactide structure)} \quad \longrightarrow \quad \left[O\text{-}\underset{\underset{O}{\|}}{C}\text{-}\underset{\underset{CH_3}{|}}{CH} \right]_n \quad \xleftarrow[-n\,H_2O]{} \quad n\ HO\text{-}\underset{\underset{CH_3}{|}}{CH}\text{-}COOH$$

lactide   poly(oxycarbonylethylidene)   lactic acid

The monomeric units $-O-CO-CH(CH_3)-$ do not change their constitution during these processes. They are also always connected in the same direction; neither peroxide groups $-O-O-$ nor diketo structures $-CO-CO-$ are formed.

An **irregular polymer**, on the other hand, is "a polymer whose molecules cannot be described by only one species of constitutional unit in a single sequential arrangement" (IUPAC). Polymers from one type of monomer can nevertheless be irregular since (a) the same constitutional units may be tied together in an irregular manner, and (b) monomeric units may isomerize before or during the polymerization. Examples:

(1) Some Ziegler-type catalysts first isomerize 2-butene, $CH_3-CH=CH-CH_3$, and then polymerize the resulting 1-butene, $CH_2=CH(C_2H_5)$, to poly(1-butene).

(2) Other monomers isomerize in part during the propagation step. For example, 4,4-dimethyl-1-pentene II polymerizes cationically at very low temperatures exclusively via the carbon-carbon double bond opening to monomeric units I. At higher temperatures, some other units III are also formed:

(2-2) $-CH_2-CH-$ $\xleftarrow{-130°C}$ $CH_2{=}CH$ $\xrightarrow{>0°C}$ $-CH_2-CH_2-CH-$
            $\quad\ \ |$ $\ \ \ |$ $\qquad\qquad\ |$
            $CH_2-C(CH_3)_3$ $CH_2-C(CH_3)_3$ $C(CH_3)_3$
                   I   II   III

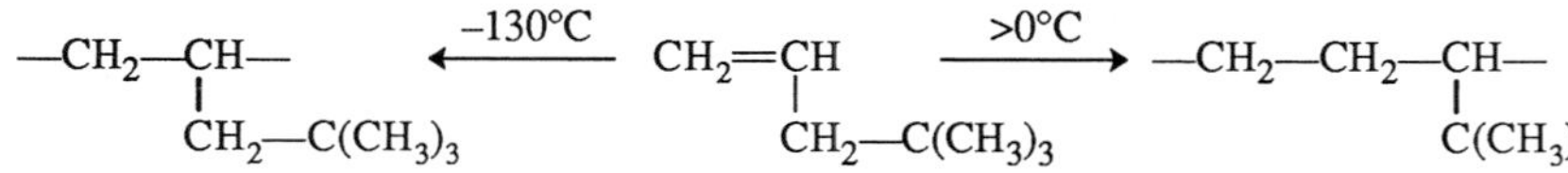

A monomer leading to units III does not exist. Polymers of this type are thus also called **phantom polymers** or **exotic polymers**. Whether such "exotic" structures are formed also depends on the initiator. Acrylamide $CH_2=CH(CONH_2)$ polymerizes radically to $+CH_2-CH(CONH_2)+_n$, but anionically to $+CH_2-CH_2-CO-NH+_n$.

The "exotic" structures are often present in small amounts only. Methacrylonitrile $CH_2=C(CH_3)(CN)$ polymerizes predominantly to $-CH_2-C(CH_3)(CN)-$ units via carbon-carbon double bonds and to a small extent to $-CH_2-C(CH_3)=C=N-$ units via the triple bond of nitrile groups. In a sense, these "exotic" units correspond to the byproducts of low-molecular weight chemistry. However, in contrast to true byproducts, they cannot be removed from the main product by purification procedures because they are part of the macromolecular structure.

(3) The same monomeric units may also be added in different ways, leading to **regio-isomerism**. Monomeric units $-CH_2-CHR-$ are usually added in **head-to-tail** position (◄◄) but to a small extent however in **head-to-head** (►◄) or **tail-to-tail** position (◄►):

$$-CH_2-CHR-CH_2-CHR- \qquad -CHR-CH_2-CH_2-CHR- \qquad -CH_2-CHR-CHR-CH_2-$$
head-to-tail head-to-head tail-to-tail

The radical polymerization of vinyl acetate (IV) leads, for example, to 1-2 % of head-to-head connections, that of vinyl fluoride (V) to 6-10 %, and that of vinylidene fluoride (VI) to 10-12 %. The polymerization of propylene oxide (VII) with diethylzinc and H_2O as initiator even delivers up to 40 % head-to-head structures.

$$CH_2=\underset{\underset{\text{IV}}{\overset{|}{OOCCH_3}}}{CH_2} \qquad CH_2{=}CHF \qquad CH_2{=}CF_2 \qquad H_2C\underset{O}{\overset{}{\diagdown\diagup}}CH(CH_3)$$

IV V VI VII

The incorporated "exotic" structures are often less stable than the "normal" ones. They act therefore as **weak links** in degradation situations. This does not mean, however, that weak links are always due to regioisomerism. Such weak links may also be produced by small proportions of impurities, for example, peroxide groups $-O-O-$ from the presence of traces of oxygen in monomers during free-radical polymerizations.

Thus, the chemical structure of a polymer molecule cannot be deduced unambiguously from the structure of its monomer molecules. An example is the polymerization of 1,4-divinylbenzene, $CH_2=CH-(1,4-C_6H_4)-CH=CH_2$. Monomer molecules with two vinyl groups are, as a rule, usually tetrafunctional, i.e., they form four chain bonds if both vinyl groups are equally reactive. This is true for free-radical polymerizations which therefore lead to branched and crosslinked poly(1,4-divinylbenzene)s (Section 2.5.10). In anionic polymerizations, however, the polymerization of the first vinyl group changes the resonance of the whole molecule and thus the reactivity of the second vinyl group. The second group does not polymerize anymore and the resulting polymer with the constitution $-[CH_2-CH(1,4-C_6H_4-CH=CH_2)]_n-$ is not crosslinked.

2.4.2 Process-Based Terms

Homopolymers are substances composed of polymer molecules derived from only one species of monomer molecules. This process-based term does not indicate the structure of the polymer, that may or may not be regular. In many cases, however, chemical experience or intuition allows one to deduce the polymer constitution from the monomer constitution.

The synthesis of **copolymers** requires two *or more* species of monomers (Chapter 12). In the past, copolymers were also called heteropolymers, interpolymers, or mixed polymers (D: Mischpolymere). The term "interpolymer" has also been used recently by Dow Chemical Co. for its brand of styrene-ethene copolymers.

For historic reasons, the term "copolymerization" is presently used exclusively for polymerizations in which two or more monomers undergo "addition polymerizations" but not if two monomers are subjected to "condensation polymerizations". The polymers resulting from the joint polymerization of ethene and propene are thus called copolymers. The polymer $+NH(CH_2)_6NH-OC(CH_2)_4CO+_n$ from the reaction of hexamethylenediamine $H_2N(CH_2)_6NH_2$ with adipic acid $HOOC(CH_2)_4COOH$, on the other hand, is considered to be the "homopolymer" polyamide 6.6 (nylon 6.6) and not a copolymer.

The term "copolymer" is used for "a polymer derived from more than one species of monomer" (IUPAC). Polymers with two or more types of constitutional or configurational monomer units are therefore not called "copolymers" if they were derived from only one species of monomers. Such polymers are sometimes called **pseudo copolymers**. Examples are certain polymers of butadiene, $CH_2=CH-CH=CH_2$, that contain both 1,4-units $-CH_2-CH=CH-CH_2-$ and 1,2-units $-CH_2-CH(CH=CH_2)-$.

Pseudocopolymers are also obtained by chemical modification of homopolymers. The partial hydrolysis of poly(vinyl acetate), $+CH_2-CH(OOCCH_3)]+_n$, for example, leads to vinyl alcohol units $-CH_2-CHOH-$ besides unchanged vinyl acetate units.

The word "copolymer" is sometimes used as a synonym for **bipolymer**, i.e., a copolymer derived from two species of monomers. Copolymers from three, four, five, ... species of monomer are called **terpolymers, quaterpolymers, quinterpolymers**, ... (see also Chapter 12 and Appendix).

Copolymers are furthermore subdivided according to the sequence of their monomer units. Both process-based and structure-based terms are used (Table 2-8). **Graft copolymers** consist of chains with monomer units that have been grafted onto or from chains with other monomer units (Section 16.3.4); this term is process-based. One talks of **comb polymers** (an undefined structure-based term), however, where such side chains were inherent in the monomer molecules.

All other relevant copolymer terms are structure-based. In chains of **alternating bipolymers**, two types of monomeric units $-a-$ and $-b-$ alternate. "Alternating bipolymer" is a special case of **periodic copolymers** that are copolymers with longer sequences of two species of monomeric units such as $+a-b-b+_n$ or $+a-a-b-b+_n$, sequences with three species of monomeric units, e.g., $+a-b-c+_n$, and so on (Table 2-8).

Graded copolymers (= **tapered copolymers**) exhibit a gradient of composition along the chain. In bipolymers, one end of the chain is enriched with "a" units and the other end with "b" units. **Block copolymers** consist of blocks of homosequences that are connected via their ends ("block polymer" if not from a copolymerization!). Multiblock copolymers with short block lengths are also called **segment(ed) copolymers**.

Statistical copolymers are copolymers with a sequence of monomeric units that is controlled by the statistics of the polymerization process. Such statistics may be Markov trials of zeroth, first, second, ... order (see also Chapter 12). Statistical copolymers with zeroth-order Markov statistics (= Bernoulli statistics) are called **random copolymers**. Some authors restrict "randomness" to those copolymers with Bernoulli statistics that also contain equal proportions of the various monomeric units.

Table 2-8 Copolymers with monomeric units a, b, and/or c. "Copolymer" is often used as a synonym of "bipolymer" although "copolymer" is a generic term.

Name	Schematic structure	Notation according to IUPAC
Bipolymer, unspecified sequence of monomer units	...a/b...	poly(A-*co*-B)
–, Markov statistics	...aabaaaabbabbbabaaababbba...	poly(A-*stat*-B)
–, Bernoulli statistics	...aaabaabbabbbbabbaababbaab...	poly(A-*ran*-B)
Alternating bipolymer	...ababababababababababab...	poly(A-*alt*-B)
Periodic bipolymer	...abbabbabbabbabbabbabbabb..	poly(A-*per*-B-*per*-B)
Periodic terpolymer	...abcabcabcabcabcabcabcabc...	poly(A-*per*-B-*per*-C)
Diblock copolymer	a................ab................b	poly(A)-*block*-poly(B)
Triblock copolymer	a..........ab..........bc..........c	poly(A)-*block*-poly(B)-*block*-poly(C)
Segmented bipolymer	(a)$_n$-(b)$_m$-(a)$_n$-(b)$_m$-(a)$_n$-(b)$_m$-(a)$_n$	many segments with small to medium *m* and *n*
Gradient bipolymer	(a)$_n$baaaaabbaaabbaabbbba(b)$_m$	-

Graft bipolymer	...a-a-a..........a-a-a..........aa...
	   | |
	   b$_m$ b$_n$
		poly(A)-*graft*-poly(B)

2.4.3 Terms for Charged Polymers

The prefixes "macro" and "poly" are used differently for charged polymers and uncharged polymers. "Macro" always refers to a large molecule. A **macroanion** is accordingly a macromolecule with *one* anionic group, for example, an anionic endgroup in so-called living styrene polymers (I) from the polymerization of styrene with monofunctional anionic initiators. Correspondingly, a **macrocation** (II) carries *one* cationic charge and a **macroradical** (III) *one* free radical. Macrodianions (-dications, -diradicals) have *two* anions (cations, radicals) per macromolecule, and so on.

Macroanions, macrodianions, etc., are *not* **polyanions** since the prefix "poly" relates here to both the presence of anions in *polymers* and the presence of *many* anions per macromolecule. The same notation applies to polycations and polyradicals.

Molecules with ionic groups may form **ion associates** with suitable counterions (Fig. 2-4). The valency of the counterions (number of electron pairs) is not decisive but the coordination number (number of nearest neighbors) is.

Ionic endgroups of linear polymers may form linear ion associates (I in Fig. 2-4) which are **supramolecular polymers**. Examples are metal salts of low-molecular weight dicarboxylic acids. In melts and highly concentrated solutions, such physical polymers exhibit similar properties to those of true chemical polymers of corresponding molecular weights, for example, high melt viscosities.

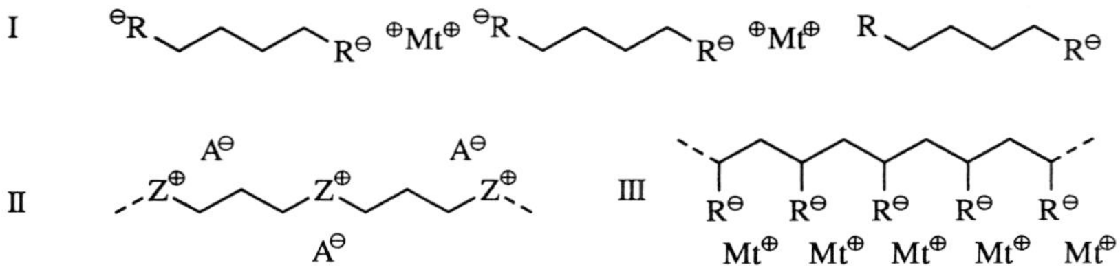

I

II   III

Fig. 2-4 I: Ion associate of linear low-molecular weight compounds with terminal anionic groups $R^\ominus$.
II: Main-chain polycation with counter-anions $A^\ominus$.
III: Side-chain polyanion with metal counter-cations $Mt^\oplus$.

Polyions contain many ionic groups either in the main chain (II in Fig. 2-4) or in substitutents (III in Fig. 2-4). In contrast to physical polymers, they maintain their macromolecular character even if the ionic groups are removed. Water-soluble polyions are called **polyelectrolytes**. This group comprises polyacids, polybases, and polysalts.

Polyacids dissociate into protons and polyanions. Examples of polyacids are poly-(orthophosphoric acid) IV, poly(vinyl orthophosphoric acid) V, poly(vinyl sulfuric acid) VI, poly(vinyl sulfonic acid) VII, and poly(acrylic acid) VIII:

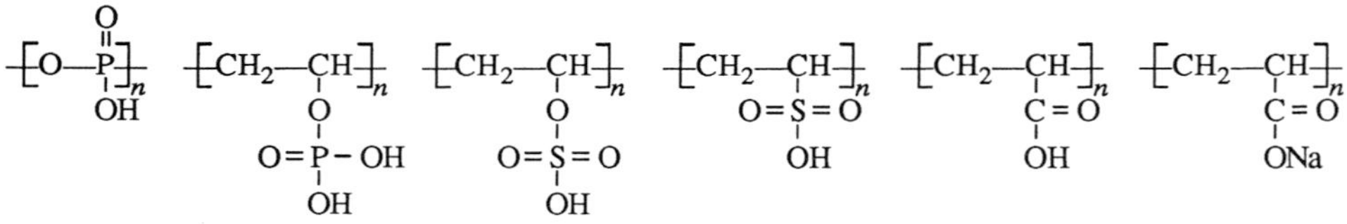

IV   V   VI   VII   VIII   IX

Salts of polyacids are called **polysalts**. The sodium salt of poly(acrylic acid) VIII is thus poly(sodium acrylate) IX (systematic name: poly(sodium-1-carboxylatoethylene)).

Ionomers are partially or completely neutralized, water-insoluble copolymers of ethene with small proportions of acidic monomers, for example, 10-15 % acrylic acid.

On ionization, **polybases** accept protons or methyl cations and convert to polycations by this *quaternization*. An example of a polybase with pro-ionic chain groups is poly-(ethylene imine) with the monomeric unit X. Poly(vinyl amine) (XI) is an example of a polybase with pro-ionic groups as substituents.

Cations are non-separable parts of chain units in **ionenes**. An example is the ionene with the monomeric unit XII which is obtained from salts of 4-vinylpyridine (XIII) by polymerization via the 1,6-position.

Polyampholytes carry many positive *and* negative charges in the same macromolecule; XIII is an example. Mixtures of polyacids and polybases are called **polyelectrolyte complexes** or **symplexes**.

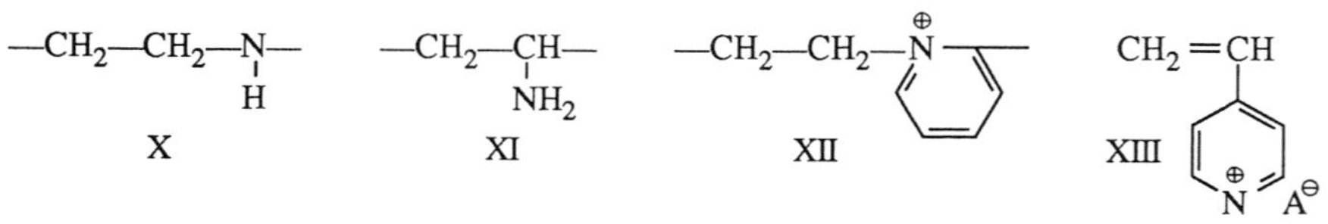

X   XI   XII   XIII

2.5   Molecular Architecture

2.5.1   Introduction

The constitution of a macromolecule is given by the constitution, number, and interconnection of its monomeric units. In the simplest cases, monomeric units are combined in open chains ("linear molecules") or closed chains ("rings"). Several open chains can be combined to branched molecules or two-dimensional or three-dimensional networks. Both the combination and arrangement of several monomeric units in a chain (Table 2-8) as well as the interconnection of several molecules (Fig. 2-5) is called **molecular architecture** (Chapter 16).

The molecular weight of a *macromolecule* is determined by the number and constitution of its monomeric units and endgroups. Special problems exist, however, for the molecular weights of synthetic *macromolecular substances* (= polymers) which are, as a rule, mixtures of macromolecules of different molecular weights. These problems are discussed in Section 3.3.

Polymer science as an interdisciplinary science uses very many different chemical and physical methods to characterize the chemical structure of polymers. Most of these methods are borrowed from neighboring fields. They are assumed to be known and will not be discussed in detail if there are no polymer specific problems. Examples are the determination of the chemical composition by analysis of chemical elements or functional groups; infrared, Raman, ultraviolet, and nuclear magnetic resonance spectroscopy (for tacticity, see Chapter 4); pyrolysis; and the various chromatographic techniques.

Size exclusion chromatography and viscometry are the most popular molecular weight methods in synthetic work; they are thus discussed in Chapter 3. All other molecular weight methods are treated in greater detail in those sections of Volume III where they follow directly from the physics of the phenomena. An example is membrane osmometry (Volume III, Chapter 10, Thermodynamics of Solutions).

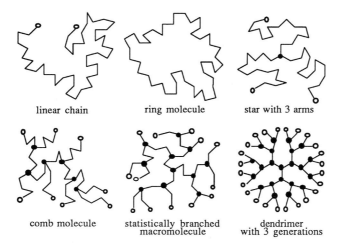

linear chain   ring molecule   star with 3 arms

comb molecule   statistically branched   dendrimer
                    macromolecule   with 3 generations

Fig. 2-5 Schematic representation of two-dimensional projections of the molecular architecture of various macromolecules with 45 monomeric units each. Monomeric units are symbolized by —, trifunctional branching points by ●, and endgroups by O.

2.5.2 Linear Chains and Rings

The simplest molecular architecture is exhibited by unbranched open or cyclic chains. **Linear chains** (p. 4) are "open" structures that always contain *two* endgroups besides their monomeric units, i.e., they are not branched (Fig. 2-5). Examples are *catena*-poly(sulfur) $+S+_n$ and poly(methylene $+CH_2+_n$. **Cyclic polymers** (**ring polymers**) are also not branched but are closed chains that do not have endgroups.

Ring molecules are *not* called macrocycles in polymer chemistry since this word is used traditionally (and inappropriately) for "large" rings of ca. 15-25 chain atoms in low-molecular weight organic chemistry. Ring polymers, on the other hand, may exhibit thousands of chain atoms per molecule. "Cyclic polymer" and "cyclopolymer" are not synonyms (Section 6.1.5). The use of "cyclolinear polymer" for linear chains with cyclic chain units (e.g., $+CH_2-Ar+_n$), confuses the issue further.

Cyclic macromolecules can be combined with themselves or with linear chains (Fig. 2-6). **Polycatenanes** (L: *catena* = chain) are chains composed of linearly intertwined rings. Contrary to conventional nomenclature, the term does not refer to polymers from monomeric catenanes or with many catenane groups (Sections 2.2.2, 16.4.2). Polycatenanes may be found as byproducts of cyclic polymers.

The stringing-up of small ring molecules on linear chains leads to **polyrotaxanes** (L: *rota* = wheel). For example, cyclodextrins (cyclic molecules with 6-12 glucose units per ring) like to form hydrogen bonds to ether groups of poly(oxyethylene) chains (PEOX), $+O-CH_2-CH_2+_n$, and slip onto the chains like pearls on a string. The chain ends are then sealed with bulky endgroups (Section 16.4.2).

Polyrotaxanes can be converted to **polymer tubes**. The reaction of cyclodextrin-poly-(ethylene oxide)-polyrotaxanes with epichlorohydrin + NaOH results in three intermolecular $-CH_2CH(OH)CH_2-$ bridges between two cyclodextrin molecules each. Stronger bases remove the bulky endgroups and the unused PEOX molecules. The resulting polymer tubes have diameters of ca. 1.5 nm (external) and 0.5 nm (internal).

Polymer tubes also exist in nature, for example, in some silicates (Volume II). Carbon black contains tube-like "buckytubes" of 0.34 nm diameter and ca. 1000 nm length in addition to the more common "buckyballs" (fullerenes) C_{60}, C_{72}, etc. (Volume II).

Fullerenes and some proteins are polymeric hollow spheres. The hollow sphere of the apoferritin molecule (G: *apo* = off, away from), a protein, has diameters of 12.2 nm (exterior) and 7.3 nm (interior). The molecules of the enzyme ferritin consist of apoferritin molecules that are filled with iron(III)hydroxide oxide phosphate (up to 4500 Fe atoms per molecule). Even bigger are bag-like protein hulls of certain bacteria.

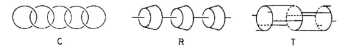

Fig. 2-6 Schematic representation of polycatenanes (C), polyrotaxanes (R), and polymer tubes (T).

2.5.3 Statistically Branched Polymers

Branched polymers contain **branch points** from which three or more subchains emanate. Depending on the relative arrangements of these subunits, star-like, comb-like, randomly branched, and dendritic polymer molecules are distinguished (Fig. 2-5).

A **subunit** extends from branching point to branching point or from a branching point to an endgroup. The subunits may be of the same size or may have a size distribution. Statistical branching usually leads to broader molecular weight distributions.

The topological concept of "branching" differs from the one of polymer chemistry. 3-Methylheptane I, for example, is a branched hydrocarbon according to the terminology of low-molecular weight organic chemistry. The corresponding poly(1-butylethylene) II, on the other hand, is considered unbranched by polymer chemistry. Similarly, the polymer III from the copolymerization of ethene with a small proportion of, e.g., 1-hexene is called in industry a *linear* low-density poly(ethylene) (LLDPE, PE-LLD):

$$CH_3-CH_2-\underset{\underset{(CH_2)_4H}{|}}{CH}-CH_3 \qquad CH_3\!\!\left[\!CH_2-\underset{\underset{(CH_2)_4H}{|}}{CH}\!\right]_{\!n}\!\!CH_3 \qquad \left[\!CH_2-CH_2\!\right]_{\!n}\!CH_2-\underset{\underset{(CH_2)_4H}{|}}{CH}\!\!\left[\!CH_2-CH_2\!\right]_{\!m}$$

$$\text{I} \qquad\qquad\qquad \text{II} \qquad\qquad\qquad\qquad \text{III}$$

Branching can also be introduced by certain higher functional monomers, some side reactions during polymerization, and by grafting (see also Chapter 16).

Short-chain branches are oligomeric subunits, **long-chain branches** polymeric ones:

$$-CH_2-\underset{\underset{CH_2CH_3}{|}}{CH}-(CH_2)_x-\underset{\underset{(CH_2)_4CH_3}{|}}{CH}-(CH_2)_y- \qquad\qquad -CH_2-\underset{\underset{(CH_2)_zH}{|}}{CH}-(CH_2)_x-\underset{\underset{(CH_2)_qCH_3}{|}}{CH}-(CH_2)_y-$$

$$\text{short-chain branched} \qquad\qquad\qquad \text{long-chain branched}$$
$$(x, y \gg 1) \qquad\qquad\qquad\qquad (x, y, z, q \gg 1)$$

The number of branching points per macromolecule and the average length and length distribution of subunits in the main chain and in branches are difficult to ascertain. The presence of *many* short-chain branches per molecule can be determined by spectroscopy or chemical means. In long-chain branched macromolecules, however, the proportion of branching units is very small compared to that of linear units. The presence and extent of long-chain branches is therefore often evaluated by indirect means, for example, by comparison of the spatial dimensions of branched and linear polymers of the same molecular weight. As one can see clearly from a comparison of star-like and linear molecules in Fig. 2-5, a macromolecule with long-chain branches has smaller spatial dimensions than a linear one with the same monomeric units (excluding branching units) and the same molecular weight. Physical properties of differently branched polymers vary widely: low-density poly(ethylene) has 1-4 % of its carbon atoms in branches and a melting temperature of ca. 110°C, the far less branched high-density poly(ethylene) one of ca. 135°C, and linear poly(methylene) one of ca. 146°C.

2.5.4 Star Molecules

Star molecules have only one branching point, the core with $f \geq 3$ arms consisting of linear chains (Fig. 2-5). The core may be a single atom (e.g., $-N<$ with $f = 3$), a low-molecular weight unit (e.g., the benzene ring with $3 \leq f \leq 6$), or even a latex particle with hundreds of connection points ($f \gg 2$). In star molecules, the functionality f is identical with both the number N_u of subunits ("arms") and the number N_{end} of endgroups.

In the simplest case, all arms consist of the same number of the same monomer units. Such molecularly uniform star *molecules* contain in each arm the same number N_{mu} of monomeric units, i.e., the arms possess equal degrees of polymerization. **Miktoarm** star polymers have arms with different constitutions (G: *miktos* = mixed). Molecularly uniform star *polymers* consist of constitutionally uniform star molecules which may or may not be uniform with respect to the different arms of the same molecule (Chapter 16.2.4).

2.5.5 Dendrimers

Star molecules with repeated star-like branching of their arms are cascade-like branched molecules with radial symmetry of repeating units around the core (Fig. 2-7). They are now mostly called **dendrimers** because of their dendritic structure (G: *dendron* = tree) but have also been named **arborols** (L: *arbor* = tree) or **cascade molecules**. (Starburst polymer® is a registered trademark for a certain class of dendrimers.)

"Radial symmetry" refers only to the chemical structure, not to the actual physical structure. Each new layer of monomer units •< is arranged "isotropically" around the central core. Dendrimers have therefore been called "iso-branched" in the past. Dendrimers are, however, not "isotropic" with respect to the spatial arrangement of repeating units. Contrary to their schematic two-dimensional representations in Fig. 2-7, they are neither spherically symmetric in three dimensions nor are their monomeric units homogeneously distributed in space (Volume III).

Each dendrimer molecule consists of a multifunctional **core** with functionality $f^* \geq 3$ that carries f^* **dendrons**. Each dendron is made up of branch units (1, 2, 3 in Fig. 2-7) that are attached to the core by one stem (distance 0-1).

On synthesis, each new layer of branch points constitutes a new (chain) **generation**. The total number of branch points per molecule is given by the functionality f^* of the core and the functionality f of the branch points (Table 2-9). The repeating **chain unit** of the dendrimer of Fig. 2-8 is $-CH_2CH_2CONHCH_2CH_2N<$. Instead of defining **chain generations** (emphasis on constitution), one can also define **branch(ing) generations** (emphasis on synthesis steps), i.e., $-NHCH_2CH_2N(CH_2CH_2CO-)_2$ for the molecule of Fig. 2-8. This amidoamine dendrimer thus has 3 chain generations with 24 endgroups $-H$ or two branching generations with 12 surface cells $-NHCH_2CH_2NH_2$.

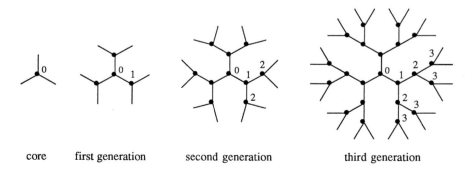

| core | first generation | second generation | third generation |

Fig. 2-7 Schematic two-dimensional representation of various generations of dendrimers with three-dimensional branch points •. A dendrimer molecule is a multidendron molecule (here: 3 dendrons).

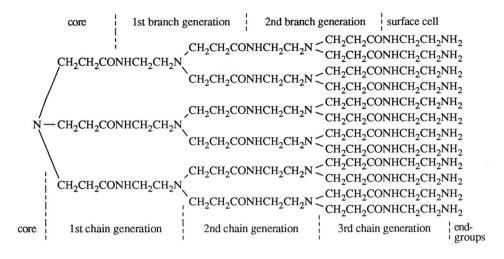

Fig. 2-8 Two branch(ing) generations and three chain generations of an amidoamine dendrimer.

All dendrimer molecules contain three different types of units: a core cell, branch cells, and surface cells (process-based teminology) or a core, linear subunits, and end-groups (structure-based terminology) (Fig. 2-9). None of these names has been proposed, recommended, or approved by IUPAC.

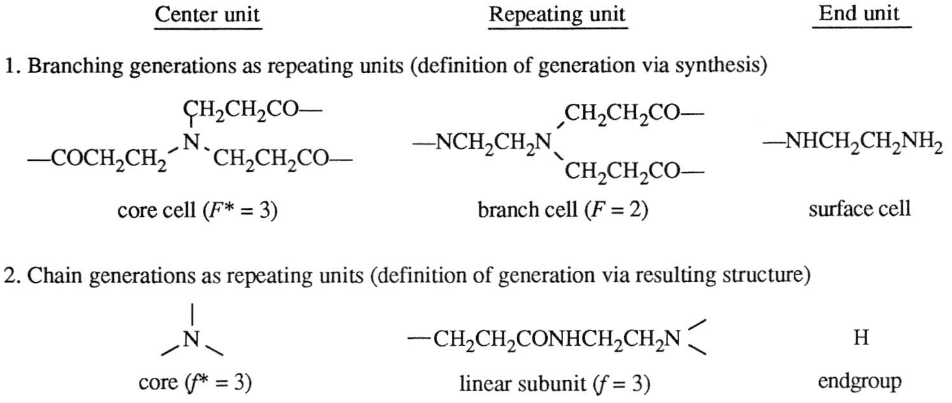

Center unit	Repeating unit	End unit

1. Branching generations as repeating units (definition of generation via synthesis)

$$\overset{\text{CH}_2\text{CH}_2\text{CO}—}{\underset{—\text{COCH}_2\text{CH}_2}{\diagdown}}\text{N}\diagdown_{\text{CH}_2\text{CH}_2\text{CO}—}$$ $$—\text{NCH}_2\text{CH}_2\text{N}\diagup^{\text{CH}_2\text{CH}_2\text{CO}—}_{\diagdown \text{CH}_2\text{CH}_2\text{CO}—}$$ $$—\text{NHCH}_2\text{CH}_2\text{NH}_2$$

core cell ($F^* = 3$)   branch cell ($F = 2$)   surface cell

2. Chain generations as repeating units (definition of generation via resulting structure)

$$\overset{|}{\underset{\diagup}{\text{N}}}_{\diagdown}$$ $$—\text{CH}_2\text{CH}_2\text{CONHCH}_2\text{CH}_2\text{N}\diagup^{\diagup}$$ H

core ($f^* = 3$)   linear subunit ($f = 3$)   endgroup

Fig. 2-9 Different definitions of center units (core cell vs. core), repeating units (branch cells vs. linear subunits), and end units (surface cells vs. endgroups).

The definition of "generation" via synthesis has the advantage of an easy visualization of the path of synthesis and the type of branching points. It has the disadvantage of not working with functionalities but with **multiplicities** F that indicate the number of units that can be added for a new generation. In Figs. 2-8 and 2-9, these multiplicities are $F^* = 3$ for the core cell but $F = 2$ for the branch cell. The structural functionalities, on the other hand, are $f^* = f = 3$. The use of structure-based terms furthermore leads to much simpler chemical names.

Table 2-9 Number N_{end} of end units and repeating units in ideal dendrimers with i chain generations and j branch generations, respectively (see Fig. 2-8). * Equals degree of polymerization, X.

Structural unit	Definition of constitutional repeating unit as	
	linear subunit	branch cell
Number of end units in a molecule (highest i or j)	$N_{end}(L) = f^*(f-1)^{i-1}$	$N_{end}(B) = F^*F^j$
Number of repeating units in a generation i or j	$N_{u,i}(L) = f^*(f-1)^{i-1}$	$N_{u,i}(B) = F^*F^j$
Total number of repeating units per molecule * (highest i or j)	$N_u(L) = f^*[(f-1)^i - 1]/[f-2]$	$N_u(B) = F^*(F^j - 1)/(F-1)$

The amidoamine dendrimer of Fig. 2-8 has a core unit with multiplicity $F^* = 3$ or functionality $f^* = 3$, repeating units of multiplicities $F = 2$ or functionalities $f = 3$, and $i = 2$ branching generations or $j = 3$ chain generations. This leads to $N_{end}(L) \equiv N_{end}(B) = 12$ endgroups (or surface cells), $N_{u,j}(L) = 12$ linear subunits or $N_{u,j}(B) = 6$ branch cells in the highest generation ($i = 2$ and $j = 3$, respectively), and a total of $N_u(L) = 21$ linear subunits or $N_u(B) = 9$ branch cells.

In amidoamine dendrimers, linear subunits are identical with repeating units. However, there are many other types of dendrimers where subunits consist of more than one repeating unit or the size of subunits increases or decreases systematically with the number of generations. An example is a dendrimer with $i = 3$ subunits in generation 1, $i = 2$ subunits in generation 2, and $i = 1$ subunit in generation 3. There is furthermore an upper limit of the possible number of generations because of the increasing crowding of end units on the "surface" of the molecule.

In other types of dendrimers, cores may not be circular or "spherically" symmetrical. An example is a "dendronized" polymer chain that carries dendrons as sidegroups:

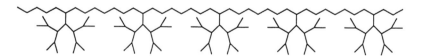

Dendrimers with a hydrophobic interior (core, low branch generations) and a hydrophilic exterior (endgroups, high branch generations) have been called **micellanes**.

2.5.6 Hyperbranched Polymers

Hyperbranched polymers consist of highly branched molecules that result from the polymerization of certain multifunctional monomers without crosslinking at higher monomer conversions if side reactions are absent. An example is a monomer of the type AB_f with $f \geq 2$ where A can only react with B but neither A with A nor B with B. Hyperbranched molecules can be envisioned as irregularly branched dendrons or as irregularly branched comb molecules. An example is one of the many different tetradecamers of AB_2 (Fig. 2-10). Hyperbranched molecules (and also dendrons, see below) with a total degree of polymerization X from f-functional monomers of the type AB_f always have $N_A = 1$ endgroup of type A and $N_B = 1 + (f-1)X$ endgroups of type B.

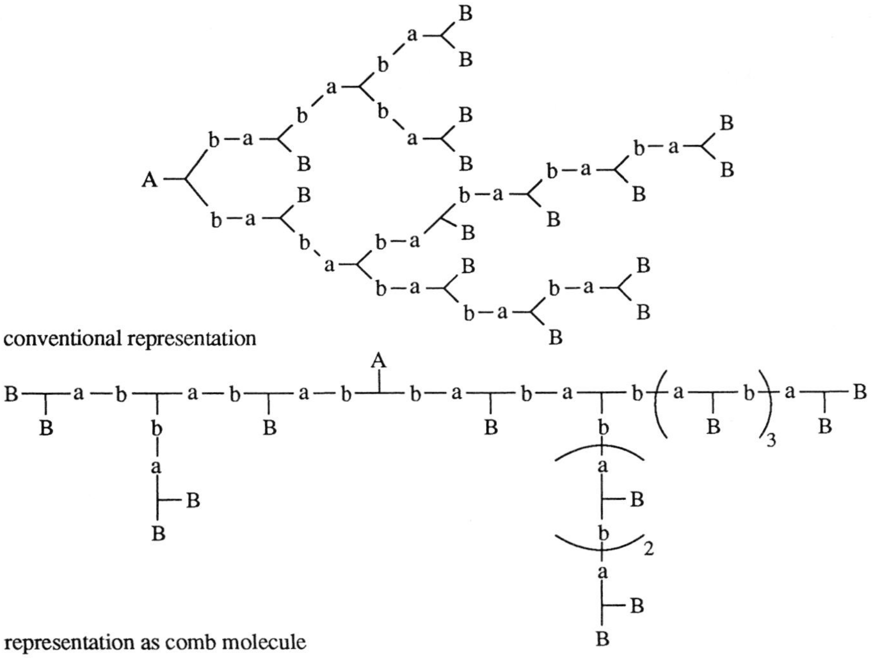

conventional representation

representation as comb molecule

Fig. 2-10 A hyperbranched tetradecamer from an AB_2 monomer ($F = 2$), showing unreacted groups A and B and reacted groups "a" and "b". This molecule has a total of $X = N_C + N_U = 14$ units, i.e., 1 core unit ($N_{core} = N_{Abb} = 1$) and 13 other monomeric units ($N_{mon} = 13$), as well as $N_A = 1$ endgroup of type A and $N_B = 15$ endgroups of type B. The 13 non-core units include $N_{abb} = 2$ branched units abb, $N_{aBb} = 7$ linear units aBb, and $N_{aBB} = 4$ terminal units aBB.

A hyperbranched *molecule* has always $N_C = 1$ **core unit** (sometimes called **seed unit** or **focal point**). In an AB_2 polymer molecule, the core unit may be either a monocore unit ABb (especially at low monomer conversions; $N_{C,Bb}$) or a dicore unit Ab_2 (especially at higher monomer conversions; $N_{C,bb}$) (Fig. 2-11) so that $(N_{C,Bb} + N_{C,bb})/N_C \equiv 1$. The molecule consists also of $N_U = N_{abb} + N_{aBb} + N_{aBB}$ non-core units that may be N_{abb} "dendritic" units abb, N_{aBb} linear units aBb, and/or N_{aBB} terminal units aBB (Fig. 2-11).

A polymer molecule from AB_3 monomers also has one core molecule (ABBb, ABbb, or Abbb) and various numbers N_{nc} of non-core units aBBb ("linear"), aBbb ("semi-branched"), and/or abbb ("dendritic"). A corresponding terminology applies to polymer molecules from monomer molecules AB_f with higher functionalities $f \geq 4$.

The total number of monomeric units N in one molecule equals the total degree of polymerization, X, of that molecule:

(2-3) $X = N_C + N_U = 1 + N_{abb} + N_{aBb} + N_{aBB}$ (AB$_2$ type)

(2-4) $X = N_C + N_U = 1 + N_{abbb} + N_{aBbb} + N_{aBBb} + N_{BBB}$ (AB$_3$ type), etc.

Inspection of simple molecules shows that the number $N_B = 1 + (f - 1)X$ of unreacted B-groups per molecule depends on the degree of polymerization and the functionality f of the core molecule ($f = 2$ for AB_2, etc.). The number of reacted b-groups in non-core units of this molecule is the sum of the products of the b-functionality f_i of the core unit

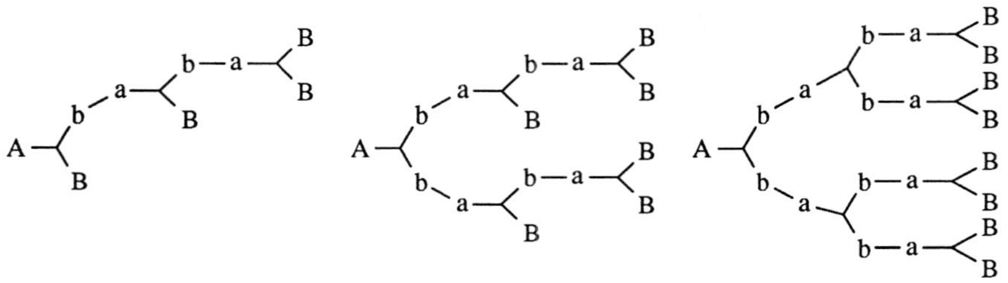

$X = 3$	$N_{abb} = 0$	$X = 5$	$N_{abb} = 0$	$X = 7$	$N_{abb} = 2$
$N_C = 1$	$N_{aBb} = 1$	$N_C = 1$	$N_{aBb} = 2$	$N_C = 1$	$N_{aBb} = 0$
$f_i = 1$	$N_{aBB} = 1$	$f_i = 2$	$N_{aBB} = 2$	$f_i = 2$	$N_{aBB} = 4$

Fig. 2-11 Molecules: Linear with monocore (left), hyperbranched with dicore (center), and perfect dendron (right). Neither core units ABb and Abb nor terminal units aBB are branching units (Fig. 2-10)!

and the number of b-group carrying units, i.e., $N_{U,b} = \Sigma_f f_i N_i$, i.e., $N_{U,b} = N_{aBb} + 2 N_{abb}$ (AB$_2$ type), $N_{U,b} = N_{aBBb} + 2 N_{aBbb} + 3 N_{abbb}$ (AB$_3$ type), etc.

Interrelationships between terminal and dendritic units depend on the number of b-groups in core units ($f_{C,b} = 1$ for ABb core units, $f_{C,b} = 2$ for Abb core units). For AB$_2$-type hyperbranched *molecules* one obtains $N_{aBB} = N_{aBb} + f_{C,b}$ and for *substances* composed of such molecules, $\Sigma_i N_{aBB,i} = \Sigma_i N_{abb,i} + \bar{f}_{C,b} \Sigma_i N_i$. For a polymer containing many molecules, one thus obtains $1 \leq \bar{f}_{C,b} \leq 2$, depending on the average degree of conversion of B-groups of the core units. For example, $f_{C,b} = 1$ for a dimer molecule (ABb–aB$_2$) but $f_{C,b} = 1$ or $f_{C,b} = 2$ for a trimer molecule (Abb(aBb)aBB *vs.* Abb(aBB)$_2$).

The **degree of branching (DB)** is actually a mole fraction x_{br}, i.e., the number of actual branching units divided by the sum of potential branching units. In an AB$_2$-type molecule, the number of actual branching units is given by N_{abb} and the number of potential branching units by $N_{abb} + N_{aBb}$. Neither the core unit (Abb or ABb) nor the terminal units (aBB) should be counted since they are not available for branching *at the given degree of polymerization* (see Figs. 2-10 and 2-11):

$$(2-5) \qquad x_{br} \equiv \frac{N_{abb}}{X - 1 - N_{aBB}} = \frac{N_{abb}}{N_{abb} + N_{aBb}} \qquad (AB_2\text{-type})$$

For AB$_3$-type molecules, two branching parameters can be defined: a degree of branching, $x_{br} \equiv (N_{abbb} + N_{aBbb})/(N_{abbb} + N_{aBbb} + N_{aBBb})$, and a degree of hyperbranching, $x_{hbr} \equiv N_{abbb}/(N_{abbb} + N_{aBbb} + N_{aBBb})$. By definition, rings should be absent.

Various "degrees of branching" have been defined by different authors. Since these authors, in part, use the same symbols but with differing meanings (the number of "dendritic" units D (= N_{abb}) may or may not include the core unit, etc.), the following expressions are written below for AB$_2$-type *molecules* with the notation used in this book:

$$(2-6) \qquad DB = (N_{abb} + N_{aBB})/(X - 1) \qquad \text{a)}$$

$$(2-7) \qquad DB = 2 (N_{abb} + 1)/[2 (N_{abb} + 1) + N_{aBb}] \qquad \text{b)}$$

$$(2-8) \qquad DB = [N_{abb} + (N_{aBB} - 2)]/[N_{abb} + N_{aBb} + (N_{aBB} - 2)] \qquad \text{c)}$$

a) Hawker-Lee-Fréchet: $N_C = 0$. b) Hölter-Burgath-Frey: core ≡ dendritic. c) Yan-Müller-Matyjaszewski: $N_C = 0$; all B-groups of the core are assumed to have reacted (Radke-Livinenko-Müller).

Table 2-10 Comparison of degrees of branching DB = x_{br} of various AB_2-type model *molecules* (for Nos. 1, 2, and 7 see Fig. 2-11, for No. 4 see Fig. 2-10). Ring formation is assumed to be absent.

No.			Number of core units				Equations for DB			
	X	f_C	N_{abb}	N_{aBb}	N_{aBB}	Type	2-5	2-8	2-6	2-7
1	3	1	0	1	1	linear, monocore	0	−1/0	0.500	0.667
2	5	2	0	2	2	linear, bicore	0	0	0.500	0.500
2a	5	1	1	1	2	hyperbranched	0.500	0.500	0.750	0.800
3	15	1	2	8	4	hyperbranched	0.200	0.333	0.429	0.428
4	16	2	2	7	4	hyperbranched	0.222	0.364	0.400	0.462
5	15	1	5	3	6	hyperbranched	0.625	0.750	0.846	0.800
6	16	2	6	2	7	hyperbranched	0.750	0.846	0.867	0.875
6a	16	2	6	1	8	hyperbranched	0.857	0.923	0.933	0.933
7	7	2	2	0	4	dendron	1	1	1	1

Table 2-10 shows that all equations give the correct numbers for dendrons (DB ≡ 1) but only Eq.(2-5) gives the right answer for the DB of the simplest type of a linear molecule (DB ≡ 0). It is therefore highly likely that Eqs.(2-6)-(2-8) overestimate DB.

For polymeric *substances*, N_{abb} and N_{aBb} in Eq.(2-5) apply to either the sum of all units in all polymer molecules ($X_i \leq 2$) or to their number averages; x_{br} is then the number-average degree of branching of the polymer. The problem here is the generally unknown average number of b-groups in core units since $1 \leq N_{C,b} \leq 2$ for individual molecules. All equations furthermore assume that no rings are formed; this would result in the loss of 1 A-group and 1-B-group per molecule. The dependence of the degree of polymerization on monomer or group conversion is discussed in Section 13.8.3.

2.5.7 Comb Molecules

Ideal comb molecules consist of chains with repeating units containing long oligomeric or polymeric side chains. An example is a poly(methyl methacrylic ester) with poly(styrene) side chains that are connected to the main chains via $-O(CH_2)_2-$ groups and terminated by $CH_2CH(CH_3)_2$:

PMMA—*comb*—PS $CH_3-\overset{\displaystyle CH_2}{\underset{\displaystyle X}{C}}-COO(CH_2)_2-(\underset{\displaystyle C_6H_5}{CH}-CH_2)_N-CH_2CH(CH_3)_2$

Such comb molecules are obtained by chain polymerization of the corresponding monomers. The monomeric units are thus also the repeating branch units. The polymer molecules are regular; all side chains have the same length.

Side chains with different lengths are created by reaction of side groups of functionalized polymers with different monomer molecules. Any variations are obviously possible. Such molecules are sometimes called **polymer brushes**.

Comb molecules also result from graft reactions to or from linear polymer molecules (Section 16.2.3). These comb molecules are irregular since both the distance between branch units and the length of side units may vary.

2.5.8 Ladder Polymers

Two organic or inorganic chains can be united through one shared chain atom to **spiro chains** and by two or more shared adjacent chain atoms to **double-strand chains**. Polymers with double-strand chains are also called **ladder polymers**. Inorganic and semi-inorganic chains can also combine to more complicated structures that may be, for example, tentatively called "pearl-string" and "double ladders" (Fig. 2-12).

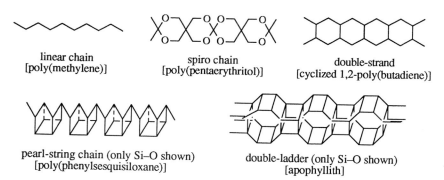

linear chain
[poly(methylene)]

spiro chain
[poly(pentaerythritol)]

double-strand
[cyclized 1,2-poly(butadiene)]

pearl-string chain (only Si–O shown)
[poly(phenylsesquisiloxane)]

double-ladder (only Si–O shown)
[apophyllith]

Fig. 2-12 Schematic representation of examples of some spiro, double-strand (ladder), pearl-string, and double-ladder polymer chains.

Some silicates have double-ladder structures, for example, apophyllith, a mineral with the composition $KCa_4Si_8O_{20}F \cdot 8\,H_2O$. The double ladder consists of two ladders that are united by four bonds per repeating unit. In each ladder, 16-membered rings alternate with 8-membered rings. Such tube-like silicates exist in many different types (Volume II). Corresponding carbon-ladder structures are unknown.

Strategies for the synthesis of organic ladder polymers are discussed in Chapter 16.3.1. Attempted syntheses, however, do not always yield the desired result. According to some authors, for example, the polymerization of the phenyl-substituted cubane analog silsesquioxane, $(C_6H_5)_8Si_8O_{12}$, does not result in the desired ladder polymer (Fig. 2-13, top right; Table 2-1, no. 8) but rather in a pearl-string polymer (Fig. 2-13, bottom right, and Fig. 2-12).

(2-9)

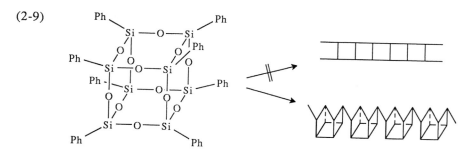

Fig. 2-13 Polymerization of $(C_6H_5)_8Si_8O_{12}$. In the formulas to the right, straight lines represent –Si–O–Si–; corners are occupied by the group $\geqslant$ Si–Ph.

2.5.9 Phyllo and Tecto Polymers

Polymers with regular two-dimensional structures are called **phyllo polymers** (IUPAC) and also **layer polymers, parquet polymers,** or **cyclomatrix polymers**. Graphite is a well-known example (Fig. 2-2). Another example is the soluble poly(silicon chloride) $[SiCl]_N$ that results from the reaction of dicalcium silicide, Ca_2Si, with iodine chloride ICl. The lattice atoms $=C<$ in graphite and $>Si-$ in $[SiCl]_N$ form planar layers that are one atom thick. Black phosphorus, on the other hand, exists in corrugated layers.

Mica, montmorrillonite, bentonite, and some other silicates have layers that are several atoms thick (Volume II): they are **sheet polymers**. A recently synthesized organic sheet polymer with molecular weights between 10^7 and 10^9 consists of chains of 22 chain atoms that are connected perpendicular to the polymeric layer plane. This sheet polymer is meltable and soluble because of its many constitutional errors (Chapter 16.3.2).

Tecto polymers such as diamond $+C+_n$, quartz $+SiO_2+_n$, and some other regularly built inorganic compounds are also called **lattice polymers**. The term "network polymers" should be avoided for tecto polymers since people associate "network" with a randomly crosslinked structure and not with a regular lattice. Lattice polymers exist exclusively in the solid state.

2.5.10 Polymer Networks

Basic Terms

Networks consist of many monomeric units or polymer chains that are irregularly interconnected by intermolecular chemical or physical bonds to "infinitely large" molecules. "Infinitely large" is a figure of speech; it is not used in the mathematical sense.

According to Avogadro, 1 mol of a substance contains ca. $6.02 \cdot 10^{23}$ molecules. If all molecules of 1000 g of a substance are united, the resulting crosslinked substance will have a molar mass of $(1000 \text{ g})(6.02 \cdot 10^{23} \text{ mol}^{-1}) = 6.02 \cdot 10^{26}$ g/mol. It is large but not "infinitely large."

Such networks result from the polymerization of monomers with three or more functionalities (exceptions: AYB_2, AXB_3, etc.) or the crosslinking reaction of polymers themselves. The latter reactions have been used by mankind from time immemorial without realizing that they have the same structural features. Materials like soft gels, pliant leather, elastic rubber, and stiff duroplastics are phenomenologically quite different but share the same molecular characteristics: crosslinks. As a consequence, crosslinking processes became known by many different names (Table 2-11).

Crosslinking processes also occur in nature; examples are coagulation (clotting, curdling) and gelation (gelling). Natural crosslinked materials include, for example, the lignins of wood, the collagen of skin, coagulated blood, gels of fruit pectins, and silicates of some rocks.

Chemically crosslinked polymers result from the intermolecular crosslinking of chains by covalent, coordinative, and/or electron-deficient bonds. The crosslinking junctions are usually trifunctional or tetrafunctional. They may consist of atoms (C, Si, N, etc.) or groups of atoms (benzene rings, etc.). Chemically crosslinked polymers are insoluble in *all* solvents.

Table 2-11 Important crosslinking processes of low-molecular weight (L), oligomeric (O), and polymeric (P) molecules to crosslinked materials (X) by chemical (c) or physical (p) bonds.

Process	Base material		Resulting material	Reaction		Bond
Drying	drying oil	→	coating	L	→ X	c
Setting	cement	→	cement stone	L	→ X	c
Hardening	resin	→	thermoset	O	→ X	c
Vulcanization	rubber	→	elastomer	P	→ X	c
Tanning	skin (collagen)	→	leather	P	→ X	c, p
Gelling	pectin solution	→	gel	P	→ X	p
Clotting	blood	→	coagulated blood	P	→ X	p
Curdling	milk	→	sour milk	P	→ X	p

Physically crosslinked polymers frequently behave like chemically crosslinked ones. Their junctions are very often composed of groups of many atoms. Examples are hydrophobic "bonds", ion domains, crystallites, microphases, and entanglements. In principle, physical networks can be de-crosslinked by appropriate solvents or a change in temperature.

Structure

A network consists of many **network chains** that extend from one chemical or physical **junction** to the next (Fig. 2-14). The type and distribution of network chains and network junctions control the mechanical properties of crosslinked polymers. Soft elastomers consist of long flexible network chains (sections between junctions) whereas stiff thermosets have very short network chains. Their rheological and thermal properties are influenced by the statistics of network junctions and the nature of the bonds.

In **ideal networks**, all junctions have the same functionality and all network chains the same number of monomeric units, leading to the same mesh sizes (Fig. 2-14). There are no dangling ends, loops, knots, or entanglements, and also no **excluded volume**, i.e., the space required by a segment that cannot be occupied by another one (see Volume III).

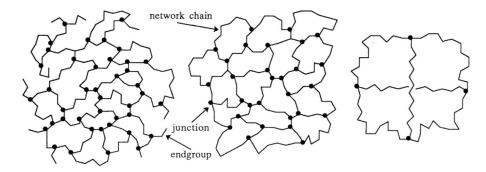

Fig. 2-14 Some networks with trifunctional network junctions or branching points.
Left: distribution of lengths of network chains and mesh sizes; dangling ends.
Center: distribution of network chains, equal mesh sizes of 14 segments each, no free ends.
Right: ideal network of equal network chains (15 segments each), equal mesh sizes, no free ends.

After a *complete* chemical crosslinking reaction, all original monomer and polymer molecules are united in a network that fills the whole reaction vessel. Crosslinking reactions are never totally complete, however, since the strongly increased viscosity of the reaction system prevents the diffusion of the remaining unreacted molecules and/or groups to reaction centers. Some monomer, oligomer and polymer molecules are thus not incorporated into the network. They can be extracted by suitable solvents.

Truly **interpenetrating networks (IPNs)** consist of two independent networks that penetrate each other without being interconnected by chemical bonds. A **semi-interpenetrating network (SIPN)** contains a network in a non-crosslinked polymer matrix.

IPNs are synthesized by swelling a crosslinked polymer poly(*cross*-a) in a monomer B that is subsequently polymerized. Monomers A and B can also be polymerized simultaneously to two independent networks poly(*cross*-a) and poly(*cross*-b) if neither reactive chain ends ~a* react with B nor reactive chain ends ~b* with A.

Since most polymers are incompatible with each other (Volumes III and IV), the goal of *complete molecular* interpenetration of the two networks is almost never obtained. During crosslinking polymerization(s), partial phase separation occurs and more or less extended domains of poly(*cross*-a) are formed in a matrix of poly(*cross*-b) and vice versa.

Semi-interpenetrating networks of poly(*cross*-a) in poly(b) are *partially crosslinked* polymeric substances. A branched polymer, on the other hand, is *not* a partially crosslinked polymer as often cited in the older literature (it can be dissolved).

Characterization

A **network chain** C extends from one trifunctional (or higher functional) crosslinking junction Y to the bifunctional monomeric unit M *before* the next junction Y' since this junction belongs already to another network chain. If these 2 junctions Y and Y' are

network chain C

connected by 4 monomer units M, then the degree of polymerization of C is $X = 5$. The mole fraction of the junctions is $x_Y = 1/5$ if the molar masses of Y and M are identical.

Chemical networks are characterized by their **degree of crosslinking** (= **crosslinking density**) which is defined as the amount-of-substance fraction, x_Y, of monomer units in junctions in all monomer units, crosslinked and uncrosslinked. This quantity equals the inverse of the number-average degree of polymerization, $\overline{X}_{n,x}$, of network chains:

(2-10) $x_Y = 1/\overline{X}_{n,x}$

Networks from the crosslinking of primary polymer molecules are characterized by a **crosslinking index** $N_{x,o}$ that indicates the number $N_{x,o}$ of network junctions per primary molecule. A **primary molecule** is a polymer molecule before crosslinking; its number-average degree of polymerization is $\overline{X}_{n,o}$. If the molecular weights of primary molecules and network chains are identical, then

$$(2\text{-}11) \qquad N_{x,o} = x_Y \frac{\overline{M}_{n,o}}{M_u} = x_Y \overline{X}_{n,o}$$

Network chains are long in elastomers but short in thermosets. Characteristic values of $\overline{X}_{n,o}$ of rubber products are 100-150 for surgical gloves, 50-80 for household gloves, 10-20 for treads of rubber tires, and 5-10 for hard rubber (Ebonite, a thermoset).

The concentration of *chemical* network junctions is usually too small to be determined by direct methods, e.g., NMR spectroscopy. Degrees of crosslinking and crosslinking indices are thus mostly determined by indirect methods, for example, mechanical properties (Volume III). Such methods deliver the *actual* concentration of all mechanically *effective* chemical *and* physical network junctions, however. The concentration of effective junctions over that of chemical junctions is increased by physical junctions such as crystalline domains and entanglements that, in part, depend on the strength and duration of the stress. A part of chemical junctions, on the other hand, may be wasted by the formation of loops and loose ends.

Lightly crosslinked chemical networks swell in solvents the more, the lower the degree of crosslinking. The swollen networks are rubbery or gel-like materials. Such **gels** may also be formed by physical networks (Volume III).

Gel particles with dimensions between 10 nm and 10 000 nm are called **microgels**. A microgel is usually soluble or dispersible. Depending on the molecule or particle size, a continuous transition exists from molecular solubility to dispersibility. For example, the intramolecularly crosslinked protein ribonuclease (124 amino acid residues) is molecularly soluble.

Literature to Chapter 2

2.2.a NOMENCLATURE OF CHEMICAL STRUCTURES (organic chemistry)
International Union of Pure and Applied Chemistry, A Guide to IUPAC Nomenclature of Organic
    Compounds. Recommendations 1993, Blackwell Sci.Publ., Oxford 1993

2.2.b NOMENCLATURE OF CHEMICAL STRUCTURES (macromolecular chemistry)
International Union of Pure and Applied Chemistry, Macromolecular Division, Commission on
Macromolecular Nomenclature:
    Compendium of Macromolecular Nomenclature, Blackwell Sci.Publ., Oxford 1991 (Purple Book)
    Nomenclature of Regular Double-Strand (Ladder and Spiro) Organic Polymers (IUPAC Recom-
        mendation 1993), Pure & Appl. Chem. **65** (1993) 1561
    Graphic Representations (Chemical Formulae) of Macromolecules (IUPAC Recommendations
        1994), Pure & Appl. Chem. **66** (1994) 2469
    Glossary of Basic Terms in Polymer Science, Pure & Appl. Chem. **68** (1996) 2287
    Source-Based Nomenclature for Non-Linear Macromolecules and Macromolecular Assemblies,
        Pure & Appl. Chem. **69** (1997) 2511
    Generic Source-Based Nomenclature for Polymers, Pure & Appl. Chem. **73** (2001) 1511
    Nomenclature of Regular Single-Strand Organic Polymers (Recommendations 2002), Pure Appl.
        Chem. **74** (2002) 1921
E.S.Wilks, Polymer Nomenclature and Structure: A Comparison of Systems Used by CAS, IUPAC,
    MDL, and DuPont. 3. Comb/Graft, Cross-Linked, and Dendritic/Hyperconnected/Star Polymers,
    J.Chem.Inf.Comput.Sci. **37** (1997) 209
E.S.Wilks, SRUs: Using the Rules, Polym.Prepr. **41**/1 (2000) 6a

2.3 ATOMIC STRUCTURE AND CHAIN FORMATION

F.G.R.Gimblett, Inorganic Polymer Chemistry, Butterworths, London 1973

A.L.Rheingold, Homoatomic Rings, Chains and Macromolecules of Main-Group Elements, Elsevier, Amsterdam 1977

J.E.Mark, H.R.Allcock, R.West, Inorganic Polymers, Prentice Hall 1992 (only polyphosphazenes, polysiloxanes, and polysilanes)

J.-M.Lehn, Supramolecular Chemistry. Concepts and Perspectives, VCH, Weinheim 1995

A.Ciferri, Ed., Supramolecular Polymers, Dekker, New York 2000

R.D.Archer, Inorganic and Organometallic Polymers, Wiley, New York 2001

J.L.Atwood, J.W.Steed, Eds., Encyclopedia of Supramolecular Chemistry, Dekker, New York 2004

2.5.2.a LINEAR CHAINS AND RINGS

G.Schill, Catenanes, Rotaxanes and Knots, Academic Press, New York 1971

J.A.Semlyen, Ed., Cyclic Polymers, Elsevier, New York 1986

E.J.Goethals, Ed., Telechelic Polymers: Synthesis and Applications, CRC Press, Boca Raton (FL) 1988

C.V.Uglea, I.A.Negulescu, Synthesis and Characterization of Oligomers, CRC Press, Boca Raton (FL) 1991

J.A.Semlyen, Ed., Large Ring Molecules, Wiley, New York 1997

2.5.2.b COPOLYMERS

J.C.Randall, Polymer Sequence Determination - Carbon 13 NMR Method, Academic Press, New York 1977

H.Inagaki, T.Tanaka, Separation and Molecular Characterization of Polymers, Dev.Polym.Charact. **3** (1982) 1

J.M.G.Cowie, Ed., Specialty Polymers, Vol. 1, Alternating Copolymers, Plenum, New York 1984

M.J.Folkes, Ed., Processing, Structure, and Properties of Block Copolymers, Elsevier, New York 1985

G.Riess, G.Hurtrez, P.Bahadur, Block Copolymers, Encycl.Polym.Sci.Eng., Wiley, New York, 2nd ed., **2** (1985) 324

N.Hadjichristidis, S.Pispas, G.A.Floudas, Block Copolymers, Wiley, New York 2003

2.5.2.c ION-CONTAINING POLYMERS

F.Oosowa, Polyelectrolytes, Dekker, New York 1971

A.Eisenberg, M.King, Ion-Containing Polymers: Physical Properties and Structure, Academic Press, New York 1977

S.Schlick, Ed., Ionomers. Characterization, Theory, and Applications, CRC Press, Boca Raton (FL) 1996

2.5.3 STATISTICALLY BRANCHED POLYMERS

P.A.Small, Long-Chain Branching in Polymers, Adv.Polym.Sci. **18** (1975) 1

Th.G.Scholte, Characterization of Long-Chain Branching in Polymers, Dev.Polym.Charact. **4** (1981) 1

W.W.Graessley, Entangled Linear, Branched and Network Polymer Systems - Molecular Theories, Adv.Polym.Sci. **47** (1982) 67

2.5.4 STAR MOLECULES

S.Bywaters, Preparation and Properties of Star-Branched Polymers, Adv.Polym.Sci. **30** (1979) 89

M.K.Mishra, S.Kobayashi, Star and Hyperbranched Polymers, Dekker, New York 1999

2.5.5 DENDRIMERS

E.S.Wilks, Nomenclature and Structure Representation for Dendritic Polymers, Polym.Prepr. **39**/2 (1998) 6

D.A.Tomalia, A.M.Naylor, W.A.Goddard III, Starburst-Dendrimere, Angew.Chem. **102** (1990) 119; Starburst-Dendrimers, Angew.Chem.Int.Ed.Engl. **29** (1990) 138

G.R.Newkome, C.N.Moorefield, F.Vögtle, Dendritic Molecules. Concepts, Syntheses, Perspectives, VCH, Weinheim 1996

P.R.Dvornic, D.A.Tomalia, Recent Advances in Dendritic Polymers, Curr.Opin.Colloid Interface Sci. **1** (1996) 221

O.A.Matthews, A.N.Shipway, J.F.Stoddart, Dendrimers–Branching out from Curiosities into New Technologies, Progr.Polym.Sci. **23** (1998) 1

A.D.Schlüter, J.P.Rabe, Dendronized Polymers, Angew.Chem. **112** (2000) 860; Angew.Chem.Int.Ed.Engl. **39** (2000) 864

G.R.Newkome, C.N.Moorefield, F.Vögtle, Dendrimers and Dendrons. Concepts, Syntheses, Applications, Wiley-VCH, Weinheim 2001

J.M.J.Fréchet, D.A.Tomalia, Eds., Dendrimers and Other Dendritic Polymers, Wiley, New York 2002

2.5.6 HYPERBRANCHED POLYMERS
M.K.Mishra, S.Kobayashi, Star and Hyperbranched Polymers, Dekker, New York 1999

2.5.7 COMB MOLECULES
N.A.Plate, V.P.Shibaev, Eds., Comb-Shaped Polymers and Liquid Crystals, Plenum, New York 1987

R.C.Advincula, W.J.Brittain, K.C.Caster, J.Rühe, Polymer Brushes, Wiley-VCH, Weinheim 2004

2.5.8-2.5.9 LADDER, PHYLLO AND TECTO POLYMERS
W.De Winter, Double Strand Polymers, Revs.Macromol.Sci. **1** (1966) 329

C.G.Overberger, J.A.Moore, Ladder Polymers, Adv.Polym.Sci. **7** (1970) 113

2.5.10 POLYMER NETWORKS
L.H.Sperling, Interpenetrating Polymer Networks and Related Materials, Plenum, New York 1981

J.E.Mark et al., Polymer Networks, Adv.Polym.Sci. **44** (1982) 1

K.Dusek, Formation and Structure of End-Linked Elastomer Networks, Rubber Chem.Technol. **55** (1982) 1

S.C.Temin, Recent Advances in Crosslinking, J.Macromol.Sci.-Revs.Macromol.Chem.Phys. C **22** (1982/83) 131

D.J.P.Harrison, W.R.Yates, J.F.Johnson, Techniques for the Analysis of Crosslinked Polymers, J.Macromol.Sci.-Revs.Macromol.Chem.Phys. C **25** (1985) 481

O.Kramer, Ed., Biological and Synthetic Polymer Networks, Elsevier Appl. Sci., New York 1988

A.Baumgärtner, C.Picot, Eds., Molecular Basis of Polymer Networks, Springer, Berlin 1989

W.Burchard, S.B.Ross-Murphy, Eds., Physical Networks, Elsevier Appl. Sci., Amsterdam 1990

L.H.Sperling, S.C.Kim, Eds., IPNs Around the World, Wiley, New York 1997

R.F.T.Stepto, Ed., Polymer Networks - Principles of Their Formation, Structure, and Properties, Blackie Academic, Glasgow, UK 1997

References to Chapter 2

[1]   O.-A.Neumüller, Römpps Chemie-Lexikon, Franckh'sche Verlagshandlung, Stuttgart, 8th ed. 1979-1988; [1a] p. 3806; [1b] p. 2642; [1c] p. 662

[2]   A.Thackray, "John Dalton". Critical Assessments of His Life and Science, Harvard University Press, Cambridge (MA) 1972

[3]   C(arl Wilhelm von) Naegeli, Die Micellartheorie, Akad.Verlagsgesellschaft, Leipzig 1928 (Ostwald's Klassiker, Neue Ausgabe, vol. 227). Naegeli lived 1817-1891.

[4]   J.Falbe, M.Regitz, Eds., Römpp Chemie Lexikon, G.Thieme Verlag, Stuttgart, 9th ed. 1989 ff., p. 650

[5]   L.Pauling, The Nature of the Chemical Bond and the Structure of Molecules and Crystals, Cornell University Press, Ithaca (NY), 3rd ed. 1960

[6]   O.Glemser, Inorganic Polymers, XXIVth International Congress of Pure and Applied Chemistry (Hamburg 1973), Butterworths, London 1974, vol **1**, p. 177

[7]   E.Grimm, Naturwiss. **27** (1938) 17

[8]   H.Staudinger, in W.Röhrs, H.Staudinger, R.Vieweg, Fortschritte der Chemie, Physik und Technik der makromolekularen Stoffe, J.F.Lehmanns Verlag, München - Berlin 1939, p. 1

[9]   R.F.Bacon, R.Fanelli, J.Am.Chem.Soc. **65** (1963) 639, Fig. 3

[10]   D.M.Gardner, K.Fraenkel, J.Am.Chem.Soc. **78** (1956) 3279, different data

[11]   A.V.Tobolsky, A.Eisenberg, J.Am.Chem.Soc. **81** (1959) 780, Fig. 4

3 Characterization of Polymers

3.1 Introduction

Organic chemists often view synthetic macromolecules as uncomplicated chemical compounds that consist of monotonous arrays of simple monomeric units. Polymers used as plastics, elastomers, and fibers do indeed have such uncomplicated structures because they are derived from simple monomers that result directly or indirectly from products obtained from fossil hydrocarbons. These monomers dominate because petroleum, natural gas, and, to some extent, coal, are the most economical feedstocks for polymers.

This picture is idealized in many respects. Plastics, elastomers, and fibers are not simply polymeric substances but polymer-based *materials*. For technological and economical reasons, they contain many adjuvants such as plasticizers, antioxidants, fillers, impact improvers, etc. Industrial polymers are also not "clean" since they always contain catalyst residues, emulsifiers, and other polymerization aids that remain after work-up but must be removed before a chemical analysis of the polymer.

The polymers themselves are not so "pure" as their idealized chemical structures suggest. During polymerization, monomeric units may be incorporated in the wrong direction (regioisomerism) or in different configurations (stereoisomerism). Side-reactions may have led to lesser or greater degrees of branching. Added regulators or polymerization conditions themselves (temperature, monomer concentration, monomer conversion, pressure) may generate different molecular weights and molecular weight distributions in both industrial processes and laboratory procedures.

Even small differences in polymer structures and/or types and proportions of adjuvants may produce large differences in physical structures and, in turn, large differences in physical properties. Examples are the various industrial poly(ethylene)s that all have the same idealized chemical structure $+CH_2-CH_2\frac{}{n}$ (Table 3-1).

Table 3-1 Physical properties of some of the 17 grades of unfilled poly(ethylene)s (PE) of Neste.
HD = high-density type (usually by insertion polymerization of ethene); LD = low-density type (usually by free-radical polymerization); LLD = very low-density type (by insertion copolymerization of ethene with several percent of 1-butene, 1-hexene, or 1-octene). The densities of PE-LDs may be lower than those of PE-LLDs since the latter are copolymers and not homopolymers.

Physical property	Physical unit	Temperature in °C	LD NCPE 1515	LLD NCPE 8030	LLD NCPE 8644	LD NCPE 3416	HD NCPE 7003	HD NCPE 7007
Density of plastic	g cm^{-3}	25	0.915	0.919	0.935	0.958	0.958	0.964
Density of melt	g cm^{-3}	225		0.787	0.794		0.814	0.819
Specific heat capacity	J kg^{-1} K^{-1}	225		3070	3150		3350	3420
Melt flow index	mL/10 min	190	18	36	9.5	0.24	3.6	8.3
Heat distortion temperature (B)	°C		41	46			84	80
Heat distortion temperature (A)	°C			39	58	50	54	
Modulus of elasticity	MPa	25	150	830	650	1350	1350	1600
Tensile strength at yield	MPa	25	8	12	17	30	30	31
Extension at yield	%	25		12	12	8.1	8.2	7.1
Notched impact strength	kJ/m^2	−30	no fracture	11	5.9	5	5.4	4.6

3.2   Chemical Composition

Since small proportions of external and internal impurities and small variations in molecular weight and molecular weight distribution may cause vast differences in the properties of polymers with the same idealized chemical structure, very many different analytical methods are employed by macromolecular science. Most of these methods are borrowed from other scientific disciplines; this book will thus discuss their polymer-relevant aspects and not their basics. Polymer-specific methods will be treated in greater detail: methods most often used in preparative polymer chemistry in this volume, all other physical methods in Volume III, Physical Structures and Properties.

3.2.1   Monomeric Units

"Chemical experience", based on the principle of smallest constitutional change in chemical reactions, often provides a reliable guess of the constitution of monomeric units. It is indeed hard to imagine that the chain polymerization of ethene, $CH_2=CH_2$, leads to other mers than $-CH_2-CH_2-$ and the polycondensation of a dicarboxylic acid, $HOOC-Z-COOH$ with a glycol $HO-Z'-OH$ to other monomeric units than $-CO-Z-CO-$ and $-O-Z'-O-$. Isomerizations of monomers before and during polymerization although not common are not rare either. The constitution of the monomeric units thus always has to be checked, preferably by more than one method since different methods may lead to different results, depending on the resolution and methodology (Table 3-2).

Elemental analysis is often not very sensitive, especially for C,H,O polymers, and copolymers thereof. A copolymer with 97 % ethylene units, $-CH_2-CH_2-$, and 3 % 1-butylene units, $-CH_2-CH(CH_2CH_3)-$, has the same elemental composition as the homopolymer of ethene $+CH_2-CH_2+_n$. Many papers thus no longer report elemental analyses. Hetero atoms such as nitrogen and chlorine are, however, easy to determine.

Wet chemical methods are useful for specific groups such as COOH, OH, CH_3CO, Cl, NCO, etc. Such groups can also be determined by many physical methods, e.g., atomic absorption spectroscopy.

Ultraviolet (UV) **spectroscopy** is important for conjugated double bonds and other groups that lead to strong absorption bands. The results may deviate from those of other methods, however, if the position of those bands depends on the sequence length of the monomeric units, for example, of styrene sequences in copolymers (Table 3-2).

Table 3-2 Composition of several styrene–methyl methacrylate copolymers by determination of the refractive index gradient dn/dc in solution, nuclear magnetic resonance spectroscopy (NMR), infrared spectroscopy (IR), elemental analysis (C, H, O), and ultraviolet spectroscopy (UV) [1].

Polymer	Percent methyl methacrylate units in copolymer by				
	dn/dc	NMR	IR	C,H,O	UV
2	72.8	73.5	74.0	74.4	78.5
4	57.0	-	53.0	58.1	57.7
6	41.5	40.2	41.0	42.2	48.5
8	24.1	24.1	23.5	23.0	28.7

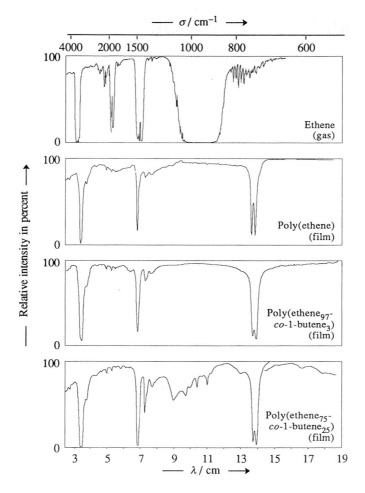

Fig. 3-1 Infrared spectra of ethene, linear poly(ethylene), and two copolymers of ethene and 1-butene. Relative intensities are presented as a function of wave length λ or wave number σ.

Infrared spectroscopy (IR) is especially suitable for polar groups but is too insensitive for the quantitative determination of small proportions of other groups. The copolymeric poly[(ethylene)$_{97\%}$–co-(1-butylene)$_{3\%}$] possesses practically the same IR spectrum as the homopolymeric poly(ethylene) (Fig. 3-1). Significant changes in the IR spectrum occur only at higher concentrations of 1-butylene units. Intensities and positions of IR bands furthermore depend considerably on the physical state. They differ for crystalline, molten, and dissolved polymers.

Nuclear magnetic resonance (NMR) spectra are fairly easy to evaluate if groups are present in sufficiently high concentrations. NMR is the most important method for the determination of configurations (tacticities) of polymer segments (Chapter 4).

All methods are relatively insensitive to the presence of small proportions of groups with structures other than the main monomeric units. The presence of these structures is however often revealed by their solution or solid state properties.

3.2.2 Endgroups

By definition, endgroups are groups located at the end of chains. Two endgroups are thus present in an unbranched (linear) macromolecule, four in a star-molecule with four arms, and none in a cyclic chain (ring molecule).

The structure of endgroups allows conclusions about certain aspects of polymer synthesis, sometimes also about molar masses and degrees of branching. The higher the molar mass of polymers with the same molecular architecture, the smaller the proportion of endgroups.

The proportion of endgroups allows one to calculate the molar mass of the polymer, provided the number of ends per molecule (the molecular architecture) and the chemical nature of *all* endgroups is known. Since the proportion of endgroups is inversely proportional to the molar mass M, endgroup determinations of *linear* polymers must relate to the **number-average molar mass**, $\overline{M}_n$, of the polymer. In general, the number-average molar mass of linear macromolecules is calculated from the amount-of-substance n_i (the "moles") of the various types i of endgroups, the number N_{end} of total endgroups per macromolecule, and the mass m (in gram) of the sample:

$$(3-1) \qquad \overline{M}_n = (N_{end}m)/(\Sigma_i\, n_i)$$

Thus, the number-average molar mass (in g/mol) is *not* the arithmetic mean of the equivalent masses of each of the various types of endgroups.

The sensitivity of molar-mass determinations via endgroups depends not only on the type of endgroup but also on the method employed. Upper limits are ca. 8000 g/mol by ^{13}C NMR, ca. 40 000 g/mol by titration, ca. 100 000 g/mol by microanalysis of iodine, ca. 200 000 g/mol if radioactive endgroups are present, and ca. 1 000 000 g/mol if the endgroups are intensively colored or are fluorescent.

3.2.3 Branches

Endgroup analysis in combination with molar-mass determination by an independent method allows one to determine the number of branches and arms, respectively, per molecule. Subsequent branching of the branches themselves falsifies the result. The direct determination of branch points by, for example, spectroscopic methods is restricted to very specific chemical structures.

The degree of branching is often estimated by comparing the spatial molecular dimension of the branched polymer with that of an unbranched one of the same composition and molar mass. Branched molecules always have smaller radii of gyration than unbranched ones. Radii can be obtained directly by scattering methods (Volume III) or indirectly from intrinsic viscosities (Section 3.6.5). Care has to be taken, however, if the polymers have broad molar mass distributions and/or are molecularly inhomogeneous.

Deviation of the experimentally determined molar mass distribution from the theoretically expected one may also indicate the presence of branching, especially in free-radical polymerizations. Such a conclusion however requires the knowledge of the termination mechanisms, for example, the ratio of termination by disproportionation to that of combination in free-radical polymerizations (Chapter 10).

3.3 Molar Mass Distributions

3.3.1 Overview

The biosynthesis of enzymes and nucleic acids, possibly also of native polysaccharides, results in molecularly uniform polymers with macromolecules of the same constitution and molar mass. Synthetic polymers and many other natural polymers are however non-uniform with regard to the degree of polymerization of their molecules. They thus have a **molar mass distribution** (= **molecular weight distribution**). Some of these polymers are not only non-uniform with respect to molecular weight but also with respect to their constitution and configuration; this additional non-uniformity may be caused by the polymerization reaction itself or by the isolation procedure.

A convention is necessary to distinguish the different "uniformities" from each other. This book uses "uniform" if all molecules have the same constitution and the same degree of polymerization, "constitutionally uniform" if only the constitution is the same but not the degree of polymerization, "molecularly uniform" if all molecules have the same degree of polymerization, and "non-uniform" if the constitution *and* the degree of polymerization vary from molecule to molecule. The literature also employs "molecularly uniform" sometimes in the sense of "constitutionally uniform" or even "uniform." For the common but incorrect use of "monodisperse" and "polydisperse" instead of "uniform" and "non-uniform," see page 44.

Distributions are described by theoretical or empirical **distribution functions**. These functions indicate the extent to which molecules with a certain property (molar mass, degree of polymerization, composition, etc.) are present in the substance. This section is concerned with distributions of degrees of polymerization and molar masses. Distributions of other properties can be treated with the same formalisms. Care has to be taken, however, to use the correct definitions of averages (Section 3.5.6).

Theories of polymerization kinetics and equilibria as well as those of polymer reactions deliver degrees of polymerization and their distributions. Experiments, on the other hand, always determine molar masses and molar mass distributions. The conversion of one into the other is possible if the composition of molecules in known; this is not always true for endgroups and for units in many hyperbranched polymers.

3.3.2 Distribution Functions

Distribution functions of properties are normalized mathematical functions that indicate the proportion of the various properties. They differ in the **continuity** of the distribution functions, the **summation** of the weighting parameters, and the type of **weighting** parameters.

The degree of polymerization X of a polymer *molecule* is a positive integer since degrees differ from each other by at least $\Delta X = 1$. Distributions of degrees of polymerization are therefore always **discontinuous** (or **discrete**) **step functions** (Fig. 3-2, left). $\Delta X = 1$ is small compared to X itself which may be in the thousands and even millions. For calculations, it can therefore be neglected and the discontinuous distribution of degrees of polymerization can be replaced by a **continuous** distribution (Fig. 3-2, right).

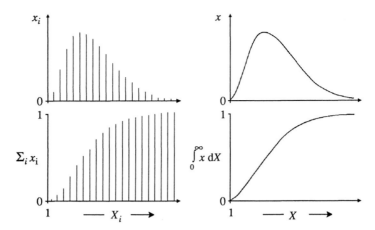

Fig. 3-2 Number distributions (expressed as mole fractions x of degrees of polymerixation, X.
Upper left: discrete differential, $x_i = f(X_i)$;   upper right: continuous differential, $x = f(X)$.
Lower left: discrete integral, $\Sigma_i x_i = f(X_i)$;   lower right: continuous integral, $_0\int^\infty x\,dX = f(X)$.

Distributions may also be **differential** or **integral (cumulative)** (Fig. 3-2). Differential distributions describe the population of a property whereas integral distributions sum all populations up to a certain property.

The statistical **weights** may be numbers, masses, or other statistical weights (see below and Volumes III and IV). Reaction mechanisms deliver **number distributions** that may be expressed as distributions of the numbers N_i of molecules of a certain property i, their amounts $n_i = N_i/N_A$ (in moles), their mole fractions $x_i = n_i/\Sigma_i n_i = n_i/n$, the corresponding number concentrations, $[N_i] = N_i/V$, or amount-of-substance concentrations ("molar concentrations"), $[n_i] = n_i/V$, where N_A = Avogadro constant and V = total volume of the solution.

Most physical methods deliver **mass distributions** ("weight distributions"). They are described by the **mass** $m_{mol,i}$ of a molecule of type i, the total mass $m_i = \Sigma_i m_{mol,i} = n_iM_i$ of all i molecules, their **mass fraction** ("weight fraction") $w_i = m_i/\Sigma_i m_i$, or their **mass concentration** ("weight concentration") $c_i = m_i/V$. Mass-statistical weights can be converted into number-statistical weights and *vice versa* if molar masses M_i are known.

Some physical methods deliver higher statistical weights or combinations of statistical weights. The **z-statistical weight** can be expressed by $z_i = m_iM_i = n_iM_i^2$, the **z-statistical weight fraction** by $Z_i = z_i/\Sigma_i z_i$,. Some mechanical properties depend on even higher statistical weights, for example, $\{z+1\}$. The notation z+1 does *not* indicate a mathematical operation since the $\{z+1\}$-statistical weight is given by $\{z+1\}_i = z_iM_i = w_iM_i^2 = n_iM_i^3$.

All expressions mentioned above apply strictly to molecularly uniform species i. In interconversions of statistical weights of non-molecularly uniform species i, the molar mass M_i is an average that corresponds to the multiplying statistical weight, as can be easily shown numerically. For example, the mass of a non-molecularly uniform species i is generally given by $m_i = n_i \overline{M}_{n,i}$ and not by $m_i = n_iM_i$, its z-statistical weight by $z_i = m_i \overline{M}_{w,i} = n_i \overline{M}_{w,i} \overline{M}_{n,i}$ and not by $z_i = m_iM_i = n_iM_i^2$, etc.

Similar rules apply to the interconversion of mole fractions, weight fractions, etc. The mass fraction of a molecularly non-uniform species i equals, for example, $x_iM_i/\overline{M}_n$ and not $x_iM_{r,i}$ (note that $M_{r,i}$ is dimensionless but M_i has the unit mass/mol).

3.3.3 Types of Distributions

There are many different types of distribution functions, most of which are named after their discoverers. In macromolecular chemistry, the following types are most important: Gaussian distribution (normal distribution, "error law"), logarithmic normal distribution (LN), generalized logarithmic normal distribution (GLN), Poisson distribution, Schulz-Zimm distribution (SZ distribution), and Kubin distribution. The number distributions (in terms of mole fractions x) of degrees of polymerization X read

$$(3\text{-}2) \qquad x(X) = \frac{1}{(2\pi)^{1/2}\sigma_n}\left[\exp\left(\frac{-(X-\overline{X}_n)^2}{2\sigma_n^2}\right)\right] \qquad \text{Gaussian}$$

$$(3\text{-}3) \qquad x(X) = \frac{1}{(2\pi)^{1/2}\sigma_n^*}\left[\exp\left(\frac{-(\ln X - \ln X_{med})^2}{2(\sigma_n^*)^2}\right)\right] \qquad \text{LN}$$

$$(3\text{-}4) \qquad x(X) = \frac{1}{(2\pi)^{1/2}\sigma_n^*}\frac{\overline{X}_n}{BX_{med}^{A+1}}X^{A-1}\left[\exp\left(\frac{-(\ln X - \ln X_{med})^2}{2(\sigma_n^*)^2}\right)\right] \qquad \text{GLN}$$

$$(3\text{-}5) \qquad x(X) = \frac{(\overline{X}_n-1)^{X-1}}{\Gamma(X)}\left[\exp(1-\overline{X}_n)\right] \qquad \text{Poisson}$$

$$(3\text{-}6) \qquad x(X) = \frac{\varsigma^{\varsigma+1}}{\overline{X}_n^{\varsigma}\,\Gamma(\varsigma+1)}\left[X^{\varsigma-1}\exp(-\varsigma X/\overline{X}_n)\right] \qquad \text{Schulz-Zimm}$$

$$(3\text{-}7) \qquad x(X) = \frac{\beta^{(\lambda+1)/\gamma}\overline{X}_n}{\Gamma[(\lambda+1)/\gamma]}\left[X^{\lambda-1}\exp(-\beta X^\gamma)\right] \qquad \text{Kubin}$$

where $\sigma_n = (\overline{X}_w\,\overline{X}_n - \overline{X}_n^2)^{1/2}$ and $\sigma_n^* = [\ln\overline{X}_w\cdot\ln\overline{X}_n - (\ln\overline{X}_n)^2]^{1/2}$ are number standard deviations, X_{med} = median ($X_{med} = \overline{X}_n$ for Gaussian distributions), $\varsigma = \overline{X}_n/(\overline{X}_w - \overline{X}_n)$ = degree of coupling of chains, and A, B, β, γ, λ = adjustable constants.

The distributions are derived in Chapters 6-16 for various types of polymerizations. Theories of thermodynamics and kinetics of polymerization always lead to number distributions since reactions depend on the number of reacting species and not on masses. However, experimental determinations of distribution functions deliver mass distributions $m(X)$, usually expressed as weight-fraction distributions, $w(X)$.

Experimental weight fractions w_i can be converted into "theoretical" mole fractions x_i by $w_i = x_i\overline{M}_{n,i}/\overline{M}_n$. The analogous conversion $w_i = x_i\overline{X}_{n,i}/\overline{X}_n$ is valid only if end-groups can be neglected. The interconversion $w_i = x_iX_i$ is wrong.

Gaussian Distribution

A Gaussian distribution describes the distribution of an occurence if all occurences are independent of each other. In mathematics, it is often called a **normal distribution**.

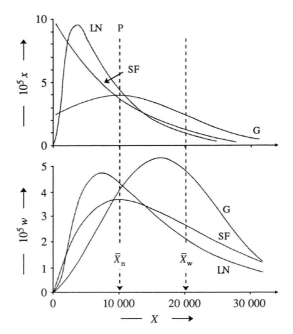

Fig. 3-3 Differential continuous number distributions of degrees of polymerization of a Gaussian distribution (G), a logarithmic normal distribution (LN), and a Schulz-Flory distribution (SF) (a special case of the Schulz-Zimm distribution, see below)), all with $\overline{X}_n = 10\,000$, $\overline{X}_w = 20\,000$, and therefore $\overline{X}_w/\overline{X}_n = 2$. Top: Distributions of mole fractions; bottom: distributions of weight fractions.

A Poisson distribution (P) with $\overline{X}_n = 10\,000$ has a ratio of weight-to-number average degrees of polymerization of only $\overline{X}_w/\overline{X}_n \approx 1.0001$. It is so narrow that it shows up in this graph as a vertical broken line at $\overline{X}_n = 10\,000$ and not as a distribution.

The Gaussian distribution is a special type of an exponential distribution. It can be written as $y = (A/\pi)^{1/2} \exp(-A^2 x^2)$ with $y = x(X)$, $A = 1/(2^{1/2} \sigma_n)$, and $x = X - \overline{X}_n$. Because $x = X - \overline{X}_n$, the most probable value of X is not at $X = 0$ but at $X = \overline{X}_n$ (Fig. 3-3). The common **error law** $y = (A/\pi)^{1/2} \exp(-A^2 x^2)$ allows negative values of x whereas the Gaussian distribution of degrees of polymerization, Eq.(3-2), is cut off at $X \leq 1$. For these formal reasons alone, Gaussian distributions cannot describe the distributions of degrees of polymerization of polymers.

Physically more serious is the fact that bond formation in polymerization, though happening at random, occurs at molecules of different degrees of polymerization. The non-exponential factors in distribution functions must therefore always contain the degree of polymerization, X, in addition to the number-average degree of polymerization, $\overline{X}_n$. This is true for Eqs.(3-4)-(3-7) but not for Eqs.(3-2) and (3-3) (see expressions for σ_n and $\sigma_n{}^*$).

Gaussian *number* distributions of degrees of polymerization, X, are symmetric about the number-average $\overline{X}_n$ (here the median) but must have a cut-off at $\overline{X}_n = 1$. Gaussian *weight* distributions are not symmetric (difficult to see in Fig. 3-3 because of the choice of parameters). The resulting weight distribution is *not* the curve that results from a formal replacement of x by w, $\overline{X}_n$ by $\overline{X}_w$, and σ_n by $(\overline{X}_z \overline{X}_w - \overline{X}_w{}^2)^{1/2}$! Note also that the maximum is neither at $\overline{X}_n$ nor at $\overline{X}_w$.

Logarithmic Normal Distribution

The logarithmic normal distribution (LN) replaces the properties X and $\overline{X}_n$ of Gaussian distributions by their natural logarithms (Eq.(3-3)). The functions $x = f(X)$ and $w = f(X)$ are thus no longer symmetric about the median (Fig. 3-3).

The LN distribution can be generalized (Eq.(3-4)). The resulting GLN distribution has two special cases that are occasionally used to describe distributions of degrees of polymerizations and molar masses, respectively. So far, neither the **Lansing-Kraemer distribution** ($A = 0$, $B = \exp[(1/2)(\sigma_w{}^*)^2]$) nor the **Wesslau function** ($A = 1$, $B = 1$) has been correlated with a reaction mechanism.

Poisson Distribution

Poisson distributions of degrees of polymerization are generated if a constant number of polymer chains starts to grow simultaneously and monomer molecules add to these chains at random and independently of the preceding steps (Chapter 8). Poisson distributions are very narrow (Fig. 3-3). Their narrowness, as measured by $\overline{X}_w/\overline{X}_n$, increases with increasing number-average degrees of polymerization:

$$(3-8) \qquad \overline{X}_w/\overline{X}_n = 1 + (1/\overline{X}_n) - (1/\overline{X}_n)^2$$

Example: A 4-year old child is twice as old as a 2-year old one. The age ratio is 2.0 and the age distribution is broad. After 70 years, these individuals are now 74 and 72. The age ratio is reduced to 1.028 and the age distribution is narrow.

Living ionic polymerizations deliver distributions that approach Poisson distributions. True Poisson distributions are not obtained because of additional effects.

Schulz-Zimm and Schulz-Flory Distributions

The formation of Poisson distributions of *degrees of polymerization* requires that the active centers initially present are *individually* preserved during the polymerization process, for example, a monomer anion as a macroanion after addition of many monomer molecules. **Schulz-Zimm distributions (SZ distributions)**, however, demand only that the *total* concentration of active centers remains constant. For example, an individual radical adds monomer molecules at random until the radical is deactivated. At the same time, new radicals are formed by decomposition of initiator molecules, chain transfer reactions, etc., so that the total radical concentration may be time-invariant.

Widths of SZ distributions are given by the **degree of coupling**, ς, of chains, i.e., the number of individually grown chains that are coupled to a dead chain. The degree of coupling is always unity for equilibrium polycondensations of bifunctional monomers (Section 13.2.5), free-radical polymerizations with termination by disproportionation (Section 10.3.5), and statistical splittings of chains to smaller molecules (Section 15.5), provided that the degrees of polymerizations are high. In this case, the SZ distribution converts to the **Schulz-Flory distribution** (SF distribution (IUPAC)) or **Flory distribution (US)**). Since the distribution applies to the thermodynamic and kinetic "standard cases" of polymerization and degradation, it is called the "**most probable distribution.**"

The function $x = f(X)$ is monotonous for SZ distributions whereas the function $w = f(X)$ shows a maximum which is located at $X = \overline{X}_n$ for SF distributions (Fig. 3-3).

Generalized Exponential Distributions

Generalized exponential distributions (**GEX distributions**) have in the exponential term the variable with some power. Such **stretched exponentials** deliver distribution functions that can be easily adopted. Often used GEX distributions include the **Kubin distribution** (Eq.(3-7)), the **Rammler-Bennett distribution**, $w = \exp(- X/X_0)^B$, and the **Weibull distribution**, $w = 1 - \exp\{(X_0 - X)^B/A]$, where A and B are adjustable constants.

The Kubin distribution is an empirical distribution that contains several other distributions as special cases (Table 3-3).

Table 3-3 Special cases of the Kubin distribution, Eq.(3-7).

Equation	Distribution	γ	λ	β	Comments
3-3	Logarithmic normal distribution	0	∞		limiting values
3-6	Schulz-Flory distribution	1	1	$1/\overline{X}_n$	
3-6	Schulz-Zimm distribution	1	ς	$\varsigma/\overline{X}_n$	
3-7	Tung distribution	γ	$\gamma - 1$		

3.4 Determination of Molar Mass Distributions

Molar mass distributions can be determined by many methods (Table 3-4). Mass spectroscopy and sedimentation equilibrium are absolute methods that need no calibration. All other methods are relative methods that require a correlation of the measured properties with molar masses that have been determined independently. Many methods are also expensive and/or time consuming (see Volume III). For these reasons, preparative and kinetic studies use only a few methods for the determination of molar mass distributions, mainly mass spectroscopy, size exclusion chromatography, and nowadays less often, preparative fractionations.

Table 3-4 Methods for the determination of molar mass distributions. A = analytical method, D = desorption from a matrix, M = in the melt, P = preparative method, S = in solution, SP = in solution with precipitant, ST = in solution with temperature variation. * See discussion in volume III.

Method	Type	State	Measured quantity
Mass spectroscopy (MALDI)	A	D	relative mass/charge
Size exclusion chromatography (SEC, GPC)	A, P	S	retention volume
Dynamic light scattering *	A	S	diffusion coefficient
Fractionation	P	SP	precipitated mass
Cloud point titration with precipitant *	A	SP	turbidity
Cloud temperature analysis *	A	ST	turbidity
Ultracentrifugation (sedimentation rate or equilibrium) *	A	S	sediment. coeff., local conc.
Field flow fractionation (FFF) *	A	S	various
Streaming birefringence *	A	S	rotational diffusion
Storage modulus, other viscoelastic properties *	A	M	various quantities

3.4.1 Mass Spectroscopy

Mass spectroscopy is the only method for the determination of molar masses and molar mass distributions that delivers a reduced molar mass, i.e., the ratio m_{mol}/N_q of the mass m_{mol} of the molecule to the number N_q of the electric charges per molecule (in the literature usually as m/z). It separates in minutes picomole to attomole quantities of charged gaseous compounds in a magnetic field.

The method has been used for years to determine the molar mass of volatile, thermally stable micromolecules. Polymer molecules are, however, neither volatile nor thermally stable so that new techniques had to be developed. These techniques (field desorption, secondary-ion mass spectroscopy, laser desorption) employ ionizations that do not require vaporization of the polymer molecules .

In *matrix-assisted laser desorption/ionization mass spectroscopy* (**MALDI-MS**), macromolecules are embedded in a matrix that strongly absorbs ultraviolet light. Such matrices are, for example 2,5-dihydroxybenzoic acid (for peptides, proteins, polysaccharides, polar synthetic polymers), indole acrylic acid (for synthetic polymers), or α-cyano-4-hydroxycinnamic acid (for proteins and synthetic polymers).

On irradiation with a UV laser, the matrix transfers its energy fast to the macromolecules that become ionized, mostly by protons, but sometimes also by alkali ions. The ionized macromolecules are desorbed and analyzed by the spectrometer. The method delivers the relative abundances (relative intensities) of singly, doubly, triply ... charged molecules as a function of the ratio M_r/N_q of relative molar masses M_r to numbers N_q of electric charges of molecules (Fig. 3-4). The latter ratio is usually called m/z in the literature.

The mass accuracy is between 0.01 % and 1 % depending on the mass analyzer used. Relative molar masses of proteins can be determined up to values of circa 1 000 000, those of synthetic macromolecules usually up to circa 100 000.

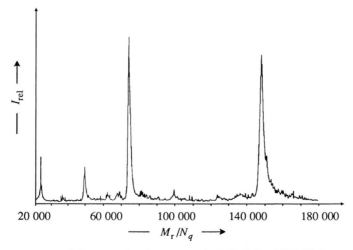

Fig. 3-4 Mass spectrum of the monoclonal mucoprotein I_gG_1 (M_r = 148 140) from *mus musculus* with signals for the single, double, and triple protonated species, using 3,5-dimethoxy-4-hydroxycinnamic acid as desorption matrix [2]. In this compound, different carbohydrate molecules are bound to the protein. By permission of the American Chemical Society, Washington, DC.

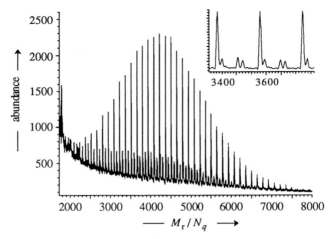

Fig. 3-5 Positive ion MALDI-TOF mass spectrum of a poly(butyl methacrylate) with a mass-average molecular weight of $\overline{M}_w$ = 3880 g/mol [3]. Primary signals are separated from each other by ΔM = 142 g/mol which corresponds to the molar mass $M_u \approx$ 142.20 g/mol of a monomeric unit. The secondary signals are always lower than the primary signals by ΔM = 72 g/mol which indicates the loss of one molecule of C_4H_8O per polymer chain (butenol, ethyl oxirane?). The insert shows an expanded view of the spectrum for $3420 \leq M_r/N_q \leq 3760$. TOF = time-of-flight. Reprinted by permission from [3]. Copyright (1995) American of the American Chemical Society.

The experimental conditions of mass spectroscopy often lead to cleavage of groups from macromolecules (Fig. 3-5) or macromolecules themselves (Fig. 3-6). The splitting-off of groups can be detected by the mass spectra themselves and the chain scissions by comparison of the mass spectra (MS) with the molar mass distribution curves obtained from size-exclusion chromatography (SEC) (next section).

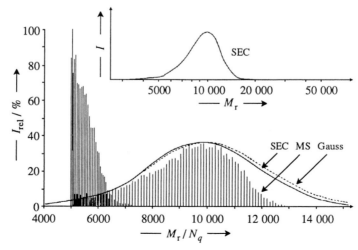

Fig. 3-6 Relative intensities I_{rel} of molecule ions of a narrow-distributed poly(styrene) (modified from [4]). The distribution curve of the single protonated molecules (N_q = 1) covers M_r/N_q values from circa 4000 to 15 000 and shows a maximum at ca. 10 000. The distances between signals correspond to the molecular weight M_r = 104.15 of a monomeric unit. The distribution curve of the double protonated species (N_q = 2) extends from ca. 7200 to lower values; it was cut off at ca. 5000.

Table 3-5 Weight averages $\overline{M}_{r,w}$ and number averages $\overline{M}_{r,n}$ of molecular weights of poly(oxyethylene)s (PEOX) in water by size-exclusion chromatography (SEC), static light scattering (LS), vapor phase osmometry (VPO), membrane osmometry (MO), and by MALDI-MS (MS) with 2,5-dihydroxybenzoic acid in water/ethanol (9:1) as matrix [5]. PEOX numbers indicate molecular weights given by the manufacturer.

| PEOX | $\overline{M}_{r,w}$ | | | $\overline{M}_{r,n}$ | | | | $\overline{M}_{r,w}/\overline{M}_{r,n}$ | |
	MS	SEC	LS	MS	SEC	VPO	MO	MS	SEC
1 470	1 500	1 443		1 462	1 382	1 300		1.02_6	1.04_4
4 250	4 018	4 059		3 957	3 958	4 100		1.01_5	1.02_6
7 100	6 865	7 043		6 655	6 874	6 600		1.03_2	1.02_5
12 600	12 492	12 320		12 426	11 843		11 500	1.00_5	1.04_0
23 000	22 190	21 228		22 115	20 040			1.00_3	1.05_9
40 000	37 039	41 500	41 400	36 842	35 500			1.00_5	1.13_7

Note that the distribution curves from mass spectroscopy (MS; Fig. 3-6) and size-exclusion chromatography (SEC; insert in Fig. 3-6) cannot be compared directly because the horizontal axis of the MS curve is directly proportional to M_r but that of the SEC curve is logarithmically proportional. This leads to an asymmetric SEC curve and a tail at lower molecular weights. The true SEC curve differs not much from that of a Gaussian distribution (Fig. 3-6, insert).

Weight average and number average molecular weights from MALDI-MS agree reasonably well with those from other methods at low molecular weights (Table 3-5). At higher molecular weights, however, weight averages are too low and number averages are too high, which feigns a too narrow polymolecularity index, $\overline{M}_{r,w}/\overline{M}_{r,n}$.

3.4.2 Size-Exclusion Chromatography

Methodology

Size-exclusion chromatography (**SEC**) is a special type of liquid chromatography. It separates molecules according to their sizes, not to their affinities to the matrix.

In SEC, a dilute polymer solution is put on top of a column that contains a packing material (matrix) with pores of 5 nm to 500 nm diameter, followed by an eluting solvent stream. Matrices should not adsorb the permeating macromolecules. The eluting solvents should have solubility parameters (see Volume III) similar to that of the matrix in order to avoid distribution effects during separation.

The matrix is either a rigid material with a relatively narrow distribution of pore sizes or a gel of a crosslinked polymer. In the latter case, the method is known as **gel permeation chromatography** (**GPC**) in synthetic polymer science and **gel filtration** in biopolymer science. Gels for organic solvents are mainly based on crosslinked poly(styrene)s, those for aqueous solutions on celluloses, crosslinked dextrans (type of polysaccharides), or crosslinked poly(acrylamide)s. Rigid matrices are usually porous glass beads.

Matrices with large pores even allow the separation of latex particles of 200 nm diameter. With this special SEC method, called **hydrodynamic chromatography**, molecules of ultrahigh molecular weight poly(ethylene)s can be separated. Such molecules are difficult to analyze by the usual SEC columns.

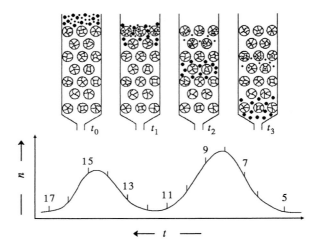

Fig. 3-7 Separation of molecules of different sizes, ● and ●, at various times t_0 (start) to t_3 during the flow through a porous matrix (top) and the resulting SEC diagram of the time dependence of differences in refractive index of solution and solvent (bottom). Numbers indicate the numbers of fractions that are proportional to the eluted volume.

The concentration of the exiting solution is measured automatically as function of time or eluent volume, for example via refractive index or ultraviolet or infrared spectroscopy. Simultaneously, viscosity or light scattering may also be measured as an indicator of molar masses of the eluted macromolecules.

The resulting SEC diagrams present the change of a concentration dependent quantity Q with time t or **elution volume** V, for example, the difference $\Delta n = n_{solution} - n_{solvent}$ in refractive indices (Fig. 3-7). $Q = f(t)$ and $Q = f(V)$ always show a maximum since first a little, then more, and finally again a little is eluted of a certain species. The maximum of the elution curve is called the **retention volume** V_R. Both retention and elution volumes are practically independent of the initial polymer concentration since SEC uses very small polymer concentrations.

The largest molecules appear first, i.e., at smaller elution volumes. This effect is assumed to be due to the molecular size: large polymer molecules have greater difficulties to squeeze into small pores hence they reside in the pores for shorter times or not at all. The residence time is shorter if the polymer is less polar than the eluting solvent.

The elution curve is broader, the wider the molecular weight distribution. Even uniform substances do not give sharp signals. This broadening of elution curves is caused by a so-called **axial dispersion** which is interpreted as the result of a distribution of residence times of identical macromolecules in differently sized pores. To correct for axial dispersions, the measured standard deviation σ of the elution curve is assumed to consist of two components, the standard deviation σ_{mol} caused by the molecular weight distribution, and the standard deviation σ_{ad} of the axial dispersion:

(3-9) $\sigma^2 = (\sigma_{mol})^2 + (\sigma_{ad})^2$

σ_{ad} is obtained if the flow of the eluted solution is reversed. The true elution curve is then obtained from σ and σ_{ad}.

Molar Mass Dependence of Retention Volume

Size-exclusion chromatography so far has defied all attempts to establish a rigorous theory of the process, mainly because the exact separation mechanism is not known. A separation of molecules by diffusion is much too slow in these small pores and a separation by flow does not seem to be the primary mechanism. All experimental data and many theoretical considerations point however to a partial exclusion of polymer molecules by the pores of the matrix as the dominant effect. The exclusion should depend on the volume of the molecules which in turn is related to the molar mass (Volume III).

The dependence of retention volume on the logarithm of molar mass shows three distinct regions, as one can see for spheroidal proteins in dilute salt solutions with a crosslinked dextran as the matrix (Pr–H$_2$O–Dex in Fig. 3-8):

- $V_R = const$ for $M \to \infty$: Very large molecules cannot enter pores; they therefore flow around and past the beads. Their retention volume must equal the interstitial volume V_I between the matrix particles ($V_R = V_I$).

- $V_R = const'$ for $M \to 0$: Very small molecules may reside in both the interstitial volume V_I and in the volume V_{pore} of the pores. The volume accessible to them is therefore $V_0 = V_I + V_{pore}$.

- $V_R = f(M)$: In the molar mass range between V_0 and V_I, larger molecules can enter the larger pores of the matrix but not the smaller ones whereas smaller molecules have access to many more pore sizes. In this range, retention volumes V_R depend linearly on the logarithm of the molar mass if the shape of the molecules and their probability to orient themselves in pores are independent of their molar mass. This is true for a series of hard (= non-deformable) spheres and approximately true for "soft" spheroidal protein molecules (Fig. 3-8). Similar *approximately* linear relationships are found for molecules with other shapes, for example, so-called random coils (Volume III) of poly(styrene) molecules (Fig. 3-8)

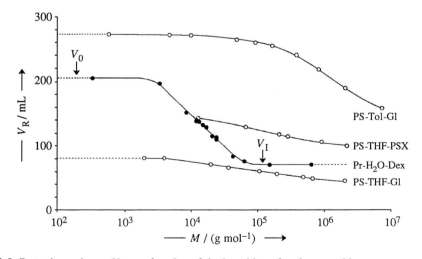

Fig. 3-8   Retention volumes V_R as a function of the logarithm of molar mass M.
  PS-THF-PSX: Poly(styrene) in tetrahydrofuran with crosslinked poly(styrene) [6].
  PS-THF-Gl: Poly(styrene) in tetrahydrofuran with porous glass beads [7].
  PS-Tol-Gl: Poly(styrene) in toluene with porous glass beads [8].
  Pr-H$_2$O-Dex: Spheroidal proteins in dilute salt solutions with crosslinked dextran [9].

Once a calibration curve $V_R = f(\lg M)$ has been established for *uniform* homologs at a certain combination of matrix, solvent, and temperature, it can be used to determine the molecular weights of unknown specimens. This calibration curve *cannot* be employed for molecules that differ in structure (constitution, configuration) even if the same matrix and the same solvent are used. This basic rule of SEC is often flagrantly violated.

The calibration should be made with molecularly uniform ("monodisperse") polymers. The unknown specimen may have a molar mass distribution, however, and the M obtained for this specimen with the help of the calibration curve is thus an average molar mass. The situation is even more complicated if the calibration was performed with specimens that are not molecularly uniform.

What, then, is the type of molar mass average that correlates with the retention volume? It is surmised that the retention volume depends on the hydrodynamic volume of the polymer molecule besides the selectivity of the matrices. The hydrodynamic volume, in turn, depends on the molecule shape (sphere, rod, random coil, etc.) and the interaction of the molecule with the solvent. Both influences are reflected by the exponent α in the relationship $[\eta] = KM^\alpha$ between intrinsic viscosity $[\eta]$ and molar mass M (see Section 3.6.5). SEC is assumed to deliver a molar mass average

$$(3\text{-}10) \qquad \overline{M}_{\text{sec}} = \sum_i m_i M_i^{1+\alpha} / \sum_i m_i M_i^\alpha \quad ; \quad m = \text{mass}$$

which leads, for example, to the following averages for Schulz-Zimm distributions of polymers with a degree of coupling of ς:

$$(3\text{-}11) \qquad \overline{M}_{\text{sec}} = \frac{\varsigma+\alpha+1}{\varsigma}\overline{M}_{\text{n}} = \frac{\varsigma+\alpha+1}{\varsigma+1}\overline{M}_{\text{w}} = \frac{\varsigma+\alpha+1}{\varsigma+2}\overline{M}_{z} = \frac{\varsigma+\alpha+1}{\varsigma+3}\overline{M}_{z+1}$$

For $\varsigma = 1$, one obtains $\overline{M}_{\text{w}}$ for spheres ($\alpha = 0$) and $\overline{M}_{z+1}$ for long rigid rods ($\alpha = 2$).

Universal Calibrations

In the simplest case, separations by SEC are controlled by the shape, size, and size distribution of the pores and the shape, size, and size distribution of the polymer molecules. These quantities must somehow be related to experimental quantities: the retention volume V_R (or, more general, the distribution coefficient K_d) and the molar mass M of the polymer molecules (or their molar mass distribution). The distribution coefficient K_d gives the fraction of the pore volume that is accessible to the polymer molecules:

$$(3\text{-}12) \qquad K_d = \frac{V_R - V_I}{V_0 - V_I}$$

The **Porath** theory leads to a distribution coefficient $K_d = K[1 - (R_{\text{eff}}/R_{\text{pore}})]^3$ for polymer molecules with an effective radius R_{eff} in uniform conical pores with radii R_{pore}. Assuming $R_{\text{eff}} = K'M^{1/2}$ (which is true for random coil molecules in so-called theta solvents (Section 5.3.2)), one arrives at Eq.(3-13) which is often used for biopolymers

$$(3\text{-}13) \qquad K_d^{1/3} = K^{1/3} - (K^{1/3}K'/R_{\text{pore}})M^{1/2} \quad ; \quad K, K' = \text{adjustable constants}$$

Another semi-empirical approach assumes the pores of the matrix to be circular cylinders with a radius R_{pore}. Polymer molecules, on the other hand, have a chromatographically effective radius R_{chr}. A separation by molecule size can be only achieved if both radii are similar: no polymer molecules can enter pores if $R_{pore} \ll R_c$ but all can do it if $R_{pore} \gg R_{chr}$.

The retention volume is determined by the interstitial volume V_I and the effective pore volume $V_{pore,eff} = fV_{pore}$. The fraction f of V_{pore} must be an exponential function of R_{chr}/R_{pore} because of the experimentally found relationship $V_P = f(\lg M)$:

$$(3\text{-}14) \qquad V_R = V_I + V_{pore,eff} = V_I + fV_{pore} = V_I + V_{pore} \exp(-R_{chr}/R_{pore})$$

The chromatographically effective radius R_{chr} is further assumed to be identical to the hydrodynamically effective Einstein radius R_h from viscosity measurements. Assuming spherical molecules with a hydrodynamic volume $V_h = (4/3)\,\pi\,R_h^3$, the Einstein radius can be calculated from the so-called intrinsic viscosity $[\eta] = (5/2)\,N_A V_h/M$ (Eq.(3-43)) which is a measure of the volume occupied by 1 gram of dissolved macromolecules ($N_A \approx 6.022 \cdot 10^{23}$ mol^{-1}; V_h in mL; M in g/mol):

$$(3\text{-}15) \qquad R_h = [3\,V_h/(4\,\pi)]^{1/3} = [(3\,[\eta]M)\,/\,(3\,\pi\,N_A)]^{1/3}$$

Insertion of $V_{pore} = V_0 - V_I$ and Eq.(3-15) into Eq.(3-14), rearrangement, and conversion into logarithms results in a relationship between the natural logarithm of the distribution coefficient K_d and the cubic root of the molar volume, $([\eta]M))^{1/3}$ (Fig. 3-9):

$$(3\text{-}16) \quad \ln K_d \equiv \ln\left(\frac{V_R - V_I}{V_0 - V_I}\right) = -\left(\frac{3}{10\,\pi\,N_A}\right)^{1/3}\frac{([\eta]M)^{1/3}}{R_{pore}} \approx \frac{-5.41 \cdot 10^{-9}\,\text{mol}}{R_{pore}}([\eta]M)^{1/3}$$

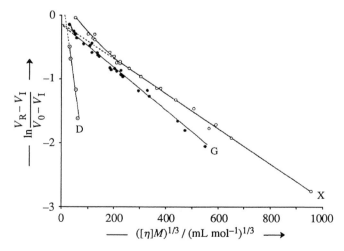

Fig. 3-9 Plots of data of Fig. 3-8 according to the Hester-Mitchell Eq.(3-16) [10].
  D: spheroidal proteins in dilute aqueous salt solutions at crosslinked dextran [9].
  G: hydrophilic synthetic polymers in dilute aqueous salt solutions at porous glass [11].
  X: hydrophobic synthetic polymers in tetrahydrofuran at crosslinked poly(styrene) [5].

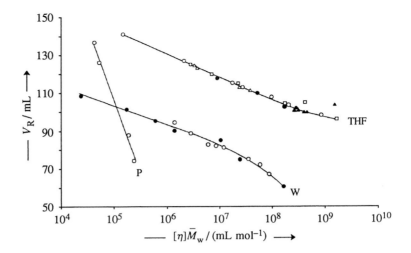

Fig. 3-10 Universal calibration of synthetic polymers.
P: Spheroidal proteins in dilute aqueous salt solutions with crosslinked dextran (data of Fig. 3-9
        for known values of [η]).
THF: Linear (O), comb-like (□), and star-like (△) poly(styrene)s, poly(phenylsiloxane)s (▲), poly-
        (butadiene)s (⊕), and poly(methyl methacrylate)s (●) in tetrahydrofuran with crosslinked
        poly(styrene) [6].
W: Dextrans (O) and poly(styrene monosulfonate)s (●) in dilute aqueous salt solutions with
        porous glass [11].

Eq.(3-16) is applicable to many polymers (Fig. 3-9). The slope of the function $\ln K_d$ = f($[\eta]M)^{1/3}$ delivers an average pore radius, for example, R_{pore} = 15.9 nm for a porous glass with pore radii ranging between 3.7 nm and 62.5 nm. The function fails, however, for low molar masses, and neither these values nor those of the linear part of $\ln K_d$ = f($[\eta/M)^{1/3}$ extrapolate to $\ln K_d = 0$ at $([\eta/M)^{1/3} = 0$ as required by Eq.(3-16).

However, K_d is unknown for many polymer-matrix-solvent-temperature combinations, since V_I is almost never accessible and V_0 only occasionally (see Fig. 3-8). Most people therefore resort to the so-called **universal calibration** by the empirical **Grubisic-Rempp-Benoit equation** that plots the retention volume V_R against the logarithm of the product of intrinsic viscosity $[\eta]$ and mass-average molar mass $\overline{M}_w$:

(3-17) $V_R = A - B \lg ([\eta]\,\overline{M}_w)$

This equation thus also assumes that SEC parameters are proportional to the molar volumes of polymers; it delivers good correlations (Fig. 3-10) except for large products of $[\eta]\,\overline{M}_w$. Since intrinsic viscosities $[\eta]$ are proportional to the ratio $\langle s^2 \rangle^{3/2}/M$ (Section 3.6; s = radius of gyration of a macromolecule), V_R should be a function of $\lg \langle s^2 \rangle^{3/2}$ that is sometimes found and sometimes not. One reason may be the neglect of corrections for the type and width of molar mass distribution since the $\langle s^2 \rangle^{3/2}$ calculated from $[\eta]$ is a number-average property (Volume III).

It may also be questionable whether the chromatographically effective radius is indeed given by a hydrodynamic radius or the radius of gyration. Cyclic polymer molecules have the same retention volumes as their linear counterparts with the same constitution and molar mass but their radii of gyration are smaller by a factor of $2^{1/2}$. Conversely, the proteins apoferritin (a hollow sphere) and ferritin (apoferritin filled with iron compounds) have the same diameters and the same retention volumes but different molar masses and radii of gyration.

3.4.3 Fractionation

Constitutionally, molecularly, and/or configurationally non-uniform polymer molecules have different solubilities (Volume III). Polymers can thus be separated by precipitation or dissolution into fractions of molecules with different structures. Such a fractionation is induced by a change in temperature or solvent/non-solvent composition.

Addition of a non-solvent to a dilute solution of a molecularly non-uniform (but constititionally and configurationally uniform) polymer causes the solution to become turbid. The solution then separates into two phases, a higher concentrated "gel phase" at the bottom and a more dilute "sol phase" on top. The gel phase is a highly viscous solution of a polymer fraction of higher molar mass than the parent polymer. Contrary to the name, it is not a highly swollen gel of chemically or physically *crosslinked* molecules.

Successive addition of non-solvent generates additional fractions with increasingly lower molar masses (**precipitation fractionation**). Alternatively, polymers in endothermal solutions may be also fractionated by successive lowering of the temperature. The latter method is however seldom used because suitable polymer-solvent-temperature combinations are rare.

Fractions deliver the differential discontinuous (discrete) molecular weight distribution. The conversion of this distribution into the integral (cumulative) distribution must consider that each fraction is not molecularly uniform but possesses a molar mass distribution itself. The molar mass of a fraction is thus an average.

In first approximation, one half of fraction 1 will have a molar mass below its molar mass average M_1 and the other half a molar mass above that average. Hence, for the calculation of the integral molar mass distribution, not the weight fraction w_1 itself has to be correlated with M_1 but a value of $w_1^* = w_1/2$. The applicable weight fraction of the second fraction is then $w_2^* = w_1 + w_2/2$, and so forth (Table 3-6). A plot of $w_j^* = f(M_j)$ delivers the discontinuous integral mass distribution that can be graphically differentiated to the continuous differential distribution. Because of the non-uniformity of the fractions, it is not permissible to go directly from a discontinuous differential distribution to a continuous differential one.

Table 3-6 Conversion of a discontinuous differential distribution of molar masses into a discontinuous integral one. Assumption: uniform fractions.

Fraction j	Molar mass $M_j/$ (g mol^{-1})	Mass fraction w_j	Cumulative fraction w_j^*
1	1500	0.0532	0.0266
2	3100	0.0740	0.0902
3	5000	0.0622	0.1583
etc.	etc.	etc.	etc.

In **dissolution fractionations**, polymers are first precipitated on a carrier. The solids are then extracted with solvent/non-solvent mixtures of increasing solvent content. The low molar mass fractions appear first. The method is used for the determination of the degree of branching of poly(ethylene)s (Volume III).

3.5 Molar Mass Averages

Distributions of properties such as degrees of polymerization, molar masses, compositions, radii of gyration, tensile strengths, etc., are often not characterized by distribution functions but only by their arithmetic, geometric, or harmonic **averages** (**means**). Degrees of polymerization and molar masses are usually arithmetic averages; they may be simple averages or complex ones. In the latter case, moments of distributions are preferably used instead of averages (Volume III).

3.5.1 Molar and Molecular Masses

Mass-related sizes of molecules can be described by various quantities that differ in their physical units.

The **molecular mass** of a molecule is the mass m_{mol} of a molecule of a substance; its physical base unit is the kilogram. An atom has correspondingly an **atomic mass** m_a. An example is the mass $m_a(^{12}C)$ of an atom of the carbon isotope ^{12}C that is the base for the **atomic mass constant** $m_u = m_a(^{12}C)/12 = 1.660\ 540\ 2 \cdot 10^{-27}$ kg. The value of the atomic mass constant equals by definition the **unified atomic mass unit** u ($m_u = 1$ u).

Mass spectroscopy (Section 3.4.1) does not determine molecular masses m_{mol} directly but the ratio m_{mol}/N_q of molecular mass to the number N_q of electric charges per molecule or molecule fragment (Fig. 3-4). The ratio m_{mol}/N_q is a **reduced molecular mass** (physical unit of a mass, e.g., kg) that is usually reported as *dimensionless* reduced relative molar mass = reduced molecular weight M_r/N_q, in literature with the symbols m/z.

The **relative molecular mass** $M_r \equiv m_{mol}/m_u$ is the *ratio* of the mass m_{mol} of a *molecule* of a substance to the atomic mass constant m_u (ISO 31-8). It is dimensionless (physical unit of unity) and traditionally called **molecular weight**. Relative molecular masses are only determined by relative chemical processes such as endgroup determinations. They are not obtained by physical methods such as osmometry or light scattering.

The **weight** $G = mg$ (in newton as physical unit) is defined as the product of mass m and acceleration g of free fall. It is permissible to use "weight" for *relative* physical quantities (because gravity cancels out) if it is used in a consistent manner. One can therefore speak of a "*mass*-average of relative molecular *mass*" = "*weight*-average of molecular *weight*" but *not* of a "*mass*-average of molecular *weight*" or a "*weight*-average of molar *mass*."

Most physical methods such as membrane osmometry, light scattering, sedimentation equilibrium, etc., determine neither molecular masses (*m*) nor molecular weights = relative molecular masses (M_r) but **molar masses** $M = m/n$ as the ratio of the mass m to the amount-of-substance n of a *substance*. It is customary to report molar masses in the physical unit g/mol since in that case "molar mass" and "molecular weight" become *numerically* identical (they are *not* identical with respect to their physical units!).

It has become customary, especially in the biosciences, to report not only reduced molecular weights (in kg) but also molar masses (in g/mol) and molecular weights (dimensionless) in "dalton" (symbol: Da). The "dalton" is by definition identical with the unified atomic mass unit (1 Da $\equiv$ 1 m_u $\equiv$ 1 u) and should therefore only be used, if at all, for reduced molecular weights. However, neither "dalton" nor its symbol "Da" are approved as physical units by the international *Conférence Générale des Poids et Mesures*.

3.5.2 Simple Averages of Molar Masses

Equilibrium methods deliver simple averages of molar masses ($\overline{M}_n$, $\overline{M}_w$, $\overline{M}_z$, etc.) since they are based on a series of simple statistical weights (Section 3.3.2). Statistical weights include not only number-based statistical weights (N, n, x) and mass-based statistical weights (m, w) but also lower and higher statistical weights such as {n–1}, z, and {z+1} and statistical weights based, for example, on lengths (Volume IV). Braces are used to indicate that the terms {n–1}, {z+1}, etc., are not mathematical operations but symbols for statistical weights; there are no special symbols for these quantities.

For molecularly *non-uniform* species i, simple statistical weights are interrelated:

$$(3\text{-}18) \qquad \{n\text{–}1\}_i \; \equiv \; (n_i)^{-1} \, (\overline{M}_{\{n\text{–}1\},i})^{-1} \; \equiv \; m_i (\overline{M}_{\{n\text{–}1\},i})^{-1} (\overline{M}_{n,i})^{-1}$$

$$(3\text{-}19) \qquad n_i \qquad \equiv n_i \qquad\qquad\quad \equiv m_i (\overline{M}_{n,i})^{-1}$$

$$(3\text{-}20) \qquad m_i \qquad \equiv n_i \overline{M}_{n,i} \qquad\qquad \equiv m_i$$

$$(3\text{-}21) \qquad z_i \qquad \equiv n_i \overline{M}_{n,i} \overline{M}_{w,i} \qquad \equiv m_i \overline{M}_{w,i}$$

$$(3\text{-}22) \qquad \{z\text{+}1\}_i \equiv n_i \overline{M}_{n,i} \overline{M}_{w,i} \overline{M}_{z,i} \equiv m_i \overline{M}_{n,i} \overline{M}_{w,i}$$

The three most important averages of molar masses are defined as

$$(3\text{-}23) \qquad \overline{M}_n \; \equiv \sum_i x_i \overline{M}_{n,i} \; = \frac{\sum_i n_i \overline{M}_{n,i}}{\sum_i n_i} \; = \frac{\sum_i m_i}{\sum_i (m_i / \overline{M}_{n,i})}$$

$$(3\text{-}24) \qquad \overline{M}_w \; \equiv \sum_i w_i \overline{M}_{w,i} \; = \frac{\sum_i m_i \overline{M}_{w,i}}{\sum_i m_i} \; = \frac{\sum_i n_i \overline{M}_{n,i} \overline{M}_{w,i}}{\sum_i n_i \overline{M}_{n,i}}$$

$$(3\text{-}25) \qquad \overline{M}_z \; \equiv \sum_i Z_i \overline{M}_{z,i} \; = \frac{\sum_i z_i \overline{M}_{z,i}}{\sum_i z_i} \; = \frac{\sum_i m_i \overline{M}_{w,i} \overline{M}_{z,i}}{\sum_i m_i \overline{M}_{w,i}} \; = \frac{\sum_i n_i \overline{M}_{n,i} \overline{M}_{w,i} \overline{M}_{z,i}}{\sum_i n_i \overline{M}_{n,i} \overline{M}_{w,i}}$$

Molecularly uniform species i have molar masses of $\overline{M}_{z,i} = \overline{M}_{w,i} = \overline{M}_{n,i} = M_i$, therefore

$$(3\text{-}26) \qquad \overline{M}_n \; \equiv \frac{\sum_i n_i M_i}{\sum_i n_i} \; = \frac{\sum_i m_i}{\sum_i (m_i / M_i)}$$

$$(3\text{-}27) \qquad \overline{M}_w \; \equiv \frac{\sum_i w_i M_i}{\sum_i w_i} \; = \frac{\sum_i n_i M_i^2}{\sum_i n_i M_i}$$

$$(3\text{-}28) \qquad \overline{M}_z \; \equiv \frac{\sum_i z_i M_i}{\sum_i z_i} \; = \frac{\sum_i m_i M_i^2}{\sum_i m_i M_i} \; = \frac{\sum_i n_i M_i^3}{\sum_i n_i M_i^2}$$

The molar masses of synthetic polymers rarely exceed several millions whereas those of some natural polymers are considerably higher, for example, the amylopectin of the easter lily (M = 2.5·10^8 g/mol) or the deoxyribonucleic acid of the lung fish (M = 6.9·10^{13} g/mol). Synthetic polymers are also always molecularly non-uniform but natural polymers may be either uniform (e.g., deoxyribonucleic acids, enzymes) or molecularly non-uniform (e.g., amylopectin); this may be either genuine or an artifact of separation procedures.

Polymers are not unique in not having a single molar mass. Even so-called "pure" substances possess molar mass distributions, albeit narrow ones. The reason is the presence of isotopes that causes atomic mass averages in most elements and molecular mass averages in almost all substances.

The number-average atomic weight (chemical atomic weight) of common hydrogen (99.985 % ^{1}H with $A_{r,i}$ = 1.007 825 035; 0.015 % ^{2}H with $A_{r,2}$ = 2.014 101 779) is $\overline{A}_{r,n}$ = 1.007 975 976 and its mass average atomic weight (= physical atomic weight) $\overline{A}_{r,w}$ = 1.008 126 64. The ratio $\overline{A}_{r,w}/\overline{A}_{r,n}$ ≈ 1.000 15 is however so small that common hydrogen appears as a "pure" substance to chemists.

3.5.3 Polymolecularity

It follows from Equations (3-18)-(3-28) that always

(3-29) $\overline{M}_{\{z+1\}} \geq \overline{M}_z \geq \overline{M}_w \geq \overline{M}_n \geq \overline{M}_{\{n-1\}}$

and that the width of a molar mass distribution is always given by the ratio of a higher molar mass average to a lower one. Most common is the (mass-number) **polymolecularity index** $Q_{w,n} \equiv \overline{M}_w/\overline{M}_n$, commonly and wrongly (p. 44) called **"polydispersity index"** **(PDI)**. Similar polymolecularity indices can be defined for other molar mass averages, for example, $\overline{M}_z/\overline{M}_w$.

The broader the molar mass distribution, the higher are the polymolecularity indices (for molecularly uniform substances: $\overline{M}_w/\overline{M}_n = \overline{M}_z/\overline{M}_w \equiv 1$). Polymolecularity indices are however not very sensitive to the width of narrow molar mass distributions. A better indicator is the **number-standard deviation**, σ_n, of the distribution, defined as

(3-30) $\sigma_n \equiv (\overline{M}_w\overline{M}_n - \overline{M}_n^2)^{1/2}$

(3-31) $Q_{w,n} \equiv \overline{M}_w/\overline{M}_n = 1 + (\sigma_n/\overline{M}_n)^2$

which is an *absolute* measure of the width of *Gaussian* distributions and a relative measure for all others. For Gaussian distributions, a value of $\overline{M}_n \pm 1\sigma_n$ corresponds always to an amount-of-substance of 68.26 %, a value of $\overline{M}_n \pm 2\sigma_n$ to one of 95.44 %, and a value of $\overline{M}_n \pm 3\sigma_n$ to one of 99.73 % (Volume III). By analogy, a mass-standard deviation $\sigma_w \equiv (\overline{M}_z\overline{M}_w - \overline{M}_w^2)^{1/2}$ may be defined.

Constant standard deviations indicate identical widths of Gaussian distributions, regardless of differences in the number average of molar mass (Fig. 3-11, left); the polymolecularity index decreases however with increasing molar mass. A constant polymolecularity index, on the other hand, signals an increase in width with increasing molar mass. The situation is quite different for Schulz-Zimm distributions. Here, polymolecularity indices are absolute measures of widths but standard deviations are relative ones. Neither Q nor σ are absolute measures of widths of all other types of distributions.

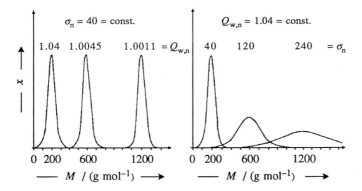

Fig. 3-11 Continuous differential Gaussian distributions of molar masses of three polymers with $\overline{M}_n/(\text{g mol}^{-1})$ of 200, 600, and 1200, and constant number-standard deviations $\sigma_n = 40$ g/mol (left) or constant polymolecularity indices $Q_{w,n} = 0.04$.

Number averages put the emphasis on the number of molecules, mass-averages on their mass, and z-averages on molecules with even higher molar masses. The fractions I-III of the polymer of Table 3-7 have, for example, the same polymolecularity index. They are also present in equal weight fractions $w_I = w_{II} = w_{III} = 1/3$ but the mole fraction of the lowest fraction I is 10 times as high as that of the highest fraction III. The situation is the opposite for the z-fractions.

Gross errors occur if the molar mass average of a polymer is calculated from the molar masses of non-uniform fractions with the help of the expressions for molar masses of *uniform* species (Equations (3-26) to (3-28)) instead of the correct Equations (3-23) to (3-25) for non-uniform ones. Such a procedure always delivers polymolecularity indices that are too low and also often wrong molar mass averages (Table 3-7, bottom).

Table 3-7 Effect of calculations on molar mass averages and polymolecularity indices of the whole polymer with the same mass fractions w_i of polymolecular fractions i that have the same polymolecularity indices $\overline{M}_z / \overline{M}_w$ and $\overline{M}_w / \overline{M}_n$. Top table: assumptions, bottom table: calculations.
(a) Correct calculations with Equations (3-23)-(3-25) for non-uniform fractions.
(b) Calculations for uniform fractions assuming $M_i = \overline{M}_{n,i}$.
(c) Calculations for uniform fractions assuming $M_i = \overline{M}_{w,i}$.

Fraction			$\dfrac{\overline{M}_{n,i}}{\text{g mol}^{-1}}$	$\dfrac{\overline{M}_{w,i}}{\text{g mol}^{-1}}$	$\dfrac{\overline{M}_{z,i}}{\text{g mol}^{-1}}$	$\dfrac{\overline{M}_{w,i}}{\overline{M}_{n,i}}$	$\dfrac{\overline{M}_{z,i}}{\overline{M}_{n,i}}$
No.	x_i	w_i	Z_i				
I	0.769	1/3	0.0625	100 000	150 000	200 000	1.5	2.0
II	0.154	1/3	0.3125	500 000	750 000	1 000 000	1.5	2.0
III	0.077	1/3	0.6250	1 000 000	1 500 000	2 000 000	1.5	2.0

Total polymer			$\dfrac{\overline{M}_n}{\text{g mol}^{-1}}$	$\dfrac{\overline{M}_w}{\text{g mol}^{-1}}$	$\dfrac{\overline{M}_z}{\text{g mol}^{-1}}$	$\dfrac{\overline{M}_w}{\overline{M}_n}$	$\dfrac{\overline{M}_z}{\overline{M}_n}$
No.	x	w	Z				
(a)	1.000	1	1.0000	$\approx 230\,769$	$= 800\,000$	$= 1\,575\,000$	≈ 3.47	≈ 6.83
(b)	1.000	1	1.0000	$\approx 230\,769$	$= 533\,333$	$= 787\,500$	≈ 2.31	≈ 1.48
(c)	1.000	1	1.0000	$\approx 346\,154$	$= 800\,000$	$\approx 1\,181\,250$	≈ 2.31	≈ 3.41

3.5.4 Complex Averages of Molar Masses

Exponent Averages

Hydrodynamic methods deliver in general so-called exponent averages of molar masses and not the simple averages $\overline{M}_n$, $\overline{M}_w$, $\overline{M}_z$, etc. The dependence of the g-average of hydrodynamic properties, $\overline{P}_{h,g}$, (diffusion coefficients, intrinsic viscosities, etc.) on the g-average $\overline{M}_{h,g}$ of the molar mass can be often described by a power expression

$$(3\text{-}32) \qquad \overline{P}_{h,g} = K_h \overline{M}_{h,g}^{\lambda}$$

where K_h and λ are empirical, molar-mass independent constants. This type of equation can be written for any relationship between properties and molar masses if the molar mass range chosen is sufficiently small. Such equations are however often applicable to astonishingly broad ranges of molar masses.

The hydrodynamic property P of a non-uniform polymer is an average $\overline{P}_{h,g}$ where g is the appropriate statistical weight (g = n, m, z, v, etc.; see p. 85 ff. and Eq.(3-63)):

$$(3\text{-}33) \qquad \overline{P}_{h,g} = \frac{\sum_i g_i P_{h,i}}{\sum_i g_i} = \frac{K_P \sum_i g_i M_i^h}{\sum_i g_i}$$

Solving Eq.(3-32) for $\overline{M}_{h,g}$ and inserting the result into Eq.(3-33) delivers the definition of the g-hydrodynamic average of the molar mass:

$$(3\text{-}34) \qquad \overline{M}_{h,g} = \left(\frac{\overline{P}_{h,g}}{K_P} \right)^{1/h} = \left(\frac{\sum_i g_i M_i^h}{\sum_i g_i} \right)^{1/h}$$

The g-hydrodynamic average $\overline{M}_{h,g}$ of the molar mass is the hth moment of the g-distribution of molar masses. It converts to a simple one-moment average of order 1 for h =1, for example, to a weight average if h = 1 and $g_i = w_i$.

An example is the so-called **viscosity-average molar mass**, $[\eta] = K_v \overline{M}_v^{\alpha}$, that is obtained from the relationship between intrinsic viscosity $[\eta]$ and molar mass (Section 3.6 and Volume III). Here, the exponent α assumes the following values in the limit of infinite molar mass: 2 for infinitely long, stiff rods with d = const., 0.764 for perturbed coils in thermodynamically good solvents, 0.5 for unperturbed coils, 0 for hard spheres, −1 for discs with constant diameter d, and −2 for stiff rods with constant length and $d \sim M$.

Compound Averages

Molar masses are sometimes obtained by the combination of two hydrodynamic quantities. An example is the Svedberg equation $M_{sd} = K_{sd} S D^{-1}$ that allows the determination of molar masses from the sedimentation coefficient S and the diffusion coefficient D of a *molecularly uniform* substance. Molar masses may also be obtained from sedimentation coefficients and intrinsic viscosities $[\eta]$ by $M_{sv} = K_{sv} S^{3/2} [\eta]^{1/2}$ or diffusion coefficients and intrinsic viscosities by $M_{dv} = K_{dv} D^{-3} [\eta]^{-1}$ (for the derivation of these equations, see Volume III).

The molar mass dependence of hydrodynamic quantities can be expressed by

(3-35) $\qquad S = K_s M^\varsigma \qquad\qquad D = K_d M^\delta \qquad\qquad [\eta] = K_v M^\alpha$

The dependence of molar masses of components i can thus be written

(3-36) $\qquad M_{sd,i} = K_{sd} S_i D_i^{-1} \qquad = K_{sd} K_s K_d^{-1} M_i^{\varsigma - \delta}$

(3-37) $\qquad M_{sv,i} = K_{sv} S_i^{3/2} [\eta]_i^{1/2} \qquad = K_{sv} K_s^{3/2} K_v^{1/2} M_i^{(3\varsigma + \alpha)/2}$

(3-38) $\qquad M_{dv,i} = K_{dv} D_i^{-3} [\eta]_i^{-1} \qquad = K_{dv} K_d^{-3} K_v^{-1} M_i^{-(3\delta + \alpha)}$

where K_s, K_d, K_v, ς, δ, and α are adjustable constants that do not depend on the molar mass. The exponents ς, δ, and α must follow the **exponent rule** since the left and right sides of each equation (3-36)-(3-38) must have the same physical units, hence:

(3-39) $\qquad 1 = \varsigma - \delta = (3\varsigma + \alpha)/2 = -(3\delta + \alpha)$

This rule is independent of any theory about shapes and interactions of molecules. The product of the constants of the right sides of Equations (3-36)-(3-38) must furthermore equal unity since Equations (3-36)-(3-38) revert to $M/M^{\varsigma - \delta} = K_{sd} K_s / K_d = 1$, $K_{sv} K_s^{3/2} K_v^{1/2} = 1$, and $K_{dv} K_d^{-3} K_v^{-1} = 1$.

All terms of Equations (3-36)-(3-38) can be measured directly: sedimentation coefficient S (Vol. III), diffusion coefficient D (Volume III), and intrinsic viscosity $[\eta]$ (Section 3.6) as well as the constants K_{sd}, K_{sv}, and K_{dv} (for the derivations, see Volume III):

(3-40) $\qquad K_{sd} = RT/(1 - \tilde{v}_2 \rho_1)$

(3-41) $\qquad K_{sv} = [(6^2/20^{1/2}) \pi N_A][\eta_1/(1 - \tilde{v}_2 \rho_1)]^{3/2}$

(3-42) $\qquad K_{dv} = [20/(6^4 \pi^2 N_A^2)][RT/\eta_1]^3$

These constants contain the molar gas constant R, kelvin temperature T, partial specific volume $\tilde{v}_2$ of the solute, density ρ_1 of the solvent, and Avogadro constant N_A.

No theoretical assumptions need to be made for S, D, $[\eta]$, K_{sd}, K_{sv}, and K_{dv}. The molar masses M_{sd}, M_{sv}, and M_{dv} of molecularly uniform polymers are therefore *absolute* molar masses. The situation is however different for the molar masses $\overline{M}_{sd}$, $\overline{M}_{sv}$, and $\overline{M}_{dv}$ of molecularly non-uniform polymers. These molar masses are also absolute because the experimental quantities required for their determinations do not need a calibration by known molar masses. However, their numerical values do depend on the molecular shape and the interaction between polymer and solvent molecules.

Intrinsic viscosities $[\eta]$ of molecularly non-uniform polymers are always mass-averages (see Section 3.6.5 and Volume III) whereas the sedimentation coefficients S and diffusion coefficients D may be obtained with different statistical weights g and g', respectively (Volume III):

(3-43) $\qquad \overline{S}_g = \sum_i g_i S_i / \sum_i g_i \;\; ; \;\; \overline{D}_{g'} = \sum_i g'_i D_i / \sum_i g_i' \;\; ; \;\; [\eta] = \sum_i w_i [\eta]_i / \sum_i w_i$

Equations (3-35)-(3-43) lead to

$$(3\text{-}44) \quad \overline{M}_{s,g;d,g'} = K_{sd}\frac{\overline{S}_g}{\overline{D}_{g'}} = \frac{K_{sd}}{K_sK_d}\left(\frac{\sum_i g_iM_i^\varsigma}{\sum_i g_i}\right)\left(\frac{\sum_i g_i'}{\sum_i g_i'M_i^\delta}\right) = \left(\frac{\sum_i g_i'}{\sum_i g_i}\right)\left(\frac{\sum_i g_iM_i^{(2-\alpha)/3}}{\sum_i g_i'M_i^{-(1+\alpha)/3}}\right)$$

$$(3\text{-}45) \quad \overline{M}_{s,g;v} = K_{sv}\overline{S}_g^{3/2}[\eta]^{1/2} = K_{sv}K_s^{3/2}K_v^{1/2}\left(\frac{\sum_i g_iM_i^{(2-\alpha)/3}}{\sum_i g_i}\right)\left(\sum_i w_iM_i^\alpha\right)^{1/2}$$

$$(3\text{-}46) \quad \overline{M}_{d,g;v} = \frac{K_{dv}}{K_d^3K_v}\left(\sum_i g_i'\right)^3\left(\sum_i g_iM_i^{-(1+\alpha)/3}\right)^{-3}\left(\sum_i w_iM_i^\alpha\right)^{-1}$$

Molar masses of molecularly non-uniform polymers by hydrodynamic measurements thus depend on the statistical weights g and g' (n, w, v, etc.) and the exponents α of the molar mass in the relationship between hydrodynamic quantities S, D, and/or $[\eta]$ and molar mass M. Fig. 3-12 shows calculations for hydrodynamic averages of molar masses calculated from weight-averages of hydrodynamic quantities.

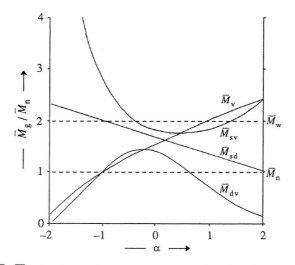

Fig. 3-12 Ratios $\overline{M}/\overline{M}_g$ of various molar mass averages as a function of exponent α of the intrinsic viscosity-molar mass relationship $[\eta] = K_v\overline{M}_v^\alpha$ [12]. $\overline{M}_g$ may be the mass-average $\overline{M}_w$, the number-average $\overline{M}_n$, or one of the hydrodynamic averages $\overline{M}_g$ that are obtained from determinations of intrinsic viscosities ($g = v$) and/or mass-average sedimentation coefficients ($g = s$) and/or diffusion coefficients ($g = d$). Calculations for a Schulz-Flory distribution ($\overline{M}_w/\overline{M}_n = 2$).

3.5.5 Degrees of Polymerization

Physical measurements always deliver averages of molar masses, M, whereas theories of polymerization are based on degrees of polymerization, X. For molecularly non-uniform polymers, both M and X are obtained as averages. The most important simple averages of degrees of polymerization are thus (see Equations (3-23)-(3-25):

$$(3\text{-}47) \qquad \overline{X}_\text{n} \equiv \sum_i x_i \overline{X}_{\text{n},i} \quad = \frac{\sum_i n_i \overline{X}_{\text{n},i}}{\sum_i n_i} \quad = \frac{\sum_i (m_i/\overline{M}_{\text{n},i}) \overline{X}_{\text{n},i}}{\sum_i (m_i/\overline{M}_{\text{n},i})}$$

$$(3\text{-}48) \qquad \overline{X}_\text{w} \equiv \sum_i w_i \overline{X}_{\text{w},i} \quad = \frac{\sum_i m_i \overline{X}_{\text{w},i}}{\sum_i m_i} \quad = \frac{\sum_i n_i \overline{M}_{\text{n},i} \overline{X}_{\text{w},i}}{\sum_i n_i \overline{M}_{\text{n},i}}$$

$$(3\text{-}49) \qquad \overline{X}_\text{z} \equiv \sum_i Z_i X_{\text{z},i} \quad = \frac{\sum_i z_i \overline{X}_{\text{z},i}}{\sum_i z_i} \quad = \frac{\sum_i m_i \overline{M}_{\text{w},i} \overline{X}_{\text{z},i}}{\sum_i m_i \overline{M}_{\text{w},i}} \quad = \frac{\sum_i n_i \overline{M}_{\text{n},i} \overline{M}_{\text{w},i} \overline{X}_{\text{z},i}}{\sum_i n_i \overline{M}_{\text{n},i} \overline{M}_{\text{w},i}}$$

The contribution of endgroups to molar masses can be neglected for high molar masses if all monomeric units have the same constitution. The various M in Equations (3-47)-(3-49) may thus be replaced by the corresponding averages of X. These equations simplify further if species i are molecularly uniform:

$$(3\text{-}50) \qquad \overline{X}_\text{n} \equiv \sum_i x_i X_i \quad = \sum_i n_i X_i / \sum_i n_i \quad = \sum_i m_i / \sum_i (m_i/X_i)$$

$$(3\text{-}51) \qquad \overline{X}_\text{w} \equiv \sum_i w_i X_i \quad = \sum_i m_i X_i / \sum_i m_i \quad = \sum_i n_i X_i^2 / \sum_i n_i X_i$$

$$(3\text{-}52) \qquad \overline{X}_\text{z} \equiv \sum_i Z_i X_i \quad = \sum_i z_i X_i / \sum_i z_i \quad = \sum_i m_i X_i^2 / \sum_i m_i X_i = \sum_i n_i X_i^3 / \sum_i n_i X_i^2$$

Numbers N_i, amounts n_i, masses m_i, etc., as statistical weights may be replaced by the corresponding statistical weight fractions, i.e., the amount-of-substance fractions ("mole fractions") x_i, mass fractions (= weight fractions) w_i, z-fractions Z_i, etc. The resulting expressions for various averages of degrees of polymerization are shown in Table 3-8.

Table 3-8 Averages of degrees of polymerization.

Average	Endgroups, non-uniform species i	No endgroups, non-uniform species i	No endgroups, uniform species i
$\overline{X}_\text{n}$	$\sum_i x_i \overline{X}_{\text{n},i}$	$\sum_i x_i \overline{X}_{\text{n},i}$	$\sum_i x_i X_i$
$\overline{X}_\text{w}$	$\sum_i w_i \overline{X}_{\text{w},i} = \dfrac{\sum_i x_i \overline{M}_{\text{n},i} \overline{X}_{\text{w},i}}{\overline{M}_n}$	$\dfrac{\sum_i x_i \overline{X}_{\text{n},i} \overline{X}_{\text{w},i}}{\overline{X}_\text{n}}$	$\dfrac{\sum_i x_i X_i^2}{\overline{X}_\text{n}}$
$\overline{X}_\text{z}$	$\sum_i Z_i \overline{X}_{\text{z},i} = \dfrac{\sum_i x_i \overline{M}_{\text{n},i} \overline{M}_{\text{w},i} \overline{X}_{\text{z},i}}{\overline{M}_\text{n} \overline{M}_\text{w}}$	$\dfrac{\sum_i x_i \overline{X}_{\text{n},i} \overline{X}_{\text{w},i} \overline{X}_{\text{z},i}}{\overline{X}_\text{w} \overline{X}_\text{n}}$	$\dfrac{\sum_i x_i X_i^3}{\overline{X}_\text{w} \overline{X}_\text{n}}$

3.5.6 Averages of Other Properties

The expressions shown above for averages of molar masses cannot be schematically applied to other properties P, for example, molecular dimensions such as radii of gyration, diffusion coefficients, sedimentation coefficients, etc. The reason is the different relationships between the various statistical weights and statistical weight fractions. For example, the weight-average $\overline{P}_w$ of a property P may be expressed by using masses m or amount-of-substances n as follows in Eq.(3-53) if the various species i are molecularly uniform. Far more complicated expressions are obtained for molecularly non-uniform species.

$$(3\text{-}53) \qquad \overline{P}_w \equiv \frac{\sum_i m_i P_i}{\sum_i m_i} = \frac{\sum_i n_i M_i P_i}{\sum_i n_i M_i} \qquad \text{but not} \qquad \overline{P}_w = \frac{\sum_i n_i P_i^2}{\sum_i n_i P_i}$$

3.5.7 Determination of Molar Masses

Polymer science employs a great number of methods for the determination of molar masses. The ranges and averages of the most important ones are shown in Table 3-9. Many of these methods are very costly with respect to instrumentation and labor. They are thus used mainly for fundamental research on relationships between structure and physical properties. The basics of their physics is discussed in Volume III.

Preparative working laboratories prefer fast and, if possible, experimentally simple routine methods. Most important here are mass spectroscopy (Section 3.4.1), size exclusion chromatography (Section 3.4.2), and dilute solution viscometry (Section 3.6). Investigations of polymerization kinetics in addition often employ methods that deliver number-average molar masses such as membrane and vapor pressure osmometry and also both static and dynamic light scattering (see Volume III).

Table 3-9 Types of averages and approximate experimentally obtainable ranges of molar masses.
  A = absolute method (no calibration necessary except for convenience), R = relative method (calibration required), E = equivalent method (constitution must be known). n = number-average, v = viscosity average (see Section 3.6.5), w = weight-average, z = z-average. Comments: the following physical properties may be obtained in addition to molar masses: second virial cooefficients A_2, radii of gyration s, frictional coefficients ξ, and/or molar mass distributions (MWD).
  § With certain assumptions; * upper range ca. 10^6 g/mol.

Method	Type	Average	Range in g/mol	Comments
Light scattering, static	A	w	> 100	A_2, s
Dilute solution viscometry	R §	v	> 200	-
X-ray or neutron small angle scattering	A	w	> 500	A_2, s
Combined sedimentation and diffusion	A	various	> 1 000	MWD, ξ
Size exclusion chromatography	R	n, w, z	> 1 000	MWD
Melt viscometry	R	w	> 1 000	-
Field flow fractionation	R	n, w, z	> 1 000	-
Membrane osmometry *	A	n	> 5 000	A_2
Ebullioscopy, cryoscopy	A	n	< 20 000	A_2
Endgroup determination by titration	E	n	< 40 000	-
Vapor pressure osmometry	A	n	< 50 000	A_2
Mass spectroscopy	A	n, w, z	< 100 000	MWD
Sedimentation equilibrium	A	w, z §	< 1 000 000	MWD
Light scattering, dynamic	R	z §	< 10 000 000	-

3.6   Viscometry

The viscometry of dilute polymer solutions delivers quantities that can be correlated with the molar mass of the polymer. The measurements are fast, experimentally simple, and inexpensive. They are still the method of choice for a quick characterization of polymers. The theoretical evaluation is not so simple, however (see Volume III).

3.6.1   Basic Terms

In the simplest case, the **dynamic viscosity** η of a material follows **Newton's law** $\eta = \sigma_{21}/\dot{\gamma}$, i.e., the ratio of shear stress σ_{21} to shear rate $\dot{\gamma} = \partial v/\partial y$, the latter being the change of flow rate v with the vertical distance y to the flow direction (Volume III). Liquids that obey Newton's law are called **Newtonian liquids**. The viscosity of these liquids does not depend on time or shear rate.

In polymer science, one is mainly interested in the **relative viscosity**, $\eta_{rel} = \eta/\eta_1$, of dilute solutions, the ratio of the Newtonian viscosity η of the solution to the viscosity η_1 of the solvent. The relative viscosity is now called **viscosity ratio** η_r by IUPAC (this new term is unnecessary since $\eta_{rel} \equiv \eta_r$ is indeed a relative quantity (Section 17.4)).

The **specific viscosity** $\eta_{sp} = (\eta/\eta_1) - 1$ is not a true specific quantity (Section 17.5). It was therefore renamed **relative viscosity increment** η_i by IUPAC. The suggested symbol η_i is however easy to confuse with η_i, the symbol for the viscosity of component i.

3.6.2   Experimental Methods

Viscosities of dilute solutions are usually measured with capillary viscometers of the Ostwald, Cannon-Fenske, or Ubbelohde type (Fig. 3-13). All types determine the time t needed for a solution of volume V and density ρ to pass between two marks in a capillary with radius R_0 and length L under a pressure $p = \rho g \Delta h$ (g = gravity). The product of time t and density ρ is proportional to the viscosity according to the **Hagen-Poiseuille law**, including the Hagenbach correction (second term on the right side):

$$(3\text{-}54) \qquad \eta = [\pi R_0{}^4/(8\ LV)]g\rho t - kV/(8\ \pi Lt)$$

The Hagenbach term corrects for the formation of eddies and the wasting of part of the potential energy by friction. k is an empirical constant.

Measured times should always exceed 100 seconds; relative viscosities should never be lower than 1.2. At flow times of (100 ± 0.1) seconds and relative viscosities of $\eta/\eta_1 = 1.2$, the error in η_{rel} is already ± 0.2 % and that in $\eta_{sp} \pm 1$ %. At $\eta_{rel} < 1.2$, anomalies are often observed; these effects are usually thought to be caused by adsorption of macromolecules on capillary walls. Relative viscosities should not exceed $\eta_{rel} \approx 2$ since then the function $\eta_{sp}/c = f(c)$ often becomes non-linear (see below).

Ostwald viscometers require exact fillings since different filling heights generate different driving pressures. In Ubbelohde viscometers, on the other hand, driving heights are always the same since the thickness of the hanging layer at the capillary exit is con-

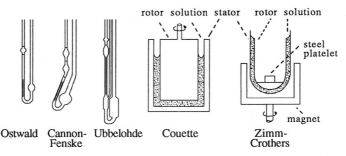

Fig. 3-13 Capillary viscometers of the Ostwald, Cannon-Fenske, and Ubbelohde type and rotary viscometers of the Couette and Zimm-Crothers design. In Couette viscometers, the rotor usually rotates in a stator and rarely around a stator.

stant. In Cannon-Fenske viscometers, the two bulbs lead to different driving pressures and thus two different shear stresses $\sigma_{21} = (pR)/(2\,L)$. The two shear gradients generated by Cannon-Fenske viscometers thus allow one to test whether the measured viscosities are indeed Newtonian.

Capillary viscosimeters usually work with shear gradients of ca. $100\ s^{-1}$. Such shear gradients are high enough to cause non-Newtonian behavior for, e.g., solutions of poly-(styrene) in benzene at 30°C if $[\eta]M \geq 4.5 \cdot 10^8\ cm^3/mol$.

Viscosities of dilute solutions of high molar mass polymers are therefore often measured with rotary viscometers since these instruments easily allow variations of shear stresses by simply changing the speed of rotation (Fig. 3-11). Rotary viscometers also create linear shear gradients between the rotor and stator if the gaps and rotational speeds are sufficiently small. Capillary viscometers, on the other hand, always generate parabolic shear gradients.

The centering of the rotor is achieved by a mechanical axis in rotary viscometers of the Couette type. A much better centering is obtained by the Zimm-Crothers viscometer. This viscometer is based on the observation that a buoyant rotor is automatically centered in the stator by the surface tension of the investigated liquid (first observed for a half-filled stemless wine glass floating in dishwasher liquid).

The rotor contains an steel platelet. The thermostated stator sits between the poles of a magnet that is attached to a motor with constant but variable speed of rotation. The coupling between the exterior magnetic field and the magnetic moment by the steel platelet generates a weak torsional momentum. With this viscometer, very low shear stresses down to 0.0004 Pa are produced and shear rates as low as $0.2\ s^{-1}$ are obtained, far lower than those in capillary viscometers (usually $1000\ s^{-1}$).

3.6.3 Concentration Dependence: Non-Electrolytes

The Newtonian viscosity η of dilute solutions or dispersions of small, hard spheres with volume V_2 in a solution of volume V can be expressed by a power series of the volume fraction $\phi_2 = V_2/V$ of the solute where η_1 = viscosity of solvent:

(3-55) $\eta = \eta_1(1 + K_1\phi_2 + K_2\phi_2^2 + ...)$

The constant K_1 has been calculated theoretically as 5/2 for unsolvated hard spheres (**Einstein**) and the constant K_2 as 6.2 (**Batchelor**). Both constants have been experimentally confirmed for carefully prepared dispersions of glass and guttapercha spheres.

In order to obtain a similar power series for other particle shapes and interactions, the volume $V_2 = N_2 V_h$ of all particles is expressed by the hydrodynamic volume V_h of *one* particle and the number N_2 of *all* particles by the number concentration $N_2/V = cN_A/M$ where c is the mass concentration of the solute. Introduction of the **reduced viscosity** (aka **viscosity number**) $\eta_{red} = (\eta - \eta_1)/(\eta_1 c)$ results in

(3-56) $\eta_{red} = (5/2)(V_h N_A/M) + K_2(V_h N_A/M)^2 c + ... = [\eta] + K_2(2\,[\eta]/5)^2 c + ...$

The quantity $[\eta] \equiv (5/2)(V_h N_A/M)$ is the **intrinsic viscosity** (IUPAC: **limiting viscosity number**); for a while, it was also called the **Staudinger index** Z_η. $[\eta]$ is usually reported in mL/g, in the United States also in 100 mL/g, and formerly in 1000 mL/g.

The names of these quantities do not correspond to their physical meanings. Neither η_{red} nor $[\eta]$ are viscosity *numbers*. $[\eta]$ is also not a intrinsic *viscosity* since it has not the physical unit Pa s of a dynamic viscosity or m^2 s^{-1} of a kinematic viscosity. Both quantities are rather specific volumes. $[\eta]$ is thus the hydrodynamic volume occupied by the unit mass of a single molecule (since $c \to 0$).

Eq.(3-56) becomes identical with the empirical **Huggins equation** if $k_H \equiv (2/5)^2 K_2$:

(3-57) $\eta_{sp}/c \equiv \eta_{red} = [\eta] + k_H[\eta]^2 c + ...$

Plots of $\eta_{sp}/c = f(c)$ are often not linear (Fig. 3-14). Many equations have thus been proposed to achieve better linearity over a greater concentration range, for example:

(3-58) η_{red} $= [\eta] + k_{SB}[\eta]\eta_{sp}$ **Schulz-Blaschke equation**

(3-59) $(\ln \eta_{rel})/c$ $= [\eta] + k_K[\eta]^2 c$ $= \eta_{inh}$ **Kraemer equation**

(3-60) $\lg \eta_{red}$ $= \lg [\eta] + k_M[\eta]c$ **Martin** or **Bungenberg-de Jong eqn.**

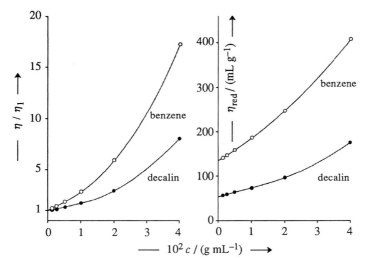

Fig. 3-14 Concentration dependence of relative viscosities (left) and reduced viscosities (right) of a poly(styrene) in benzene or decalin at 25°C [13].

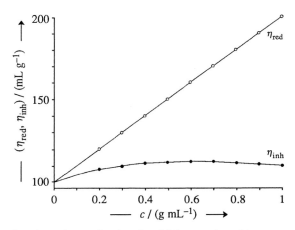

Fig. 3-15 Concentration dependence of reduced and inherent viscosities. η_{red} was assumed to be a linear function of c with $[\eta] = 100$ mL/g and $k_H = 1$; η_{inh} was recalculated from these data.

In these equations, k_H, k_{sB}, $k_K = k_H - (1/2)$, and k_M are empirical constants. The Huggins constant k_H usually adopts values of 0.3-0.8 for neutral, flexible, linear macromolecules. More rigid macromolecules show higher values.

In order to determine intrinsic viscosities more accurately, reduced viscosities η_{red} and **inherent viscosities** (IUPAC: **logarithmic viscosity numbers**) $\eta_{inh} \equiv \eta_{ln} = (\ln \eta_{rel})/c$ are sometimes plotted in one diagram and extrapolated linearly to a common intercept which is taken as the intrinsic viscosity (Fig. 3-15). Such an extrapolation can lead to gross errors. If, for example, η_{red} is a linear function of c, then η_{inh} cannot have a linear concentration dependence except for $k_H = 1/3$ which can be shown as follows. Inherent viscosities $\ln \eta_{rel} = \ln (1 + \eta_{sp})$ can be developed in a Taylor series so that $\eta_{inh} = [\eta] + (k_H - (1/2))[\eta]^2 c + ((1/3) - k_H)[\eta]^3 c^2 + \ldots$ Higher terms cannot be neglected since the Taylor series converges very slowly. For very small relative viscosities ($\eta_{rel} < 1.2$), one obtains $k_K \approx k_H + [(1/3) - k_H[\eta]c]$.

Intrinsic viscosities are sometimes calculated from data at a single concentration with the help of the **Solomon-Ciuta equation** $[\eta] = [2 (\eta_{sp} - \ln \eta_r)]^{1/2}/c$. The derivation of this equation indicates that it can only be valid for $k_H = 1/3$.

3.6.4 Concentration Dependence: Polyelectrolytes

The concentration dependence of reduced viscosities of polyelectrolyte solutions differs strongly from that of its nonionic counterparts (Fig. 3-16). Reduced viscosities of polyelectrolyte solutions without added salt are very high at small polymer concentrations. With increasing polymer concentration, they pass through a maximum, and then decrease strongly until they approach an asymptote. These effects decrease with increasing concentrations of added salt(s); at high salt concentrations, solutions of polyelectrolytes behave like those of neutral polymers.

This behavior is not only found for polyacids and polybases but also for polyampholytes despite the charge compensation, . An example is a copolymer with equal amounts of $-CH_2-CH(CONH_3)^{\oplus}-$ and $-CH_2-CHSO_3^{\ominus}$ units.

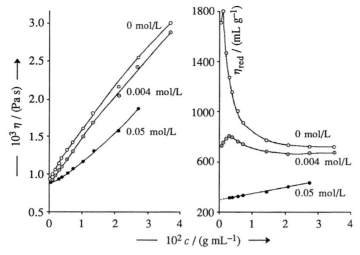

Fig. 3-16 Concentration dependence of viscosities η (left) and reduced viscosities η_{red} (right) of a sodium pectate in different concentrations of aqueous NaCl solutions at 27°C [14].

These effects are explained as follows. Polyelectrolytes dissociate more strongly at higher dilutions. In polysalts such as sodium pectate or the sodium salt of poly(acrylic acid), sodium counterions form a cation sphere around the negatively charged polyacid molecules. The thickness of that ion sphere is much larger than the diameter of the coiled polyelectrolyte molecules.

The carboxylic acid groups $COO^{\ominus}$ of the polyacid molecules repel each other which stiffens the chain and increases the viscosity. At medium polyelectrolyte concentrations, some counter ions are in the interior of the polymer coils and some on the outside. At very high polyelectrolyte concentrations, the concentration of counterions is higher in the interior than at the exterior. The resulting osmotic effect causes more water to enter the coils that expands them. The osmotic effect thus dominates at high polyelectrolyte concentrations and the electrostatic effect at low ones.

Addition of salt increases the ion strength of the exterior relative to the interior and decreases the thickness of the ion cloud. Both effects decrease coil diameters and thus also reduced viscosities η_{red}.

The concentration dependence of polyelectrolyte solutions for concentrations greater than the maximum of η_{red} is often described by the **Fuoss equation** for the dependence of inverse reduced viscosities on the square root of polymer concentration:

$$(3-61) \qquad 1/\eta_{red} = A_{FS} + K_{FS}c^{1/2}$$

A_{FS} and K_{FS} are empirical constants. The intercept A_{FS} is often wrongly identified with the inverse of intrinsic viscosity ($A_{FS} = 1/[\eta]$) because experiments seem to "confirm" the validity of the Fuoss equation. Modern scaling theories predict however quite different and more complicated dependencies of η, η_{sp}, and η_{red} on polyelectrolyte and salt concentrations for the various concentration regimes (Volume III). It seems that the Fuoss-type behavior of polyelectrolyte solutions is caused by non-Newtonian behavior, i.e., a strong shear rate dependence of viscosities.

3.6.5   Intrinsic Viscosities and Molar Masses

Averages

According to Eq.(3-56), intrinsic viscosity $[\eta]$ is controlled by the ratio V_h/M. The molar mass M can thus be obtained from intrinsic viscosity if the hydrodynamic volume V_h varies systematically with the molar mass. The function $V_h = f(M)$ can often be represented by a power equation, the **Kuhn-Mark-Houwink-Sakurada equation (KMHS equation)**, where K_v and α are constants that have to be determined empirically for a given combination of polymer, solvent, and temperature:

$$(3\text{-}62) \qquad [\eta] = K_v M^\alpha \qquad\qquad \text{KMHS equation}$$

The intrinsic viscosity of a molecularly non-uniform ("polydisperse") polymer is the weight-average $\overline{[\eta]}_w$ of the intrinsic viscosities $[\eta]_i$ of all species i. According to Eq.(3-57), $\eta_{sp} = [\eta]c$ for $c \to 0$. Experiments show that specific viscosities η_{sp} of nonionic polymer homologs are additive, i.e., $\eta_{sp} = \Sigma_i \eta_{sp,i} = \Sigma_i [\eta]_i c_i$. Introducing this expression into $\eta_{sp} = [\eta]c$ and setting $c = \Sigma_i c_i$ und $w_i = c_i/c$ shows that the intrinsic viscosity is indeed a weight average: $[\eta] = [\Sigma_i [\eta]_i c_i]/c = \Sigma_i w_i[\eta]_i \equiv \overline{[\eta]}_w$.

The molar mass in Eq.(3-62), on the other hand, is an exponent average (Eq.(3-33)) with $K_P = K_v$, $g = w$, $h = \alpha$. Solving Eq.(3-62) for the molar mass and setting first $[\eta] = \Sigma_i w_i[\eta]_i$, and then $[\eta]_i = K_v M_i^\alpha$ for each species i, shows that M is the **viscosity-average molar mass**:

$$(3\text{-}63) \qquad M = \{[\eta]/K_v\}^{1/\alpha} = \{(\Sigma_i w_i[\eta]_i)/K_v\}^{1/\alpha} = \{\Sigma_i w_i M_i^\alpha\}^{1/\alpha} \equiv \overline{M}_v$$

The viscosity-average molar mass $\overline{M}_v$ is identical with the mass-average molar mass $\overline{M}_w$ for $\alpha = 1$. It is smaller than $\overline{M}_w$ for $\alpha < 1$ and larger for $\alpha > 1$.

Molar masses calculated with Eq.(3-62) are true viscosity averages only if the calibration was performed with uniform polymers or with polymers with known true viscosity-average molar masses. All other calibrations deliver undefined molar mass averages.

Measurements on polymers with different widths of molar mass distributions result in a set of parallel lines in the plot of $\lg [\eta] = f(\lg M)$ (Fig. 3-17). In order to obtain the sought-for dependence of intrisic viscositiy on viscosity-average of molar mass, a correction factor q_{KMHS} for the polymolecularity must be introduced (Table 3-10):

$$(3\text{-}64) \qquad [\eta] = K_v \overline{M}_v^\alpha = K_v q_{KMHS} \overline{M}_g^\alpha$$

Table 3-10 Polymolecularity correction factors. $\varsigma = [(\overline{M}_w/\overline{M}_n) - 1]^{-1}$. α = exponent in Eq.(3-64).

	Schulz-Zimm distributions	Logarithmic normal distributions
if $\overline{M}_g = \overline{M}_w$	$q_{KMHS} = \dfrac{\Gamma(\varsigma+\alpha+1)}{(\varsigma+1)^\alpha \, \Gamma(\varsigma+1)}$	$q_{KMHS} = (\overline{M}_w/\overline{M}_n)^{(\alpha^2-\alpha)/2}$
if $\overline{M}_g = \overline{M}_n$	$q_{KMHS} = \dfrac{\Gamma(\varsigma+\alpha+1)}{\varsigma^\alpha \, \Gamma(\varsigma+1)}$	$q_{KMHS} = (\overline{M}_w/\overline{M}_n)^{(\alpha^2+\alpha)/2}$

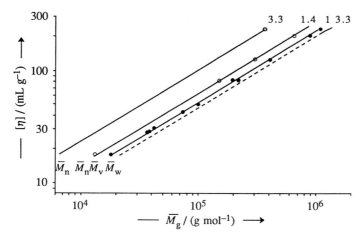

Fig. 3-17 Dependence of the logarithms of intrinsic viscosities of poly(vinyl acetate)s with different polymolecularity ratios in butanone at 25°C on different molar mass averages $\overline{M}_g$. Points are experimental, lines calculated. Data points for mass-averages are not shown since they all fall on the dotted line. Number-average data are shown for $\overline{M}_w / \overline{M}_n = 1.4$ and 3.3 only. Viscosity-average molar masses $\overline{M}_v$ have been calculated from number-average and mass-average data with $\alpha = 0.62$.

Table 3-11 shows some correction factors for unperturbed coils ($\alpha = 1/2$), perturbed flexible coils ($\alpha = 0.764$ (Volume III)), infinitely long and stiff rods ($\alpha = 2$), and some other α values ($\alpha = 0.6$, $\alpha = 1.5$, etc.) without physical meaning, all for Schulz-Zimm or logarithmic normal distributions and for calibrations of Eq.(3-62) with mass-average or some number-average molar masses. The corrections for Schulz-Zimm distributions with $\overline{M}_w/\overline{M}_n = 2$ vary between 6 % (if $\alpha = 1/2$) and 25 % (if $\alpha = 2$). They are much larger for logarithmic normal distributions with the same width of distribution: 8.3 % (if $\alpha = 1/2$) and 100 % (if $\alpha = 2$). Far larger corrections apply if calibrations are performed with number-average molar masses instead of viscosity-average ones.

Table 3-11 Polymolecularity correction factors q_{KMHs} for $\overline{M}_g = \overline{M}_w$ [15].

$\overline{M}_w/\overline{M}_n$	Schulz-Zimm distributions and $\alpha =$					Logarithmic normal distributions and $\alpha =$				
	0.5	0.6	0.764	1.5	2	0.5	0.6	0.764	1.5	2
					q_{KMHS} for $\overline{M}_g = \overline{M}_w$					
1.1	0.989	0.990	0.992	1.034	1.091	0.988	0.989	0.991	1.036	1.100
1.3	0.971	0.972	0.980	1.083	1.231	0.968	0.969	0.977	1.103	1.300
1.5	0.959	0.961	0.971	1.119	1.333	0.951	0.953	0.964	1.164	1.500
2.0	0.940	0.943	0.958	1.175	1.250	0.917	0.920	0.939	1.297	2.000
3.0	0.921	0.926	0.946	1.228	1.667	0.872	0.876	0.906	1.510	3.00
5.0	0.907	0.912	0.936	1.270	1.800	0.818	0.824	0.865	1.829	5.00
10.0	0.896	0.903	0.929	1.300	1.900	0.750	0.759	0.813	2.37	10.0
30.0	0.845	0.897	0.925	1.320	1.968	0.654	0.665	0.736	3.29	30.0
					q_{KMHS} for $\overline{M}_g = \overline{M}_n$					
2.0	1.329	1.430	1.627	3.32	6.00	1.214	1.395	1.595	3.67	8.00
5.0	2.028	2.400	3.20	14.2	45.0	1.569	2.165	2.96	20.4	125

Effect of Molecular Shape

Intrinsic viscosities measure the hydrodynamic volume per mass (p. 95). The hydrodynamic volume depends on the macroconformation (the "shape") of the polymer molecule and, by implication, on the density distribution of its segments, as well as the interaction of the molecule segments with themselves and the solvent. Fig. 3-18 shows typical $\lg [\eta] = f(\lg \overline{M}_w)$ curves for various types of molecules and interactions.

Strong intramolecular interactions lead molecules to adopt compact macroconformations with little, if any, interactions with solvents. These systems behave as dispersions of "hard" bodies. It follows from Eq.(3-56) for **hard spheres** with $V_h = (4/3)\,\pi\,R_{sph}^3 = m_2/\rho_2 = M/(N_A\rho_2)$ that $[\eta] = 5/(2\,\rho_2)$: the intrinsic viscosity of hard spheres depends only on the bulk density ρ_2 of the sphere = polymer molecule.

Such a behavior is shown by the hyperbranched polymers of the α,ε-lysine derivative $NH_2CH\{(CH_2)_4NHCOOC(CH_3)_3\}COOH$ (Fig. 3-19). Their intrinsic viscosity is independent of the molar mass and only slightly higher than the theoretical one of $[\eta]/(mL\ g^{-1}) = 2.12$ which indicates a slight swelling of the spherical molecules by the solvent.

Hydrodynamic volumes $V_h = Q^3 s^3$ can be expressed by mean-square radii of gyration, $\langle s^2 \rangle$, where Q depends on the type of entity (hard sphere, long rigid rod, random coil, etc.); $Q = (5/3)^{1/2}$ for hard spheres (see Volume III). Eq.(3-56) can thus be written

$$(3\text{-}65) \qquad [\eta] = (5\,N_A Q^3/2)[\langle s^2 \rangle^{3/2}/M] = \Phi(\langle s^2 \rangle^{3/2}/M)$$

where Φ is a shape and interaction dependent constant. The radius s of long, thin ($L \gg R$), rigid, circular rods is given by $\langle s^2 \rangle^{1/2} = L/12^{1/2}$ (Volume III). Since the length L is proportional to the molar mass M, it follows that the intrinsic viscosity of such a **rigid rod** is proportional to the square of the molar mass. An example is imogolite, an aluminum silicate $SiO_2 \cdot Al_2O_3 \cdot 2\ H_2O$ (Fig. 3-18, IM).

The proportionality $[\eta] \sim M^2$ breaks down for smaller molar masses since the contribution of the particle radius to the value of the radius of gyration can no longer be neglected (imogolite in Fig. 3-18). Very long rods are furthermore not infinitely stiff (cf. the behavior of short and long garden hoses). As a result, the logarithm of intrinsic viscosity becomes an S-shaped function of the logarithm of molar mass (see poly(hexyl isocyanate) in Fig. 3-18).

The calculation of the proportionality factor Φ for the various molecule structures, shapes, and interactions requires elaborate theories (Volume III). The results are given here in qualitative terms only.

Chain segments of thin, flexible chains can adopt many different microconformations (Chapter 5) while the chains adopt the macroconformations of kidney-shaped **random coils**. Infinitely thin chain segments can cross over each other and the coil is **unperturbed**. Chain segments of finite thicknesses can also be in this state if attraction and repulsion just compensate each other (**theta state**). Coils are **perturbed** and expanded from their unperturbed state if repulsion dominates. In both cases, the mean-square radius of gyration depends on the power $2v$ of the molar mass, i.e., $\langle s^2 \rangle = K_s M^{2v}$. Insertion of this expression into Eq.(3-65) and comparison with Eq.(3-64) shows that the exponents v and α are related via $\alpha = 3\,v - 1$:

$$(3\text{-}66) \qquad [\eta] = \Phi\langle s^2 \rangle^{3/2}/M = \Phi K_s^{3/2} M^{3v-1} = K_v M^\alpha$$

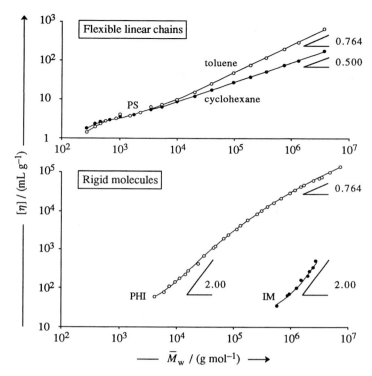

Fig. 3-18 Intrinsic viscosity as function of weight-average molar mass. Numbers indicate α.
Top: Flexible chains of poly(styrene) (PS) in the thermodynamically good solvent toluene at 15°C (α = 0.764; perturbed coils) and in the theta solvent cyclohexane at 34°C (α = 0.500; unperturbed coils) [16].
Bottom: Rigid molecules of poly(hexyl isocyanate)s (PHI) in hexane at 25°C [17] and imogolite (IM) in dilute acetic acid (+ 0.02 wt% NaN₃; pH = 3) at 30°C [18].

Theory predicts and experiment confirms for sufficiently high molar masses that v = 1/2 and α = 1/2 for unperturbed coils of **flexible linear macromolecules** and v = 0.588 and α = 0.764 for perturbed ones (PS in Fig. 3-18). The exponents of lg [η] = f(lg M) curves deviate from these values for molar masses of less than 10^3-10^4 g/mol for two reasons: (a) the chains are too short to adopt the theoretically assumed statistics of their segment distribution and (b) endgroups cannot be neglected.

Values of v = 0.588 and α = 0.764 are also obtained for the high-molar mass range of **stiff linear** macromolecules which form **worm-like chains** in solution (PHI in Fig. 3-16). The whole function lg [η] = f(lg M) of these molecules is, however, S-shaped. Their low-molar mass molecules behave like rigid rods. Since the rods are short, they do not follow $\langle s^2 \rangle^{1/2} = L/12^{1/2}$ for very long rods of the same diameter. With increasing length of the rods, the molecules become more flexible although their segments remain stiff. As a result, the values of α = 2 and v = 1 of truly rigid rods are never approached. The molecules behave instead more and more like perturbed coils. The result is an S-shaped form of the lg [η] = f(lg M) curves. At high molar masses, the values of α = 0.764 and v = 0.588 of perturbed chains are observed albeit with far higher values of [η] than those of flexible macromolecules of the same molar mass.

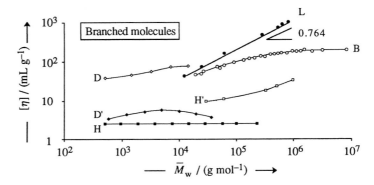

Fig. 3-19 Intrinsic viscosity as function of weight-average molar mass for different branched molecules. Linear poly(ethylene) (curve L) is used for a comparison. Numbers indicate α.
    B   = Statistically branched low-density poly(ethylene)s in tetralin at 120°C [19].
    D   = Polyamidoamine dendrimers in 0.1 mol/L citric acid in water at 25°C [20]. $[\eta] \times 10$.
    D' = Dendrimers with 3,5-dioxybenzylidene units in tetrahydrofuran at 30°C [21].
    H   = Hyperbranched poly(α,ε-lysine)s in *N,N*-dimethylformamide at 25°C; $\rho = 1.18$ g/mL [22].
    H' = Hyperbranched polymers of 3,5-diacetoxybenzoic acid in tetrahydrofuran at 25°C [23].
    L   = Linear high-density poly(ethylene)s in tetralin at 120°C [19].

Different effects are observed for branched macromolecules (Fig. 3-19). **Randomly
branched** polymers (Fig. 3-19, B) at low molar masses behave approximately like linear
macromolecules. Random branching during polymerization (Section 10.4.4) leads to an
increase of the number of branches per molecule with increasing molar mass. The shape
of the dissolved macromolecules becomes more "rounded" at higher molar masses (see
Fig. 2-5). At very high molar masses, such molecules approach $\alpha = 0$ of true spheres.
Because of $[\eta] = 5/(2\,\rho_2)$ (p. 100), the densities of these sphere-like structures are, however, much lower than the bulk density of the polymers themselves.

Dendritic and hyperbranched polymers are very highly branched and have accordingly very low intrinsic viscosities that are even lower than those of randomly branched
polymers of comparable molar mass (Fig. 3-19, bottom). Generalizations about the
shape of their $\lg [\eta] = f(\lg M)$ functions are however difficult because of lack of data.

The intrinsic viscosities of all but one of the investigated **dendrimers** increase slightly
with increasing molar mass (or number of generations). They finally seem to become
constant for PAMAM dendrimers (Fig. 3-19, D) and also for other types of dendrimers.
Since intrinsic viscosities of higher generations of all these dendrimers have not been
determined, it remains unclear whether the $[\eta]$ of higher generation dendrimers are indeed independent of M or pass through a maximum (Fig. 3-19, D').

Hyperbranched polymers with strong intersegmental attractions (Fig. 3-19, H) show
intrinsic viscosities that are independent of the molar mass and indicative of small, slightly swollen compact spheres (p. 100). Hyperbranched polymers with weaker intersegmental interactions exhibit larger intrinsic viscosities (Fig. 3-19, H'). The function $\lg [\eta] =
f(\lg \overline{M}_w)$ is curved. No conclusions about the shape of this function can be drawn, how-
ever, since there is a systematic increase of the polymolecularity index $\overline{M}_w/\overline{M}_n$ from 6.3
at $\overline{M}_w = 30\,000$ g/mol to 30 at $\overline{M}_w = 498\,000$ g/mol. Such systematic changes cause
dramatic shifts in the intrinsic viscosity–molar mass plots if weight-average molar masses
are used instead of the correct viscosity averages (see Fig. 3-17 and Table 3-11).

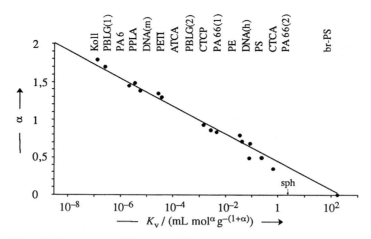

Fig. 3-20 Relationship between α and lg K_v for different polymers.
ATCA = amylose tricarbanilate in acetone (20°C); CTCA = cellulose trioctanate in DMF (140°C); CTCP = cellulose tricarbanilate in acetone (25°C); DNA = Deoxyribonucleic acids of high (h) or medium (m) molar mass in 0.11 mol/L NaCl in water (20°C); Koll = collagen in citrate buffer (24.8°C); PA 6 = poly(ε-caprolactam) in SbCl$_3$ (25°C); PA 66 = poly(hexamethylene adipamide) in (1) 90 % formic acid and (2) 90 % formic acid + 2.2 mol/L KCl (25°C); PBLG(1) = poly(γ-benzyl-L-glutamate) in DMF (25°C), PBLG(2) in dichloroacetic acid (25°C); PE = poly(ethylene) in tetralin at 130°C; PETI = poly(α-phenylethylisocyanate) in THF (30°C); PPLA = poly(D,L-phenylalanine) in CHCl$_3$ (25°C); PS = poly(styrene) in cyclohexane (34.5°C); br-PE = highly branched poly(ethylene) in tetralin (130°C); sph = hard spheres.

The K_v and α values of linear polymers are interrelated by the empirical equation α = $A - B$ lg K_v (Fig. 3-20). The slight scattering of data may be caused by the different bulk densities of polymers, thicknesses of polymer chains, and experimental errors.

Historical Notes to Chapter 3

Size-Exclusion Chromatography

G.H.Lathe, C.R.Ruthven, Biochem.J. **62** (1956) 665 (separation of proteins at starch granules)
J.Porath, P.Flodin, Nature **183** (1959) 1657 (gel filtration of water-soluble polymers)
J.C.Moore, J.Polym.Sci. **A 2** (1964) 835 (gel chromatography in organic solvents)

Viscosity of Dilute Solutions

H.Staudinger, W.Heuer, Ber.Dtsch.Chem.Ges. **B 63** (1930) 222
Experiments of Staudinger and his coworkers showed that specific viscosities $\eta_{sp} \equiv (\eta - \eta_1)/\eta_1$ of macromolecules depend on molecular weights M_r. Staudinger investigated macromolecules with relatively low molar masses that had relatively small concentration dependences of $\eta_{sp} = f(c)$. He thus used η_{sp}/c instead of [η]. Macromolecular chains in solution were envisioned as rigid rods where each of the N_u monomeric units with molecular weight $M_{u,r}$ made the same viscosity contribution to the molecular weight M_r of the whole macromolecule. Staudinger's empirical viscosity law was thus $\eta_{sp}/c = K_v N_u M_{u,r} = K_v M_r$.

K.H.Meyer, Kolloid-Z. **53** (1930) 8
    Linear macromolecules cannot be rigid rods since their shape is controlled by the free rotation of chain units around chain bonds.

W.Haller, Kolloid-Z. **56** (1931) 257
    The author postulates a free rotation of chain units around chain bonds (correct) as well as a deformation of valence bonds (incorrect because it would require very high energies). These two factors generate coil-like, solvated macromolecules that are caused to rotate by viscous flow. Assuming no variation of coil density with coil radius (not true), the viscosity law is calculated as $\eta_{sp}/c = K_v M_r^{2/3}$ instead of Staudinger's $\eta_{sp}/c = K_v M_r$.

L.Onsager, Phys.Rev. **A 40** (1932) 1028
    Staudinger's law $\eta_{sp}/c = K_v M_r$ is probably not caused by the presence of rod-like molecules (correct) but by the slipping of molecules past each other (incorrect).

E.Guth, H.Mark, Ergeb.Exakt.Naturwiss. (1933) 113
    This paper emphasizes that hydrodynamics delivers only relationships between specific viscosities and molecular dimensions. Additional assumptions are necessary for the dependence of dimensions on molecular weights.

W.Kuhn, Kolloid-Z. **68** (1934) 2
    Chain molecules are not rigid rods since that would imply $\eta_{sp}/c \sim M_r^8$. Such molecules rather form bean-shaped coils. The finite thickness of chains leads to an excluded volume that expands the coil and leads to $\eta_{sp}/c \sim M_r^\alpha$. The exponent is estimated as $\alpha = 0.84$.

E.Guth, H.Mark, Monatsh.Chem. **65** (1934) 93
    Development of the random-coil model of infinitely thin chain molecules (simultaneously with W.Kuhn). The hydrodynamic volume of a random coil is far greater than the volume of the dry molecule which leads to the high viscosity of dilute polymer solutions.

H.Mark, in E.Saenger, Ed., Der feste Körper (lecture at the 50th anniversary of the Physical Society of Zurich (1937)), Hirzel, Leipzig 1938
    The viscosity number η_{sp}/c is proportional to the α power of the molecular weight ($\eta_{sp}/c = K_v M_r^\alpha$). K_v and α depend on the system. The paper is not easily accessible and was apparently never included in Chemical Abstracts and Chemisches Zentralblatt.

I.Sakurada, Proceedings of the Congress of Nippon Kagaku Seni Kenkyu-sho **5** (1940) 33, see T.Saegusa, Makromol.Symp. **98** (1995) 1199
    Finds $\eta_{sp}/c = K_v M_r^\alpha$.

R.Houwink, J.Prakt.Chem. **157** (1941) 15
    Finds $\eta_{sp}/c = K_v M_r^\alpha$.

    In subsequent years, many new and sophisticated hydrodynamic models were developed for random coils

    J.G.Kirkwood, J.Riseman, J.Chem.Phys. **16** (1948) 565
    P.E.Rouse, J.Chem.Phys. **21** (1953) 1272: free-draining coils show $\alpha = 1$.
    B.H.Zimm, J.Chem.Phys. **24** (1956) 269: non-draining coils show $\alpha = 1/2$.
    P.J.Flory, Principles of Polymer Chemistry, Cornell University Press, Ithaca, New York 1953,
        p. 622: coils in good solvents have an upper limit of 4/5 for α.

and subsequently for perturbed coils, worm-like chains, etc. (see Volume III).

Literature to Chapter 3

3.1 INTRODUCTION: Nomenclature, Comprehensive Works
IUPAC, Commission on Macromolecular Nomenclature, Compendium of Macromolecular Nomen-
    clature, Blackwell, London 1991 ("Purple Book")
G.Allen, J.C.Bevington, Eds., Comprehensive Polymer Science; C.Booth, C.Price, Eds., vol. 1,
    Polymer Characterization, Pergamon Press, Oxford 1989

3.1 INTRODUCTION: General Analytical Methods
J.Urbanski, W.Czerwinski, K.Janicka, F.Majewska, H.Zowall, Handbook of Analysis of Synthetic
    Polymers and Plastics, Wiley, New York 1977
L.S.Bark, N.S.Allen, Eds., Analysis of Polymer Systems, Appl.Sci.Publ., Barking, Essex 1982
J.Mitchell, Jr., Ed., Applied Polymer Analysis and Characterization, Hanser, Munich 1987
E.Schröder, G.Müller, K.-F.Arndt, Polymer Characterization, Hanser, Munich 1989
T.R.Crompton, Analysis of Polymers, Pergamon, Oxford 1989
J.R.White, D.Campbell, Polymer Characterization. Physical Techniques, Chapman and Hall, New
    York 1989
J.I.Kroschwitz, Ed., Polymers: Characterization and Analysis, Wiley, New York 1990 (reprints of
    contributions to Encyclopedia of Polymer Science and Engineering)
H.G.Barth, J.W.Mays, Eds., Modern Methods of Polymer Characterization, Wiley, New York 1991
B.J.Hunt, M.I.James, Eds., Polymer Characterisation, Blackie Academic, Glasgow 1993
R.A.Pethrick, J.V.Dawkins, Modern Techniques for Polymer Characterisation, Wiley, London 1999
B.Stuart, Polymer Analysis, Wiley, New York 2002
R.F.Brady, Jr., Ed., Comprehensive Desk Reference of Polymer Characterization and Analysis,
    ACS Books, Washington (DC) 2003

3.2 CHEMICAL COMPOSITION: Special Methods
Infrared and Raman Spectroscopy
W.Klöpffer, Introduction to Polymer Spectroscopy, Springer, Berlin 1984
R.A.Nyquist, The Infrared Building Blocks of Polymers, Sadtler, Philadelphia 1989
D.I.Bower, W.F.Maddams, The Vibrational Spectroscopy of Polymers, Cambridge University Press,
    Cambridge 1989
D.O.Hummel, Atlas of Polymer and Plastics Analysis, VCH, Weinheim, 3rd ed., 4 vols. 1991-2000
P.J.Hendra, J.K.Agbenyega, The Raman Spectra of Polymers, Rapra Technol., Shawbury (UK) 1993
M.Buback, H.P.Vögele, FT-NIR Atlas, Rapra Technol., Shawbury (UK) 1993
A.H.Fawcett, Ed., Polymer Spectroscopy, Wiley, New York 1996
G.Zerbi, H.W.Siesler, I.Noda, M.Tasumi, S.Krimm, Modern Polymer Spectroscopy, Wiley-VCH,
    Weinheim 1999
J.L.Koenig, Spectroscopy of Polymers, Elsevier, New York 1999
P.A.Mirau, A Practical Guide to Understanding the NMR of Polymers, Wiley, New York 2005

Nuclear Magnetic Resonance and Electron Spin Resonance Spectroscopy
B.Ranby, J.F.Rabek, ESR Spectroscopy in Polymer Research, Springer, Berlin 1977
Q.T.Pham, R.Petiaud, H.Waton, Proton and Carbon NMR Spectroscopy of Polymers, Wiley,
    New York 1983 (2 vols.)
R.A.Komoroski, Ed., High Resolution NMR Spectroscopy of Synthetic Polymers in Bulk,
    VCH, Weinheim 1986
R.R.Ernst, G.Bodenhausen, A.Wokann, Principles of Nuclear Magnetic Resonance in One and Two
    Dimensions, Oxford University Press, New York 1987
B.Schrader, Infrared and Raman Spectroscopy, VCH, Weinheim 1995
F.A.Bovey, P.A.Mirau, NMR of Polymers, Academic Press, San Diego 1997
I.Ando, T.Asakura, Solid State NMR of Polymers, Elsevier Science, Amsterdam 1998
R.N.Ibett, Ed., NMR Spectroscopy of Polymers, Blackie Academic, London 1993

Gas Chromatography
V.G.Berekin, V.R.Alishoyev, I.B.Nemiroskaya, Gas Chromatography of Polymers, Elsevier,
    Amsterdam 1977
S.A.Liebman, E.J.Levy, Pyrolysis and GC in Polymer Analysis, Dekker, New York 1986

Other Chromatographic Methods
H.Inagaki, Polymer Separation and Characterization by Thin-Layer Chromatography,
    Adv.Polym.Sci. **24** (1977) 189
J.Janca, Field-Flow Fractionation, Dekker, New York 1987

3.2.2 END GROUP DETERMINATIONS
S.R.Palit, B.M.Mandal, End-Group Studies Using Dye Techniques, J.Macromol.Sci.Rev. C **2**
    (1968) 225
R.G.Garmon, End Group Determinations, Techniques and Methods of Polymer Evaluation
    4/1 (1975) 31

3.3 MOLAR MASS DISTRIBUTIONS
S.R.Rafikov, S.Pavlova, I.I.Tverdokhlebova, Determination of Molecular Weights and Polydispersity
    of High Polymers, Akad.Wiss.USSR, Moscow 1963; Israel Program of Scientific Translation,
    Jerusalem 1964
J.Aitchison, J.A.C.Brown, The Lognormal Distribution, Cambridge University Press, Cambridge
    1969
L.H.Peebles, Molecular Weight Distributions in Polymers, Interscience, New York 1971
H.-G.Elias, Polymolecularity and Polydispersity in Molecular Weight Determinations,
    Pure Appl.Chem. **43**/1-2 (1975) 115
S.R.Holding, E.Meehan, Molecular Weight Characterization of Synthetic Polymers (Rapra Review
    Report), Plastics Design Library, Norwich (NY), 1997

3.4.1 MASS SPECTROSCOPY
G.Siuzdak, Mass Spectrometry for Biotechnology, Academic Press, London 1996
G.Montaudo, R.P.Lattimer, Mass Spectroscopy of Polymers, CRC Press, Boca Raton (FL) 2002
H.Pasch, W.Schnepp, MALDI-TOF Mass Spectroscopy of Synthetic Polymers, Springer, Berlin
    2003
E. de Hoffmann, V.Stroobant, Mass Spectrometry. Principles and Applications, Wiley, Chichester,
    2nd ed. 2002
J.H.Gross, Mass Spectrometry: A Textbook, Springer, Berlin 2004

3.4.2 SIZE-EXCLUSION CHROMATOGRAPHY
W.W.Yan, J.J.Kirkland, D.D.Bly, Modern Size-Exclusion Liquid Chromatography, Wiley, New
    York 1979
G.Glöckner, Polymercharakterisierung durch Flüssigkeitschromatographie, VEB Dtsch.Vlg.Wiss.,
    Berlin 1980, Hüthig, Heidelberg 1982; Polymer Characterization by Liquid Chromatography,
    Elsevier, Amsterdam 1986
C.G.Smith, N.E.Skelly, C.D.Chow, R.A.Solomon, Chromatography: Polymers, CRC Press,
    Boca Raton, FL 1982
B.G.Belenkii, L.Z.Vilenchek, Modern Liquid Chromatography of Macromolecules, Elsevier, New
    York 1983
J. Janca, Ed., Steric Exclusion Chromatography of Polymers, Dekker, New York 1984
B.J.Hunt, S.Holding, Eds., Size Exclusion Chromatography, Blackie, Glasgow 1989; Chapman and
    Hall, New York 1990
C.S.Wu, Handbook of Size Exclusion Chromatography, Dekker, New York 1995
M.Potschka, Inverse Size Exclusion Chromatography and Universal Calibration, Macromol.Symp.
    110 (1996) 121
H.Pasch, B.Trathnigg, HPLC of Polymers, Springer, Berlin 1997
R.W.A.Oliver, Ed., HPLC of Macromolecules, Oxford Univ. Press, New York, 2nd ed. 1998
    (biological macromolecules)
S.Mori, H.G.Barth, Size Exclusion Chromatography, Springer, Heidelberg 1999
K.M.Gooding, F.E.Regnier, Eds., HPLC of Biological Macromolecules, Dekker, New York 2002

3.5 MOLAR MASS AVERAGES
H.-G.Elias, R.Bareiss, J.G.Watterson, Mittelwerte des Molekulargewichtes und anderer Eigenschaf-
    ten, Adv.Polym.Sci.-Fortschr.Hochpolym.Forschg. **11** (1973) 111
R.E.Bareiss, Polymolecularity Correction Factors, in J.Brandrup, E.H.Immergut, E.Grulke, Eds.,
    Polymer Handbook, Wiley, New York, 4th ed. 1998

3.5.7 DETERMINATION OF MOLAR MASSES

R.U.Bonner, M.Dimbat, F.H.Stross, Number Average Molecular Weights, Interscience, New York 1958

P.E.Slade, Jr., Polymer Molecular Weights, Dekker, NY 1975 (2 vols.)

N.C.Billingham, Molar Mass Measurements in Polymer Science, Halsted Press, New York 1977

A.R.Cooper, Ed., Determination of Molecular Weight, Wiley 1990

A.K.Bledzki, T.Spychaj, Molekulargewichtsbestimmungen von hochmolekularen Stoffen, Hüthig und Wepf, Basel 1991 (Übungsaufgaben)

3.6 VISCOMETRY

M.Kurata, W.H.Stockmayer, Intrinsic Viscosities and Unperturbed Dimensions of Long Chain Molecules, Fortschr.Hochpolym.Forschg.-Adv.Polym.Sci. **3** (1961/64) 196

H.van Oene, Measurement of the Viscosity of Dilute Polymer Solutions, in Characterization of Macromolecular Structure, Natl.Acad.Sci.U.S., Publ. 1573, Washington (DC) 1968

H.Yamakawa, Modern Theory of Polymer Solutions, Harper and Row, New York 1971

M.Bohdanecky, J.Kovar, Viscosity of Polymer Solutions, Elsevier, Amsterdam 1982

R.E.Bareiss, Polymolecularity Correction Factors, in J.Brandrup, E.H.Immergut, E.Grulke, Eds., Polymer Handbook, Wiley, New York, 4th ed. 1998

W.-M.Kulicke, C.Clasen, Viscosimetry of Polymers and Polyelectrolytes, Springer, Heidelberg 2004 (laboratory handbook)

References to Chapter 3

[1]   H.-G.Elias, U.Gruber, Makromol.Chem. **86** (1965) 168, Table 2

[2]   R.C.Beavis, B.T.Chait, Anal.Chem. **62** (1990) 1836, Fig. 2b

[3]   P.O.Danis, D.E.Karr, W.J.Simonsick, Jr., D.T.Wu, Macromolecules **28** (1995) 1229, Fig. 1

[4]   K.Rollins, J.H.Scrivens, M.J.Taylor, Rapid Commun.Mass Spectrom. **4** (1990) 355

[5]   U.Bahr, A.Deppe, M.Karas, F.Hillenkamp, U.Giessmann, Anal.Chem. **64** (1992) 2866, Table I

[6]   Z.Grubisic, P.Rempp, H.Benoit, J.Polym.Sci.-Polym.Lett. Ed. **5** (1967) 753, Table I

[7]   A.L.Spatorico, J.Appl.Polym.Sci. **19** (1975) 1601, Table I

[8]   A.R.Cooper, E.M.Barrall, II, J.Appl.Polym.Sci. **17** (1973) 1253, Table III

[9]   P.Andrews, Biochem.J. **91** (1964) 222, Tables 1 and 3

[10] R.D.Hester, P.H.Mitchell, J.Polym.Sci.-Polym.Chem.Ed. **18** (1980) 1767

[11] A.L.Spatorico, G.L.Beyer, J.Appl.Polym.Sci. **19** (1975) 2933, Tables I and IV

[12] H.-G.Elias, R.Bareiss, J.W.Watterson, Adv.Polym.Sci. **11** (1973) 111, Fig. 4 (modified)

[13] D.J.Streeter, R.F.Boyer, Ind.Eng.Chem. **43** (1951) 1790, Table I

[14] D.T.F.Pals, J.J.Hermans, Rec.Trav. **71** (1952) 433, Table I

[15] R.E.Bareiss, Polymolecularity Correction Factors, in J.Brandrup, E.Immergut, E.A.Grulke, Eds., Polymer Handbook, Wiley-Interscience, New York 1999, p. VIII/215 (amended)

[16] F.Abe, Y.Einaga, H.Yamakawa, Macromolecules **26** (1993) 1891, Tables I and III

[17] H.Murakami, T.Norisuye, H.Fujita, Macromolecules **13** (1980) 345, Tables I and II

[18] N.Donkai, H.Inagaki, K.Kajiwara, H.Urakawa, M.Schmidt, Makromol. Chem. **186** (1985) 2623, Table 1

[19] R.Kuhn, H.Krömer, G.Rossmanith, Angew.Makromol.Chem. **40/41** (1974) 361, Fig. 3

[20] P.R.Dvornic, S.Uppuluri, in J.M.J.Fréchet, D.A.Tomalia, Eds., Dendrimers and Other Dendritic Polymers, Wiley, New York 2001; S.Uppuluri, Ph.D. Thesis, Michigan Technological University, Houghton (MI), 1997

[21] T.H.Mourey, S.R.Turner, M.Rubinstein, J.M.J.Fréchet, C.J.Hawker, K.L.Wooley, Macromolecules **25** (1992) 2401, Table I, SEC data of P2

[22] S.M.Aharoni, N.S.Murthy, Polym.Commun. **24** (1983) 132, recalculated from Table 1

[23] R.S.Turner, B.I.Voit, T.H.Mourey, Macromolecules **26** (1993) 4617, Table II

4   Configuration

4.1   Stereoisomers

4.1.1   Historical Development

In 1869, E.Paternò presented the first tetrahedral model of the carbon atom albeit in a rarely read journal. Five years later, Jacobus Hendricus van't Hoff in the Netherlands and Joseph Achille Le Bel in France concluded simultaneously that the four valences of carbon atoms $C^1R^2R^3R^4R$ are arranged "asymmetrically". van't Hoff also showed that a molecule composed of i stereogenic "asymmetric" carbon atoms and j unsymmetrically substituted carbon-carbon double bonds forms a total of 2^i2^j stereoisomers, i.e., 2^i optical isomers (**configurational stereoisomers**) and 2^j geometrical isomers (**torsional stereoisomers**). Stereoisomers are either **enantiomers** *or* **diastereomers** (Section 4.1.4).

Polymer molecules with monomeric units –CHR–, –CRR'–, –CH$_2$CHR–, –ZCH$_2$CHR–, etc., can thus, in principle, form many different stereoisomers. Hermann Staudinger was the first to point out in 1929 that the non-crystallizability of poly(indene) and vinyl polymers of the type $+CH_2-CHR+_n$ is probably caused by the random arrangement of different diastereomeric units in the chain. No experimental proof was given, however.

Staudinger's depiction of stereoisomerism of poly-(indene)s (today called "erythro" and "threo", see Section 4.2.5).

Different configurational structures were also invoked in 1942 by Maurice L. Huggins in order to explain why different intrinsic viscosity-molecular weight relationships were obtained for polymers of the same monomer if the polymerization took place at different temperatures. Again, no experimental proof was available.

In 1948, Calvin E. Schildknecht and coworkers observed that poly(vinyl isobutyl ether)s from the polymerization by BF$_3$ in liquid propane at –70°C were amorphous and non-crystallizable whereas those from the polymerization by a complex of BF$_3$ and ether were crystalline. They correctly attributed this behavior not only to a difference in stereochemical structures but also to the specific type of stereochemistry (what we now call isotactic and syndiotactic). Unfortunately, the assignments were mixed up because they were based on "chemical intuition" since no experimental methods for the determination of configurations were available.

These observations did not raise much interest, however, probably for two reasons: the initiator systems could not be used for the controlled stereochemical synthesis of other monomers, and the polymers were not of sufficient industrial interest. The situation changed after Guilio Natta and coworkers found that the newly discovered Ziegler catalysts from AlR$_3$ + TiCl$_4$ for the low-pressure, room temperature polymerization of ethene also convert 1-olefins to stereoregular, and often crystallizing, polymers. Modified Ziegler catalysts later allowed the polymerization of many other types of monomers. Today many different stereoregular polymers are known.

4.1.2 Configurational and Conformational Isomers

Molecules with the same number of identical atoms in different arrangements are called **isomers** in chemistry (in physics, "isomer" refers to nuclides in different excited states (nuclear isomerism) or to the spin reversal in H_2 or He (ortho-para isomerism) (G: *isos* = equal, *meros* = part)). The number N of isomers increases dramatically with the number of atoms per molecule so that macromolecules can exist, in principle, in a very great number of isomers. An example is the series of alkanes, C_nH_{2n+2}:

propane	C_3H_8	$N =$	1
decane	$C_{10}H_{22}$	$N =$	75
dicosane (icosane)	$C_{20}H_{42}$	$N =$	366 319
triacontane	$C_{30}H_{62}$	$N =$	4 111 840 763
tetracontane	$C_{40}H_{82}$	$N =$	62 491 178 805 831 etc.

Isomers are subdivided into constitutional isomers and stereoisomers. **Constitutional isomers** differ in the succession of atoms (Chapter 2). **Stereoisomers**, on the other hand, have the same succession of atoms but different arrangements of these atoms in space. An isomer is thus *either* a constitutional isomer *or* a stereoisomer; it can never be both.

Historically, stereoisomers are either divided into configurational and conformational isomers according to their **energy barriers** (Chapter 5) or into enantiomers and diastereomers (Section 4.1.4) according to their symmetry properties (Section 4.1.3).

Configurational isomers have a "high" energy barrier for the conversion of one isomer into another, **conformational isomers** a "low" energy barrier. A high energy barrier allows separation of configurational isomers into their pure compounds. The low energy barrier of conformational isomers, on the other hand, leads to the rapid conversion of these isomers into each other; conformational isomers cannot be separated. The dividing line between configurational and conformational isomers is not sharp, however.

Life-times of several hours (i.e., more than ca. 10^4 seconds ≈ 2.8 h) are required for separating and purifying isomers. Since the **isomerization** of a compound A is a unimolecular reaction with a rate of $-d[A]/dt = k[A]$, rate constants k must be at least $k = 10^4$ s^{-1}. According to the **Eyring equation** for rate processes

$$(4-1) \qquad k = K(k_BT/h) \ \exp(-\Delta G^{\ddagger}/RT)$$

(k_B = Boltzmann constant, R = molar gas constant; h = Planck constant), an isomerizations with $k = 10^4$ s^{-1} requires a Gibbs activation energy of $\Delta G^{\ddagger} \approx 49.8$ kJ/mol for $K = 1$ and room temperature (23°C = 296.15 K) (see also Table 4-1).

Separations of isomers are successful only if the molecules cannot isomerize appreciably within the time necessary for separation and work-up of isomers. Slow isomerizations thus require very small rate constants k which necessitates high activation energies and/or low temperatures. For example, in order to convert 5 % of an isomer into another isomer, only $t_{0.05} = 0.05/k = 2061$ s ≈ 34 min are necessary for reaction with a Gibbs activation energy of $\Delta G^{\ddagger} = 100$ kJ/mol at a temperature of $T = 300$ K.

No distinction is necessary between configurational and conformational isomers if one is not interested in the isolation of isomers (like chemists) but only in the spatial arrangement of atoms of a molecule regardless of the bonds between atoms (like physicists). In physics, the statistics of conformations is thus often called "configurational

Table 4-1 Rate constants k and times $t_5 \approx 0.05/k$ for a 5 % isomerization at selected temperatures.

$\dfrac{\Delta G^{\ddagger}}{\text{kJ mol}^{-1}}$	$\dfrac{k}{\text{s}^{-1}}$ for $T =$			$\dfrac{t_5}{\text{s}}$ for $T =$		
	100 K	300 K	500 K	100 K	300 K	500 K
10	$1.3 \cdot 10^7$	$1.1 \cdot 10^{11}$	$9.4 \cdot 10^{11}$	$4 \cdot 10^{-9}$	$4 \cdot 10^{-13}$	$5 \cdot 10^{-14}$
25	$1.8 \cdot 10^{-1}$	$2.8 \cdot 10^8$	$2.6 \cdot 10^{10}$	$3 \cdot 10^{-1}$	$2 \cdot 10^{-10}$	$2 \cdot 10^{-12}$
50	$1.6 \cdot 10^{-14}$	$1.2 \cdot 10^4$	$6.2 \cdot 10^7$	$3 \cdot 10^{12}$	$4 \cdot 10^{-6}$	$8 \cdot 10^{-10}$
100	$1.2 \cdot 10^{-40}$	$2.4 \cdot 10^{-5}$	$3.7 \cdot 10^2$	$4 \cdot 10^{38}$	$2 \cdot 10^3$	$1 \cdot 10^{-4}$

statistics". The statistics of the spatial arrangement of atoms is however not the statistics of relative configurations (tacticity, Section 4.3) but, in the language of chemists, *conformational statistics*.

The classical division of stereoisomers into configurational and conformational isomers is based on the lifetime of isomers. The distinction between configurational and conformational isomers is therefore based on the speed of separation and thus also on the skill of the experimentalist (in preparative work) or the speed of the instrument (in analytical work). It becomes nonsensical for "fast" methods. For example, conformational isomers with lifetimes of microseconds appear as configurational isomers if they are investigated by infrared, Raman, or electron spin resonance spectroscopy. As a consequence, these time scales make it difficult to distinguish between configurational and conformational isomers.

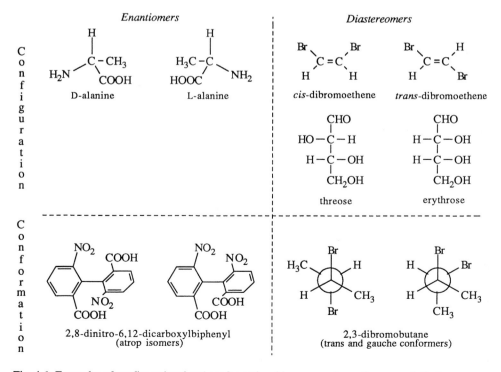

Fig. 4-1 Examples of configurational and conformational isomers and enantiomers and diastereomers.

Conformational isomers are defined by IUPAC as stereoisomers that result from the rotation around single bonds. These rotations usually require low activation energies. According to this definition, **configurational isomers** can be interconverted only by a cleavage of chemical bonds which requires high activation energies.

Some examples are shown in Fig. 4-1. Both alanines, both 1,2-dibromoethenes, and the pair threose-erythrose are configurational isomers since the transformation of one isomer into the other requires the breaking and reforming of chemical bonds. The two isomers of 2,3-dibromobutane, on the other hand, are clearly conformational isomers since they can be interconverted by rotation around the $C^2–C^3$ bond.

The two isomeric 2,8-dinitro-6,12-dicarboxyl biphenyls are also conformational isomers since they result from the rotation around the central phenyl-phenyl bond. The rotation requires, however, such a high activation energy that the two isomers can be separated and isolated. These compounds thus are conformational isomers according to bond classification but configurational isomers according to energy classification. This isomerism is thus caused by a hindred rotation around single bonds; it is called **atropisomerism**. The often considerable optical activity of atropisomers disappears at higher temperatures because of the more easy rotation around the central axis.

The two alanines and the two atropisomers are enantiomers but threose/erythrose, the two dibromoethenes, and the two 2,3-dibromobutanes, are diastereomers (Section 4.1.4).

4.1.3 Symmetry

Group theory describes the symmetry of small molecules by three simple and two complex symmetry elements and the corresponding five symmetry operations. Chain molecules require in addition translation as a symmetry operation of the symmetry element "identity" and two additional complex symmetry operations (Table 4-2).

The three simple symmetry elements of small molecules are equivalent to three "dimensionalities": zero-dimensional (**point** or **center**), one-dimensional (**line** or **axis**), and two-dimensional (**area** or **plane**). The corresponding symmetry operations are the **inversion** about a center (= inversion center), the **rotation** about an axis (= proper or simple axis of rotation), and the **reflection** at a plane (= mirror plane of symmetry).

Table 4-2 Symmetry elements and symmetry operations.

Symmetry element	Symbol	Symmetry operation
Simple		
Identity	-	Translation as repetition of identity
Symmetry center	i	Inversion around a point
Symmetry axis (proper axis of rotation)	C	Rotation around an axis
Symmetry plane (mirror plane of symmetry)	σ	Reflection at a plane cutting through the object
Complex		
Rotatory inversion axis	-	Combination of rotation and inversion
Mirror axis	S	Combination of rotation and reflection
Screw axis	-	Combination of translation and rotation
Glide plane	-	Combination of translation and reflection

The two complex symmetry elements and operations of small molecules combine either rotation and inversion or rotation and reflection (Table 4-2). The rotation-reflection axis is also called **mirror axis, improper axis,** or **alternating axis.** Molecules with symmetry axes but without symmetry centers, symmetry planes, and rotation-reflection axes are said to be **dissymmetric.**

Chain molecules have in addition the simple symmetry element of **identity** and the corresponding simple symmetry operation of **translation** as repetition of identity. Translations do not have any mirror symmetries. The additional element leads to two additional complex symmetry elements. For example, the combination of translation and rotation generates molecules with screw axes with the macroconformation (shape) of a helix. Such molecules are *per se* dissymmetric. On the other hand, the combination of translation and reflection at a plane containing the translation axis leads to a mirror (reflection) plane with the corresponding mirror symmetry.

Symmetry operations on small molecules cause the object to change its position but not its appearance. At least one point of the object maintains its original position so that the object can be decribed by so-called **point groups** or **symmetry groups** as groups of symmetry elements. The most important symmetry groups are shown in Table 4-3; examples are depicted in Fig. 4-2. Further subdivision of symmetry groups with respect to the foldedness of axes leads to 32 symmetry groups (= crystal classes).

Table 4-3 Definition of important symmetry groups. * = Chiral symmetry groups.

Symmetry group	Symbol	Example
No symmetry axis		
One-fold axis of rotation	$*C_1$	L-alanine (C_1)
One inversion center	C_i or S_2 or i	2,3-dibromobutane ($C_i = S_2$)
One mirror plane	C_s or S_1 or σ	vinyl chloride (C_s)
One n-fold symmetry axis		
n-fold axis of rotation (n > 1)	$*C_n$	1,3-dibromoallene (C_2)
n-fold axis of rotation and perpendicular to this axis a symmetry plane	$C_{nh} = C_n + \sigma_h$	*trans*-1,2-dibromoethene (C_{2h})
n-fold axis of rotation with n symmetry planes	$C_{nv} = C_n + n\,\sigma_v$	H_2O (C_{2v}), CH_3–$CHBr_2$ (C_{2v}), $CHCl_3$ (C_{3v}), $CH\equiv CCl$ ($C_{\infty v}$)
n-fold mirror axis with n ≤ 4 (even only)	S_n	tetramethylspiropyrrolidinium
One n-fold and n 2-fold symmetry axes		
One n-fold axis of rotation, perpendicular to it n 2-fold axes of rotation	$*D_n = C_n + n\,C_2$	double-bridged biphenyl (D_2)
One n-fold axis of rotation, n 2-fold axes of rotation, and n vertical mirror planes	$D_{nd} = C_n + n\,C_2 + n\,\sigma_d$	allene (D_{2d})
One n-fold axis of rotation, n 2-fold axes of rotation, n vertical mirror planes, and one horizontal mirror plane	$D_{nh} = C_n + n\,C_2 + n\,\sigma_v + n\,\sigma_v + \sigma_h$	ethene (D_{2h}), benzene (D_{6h}), acetylene ($D_{\infty h}$)
Several n-fold symmetry axes with n > 2		
Tetrahedron	$T_d = 4\,C_3 + 3\,C_2 + 6\,\sigma$	methane (T_d)
Octahedron	$O_h = 3\,C_4 + 4\,C_3 + 6\,C_2 + 9\,\sigma$	chromohexacarbonyl (O)
Sphere	K = all symmetry elements	

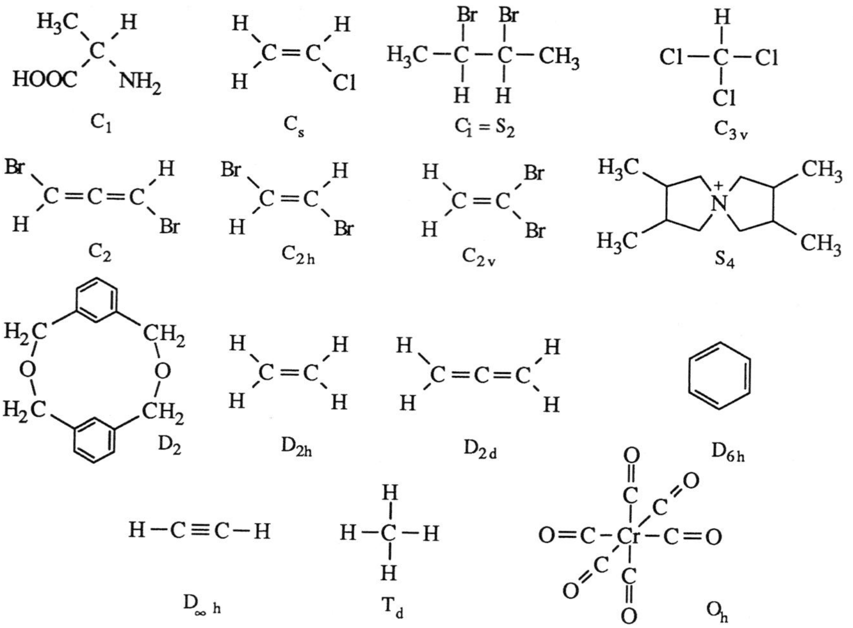

C_1 C_s $C_i = S_2$ C_{3v}

C_2 C_{2h} C_{2v} S_4

D_2 D_{2h} D_{2d} D_{6h}

$D_{\infty h}$ T_d O_h

Fig. 4-2 Examples of symmetry groups (for designations, see Table 4-3).

The following symbols are used in the **Schönflies notation**: T = *tetrahedron*, O = *octa-hedron*, S = *sphenoid mirror axis*; G: *sphen* = wedge), C = *cyclus* (= axis of rotation; L: *cyclus*, from G: *kyklos* = circle), and D = *digyre* (= 2-fold axis; G: *di* = two; L: *gyros*, from G: *gyros* = circle), perpendicular to which are other 2-fold axes.

An axis of rotation is n-fold (symbol C_n) if the object is n times congruent after a complete rotation of 360°. Axes of rotation may be imaginary, i.e., they do not need to pass through an atom or an atom group. The axis of rotation with the highest n is called the **main axis** (of rotation). It serves as the reference for all other symmetry elements.

Reflections and their corresponding mirror planes are often designated by σ. Reflections at a plane perpendicular to the main axis (i.e., *horizontal* plane) carry the symbol σ_h. Mirror planes parallel to the main axis (i.e., *vertical* planes) have the symbol σ_v. Since the mirror plane halves certain angles (it makes them *dihedral*), σ_v is often replaced by σ_d.

4.1.4 Enantiomers and Diastereomers

Stereoisomers are subdivided according to their *symmetry* into enantiomers and dia-stereomers. Enantiomers relate to each other like image and mirror image (G: *enantios* = opposite) but diastereomers do not (Fig. 4-1). Two stereoisomers are therefore always *either* enantiomers *or* diastereomers but never both.

The image/mirror image property makes enantiomers "chiral": the two stereoisomers relate to each other like the left and right hand (G: *cheir* = hand). **Chirality** is characterized by the absence of symmetry centers, symmetry planes, and mirror axes. Symmetry axes may be present, however. Because of this special feature, symmetry axes are distinguished from other symmetry elements.

Molecules are called **antipodes** if they relate to each other like image and mirror image (G: *anti* = opposite, against; *pous* = foot). They are assumed to be energetically equivalent.

In nature, one antipode usually dominates; for example, L-molecules of α-amino acids or D-molecules of sugars. Literature offers four hypotheses for this finding:

(1) The combination of three macroscopic orientational factors (e.g., electric field + surface interaction + directional transport) creates a chiral effect on reactant molecules, thus leading to an absolute asymmetric synthesis.

(2) One of the antipodes is preferentially synthesized at a solid, chiral surface.

(3) Both antipodes are synthesized in equal amounts by nature but circularly polarized light destroys one antipode. Such light is generated if unpolarized light is scattered at non-spherical entities that are oriented in a strong magnetic field. These fields were present in the early periods of earth.

(4) Violation of parity in the subatomic range causes very small differences in the energy content of the two antipodes, e.g., $\Delta E \approx 10^{-14}$ J/mol for D- and L-alanine. Both antipodes are synthesized in equal amounts by nature but the less stable antipode is destroyed in geological time periods by the usual chemical reactions.

Chiral molecules are always **optically active**, i.e., they rotate the plane of polarized light. Chiral molecules are also dissymmetric since n-fold mirror axes are absent. They belong to the symmetry groups C_1, C_n, and D_n (Table 4-3).

One of these symmetry groups, C_1, also has only a 1-fold axis of rotation but no axes with $n \geq 2$. Molecules with the symmetry group C_1 are called **asymmetric**. Such an asymmetry is found if 4 different ligands are tetrahedrally arranged about a central atom such as C, Si, P^+, or N^+. The "asymmetric central atom" (= "chiral atom" of old literature) in now called a **stereogenic center** since it is neither asymmetric nor chiral.

Atoms with four different ligands are not asymmetric (stereogenic) if all four ligands reside in a symmetry plane. According to theoretical calculations, such molecules are the hypothetical molecules 1,2-dilithium methane, Li_2CH_2 (planar), and 1,1-dilithium ethene, $Li_2C=CH_2$ (orthogonal).

Mixtures of equal amounts of two antipodes are called **racemates** (first known example: D,L-tartaric acid (from G: *tartaron* = tartar = winestone = potassium hydrogen tartrate) = uvic acid (L: *uva* = grape) = racemic acid (L: *racemys* = grape, grape juice)) (see p. 7). The chirality of racemates is externally compensated; racemates are not optically active. Chiralities can, however, be internally compensated if a molecule has two centers with opposite chirality (example: meso-tartaric acid, p. 7). Such molecules are called **meso compounds**. They are achiral and also not optically active.

Diastereomers, on the other hand, do not exist as mirror and mirror image. They can be achiral as well as chiral. For example, the diastereomers *cis*- and *trans*-1,2-dibromoethene are achiral but the diastereomers threose and erythrose are chiral.

Epimers are a subclass of diastereomers. They have several asymmetry centers, all of which with the same structure except one.

Pseudoasymmetric atoms possess only a few properties of asymmetric atoms. An example is trihydroxyglutaric acid (Fig. 4-3) whose center carbon atom may be studded with two like configured (I-1 and I-2) or two unlike configured ligands (I-3 and I-4).

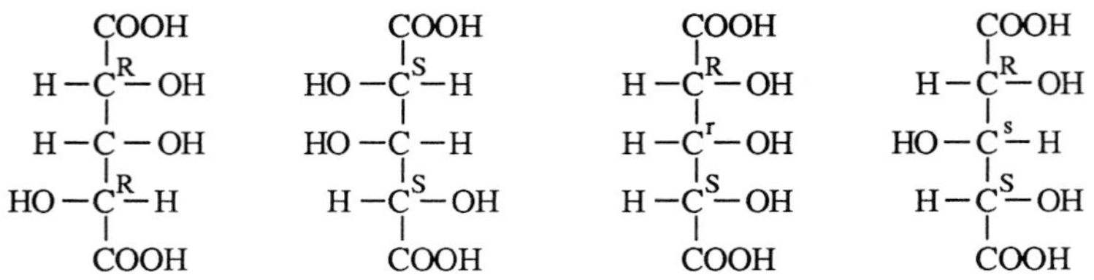

$$
\begin{array}{cccc}
\text{COOH} & \text{COOH} & \text{COOH} & \text{COOH} \\
\text{H}-\overset{R}{\text{C}}-\text{OH} & \text{HO}-\overset{S}{\text{C}}-\text{H} & \text{H}-\overset{R}{\text{C}}-\text{OH} & \text{H}-\overset{R}{\text{C}}-\text{OH} \\
\text{H}-\text{C}-\text{OH} & \text{HO}-\text{C}-\text{H} & \text{H}-\overset{r}{\text{C}}-\text{OH} & \text{HO}-\overset{s}{\text{C}}-\text{H} \\
\text{HO}-\overset{R}{\text{C}}-\text{H} & \text{H}-\overset{S}{\text{C}}-\text{OH} & \text{H}-\overset{S}{\text{C}}-\text{OH} & \text{H}-\overset{S}{\text{C}}-\text{OH} \\
\text{COOH} & \text{COOH} & \text{COOH} & \text{COOH} \\
\text{I-1} & \text{I-2} & \text{I-3} & \text{I-4} \\
\end{array}
$$

Enantiomers	Diastereomeric meso compounds

Fig. 4-3 Enantiomeric and diastereomeric meso (pseudoasymmetric) forms of D-arabo trihydroxyglutaric acid (in Fischer projection, see Section 4.1.5). For the R,S system, see Section 4.1.6.

In the first case (I-1 and I-2), the center carbon atom is achiral but the whole molecule is chiral since it does not possess any symmetry centers, symmetry planes, or mirror axes. In the second case (I-3 and I-4), the center carbon atom is by definition asymmetric since it possesses four different substituents in a tetrahedral arrangement. In contrast to a true chirality center, it is however dissected by a symmetry plane. A change of positions of substituents in I-3 and I-4 does not result in enantiomers but in two different meso structures.

Such carbon atoms are thus called **pseudoasymmetric**. They are always found in compounds of the type CABE∃ where A and B are two different achiral substituents and E and ∃ two constitutionally identical but image/mirror image like (enantiomorphic) substituents. Pseudoasymmetric carbon atoms are marked by the small letters r and s, truly asymmetric ones by capital letters R and S (see Fig. 4-3).

Molecules of the type CAABD become chiral if one of the two identical achiral substituents A is replaced by a new substituent E. Arrangements such as CAABD are therefore called **prochiral** or **potentially chiral**. These molecules possess symmetry centers or symmetry planes but no symmetry axis through the prochiral carbon atom C.

The two achiral (identical) substituents A and A react, of course, equally fast with achiral reagents but they react with different speeds with chiral reagents. One sees this if the two identical substituents A are tagged as A' and A" (Fig. 4-4). In this case, the triangle DBA", as viewed from A' is the mirror image of the triangle DBA' as viewed from A". If the seniority of substituents decreases in the order A-B-D, then the sequence A"→B→D in clockwise direction is called Re (L: *rectus* = right, lucky). The counterclockwise arrangement A"←B←D is called Si (L: *sinister* = left, unlucky, evil).

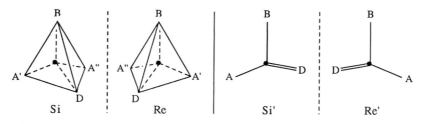

Fig. 4-4 Three-dimensional (left) and two-dimensional (right) prochiral molecules. Examples:

Si and Re:	bromochloromethene	CH_2BrCl	with A' = A" = H	B = Br	D = Cl
Si' and Re':	propene	$CH_3–CH=CH_2$	with A = CH_3	B = H	D = CH_2
	acetaldehyde	$CH_3–CH=O$	with A = CH_3	B = H	D = O

Since the two triangles DBA" and DBA' of the tetrahedral CA'A"BD are planar, the same considerations also apply to the planar, trigonal arrangement of A, B and D in compounds CABD (Fig. 4-4). Monomers like acetaldehyde and propene have thus front sides and backsides. They are prochiral or two-dimensionally chiral.

4.1.5 Stereostructures

Three-dimensional arrangements of atoms in space need to be pictured two-dimensionally on planar information carriers such as book pages or computer screens. Perspective drawings like that of the molecule CABDH in Fig. 4-5 (I) are not always practical. Such a molecule however can also be depicted in another way, i.e., by putting C (as ●) and B and D into the paper plane. Because of the tetrahedral structure of the molecule, the substituent A must then be above the paper plane and the substituent H, below (Fig. 4-5, II). Bonds in the paper plane are symbolized by lines, above the paper plane by wedges, and below paper planes by dotted lines.

Fischer projections, on the other hand, put the plane of projection in such a way through the asymmetric carbon atom that two substituents are above and the other two substituents are below the paper plane. (Fig. 4-5, III). The two lower substituents take the vertical positions and the two upper substituents the horizontal positions (Fig. 4-5, IV). The upper, lower, and in-plane positions are no longer represented by special symbols (Fig. 4-5, V).

Molecules with two or more centers of chirality, for example, CABD–CKLM, can also be represented in two dimensions (Fig. 4-5, bottom). The different representations are based on certain conformations (Chapter 5). Fischer projections always show molecules in cis conformations ("synperiplanar" in organic chemistry). The Newman, wedge, and sawhorse representations, on the other hand, depict molecules in the trans conformation ("antiperiplanar"). Fischer projections of macromolecules are not very useful, however, since polymer chains rarely exist in cis macroconformations.

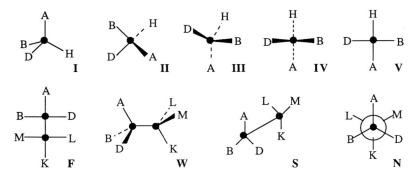

Fig. 4-5 Representation of three-dimensional molecules on two-dimensional carriers of information.
  Top: The perspective (three-dimensional) picture of the molecule CABDH (**I**) can be represented two-dimensionally by wedges (bonds above the plane), dotted lines (bonds below paper plane), and solid lines (bonds in the plane) (**II, III**). The two-dimensional Fischer projection (**V**) corresponds to the "spatial" representation **III** and the pseudo-spatial representation **IV**. ● Stereogenic carbon atom.
  Bottom: CABD–CKLM in cis conformation in the Fischer projection (**F**) and in trans conformation in the Newman projection (**N**) and in the wedge (**W**) and sawhorse (**S**) representations.

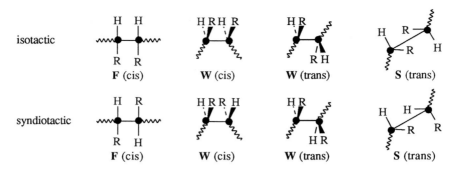

Fig. 4-6 Stereo representations of chain structures. ∿∿ Chain segments. Fischer projections (**F**) always show the cis conformation whereas wedge (**W**) and sawhorse (**S**) projections depict cis or trans conformations. Top: isotactic structures; bottom: syndiotactic structures. R = substituent.

Whether substituents R of two adjacent chain units –CHR– appear in stereo projections on the same side or on opposite sides depends on the relative configuration of the substituents as well as the relative conformation of the chain segments ∿∿ on both sides of the diad –CHR–CHR–.

The diad is called **isotactic** if the two carbon atoms of the diad have the same *relative* configuration of substituents (G: *isos* = equal; *taktikos* = ordered). Diads with non-identical relative configurations are called **syndiotactic** because these configurations repeat after every second chain unit (G: *syn* = together; *dios* = two).

Fig. 4-6 shows stereo projections of isotactic and syndiotactic diads –CHR–CHR–. Fischer projections of polymer chains are turned by 90° (Fig. 4-6, **F**). In Fischer projections, substituents R of *isotactic* diads are "on the same side" since such projections always depict chains in cis conformations. The same is true for wedge representations of cis conformations (Fig. 4-6, **W** (cis)). The two substituents appear "on the other side" if trans conformations are represented as wedge or saw-horse projections (Fig. 4-6, **W** (trans) and **S** (trans)). For syndiotactic diads, the situation is reversed: substituents are on different sides for cis conformations but on the same side for trans conformations. It is important to note that this discussion is concerned with the *relative* configuration of subsequent units and not with the absolute configuration (see Sections 4.1.6 and 4.2.2).

The stereo representations of Fig. 4-6 show the configurations of substituents around central atoms (i.e., chain atoms) but not their true spatial positions. Cis conformations of polymer chains are very rare (Chapter 5) so that representions of trans conformations are more realistic. However, trans conformations are not necessarily the conformations with the lowest energy.

It must be noted that the use of –CHR–CHR– is reserved for polymer chains with unspecified relative configurations.

4.1.6   D,L and R,S Systems

The absolute configuration of ligands at a stereogenic center can be described by various conventions. Most popular are the older D,L-system and the newer R,S-system. Both systems are based on definitions.

The **D,L-system** uses the dextrorotatory (+)-glyceraldehyde as a reference molecule; this molecules turns the plane of polarized light to the right (L: *dexter* = right, happy) (see Volume III for optical activity). The molecule is thus assigned a D-configuration. Correspondingly, the levorotatory (–)-glyceraldehyde is attributed the L-configuration (L: *laevus* = left). Optical activity (–, +) and configuration (L, D) are not necessarily correlated, however. There are dextrorotatory L-compounds, for example, L-(+)-alanine.

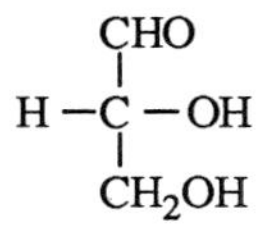

$$
\begin{array}{c}
CHO \\
| \\
H-C-OH \\
| \\
HO-C-H
\end{array}
$$

CHO	H−C−OH	COOH	COOH	COOH
H−C−OH	H−C−OH	H−C−NH$_2$	H$_2$N−C−H	H$_2$N−C−H
CH$_2$OH	CH$_2$OH	CH$_2$OH	CH$_2$OH	CH$_3$
D-(+)-glyceraldehyde	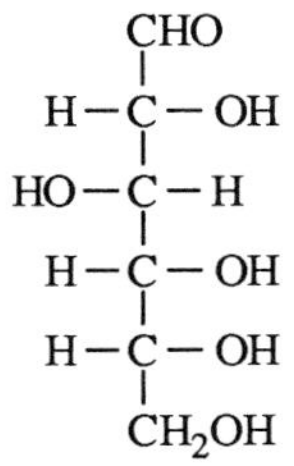 D-(+)-glucose	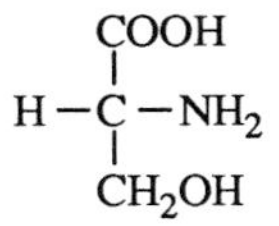 D-(+)-serine	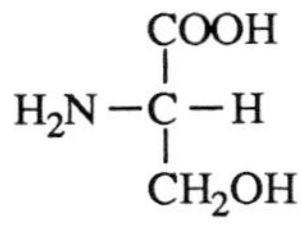 L-(–)-serine	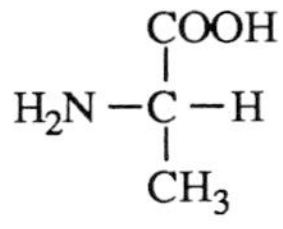 L-(+)-alanine

By definition, the substituent OH of glyceraldehyde is placed on the right side of the molecule in Fischer projections. In sugar molecules with several centers of chirality, the center which is farthest away from the carbon atom with the highest oxidation stage (CHO in glucose) is decisive for the assignment of D or L configurations. The dextrorotatory (+)-glucose is thus assigned the D-configuration. In α-amino acids, the levorotatory L-(–)-serine serves as the reference substance, however. For this reason, the R,S-system, and not the D,L-system, is used for most organic compounds. The D,L-system is, however, still employed for naturally occurring molecules.

The **R,S-system** has a topological foundation; it is therefore independent of reference molecules. It orders the ligands of an enantiogenic center in a somewhat arbitrary sequence that is then used to assign the R or S configuration. The ligands are given different seniorities that are based on the position in the periodic system of elements of the atom that is directly connected to the enantiogenic center. In this order of seniorities, iodine is assigned the highest seniority, a lone electron pair the lowest:

I, Br, Cl, HSO$_3$, HS, F,
C$_6$H$_5$COO, CH$_3$COO, HCOO, C$_6$H$_5$O, C$_6$H$_5$CH$_2$O, C$_2$H$_5$O, CH$_3$O, HO,
NO$_2$, NO, (CH$_3$)$_3$N$^+$, (C$_2$H$_5$)$_2$N, ((CH$_3$)$_2$NH)$^+$, CH$_3$NH, (NH$_3$)$^+$, NH$_2$,
CCl$_3$, COCl, CF$_3$, COOCH$_3$, COOH, CONH$_2$, C$_6$H$_5$CO, CH$_3$CO, CHO, CR$_2$OH, CH$_2$OH,
(C$_6$H$_5$)$_3$C, C$_6$H$_5$, (CH$_3$)C, C$_6$H$_{11}$, CH$_2$=CH, (CH$_3$)$_2$CH, C$_6$H$_5$CH$_2$, (CH$_3$)$_2$CHCH$_2$,
C$_6$H$_{13}$, C$_5$H$_{11}$, C$_4$H$_9$, C$_3$H$_7$, C$_2$H$_5$, CH$_3$, Li, D, H, lone electron pair.

A chirality center is then viewed from the side that is opposite to the ligand with the lowest seniority. The chirality center is assigned the symbol R (L: *rectus* = right) if the remaining ligands follow each other clockwise with decreasing seniority. The symbol S (L: *sinister* = left) is used if the seniority of ligands decreases counter-clockwise.

The D-(+)-glyceraldehyde is therefore the [R]-glyceraldehyde since OH > CHO > CH$_2$OH > H and the L-(–)-serine the [R]-serine since NH$_2$ > COOH > CH$_2$OH > H. The R,S-system is unambiguous but formal. Similar compounds may be assigned different configurations since ligands have different seniorities. Examples are [S]-alanine and [R]-

trifluoroalanine since CH_3 has a lower seniority than CF_3 relative to COOH:

$$
\begin{array}{cccc}
\text{CHO} & \text{COOH} & \text{NH}_2 & \text{NH}_2 \\
| & | & | & | \\
\text{H}-\text{C}-\text{OH} & \text{H}_2\text{N}-\text{C}-\text{H} & \text{CF}_3-\text{C}-\text{H} & \text{CH}_3-\text{C}-\text{H} \\
| & | & | & | \\
\text{CH}_2\text{OH} & \text{CH}_2\text{OH} & \text{COOH} & \text{COOH}
\end{array}
$$

| [R]-glyceraldehyde | [R]-serine | [R]-trifluoroalanine | [S]-alanine |
| D-(+)-glyceraldehyde | L-(–)-serine | | D-(–)-alanine |

4.2 Polymers with Ideal Tacticities

4.2.1 Definitions

In macromolecular chemistry, one distinguishes configurational base units from configurational repeating units and stereorepeating units. A **configurational base unit** is a constitutional repeating unit with a defined configuration at a stereogenic central atom or several of such atoms. Configurational base units are identical with **configurational repeating units** in *regular* polymer molecules.

For example, poly(methylmethylene) I has one type of constitutional repeating unit, $-CH(CH_3)-$, that may exist in two possible types of configurational repeating units, Ia and Ib. Poly(propylene), on the other hand, has two possible types of constitutional repeating units, $-CH(CH_3)-CH_2-$ (II) and $-CH_2-CH(CH_3)-$ (III), each with two configurational repeating units, IIa + IIb and IIIa + IIIb, respectively.

The configurational unit Ib is the mirror image of Ia; the two units Ia and Ib are therefore enantiomers. The pair IIa and IIb is also enantiomeric and so is the pair IIIa and IIIb. The pairs IIa + IIIa and IIb + IIIb are however diastereomers since they each have two different constitutional base units. This classification of IIa + IIb and IIIa + IIIb as pairs of enantiomers and IIa + IIIa and IIb + IIIb as pairs of diastereomers refers to the *units* shown below and *not* to the infinitely long chains derived from these units and groups of units, respectively.

$$
\begin{array}{cccccc}
\text{H} & \text{CH}_3 & \text{H} & \text{CH}_3 & \text{CH}_3 & \text{H} \\
| & | & | & | & | & | \\
-\text{C}- & -\text{C}- & -\text{C}-\text{CH}_2- & -\text{C}-\text{CH}_2- & -\text{CH}_2-\text{C}- & -\text{CH}_2-\text{C}- \\
| & | & | & | & | & | \\
\text{CH}_3 & \text{H} & \text{CH}_3 & \text{H} & \text{H} & \text{CH}_3
\end{array}
$$

| Ia | Ib | IIa | IIb | IIIa | IIIb |

Stereorepeating units are configurational repeating units that have defined configurations of *all* stereogenic centers. All units IIa, IIb, IIIa, and IIIb are such stereorepeating units. The unit $-CH(CH_3)-CH_2-$ is a sterically undefined configurational repeating unit.

A **stereoregular polymer** is a polymer whose molecules consist of only one type of stereorepeating units that are always connected in the same way. If the steric repeating units of such molecules are replaced by configurational repeating units, **tactic polymers** are obtained (G: *taktikos* = order, arrangement). A stereoregular polymer is always tactic but a tactic polymer is not necessarily stereoregular since not all centers of stereoisomerism need to be defined in the latter case.

Poly(2-pentene) IVa is thus a tactic polymer with regard to the configuration of the chain atom that carries the ethyl group. It is not a stereoregular polymer, however, since the configuration of the –CH(CH₃)– unit remains undefined. The polymer IVb is also tactic and not stereoregular. The polymer IVc is however stereoregular and therefore also tactic.

$$
\left[\begin{array}{c} \text{H} \\ | \\ \text{C}-\text{CH(CH}_3) \\ | \\ \text{C}_2\text{H}_5 \end{array}\right]_n \qquad \left[\begin{array}{c} \text{H} \\ | \\ \text{CH(C}_2\text{H}_5)-\text{C} \\ | \\ \text{CH}_3 \end{array}\right]_n \qquad \left[\begin{array}{c} \text{C}_2\text{H}_5\ \ \text{H} \\ |\qquad| \\ \text{C}\ \ -\text{C} \\ |\qquad| \\ \text{H}\ \ \ \text{CH}_3 \end{array}\right]_n
$$

IVa   IVb   IVc

The three most simple steric repeating units of poly(propylene) contain one, two, or four monomeric units:

$$
\begin{array}{ccc}
\text{H} & \text{H}\qquad\text{CH}_3 & \text{H}\qquad\text{H}\qquad\text{CH}_3\qquad\text{CH}_3 \\
| & |\qquad\ | & |\qquad\ |\qquad\ |\qquad\ | \\
-\text{C}-\text{CH}_2- & -\text{C}-\text{CH}_2-\text{C}-\text{CH}_2- & -\text{C}-\text{CH}_2-\text{C}-\text{CH}_2-\text{C}-\text{CH}_2-\text{C}-\text{CH}_2- \\
| & |\qquad\ | & |\qquad\ |\qquad\ |\qquad\ | \\
\text{CH}_3 & \text{CH}_3\qquad\text{H} & \text{CH}_3\qquad\text{CH}_3\qquad\text{H}\qquad\text{H}
\end{array}
$$

IT   ST   HT

The repetition of these units results in isotactic (IT), syndiotactic (ST), or heterotactic (HT) polymer chains. It does not matter whether IIa or IIIa is chosen as the simplest configurational repeating unit IT since infinitely long poly(propylene) molecules from IIa differ from those of IIIa only by the orientation of the repeating units. Chains with IIa units are thus not enantiomeric to chains of IIIa although the corresponding configurational repeating units themselves are enantiomeric. The same reasoning applies to the chains from units of the pair IIb and IIIb.

An **isotactic repeating unit** (IT) = **isotactic unit** (it) consists of a single constitutional repeating unit. Isotactic polymer *molecules* thus have only identical units (G: *isos* = equal).

The **syndiotactic unit** (st) is identical with the **syndiotactic repeating unit** (ST). It consists of two enantiomeric configurational repeating units. The regular repetition of identical *pairs* of these units leads to syndiotactic polymer molecules (G: *syn* = together; *dios* = two).

A **heterotactic unit** (ht) consists of three monomeric units but a **heterotactic repeating unit** (HT) has four monomeric units. The repetition of a heterotactic repeating unit HT generates a heterotactic chain $+\text{HT}\xrightarrow{}_n$ = $+\text{IT-}alt\text{-ST}\xrightarrow{}_n$ in which heterotactic triads ht = it-st are connected in head-to-tail position. Adjacent heterotactic triads are diastereomeric. The repetition of heterotactic triads ht, composed of three monomeric units, does not lead to a heterotactic chain $+\text{IT-}alt\text{-ST}\xrightarrow{}_n$ but to $+per(\text{IT-ST-ST})\xrightarrow{}_n$.

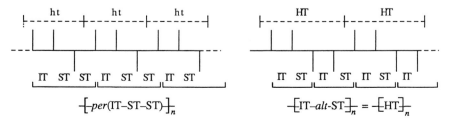

Isotactic poly(propylene)s $-[-IT-]_{\overline{n}}$, syndiotactic poly(propylene)s $[-ST-]_{\overline{n}}$ and heterotactic poly(propylene)s $-[-HT-]_{\overline{n}}$ (see above) contain one, two, or four configurational monomeric units in each configurational repeating unit. Each of these units possesses a single center of stereosymmetry. Such tactic polymer molecules are called **monotactic.**

The tactic poly(2-pentene) molecule (IVc), on the other hand, contains two defined centers of stereoisomerism per configurational monomeric unit; it is a **ditactic polymer molecule. Tritactic polymer molecules** have three defined centers of stereoisomerism per configurational repeating unit.

Atactic polymers are regular polymers like isotactic or syndiotactic ones. By definition, they contain all types of configurational units in equal amounts but with random distribution from molecule to molecule (Section 4.3). Literature does not use "atactic" rigorously, however, but rather like "not completely tactic" or "not predominantly tactic".

Stereoblock polymers are similarly defined as constitutional block polymers: each block has a different configurational make-up but all blocks contain the same type of constitutional monomeric units.

Tactic block polymers relate to stereoblock polymers like tactic polymers to stereoregular polymers: all centers of stereoisomerism are defined in stereoblock polymers but not in tactic block polymers. **Stereoblock *copolymers*** and **tactic block *copolymers*** are derived from two or more constitutionally different monomers!

4.2.2 Relative and Absolute Configurations

The **absolute configuration** of organic chemistry refers to the configuration of *each* center of stereoisomerism relative to its substituent with the lowest seniority. **Relative configurations** consider the configuration relative to the preceeding one.

The relative configurations in isotactic polymer molecules are identical if one moves along the chain. The ligands H, R and chain $(\sim\!\!\sim\!\!\sim)$ always follow in the same direction, for example, clockwise in the W(cis) representation of $-[-CHR-]_{\overline{n}}$ in Fig. 4-6. The same is true for poly(α-amino acids) $-[-NH-*CHR-CO-]_{\overline{n}}$ with 100 % D-units *or* 100 % L-units. The same poly(α-amino acids) with alternating D and L units are however syndiotactic. The *C atoms of $-[-NH-*CHR-CO-]_{\overline{n}}$ chains are stereogenic since their four ligands NH, H, R, and CO are all different and not in a plane. Poly(D-α-amino acid)s and poly(L-α-amino acid)s are therefore optically active but poly[(D-*alt*-L)-α-amino acid]s are not.

Poly(propylene)s do not possess D and L units, however, since two of the four ligands of ^{0}C in their constitutional units $-CH_2-^{0}CH(CH_3)-CH_2-$ are identical. The prochiral units $-^{0}CH(CH_3)-$, however, behave like chiral units with respect to tacticity.

Relative configurations of isotactic polymer molecules are independent of the direction of travel along the chain (from left to right or right to left). The situation is quite different for absolute configurations. 2,4-Dichloropentane $CH_3-CHCl-CH_2-CHCl-CH_3$, for example, has two stereogenic carbon atoms (numbers 2 and 4) with 4 ligands each. The ligands have the seniorities $Cl > CH_2-CHCl-CH_3 > CH_3 > H$. The molecule I in Fig. 4-7 thus has the configurational sequence RS. It is a meso compound since the stereogenic central atoms have opposite configurations.

Molecule I can be converted to molecule III by a rotation of 180°; I and III are thus identical. Molecules II and IV, on the other hand, are not interconvertible by a rotation. They are thus enantiomers; their 1:1 mixture is racemic.

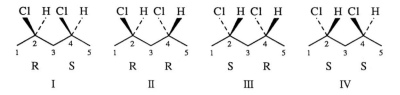

Fig. 4-7 The four possible configurations of 2,4-dichloropentane.

The situation is different for polymer chains. An example is the hepta(propylene) molecule $(CH_3)_2CH[CH_2CH(CH_3)]_7CH_2CH(CH_3)_2$ with one isopropyl and one isobutyl endgroup (Fig. 4-8). Its isotactic heptamer Ia begins with an S-configuration, its isotactic heptamer Ib with an R-configuration. Ia and Ib are thus enantiomers.

If however the contributions by endgroups are negligible (i.e., endgroups practically identical and/or infinite chain length), then Ia and Ib each possess a mirror plane. One half of the molecule has an R-configuration, the other half an S-configuration. The "infinitely long" molecules Ia and Ib are meso compounds and not chiral.

The absolute configurations R and S alternate in the two syndiotactic heptamers IIa and IIb. A very long syndiotactic chain can thus not be optically active. The infinitely long IIa and IIb are not enantiomers but the same single molecule. Syndiotactic polymers are thus not "racemic" and the symbol "rac" for st-polymers is unjustified ("meso" for isotactic polymers is justified, however, see above).

The use of "m" for the mole fraction of isotactic units and that of "r" for the mole fraction of syndiotactic units is however very common. These symbols are not only superfluous (in contrast, "i" and "s" need no translation) but wrong. IUPAC recently saw the light and is now sheepishly recommending "racemo" instead of "racemic".

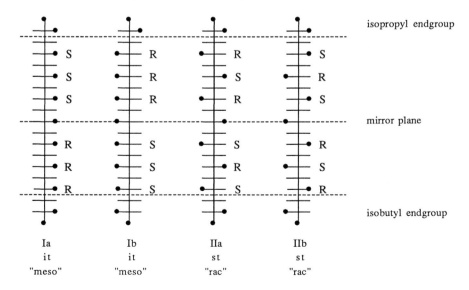

Fig. 4-8 Fischer projections of the configurations of α-isopropyl-ω-isobutyl-hepta(propylene). R and S are defined as absolute configurations. ● Methyl group. The designations "meso" and "rac" refer to infinitely long chains (negligible endgroups).

4.2.3   Presentation of Relative Configurations

Substituents R of isotactic polymer molecules are always "on the same side" in Fischer projections and other stereoprojections based on cis-conformations of chains (Fig. 4-9), regardless of the number of chain atoms per monomeric unit.

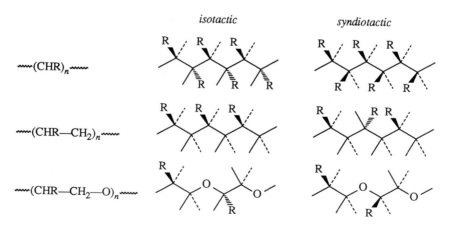

Fig. 4-9 Fischer projections (projections of cis conformations on a plane) of segments of isotactic (left) and syndiotactic (right) polymer molecules. From top to bottom for R = CH$_3$: poly(methyl-methylene), poly(propylene), and poly(oxypropylene).

For stereoprojections of chains in trans-conformation (i.e., zig-zag chains), the "same side" rule applies only to monomeric units with even numbers of chain atoms, for example, $+CHR-CH_2+_n$ or $+CHR-O+_n$ (Fig. 4-10). Substituents R are on opposite sides of the chains in stereoprojections of polymer chains with one chain atom per monomeric unit, such as $+CHR+_n$, with three chain atoms per monomeric unit $(+CHR-CH_2-O+_n$ or $+NH-CHR-CO+_n$, with five chains atoms per monomeric unit, etc.

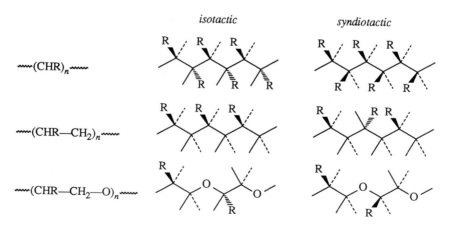

Fig. 4-10 Segments of isotactic (left) and syndiotactic (right) polymer chains in hypothetical all-trans-conformations. From top to bottom for R = CH$_3$: poly(methylmethylene), poly(propylene), and poly(oxypropylene). Compare the relative positions of substituents R in chains with all-trans conformations (Fig. 4-10) with those in all-cis conformation (Fig. 4-9).

4.2.4 Hemitactic Polymers

Hemitactic polymer chains contain two alternating types of stereogenic atoms (Fig. 4-11). The configuration of the first, third, fifth ... central atom is always exactly defined whereas those of the second, fourth, sixth ... central atom are random. Only one-half of the central atoms is therefore tactic (G: *hemi* = half). According to the tacticity of the first, third, fifth ... central atom, one distinguishes **hemi-isotactic** (hemi-it) from **hemi-syndiotactic** (hemi-st) polymer molecules.

The conversion of undefined central atoms into defined ones transforms the hemi-isotactic polymer molecules into either isotactic or syndiotactic ones (Fig. 4-11). Hemi-syndiotactic polymer molecules, on the other hand, are either transformed into hetero-tactic moleculess with the sequence –i-*per*-s– or into bi-heterotactic molecules with the sequence i-*per*-i-*per*-s-*per*-s– (Fig. 4-11).

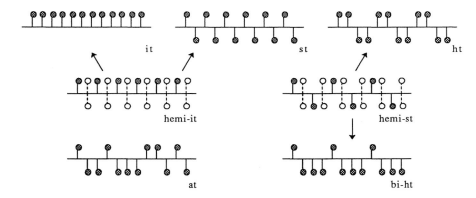

Fig. 4-11 Schematic representation of segments of isotactic (it), syndiotactic (st), heterotactic (ht), hemi-isotactic (hemi-it), hemi-syndiotactic (hemi-st), atactic (at) (Section 4.3.1), and bi-heterotactic (bi-ht) chains. Substituents in exactly defined configurations are characterized by ⊗, those in arbitrary configurations by O.

4.2.5 Ditactic Polymers

Ditactic polymer molecules have two centers of stereoisomerism per constitutional monomeric unit, tritactic three, etc. Ditactic polymers result from the polymerization of 1,2-disubstituted ethene derivatives, for example, 2-pentene:

(4-2) $CH_3-CH_2-CH=CH-CH_3 \longrightarrow +CH(C_2H_5)-CH(CH_3)+_n$

The two stereogenic carbon atoms * in the monomer unit $-*CH(C_2H_5)-*CH(CH_3)-$ may have the same configuration (**erythro** unit; in analogy to erythrose (Fig. 4-1); G: *erythros* = red) or different ones (**threo** unit; in analogy to threose; unknown word origin). Diads composed of two monomeric units may be furthermore isotactic or syndiotactic. There are thus four possible configurations (Fig. 4-12). This erythro-threo nomenclature applies to all units containing directly adjacent stereogenic centers.

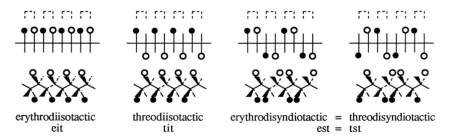

| erythrodiisotactic | threodiisotactic | erythrodisyndiotactic = threodisyndiotactic |
| eit | tit | est = tst |

Fig. 4-12 The four possible ditactic configurations of a segment $-[CH(CH_3)-CH(C_2H_5)]_4-$ of poly(2-pentene) with four monomeric units. CH_3 is symbolized by ● and CH_2H_5 by O. Top: Fischer projections (cis-conformations); bottom: stereoprojections (trans-conformations).

The disyndiotactic poly(2-pentene) with the mer $-CHR-CHR-$ does not have erythro and threo configurations as can be seen by a rotation of the chain direction by 180°. The situation is different for polymers with the segment $-[CH_2-CHR-CHR'-CH_2]_2-$:

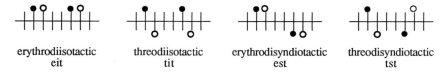

| erythrodiisotactic | threodiisotactic | erythrodisyndiotactic | threodisyndiotactic |
| eit | tit | est | tst |

The substituents R and R' of erythrodiisotactic configurations (eit) are always "on the same side" in Fischer projections but on different sides in wedge projections. Newman projections of the eit configuration in cis conformation show correspondingly R above R' and H above H (not shown).

Polymers with rings as stereogenic centers can be obtained by polymerization of the double bonds of unsaturated rings. The ring atoms bound directly to the ring atoms acting as chain atoms can here be treated like substituents. Similarly to poly(2-pentene), poly(1,3-cyclopentene) can thus form four different configurations (Fig. 4-13). These polymers show special features for those bonds where the ring enters or exits the chain: they have a cis conformation in erythro polymers and a trans conformation in threo polymers.

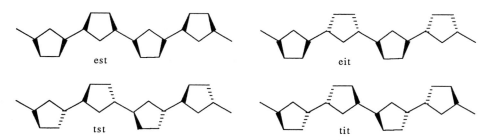

Fig. 4-13 The four relative configurations of the ditactic poly(1,3-cyclopentene)s.

The erythro-threo nomenclature applies to directly adjacent stereogenic centers with different constitutions. A meso-racemo nomenclature is used however for constitutionally identical stereogenic centers that are either directly adjacent or connected via symmetric groups Z such as $-O-$, $-CH_2-$, $-CH_2-CH_2-$, and $-CH_2-CHR-CH_2-$ (Fig. 4-14).

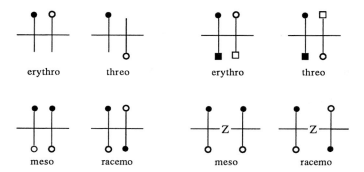

erythro   threo   erythro   threo

meso   racemo   meso   racemo

Fig. 4-14 Comparison of erythro-threo and meso-racemo nomenclatures for different ditactic polymers. Z = symmetric connecting group; ●, O, □, ■ = substituents with different constitutions.

4.2.6 Geometric Isomers

Polymer molecules with different **torsional stereoisomerisms = geometric isomerisms** are also tactic. Depending on the configurational arrangement of chain sections relative to the double bond in the chain, cis-tactic (ct) and trans-tactic (tt) structures are distinguished. Cis configurations correspond to E isomers of the R,S nomenclature, trans configurations to Z isomers.

1,4-cis (ct) (E)   1,4-trans (tt) (Z)   1,2 (it or st)   3,4 (it or st)

Examples are the 1,4-poly(diene)s $+CH_2-CR=CH-CH_2+_n$ that result from the polymerization of 1,3-dienes $CH_2=CR-CH=CH_2$ via polymerization of the two double bonds. R is H in 1,3-butadiene, CH_3 in isoprene, Cl in chloroprene, etc. Natural rubber (cis) and balata and guttapercha (both trans) are natural polymers. The polymerization of a single double bond of 1,3-dienes leads to 1,2-poly(diene)s that may be isotactic, syndiotactic, or atactic.

4.3 Polymers with Non-ideal Tacticities

4.3.1 Overview

The smallest configurational entities of polymers with configurational tacticity are the diads composed of two monomeric units. These diads may be either isotactic (symbol i; in the literature often as m) or syndiotactic (symbol s; in literature as r) (see p. 123).

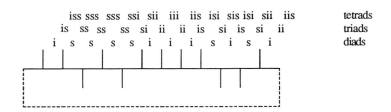

Fig. 4-15 Tactic diads, triads, and tetrads of a cyclic dodecamer.

Each monomeric unit belongs to 2 tactic diads (dyads), 3 tactic triads, 4 tactic tetrads, 5 tactic pentads, etc. A cyclic dodecamer with 12 monomeric units thus has 12 tactic diads, 12 tactic triads, 12 tactic tetrads, etc. (Fig. 4-15). Simple **ideal-tactic** polymers have only one type of diads, one type of triads, one type of tetrads, etc., for example, only isotactic diads (i), isotactic triads (ii), isotactic tetrads (iii), etc., in isotactic polymers. The corresponding amount-of-substance fractions ("mole fractions") are $x_i = x_{ii} = \ldots = 1$. The mole fractions of ideal syndiotactic polymers are similarly $x_s = x_{ss} = x_{sss} = \ldots = 1$.

The other limiting case is characterized by the maximum possible number N_J of types per J-ad (diad, triad, tetrad, etc.). In this case, two different diads (i, s), four different triads (ii, is, si, ss), eight different tetrads (iii, iis, isi, sii, ssi, sis, iss, sss), etc., are possible:

(4-3) $N_J = 2^{J-1}$;   J = number of monomeric units per J-ad

In the limiting case of **ideal atactic polymers** (at*), each of the various J-ads is present in equal amounts ($x_i = x_s$; $x_{ii} = x_{is} = x_{is} = x_{ss}$; etc.).

Based solely on the proportions of diad fractions ($x_i = x_s = 1/2$) and triad fractions ($x_{ii} = x_{is} = x_{si} = x_{ss} = 1/4$), the dodecamer of Fig. 4-15 would be called an ideal-atactic compound. Although it has the tetrad equalities $x_{iii} = x_{iss} = x_{sis} = x_{ssi}$ and $x_{iis} = x_{isi} = x_{sii} = x_{sss}$, it is not a truly ideal-atactic polymer because of the inequalities $1/12 = x_{iii} \neq x_{sss} = 1/6$, $1/12 = x_{iss} \neq x_{sii} = 1/6$, etc. Correct classifications of polymers as ideal-atactic, Markov 1st order, Markov 2nd order, etc. require the determination of higher J-ads (pentads, etc.). The proportion of higher J-ads may often be obtained from nuclear magnetic resonance spectroscopy, the most important method for the determination of tacticity.

NMR spectroscopy does not detect inverse J-ads separately but only their sums. For example, x_{is} and x_{si} are measured together as $x_{is+si} \equiv x_{|is|}$. The number N_J^* of J-ads measured as one type by NMR is thus reduced to

(4-4) $N_J^* = 2^{J-2} + 2^{(J-L)/2}$

for ideal atactic (at) polymers. L = 2 applies to even numbers of J and L = 3 to odd numbers of J. Table 4-4 compares the number N_J of types of J-ads in ideal-atactic polymers as they can be obtained from NMR (at$_{NMR}$) with those that are actually present (at), i.e., in cases where directions can be distinguished from each other (is vs. si, etc.).

4.3.2   J-Ads

The preceding discussion of stereoregular and tactic polymers dealt with ideal structures. Real polymers are always irregular, however; they do not posses perfect steric or

Table 4-4 Numbers of types of diads, triads, tetrads, etc. per type of J-ad in isotactic (it), syndiotactic (st), heterotactic (ht), bi-heterotactic (bi-ht), hemi-syndiotactic (hemi-st), hemi-isotactic (hemi-it), and ideal-atactic (at$_{NMR}$) vinyl polymers as compared to those detected by NMR in ideal-atactic polymers (at) [1].

J-ad		Number of types of J-ads in polymers of type							
Type	number of mers	it	st	ht	bi-ht	hemi-st	hemi-it	at$_{NMR}$	at
Diad	2	1	1	2	2	2	2	2	2
Triad	3	1	1	1	3	3	3	3	4
Tetrad	4	1	1	2	2	4	4	6	8
Pentad	5	1	1	1	3	6	7	10	16
Hexad	6	1	1	2	2	8	8	20	32
Heptad	7	1	1	1	3	13	14	36	64
Octad	8	1	1	2	2	16	16	72	128
Nonad	9	1	1	1	3	25	28	136	256
Decad	10	1	1	2	2	32	32	272	512
Undecad	11	1	1	1	3	51	54	528	1024

tactic structures. The proportions and sequences of steric and configurational diads, triads, etc., are thus described by suitable statistical parameters.

The sum of amount-of-substance fractions ("mole fractions") of both diads is unity:

$$(4\text{-}5) \qquad x_i + x_s \equiv 1$$

The mole fraction of isotactic diads in compounds with chiral monomeric units (e.g., α-amino acid residues) is the sum of the mole fractions of DD and LL diads. The mole fraction of syndiotactic diads equals the sum of the mole fractions of DL and LD diads:

$$(4\text{-}6) \qquad x_i = x_{DD} + x_{LL}$$

$$(4\text{-}7) \qquad x_s = x_{DL} + x_{LD}$$

The sums of the mole fractions of triads and tetrads are thus

$$(4\text{-}8) \qquad x_{ii} + x_{is} + x_{si} + x_{ss} \equiv 1 \qquad\qquad \text{(triads)}$$

$$(4\text{-}9) \qquad x_{iii} + x_{iis} + x_{isi} + x_{sii} + x_{iss} + x_{sis} + x_{ssi} + x_{sss} \equiv 1 \quad \text{(tetrads),} \quad \text{etc.}$$

The various types of J-ads are interrelated since each J-ad is composed of other types of J-ads: diads can be expressed by triads, triads by tetrads, etc. (Table 4-5). For example, isotactic diads i can be found in isotactic triads ii and in the two types of heterotactic triads, is and si. By definition, $x_{|is|} \equiv x_{is} + x_{si}$, $x_{|iis|} \equiv x_{iis} + x_{sii}$, etc.

Atactic and imperfect isotactic, syndiotactic, hemitactic, etc. polymers contain isotactic sequences I composed of one (i), two (ii), three (iii), etc., diads. In such polymers, an isotactic sequence starts at a heterotactic triad ...si... and ends at a heterotactic triad ...is... The number-average degree $\overline{X}_{I,n}$ of isotactic sequence lengths is thus

$$(4\text{-}10) \qquad \overline{X}_{I,n} = 2\, x_i/(x_{is} + x_{si}) = 2\, x_i/x_{|is|}$$

Table 4-5 Relationships between the mole fractions of diads, triads, tetrads, and pentads.

Diads = f(triads)

$x_i \ = x_{ii} + (1/2)\, x_{|is|}$

$x_s \ = x_{ss} + (1/2)\, x_{|is|}$

Triads = f(tetrads)

$x_{ii} \ = x_{iii} + (1/2)\, x_{|iis|}$

$x_{ss} \ = x_{sss} + (1/2)\, x_{|iss|}$

$x_{|is|} \ = x_{|iis|} + 2\,(x_{isi} + x_{sis}) + x_{|iss|}$

Tetrads = f(pentads)

$x_{iii} \ = x_{iiii} + (1/2)\, x_{|iiis|}$

$x_{sss} \ = x_{ssss} + (1/2)\, x_{|ssss|}$

$x_{isi} \ = (1/2)(x_{|isis|} + x_{|iisi|})$

$x_{sis} \ = (1/2)(x_{|isis|} + x_{|siss|})$

$x_{ssi} \ = x_{|isss|} + 2\, x_{|issi|} = x_{|iiss|} + x_{|siss|}$

$x_{|iis|} \ = x_{|iiis|} + 2\, x_{siis} = x_{|iisi|} + x_{|iiss|}$

Tetrads = f(tetrads)

$x_{|iis|} + 2\, x_{sis} \ = x_{|iss|} + 2\, x_{isi}$

Pentads = f(pentads)

$x_{|iiis|} + 2\, x_{siis} \ = x_{|iisi|} + x_{|iiss|}$

$x_{|isss|} + 2\, x_{issi} \ = x_{|sssi|} + x_{|ssii|}$

The number-average degree of syndiotactic sequence lengths is $\bar{X}_{S,n} = 2\, x_s/x_{|is|}$. The number-average degree of *all sequences* is thus

(4-11) $\bar{X}_{seq,n} = (1/2)\,[\,\bar{X}_{I,n} + \bar{X}_{S,n}\,] = 1/x_{|is|}$

4.3.3 Experimental Methods

Tacticities are determined by absolute or relative methods. Absolute methods do not require calibrations. Examples are nuclear magnetic resonance (NMR), X-ray spectroscopy, infrared spectroscopy (all below), and the determination of optical activity (see Volume III). Relative methods comprise the determinations of the degree of crystallinity, solubility, glass and melting temperatures, and some chemical reactions (Table 4-6).

X-ray Spectroscopy

The positions and intensities of reflections in X-ray diagrams allow conclusions about the distances of atoms in crystals and thus about configurations. Model compounds are not required. This expensive method is used to calibrate relative methods with well crystallizing compounds of high steric purity (see Volume III).

Table 4-6 Determination of tacticities of solid (S) or dissolved (D) polymers. *See Volume III.

State of matter	Method	Determination of tacticity by		
		presence	type	proportion
S	Glass or melting temperature *	sometimes	no	sometimes
S	X-ray spectroscopy *	if crystalline	yes	no
S	Infrared spectroscopy	yes	sometimes	diads only
S	NMR spectroscopy	yes	yes	
D	NMR spectroscopy	yes	in principle	diads to pentads
D	Optical activity *	if chiral	yes	yes
D	Solubility *	yes	no	sequences
D	Chemical reactions	sometimes	no	yes

Nuclear Magnetic Resonance Spectroscopy

Nuclear magnetic resonance spectroscopy (NMR) of dissolved polymers is the most important method for the determination of the configuration and degree of tacticity because it can be applied to both crystallizing and non-crystallizing polymers. Tacticities can be ascertained because the chemical shift of signals from ^1H atoms ("protons"), ^{13}C atoms, ^{19}F atoms, etc., depends on the configuration of the main chain. The method is absolute but is used as a relative one in one-dimensional NMR spectroscopy because of technical problems.

An example is the analysis of proton NMR spectra of poly(methyl methacrylate)s (PMMA) of various tacticities. The monomeric units $-CH_2-CH(CH_3)(COOCH_3)-$ of this polymer (see also next page) contain methylene protons in $-CH_2-$, α-methyl protons in $-CH_3$, and methyl ester protons in $-COOCH_3$. The different chemical environments of these three types of protons leads to three regions of chemical shifts in NMR spectra (Fig. 4-16). These can be assigned to the proton type by comparison with the chemical shifts of the protons in methyl pivalate, $(CH_3)_3C-COOCH_3$, as a model compound.

Methyl ester protons always give only one signal at the same position, regardless of the tacticity of the specimens (not shown in Fig. 4-16), probably because the distance between protons of two methyl ester groups is too large to affect their chemical shifts. They can thus not be used for the determination of tacticities.

The two methylene protons of a monomeric unit in st-PMMA are in chemically equivalent environments since each is flanked by one α-methyl and one methyl ester group. The two methylene groups in it-PMMA are not chemically equivalent: one proton is surrounded by two α-methyl groups and the other one by two methyl ester groups.

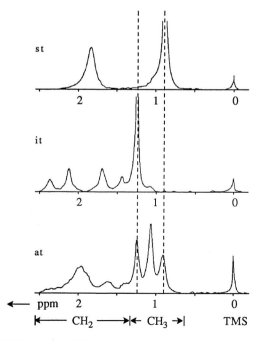

Fig. 4-16 Part of the 60 MHz proton NMR spectra of isotactic (it), syndiotactic (st), and atactic (at) PMMA. $COOCH_3$ signals are not shown. TMS = Reference signal from tetramethylsilane $Si(CH_3)_4$.

st-PMMA   it-PMMA

The actual conformation is irrelevant since one observes only the time average of the positions. The two equivalent methylene protons of st-PMMA thus deliver a single proton resonance signal whereas the chemically non-equivalent methylene protons of it-PMMA give rise to an AB quartet (Fig. 4-16). These features can be seen clearly in the spectra of isotactic and syndiotactic PMMA. They overlap however in atactic PMMA which makes it difficult to resolve the relative proportions of contributions by isotactic and syndiotactic units in such low-field experiments.

The situation is much better for the signals from α-methyl protons. Atactic PMMAs show three signals. One corresponds to the signal from α-methyl protons from truly syndiotactic PMMA and another one to the signal from isotactic PMMA. The third signal is centered between these two. One concludes that the three signals come from the three types of triads: syndiotactic (ss), isotactic (ii), and heterotactic (ht = is + si). The relative areas of the signals indicate the relative proportion of the triad types.

Poly(methyl methacrylate), $+CH_2-C(CH_3)(COOCH_3)+_n$, thus does not deliver the fractions x_{is} and x_{si} separetely but only their sum $x_{|is|}$. The reason is that the stereogenic C atom is surrounded by only three different types of ligands: CH_3, $COOCH_3$, and CH_2 (one each on opposite sides). However, in order to distinguish the triads is and si by NMR spectroscopy, all four of the adjacent ligands of the stereogenic atom must be different. An example is stereogenic *C of poly(oxypropylene), $+O-*CH(CH_3)-CH_2+_n$ with its four different ligands O, H, CH_3, and CH_2.

The reduction of the types of ligands to three from four also lowers the number of detectable triads from 4 (ii, is, si, ss) to 3 (ii, ss, and |is| ≡ (is + si)). The number of detectable tetrads decreases correspondingly from 8 to 6, that of pentads from 16 to 10, etc. (Table 4-4, at$_{NMR}$ vs. at).

In general, signals from protons of dissolved polymers are broader than signals of dissolved low-molecular weight model compounds. This difference is due to the different environments of NMR-active groups. In dilute solutions of low-molecular weight compounds, these groups are separated from each other since they reside in different molecules. The higher the molecule concentration, the smaller the distance between the groups. The orientation of the nuclei of the groups increases and so does their magnetic interaction: the signals broaden. The broadening is also greater at lower temperatures since the residence time of the orientations is larger.

The situation is different for polymer molecules since their NMR-active nuclei are bound together in a chain. The distance between adjacent nuclei is constant and cannot by reduced by lowering the polymer concentration: dilution does not increase the sharpness of signals as long as intermolecular associations are absent. The sharpness of signals also does not depend on the molar mass of random-coil-forming polymers. Sharper signals from polymer solutions are therefore only obtained from measurements at elevated temperatures.

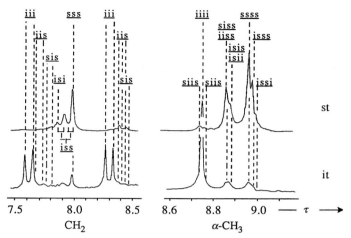

Fig. 4-17 220 MHz proton resonance spectra of a predominantly syndiotactic poly(methyl methacry-late) (top) and a predominantly isotactic one (bottom), both in ca. 0.1 g/mL solution in chlorobenzene at 135°C. Left: tetrad signals from β-methylene groups. Right: pentad signals from α-methyl protons.

Tetrads and pentads are accessible by higher strengths of magnetic fields which lead to larger chemical shifts of signals and better resolutions of spectra (Fig. 4-17). Such measurements can be made with frequencies of 220-900 MHz using supraconduc-ting magnets that are cooled by liquid helium.

Higher J-ads are also obtainable from ^{13}C spectra because chemical shifts from ^{13}C are larger than those of ^1H (up to 250 ppm from ^{13}C vs. up to 10 ppm from ^1H). An example is the spectrum of a fraction of a poly(propylene) (Fig. 4-18)

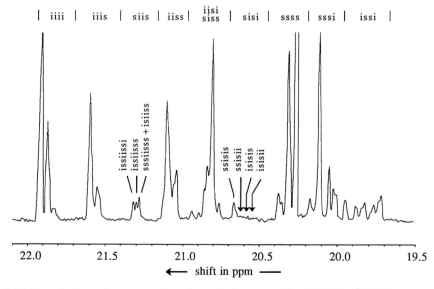

Fig. 4-18 Pentads, heptads, and nonads in the methyl region of the 150 MHz ^{13}C NMR spectrum of the pentane-soluble fraction of a poly(propylene). Arrows indicate expected signals that were not ob-served. Reprinted with permission from [2]. Copyright (1997) American Chemical Society.

Very many sophisticated NMR techniques are now available. Examples are double-resonance techniques to resolve spin-spin couplings of adjacent CH_2 and CH groups in polymers of the type $+CH_2-CHR+_n$ use of spin-spin and spin-lattice relaxation times for the determination of tacticities, two-dimensional D-NMR spectroscopy for absolute assignments, ^{13}C NMR spectroscopy of solid polymers, and many more.

Infrared Spectroscopy

The concentration of diads (and diads only) is often obtained from infrared spectroscopy, using oligomers or polymers of known configuration for assignments. In some cases, absorption frequencies can be calculated for several configurations. Different configurations also often generate different deformational vibrations of CH_2 and CH groups, and sometimes also of the amide I band of poly(α-amino acid)s.

The proportion of the various diads may also be obtained from the so-called crystalline bands since polymers with different stereoregularities crystallize to different extents and IR spectra are sensitive to crystallinity between 670 cm^{-1} and 1000 cm^{-1}. This method for the determination of tacticities is questionable, however, since "crystallinity" is not an inherent property of a substance but a property of a material (Volume III).

Crystallinity and Solution Properties

Polymers with greater tacticities are often more crystalline than those with lower ones. Hints of different crystallinities (and thus of different stereoregularities if the specimens were prepared in the same manner) are thus often indicated by different melting and glass temperatures (volume III).

Crystallinity *per se* is not an indication of tacticity, however, since certain atactic polymers also crystallize. For example, at-poly(vinyl acetate), $+CH_2-CH(OOCCH_3)+_n$, does not crystallize but its saponification product poly(vinyl alcohol), $+CH_2-CHOH+_n$, does. The crystallizable, isotactic poly(styrene) can be converted into the non-crystallizable poly(p-iodostyrene) and the non-crystallizable poly(p-lithium styrene) and, without reversal of configurations, back into the crystallizable isotactic poly(styrene):

Tacticity, crystallinity, and solution properties do not always go hand in hand, both with respect to *qualitative* differences (different types of tacticity) and *quantitative* differences (different degrees of tacticity). A greater tacticity leads in general to a higher crystallinity and thus to a lower solubility. Sometimes it is just the reverse, however: isotactic poly(2-hydroxyethyl methacrylate) is crystalline and dissolves in water but its non-crystalline atactic counterpart is not water soluble.

Molecules with higher degrees of tacticity can be separated from those with lower ones by their different solubility if they do not differ in the type of tacticity. Isotactic poly(propylene), for example, is characterized in industry by an **isotacticity index** that indicates the percentage of poly(propylene) that is insoluble in boiling xylene. This index does not measure the average tacticity of the polymer but the fraction of those polymer molecules whose degree of tacticity is just sufficient for their incorporation in those crystalline regions that cannot be dissolved under test conditions. The isotacticity index thus depends also on the morphology (and thus the thermal history) of the specimen and the interaction between polymer segments and solvent molecules.

Solubilities of partially crystalline polymers furthermore depend on molar masses. Highly soluble fractions of high-molecular weight atactic molecules may thus contain low-molecular weight fractions of stereoregular molecules and *vice versa*.

Different tacticities may also be recognized by **cloud-point titrations** where dilute polymer solutions of different concentrations are titrated with a precipitant until the beginning phase separation is indicated by an onset of cloudiness of the solution (Volume III). Similarly, the temperature may be lowered until the solution becomes cloudy. An onset of clouding upon temperature increase is relatively rare, however.

Historical Notes to Chapter 4

Stereochemistry

E.Paternò, Giornale di Scienze Naturali ed Economiche di Palermo **5** (1869) 117
    First tetrahedral model of the carbon atom.

J.H. van't Hoff, Arch.néerl.sci. exactes et naturelles **9** (1874) 445
J.A. Le Bel, Bull.Soc.Chim.France **22** (1874) 337
    Concluded simultaneously that the four valences of a carbon atom lead to an "asymmetric carbon atom" if all four univalent substituents are different, e.g., in $CR^1R^2R^3R^4$. The asymmetric carbon atom was defined as one that has no plane of symmetry (J.H. van't Hoff, La Chimie dans l'Espace, P.M.Bazendijk, Rotterdam 1875; ibid, Die Lagerung der Atome im Raume, Vieweg, Braunschweig 1877). In 1901, van't Hoff received the first Nobel prize ever albeit for the discovery of the laws of chemical dynamics and osmotic pressure in solution.
    Le Bel, on the other hand, favored a squarish pyramid for $CR^1R^2R^3R^4$, with carbon at the apex. He denied the existence of a tetrahedral carbon, advanced by van't Hoff, except for special cases.

J.K.O'Loane, Optical Activity in Small Molecules, Nonenantiomorphic Crystals, and Nematic Liquid Crystals, Chem.Rev. **80** (1980) 41
    This paper is an excellent review of the van't Hoff–Le Bel discoveries and controversies, optical activity shown by crystals and molecules, asymmetry, dissymmetry, etc.

A.G.Ogston, Nature **162** (1948) 963
    First recognition of prochirality. See also R.Bentley, Ogston and the Development of Prochirality Theory, Nature **276** (1978) 673.

Tacticity of Polymers

H.Staudinger, A.A.Ashcroft, M.Brunner, H.A.Bruson, S.Wehrli, Helv.Chim.Acta **12** (1929) 944
H.Staudinger, Die hochmolekularen organischen Verbindungen, Springer, Berlin 1932, p. 113
    The inability of polymers to crystallize is caused by a random arrangement of steric units.

M.L.Huggins, J.Am.Chem.Soc. **66** (1944) 1991
    Hypothesized that the different intrinsic viscosity-molecular weight relationships obtained for constitutionally identical vinyl polymers from various polymerization procedures are caused by different configurational structures.

C.E.Schildknecht, S.T.Gross, H.R.Davidson, J.M.Lambert, A.O.Zoss, Ind.Eng.Chem. **40** (1948) 2104
    Experimental observation that different polymerization procedures lead indeed to polymers with different physical properties that are assumed to be caused by different steric structures.

K.Ziegler, Angew.Chem. **64** (1952) 323; K.Ziegler, E.Holzkamp, H.Breil, H.Martin, Angew.Chem. **67** (1955) 541; K.Ziegler, Angew.Chem. **76** (1964) 545 (Nobel lecture)
    New types of catalysts allow the polymerization of ethene at room temperature and low pressures.

G.Natta, Acc.Naz.Lincei.Mem. **4** [sez. 2] (1955) 61; G.Natta, P.Pino, P.Corradini, F.Danusso, E.Mantica, G.Mazzanti, G.Moraglio, J.Am.Chem.Soc. **77** (1955) 1708; G.Natta, Angew.Chem. **76** (1964) 553 (Nobel lecture)
    Ziegler's catalysts allow the polymerization to polymers with regular steric structures.

Literature to Chapter 4

NOMENCLATURE
IUPAC Commission on Nomenclature of Organic Chemistry, 1974 Recommendations, Section
    E, Fundamental Stereochemistry, Pure Appl.Chem. **45** (1976) 11
International Union of Pure and Applied Chemistry, Macromolecular Division, Commission on
    Macromolecular Nomenclature, Compendium of Macromolecular Nomenclature, Blackwell
    Scientific, Oxford 1991

GENERAL OVERVIEWS
E.L.Eliel, Stereochemistry of Carbon Compounds, McGraw-Hill, New York 1962
K.Mislow, Introduction to Stereochemistry, Benjamin, New York 1965
G.Natta, M.Farina, Stereochemistry, Longman, New York 1972
B.Testa, Principles of Organic Stereochemistry, Dekker, New York 1979
M.Nógrádi, Stereochemistry. Basic Concepts and Applications, Pergamon Press, Oxford 1981

GROUP THEORY
F.A.Cotton, Chemical Applications of Group Theory, Interscience, New York 1963
J.D.Donaldson, S.D.Ross, Symmetry in the Stereochemistry, Intertext Books, London 1972
R.L.Flurry, Symmetry Groups. Theory and Chemical Applications, Prentice-Hall, Englewood
    Cliffs, NJ 1980

CONFIGURATIOßN and TACTICITY
G.Natta, F.Danusso, Stereoregular Polymers and Stereospecific Polymerizations, Pergamon,
    Oxford 1967
A.D.Ketley, Ed., The Stereochemistry of Macromolecules, Dekker, New York 1967 (3 vols.)
F.A.Bovey, Polymer Conformation and Configuration, Academic Press, New York 1969
E.Sélégny, Ed., Optically Active Polymers, Reidel Publ., Dordrecht 1979
J.L.Koenig, Chemical Microstructure of Polymer Chains, Wiley, New York 1980

INFRARED SPECTROSCOPY
S.Krimm, Infrared Spectra of High Polymers, Fortschr.Hochpolym.Forschg. **2** (1960) 51

NUCLEAR MAGNETIC RESONANCE SPECTROSCOPY
F.A.Bovey, High Resolution NMR of Macromolecules, Academic Press, New York 1972
A.E.Tonelli, NMR Spectroscopy and Polymer Microstructure, VCH Weinheim 1989
H.Friebolin, Basic One- and Two-Dimensional NMR Spectroscopy, Wiley-VCH, Weinheim,
    2nd ed. 1993

References to Chapter 4

[1]   M.Farina, G.Di Silvestri, P.Sozzani, B.Savaré, Macromolecules **18** (1985) 923, Table I and
        other information
[2]   V.Busico, R.Cipullo, G.Talarico, A.L.Segre, J.C.Chadwick, Macromolecules **30** (1997) 4786

5 Conformation

5.1 Fundamentals

In molecules with defined constitution and configuration, rotation of atoms or groups of atoms around single bonds generates species with different spatial arrangements of atoms. In Fig. 5-1, the "chain" atom 4 moves on a circle around an axis that is a fictitious extension of the bond between chain atom 2 and chain atom 3. The position of the circle is defined by the bond length $b_{3\text{-}4}$ and the valence angle τ of the chain atoms. The chain atom 4 may, in principle, reside in an infinite number of positions along this path.

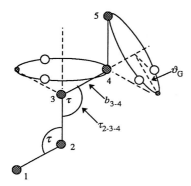

Fig. 5-1 Rotation of the chain atom 4 around the extension of the chain bond 2-3 and that of the chain atom 5 around the extension of the chain bond 3-4. All chain atoms ● are in trans position. Cis positions are marked by small ⊗.

Shown also are the two gauche positions O with the conformational angle $\vartheta_G = \pm 60°$ (organic-chemical convention) but not the corresponding anti positions ($\vartheta_A = \pm 120°$) (see below).

Because of attraction and repulsion forces between substituents of chain atoms (not shown in Fig. 5-1), only a few of the many possible positions are energetically distinguished, however. Ethane has two such positions: maximum repulsion of hydrogen atoms 1 and 6 in cis position, and minimum repulsion in trans position (Fig. 5-2). These distinguished spatial locations are called conformations; they cannot be made congruent. In organic chemistry, the term **conformation** refers to the spatial arrangement about a *single* bond. In **polymer chemistry**, one distinguishes between **microconformations (local conformations)** about single bonds and **macroconformations** of the whole molecule, the latter is called **configuration** in physics (see p. 110).

"Conformation" is a neolatin word (L: *forma* = shape; old Latin: *com* = equal, similar. If followed by consonants, *com* becomes *con* except for b, p and m).

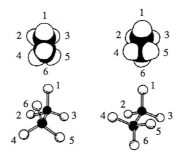

Fig. 5-2 Distinguished conformations of ethane (see also Table 5-1). Top: space-filling models; bottom: ball-and-stick models. Left: ecliptic (cis, synperiplanar); right: staggered (trans, antiperiplanar).

Small molecules with energetically distinguished spatial positions of their ligands can be considered defined species. They are called **conformers, conformational stereoisomers**, or **conformational isomers**, and, if open-chained, also **rotational isomers** or **rotamers**. In polymer chemistry, the same terms apply to local conformations.

This classic definition of **conformational stereoisomerism** is sometimes extended by inclusion of the so-called **torsional stereoisomerism** which results from the rotation of ligands about double or partial double bonds. An example is the **geometric isomerism** about the double bonds of cis and trans-polydienes, $\{CH_2–CR=CH–CH_2\}_{\overline{n}}$ (R = CH_3, Cl, etc.; Section 4.2.6). Peptide bonds form partial double bonds because of the equilibrium $–NH–CO– \rightleftarrows –N=C(OH)–$. Peptides may thus exist in cis and trans conformations.

A macromolecule contains many chain bonds and thus many microconformations. The sequence of these local conformations determines the **macroconformation** (or **molecular conformation**) of the molecule, i.e., its shape and spatial size (Volume III). If, for example, *all* microconformations about chain bonds of a linear chain are trans, then the macroconformation of the whole macromolecule is that of a zig-zag chain. The macroconformation of the chain is a spiral (helix) if trans and gauche(plus) or trans and gauche(minus) microconformations (see Section 5.2) alternate. The overall outer shape of the chain is a rod in the latter cases.

The macroconformation of polymer molecules and ensembles of polymer molecules as well as their static and dynamic properties can be simulated mathematically by **molecular modeling** if the potential functions and the potential energies of the various types of microconformations are known (volume III). The potential energy is obtained by the addition of the energies that are required to change valence angles, cause torsions about valence bonds, etc., using various force fields. In low-molecular weight chemistry, molecular modeling is very important for the calculation of spectra, three-dimensional molecule structures, and chemical reactivities, especially for the development of pharmaceutically active compounds.

In polymer science, one is especially interested in the effect of macroconformation on physical properties such as mechanical stiffness and strength. Methods for molecular modeling comprise Monte Carlo processes, Brownian dynamics, molecular mechanics, and molecular dynamics. All methods use force fields to calculate the distribution of probabilities of the various macroconformations (Volume III). All methods are limited by the capacity of computers.

5.2   Microconformations

5.2.1   Definitions

The geometry of simple molecules of the type A–B–D is completely defined by the two bond lengths $b_{A–B}$ and $b_{B–D}$ and the bond angle $\tau_{A\text{-}B\text{-}D}$ (valence angle) between the bonds A–B and B–D. An example is the methane molecule $H–C(H)_2–H$ with A = D = H and B = $C(H)_2$. This molecule has the shape of a regular tetrahedron (Fig. 4-4, left) with equal side lengths = bond lengths of $b_{C–H}$ = 0.1094 nm and equal tetrahedral angles = bond angles of $\tau_{H–C–H}$ = 109°28'.

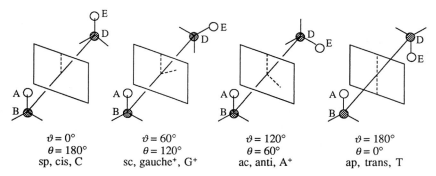

$$\vartheta = 0°$$
$$\theta = 180°$$
sp, cis, C

$$\vartheta = 60°$$
$$\theta = 120°$$
sc, gauche⁺, G⁺

$$\vartheta = 120°$$
$$\theta = 60°$$
ac, anti, A⁺

$$\vartheta = 180°$$
$$\theta = 0°$$
ap, trans, T

Fig. 5-3 Definition of torsional angles ϑ (organic-chemical convention) and θ (macromolecular convention) as projection - - - - of the bonds A–B and D–E (both ⊗—O) on a plane that is normal to the bond B-D (⊗—⊗). Angles ⌐⌐⌐ indicate the organic-chemical designations ϑ.

   Torsional angles are measured from –180° to 180° and not from 0° to 360°. The torsional angle ϑ is defined as positive if the bond A–B needs the smaller of the two possible rotations (clockwise and counter-clockwise) in order to become congruent with the bond D–E if the molecule A–B–D–E is viewed in the direction of the central bond B–D.

   Ethane $H-{}^{1}CH_2-{}^{2}CH_2-H$, on the other hand, is a molecule of the type A–B–D–E. The geometry of such a molecule is described completely if one knows not only the three bond lengths b_{A-B}, b_{B-D} and b_{D-E} and the two bond angles τ_{A-B-D} and τ_{B-D-E} but also the **torsional angle (= conformational angle, rotational angle, dihedral angle)**. The torsional angle is determined by projecting the bonds A–B and D–E on a plane that is normal to the bond B-D. The resulting angle between these two projections is the torsional angle (Fig. 5-3). It describes the relative spatial position of the "bound" substituents to the "unbound" ones: if the three hydrogen atoms of ${}^{1}C$ of ethane are considered "bound", then the hydrogen atoms at ${}^{2}C$ are "unbound" (to ${}^{1}C$) and vice versa.

   Torsional angles are defined, named, and measured differently in organic chemistry and polymer science (Fig. 5-3, Table 5-1). According to IUPAC, one can either use organic chemical designations (synclinal, anticlinal, synperiplanar, antiperiplanar or polymer designations (gauche, anti, cis, trans). The organic-chemical convention assigns the angle $\vartheta = 0°$ to the synperiplanar position (cis) and the angle $\vartheta = \pm 180°$ to the antiperiplanar position (trans) (Fig. 5-3). Practically all polymer scientists however use the complementary angle $\theta = 0°$ as the torsional angle for the trans conformation because (a) trans conformations are often the most stable conformations and (b) cis conformations of chains (—O— replaced by —⊗— in Fig. 5-3) are difficult to represent graphically.

Table 5-1 Nomenclature of conformations.

Polymer science			Organic chemistry			Old terms
Name	Symbol	Torsional angle θ	Name	Symbol	Torsional angle ϑ	
cis	C	± 180°	synperiplanar	sp	0°	ecliptic, parallel, planar-syn
gauche	G	± 120°	synclinal	sc	± 60°	gauche-staggered, skew, skew-syn
anti	A	± 60°	anticlinal	ac	± 120°	part ecliptic, part covered, skew-anti
trans	T	0°	antiperiplanar	ap	± 180°	trans-staggered, anti-parallel

Ethane has only two distinguished microconformations (Fig. 5-2). The three hydrogen ligands of each carbon atom here are either staggered with respect to the three hydrogen ligands of the other carbon atom in the trans conformation (antiperiplanar, ap) or, parallel to them, in the cis conformation (synperiplanar, sp). A rotation of one methyl group by 60° around the C-C axis converts the trans conformation into the cis conformation and vice versa. The total rotation of one methyl group by 360° thus creates three identical trans conformations and three identical cis conformations.

The trans conformation belongs to the symmetry group D_{2d}, and the cis conformation to the symmetry group D_{3h} (Section 4.1.3). Both conformations are thus achiral (no image-mirror image). Conformations between the distinguished ones, i.e., at torsional angles other than 0° and ±60° (not shown in Fig. 5-1), belong to the symmetry group D_3. They are chiral.

The distance between bound and unbound hydrogen atoms is shorter in the cis conformation than in the trans conformation. Hence, the trans conformation of ethane is energetically favored because the hydrogen ligands repel each other, which leads to a steric hindrance in the cis conformation. The cis conformation thus corresponds to an energy maximum and the trans conformation to an energy minimum (Fig. 5-4).

In *butane* $^1CH_3-^2CH_2-^3CH_2-^4CH_3$, each of the two central carbon atoms ^2C and ^3C carries 2 hydrogen ligands and 1 methyl ligand. The three identical trans conformations of ethane are replaced in butane by one trans conformation (0°) and two gauche (sc) conformations (+60°, –60°) (F: *gauche* = left, awkward) and three identical cis conformations by one cis (±180°) and two anti (ac) conformations (+ 120°, –120°).

The trans conformation has planar symmetry (symmetry group C_{2h}). The two gauche conformations are chiral and thus enantiomers; they belong to the symmetry group C_2.

The different repulsions CH_3/CH_3, CH_3/H, and H/H lead to a deep energy minimum at 0°, a high maximum at 180°, two lesser minima at ±120°, and two lesser maxima at ±60° (Fig. 5-4).

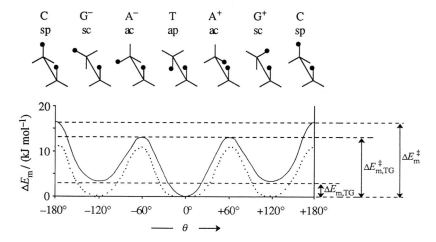

Fig. 5-4 Ideal conformations (top) and rotational barriers (bottom) around the C–C bond of ethane CH_3-CH_3 (dotted curve) and around the central C–C bond of butane $CH_3CH_2-CH_2CH_3$ (solid curve) as a function of the torsional angle θ (macromolecular convention for C, G, A, and T). • is a hydrogen atom in ethane and a CH_3 group in butane. The energies to the right refer to butane.

The trans conformation produces the deepest energy minimum because the methyl groups are farthest away from each other (least repulsion). They are nearest to each other in the cis conformation, which results in the highest energy maximum (highest repulsion). The gauche and anti conformations are intermediate between these, generating two less deep minima (G^+, G^-) and two less high maxima (A^+, A^-). The energy difference between the lowest energy minimum (T) and the next higher energy minimum (G^+ or G^-) is called the **conformational energy** or **potential energy**. In butane, it is $\Delta E_{m,TG} = \Delta E_m^\ddagger - \Delta E_{m,TG}^\ddagger \approx 2.9$ kJ/mol (Fig. 5-4).

The transition from the trans conformation to the gauche conformation via the energy maximum at the anti conformation requires an activitation energy ($\Delta E_{m,TG}^\ddagger = 13$ kJ/mol for butane). The total activation energy for the transition from the lowest energy minimum (trans) via the highest energy maximum (cis) to trans is called the **rotational barrier** or **potential barrier**; it is $\Delta E_m^\ddagger \approx 15.9$ kJ/mol for butane.

Because of interactions between substituents, the threefold rotation potentials of *substituted butanes* do not always show energy extrema at the ideal positions of $0°$, $\pm 60°$, $\pm 120°$, and $\pm 180°$. These non-ideal conformations however carry the same designations if the non-ideal ones deviate by no more than $\pm 30°$ from the ideal ones. A conformation with a torsional angle of $+10°$ instead of $0°$ is therefore also called "trans". Enantiomorphic conformations with unknown signs are designated $G/\overline{G}$, $A/\overline{A}$, $C/\overline{C}$, and $T/\overline{T}$. The last two symbols are used only if the torsional angles are not exactly $180°$ (C) or $0°$ (T).

Butane $^1CH_3-^2CH_2-^3CH_2-^4CH_3$ has only one central bond $^2C-^3C$ and therefore only one conformational monads. The unbound H atoms are nearer to the bound H atoms in monads C and $A^\pm$ than in monads $G^\pm$ and T. The resulting repulsion makes C and $A^\pm$ less probable conformations; they can be neglected for higher alkanes. They may be present however in other molecules (Section 5.3.2).

Pentane $^1CH_3-^2CH_2-^3CH_2-^4CH_2-^5CH_3$ possesses **conformational diads** composed of two conformational monads at $^2C-^3C$ and $^3C-^4C$ (Fig. 5-5). Each of these monads may exist in one trans (T) or one of the two gauche conformations (G^+ or G^-). Pentane thus exhibits four different types of conformational diads: TT, $TG^+ = TG^- = G^-T = G^+T$, $G^+G^+ = G^-G^-$, and $G^+G^- = G^-G^+$. Conformational calculations of carbon chains usually assume the absence of $G^+G^- = G^-G^+$ because of the high potential energies for these very sterically hindered conformations.

Ethane, propane, butane, pentane, etc., have threefold rotation potentials whereas 1,4-phenylene groups and sulfur chain atoms in *catena*-sulfur have twofold ones.

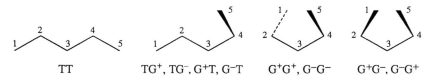

| TT | TG^+, TG^-, G^+T, G^-T | G^+G^+, G^-G^- | G^+G^-, G^-G^+ |

Fig. 5-5 Conformational diads of pentane $^1CH_3-^2CH_2-^3CH_2-^4CH_2-^5CH_3$.

5.2.2 Rotational Barriers

Rotational barriers are caused by the difference between repulsive and attractive forces. In ethane, the total attraction energy E_{ne} stems from the energy between the (negative) binding electrons (index e) of the C–H bonds and the (slightly positive) hydrogen atoms (index n; for nucleus) that are not bonded to these bonds. The total repulsion energy of ethane is composed of three contributions: the energy E_{nn} from the nuclei of non-bonded hydrogen atoms (both positive), the energy E_{ee} between the binding electrons of two C–H bonds (both negative), and the kinetic energy E_{kin} of the electrons (all negative). The energy E_b between the binding electrons of the carbon atoms need not be considered, however, since it is prevalent only if the 4f state participates in the binding (in that case, the corresponding molecular orbitals are not cylinder-symmetrical to the C–C bond). The potential barrier is then given by the sums of all these energy contributions, i.e., by $\Delta E^{\ddagger} = E_{ne} - (E_{nn} + E_{ee} + E_{kin})$.

For the attractive forces as a function of the torsional angle, the ethane molecule shows a curve with three identical maxima and minima. The same curve is obtained for the repulsive forces. The phases of the attraction energy and repulsion energy are shifted by 120°, however. The difference between the maximum and the minimum of the energy is $\Delta E_{attr} = 82.5$ kJ/mol for the attraction and $\Delta E_{rep} = 93.8$ kJ/mol for the repulsion. The molar potential barrier of ethane is thus $\Delta E_m^{\ddagger} = (93.8 - 82.5)$ kJ/mol $= 11.3$ kJ/mol (see Fig. 5-4 and Table 5-2). This activation energy is relatively small so that the conformational isomers are interconverted relatively fast. Torsional isomers have much higher potential barriers, which allows their separation.

The energy for interconversion is delivered by the collision of two molecules. However, only a part of thermal energy (ca. $RT/2 \approx 1.24$ kJ per mole per degree of freedom) is transferred to the other molecule. Because of the Maxwell-Boltzmann energy distribution, the energy for surmounting the potential barrier is only transferred by a small fraction of collisions. Most collisions cause only vibrations of no more than ±20° around the potential minima. At normal temperatures, the majority of molecules thus remain in conformations with a minimum of potential energy and one can thus treat the bonds (and small molecules) as if they exist only in discrete rotational states, i.e., as distinct **conformers**.

The transition from one conformation into another determines the flexibility of the molecule. A microconformer is **dynamically flexible** if the rotational barrier for the trans-gauche conversion is smaller than circa RT (i.e., $\Delta E_{m,TG}^{\ddagger} < 2.48$ kJ/mol at $T = 298.15$ K (25°C)). The conversion of trans into gauche (or *vice versa*) is 100 % within a time of $t = 4.4 \cdot 10^{-13}$ s at a rotational barrier of $\Delta E_{m,TG}^{\ddagger} = RT$.

Microconformers are **statically flexible** if the molar conformational energy is smaller than RT ($\Delta E_{m,TG} < RT$ for butane). Since macroconformers consist of many microconformers, one has to distinguish between local and global flexibilities. A chain segment may be considered rigid with respect to several monomeric units (microconformers) but the whole chain may be flexible (see the molar mass-dependent behavior of poly(hexylisocyanate) in Fig. 3-18).

The static flexibility of a polymer chain can be described by a **persistence length** L_{pers} (L: *persistere* = to persist) that is then compared to the *conventional* **contour length** r_{cont}. The conventional contour length is the maximum length of a chain that is *physically* possible. In many cases, it is the **end-to-end distance** of a rigid zig-zag chain. The conventional contour length is not identical with the *historical* contour length (= geometric contour length) which is obtained by following the contour of a chain. In a zig-zag chain with 1000 chain bonds of length L_b, $L_{cont,hist} = N_b L_b = 1000\,L_b$.

The temperature dependence of the persistence length L_{pers} depends on the conformational energy $\Delta E_{m,TG}$ and a system-dependent constant L_0:

(5-1) $L_{pers} = L_0 \exp [\Delta E_{m,TG}/RT]$; $L_0 = (0.1\text{-}0.9)$ nm

The chain is flexible if the persistence length is smaller than r_{cont} but greater than L_0 ($r_{cont} > L_{pers} > L_0$). The **flexibility of a chain** with the degree of polymerization, X, is thus determined by the ratio of persistence length and conventional contour length, for example, for a chain with trans and gauche microconformations:

(5-2) $L_{pers}/r_{cont} = (1/X) \exp [\Delta E_{m,TG}/(RT)]$

In order to be totally stiff ($L_{pers}/r_{cont} = 1$), a chain with $X = 10^4$ at $T = 298.15$ K (25°C) must therefore possess a $\Delta E_{m,TG} = 22.8$ kJ/mol.

5.2.3   Effect of Constitution

Rotational barriers and conformational energies of single bonds depend on the length of the bond, the size of the substituents, the number of bound ligands, and the interaction between ligands (Table 5-2). Rotational barriers are greater in torsional isomers than in conformational isomers (e.g., double bonds versus single bonds): ethane H_3C-CH_3 has a rotational barrier of $\Delta E^{\ddagger} = 11.3$ kJ/(mol ⯈C–C⯇ bond) but ethene $H_2C=CH_2$ one of 272 kJ/(mol >C=C< bond).

The rotational barrier decreases with increasing bond length in compounds of similar constitution (Table 5-2). An example is the series H_3C-CH_3, H_3C-SiH_3, $H_3Si-SiH_3$ where the bond length of the central bond increases (0.154 nm $\rightarrow$ 0.193 nm $\rightarrow$ 0.234 nm) whereas the rotational barrier decreases (12.1 kJ/mol $\rightarrow$ 7.1 kJ/mol $\rightarrow$ 4.2 kJ/mol).

Table 5-2 Rotational barriers $\Delta E^{\ddagger}$ and bond lengths L of single bonds (indicated by –).

Compound	$\dfrac{\Delta E_m^{\ddagger}}{\text{kJ mol}^{-1}}$	$\dfrac{L}{\text{nm}}$	Compound	$\dfrac{\Delta E_m^{\ddagger}}{\text{kJ mol}^{-1}}$	$\dfrac{L}{\text{nm}}$
SiH_3-SiH_3	4.2	0.234	CCl_3-CCl_3	42.0	0.154
SiH_3-CH_3	7.11	0.193	CF_3-CF_3	16.40	0.154
CH_3-CH_3	11.30	0.154	CF_2H-CH_3	13.3	0.154
$CH_3-CH_2CH_3$	14.22	0.154	CFH_2-CH_3	13.75	0.154
$CH_3-CH(CH_3)_2$	16.31	0.154			
$CH_3-C(CH_3)_3$	19.67	0.154	$\sim CH_2-SCH_2CH_2\sim$	8.8	0.181
$CH_3-CH=CH_2$	8.35	0.154	CH_3-SH	1.86	0.181
$\sim CH_2-CH_2COCH_2\sim$	9.6	0.154	CH_3-OH	1.56	0.144
$\sim CH_2-COCH_2CH_2\sim$	3.4	0.154	CH_3-OCH_3	11.00	0.143
$\sim CH_2-COOCH_2\sim$	2.1	0.154	$\sim CH_2-OCOCH_2\sim$	5.0	0.143
$CH_3-C_6H_5$	3.35		C_6H_5-OH	13.0	
$CH_3-C\equiv CCH_3$	2.11		$\sim CH_2-NHCH_2CH_2\sim$	13.8	0.147
$CH_3CH_2-CH_2CH_3$	15.89	0.154	CH_3-NH_2	8.28	0.147
$(CH_3)_2CH-C(CH_3)_3$	29.2	0.154	CH_3-NO_2	0.025	

Rotational barriers are lower, the smaller the number of nonbonded ligands. The rotational barrier of H_3C-CH_3 is 12.1 kJ/mol but that of H_3C-OH only 1.6 kJ/mol although CH_3 and OH have almost the same volume. $\sim CH_2-O\sim$ and $\sim CH_2-CO\sim$ have considerably lower rotational barriers than $\sim\sim CH_2-CH_2\sim\sim$. The chains of polyethers and polyesters are thus more flexible than those of polyalkanes. Polyethers and polyesters are therefore used industrially as polymeric plasticizers.

Rotational barriers increase with increasing size of nonbonded ligands. Examples are

$$CH_3-CH_3 \quad < \quad CH_3-CH_2CH_3 \qquad < \quad CH_3-CH(CH_3)_2 \quad < \quad CH_3-C(CH_3)_3$$
$$O=CHCH_3 \quad < \quad CH_2=CHCH_3 \qquad < \quad CH_2=C(CH_3)_2$$
$$CH_3-CH_3 \quad < \quad (CH_3)_2CH-C(CH_3)_3 \quad < \quad CCl_3-CCl_3$$
$$CH_3-OH \quad < \quad C_6H_5OH$$

The spatial requirements of ligands are controlled by van der Waals radii and not by atomic radii. Not only do repulsive forces (steric effects) between bound and unbound ligands have to be considered but also the influence of polarity, even for hydrocarbons. For example, all angles $\angle_{H-C-H}$ of methane have the exact dihedral angle $\tau_{H-C-H} = 109°28'$ of a regular tetrahedron. The bond angle $\angle_{C-C-C}$ of crystalline poly(ethylene) is however enlarged to $\tau_{C-C-C} = 111.5°$ because of the different space requirements of carbon and hydrogen atoms and the different polarities of C–H and C–C bonds.

Bound and unbound hydrogen atoms repel each other in aliphatic hydrocarbons. This "steric" reason causes the chains in crystalline poly(ethylene) to adopt all-trans conformations (zig-zag chains).

If, however, the adjacent chain atom has electronegative groups or non-bonded electron pairs as ligands, then the attraction between nucleus and electrons may become so large in one of the possible conformations that the balance between attractive and repulsive forces is changed. In *polar* environments, such chains try to obtain the conformations with the maximum number of gauche interactions between adjacent electron pairs and/or electronegative ligands (**gauche effect**). At room temperature, the chains of crystalline poly(oxymethylene) $+O-CH_2+_n$ thus exist in an all-gauche conformation.

Gauche effects are not produced by internal effects of microconformations *per se* but by polar or polarizable environments. For example, the most stable conformation of $CH_3O-CH_2-CH_2-OCH_3$ is TTT according to *ab initio* calculations. Experiments show, however, that this conformation exists only in crystals at $T < 34$ K (frozen-in rotations, etc.) and in the gas phase (no interactions with other molecules). In crystals at $T = 40$ K, in melts, and in solution, the most stable conformation is TGT.

Special effects are observed for molecules that contain rings as part of the chain structure. The inversion of 4-membered rings requires only relatively low activation energies (Table in Fig. 5-6). Much higher activation energies are necessary for chair-boat inversions of 6-membered rings. Thus, the cyclohexanediyl unit and its derivatives behave in chains as relatively rigid entities.

High flexibilities of chains are thus caused by several factors: large bond lengths between chain atoms, many competing microconformations resulting from absent or identical ligands at chain atoms, and small differences in the potential energies of gauche and trans conformations. All three factors are present in poly(dimethylsiloxane)s with the repeating unit $-O-Si(CH_3)_2-$: a fairly large length of the Si–O bond ($b_{Si-O} = 0.164$ nm),

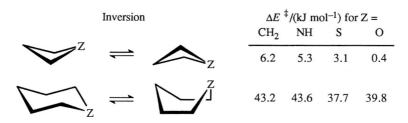

Fig. 5-6 Activation energies for the inversion of rings with different ring atoms Z.

an unsubstituted (–O–) and a symmetrically substituted (–Si(CH3)2–) chain atom, and polarized chain bonds –Si–O–. The high flexibility of the chains leads to a low glass temperature (freezing-in of the "amorphous" state of melts) of these polymers that are therefore highly viscous liquids, even at very high molar masses.

5.3 Macroconformations

The macroconformation (shape) of a polymer molecule is controlled by its constitution and macroconfiguration, the type and sequence of its microconformations (intra-molecular effects) as well as the interaction of the molecule with its environment (inter-molecular effects). Because of these intermolecular effects, macroconformations of polymer molecules may differ in crystalline states, solid amorphous states, melts, and in various solvents (Fig. 5-7).

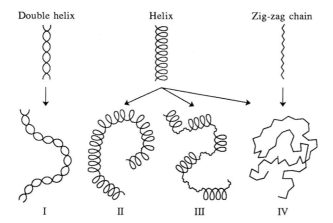

Fig. 5-7 Planar projection of characteristic shapes (macroconformations) of linear polymer molecules in crystalline states (top) and dilute solutions (bottom).
I   = Wormlike chain of the double helix of deoxyribonucleic acids in dilute salt solutions (25°C).
II   = Wormlike chain of helical low-molecular weight poly(γ-benzyl-L-glutamate) in DMF (25°C).
III = Random coils with helical and non-helical coil segments. Examples: poly(oxyethylene) in water at 25°C; at-poly(methyl methacrylate) in acetonitrile at 44°C.
IV = Random coils. Examples: poly(ethylene) in xylene at 160°C; poly(γ-benzyl-L-glutamate) in di-chloroacetic acid at 25°C; at-poly(methylmethacrylate) in acetone at 25°C.

For example, all-trans conformations may be stabilized in crystals only by the pack-ing of the chains (example: poly(ethylene)) or, in addition, by *inter*molecular interac-tions (example: hydrogen bonds in polyamide 6). All *inter*molecular interactions may survive in some molecules in certain solvents (example: desoxyribonucleic acids in dilute salt solution) although these cases are relatively rare. *Intra*molecular interactions such as hydrogen bonds, dipole-dipole interactions, and van der Waals interactions, on the other hand, may be or may be not present in solution, depending on the interaction of the polymer with the solvent. A helical molecule may thus form in solution either a worm-like chain, a random coil with "stiff" helical sections, or a truly random coil.

5.3.1 Crystalline Polymers

Although some polymers may crystallize (volume III), perfect crystallinity is almost never obtained because of imperfect polymer structures as well as kinetic effects. Poly-mers are therefore only **partially crystalline (semi-crystalline)**.

Equilibrium States

In equilibrium, polymer chains pack in crystals as tight as their constitution, configu-ration, and *intra* and *inter*molecular interactions allow. The smallest possible distance between chains is controlled by the van der Waals radii of atoms if *inter*molecular repul-sion dominates. One can calculate, for example, from the bond length $b_{C-C} = 0.154$ and bond angle $\angle_{C-C-C} = 111.5°$ of carbon atoms in poly(ethylene) $+CH_2-CH_2\frac{}{}n$ that the maximum distance between unbound hydrogen atoms of chains in an all-trans confor-mation is 0.254 nm. This distance is smaller than the sum of van der Waals radii of adja-cent hydrogen atoms (0.264 nm). Therefore, in the crystalline equilibrium state, poly-(ethylene) molecules adopt the macroconformation with the lowest energy, the all-trans conformation of a zig-zag chain.

The methyl groups of isotactic poly(propylene), $+CH_2-CH(CH_3)\frac{}{}n$, require more space than the hydrogen ligands of poly(ethylene). As a consequence, an all-trans con-formation of the chain would result in a sum of van der Waals radii that is greater than 0.254 nm. Because of this space requirement, every other chain bond thus avoids the trans microconformation and adopts the energetically next best one, i.e., gauche. All gauche conformations of a single chain must however have the same type of gauche conformations (...TG$^+$... or ...TG$^-$...) since sequences such as ...TG$^+$TG$^-$TG$^+$TG$^-$... are sterically impossible. The sequence of like conformational diads forces the chain to adopt the shape of a helix (Fig. 5-8). Since gauche positions differ from trans positions by 120°, 3 repeating units are required for 1 complete rotation of 360°: the chain adopts the macroconformation of a 3_1 helix (G: *hélix* = spiral, screw). Macroconformations ...TG$^+$TG$^+$... and ...TG$^-$TG$^-$... are energetically equivalent, however, and crystalline it-poly(propylene) thus contains equal proportions of these two macroconformations.

The helix expands with increasing size of the groups *near* the chain atoms; this effect is caused by larger bond angles between chain atoms as well as an increase in the number of chain atoms per turn. For example, the bond angles $\angle_{C-C-C}$ of chain atoms are 111.5° in the all-trans chain of poly(ethylene), 114° in the 3_1 helix of it-poly(pro-

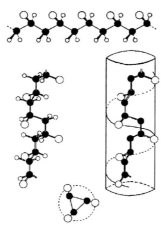

Fig. 5-8 Section of the hypothetical all-trans macroconformation (top) and the helical $(TG^+)_n$ macro-conformation (center left and right) of isotactic poly(propylene).
Center left: spatial arrangements of carbon atoms (●), methyl groups (O), and hydrogen atoms (o). Center right: the same chain in the cylinder that forms the envelope around the methyl groups in the chain direction. This helix is left-handed since the screw turns counter-clockwise away from the observer if it is viewed in the direction of the cylinder axis

pylene), and 116° in the 3_1 helix of it-poly(styrene). The torsional angles in all three polymers stay constant at 0° and ±120°, however.

Decreasing distances of large R groups to main chains require more chain atoms per helix turn and stronger deviations of torsional angles from the ideal values of 0° and 120°. Examples are the substituents R of isotactic poly(1-olefin)s $+CH_2-CHR+_n$.

substituents R	$-CH_2-CH_2-CH(CH_3)(C_2H_5)$	$-CH_2-CH(CH_3)_2$	$-CH(CH_3)_2$
polymer chains	3_1 helix	7_2 helix	4_1 helix
torsional angles	±0°, +120°	−13°, +110°	−24°, +96°

Macroconformations are also effected by electrostatics (p. 144), for example, the gauche effect causing the all-gauche conformation of poly(oxymethylene). Electrostatic effects may also lead to anti, and even cis, conformations in polymer chains. An example is the macroconformation $+A^-TA^+T+_n$ of crystalline 1,4-*trans*-poly(butadiene).

Non-Equilibrium States

Large crystals of polymers are rare. Examples are the molecular lattices of spherical proteins in which each lattice site is occupied by a protein molecule. These protein crystals contain plenty of hydrating water that allows doping of protein crystals by heavy metal ions for the analysis of their crystal structure by X-ray analysis.

It seems that only one case of large crystals of *synthetic* polymers is known: centimeter-sized crystals of poly(oxy-2,6-diphenyl-1,4-phenylene), grown from tetrachloro-ethane solutions. These crystals contain up to 35 % of solvent, however.

Most crystallizable polymers do not form crystals but only **crystalline regions** that may vary in size from 10-80 nm for crystallites to centimeter-size for larger morpholog-ical units such as spherulites (Volume III). The crystalline regions are embedded in non-

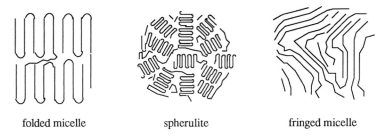

folded micelle   spherulite   fringed micelle

Fig. 5-9 Models for semi-crystalline polymers. Depending on the chemical structure, lines represent chain sections in either zig-zag or helical conformation.

crystalline regions or *vice versa*. The existence of crystalline regions in such **semi-crys-talline polymers** can be detected by thermal analysis, their structure by X-ray studies.

The coexistence of crystalline and non-crystalline regions in semi-crystalline poly-mers is visualized by models (Fig. 5-9). The model of a **fringed micelle** assumes that a polymer chain runs through several crystallites and that each crystallite contains chain segments of various polymer molecules. This **2-phase model** envisions a sharp border between perfect crystalline regions and completely amorphous ones. Such fringed mi-celles may result from the crystallizations of melts or concentrated solutions.

In **folded micelles**, a polymer chain folds itself into many parallel, straight sections that are bridged by more or less perfect loops composed of a few monomeric units and occasionally to another row of folds (Fig. 5-9). The crystallization of polymers from di-lute solution results in so-called **polymer single crystals** that consist of one layer of such fold structures. The crystallization from melts delivers **lamellar structures** composed of several fold structures that are stacked upon each other.

Depending on polymer structure and crystallization conditions, many other morphol-ogies may be generated, for example, **spherulites** (G: *sphaira* = ball, *lithos* = stone) (Fig. 5-9). These physical structures contain unordered and less ordered regions; they are never 100 % crystalline. Solid polymers are therefore characterized by their **degree of crystallinity** that is based on the 2-phase model. The degree of crystallinity varies with the method of determination (density, infrared, X-ray, etc.) since each method responds differently to different degrees of order (see Volume III).

5.3.2 Polymer Solutions

A statistical sequence of the different types of microconformations leads to many dif-ferent macroconformations and thus to statically flexible polymer chains. Six different but energetically equal types of microconformations cause a poly(methylene) chain of degree of polymerization $X = 30$ to adopt a total of $6^{X-2} = 6^{28} \approx 6.14 \cdot 10^{21}$ different macroconformations (cf. Fig. 5-10). Each microconformation, and thus each macrocon-formation, exists however for only a very short period of time. Since a polymer contains many polymer chains and each polymer chain is in a different macroconformation at any given moment, the observed characteristic chain dimension is thus a *temporal* and *spatial* average of all macroconformations. Such a dimension is the average end-to-end distance of a linear chain or the average radius of gyration.

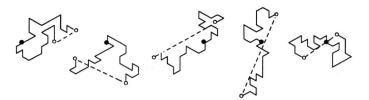

Fig. 5-10 Five of the $6^{28} = 6.14 \cdot 10^{21}$ possible two-dimensional macroconformations of a chain with the degree of polymerization $X = 30$ and six possible, energetically equal microconformations. The directions of single bonds were determined by rolling a dice, leaving out two successive congruent bonds. ● = central atom, O = chain end. The broken line indicates the end-to-end distance.

Isolated, *flexible* chain molecules thus adopt the shapes of **random coils**; these coils are *not* spherical entities (Fig. 5-10; see also Volume III). A random coil may be either unperturbed or perturbed. Chains of real coils have a finite thickness. Two chain segments crossing each other in space cannot occupy the same volume element. The **excluded volume** resulting from this effect causes the random coil to expand and take up a larger volume. The random coil is said to be **perturbed**.

Infinitely thin chains can of course cross each other without perturbing the chain. However, such infinitely thin chains do not exist. Nevertheless, chains may behave *as if* they are **unperturbed** which is the case if the segment-segment and segment-solvent interactions just compensate each other in such a way that the chain *appears* as infinitely thin. Such **phantom chains** are created by certain thermodynamically bad solvents at the so-called **theta temperature** (Volume III).

5.3.3   Melts

In melts, the thermal energy is large enough to allow rotations about chain bonds, which creates irregular sequences of microconformations that vary with time. Chain molecules in melts thus adopt the macroconformation of random coils, regardless of their macroconformation in the solid (crystalline or amorphous) states.

In such melts, one can estimate from the densities of melts and their frozen-in states (Volume III) that the random coils must interpenetrate each other. Coils in melts tend to expand in order to gain entropy but this expansion is prevented by the presence and volume requirements of the other coils. *Intra*molecular forces are the same as *inter*molecular ones since all coils have the same constitution and configuration. Expansion and compression of coils thus balance each other: polymer molecules in melts adopt their unperturbed dimensions.

The interpenetration of random coils is the stronger, the more flexible the polymer chains and the larger the molecular weight. At certain **molecular weights**, M_{ent}, chains become so much intertwined (**entangled**) that they offer additional resistances to flow and diffusion. For example, the molecular weight dependence of melt viscosity η changes from $\eta \sim M^1$ at $M < M_{ent}$ to $M \sim M^{3.4}$ at $M > M_{ent}$ (Volume III). This transition is assumed to play a role in free-radical polymerizations to high monomer conversions.

Macroconformations of *non-crystallizable* polymers are frozen-in if their melts are cooled below the **glass temperature**, which is the temperature at which the elastomeric

melt becomes a "solid", "glass-like" (i.e., **amorphous**) body. Similarly to melting temperatures (Fig. 1-1), glass temperatures increase with increasing molecular weights until
they finally become independent of the molecular weight. On heating, amorphous polymers transform from a glassy mass into a **rubber-like state** (at high molecular weights)
or a viscous melt (at lower molecular weights).

Amorphous states are also obtained by a fast cooling of melts of *crystallizable* macromolecules below their glass temperatures since the chain segments have not time to arrange themselves in a crystal lattice. Crystallization occurs if the melts of crystallizable
polymers are cooled slowly but even here the times are too short to disentangle the
chains and reorder the chain segments *completely* in crystal lattices. A 100 % crystallinity of polymers is practically never obtained: all "crystalline" polymers are, in reality,
only **semicrystalline** albeit to various degrees.

Chains with rigid segments that are separated by flexible segments may show a different thermal behavior. On heating, their flexible segments become mobile but the rigid
segments remain stiff and arrange themselves in ordered, crystal-like domains. The
polymer forms a **thermotropic liquid crystal** (Volume III). **Lyotropic liquid-crystalline
states** are correspondingly formed in solution.

Historical Notes to Chapter 5

G.H.Christie, J.Kenner, J.Chem.Soc. **121** (1922) 614
First cleavage of the racemate of an atropisomer (dissymmetry of molecule caused by hindrance of
free rotation), 2,8-dinitro-6,12-dicarboxylbiphenyl (see Fig. 4-1). No free rotation exists around
carbon-carbon bonds if the carbon atoms carry large substituents with strong interactions.

W.H.Haworth, Bull.Soc.Chim. **45** (1929) 1; s.a. M.Stacey, Chem.Soc.Rev. **2** (1973) 145
Proposed to call the distinguished spatial position of a substituent in molecules with hindred rotation around single bonds a "conformation".

A.Müller, Proc.Roy.Soc. A **114** (1927) 542; W.Hengstenberg, Z.Kristallogr. **67** (1928) 437
Long-chain paraffins crystallize in all-trans conformation (zig-zag chain). Confirmation for poly-
(ethylene) by C.W.Bunn, Trans.Faraday Soc. **35** (1939) 482.

W.Hengstenberg, Ann.Physik **84** (1927) 245
Poly(oxymethylene) does not crystallize in zig-zag chains.

E.Sauter, Z.physik.Chem. **B 18** (1932) 417
The chains of poly(oxymethylene)s form "helicogyres" in the crystalline state; the helicogyres are
shown two-dimensionally as "bathtubs". A later paper presented space-filling models of helicogyres
that clearly showed the helix structure (E.Sauter, Z.physik.Chem. **B 21** (1933) 161, 186 (Fig. 2).

H.Lohmann, in H.Staudinger, Die hochmolekularen organischen Verbindungen. Kautschuk und Cellulose, Springer, Berlin 1932, S. 287-332
Suggested a "meander structure" for poly(oxymethylene), based on X-ray data (from G = *maiandros*,
a strongly winding river in Phrygia, an ancient country in west-central Asia Minor. The river is now
called "Menderes" in Turkish). According to Sauter (l.c. 1933), the meander structure is not two-dimensionally plane but "strongly wound" with "screw axes" (i.e., a helix!).

bathtub representation of poly(oxymethylene)   meander form of poly(oxyethylene)

C.S.Hanes, New Phytologist **36** (1937) 101
   Suggested that amylose crystallizes in spirals; experimentally confirmed by R.E.Rundle, R.R.Baldwin, J.Am.Chem.Soc. **65** (1943) 554.

C.W.Bunn, Proc.Roy.Soc. **180** (1942) 67
   Postulated a helical conformation for the main chain of vinyl polymers with sterically equivalent monomeric units. Experimental confirmation for poly(1-olefin)s by G.Natta, Acc.Naz.Lincei.Mem. **4** (sez. 2) (1955) 61; -, J.Polym.Sci. **16** (1955) 143; -, Makromol.Chem. **16** (1955) 213; -, Acc.Naz.Lincei.Mem. **4** (sez. 2) (1955) 73; G.Natta, P.Corradini, J.Polym.Sci. **20** (1956) 251; G.Natta, P.Pino, P.Corradini, F.Danusso, E.Mantica, G.Mazzanti, G.Moraglio, J.Am.Chem.Soc. **77** (1955) 1708.

M.L.Huggins, Chem.Revs. **32** (1943) 195
   Suggested helical structures for poly(α-amino acid)s and proteins; experimental confirmation by L.Pauling, R.B.Corey, H.R.Branson, Proc.Natl.Acad.Sci. (USA) **37** (1951) 205.

J.D.Watson, F.H.C.Crick, Nature [London] **171** (1953) 737, 964
   Proposal of the double helix of deoxyribonucleic acids based on experimental results of various authors.

G.N.Ramachandran, G.Kartha, Nature **176** (1955) 593
   Experimental proof of the triple helix structure of collagen.

Literature to Chapter 5

5.1 FUNDAMENTALS (definitions)
IUPAC Commission on Nomenclature of Organic Chemistry, 1974 Recommendations, Section E, Fundamental Stereochemistry, Pure Appl.Chem. 45 (1976) 11
International Union of Pure and Applied Chemistry, Macromolecular Division, Commission on Macromolecular Nomenclature, Compendium of Macromolecular Nomenclature, Blackwell Scientific, Oxford 1991

5.2 MICROCONFORMATIONS
S.Mizushima, Structure of Molecules and Internal Rotation, Academic Press, New York 1956
E.L.Eliel, Stereochemistry of Carbon Compounds, McGraw-Hill, New York 1962
M.Hanack, Conformation Theory, Academic Press, New York 1965
W.Orville-Thomas, Ed., Internal Rotation in Molecules, Wiley, New York 1974
E.L.Eliel, N.L.Allinger, S.J.Angyal, G.A.Morrison, Conformational Analysis, Am.Chem.Soc., Washington, D.C., 1981

5.3 MACROCONFORMATIONS (see also Volume III)
M.V.Volkenstein, Configurational Statistics of Polymer Chains, Interscience, New York 1963
T.M.Birshtein, O.B.Ptitsyn, Conformation of Macromolecules, Interscience, New York 1966
P.J.Flory, Statistical Mechanics of Chain Molecules, Interscience, New York 1966; reprinted by Hanser, Munich 1989
A.J.Hopfinger, Conformational Properties of Macromolecules, Academic Press, New York 1973
W.L.Mattice, U.W.Suter, Conformational Theory of Large Molecules, Wiley, New York 1994
M.Rehahn, W.L.Mattice, U.W.Suter, Eds., Rotational Isomeric State Models in Macromolecular Systems, Adv.Polym.Sci. **131/132** (1997) 1

6 General Aspects of Polymerization

6.1 Basic Terms

6.1.1 Prerequisites

Polymer molecules are synthesized either from monomer molecules by **polymerizations** (Chapters 7-14 and 16) or from macromolecules of other constitution and/or configuration by **polytransformations** (Chapters 15 and 16). For polyreactions to happen, certain chemical, thermodynamic, and mechanistic conditions must be fulfilled.

Polymerizations are *chemically* possible only if monomer molecules are at least bifunctional (Section 6.1.4). The functionality depends not only on the chemical structure of the molecule but also on the reaction partner and the synthesis conditions.

Thermodynamics requires a negative or zero Gibbs polymerization energy. Polymerizations can thus only be conducted within a certain temperature interval (Chapter 7).

Mechanistically, two conditions must be fulfilled. Monomer molecules must be activated easily and the rate of the polymerization reaction of monomer molecules must be much larger than the sum of all rates that may block the reactive positions.

6.1.2 Historical Development

Depending on the emphasis put on one or the other aspect, polymerizations are subdivided into various groups. Most common are subdivisions according to the

- *origin* of monomers (natural vs. synthetic);
- *chemical structure* of monomers (vinyl polymerizations, diene polymerizations, ring opening polymerizations, etc.);
- *relative compositions* of monomer molecules and monomeric units;
- *constitution of resulting polymers* (linear, branching, crosslinking, isomerizing, etc., polymerizations, cyclopolymerizations, etc.);
- *stereocontrol* of polymerizations (stereospecific, enantioselective, etc.);
- *presence* or *absence of low-molecular weight reaction products* (chain polymerization and polyaddition versus polycondensation and polyelimination);
- *type of start of polymerization* (thermally, catalytically, enzymatically, photochemically, electrochemically, by initiators, etc.);
- *nature of active species* (anionic, cationic, free-radical, etc.);
- *mechanism* of polymerization (in equilibrium, by addition, by insertion, etc.; living, controlled, uncontrolled, etc.);
- *reaction medium* (in bulk, solution, emulsion, suspension; by precipitation; in mesophases, crystals, inclusion compounds, etc.);
- *state of matter* during polymerization (homogeneous, hetereogeneous, gaseous state, liquid state, plasma, solid state, etc);
- *employed strategy* (stepwise, divergent, convergent, etc.).

Historically, the fight was over the question whether the combination of low-molecular weight molecules to form larger entities is a chemical reaction (in modern technical parlance: a polymerization) or a physical process (today: association or aggregation). Even today, the physical coupling of so-called subunits (chemical molecules) to larger entities (quaternary structures, associates) is called a "polymerization" in biochemistry.

In the early years of the 20th century, chemical polymers were often confused with physical ones and vice versa (Section 1.3). Hermann Staudinger thus suggested in 1920 that only those substances whose atoms are held together by "primary valences" (= covalent bonds) should be called "polymerization products" but not those whose atoms are bonded by "secondary valences" (hydrogen bonds, dipole-dipole interactions, van der Waals bonds). It followed that "polymerization processes in a wider sense are all processes where two or more molecules unite to a product with the *same composition*" (my emphasis) "but higher molecular weight". He subdivided these processes into two classes:

– *Genuine polymerization processes* "where the polymerization product still has the same type of binding of atoms as the monomeric body". Examples included "the conversion of styrene into meta-styrene, of isoprene into rubber, of formaldehyde into paraformaldehyde" (all polymerizations in the modern sense) and also "the formation of hexaphenylethane from triphenylmethane, of paraldehyde from acetaldehyde, of cyclobutanedione derivatives from ketenes" (all *not* polymerizations in the modern sense).

– *Non-genuine* (G: "unecht" = untrue, false, fake) or *condensating polymerization processes* "which are accompanied by a more or less pronounced shift of atoms, where the polymerization products thus no longer exhibit the original type of binding of atoms as the monomeric compound". He included in this class "the aldol polymerization, the polymerization of formaldehyde ..., the formations of distyrene from styrene, diisobutylene from isobutylene, diacrylic ester from acrylic ester ..." (mostly not polymerizations in the modern sense). It may have been the choice of the word "unecht" that prevented Staudinger from studying condensation polymerizations since "unecht" in German also has the connotation of "false", "fake", "improper", and "spurious".

The two classes of polymerization processes were redefined in 1929 by Wallace Hume Carothers by dropping the requirement that a polymer and its monomer must have the same atoms in the same proportion in order to be considered a polymerization product. Instead, Carothers concentrated on the constitution of the *resulting* polymer molecules:

– "*Addition* or *A polymers*. The molecular formula of the monomer is identical with that of the structural unit". As examples he mentioned natural rubber, poly(styrene), poly(oxymethylene), and paraacetaldehyde (now called paraldehyde).

– "*Condensation* or *C polymers*. The molecular formula of the monomer differs from that of the structural unit". His examples were poly(ethylene glycol), cellulose, and silk fibroin. However, poly(ethylene glycol) is a C polymer only if it is synthesized from ethylene glycol by removal of water. It is an A polymer if it is prepared by the ring-opening polymerization of ethylene oxide with a trace of water:

$$(6\text{-}1) \quad n\ HO{-}CH_2{-}CH_2{-}OH \qquad\qquad\qquad n\ H_2C{-\!-}CH_2$$

$$\searrow\ -(n-1)\ H_2O \qquad +H_2O\ \nearrow \qquad\qquad \underset{O}{\diagdown\!\diagup}$$

$$HO(CH_2{-}CH_2{-}O)_n OH$$

C polymerization   A polymerization

Carothers further distinguished "condensation polymerizations" (C polymerizations) as processes leading to C-polymers from "addition polymerizations" (A polymerizations) as those resulting in A-polymers. These definitions however mix a process-related characteristic (release of a low-molecular weight product, such as water in Eq.(6-1)), with a constitution-related one (identity or non-identity of the composition of monomer molecules and monomeric units). Both characteristics are however not necessarily identical.

Eight years later, Otto Bayer et al. showed that the reaction of diisocyanates with diols proceeds like an A polymerization but without the release of low-molecular weight molecules. Similarly to C polymerizations, however, the molecular structure of the resulting polyurethanes differs from that of the monomers:

(6-2) n O=C=N–Z–N=C=O + n HO–Z'–OH

$\longrightarrow$ O=C=N–Z–NH–CO–(O–Z'–O–CO–NH–Z–NH–CO)$_{n-1}$O–Z'–OH

Bayer called this type of polymerization a "polyaddition". This term caused a monumental confusion in English speaking countries since "polyaddition" sounds similar to "addition polymerization". The reason for the choice of words and the resulting linguistic confusion is a historic difference in terminology.

In Germany, scientists followed Staudinger's lead. "Polymerization" there referred exclusively to what Carothers called "addition polymerization", a term that never caught on in Germany. The German Otto Bayer thus felt free to call his "diisocyanate addition process" a "polyaddition". In German, "polymerization" thus refers to only one of the four basic types of monomer-to-polymer conversion (see below) whereas it covers all four of them in English.

The fourth type of polymerization is similar to the A type with respect to the preservation of the composition of the monomer unit. Like the C type, low-molecular weight molecules are released during the polymerization. It has no established name (see below). This type of reaction is prevalent in certain types of biochemical polymerizations. An example in synthetic polymer chemistry is the polymerization of *N*-carboxy anhydrides of α-amino acids (Leuchs anhydrides) to poly(α-amino acid)s:

(6-3)

$$\begin{array}{c} R \\ | \\ H-C----N \\ | \quad\quad | \\ C \quad C \\ O \quad O \quad O \end{array} \quad \xrightarrow[-CO_2]{} \quad \begin{array}{c} R \\ | \\ N-C-C \\ | \quad | \quad || \\ H \quad H \quad O \end{array}$$

6.1.3   Classifications

Polymerization reactions are not only distinguished by the equality/inequality of composition of monomer molecules and monomeric units but also by the type and rate of growth processes. In the discussion that follows, one has to remember that the term "chain" is often used in polymer science without further specification as "polymer chain" (a molecule) or "kinetic chain" (a process), trusting that the meaning is revealed by the context. The polymer science use of the terms "chain(-growth) polymerization" and "step(-growth) polymerization" also differs from that of general chemical kinetics.

In general chemical kinetics, a **gross chemical reaction** describes the stoichiometry of a chemical reaction from starting to final molecules without presentation of intermediates. Such a reaction may proceed as unimolecular or bimolecular **elementary reaction** (termolecular reactions are very rare) or as a **composite reaction (complex reaction)** of more than one elementary reaction. The intermediates of composite reactions may be detected by analytical means if they are too short-lived or preparatively isolated if they are long-lived. The elementary reactions of a composite reaction may either proceed **simultaneously (in parallel, concurrently)** or in **succession (in steps)**.

It is clear from Eq.(6-1) that the polymerization of monomer molecules to polymer molecules cannot be an elementary reaction. The elementary reactions of a C polymerization *or* an A polymerization must proceed in succession, that is *in steps* (in succession) in the terminology of general chemical kinetics, where a chain reaction is presented as a typical example of a step reaction.

However, the term "step-wise" ("step-growth") became somehow exclusively associated with C polymerizations in polymer chemistry where it is contrasted with the "chain-growth" of A polymerizations (in the older literature, this "chain-growth" is correctly (in the sense of kinetics) identified as a special kind of step-reaction). "Step-growth" probably owes its name to the fact that C polymerizations are usually so slow, even in the melt, that intermediates (polymer molecules with lower molecular weight) can be isolated, e.g., by lowering the polymerization temperature, similarly to the "steps" in preparative low-molecular weight organic chemistry.

In principle, the faster A polymerizations can also be stopped if the temperature is lowered considerably, for example, fast anionic polymerizations by very fast cooling to the temperature of liquid nitrogen. The success of this procedure thus depends on the velocity of the experimental procedure and this, in turn, on the skill of the experimentalist. Physical definitions should never be based on skill, however.

"Chain-growth" and "step-growth" polymerizations differ, however, in another aspect although it is not reflected in the respective designations. In "chain-growth polymerizations", a polymer chain grows exclusively by a reaction between monomer molecules and a reactive site on the polymer chain that stays there during the entire growth of the polymer chain. In "step-growth polymerizations", the growth of polymer chains proceeds by reactions between molecules of all sorts of degrees of polymerization ($X = 1, 2,$ ... $N-1, N$). In contrast to "chain-growth polymerizations", the active molecule site in "step-growth polymerizations" has also to be activated in every "step".

A new IUPAC terminology tries to overcome the semantic and scientific problems associated with the older terms presented above. This series of books will use the following terminology: Monomer molecules carry the symbol M while polymer molecules with the symbols P_i, P_j, P_{j+1}, etc., are defined as molecules with *two or more* monomeric units (i, j $= 2, 3, ..., N-1, N$). The molecules M, P_i, P_j, etc., are collectively called **reactants** R. Low-molecular weight by-products are designated **leaving molecules** (symbol L).

According to IUPAC, one can distinguish two types of polymerization with respect to the formation or absence of leaving molecules and two types of polymerization with respect to the participating species (P + M versus R + R) (Table 6-1). This table also shows older and conventional English terms. Because of the still existing confusion regarding the words "addition" (addition polymerization, polyaddition) and "polymerization", especially in the older literature, the table also includes the corresponding German words.

Table 6-1 Types and names of polymerization reactions. M = monomer molecule, P = polymer molecule (degree of polymerization of 2 or higher), R = reactant molecule (monomer or polymer molecule), L = leaving molecule. "Chain" in IUPAC terminology refers to a *kinetic* chain.

	Classification (older classification)	
	Polymer-monomer reaction (Chain-growth polymerization)	Reactant-reactant reaction (Step-growth polymerization)

Without formation of leaving molecules (byproducts)

	$P_i + M \longrightarrow P_{i+1}$	$R_m + R_n \longrightarrow R_{m+n}$
English, IUPAC	chain polymerization	polyaddition
English, conventional	chain-growth polymerization, addition polymerization	no name
English, historical	A polymerization	no name
German, historical	Polymerisation, Additionspolymerisation	Polyaddition
This book	**chain polymerization**	**polyaddition**

With formation of leaving molecules (byproducts)

	$P_i + M \longrightarrow P_{i+1} + L$	$R_m + R_n \longrightarrow R_{m+n} + L$
English, IUPAC	condensative chain polymerization	polycondensation
English, conventional	no name	polycondensation, condensation polymerization, step-growth polymerization
English, historical	no name	C polymerization
German, historical	no name	Polykondensation
This book	**polyelimination**	**polycondensation**

Polymerization will be used as the general term for all reactions leading from monomers to polymers. The IUPAC names "**chain polymerization**", "**polyaddition**", and "**polycondensation**", as defined, present no linguistic problem. "Condensative chain polymerization" is not a very good choice, however. If "chain polymerization" is the generic term and if there are two subclasses, then *both* subclasses have to be distinguished by adjectives. An example would be "condensative chain polymerization" vs. "additive chain polymerization". Since such names are unwieldy and since the latter term creates new confusion, it is preferable to give "condensative chain polymerization" a new, shorter name. This book will thus use "**polyelimination**" for that type of polymerization.

The various types of polymerization differ characteristically in the increase of the number-average degree of polymerization of polymers with the **extent of reaction**, p, of functional groups (Fig. 6-1). It is useful to distinguish the extent of reaction from the **conversion of monomer molecules**, u, and the **yield of polymer**, y (see Section 13.2.3).

In conventional polycondensations of bifunctional monomer molecules AB or AA + BB, reactants combine at random, for example, 2 AB $\rightleftarrows$ AbaB, 2 AbaB $\rightleftarrows$ AbabaB, AbaB + AB $\rightleftarrows$ AbabaB, 2 AbabaB $\rightleftarrows$ AbababaB, AbabaB + AB $\rightleftarrows$ AbababaB, etc. The number-average degree of polymerization increases only slowly with increasing extent of reaction of functional groups A or B (Fig. 6-1). Higher number-average degrees of polymerization are only obtained at very high degrees of reaction, $p > 0.98$ (Chapter 13).

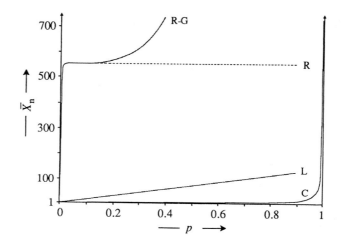

Fig. 6-1 Number-average degree of polymerization of polymers as a function of the extent of reaction, p, of monomers in various polymerizations.
C: bifunctional equilibrium polycondensation or polyaddition (independent of catalyst).
L: living chain polymerization. Example: anionic polymerization of styrene (S) by the mono-functional initiator butyl lithium (I) at an initial molar ratio $[S]_o/[I]_o = 141{:}1$.
R: free-radical chain polymerization. Example: polymerization of styrene in bulk by *N,N-azobis*isobutyronitrile (AIBN) as initiator; initial molar ratio 141:1. Each AIBN molecule generates two radicals. Only a part of these radicals initiates the polymerization, see Section 10.2.3.
R-G: free-radical chain polymerization with autoacceleration or gel effect (schematically).

Polycondensations and polyadditions are often *catalyzed*. The catalyst activates a functional group but is split off after the reaction of two functional groups. Catalysts thus do not become part of the polymer structure. However, they may be incorporated if they are used in high concentrations. For example, a sulfuric acid ester is formed as an endgroup if sulfuric acid is used as polyesterification catalyst.

Chain polymerizations and polyeliminations are *initiated*. Initiator molecules dissociate and the resulting polymerization-active fragments I* initiate the polymerization by adding one monomer molecule M each. The resulting **activated monomer molecules** I–M* add other monomer molecules to give I–M–M*, and so on. The further fate of the activated polymer chain depends on the initiator which may furnish initiating anions, initiating cations, or free radicals.

Anionic chain polymerizations very often remain "living" during the course of the experiment. An example is the anionic polymerization of styrene by butyl lithium:

(6-4) $C_4H_9^{\ominus} \,//\, Li^{\oplus} \;+\; n\; CH_2{=}CH(C_6H_5) \;\longrightarrow\; C_4H_9[CH_2{-}CH(C_6H_5)]_n^{\ominus} \,//\, Li^{\oplus}$

C_4H_9Li dissociates in suitable solvents into butyl anions and lithium cations. Butyl anions add styrene molecules, giving "**monomer anions**", $C_4H_9{-}CH_2{-}^{\ominus}CH(C_6H_5)$ and then macroanions, $C_4H_9{+}CH_2{-}CH(C_6H_5){+}_n^{\ominus}$, until all monomer is polymerized. Addition of new monomer results in further polymerization. All C_4H_9Li molecules are completely dissociated at the beginning of the experiment, all butyl anions immediately add monomer molecules, all monomer anions grow simultaneously with the same rate, etc. At any given time (or extent of reaction), the number of monomer molecules added per initiator anion is therefore the same for all growing chains: the number-average degree of polymerization increases linearly with the extent of reaction of double bonds (Fig. 6-1). Living polymerizations are not only obtained in some anionic polymerizations (Section 8.3) but also in several cationic (Section 8.4) and coordination polymerizations (Chapter 9), and even in some special radical polymerizations (Chapter 10).

The situation is quite different for conventional free-radical polymerizations. Here, an initiator such as dibenzoyl peroxide, $C_6H_5COO-OOCC_6H_5$ (BPO), dissociates into two radicals $C_6H_5COO^\bullet$ (and further into $C_6H_5^\bullet$). The radicals $I^\bullet$ add monomer molecules M and become **monomer radicals** $I-M^\bullet$ and then **polymer radicals** $I-M_2^\bullet$, $I-M_3^\bullet$, etc. All BPO molecules do not dissociate simultaneously, however, and polymer radicals are thus generated successively. However, the rate constants $k_p/(L \text{ mol}^{-1} \text{ s}^{-1})$ of the bimolecular propagation steps $I-M_N^\bullet + M \rightarrow I-M_{N+1}^\bullet$ are much larger ($10-10^5$) than the rate constants of polycondensations $M_i + M_j \rightarrow M_{i+j}$ ($10^{-5}-10^{-2}$).

The degree of polymerization thus becomes very large at very small extents of reaction (Fig. 6-1, R). The growing polymer radicals will however react with other radicals or with other molecules, which either terminates the polymer chain or the kinetic chain (see Chapter 10), all while new initiator radicals are being formed. Finally, a steady state between radical formation and removal is established and the number-average degree of polymerization becomes constant, unless of course additional reactions or physical phenomena are present.

For example, the viscosity of the reaction mixture may become so high during polymerization in the melt that two polymer radicals have difficulty finding each other. The average life-time of radicals will become larger than in the steady state which allows them to add more monomer molecules before being terminated. This effect leads to a strong increase in the degree of polymerization (Fig. 6-1, R-G).

6.1.4 Functionality

Monomer molecules must be at least bifunctional in order to polymerize. The **functionality** of a *monomer molecule* is defined as the number of covalent bonds that the molecule can form with other reactants (monomer or polymer molecules) at the existing reaction conditions. A monomer molecule ($X = 1$) with the functionality $f_{mon} = 1$ can form only dimer molecules ($X_R = 2$) with a functionality of $f_R = 0$. A bifunctional monomer leads to linear bifunctional polymer molecules with $f_R = 2$ in intermolecular reactions but to rings ($f_R = 0$) in intramolecular reactions.

The *maximum* functionality f_R of a reactant molecule with a degree of polymerization X_R obtained from like monomer molecules with the functionality f_{mon} is

(6-5) $\qquad f_R = 2(f_{mon} - 1) + (X_R - 2)(f_{mon} - 2) = 2 + X_R(f_{mon} - 2)$

A hectamer ($X_R = 100$) from a trifunctional monomer molecule ($f_{mon} = 3$) will thus have $f_R = 102$ functional endgroups. The reaction of two reacting groups of each hectamer molecule with one group each of two other hectamer molecules suffices to crosslink *all* hectamer molecules of a substance to one giant macromolecule. A *totally* **crosslinked** *substance* consists of *one* giant molecule. A *partially* crosslinked *substance* contains one or more giant molecules and soluble molecules of far lower molecular weight.

The *intramolecular* reaction of *two* side-groups of the same reactant molecule removes two functional groups and creates one ring (one cyclic molecule). One *intermolecular* ring forms if two groups each from two molecules react but here the reaction
of the first two groups of different molecules is intermolecular so that only the reaction
of the second pair counts as intramolecular. A molecule with N_{cycl} rings has thus 2 N_{cycl}
functional groups less than a corresponding linear or branched molecule. The functionality of linear, branched, hyperbranched, dendritic, or crosslinked molecules is thus

$$(6\text{-}6) \qquad f_R = 2 + X_R(f_{mon} - 2) - 2\,N_{cycl}$$

The functionality of a *molecule* may be zero or any positive *whole* number. The
functionality of a *substance*, on the other hand, may also be a positive *fractional* number
since the substance may contain molecules of different functionalities and the functionality of a monomer molecule may change as a result of the reaction of some of its functional groups.

Linear (unbranched) chains are formed if the functionality of the monomer *molecules* equals exactly two. Bifunctionality may be produced by a single chemical group
such as the isocyanate group in the polymerization of monoisocyanate molecules via the
N=C double bond with bases as initiators:

$$(6\text{-}7) \qquad R-N{=}C{=}O \longrightarrow \quad \text{\small w}N-C\text{\small w}$$
$$\qquad\qquad\qquad\qquad\qquad\qquad\quad |\quad ||$$
$$\qquad\qquad\qquad\qquad\qquad\qquad R\quad O$$

Bifunctionality may also arise from the reaction of two monofunctional groups per
monomer molecule, for example, in the reaction of diisocyanates with diols according to
Eq.(6-2). The resulting urethane groups may however react further with isocyanate
groups to form allophanate groups. This reaction generates branched polymers:

$$(6\text{-}8) \qquad \text{\small w}N{=}C \;+\; \text{\small w}NH-C-O\text{\small w} \longrightarrow \text{\small w}NH-C-N-C-O\text{\small w}$$

Hence, depending on reaction conditions, the isocyanate group may exhibit different
functionalities: bifunctional in chain polymerizations (Eq.(6-5)), monofunctional in
polyadditions (Eq.(6-2)), and half-functional in the allophanate reaction (Eq.(6-8)).
"Functionality" is thus not an absolute property of a molecule but a relative one that depends on the type of the reaction partner and the reaction conditions.

p-Divinylbenzene $CH_2{=}CH-(p\text{-}C_6H_4)-CH{=}CH_2$ (DVB) has a formal functionality of
4 since each of the two vinyl groups has a functionality of 2. Accordingly, branched
and, at larger monomer conversions, crosslinked polymers result from free-radical polymerizations. However, the polymerization of the first vinyl group of a DVB monomer
molecule will leave the second vinyl group with a somewhat different reactivity since the
resonance structure of the monomeric unit differs from that of the monomer molecule.
The second vinyl group does not polymerize in certain ionic polymerizations and the resulting polymer molecules are linear. Examples are the anionic polymerization of DVB
by lithium diisopropylamide in tetrahydrofuran and the cationic polymerization of DVB
by acetyl perchlorate in methylene chloride. Functionalities may also be lost by cyclopolymerizations (Section 6.1.5).

6.1.5   Cyclopolymerization

Cyclopolymerizations are "polymerizations in which the number of cyclic structures in the constitutional units of the resulting macromolecules is larger than in the monomer molecules" (IUPAC 1996). **Cyclopolymers** result from addition polymerizations with alternating intramolecular and intermolecular functional group reactions if the newly formed cyclic structures are relatively free of strain as in 5- and 6-membered carbon rings. Such structures can be obtained from monomer molecules with two carbon-carbon double bonds that are in the 1,5 or 1,6 position. Examples are

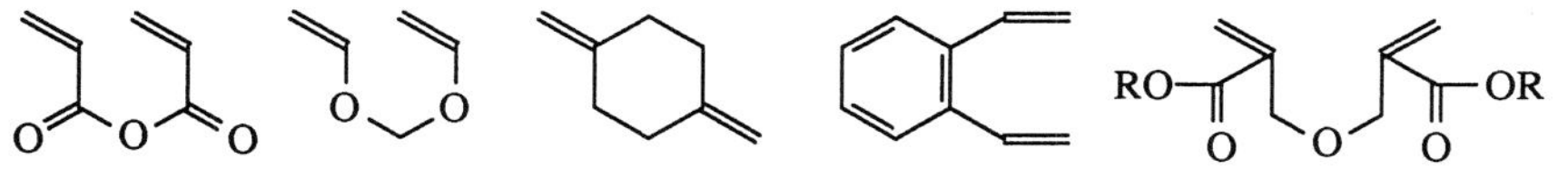

| acrylic anhydride | divinyl-formal | *p*-dimethylene cyclohexane | *o*-divinyl-benzene | dialkyl methacrylates bridged by an ether group (EAMA) |

The polymerization of acrylic anhydride may result in monomeric units with acrylic groups in the side chain (I) which may then react further to branch and even crosslink polymers. By back-biting, six-membered (II) or five-membered (III) rings are formed. The proportion of III increases with higher temperatures, lower concentrations, and increasing polarity of solvents. The polymerization of *p*-dimethylenecyclohexane (DCH) also delivers 5-membered rings. In the following formulas, the symbol * indicates an active center (e.g., a radical) and the symbol ⌇⌇ a polymer chain:

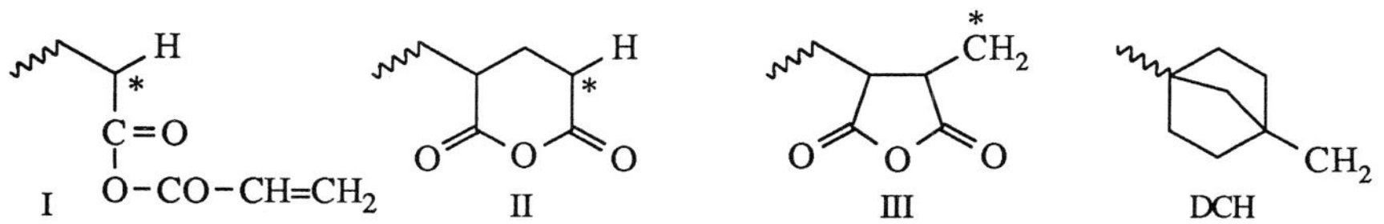

The proportion f_{ring} of intramolecular cyclizations depends on the relative rates of ring formation (R_{ring}) and linear propagation (R_{chain}); it is almost never 100 %:

(6-9) $f_{ring} = R_{ring}/(R_{ring} + R_{chain})$

Ring formation, $P_i^* \rightarrow$ cyclo-P_i, is a unimolecular reaction; its rate is proportional to the concentration $[P_i^*]$ of growing molecules in kinetically controlled processes. Chain formation, $P_i^* + M \rightarrow P_{i+1}^*$, on the other hand, is a bimolecular reaction:

(6-10) $R_{ring} = k_{ring}[P_i^*]$; $R_{chain} = k_{chain}[P_i^*][M]$

Introduction of Eqs.(6-10) into Eq.(6-9) results in

(6-11) $1/f_{ring} = 1 + (k_{chain}/k_{ring})[M]$

Rings are thus preferably formed at low monomer concentration (**Ruggli-Ziegler dilution principle**). The ratio k_{ring}/k_{chain} indicates the molar monomer concentration at which chains and rings are formed with equal probability.

Table 6-2 Observed ratios k_{ring}/k_{chain} of rate constants at temperatures T and differences of activation energies for the cyclopolymerization of various monomers.

Monomer	Solvent	Initiation	$\dfrac{T}{°C}$	$\dfrac{k_{ring}/k_{chain}}{mol\ L^{-1}}$	$\dfrac{E^{\ddagger}_{ring} - E^{\ddagger}_{chain}}{kJ\ mol^{-1}}$
Acrylic anhydride	cyclohexanone	radical	60	5.9	10.1
Methacrylic anhydride	DMF	radical	60	2.4	10.9
Methacrylic anhydride	cyclohexanone	radical	60	45.0	-
Divinyl formal	benzene	radical	60	130.0	10.9
o-Divinylbenzene	benzene	radical	50	2.7	-
o-Divinylbenzene	-	radical	50	2.1	-
o-Divinylbenzene	toluene	AlCl$_3$	0	4.9	-
o-Divinylbenzene	toluene	BF$_3$-O(C$_2$H$_5$)$_2$	0	0.7	-
o-Divinylbenzene	CCl$_4$	BF$_3$-O(C$_2$H$_5$)$_2$	0	1.4	-
EAMA, R = methyl (p. 161)	toluene	radical	80	16.7	11.6
EAMA, R = t-butyl (p. 161)	toluene	radical	80	87.0	20.5

The ratio k_{ring}/k_{chain} can be estimated from the probabilities of chain and ring formation. The probability of ring formation is obtained from the distribution of probabilities of the end-to-end distances of very short chains (Volume III). The probability of chain formation is calculated from the average distance of two double bonds in the polymerizing melt or solution and from their largest distance that may still result in a reaction. Assuming a random distribution of monomer molecules and no steric hindrance during the reaction, a ratio of k_{ring}/k_{chain} = 1.11 mol/L is obtained. According to Eq.(6-11), a monomer concentration of [M] = 0.01 mol/L should thus result in a ring formation of 99.1 %. At a monomer concentration of ca. 7.43 mol/L (corresponds to the polymerization of 1,6-dienes in bulk), a cyclization of only ca. 13 % is expected. However, far higher ratios k_{ring}/k_{chain} and thus far higher degrees of cyclization (up to 90-100 %) are observed experimentally for molar concentrations of 1-8 mol/L (Table 6-2).

These high degrees of cyclization are caused by much shorter distances between double bonds than assumed by simple probability calculations. This view is supported by the experimental finding of a strong bathochromic shift of the UV absorption of double bonds which indicates their interaction in the transition state.

The degree of cyclization is strongly affected by the thermodynamic quality of the solvent. Cyclopolymers are soluble even at high degrees of monomer conversion.

6.2 Mechanism and Kinetics

6.2.1 Activation of Monomers

Monomer molecules can unite to form polymer molecules by opening of multiple bonds, cleavage and reassembling of σ bonds, or saturation of coordinatively unsaturated groups. In general, all these processes need external activation. Very few monomers polymerize by self-catalysis of their functional groups (Section 13.4.2), thermal polymerization (Section 10.2.1), or charge transfer between different species (Section 12.5.3).

External activations are provided by *small* proportions of catalysts or initiators. The **catalyst** and the reactant molecule (monomer, polymer) form an intermediary compound that reacts faster with another reactant molecule than the plain reactants. The catalyst is split off after *each* reaction. In the classic case, the catalyst remains unchanged; it does not appear in the reaction scheme (see, for example, the polycondensation of ethylene glycol, Eq.(6-1)). At high catalyst concentrations, however, catalyst fragments may become part of the reaction product. An example is the sulfuric ester endgroup in the sulfuric acid catalyzed polyesterification of diols and dicarboxylic acids.

Catalysts prefer to react with heteroatoms since such atoms contain non-binding electron pairs or electron-pair gaps. Most polycondensations and polyadditions are therefore easily catalyzed since the reactants almost always contain heteroatoms.

Initiators are non-classical catalysts; they are not recovered after their reaction. They act either by addition to (type A) or by coordination with (type C) a monomer molecule.

Type A initiator molecules I-J either dissociate spontaneously into an initiating ion and a non-initiating counterion ($I^{\ominus} + J^{\ominus}$; $I^{\ominus} + J^{\oplus}$) or by the action of heat, radiation, or an activator into two initiating radicals $I^{\bullet}$ and $J^{\bullet}$. The initiating species I* (radicals $I^{\bullet}$, $J^{\bullet}$, anion $I^{\ominus}$, cation $I^{\oplus}$) forms a *covalent* bond with a monomer molecule. The resulting activated species I–M* adds more molecules to give I–M_2*, I–M_3*, etc., whereby the active center moves to the end of the growing chain to give macroradicals $I(M_{N-1})M^{\bullet}$ in **free-radical polymerizations** (Chapter 10), macroanions $I(M_{N-1})M^{\ominus}$ in **anionic polymerizations** (Section 8.3), and macrocations $I(M_{N-1})M^{\oplus}$ in **cationic polymerizations** (Section 8.4). In these **chain polymerizations** ("addition polymerizations"), growing chain ends remain active until they are terminated by reactions with other molecules.

Type C initiators form *coordination* compounds with monomers, either $I^{\delta+}M/J^{\delta-}$ or $I^{\delta-}M/J^{\delta+}$. The chains grow by *insertion* of monomer molecules into the $I^{\delta+}$–M or $I^{\delta-}$–M bond, respectively (Chapter 9). In some of these **coordination polymerizations** (emphasis on the initiation reaction) or **insertion polymerizations** (emphasis on the propagation reaction), the reactive chain end ($I^{\delta+}$, $I^{\delta-}$) is split off spontaneously by dissociation or by reaction with another molecule *after* the chain has been formed. Because the initiator fragment is no longer part of the chain, such polymerizations are also called *catalyzed*. Other than in truly catalyzed reactions, however, the "catalyst" is not split off after *each* reaction between the growing species and *one* monomer molecule.

The **polymerizability** of a monomer molecule, i.e., whether it can be activated and polymerized, depends on the polarization of the bond or group, the resonance stabilization, and the steric hindrance by substituents. Polarization may be caused by heteroatoms in multiple bonds (e.g., O=C<) or in partial double bonds (e.g., –NH–CO– ↔ –N=C(OH)–) or by electron-attracting or electron-withdrawing substituents.

Heterocycles are polymerized easily by initiator ions if (a) the polymerization is allowed thermodynamically (Chapter 7) and (b) the reaction between initiator ion I* and monomer molecule M results in a molecule I–M* (and later to I-M_2* ... I–M*, etc.) with an active center –M* similar to that of I*. Unsaturated monomers with heteroatoms in substituents often do not polymerize ionically since condition (b) is not fulfilled because initiator ions react preferably with the substituents.

Initiator ions also polymerize monomer molecules with polarized double bonds. *Formaldehyde* has a negative partial charge on the oxygen atom and a positive partial charge on the carbon atom. An initiator anion can therefore attack the carbon atom, which leads to a new resonance-stabilized anion. An initiator cation, on the other hand, attacks the oxygen atom and forms a new resonance-stabilized cation:

$$(6\text{-}12) \quad \overset{\delta^-}{O}{=}\overset{\delta^+}{CH_2} \xrightarrow{\ +\ R^\oplus\ } \left[R{-}\overset{\oplus}{O}{=}CH_2 \longleftrightarrow R{-}O{-}\overset{\oplus}{C}H_2 \right] \xrightarrow{\ +\ HCHO\ } R{-}O{-}CH_2{-}O{-}\overset{\oplus}{C}H_2$$

Substituents in monomers with carbon-carbon double bonds act similarly:

$$\overset{\beta}{\underset{\delta^-}{C}H_2}{=}\overset{\alpha}{\underset{\delta^+}{C}H} \qquad \overset{\beta}{\underset{\delta^+}{C}H_2}{=}\overset{\alpha}{\underset{\delta^-}{C}H}$$

$$\qquad\quad \text{D} \qquad\qquad\qquad \text{A}$$

Propene $CH_2{=}CH(CH_3)$, for example, carries an electron-donating methyl group (D = CH_3) in the α-position. An initiating anion $I^\ominus$ can neither attack the partially negative β-carbon atom nor the sterically hindered α-carbon. Propene can thus only be polymerized by initiator cations or initiator radicals. Such polymerizations are prone to strong side reactions; they result in branched, low-molecular weight polymers (Volume II). High molecular weight poly(propylene) is obtained by coordination polymerization.

Acrylic esters have an electron-withdrawing (electron-accepting) substituent (A = COOR). They polymerize only anionically but not cationically.

Alkyl vinyl ethers have an electron-donating ether group (D = OCH_3); they are strongly resonance stabilized according to $[CH_2{=}CH{-}O{-}R \leftrightarrow {}^\ominus CH_2{-}CH{=}O^\oplus{-}R]$. Their bulk polymerization by initiator cations from, e.g., $BF_3\cdot O(C_2H_5)_2$ as initiator is fast and violent (**flash polymerization**). Radical polymerizations, on the other hand, are very slow because the growing macroradicals are not resonance stabilized.

Table 6-3 compares the action of some types of initiation.

Monomers for initiated polymerizations may have one type or more than one type of polymerizable groups. Such groups may be double bonds (carbon-carbon, carbon-oxygen, etc.) or heteroatoms containing groups in rings (cyclic amides, oxides, etc.). An exclusive reaction of a monomer group is possible only if all three factors work in unison.

Table 6-3 Initiation of some monomers by various initiators. A = electron-accepting group, D = electron-donating group. ( ) low molecular weight.

Monomer	Substituent	Initiation			
		Free-radical	Cationic	Anionic	Coordination
Ethene	-	+	(+)	-	+
Propene	D	-	(+)	-	+
Isobutene	D	-	+	-	-
Vinyl ether	D	(+)	+	-	+
Vinyl ester	D	+	-	-	-
Acetaldehyde	D	-	+	+	-
Vinyl chloride	A	+	-	+	+
Acrylic ester	A	+	-	+	+
Styrene	D,A	+	+	+	+
Formaldehyde	-	-	+	+	-
Tetrahydrofuran	-	-	+	-	-

So-called **functional monomers** have two or more *different* functional groups and can thus polymerize by several possible types of mechanisms. An example is the polymerization of *2-vinyl oxazoline* (I) by cations, anions, or free radicals:

(6-13)

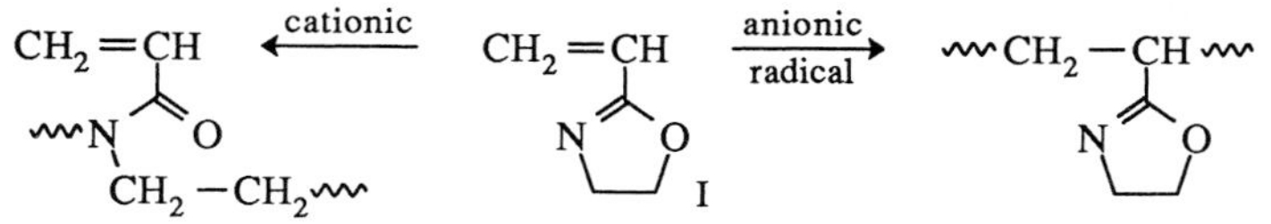

Diketene (III), the dimer of ketene (II), offers even more possibilities. γ-Rays initiate the addition polymerization of the vinyl group (IV) as well as the **ring-opening polymerization** of the β-lactone to a polyester, resulting in a **pseudo-copolymer** with units IV and V. A polyketone VI is obtained by the polymerization of III with $HgCl_2$ or $Al(OC_3H_7)_3$ as initiator. Lewis acids or $ZnCl_2$, on the other hand, result in polyketones (VI) or their polyenols (VII):

(6-14)

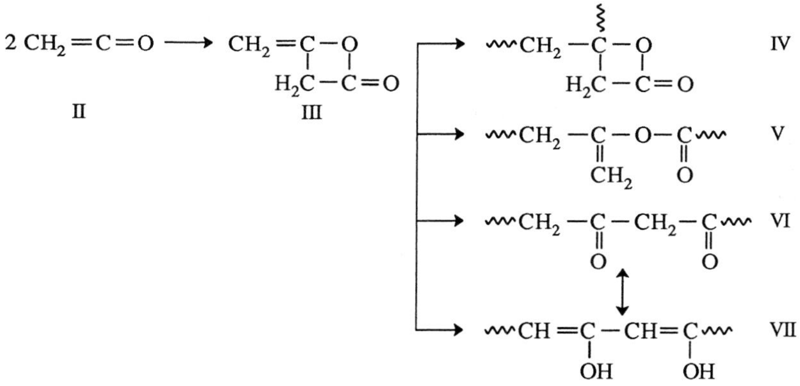

Some monomers do not polymerize *per se* but only after an isomerization by the catalyst, especially by catalysts that contain a transition metal. $Al(C_2H_5)_3/TiCl_3/Ni(acetyl-acetone)$ isomerizes *3-heptene* (VIII) first to 2-heptene (IX) and further to 1-heptene (X) before the latter polymerizes to poly(1-heptene) (XI):

(6-15)

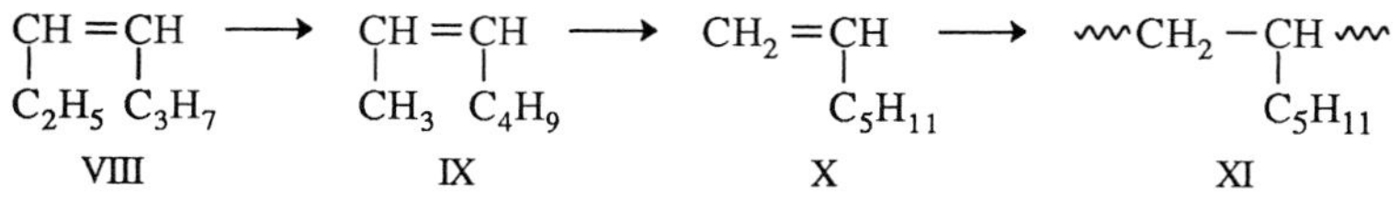

Isomerization and disproportionation of monomers have also been observed for other monomer–catalyst systems, for example:
– *Propene* $CH_2=CH(CH_3)$ disproportionates in presence of carrier-supported MoO_3 to ethene $CH_2=CH_2$ and 2-butene $CH_3–CH=CH–CH_3$; only ethene polymerizes;
– *1,4-dihydronaphthalene* isomerizes in the presence of sodium naphthalene to 1,2-dihydronaphthalene which subsequently polymerizes anionically;
– *vinyl benzoate* (XII) isomerizes under the action of light to the enol (XIII) which is in equilibrium with its ketoaldehyde (XIV). The latter disproportionates into methylphenyl ketone (XV) and carbon monoxide (XVI) (Eq.(6-16)). XIII-XVI do not polymerize.

(6-16) $\begin{array}{c} CH_2 \\ \| \\ C-H \\ | \\ O-C-C_6H_5 \\ \| \\ O \quad XII \end{array}$ $\longrightarrow$ $\begin{array}{c} CH-CO-C_6H_5 \\ \| \\ C-H \\ | \\ OH \\ \\ XIII \end{array}$ $\rightleftharpoons$ $\begin{array}{c} CH_2-CO-C_6H_5 \\ | \\ C-H \\ \| \\ O \\ \\ XIV \end{array}$ $\longrightarrow$ $\begin{array}{c} CH_3-CO-C_6H_5 \\ + \\ CO \\ \\ XVI \end{array}$ XV

Some polymers isomerize even after their synthesis. It is therefore advisable to conduct unknown or little known polymerizations in the dark.

6.2.2   Elementary Reactions

Biosyntheses of nucleic acids and enzymes are controlled by matrices. Such **matrix-controlled polymerizations** deliver uniform macromolecules whose chemical structures are either exact copies of the matrix (nucleic acids) or follow the instructions conveyed by the matrix (enzyme molecules).

Biosyntheses of polysaccharides and all nonbiological syntheses are **time-controlled** polymerizations. The growth of an individual chain (the propagation reaction) is halted by either thermodynamics (**equilibrium polymerizations**) or kinetics and in the latter case by either complete consumption of monomer molecules (**living polymerizations**) or by intrinsic or extrinsic deactivations.

"Chemical experience" is often not a good guide to discern the nature of the propagation step and the resulting polymer constitution. **Regioselectivities** are not always 100 % (p. 45) since they decrease with decreasing atomic radii of the substituents.

The polymer constitution may be sometimes even completely different from the one predicted by chemical experience. Acrylamide $CH_2=CH(CONH_2)$ does polymerize free-radically to poly(acrylamide) $+CH_2-CH(CONH_2)+_n$ as may be expected. However, its polymerization by strong bases leads to poly(β-alanine) $+CH_2-CH_2-CO-NH+_n$. The polymerization of 1-olefins $CH_2=CH(CH_2)_nCH_3$ by nickel(0)cyclooctadiene/bis(trimethylsilyl)amino*bis*(trimethylsilylimino)phosphorane does not proceed via the 1,2-carbons to poly(1-olefin)s $+CH_2-CH\{(CH_2)_nCH_3\}+_n$ but by hydride shift via the 2,ω-position to $+CH(CH_3)-(CH_2)_{n+1}+_n$, a **phantom polymerization**.

A polymerization leads only to high-molecular weight compounds if deactivation reactions are absent or small. The rate of propagation must always be larger than the sum of rates of all reactions that may terminate individual polymer chains. The possible deactivation mechanisms can be subdivided into termination, transfer, isomerization, and complexation reactions.

Termination is the deactivation of an active polymerization center (radical, anion, cation, etc.) without formation of a new active center. Terminations may be internal (e.g., by reorganization of the active center in certain ionic polymerizations) or external by reaction with other molecules in the polymerization system (e.g., radicals in free-radical polymerizations), impurities, additives, etc. Termination reactions deprive macromolecules of the chance to grow further; they lower the degree of polymerization.

Termination reactions also decrease the rate of polymerization since they reduce the concentration of active centers. For example, the concentration of free radicals is only ca. 10^{-8} mol/L to 10^{-9} mol/L in the **steady state** where the production and elimination of radicals just balance each other (Chapter 10).

The small concentration of active centers is one of the reasons why free-radical polymerizations are usually slower than ionic polymerizations. Because of their equal charges, two growing macroions cannot undergo a mutual deactivation. Their unimolecular termination (see Chapter 8) requires, however, a much higher activation energy. The concentration of growing macroions of ca. 10^{-2} mol/L to 10^{-3} mol/L is thus much larger than that of growing macroradicals at the usual experimental conditions.

Transfer reactions also deactivate growing chains, for example:

(6-17) $\sim\sim CH_2$—$^\bullet CHCl + RCl \longrightarrow \sim CH_2$—$CHCl_2 + R^\bullet$

Low-molecular weight transfer agents RCl (monomers, solvents, initiators, etc.) reduce the degree of polymerization. They do not markedly decrease the polymerization rate, however, if the new species $R^\bullet$ and the old species $\sim\sim CH_2$–$^\bullet CHCl$ have about the same reactivity. Transfer reactions can also be unimolecular, for example, β-eliminations such as

(6-18) $\sim\sim CHCl$—$CHCl$—$^\bullet CHCl \longrightarrow \sim\sim CHCl$—$CH{=}CHCl + {}^\bullet Cl$

Spontaneous isomerizations of growing macroions with their low-molecular weight counterions to inactive compounds also deactivate chains (Chapter 8).

Complex formation between monomer and initiator can prevent polymerizations below certain *kinetic* floor temperatures (Chapter 7).

6.2.3   Distinction Between Mechanisms

It is not always easy to determine the mechanism of a polymerization. For example, Ziegler catalysts composed of a transition-metal compound of groups 4-10 (e.g., $TiCl_4$) and an organometallic compound of groups 1, 2, or 13 (e.g., AlR_3) usually lead to coordination polymerizations (Chapter 9). However, the system phenyltitaniumtriisopropoxide + aluminumtriisopropoxide initiates the free-radical polymerization of styrene.

The system borontrifluoride + cocatalyst usually initiates a cationic polymerization (Section 8.4). With diazomethane as monomer, however, the same system leads to a free-radical polymerization via intermediately generated boron alkyls as initiators.

Also, the action of an initiator depends not only on the monomer but also on the medium. Iodine $I^\oplus[I_3]^\ominus$ initiates the cationic polymerization of vinyl ethers. However, iodine complexes $[DI]^\oplus[I_3]^\ominus$ with D = benzene, 1,4-dioxane, or some monomers cause the anionic polymerization of 1-oxa-4,5-dithiacycloheptane.

Several criteria must thus be employed to determine the action of initiators and the mechanism of polymerization. Such criteria usually include variation of temperature, solvent, additives, and monomers.

The *temperature dependence* of polymerization rates is often an unreliable criterion. Polyinsertions and ionic polymerizations are fast at low temperatures, as are some free-radical polymerizations.

The variation of the *solvent* lets one draw the following conclusions. A free-radical polymerization practically does not depend on the relative permittivity ("dielectric constant") of the solvent. "Ions" are more likely to initiate true ionic polymerizations if the solvent is very polar; decreasing polarity favors polyinsertions, however.

A polymerization in the presence of oxygen group-containing solvents is unlikely to be cationic because of the formation of oxonium salts; the cationic polymerization of oxygen-containing monomers is possible, however. Thus, alkyl vinyl ethers can be polymerized cationically in divinyl ethers but not olefins,

Anionic polymerizations of monomers M in the presence of alkyl halogenides RCl are practically impossible since the counterions $Mt^{\oplus}$ react with RCl:

$$(6\text{-}19) \qquad \sim M....M^{\ominus}\ Mt^{\oplus} + RCl \longrightarrow \sim M....M—R + MtCl$$

The mechanism of a polymerization can sometimes also be evaluated from the action of an *additive*. Added 1,1-diphenyl-2-picrylhydrazide catches radicals; it stops free-radical polymerizations (Section 10.4). Benzoquinone is also an inhibitor for free-radical polymerizations. However, it reacts also with cations and is therefore no help in distinguishing free-radical and cationic polymerizations.

Ionic polymerizations can be discerned by tritium (T) or ^{14}C labeled methanol:

$$(6\text{-}20) \qquad \sim M^{\ominus} + CH_3OT \longrightarrow \sim MT + CH_3O^{\ominus}$$

$$(6\text{-}21) \qquad \sim M^{\oplus} + {}^{14}CH_3OH \longrightarrow \sim M—O—{}^{14}CH_3 + H^{\oplus}$$

A ^{14}C radioactive polymer must have been produced by a cationic polymerization. The absence of tritium in the dead polymer, when working with tritiated methanol, CH_3OT, is proof of the absence of an anionic polymerization. An inactive polymer from an ionic polymerization is, however, neither proof of the absence of a cationic polymerization nor of the presence of an anionic one since the alkoxide ion may also remove a hydrogen atom from the growing macrocation:

$$(6\text{-}22) \qquad \sim CH_2—{}^{\oplus}CHR + CH_3O^{\ominus} \longrightarrow \sim CH{=}CHR + CH_3OH$$

Some monomers polymerize exclusively by a certain mechanism, for example, poly(isobut(yl)ene) polymerizes only cationically but not anionically or free-radically. An initiator for isobutene is therefore most likely also a cationic initiator for other monomers. Acrylates and methacrylates, on the other hand, do not polymerize cationically but anionically and free-radically. Cyclic sulfides and oxides cannot be polymerized by free radicals.

The initiator action can also be evaluated by copolymerization (Table 6-4). Alternatively, multifunctional monomers may be used. For example, 2-vinyl oxyethylene methacrylate is polymerized by cations (vinyl), anions (acrylate), or radicals (both groups).

Table 6-4 Monomer pairs for the evaluation of the initiator action.

Monomers		Polymers resulting from the initiation by		
I	II	Cations	Radicals	Anions
Styrene	methyl methacrylate	poly(styrene)	poly(S-*stat*-MMA)	poly(methyl methacrylate)
Isobutene	vinyl chloride	poly(isobutene)	poly(IB-*alt*-VC)	-
Isobutene	vinylidene chloride	poly(isobutene)	poly(IB-*alt*-VDC)	poly(vinylidene chloride)

6.2.4 Analysis of Polymerization Mechanisms

Polymerization mechanism can be analyzed by two different approaches:

1. The *kinetic method* sets up a reaction scheme with the probable elementary reactions. The corresponding differential equations are integrated. The method is flexible and delivers directly in many cases the absolute rate constants of elementary reactions. It has the disadvantage that each type of polymerization requires a new set of assumptions and equations.

2. The *statistical method* looks at the probability of a reaction step relative to all competing steps. This approach delivers generalized equations that can be applied to many more different chemical reactions than the kinetic method. It is therefore expedient to discuss the results of the statistical method in Section 6.3 of this chapter but those of the kinetic approach in Chapters 8-13.

The rate of polymerization is not only determined by that of the propagation reaction but also by the rates of initiation, termination, transfer, etc., reactions and furthermore by equilibration reactions between initiator molecules and the actual initiating molecules. The observed overall rate of polymerization, R_{gross}, is thus often not identical to the rate of propagation, R_p. It is rather a function of R_p and the rate of formation of the active species which in turn depends on the concentration [I] of the initiator. In general, one can write for an initiated polymerization

$$(6\text{-}23) \qquad R_{gross} = k[I]^m[M]^n$$

where [I] is the *instantaneous* amount-of-substance concentration of the initiator and [M] that of the monomer. The overall rate constant k contains the rate constants of several elementary reactions. Exponents m and n may adopt different values.

Propagation reactions are in most cases bimolecular reactions of a growing species P* and a monomer M. The rate R_p of the propagation reaction is given by

$$(6\text{-}24) \qquad R_p = [M] \, \Sigma_r \, k_{p,r}[P^*]_r$$

if back reactions can be neglected. In order to obtain the rate constant $k_{p,r}$, the relationship between Eqs.(6-23) and (6-24) as well as the number r and the concentrations [P*] of the various types of propagating species must be known.

In free-radical polymerizations, only one type of active species is present. The concentration $[P^\bullet] = [P^*]$ of the growing macroradicals is, however, too low to be determined directly. It is often eliminated mathematically (Chapter 10).

In ionic polymerizations, several types of propagating species often exist: free ions, ion pairs, ion associates, etc. (Chapter 8). The concentration of the various species can often be obtained from electrical conductivities, spectroscopic investigations, or stoppage of the growth of macrocations by bases and of macroanions by acids.

Kinetic treatments are simplified greatly by assuming that the reactivity of the growing species (macroradical, macrocation, macroanion, etc.) is independent of the size of the reacting molecule (**principle of equal chemical reactivity**). Based on early experiments with low-molecular weight compounds (p. 455), it is usually assumed that this is true if the degree of polymerization exceeds values of ca. 2 or 3.

Early opposition to the principle of equal chemical reactivity pointed out that rate constants should decrease since molecules become less mobile with increasing molecular weight of the reacting species. However, the mobility of the *molecule* is not decisive but that of the reacting group carrying *segment*. Examples are crosslinking reactions where endgroups become more and more immobilized with increasing network density. It is true that the number of collisions (and thus the probability of reactions) decreases with the increase of the viscosity of the system that is caused by the increasing concentration and molecular weight of polymers. However, higher viscosities also increase contact times which counterbalances the decrease of the number of collisions. The steric factor is also not affected by the size of the molecule since the molecule cannot distinguish whether its reacting segment is intramolecularly or intermolecularly screened.

The principle of equal chemical reactivity does not apply if the functional groups are only formally separated from each other. For example, the two vinyl groups of divinyl-benzenes, $CH_2=CH-C_6H_4-CH=CH_2$, are part of a resonance system. The reaction of one vinyl group decreases the resonance stability of the molecule and thus the reactivity of the second vinyl group. The reactivity decreases also if conjugated bonds are formed; an example is the polymerization of acetylene $CH\equiv CH$ to poly(vinylene) $\{CH=CH\}_n$.

6.2.5 Activation Parameters

Two theories are usually used to describe the temperature dependence of rate constants of elementary reactions. According to **collision theory**, rate constants are controlled by the number of successful collisions of molecules with sufficiently high energies. The temperature dependence of the rate constant k can be expressed by

$$(6\text{-}25) \qquad k = R_{coll}f_{ster} \exp(-\Delta E^{\ddagger}/RT) = A^{\ddagger} \exp(-\Delta E^{\ddagger}/RT)$$

R_{coll} is the collision rate at unit concentration (gases: $R_{coll} \approx 10^{10}$ molecules per second). The steric factor f_{ster} indicates the fraction of successful collisions (ca. 0.01–1); it measures the probability of the reaction. Neither R_{coll} nor f_{ster} are easy to determine individually; their experimentally accessible product $A^{\ddagger} = R_{coll}f_{ster}$ is called the **pre-exponential factor**. The **Boltzmann factor** $\exp(-\Delta E^{\ddagger}/RT)$ is a measure of the number of molecules that possess a sufficiently high energy for the reaction. It contains the **apparent** or **Arrhenius activation energy**, $\Delta E^{\ddagger}$.

The **transition state theory** assumes that the transition state can be described by an equilibrium constant $K^{\ddagger}$ and a universal frequency factor $k_B T/h$ where k_B = Boltzmann constant, h = Planck constant, R = gas constant, and $\Delta G^{\ddagger}$ = **Gibbs activation energy**:

$$(6\text{-}26) \qquad k = (k_B T/h)K^{\ddagger} = (k_B T/h) \exp(-\Delta G^{\ddagger}/RT)$$

Introduction of the 2nd Law of Thermodynamics, $\Delta G = \Delta H - T\Delta S$, into Eq.(6-26) leads to the **Eyring equation:**

$$(6\text{-}27) \qquad k = (k_B T/h) \exp(\Delta S^{\ddagger}/R) \exp(-\Delta H^{\ddagger}/RT)$$

Comparison of Eqs.(6-25) and (6-27) indicates the relationship between activation enthalpy and Arrhenius activation energy on one hand and activation entropy and pre-exponential factor on the other:

$$(6\text{-}28) \qquad \Delta H^{\ddagger} = \Delta E^{\ddagger} - RT \qquad ; \qquad \Delta S^{\ddagger} = R\,[\ln A^{\ddagger} - \ln(k_B T/h) - 1]$$

6.3 Statistics of Polymerizations

Statistics subdivides polymerization reactions into single and multiple mechanisms. **Single mechanisms** are characterized by a single propagation state. Examples are free-radical homopolymerizations and some copolymerizations.

Multiple mechanisms include two subclasses. In *concurrent states*, two or more propagation states are present simultaneously. An example is the presence of two different catalytic centers in heterogeneous Ziegler-Natta polymerizations (Chapter 9). In *consecutive states*, two or more propagation states are continuously interconverted. An example is the equilibria between free macroions and various ion pairs and ion associates in ionic polymerizations (Chapter 8).

6.3.1 Basic Terms

Statistics of polymerization looks at the probability of interconnecting units. A "unit" may be a monomeric unit, a diad of two monomeric units, etc. Examples are the two constitutionally different monomeric units of acrylonitrile and styrene in their bipolymerization (Chapter 12) or the monomeric units of D- and L-propylene oxide in their stereoselective polymerization (see below). In stereostatistics, the smallest statistical units may be configurational diads, for example, the isotactic and syndiotactic diads composed of two propylene units.

The smallest statistical units of a polymer chain will be designated "a" and "b". The mole fraction x_a may thus be the amount-of-substance fraction of "a" units that were generated from monomer molecules A in copolymerizations of A and B. It may also be the mole fraction of isotactic diads "i" in, e.g., vinyl polymers, so that $x_a = x_i$.

The following discussion is restricted to the most important case of binary mechanisms, i.e., polymerizations in which the controlling unit can react in only two ways. Such mechanisms are subdivided according to the range of action, i.e., the number of controlling units at the growth center, into **Markov trials** of zeroth, first, ... *n*th order. In chain polymerizations, growth centers may be either active chain ends (radicals, etc.) in addition polymerizations or active complexes in insertion polymerizations.

Markov trials are of zeroth order if the last unit (i.e., zero unit) does not control the addition of a new unit. These truly **random trials** or **Bernoulli trials** are a special case of **statistical trials** which comprise, in addition to Bernoulli trials, all Markov trials of higher order and all non-Markov trials. The addition step is controlled by one unit in first-order Markov trials and by two units in second-order Markov trials. Higher order Markov trials are not known for polymerizations.

The influence of the controlling units is expressed as a **conditional probability** of the connecting step. In **Markov first-order trials**, one has to consider four conditional probabilities, i.e., the conditional probabilities of adding A or B to a center "a" and those of adding A or B to a center "b": $p_{a/A}$, $p_{a/B}$, $p_{b/A}$, and $p_{b/B}$. The sum of conditional probabilities at each center must be unity, i.e.,

(6-29) $\qquad p_{a/A} + p_{a/B} \equiv 1 \quad ; \quad p_{b/B} + p_{b/A} \equiv 1 \qquad$ **(Markov first-order trials)**

(6-30) $\qquad p_{a/A} + p_{a/B} = p_{b/B} + p_{b/A}$

In **second-order Markov trials**, eight different conditional probabilities have to be considered since there are four different controlling units of two monomeric units each: $p_{aa/A}$, $p_{aa/B}$, $p_{ba/A}$, $p_{ba/B}$, $p_{ab/A}$, $p_{ab/B}$, $p_{bb/A}$, and $p_{bb/B}$. Again, the sum of probabilities for each controlling unit must be unity: $p_{aa/A} + p_{aa/B} \equiv 1$, $p_{ba/A} + p_{ba/B} \equiv 1$, $p_{ab/A} + p_{ab/B} \equiv 1$, and $p_{bb/A} + p_{bb/B} \equiv 1$.

In **Bernoulli trials (zeroth order Markov trials)**, on the other hand, only two conditional probabilities exist (p_A and p_B) since the nature of the last controlling unit (a or b) is irrelevant: $p_A = p_{a/A} = p_{b/A}$ and $p_B = p_{b/B} = p_{a/B}$; thus $p_A + p_B \equiv 1$. At a controlling center, the conditional probability of the homopropagation equals that of the cross-propagation whereas in Markov trials it does not (Table 6-5).

Depending on the conditional probability of the homopropagations, both Bernoulli and Markov trials can be subdivided into symmetric and asymmetric mechanisms. Symmetric Bernoulli mechanisms are known in copolymerizations of constitutionally different monomers as "ideal azeotropic copolymerizations", in copolymerizations of configuratively different but constitutionally identical monomerss occasionally as "random-flight polymerizations", and in the stereocontrol of polymerizations of non-chiral monomers as "ideal-atactic polymerizations."

Conditional probabilities are not identical with the probabilities of finding a certain unit in a polymer, the latter being the mole fractions x_j of monads, x_{jk} of diads, x_{jkl} of triads, etc. The mole fractions of diads "a" and "b", triads aa, ab, ba, and bb, etc., are

(6-31) $x_a + x_b \equiv 1$ monads

(6-32) $x_{aa} + x_{ab} + x_{ba} + x_{bb} \equiv x_{aa} + x_{|ab|} + x_{bb} \equiv 1$ diads

(6-33) $x_{aaa} + x_{|aab|} + x_{aba} + x_{|abb|} + x_{bab} + x_{bbb} \equiv 1$ triads

Experiments often do not distinguish between diads ab and ba. The sum of their mole fractions x_{ab} and x_{ba} is thus symbolized by $x_{|ab|}$. By analogy, one can write $x_{|aab|} \equiv x_{aab} + x_{baa}$ and $x_{|abb|} = x_{abb} + x_{bba}$ for triads.

Table 6-5 Conditional probabilities p of Bernoulli and Markov trials and resulting mole fractions x for polymerizations to infinitely long chains.

| | Bernoulli | | Markov first order | |
	symmetric	asymmetric	symmetric	asymmetric
Assumptions	$p_{a/A} \equiv p_{b/B}$	$p_{a/A} \neq p_{b/B}$	$p_{a/A} \equiv p_{b/B}$	$p_{a/A} \neq p_{b/B}$
	$p_{a/B} \equiv p_{b/B}$	$p_{a/B} \equiv p_{b/B}$	$p_{a/B} \neq p_{b/B}$	$p_{a/B} \neq p_{b/B}$
	$p_{b/A} \equiv p_{a/A}$	$p_{b/A} \equiv p_{a/A}$	$p_{b/A} \neq p_{a/A}$	$p_{b/A} \neq p_{a/A}$
Results	$p_{a/A} = p_A$	$p_{a/A} = p_A$		
	$p_{a/B} = p_{b/A} = 1/2$	$p_{a/B} \neq p_{b/A}$	$p_{a/B} = p_{b/A}$	$p_{a/B} \neq p_{b/A}$
Mole fractions	$x_a = 1/2$	$x_a = p_a \neq p_b$	$x_a = 1/2 = x_b$	$x_a = p_{b/A}/(p_{b/A} + p_{a/B})$
	$x_{aa} = 1/4$	$x_{aa} = x_a^2$	$x_{aa} = (1/2)\,p_{a/A}$	$x_{aaq} = x_a p_{a/A}$
	$x_{|ab|} = 1/2$	$x_{|ab|} = 2\,x_a(1 - x_a)$	$x_{|ab|} = 1 - p_{a/A}$	$x_{|ab|} = x_a p_{a/B} + x_B p_{b/A}$

6.3.2 Bernoulli and Markov Trials

The following derivations relate to infinitely long chains. Effects of initiating reactions and endgroups are thus not considered. For infinitely long chains, the probability of finding an "ab" link must equal the probability of finding a "ba" link. These probabilities are given by the products of the mole fractions and conditional probabilities:

$$(6\text{-}34) \qquad x_a p_{a/B} = x_b p_{b/A}$$

In a binary *first-order Markov trial*, the combination of Eqs.(6-31) and (6-34) delivers the following expression for the mole fractions of units "a" and "b":

$$(6\text{-}35) \qquad x_a = p_{b/A}/(p_{b/A} + p_{a/B}) = (1 - p_{b/B})/(2 - p_{a/A} - p_{b/B}) = 1 - x_b$$

Correspondingly, the expressions for the mole fractions of diads are

$$(6\text{-}36) \qquad x_{aa} \quad \equiv x_a p_{a/A} \qquad\qquad = x_a(1 - p_{a/B})$$

$$(6\text{-}37) \qquad x_{bb} \quad \equiv x_b p_{b/B} \qquad\qquad = x_b(1 - p_{b/A})$$

$$(6\text{-}38) \qquad x_{|ab|} \equiv x_a p_{a/B} + x_b p_{b/A}$$

In the asymmetric case, the two conditional probabilities $p_{a/B}$ and $p_{b/A}$ for the crossover are unequal. They can be calculated from the experimentally determined mole fractions with the help of Eqs.(6-36)-(6-38). It follows from Eqs.(6-29) and (6-30) for $p_{a/A} > p_{b/A}$ that $p_{b/B} > p_{a/B}$. A tendency to form long "a" segments is thus accompanied by a tendency to generate long "b" segments.

In the symmetric case, the condition $p_{a/A} = p_{b/B}$ requires $p_{a/B} = p_{b/A}$ (Table 6-5). It follows from Eq.(6-35) that $x_a = 1/2 = x_b$ and from Eqs.(6-36)-(6-38) that $x_{aa} = x_{bb} = (1/2) p_{a/A}$ and $x_{|ab|} = 1 - p_{a/A}$. Depending on the conditional probability $p_{a/A}$, the mole fraction x_{aa} varies between 0 at $p_{a/A} = 0$ and 1/2 at $p_{a/A} = 1$. In the same range, $x_{|ab|}$ decreases to 0 from 1. The copolymerization of A and B monomers according to a symmetric Markov first-order trial thus always delivers equal fractions of "a" and "b" units whereas the mole fractions of diads may vary.

The *asymmetric Bernoulli trial* is another special case (Table 6-5). It follows from the definitions that $p_{a/A} = p_{b/A} = p_A$ and thus

$$(6\text{-}39) \qquad x_a = p_A \neq p_B = x_b$$

and for the mole fractions of aa, |ab|, and bb units:

$$(6\text{-}40) \qquad x_{aa} \ = x_a^2 \qquad\qquad = (1 - x_b)^2$$

$$(6\text{-}41) \qquad x_{|ab|} = x_a x_b + x_b x_a \ = 2 \, x_a(1 - x_a)$$

$$(6\text{-}42) \qquad x_{bb} \ = x_b^2 \qquad\qquad = (1 - x_a)^2$$

Asymmetric Bernoulli trials are thus characterized by a single parameter, for example x_a or x_b (Fig. 6-2, dotted curves).

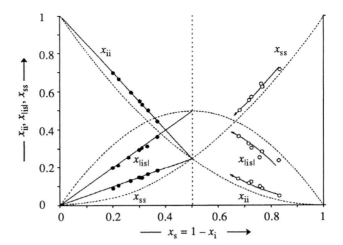

Fig. 6-2 Mole fractions x_{ii}, x_{lisl} and x_{ss} of tactic triads as a function of the mole fraction x_s of syndiotactic diads. The dotted lines ····· correspond to an asymmetric Bernoulli trial.
    Left of broken line: Polymerization of the prochiral methyl vinyl ether $CH_2=CH(OCH_3)$ by $Al_2(SO_4)_3 \cdot H_2SO_4$ in toluene (●). Solid lines —— correspond to a polymerization with an enantiomorphic catalyst (see below). The dotted vertical line - - - indicates the maximum syndiotacticity that can be obtained by such a catalyst. The diad tacticity is always $0 \le x_s \le 1/2$.
    Right of broken line: Free-radical polymerization of 9.1 mol% methyl methacrylate in acetonitrile (○) at temperatures from −5°C to +120°C. The polymerization follows a first-order Markov trial. Diad tacticities x_s may, in principle, range from 0 to 1.

All links are equally probable in *symmetric Bernouilli mechanisms* and thus

(6-43) $x_a = x_b = 1/2$

(6-44) $x_{aa} = x_{bb} = 1/4 = (1/2)\, x_{labl}$

Such a mechanism, for example, leads to an ideal-atactic polymer with an ideal-statistical (= truly random) distribution of diads, triads, etc. The term "atactic" is, however, rarely used in this exact meaning but loosely in the sense of "not appreciably tactic".
    Bernoulli and Markov mechanisms can be distinguished from each other by (a) comparison of experimentally determined tacticities of diads, triads, tetrads, etc., with conditional probabilities calculated for a certain type of trial (Table 6-6), or (b) less clear-cut, graphically as shown in Fig. 6-2. A deviation from one *assumed* trial, for example, a Bernoullian, is however no proof of the presence of another binary trial, for example, a first-order Markovnian. It only excludes the assumed trial. It may also be that non-binary trials are operating. Such distinctions require the experimental determination of higher J-ads, e.g., pentads.
    For example, the stereocontrol of the free-radical polymerization of methyl methacrylate follows a first-order Markov trial but this cannot be deduced from the Bernoullian lines in Fig. 6-2. A certain statistic may furthermore not operate under all experimental conditions. For example, the free-radical polymerization of glycidyl methacrylate was Bernoullian ($p_{s/s} = 0.97$, $p_{i/s} = 1.00$) at −78°C but not at +60°C ($p_{s/s} = 0.77$, $p_{i/s} = 0.86$).

Table 6-6 Comparison of various statistical models. η = Block character (p. 176).

Type	Values for			
	x_{ii}	x_{ss}	$x_{|isi|}$	η
One mechanism				
Bernoulli, symmetric	1/4	1/4	1/2	1
Bernoulli, asymmetric	x_i^2	x_s^2	$2\,x_i x_s$	1
Markov first-order, symmetric	$-(1/2)\,p_{|isi|}$	$x_s - (1/2)\,p_{|isi|}$	$p_{|isi|}$	$2\,x_i x_s / p_{|isi|}$
Markov first-order, asymmetric	$x_i(1 - p_{|isi|})$	$1 - x_i(1 + p_{|isi|})$	$2\,x_i p_{|isi|}$	$x_s / p_{|isi|}$
Two mechanisms				
Bernoulli, symmetric	x_i^2	x_s^2	$2\,x_i x_s$	1
Bernoulli, asymmetric	$(1/2)(3\,x_i - 1)$	$(1/2)\,x_s$	x_s	$2\,x_i$

6.3.3 Enantiomorphic Site-Control

Copolymerizations of D- and L-monomers present a special problem. Isotactic diads of monomeric units here may be DD or LL diads and syndiotactic diads either DL or LD diads. According to Eq.(6-37), one would expect for an *asymmetric first-order Markov trial* with respect to monomeric units

$$(6\text{-}45) \qquad x_i \equiv x_{LL} + x_{DD} = x_L p_{L/L} + x_D p_{D/D}$$

or, with Eqs.(6-29) and (6-30),

$$(6\text{-}46) \qquad x_i = 1 - 2\,p_{L/D} p_{D/L}/(p_{L/D} + p_{D/L}) = 1 - x_s$$

Similar equations can be derived for the mole fractions of the types of configurational triads:

$$(6\text{-}47) \qquad x_{ii} = x_D(p_{D/D})^2 + x_L(p_{L/L})^2 = 1 + p_{L/D} p_{D/L} - 4\,p_{L/D} p_{D/L}/(p_{L/D} + p_{D/L})$$

$$(6\text{-}48) \qquad x_{ss} = x_D p_{D/L} p_{L/D} + x_L p_{L/D} p_{D/L} = p_{L/D} p_{D/L}$$

$$(6\text{-}49) \qquad x_{|isi|} = x_D p_{D/L} p_{L/L} + x_L p_{L/D} p_{D/D} + x_L p_{L/L} p_{L/D} + x_D p_{D/D} p_{D/L}$$
$$= 4\,p_{L/D} p_{D/L}/(p_{L/D} + p_{D/L}) - 2\,p_{L/D} p_{D/L}$$

Polymerizations of chiral or prochiral monomers by *asymmetric Bernoulli mechanisms* are known as **polymerizations with enantiomorphic site-control**. The linking of the new monomeric unit here is independent of the type of the preceding units, i.e., $p_{D/L} = p_{L/L}$, $p_{L/D} = p_{D/D}$, and $p_{D/L} \neq p_{L/D}$. Introduction of these relations into Eqs.(6-46)-(6-49) and combination with Eqs.(6-29) and (6-30) delivers

$$(6\text{-}50) \qquad x_i = 1 - 2\,p_{L/D}(1 - p_{L/D}) = 1 - x_s$$

$$(6\text{-}51) \qquad x_{ss} = p_{L/D}(1 - p_{L/D}) \qquad = (1/2)\,x_s \qquad = x_{ss}$$

$$(6\text{-}52) \qquad x_{|isi|} = 2\,p_{L/D}(1 - p_{L/D}) \qquad = x_s \qquad = 2\,x_{ss}$$

$$(6\text{-}53) \qquad x_{ii} = 1 - 3\,p_{L/D}(1 - p_{L/D}) = 1 - (3/2)\,x_s = 1 - 3\,x_{ss}$$

In polymerizations with enantiomorphic site control, i.e., asymmetric Bernoulli trials, the mole fraction $x_{|is|}$ of heterotactic triads should thus always be twice the mole fraction x_{ss} of syndiotactic triads (Fig. 6-2, left). In regular asymmetric Bernoulli trials, this is only true for $x_i = x_s = 1/2$ (dotted line in Fig. 6-2, left).

Polymerizations with enantiomorphic site control restrict the maximum proportion of syndiotactic diads and triads. As indicated by Eqs.(6-50)–(6-53), fractions of syndiotactic diads cannot exceed $x_s = 1/2$. Such polymerizations always lead to predominantly isotactic polymers. In enantiomorphic site-controlled polymerizations, configurational mistakes are furthermore always of the type ...issi... (or ...siis...) instead of the mistakes ...isi... (or ...sis...) in end-controlled polymerizations.

A linking that is independent of the last unit (e.g., $p_{D/L} = p_{L/L}$) but different for the two antipodes (e.g., $p_{D/L} \neq p_{L/D}$) indicates the presence of two chemically different active sites (here: D* and L*). One site prefers the linking of D-monomers, the other one that of L-monomers. The probability of adding a D-monomer to a D*-site may be greater than the probability of adding an L-monomer to an L*-site, i.e., $p_{D/D} > p_{L/L}$ for D*-sites and $p_{L/L} > p_{D/D}$ for L*-sites. The enantiomorphic site control for the polymerization of racemic monomers thus predicts the formation of two types of chains, one rich in D-units and the other rich in L-units, with opposite microconfigurations and opposite optical activities but present in equal amounts.

Examples of such polymerizations with enantiomorphic catalysts are the polymerization of [RS]-4-methyl-1-hexene by Al(i-C$_4$H$_9$)$_3$/TiCl$_4$ as catalyst, of racemic methyloxirane (= propylene oxide) by zinc alkoxides, and of the N-carboxy anhydride of racemic leucine by racemic α-methylbenzylamine. In the first two cases, the resulting optically inactive polymers could be separated chromatographically into two optically active polymers with opposite signs of optical rotation.

The enantiomorphic catalyst model is also followed by the polymerization of prochiral propene by, e.g., ZrCl$_4$/(C$_2$H$_5$)$_3$Al to predominantly isotactic poly(propylene). The polymerization of propene by VCl$_4$/(C$_2$H$_5$)$_2$AlCl/anisole, on the other hand, results in syndiotactic poly(propylene). Here, the dependence of triad fractions on diad fractions adheres neither to the enantiomorphic catalyst model nor to the simple Bernoulli trial.

The latter polymerization can be described by a *symmetric Markov trial*, however. This model assumes equal conditional probabilities for the formation of homotriads, i.e., $p_{i/i} = p_{s/s}$. The mole fraction x_{ss} of syndiotactic triads is calculated from Eqs.(6-29), (6-30), and (6-38) as well as the universal relationships $x_i + x_s \equiv 1$ and $x_i = x_{ii} + (1/2)(x_{is} + x_{si}) = x_{ii} + (1/2) x_{|is|}$ (Table 4-5):

$$(6-54) \qquad x_{ss} = x_s + (1/2)(p_{s/s} - 1) = x_s - (1/2) + (1/2)\, p_{s/s}$$

The model predicts a linear relationship between x_{ss} and x_s, with a slope of 1 and a negative intercept, all found experimentally. For $p_{i/i} = p_{s/s} = 1$, it also predicts infinitely long stereoblocks, while $p_{i/i} = p_{s/s} = 0$, on the other hand, leads to $p_{i/s} = p_{s/i} = 1$ and thus $x_{|is|} = 1$, i.e., a completely heterotactic polymer.

The various statistical models also differ in the expressions for the **block character** η = $x_{|is|}/(2\, x_i x_s)$ and the **average sequence length** $\overline{N}_{seq} = 1/x_{|is|}$, i.e., the inverse of the mole fraction $x_{|is|}$ of all heterotactic triads since an it-sequence ...iii... is terminated by a heterotactic triad "is" and an st-sequence ...sss... by a heterotactic triad "si".

6.3.4 Rate Constants

The rate constants of the four different propagation reactions i/i, i/s, s/i, and s/s of first-order Markov trials can be calculated with the help of experimental diad and triad fractions. In infinitely long chains, a steady state exists for each type of diads since the termination of an isotactic sequence by a cross-reaction is followed by the appearance of a syndiotctic sequence and vice versa:

(6-55) $d[P_i{}^*]/dt = R_{s/i} - R_{i/s} = k_{s/i}[P_s{}^*][M] - k_{i/s}[P_i{}^*][M] = 0$

(6-56) $d[P_s{}^*]/dt = R_{i/s} - R_{s/i} = k_{i/s}[P_i{}^*][M] - k_{s/i}[P_s{}^*][M] = 0$

It follows that $[P_s{}^*]/[P_i{}^*] = k_{i/s}/k_{s/i}$. The *instantaneous* mole fraction of isotactic diads is given by

(6-57) $(x_i)_{inst} = [P_i{}^*]/([P_i{}^*] + [P_s{}^*]) = k_{s/i}/(k_{s/i} + k_{i/s})$

The mole fraction of isotactic diads in the *dead* polymer is calculated as follows. The change of the amount of isotactic diads, n_i, with time is given by

(6-58) $dn_i/dt = k_{i/i}[P_i{}^*][M] + k_{s/i}[P_s{}^*][M]$

and the polymerization rate by

(6-59) $- d[M]/dt = k_p[P^*][M] = K[M]$; $K = k_p[P^*]$

where $[P^*]$ = mole fraction of all growing isotactic and syndiotactic chain ends. K is a constant in living polymerizations and in the steady state of non-living ones. The change of the amount n_i of isotactic diads with the change of monomer concentration can be calculated from Eq.(6-57), Eq.(6-58), and $[P_s{}^*]/[P_i{}^*] = k_{i/s}/k_{s/i}$:

(6-60) $- dn_i/d[M] = (k_{i/i} + k_{i/s})[P_i{}^*]/K$

Integration of Eq.(6-60) from 0 to n_i, or $[M]_0$ to $[M]$, respectively, leads to

(6-61) $n_i = (k_{i/i} + k_{i/s})[P_i{}^*]([M]_0 - [M])/K$

Introduction of $[P_s{}^*]/[P_i{}^*] = k_{i/s}/k_{s/i}$, the mole fraction of isotactic diads, and the corresponding expressions for syndiotactic diads results in

(6-62) $x_i = \dfrac{n_i}{n_i + n_s} = \dfrac{k_{s/i}(k_{i/i} + k_{i/s})}{k_{s/i}(k_{i/i} + k_{i/s}) + k_{i/s}(k_{s/s} + k_{s/i})}$

Comparison of Eqs.(6-62) and (6-57) shows that the instantaneous mole fraction of isotactic diads, i.e., those at the growing chain end, equals the mole fraction of isotactic diads in the dead polymer only if the following relationship is true:

(6-63) $k_{i/i} + k_{i/s} = k_{s/s} + k_{s/i}$

With these equations, individual rate constants can be calculated, for example, $k_{i/i}$:

$$(6-64) \quad k_p = k_{i/i}x_i\left[1+\left(2+\frac{x_{ss}}{x_s-x_{ss}}\right)\left(\frac{x_i-x_{ii}}{x_{ii}}\right)\right] = k_{i/s}x_i\left[2+\frac{x_{ss}}{x_s-x_{ss}}+\frac{x_{ii}}{x_i-x_{ii}}\right]$$

In a similar manner, expressions for various rate constants of other statistical trials can be derived. Table 6-7 shows ratios of rate constants that can be calculated from ratios of J-ads, assuming the validity of Eq.(6-63). For example, the ratio of mole fractions of isotactic and syndiotactic diads delivers the ratio of rate constants of isotactic and syndiotactic homopropagations for Bernoulli trials, but the ratio of the i/s and s/i cross-propagations for first-order Markov trials.

Table 6-7 Ratios of rate constants that can be calculated from diad and triad concentrations.

J-Ads	Bernoulli trials	Markov 1st order trials	Markov 2nd order trials
x_i/x_s	k_i/k_s	$k_{s/i}/k_{i/s}$	$[k_{ss/i}(k_{si/i}+k_{ii/s})]/[k_{ii/s}(k_{is/s}+k_{ss/i})]$
x_{ii}/x_{ss}	$(k_i/k_s)^2$	$(k_{i/i}k_{s/i})/(k_{s/s}k_{i/s})$	$(k_{ss/i}k_{si/i})/(k_{ii/s}k_{is/s})$
x_{ii}/x_{is}	k_i/k_s	$k_{i/i}/k_{i/s}$	
x_{iii}/x_{sss}	$(k_i/k_s)^3$	$[(k_{i/i})^2k_{s/i}]/[(k_{s/s})^2k_{i/s}]$	$(k_{ss/i}k_{si/i}k_{ii/i})/(k_{ii/s}k_{is/s}k_{ss/s})$

6.3.5   Activation Constants

The Arrhenius equations (6-25) for rate constants k_J, conditional probabilities p_J, or mole fractions x_J of the various J-ads allows one to calculate apparent activation energies and pre-exponential factors or, with Eq.(6-28), also activation enthalpies and activation entropies. These quantities have simple meanings for Bernoulli trials but not for all other kinds of trials because the latter assume more than one type of propagation reactions.

The rate constants of the various propagation steps can be calculated from the gross rate constant k_p if the type of trial is known (see Eq.(6-64)). In general, however, only the temperature dependence of *ratios* of rate constants is discussed. For first-order Markov trials, such ratios are (see also Table 6-7):

$$\frac{k_{i/i}}{k_{i/s}}=\frac{2\,x_{ii}}{x_{|is|}} \qquad \frac{k_{i/i}}{k_{s/i}}=\frac{2\,x_s x_{ii}}{x_i x_{|is|}} \qquad \frac{k_{i/i}}{k_{s/s}}=\frac{x_s x_{ii}}{x_i x_{ss}}$$

$$\frac{k_{i/s}}{k_{s/i}}=\frac{x_s}{x_i} \qquad \frac{k_{i/s}}{k_{s/s}}=\frac{x_s x_{|is|}}{2\,x_i x_{ss}} \qquad \frac{k_{s/i}}{k_{s/s}}=\frac{x_{|is|}}{2\,x_{ss}}$$

The ratio k_a/k_b may refer to various stereocontrols in homopolymerizations (e.g., $k_{i/i}/k_{i/s}$ or $k_{i/i}/k_{s/i}$). The temperature dependence of these ratios is described by Eq.(6-27):

$$(6-65) \quad \frac{k_a}{k_b} = \exp\left[\frac{\Delta S_a^{\ddagger}-\Delta S_b^{\ddagger}}{R}\right]\exp\left[\frac{-(\Delta H_a^{\ddagger}-\Delta H_b^{\ddagger})}{RT}\right]$$

Very often, a linear relationship is observed betwen the differences of activation enthalpies and differences of activation entropies:

(6-66) $[\Delta H_a^{\ddagger} - \Delta H_b^{\ddagger}] = \Delta\Delta H_o^{\ddagger} + T_o [\Delta S_a^{\ddagger} - \Delta S_b^{\ddagger}]$

Such **compensation effects** are observed for polymerizations of the same monomer by the same initiator at constant temperature but in various solvents or for a systematic variation of the monomer structure at constant polymerization conditions. The slope constant T_o of Eq.(6-66) must have the physical unit of a temperature. At this **compensation temperature**, no effect of the solvent is found for the stereocontrol of polymerization. An example is the free-radical polymerization of methyl methacrylate at 80°C.

6.4   Polymerization of Chiral Monomers

6.4.1   Definitions

The polymerization of configuratively different but constitutionally identical monomers such as D- and L-monomers is a special case of a copolymerization. Such polymerizations are subdivided into stereospecific and stereoselective ones.

Low-molecular weight *organic chemistry* considers a reaction as stereospecific if a stereoisomer is converted 100 % into another stereoisomer. A reaction is called stereoselective if one of the new stereoisomers is formed in excess. Stereospecificity is thus the limiting case of stereoselectivity.

Polymer chemistry defines "stereospecific" and "stereoselective" differently. A **stereospecific polymerization** converts a monomer into a *tactic polymer*. A **stereoselective polymerization**, on the other hand, selects from a mixture of stereoisomeric monomer molecules only one of the stereoisomers. Stereoselective polymerizations comprise the subclasses of enantiosymmetric, enantioasymmetric, diastereosymmetric, and diastereoasymmetric polymerizations.

For example, the polymerizations of propene to isotactic or syndiotactic poly(propylene) are both stereospecific but not stereoselective since propene is not stereoisomeric. The situation is different for the polymerization of racemic propylene oxide:
- The exclusive polymerizations of the D-monomer of this racemate to poly(D-propylene oxide) or its L-monomer to poly(L-propylene oxide) are both stereospecific (with regard to polymer molecules) and stereoselective (with regard to monomer molecules).
- The polymerization of the same D,L-racemate to poly(D-alt-L-propylene oxide) is stereospecific but not stereoselective.
- The exclusive polymerization of the L-monomer of the racemate to a poly(L-propylene oxide) with a hypothetical random distribution of head-to-tail, head-to-head, and tail-to-tail links would be stereoselective but not stereospecific.
- The polymerization of D,L-propylene oxide to a mixture of poly(L-propylene oxide) and poly(D-propylene oxide) is stereospecific. It is also stereoselective but only with regard to the monomer and the resulting poly(L-propylene oxide) and poly(D-propylene oxide) *molecules* and not with regard to the *substance* poly(propylene oxide).

6.4.2 Chiral Monomers

The stereogenic center of a chiral monomer molecule may be either in the substituent or in the part that will incorporate into the main chain. An example of the former is a chiral α-olefin such as 4-methyl-1-hexene; examples of the latter are cyclic compounds such as 2,3-dimethylthiirane or methyloxirane (= propylene oxide). These chiral monomers may or may not lead to chiral polymer molecules, depending on polymerization conditions and the constitution of the monomer. A polymer composed of chiral polymer molecules may be optically active if all or an excess of its molecules are of one type of chirality, or not optically active at all if a racemate was formed.

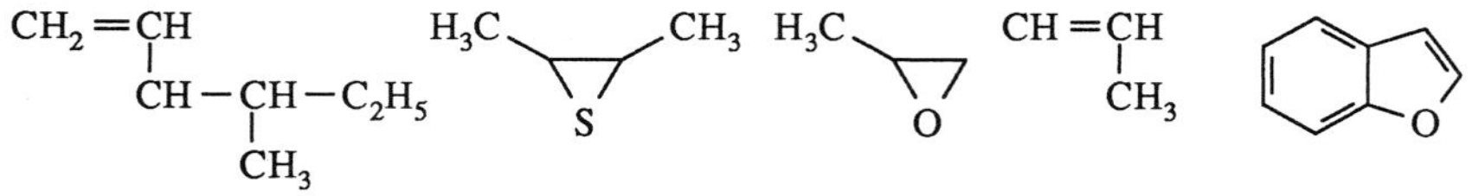

CH₂=CH
  |
  CH−CH−C₂H₅
      |
      CH₃

4-methyl-1-hexene   2,3-dimethyl thiirane   methyloxirane   propene   benzofuran

Chiral polymer molecules may also be obtained from prochiral monomer molecules such as ring-shaped molecules with a symmetry plane (e.g., benzofuran) or ethene derivaties with a prochiral double bond (e.g., propene).

Polymerizations of monomers with *chiral substituents* are, by definition, not stereospecific if the absolute configuration is maintained. An optically active monomer here is just converted into the corresponding optically active polymer. An example is the polymerization of [R]-4-methyl-1-hexene.

Cyclic monomer molecules with *stereogenic centers in the ring* deliver optically active macromolecules only if the polymerization is stereospecific; these macromolecules have only one type of configurational repeating unit. For example, the stereospecific polymerization of the optically active L-propylene oxide via O–CH₂ bond opening leads to optically active poly(L-propylene oxide) without change of the configuration of the monomer molecule and the monomeric unit. An attack at the O–CH(CH₃) bond, however, results in D-propylene oxide units and, if exclusive, in poly(D-propylene oxide).

The polymerization of monomers with *two stereogenic centers* may lead to polymers that are optically active (from the cis monomer) or not (from the trans monomer). An example is 2,3-dimethyl thiirane.

6.4.3 Enantioasymmetric Polymerizations

Polymerizations of mixtures of enantiomers are called enantioasymmetric if only one of the enantiomers is polymerized. Both complete preservation or complete inversion of the monomer configuration lead to optically active, stereoregular polymers with a single type of configurational repeating units. The remaining monomer is enriched in the other type of enantiomer. The polymerization stops at a monomer conversion of 50 % since all of the polymerizable enantiomer molecules have been used up.

Ideal enantioasymmetric polymerizations with exclusive polymerization of one of the two types of enantiomeric monomer molecules have not been found as yet. Known, however, are **partially enantioasymmetric polymerizations** in which one enantiomer type reacts faster with a chiral catalyst than the other one. The faster reacting enantiomer is almost completely polymerized before the slower reacting one starts to polymerize. The monomer is finally completely exhausted. Such partially enantioasymmetric polymerizations have been formerly called "stereoselective" or "asymmetric-selective".

An example is the anionic polymerization of racemic α-methyl benzyl methacrylate $CH_2=C(CH_3)(CO-O-CH(CH_3)-C_6H_5)$ (Fig. 6-3). At the beginning, practically only the [S]-enantiomer is polymerized. The polymerization of the [R]-enantiomer starts at an [S]-conversion of $u_R = u_S \approx 30$ % ($= y_p = 30$ %). The polymerization of the [S]-enantiomer is practically complete at $y_p = 67$ %.

The triad isotacticity of the polymer is almost $x_{ii} = 100$ % at $y_p = 0$ but decreases linearly to $x_{ii} \approx 90$ % at $y_p = 67$ %. It then increases again but does not attain $x_{ii} = 100$ % at $y_p = 100$ %. The deviations from complete isotacticity are caused by the formation of heterotactic triads (is + si); the fraction of syndiotactic triads is very small, if any.

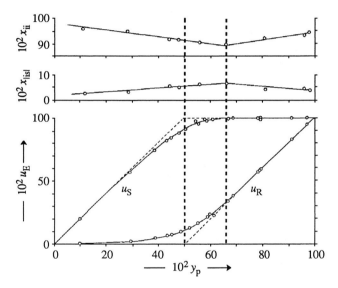

Fig. 6-3 Mole fractions x_{ii} of isotactic triads and x_{isi} of heterotactic triads as well as conversions u_E of the [S]-enantiomer (u_S) and [R]-enantiomer (u_R) as function of the polymer yield y_p of the polymer. Polymerization of [R,S]-α-methyl benzyl methacrylate by the homogeneous initiator system cyclohexyl magnesium chloride plus (–)-sparteine (structure p.185) in toluene at –78°C. Data of [1].

The partial enantioasymmetric polymerization of mixtures of antipodes constitutes a copolymerization of configurationally different but constitutionally identical monomers. Such copolymerizations follow first-order Markov statistics. The two active centers $\sim r*$ and $\sim s*$ react with the two enantiomers R and S in four propagation reactions r/R, r/S, s/R, and s/S. The propagation rates are $(R_p)_{jJ} = k_{jJ}[P_j*][J]$ (where $j = \sim r*, \sim s*$ and $J = R, S$) if the propagation step is of first order with respect to both monomer and active center. The relative change of monomer concentrations is then

(6-67)
$$\frac{d[R]}{d[S]} = \frac{k_{sR} + k_{rR}([P_r*]/[P_s*])}{k_{sS} + k_{rS}([P_r*]/[P_s*])}\left(\frac{[R]}{[S]}\right)$$

This equation reduces to the ideal copolymerization equation (Section 12.2.3) in two special cases with different meanings of the parameter Q:

(6-68) $d[R]/d[S] = Q([R]/[S])$

1. In a steady-state polymerization, both cross-propagation steps (~r + S and ~s + R) are non-zero and of the same rate ($R_{rS} = R_{sR} \neq 0$). The classic condition $k_{rR}/k_{rS} = k_{sR}/k_{sS}$ for ideal copolymerizations leads here to $Q = k_{rR}/k_{sS}$, i.e., the parameter Q equals the ratio of the rate constants of the two homosteric propagation steps.

2. If cross-propagations are absent ($R_{rS} = R_{sR} = 0$) and the rate constants of the two homopropagation steps are equal ($k_{rR} = k_{sS}$), then Q will equal the ratio of the concentrations of the two active centers ($Q = [P_r^*]/[P_s^*]$) and not the ratio of the two homosteric propagation steps.

Q will be constant in both cases. Eq.(6-68) can thus be integrated to

(6-69) $[R]/[R]_o = ([S]/[S]_o)^Q$

The monomer conversion u can be calculated from the actual and initial molar concentrations of R and S as

(6-70) $u = 1 - \left(\dfrac{[R]-[S]}{[R]_o + [S]_o} \right)$

and the optical activity α of the monomer that remains after a monomer conversion u from the actual concentration of R and S and the optical activity α_R of the pure R antipode:

(6-71) $\alpha = \alpha_R \left(\dfrac{[R]-[S]}{[R]+[S]} \right)$

Equations (6-69)-(6-71) can be united to give

(6-72) $(1-u)^{Q-1} = \dfrac{1+(\alpha/\alpha_R)}{[1-(\alpha/\alpha_R)]^Q} \left(\dfrac{2^{Q-1}[S]_o^Q}{[R]_o \{[R]_o + [S]_o\}^{Q-1}} \right)$

which reduces for a racemic mixture of enantiomers to

(6-73) $(1-u)^{Q-1} = \dfrac{1+(\alpha/\alpha_R)}{[1-(\alpha/\alpha_R)]^Q}$

An example is the enantiosymmetric polymerization of [R,S]-methylthiirane by the optically active catalyst system Zn(C$_2$H$_5$)$_2$ + [R](–)-3,3-dimethyl-1,2-butanediol (Fig. 6-4). This polymerization is "living", i.e., without termination reaction and without steady state but with constant concentration of the active centers. The factor Q is thus the ratio of concentrations of active species; it was found to be $Q = [P_r^*]/[P_s^*] = 0.43$. The polymerization of the racemic monomer delivers an optically left-rotating polymer.

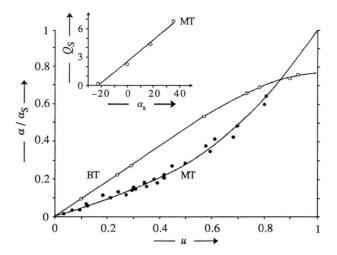

Fig. 6-4 Relative optical rotation α/α_S of the unreacted monomer as a function of the monomer conversion u during the polymerization of [R,S]-methylthiirane ($\bullet$, MT) and of [R,S]-t-butylthiirane (O, BT) by zinc diethyl plus [R](–)-3,3-dimethyl-1,2-butanediol (1:1). Data from [2].

Solid lines: calculations for a first-order reaction with respect to the monomer for MT ($Q = 0.430$) and a second-order reaction for BT ($Z = 0.123$). All polymerizations deliver optically left-rotating monomeric units, i.e., $Q_S = 1/Q \approx 2.33$ and $Z_S = 1/Z \approx 8.13$. See Eq.(6-74) for Z.

Insert: Apparent values of Q_S for the polymerization of non-racemic mixtures of [R] and [S]-methyl thiirane as a function of the optical activity of the initial monomer mixture as a measure of the relative concentration of the antipodes.

The interpretation of Q as a ratio of concentrations (and not as a ratio of rate constants) easily explains why Q_S values of polymerizations of non-racemic mixtures of antipodes increase linearly with the concentration of one antipode (insert of Fig. 6-4). The chiral initiator causes a higher concentration of one type of active species as a result of the higher concentration of one antipode. The interpretation of Q as a ratio of two rate constants, on the other hand, would require that the excess of one antipode changes the chirality of the polymerization-active centers which is fairly improbable.

Enantiosymmetric polymerizations have also been observed for the polymerizations of methyl oxirane (propylene oxide) and the N-carboxy anhydride (Leuchs anhydride) of γ-benzyl-L-glutamate. As expected, the observed stereoselectivities depended strongly on the nature of the chiral center of the initiator.

The polymerization of racemic t-butylthiirane is also enantiosymmetric but does not follow Eq.(6-73) which assumes a first-order reaction with regard to the monomer. If however the propagation takes place only at those centers which were already complexed by the monomer molecules, then the reaction would be of second order with respect to the monomer. The relative monomer conversion is now a function of the square of the ratio of monomer concentrations and Eq.(6-68) converts to

(6-74) $d[R]/d[S] = Z([R]/[S])^2$

where the proportionality constant is now called Z in order to distinguish it from the proportionality constant Q of Eq.(6-68).

For polymerization of racemates, Z is obtained from the monomer conversion u and the optical activities α_o of one of the antipodes and α of the remaining monomer:

$$(6\text{-}75) \qquad \frac{1}{(1-u)(1+(\alpha/\alpha_o))} = 1 - Z + \frac{Z}{(1-u)(1-(\alpha/\alpha_o))}$$

In such polymerizations, Z is independent of the optical purity of the starting monomer since no new active center is formed for initiation reactions at already complexed centers. The ratio of concentrations of active centers is thus independent of the ratio of initial monomer concentrations. Complexation equilibria vary strongly with temperature, however, and so does Z.

Ideal enantioasymmetric polymerizations of enantiomeric cyclic monomers with two chiral centers per ring lead to optically inactive, stereoregular polymers. Such a monomer is *trans*-2,3-dimethylthiirane (the corresponding cis compound is meso). Its polymerization is, however, only partially enantioasymmetric.

The enantioasymmetric polymerization of racemic 1-olefins, on the other hand, delivers optically active polymers. These polymers are not necessarily stereoregular since their molecules may contain both isotactic and syndiotactic diads. An example is the polymerization of [R]-4-methyl-1-hexene.

6.4.4   Other Stereoselective Polymerizations

Enantiosymmetric polymerizations of mixtures of enantiomers deliver two types of polymer molecules with opposite absolute configurations. A racemic monomer ([R]/[S] $\equiv 1$) leads to equal amounts of poly(R) and poly(S). Enantiosymmetric polymerizations thus deliver polymer *molecules* that are both stereoregular and optically active. The *polymer*, however, is internally compensated and thus optically inactive.

In principle, polymerizations can proceed to 100 % monomer conversion. At incomplete conversion, the remaining monomer is not optically active, in contrast to ideal and partial enantioasymmetric polymerizations.

Diastereosymmetric and **diastereoasymmetric** polymerizations are defined for diastereomers in the same way as enantiosymmetric and enantioasymmetric polymerizations are for enantiomers. An example of a diastereosymmetric polymerization is the stereospecific polymerization of [S]-*s*-butyl-[R,S]-oxirane with optically inactive initiators. The polymerization of the same monomer with optically active initiators is diastereoasymmetric. The latter polymerization is, however, not ideal diasteroasymmetric since the initiation reactions are not exclusive.

6.4.5   Enantiogenic Polymerizations

Enantiogenic polymerizations are polymerizations with preservation of the absolute configuration of prochiral monomers to polymers with macromolecules consisting of only one type of configurational repeating unit. Formerly, such polymerizations were called "asymmetric".

All known polymerizations of this type are only partially enantiogenic. This class includes polymerizations of monomers by chiral initiators, for example, of

- *cis*-2,3-dimethylthiirane (meso compound) to optically active polymers with exclusively R,R or S,S diads;

- benzofuran to optically active polymers with either R or S units;

- 1,4-pentadiene by 1,4-addition to optically active polymers with only one type of configurative repeating units.

In the ideal case, enantiogenic polymerizations of prochiral monomers with non-chiral initiators lead to polymer molecules with identical configurational repeating units. These polymers are optically inactive since the polymerization delivers equal amounts of left and right winding helices. An example is the polymerization of propene to isotactic poly(propylene) (see also Section 4.2.2 and Volume II).

However, the enantiogenic polymerization of prochiral monomers does lead to optically active polymers if the polymer molecules prefer a certain type of helix. An example is the polymerization of triphenylmethyl methacrylate, $CH_2=C(CH_3)COOC(C_6H_5)_3$, by $C_6H_5CH(CH_3)N(C_6H_5)Li$ as initiator (gives helices with chiral endgroups) or by the initiator system C_4H_9Li plus (–)-sparteine (gives helices without chiral endgroups). Stiff helices with identical screw senses are also formed by poly(isocyanide)s = poly(isonitrile)s, for example, $\{C(=N-C(CH_3)_3)\}_{\overline{n}}$. Such polymers are **atropisomers**.

Free-radical copolymerization of methyl methacrylate (MMA) with *p*-vinyl phenyl boronic ester (PVPB) of D-mannitol results in high-molecular weight copolymers that are optically active after removal of the mannitol. The optical activity is highest for a content of 16 % boronic acid groups. The MMA monomeric units are probably incorporated between PVPB units that are in the 1,2 and 5,6 positions of the mannitol. Similar "asymmetrically induced" enantiogenic polymerizations have been observed for the free-radical copolymerization of styrene with (–)-menthyl maleate and the free-radical polymerization of *N*-vinyl carbazole in the presence of (+)-D-camphor sulfonic acid.

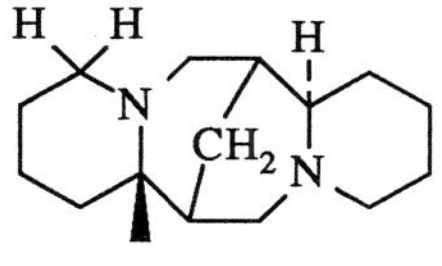

sparteine 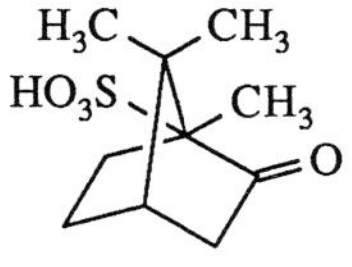   camphor-10-sulfonic acid

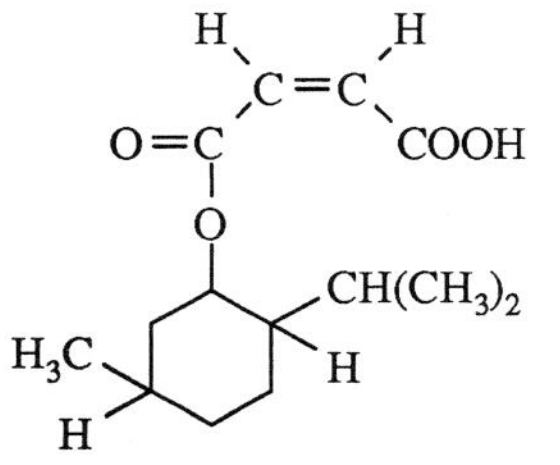

menthyl maleate 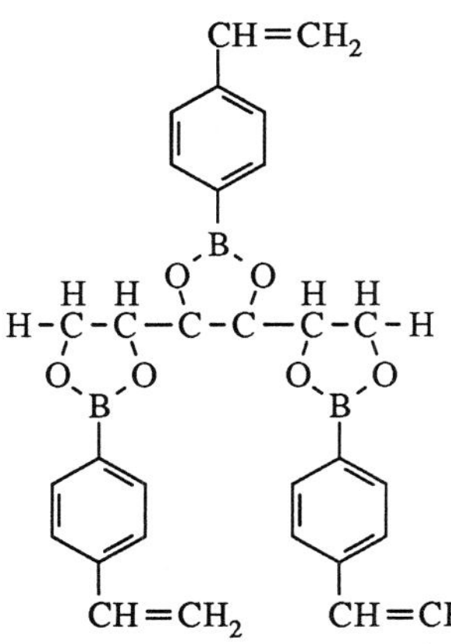   D-mannitol-*p*-vinyl phenyl boronic acid ester

6.5   Polymerization Experiments

6.5.1   Proof and Quantitative Determination of Polymer Formation

Purification

Most commercially available monomers are stabilized by polymerization inhibitors in order to prevent premature polymerization. For polymerizations, industry usually does not remove the small concentrations of such stabilizers but overrides their action by an increase of the initiator concentration.

Laboratory experiments require careful purification of monomers and labware as well as rigorous experimental conditions. A few parts per million of oxygen is sufficient to inhibit the free-radical polymerization of styrene. Traces of water prohibit anionic polymerizations. One percent of a monofunctional impurity limits the number-average degree of polymerization of polymers from polycondensations to 100. Compounds with functionalities of three and higher must be excluded if linear polymers are desired.

At least the final purification operation and the polymerization itself should take place in the absence of water and light (if an effect is suspected). Light may cause the formation of radicals from the monomer and/or solvent which may either start an unintended polymerization, terminate growing polymer chains, or degrade polymers.

It is also advisable to conduct ionic chain polymerizations in flame-dried quartz vessels since the surface groups of glass containers may interfere with the polymerization. Polymerizations should also take place in a nitrogen, helium, or argon atmosphere.

Investigations of kinetics of chain polymerizations benefit from a prepolymerization (to get rid of impurities) in which the monomer is polymerized by the same initiator or catalyst as in the main experiment. After a monomer conversion of ca. 20 %, the remaining monomer is distilled into a reaction vessel that already contains the initiator.

Absolute purification criteria for monomers are difficult to define since most analytical methods are not sensitive enough to determine traces of impurities. One can be reasonably sure about the purity of the monomer and the apparatus, however, if the polymerization kinetics is reproducible, especially if the monomer has been purified by several methods.

Course of Polymerization

The course of a polymerization can be followed by the formation of the polymer, the disappearance of the monomer(s), or the formation of a byproduct. Ideally, all three methods should be applied, using gravimetry, spectroscopy, and the like.

Less unequivocal is the use of viscometry. The increase of viscosity with time does indicate the formation of polymer. The quantitative evaluation of data is difficult, however, since viscosity depends not only on polymer concentration (and thus on polymer yield) but also on polymer molecular weight and interaction with the solvent (if any).

Indirect methods allow one to follow the course of polymerization continuously without isolation of the polymer. Especially important is **dilatometry** (L: *dilatare* = to enlarge, extend) which measures the change of volume of the polymerizing melt or solution. A dilatometer consists of a calibrated tube of ca. 3 mm diameter and a reaction vessel of 4-8 mL volume which already contains the initiator. The monomer is distilled under nitrogen into the reaction vessel which is then inserted into the temperature bath.

The monomer conversion u is calculated from the volumes V and V_0 of the polymerizing melt at times t and t_0 and the partial specific volumes $\tilde{v}_{mon}$ and $\tilde{v}_{pol}$ of monomer and polymer:

(6-76)
$$u = \left(\frac{V_0 - V_t}{V_0}\right)\left(\frac{\tilde{v}_{mon}}{\tilde{v}_{mon} - \tilde{v}_{pol}}\right)$$

For solutions, the volume of the monomer is calculated from the difference of volumes of solution and solvent if volumes are additive. In many cases, this is not true. One can, however, often assume an additivity of the weighted densities.

In polymerizations, intermolecular van der Waals, dipole-dipole, and hydrogen bond interactions between monomer molecules are replaced by intramolecular covalent bonds between monomeric units. Since atomic distances are 0.3-0.5 nm for van der Waals bonds but only 0.14-0.19 nm for covalent bonds, polymerizing liquids *usually* contract. Contractions are larger, the smaller the monomer molecules since more van der Waals bonds per unit mass are eliminated.

The larger the substituents, the smaller is thus the *contraction* during chain polymerization: 66 % for ethene $CH_2=CH_2$, 34 % for vinyl chloride $CH_2=CHCl$, and only 14 % for styrene $CH_2=CHC_6H_5$. The chain polymerization of the small ring of ethylene oxide is accompanied by a contraction of the melt of 23 %, that of the larger ring of tetrahydrofuran by 10 %, and that of the even larger ring of octamethylcyclotetrasiloxane by only 2 %. In polycondensations, the contraction is smaller, the smaller the molecules of the leaving byproduct: 22 % for the release of water in the polycondensation of hexamethylene diamine with adipic acid but 66 % for that of octanol from the polycondensation of hexamethylene diamine with dioctyl adipate.

In rare cases, an *expansion* is observed. Polymers have larger specific volumes than their monomers if the polymerization reduces the *total number* of chemical bonds, σ-bonds are replaced by π-bonds, and/or crystalline monomers are converted into less dense amorphous polymers. An example is the polymerization of norbornene spiroorthocarbonate where 12 σ-bonds of the monomer molecule are replaced by 10 σ-bonds and 1 π-bond (C=O) of the monomeric unit:

(6-77)

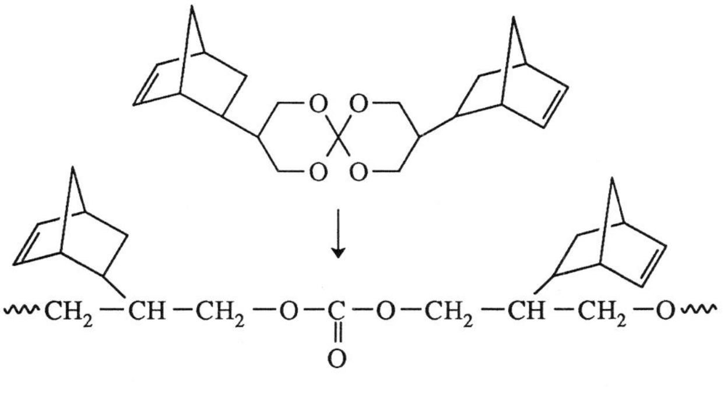

Many other experimental techniques are also employed for polymerization monitoring, most of them either for measuring property changes of the reaction mixture (e.g., refractive index) or of the polymer (e.g., endgroups, molecular volume, etc.).

The disappearance of the monomer(s) is far less often monitored since monomers may be removed not only by reactions other than polymerization but also because they are not easy to remove from the highly viscous melt or solution. The monomer content can be determined directly by size-exclusion chromatography if the polymerization took place in solution and by gas chromatography after the polymer has been removed from the polymerization system (but see below).

6.5.2 Isolation and Purification of Polymers

Polymerizations are stopped by the addition of terminating moieties or by rapid lowering of the polymerization temperature. **Distillation** may be used to remove the monomer and solvent (if any) from the polymer but this procedure has many disadvantages. The high viscosity of the remaining polymer prevents the complete removal of low-molecular weight materials. Distillation also does not eliminate catalyst and initiator residues which may cause post-reactions. Too high distillation temperatures may also degrade the polymer.

Polymers are therefore usually separated from residual monomers, solvents, catalysts, etc., by **precipitation**. Polymer solutions are further diluted and molten polymers are dissolved to 1-5 % solutions. A thin stream of this solution is then sprayed or slowly poured into a ca. 10-fold surplus of a vigorously stirred liquid precipitant. The precipitant should be neither too weak (results in incomplete precipitation) nor too strong (leads to an inclusion of monomer and/or initiator (see below)). Precipitations should take place at low temperatures since a precipitation above the glass temperature of the polymer will lead to a sticky material.

The precipitated polymer is usually amorphous or semicrystalline; crystallizable polymers are rarely obtained as crystallites. The precipitate also always contains solvent. The removal of the solvent by **drying** the precipitate **at elevated temperatures** is usually incomplete since the diffusion of the solvent from the interior to the surface is more and more impeded by increasing viscosity until it finally comes to a stop at temperatures below the glass temperature of the polymer. The **solvent inclusion** can be considerable: up to 20 % of carbon tetrachloride and up to 2.5 % of butanone in poly(styrene) and up to 10 % of N,N-dimethylformamide in poly(acrylonitrile). Inclusions decrease if the solvent contains a non-solvent that can be distilled off azeotropically with the solvent.

A far better method is **freeze-drying**. The polymer is dissolved, the solutions are shock-frozen in liquid air, and the solvent is then removed by sublimation at temperatures below the glass temperature of the polymer. Suitable freeze-drying solvents are water, benzene, p-dioxane, and formic acid. Freezing must be fast since slow freezing would produce large solvent crystallites that can cleave the less polymer chains. Polymer degradations resulting from this effect are observed especially for high-molecular weight polymers, probably because of strong entanglements of polymer chains.

Polymers can usually not be purified by **recrystallization** since (a) many polymers do not crystallize and (b) crystallizing polymers lead to trapped solvents and impurities. For example, the slow evaporation of solvent from solutions of poly[oxy-(2,6-diphenyl-1,4-phenylene)] in tetrachloroethane generates centimeter-sized crystals that contain ca. 30 % solvent. Also, large water contents are always present in protein single crystals that are obtained from aqueous solutions.

Some polymers can be purified by **extraction** with swelling solvents. **Dialysis** is suitable for water-soluble polymers, **electrophoresis** for charged polymers. Emulsions of polymers can be broken by freezing and thawing, by the addition of acids or bases, by boiling, or by the addition of electrolytes. Monomers and oligomers can be furthermore removed from polymers by **chromatography with supercritical fluids**.

Historical Notices to Chapter 6

Systematics of Polymerizations

H.Staudinger, Ber.Dtsch.Chem.Ges. **53** (1920) 107
  Distinguishes "genuine" from "non-genuine" polymerizations, the first being characterized by the same structure of monomer molecules and monomeric units in polymers.

W.H.Carothers, J.Am.Chem.Soc. **51** (1929) 2548
  Distinguishes "addition polymers" from "condensation polymers".

O.Bayer, H.Rinke, W.Siefken, L.Orthner, H.Schild, DRP 728 981 (1937);
O.Bayer, Das Diisocyanat-Polyadditionsverfahren (Polyurethane), Angew.Chem. **59** (1947) 257
  Discovery of polyaddition.

Tacticities

F.A.Bovey, G.V.D.Tiers, J.Polym.Sci. **44** (1960) 173
R.L.Miller, J.Polym.Sci. **46** (1960) 303; SPE Transact. **3** (1963) 123
  Bernoulli statistics of tacticities.

R.A.Shelden, T.Fueno, T.Tsunetsugu, J.Furukawa, J.Polym.Sci. (Letters) **3** (1965) 23
  Model of enantiomorphic site control.

Literature to Chapter 6

6.1.a MONOMERS
–, The Brandon Worldwide Monomer Reference Guide and Sourcebook, Brandon Associates, Merrimack (NJ), 3rd ed. 1993 (data on ca. 3800 monomers incl. glass temperatures of their homopolymers)

6.1.b HANDBOOKS
See Literature to Chapter 1

6.1.c TEXTBOOKS
R.W.Lenz, Organic Chemistry of Synthetic High Polymers, Interscience, New York 1967
P.Rempp, E.Merrill, Polymer Syntheses, Hüthig und Wepf, Basle, 2nd ed. 1991
D.Braun, H.Cherdron, H.Ritter, Polymer Synthesis. Theory and Praxis, Springer, Berlin, 3rd ed. 2001
G.Odian, Principles of Polymerization, Wiley, Hoboken (NJ), 4th ed. 2004

6.1.d GENERAL MONOGRAPHS
R.H.Yocum, E.B.Nyquist, Functional Monomers, Dekker, New York, 2 vols. 1973
S.R.Sandler, W.Karo, Polymer Syntheses, Academic Press, New York 1974-1980 (3 vols.); 2nd ed., vol. 1 (1992)
R.N.Haward, Ed., Developments in Polymerization, Appl.Sci.Publ., London 1979, 2 vols.
E.J.Goethals, Ed., Telechelic Polymers: Synthesis and Applications, CRC Press, Boca Raton (FL) 1988

J.P.Ebdon, Newer Methods of Polymer Synthesis, Routledge, Chapman and Hall, New York 1991
J.R.Ebdon, G.Eastmond, Ed., New Methods of Polymer Synthesis, Chapman and Hall, London 1996
K.Hatada, T.Kitayama, O.Vogl, Ed., Macromolecular Design of Polymeric Materials, Dekker, New
    York 1996
K.Takemoto, R.M.Ottenbrite, M.Kamachi, Ed., Functional Monomers and Polymers, Dekker, New
    York, 2nd ed. 1998
A.-D.Schlüter, Ed., Synthesis of Polymers, Wiley-VCH, Weinheim 1999
H.R.Kricheldorf, O.Nuyken, G.Swift, Eds., Handbook of Polymer Synthesis, Dekker, New York,
    2nd ed. 2005

6.1.e RING-OPENING POLYMERIZATION (see also Chapter 9)
K.C.Frisch, S.L.Regen, Ring-Opening Polymerization, Dekker, New York 1969
K.C.Frisch, Ed., Cyclic Monomers, Wiley-Interscience, New York 1972
S.Penczek, Ed., Polymerization of Heterocycles (Ring-Opening), Pergamon Press, Oxford 1976
N.C.Billingham, Recent Developments in Ring-Opening Polymerisation, Dev.Polym. **1** (1979) 147
K.J.Ivin, T.Saegusa, Ed., Ring-Opening Polymerization, Elsevier Appl.Sci. Publ., New York 1984
V.Dragutan, A.T.Balaban, M.Dimonic, Olefin Metathesis and Ring-Opening Polymerization of
    Cyclo-Olefins, Wiley, New York 1985

6.1.f RING-FORMING POLYMERIZATION (see also Chapter 9)
R.J.Cotter, M.Matzner, Ringforming Polymerizations, Academic Press, New York 1969, 2 vols.

6.1.5 CYCLOPOLYMERIZATION
G.B.Butler, Cyclopolymerization and Cyclocopolymers, Dekker, New York 1992
L.J.Mathias, Cyclopolymerization Overview: Recent Developments Leading to Molecular Control of
    Cyclization Efficiency, Trends in Polymer Science **4** (1996) 330

6.3 STATISTICS OF POLYMERIZATION
G.Lowry, Ed., Markov Chains and Monte Carlo Calculations in Polymer Science, Dekker,
    New York 1970
J.L.Koenig, Chemical Microstructure of Polymer Chains, Wiley, New York 1980

6.3.-6.4 STEREOCONTROL
IUPAC, Terminology for Polymerizations Involving Chiral Monomers or Resulting in Optically
    Active Monomers, Recommendations 1984
T.Tsuruta, Stereoselective and Asymmetric-Selective (or Stereoelective) Polymerizations,
    J.Polym.Sci. D (Macromol.Revs.) **6** (1972) 179
R.W.Lenz, F.Ciardelli, Eds., Preparation and Properties of Stereoregular Polymers, Reidel,
    Dordrecht 1980
J.Steinke, D.C.Sherrington, I.R.Dunkin, Imprinting of Synthetic Polymers Using Molecular
    Templates, Adv.Polym.Sci. **123** (1995) 81

6.5 POLYMERIZATION EXPERIMENTS
S.H.Pinner, A Practical Course in Polymer Chemistry, Pergamon Press, New York 1961
C.G.Overberger, Ed., Macromolecular Syntheses, Wiley, New York, Bd. 1 (1963), vol. 2 ff., various
    editors
E.M.McCaffery, Laboratory Preparation for Macromolecular Chemistry, McGraw-Hill, New York 1970
E.A.Collins, J.Bares, F.W.Billmeyer, Experiments in Polymer Science, Wiley, New York 1973
D.Braun, H.Cherdron, W.Kern, Praktikum der makromolekularen Chemie, Hüthig, Heidelberg,
    3.Aufl. 1979; Practical Macromolecular Organic Chemistry, Harwood Academic Publ., Chur 1984
R.K.Sadhir, R.M.Luck, Expanding Monomers: Synthesis, Characterization, and Applications, CRC
    Press, Boca Raton (FL) 1992
S.R.Sandler, W.Karo, J.Bonesteel, E.M.Pearce, Polymer Synthesis and Characterization. A Labora-
    tory Manual, Academic Press, San Diego (CA) 1998
S.R.Sandler, W.Karo, Sourcebook of Advanced Polymer Laboratory Preparations, Academic Press,
    San Diego (CA) 1998
G.-W.Oetjen, Freeze-Drying, Wiley-VCH, Weinheim 1999
W.R.Sorensen, F.Sweeney, T.W.Campbell, Preparative Methods of Polymer Chemistry, Wiley, New
    York 2001

References to Chapter 6

[1]   Y.Okamoto, K.Ohta, H.Yuki, Macromolecules **11** (1978) 724, Table 1, Fig. 2
[2]   M.Sepulchre, N.Spassky, P.Sigwalt, Israel J.Chem. **15** (1976/77) 33, Fig. 1, Table 5

7 Thermodynamics of Polymerization

7.1 Overview

Polymerizations are thermodynamically controlled by the **Gibbs polymerization energy**, ΔG_p, which indicates the difference between the Gibbs energies of **products** (= end products) and **educts** (= starting materials; L: *educere* = to bring out, evolve (*ex* = out, *ducere* = to lead)) ("**substrates**" in the biosciences). The Gibbs polymerization energy is given by the difference of the enthalpy term, ΔH_p, and the entropy term, $T\Delta S_p$, or by an expression that contains the equilibrium constant K for the *complete* equilibrium between *all* participants:

$$(7\text{-}1) \qquad \Delta G_p = \Delta H_p - T\Delta S_p = - RT \ln K$$

Many (but not all!) polycondensations and polyadditions are so slow that each "step" leads to such a complete equilibrium. An example is the polycondensation of stoichiometric amounts of hexamethylene diamine, $H[NH(CH_2)_6NH]H$, (symbol: H-B-H) and adipic acid, $HO[OC(CH_2)_4CO]OH$ (symbol: HO-A-OH) whereby leaving molecules $H–OH = H_2O$ are produced. In the first step, a dimer H–B–A–OH with a degree of polymerization of $X = 2$ is generated. In the second step, this dimer can react with either of the two types of monomers and generate two different trimers ($X = 3$). Three different reactions generate a tetramer ($X = 4$), etc.:

$$(7\text{-}2)$$

H-B-H	+ HO-A-OH	⇄ H-B-A-OH	+ H_2O	;	$X = 2$
H-B-H	+ HO-A-B-H	⇄ H-B-A-B-H	+ H_2O	;	$X = 3$
HO-A-OH	+ H-B-A-OH	⇄ HO-A-B-A-OH	+ H_2O	;	$X = 3$
HO-A-OH	+ H-B-A-B-H	⇄ HO-A-B-A-B-H	+ H_2O	;	$X = 4$
H-B-H	+ HO-A-B-A-OH	⇄ H-B-A-B-A-OH	+ H_2O	;	$X = 4$
H-B-A-OH	+ H-B-A-OH	⇄ H-B-A-B-A-OH	+ H_2O	;	$X = 4$

- -

N H-B-H	+ N HO-A-OH	⇄ H(B-A)$_N$OH	+ $(2N - 1) H_2O$;	$X = 2N$

Some polycondensations and polyadditions are so fast, however, that the polymerization itself does not lead to a complete equilibrium. Keeping the polymerization system for a long time at a certain temperature, however, may promote **exchange reactions** between segments of different polymer molecules or between a segment and a chain end. Examples of exchange reactions between chain segments of two different molecules are transesterification of polyesters and the so-called **equilibration** of polysiloxanes:

$$(7\text{-}3) \qquad \text{~~}[SiR_2-O]_i \qquad [SiR_2-O]_m\text{~~} \qquad \text{~~}[SiR_2-O]_i-[SiR_2-O]_m\text{~~}$$
$$+ \qquad \rightleftharpoons \qquad +$$
$$\text{~~}[O-SiR_2]_j \qquad [O-SiR_2]_n\text{~~} \qquad \text{~~}[O-SiR_2]_j-[O-SiR_2]_n\text{~~}$$

The so-called **transamidation** of polyamides, on the other hand, proceeds mainly between a polymer segment and an amine endgroup:

$$(7\text{-}4) \qquad \text{~~}Z{-}CO{-}NH{-}Z'\text{~~} \qquad \text{~~}Z{-}CO \quad + \quad H_2N{-}Z'\text{~~}$$
$$+ \qquad \rightleftharpoons$$
$$\text{~~}Z''{-}NH_2 \qquad \text{~~}Z''{-}NH$$

In polycondensations and polyadditions, endgroups (e.g., Eq.(7-2)) or center groups (e.g., Eqs.(7-3) and (7-4)) are active or can be activated by catalysts at any time. In principle, equilibria are always established though often at very long times.

In chain polymerizations and polyeliminations, however, such equilibria can only be established between monomer molecules M and *active* chains $\sim\sim M_i^*$ where $i = 1, 2, \ldots \infty$ and * is the active center at either a growing chain end ($* = \cdot, \ominus, \oplus$) or in a coordination catalyst. The active center must be permanently fixed at an individual polymer chain. It should neither be terminated nor transferred to other molecules. Only "**living polymerizations**" without such chain termination or transfer allow one to study polymerization equilibria (see also Chapter 8).

Most polymerization equilibria of chain polymerizations and polyeliminations are, however, not *complete* equilibria between *all* participants. Exchange reactions between chains of different degrees of polymerization are often very slow. They are completely absent in all-carbon chains such as $\sim CH_2-CHR\sim$ where polymerization equilibria can be established only between growing polymer chains $\sim\sim M_i^*$ and monomer molecules M:

(7-5) $\sim M_i^* + M \rightleftarrows \sim M_{i+1}^*$ but not $\sim M_i\sim + \sim M_j\sim \rightleftarrows \sim M_{i-k}\sim + \sim M_{j+k}\sim$

Exchange reactions are known, however, for polymeric sulfur, $\cdot S_i\cdot$, and for some heterochains from ring-opening polymerizations.

The position of the equilibrium is independent of the path leading to it. Polymerization equilibria can thus be achieved by polymerization or depolymerization and by ionic or free-radical mechanisms. Polymer growth by polymeric monoradicals is, however, always accompanied by deactivation (termination, transfer) by other radicals, which removes the polymer radical irreversibly. Polymerization equilibria are therefore usually studied for living ionic polymerization or insertion polymerizations.

The thermodynamic ability to polymerize decreases in most cases (but not all) with increasing temperature (Section 7.3). The inability of a monomer to polymerize may therefore have not only kinetic reasons (very slow rate, strong competing reactions) but also thermodynamic ones. To increase the temperature in the hope to get a polymerization to go may be just the wrong approach.

The quantitative description of polymerization equilibria requires careful definitions of all reactants and types of equilibrium (Section 7.2.1), especially if other equilibrium reactions are present in addition to the propagation equilibrium (for example, chain-ring equilibria). Equilibria may be furthermore thermodynamically non-ideal.

7.2   Polymer-Monomer Equilibria

7.2.1   Simple Cases

There are three simple types of polymerization equilibria: direct reaction of monomer molecules to polymer molecules (Type I), polymerization of activated monomer molecules (Type II), and initiation followed by monomer addition (Type III). Most other polymerization equilibria are very complex.

Type I

Monomer molecules M_1 are directly converted into polymer molecules M_2, M_3, ... In the simplest case, equilibrium constants $K_2 = [M_2]/[M_1]^2$, $K_3 = [M_3]/([M_2][M_1])$, etc., are independent of the degree of polymerization of the polymer molecule ($K_2 = K_3 = ...$ $= K_\infty$). An example is the ring-expansion polymerization of 1,3-dioxolane:

(7-6) $\qquad$ $\qquad$ etc.

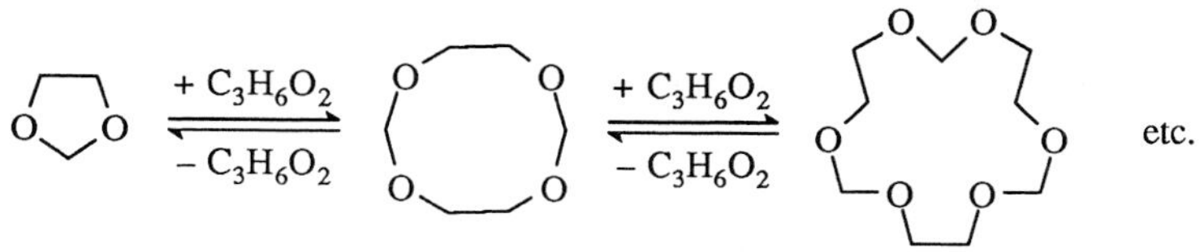

Competing with ring-expansion is often ring-opening, so that two simultaneous equilibrium reactions have to be considered.

Type II

Monomer molecules M_1 are first activated, for example, to biradicals $^\bullet M^\bullet$ or zwitterions $^\oplus M^\ominus$ (both symbolized by $^*M_1^*$) before they react further with non-activated monomer molecules to form $^*M_2^*$, $^*M_3^*$, etc. Equilibrium constants $K_1 = [^*M_1^*]/[M_1]$ of activation differ from equilibrium constants $K_i = [^*M_i^*]/([^*M_{i-1}^*][M_1])$, etc. ($i \geq 2$), of propagation.

An example is the ring-opening of *cyclo*-octa(sulfur) to the corresponding biradical which then reacts with non-activated *cyclo*-octa(sulfur) molecules (the actual formation of polymeric sulfur is accompanied by more complicated equilibria (p. 37):

(7-7)

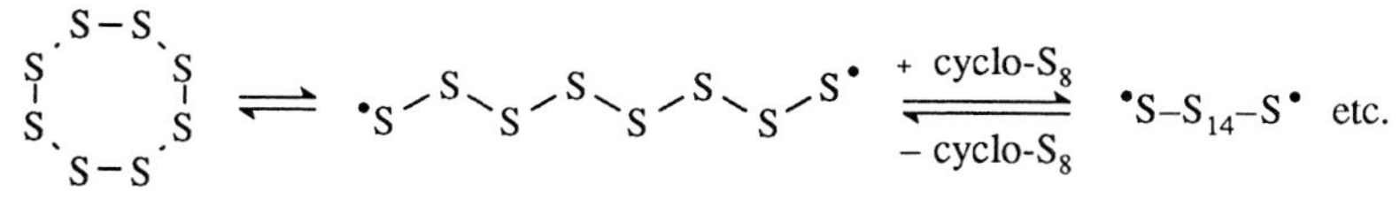

The formally similar polymerization of *p*-cyclophane to poly(*p*-xylylene) via *p*-xylylene and the corresponding biradical

(7-8)

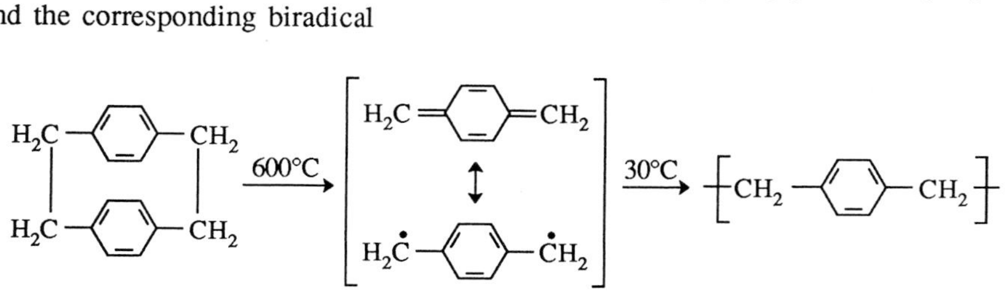

leads to another thermodynamic equilibrium, however, since *two* biradicals are formed from *p*-cyclophane, and the biradicals probably do *not* react with other *p*-cyclophane molecules but recombine with other biradicals. Another type of equilibrium would result if the biradicals react with monomer molecules since then only biradicals with an uneven number of monomer units per molecule would be formed (i.e., $^\bullet M_1^\bullet$, $^\bullet M_3^\bullet$, ..., but not $^\bullet M_2^\bullet$, $^\bullet M_4^\bullet$, ... where $M \equiv CH_2-(p-C_6H_4)-CH_2$). However, termination reactions exist in addition to Eq.(7-8); the polymerization is thus not an *equilibrium* polymerization.

Type III

Initiator molecules IJ react with monomer molecules M to form IMJ that adds further monomer molecules to give IM_2J, IM_3J, ... Equilibrium constants differ for initiation, K_1 = [IMJ]/([IJ][M]), and propagation, K_2 = $[IM_2J]$/([IMJ][M]); K_3 = $[IM_3J]$/($[IM_2J]$[M]), etc., ($K_2 = K_3 = ... = K_\infty$). An example is the ring-opening polymerization of tetrahydrofuran with acetic anhydride as initiator and $H[SbF_6]$ as catalyst:

(7-9)

$$\underset{O}{\triangle} \xrightarrow[H[SbF_6]]{+ (CH_3CO)_2O} CH_3COO[(CH_2)_5O]OCCH_3$$

$$\xrightarrow[H[SbF_6]]{+ C_4H_8O} CH_3COO[(CH_2)_5O]_2OCCH_3 \quad \text{etc.}$$

7.2.2 Degrees of Polymerization

Type I-III polymerizations differ in the types of species in equilibrium. Hence, they deliver different expressions for the number-average degree of polymerization, $\overline{X}_n$, the reduced monomer concentration, $([M]_0 - [M])/[M] = ([M]_0/[M]) - 1$, and the reduced initiator concentration, $([IJ]_0 - [IJ])/[IJ] = ([IJ]_0/[IJ]) - 1$ (Table 7-1). The expressions in Table 7-1 count IMJ and *M* as polymer molecules and assume identical equilibrium constants for propagation equilibria ($K_2 = K_3 = ... = K_\infty$). For the derivations, see Appendix A-7.

Table 7-1 Three simple polymerization equilibria (Types II and III [1]; Type I [2, 3]). $\overline{X}_n \geq 1$.

	Type I	Type II	Type III
Initiator	-	-	IJ
Monomer	M_1	M_1	M_1
Polymers	M_2, M_3, etc.	*M_1*, *M_2*, etc.	IMJ, IM_2J, etc.
$\overline{X}_n$	$1 + \{1/(1 - K_\infty[M])\}$	$1/(1 - K_\infty[M])$	$1/(1 - K_\infty[M])$
$([M]_0/[M]) - 1$	$(K_1/K_\infty)(1 - K_\infty[M])^{-2} - (K_1/K_\infty)$	$K_1(1 - K_\infty[M])^{-2}$	$[IJ]K_1(1 - K_\infty[M])^{-2}$
	$= (K_1/K_\infty) \overline{X}_n (\overline{X}_n - 2)$	$= K_1 \overline{X}_n^2$	$= K_1[I-J] \overline{X}_n^2$
$([I-J]_0/[I-J]) - 1$	-	-	$K_1[M] \overline{X}_n$

The expressions for the equilibrium concentration of the monomer, [M], and the number-average degree of polymerization, $\overline{X}_n$, allow the calculation of the equilibrium constant, K_∞, of the propagation equilibrium. An example is the polymerization of tetrahydrofuran, (Eq.(7-9)), that can be treated as a Type III equilibrium. The equilibrium concentration of the monomer increases here with increasing number-average degree of polymerization of the polymer; it becomes constant at $\overline{X}_n \approx 10$ (Fig. 7-1, top). Equilibrium constants K_i were calculated from [M] and $i = \overline{X}_n = 1/(1 - K_i[M])$. They decrease first and then become constant at $\overline{X}_n \approx 3$. The calculated product $K_i[M]$ is also not constant but increases with $\overline{X}_n$, reaches the theoretical value of $K_i[M] = 1$ at $\overline{X}_n \approx 15$, but continues to increase to $K_i[M] = 1.153$ at $\overline{X}_n = i = 21.05$.

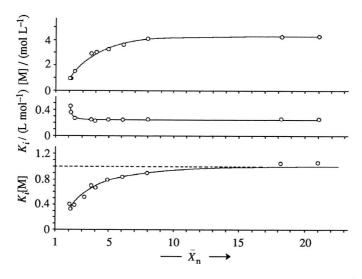

Fig. 7-1 Equilibrium monomer concentrations [M] (top), equilibrium constants $K_i = K_2 = K_3 = ...$ calculated from $\overline{X}_n = (1 - K_i[M])^{-1}$ (center), and calculated products $K_i[M]$ (bottom) as a function of the experimental number-average degree of polymerization for the equilibrium polymerization of tetrahydrofuran with HSbF$_6$/acetic anhydride in CH$_2$Cl$_2$ at 10°C. Data from reference [4].

The deviations of the data from the predictions of the simple Type III case equilibrium $K_2 = K_3 = K_4 = ... K_\infty$ are probably caused by an equilibrium constant $K_2 > K_3 = K_4 = ... = K_\infty$ (Type IIa). Instead of $\overline{X}_n = (1 - K_\infty[M])^{-1} \to K_\infty[M] = (\overline{X}_n - 1)/\overline{X}_n$ (Table 7-1, Type III), one now has $K_\infty[M] = (\overline{X}_n - 2)/(\overline{X}_n - 1)$ (Type IIa). A plot of $K_\infty[M]$ as a function of $(\overline{X}_n - 2)/(\overline{X}_n - 1)$ should pass through the origin at $(\overline{X}_n - 2)/(\overline{X}_n - 1) = 0$ (corresponds to $\overline{X}_n = 2$) and should have a slope of unity if Type IIIa applies. The experimental data (not shown) do give a straight line with a slope of 1.08 (correlation coefficient 0.980) for experimental values of $\overline{X}_n > 2.5$.

Values of $K_i[M] > 1$ must be due to experimental error since $K_i[M]$ and $K_\infty[M]$ can only approach, but never exceed, a value of unity. The total concentration of monomeric units = initial concentration of monomer molecules, [M]$_0$, is given by the sum of the concentrations of unreacted monomer molecules and monomeric units in polymer molecules, for example, for Type I:

(7-10) $[M]_0 = [M]_{free} + [M]_{bound} = [M] + \sum_{i=2}^{\infty} i[M]$

Introduction of $K_i = [M_{i+1}]/([M_i][M])$ delivers with $K_i = K_1$ for $i = 1$ and $K_i = K_\infty$ for $i \geq 2$:

(7-11) $[M]_0 = [M]\{1 + K_1[M] \sum_{i=2}^{\infty} i(K_\infty[M])^{i-2}\}$

The degree of polymerization, $i = X_i$, can, in principle, adopt any value from 2 to infinity. Since [M]$_0$ can only be a finite value, it follows that $K_\infty[M] < 1$ because the sum in Eq.(7-11) would diverge for $K_\infty[M] \geq 1$. The same is true for Type II and III equilibria.

Weight-average degrees of polymerization, $\overline{X}_w$, can be calculated from Eq.(3-51), setting $X_i = i$ and $n_i \sim [M_i]$. Types II and III deliver identical expressions for $\overline{X}_w$ that are also identical with the one derived for reactants from linear polycondensations in equilibrium (Eq.(13-47)):

$$(7\text{-}12) \qquad \overline{X}_w = \frac{1 + K_\infty[M]}{1 - K_\infty[M]} = 2\overline{X}_n - 1$$

Type I equilibrium leads to a more complicated interrelationship between weight and number-average degree of polymerization. For $\overline{X}_n > 10$, it reduces to the simple relationship $\overline{X}_w \approx 2\,\overline{X}_n - 2$ with less than 5 % error for $X > 10$.

7.2.3 Distributions of Degrees of Polymerization

Mole and weight fractions of polymer molecules of different degrees of polymerizations, $X_i \equiv i$, can be calculated from the equations for polyadditions (Chapter 13) since polymerization equilibria are independent of the path leading to polymer molecules. The mole fraction of a polymer i is thus $x_i = (1 - p_i)p_i^{i-1}$ and the weight fraction $w_i = i(1 - p_i)^2 p_i^{i-1}$ where p_i is the probability of finding molecules in a polymer with the degree of polymerization i. Summation delivers for *complete* equilibria:

$(7\text{-}13\text{-}I) \quad p_i = [M_i]/\Sigma_i\,[M_i] \qquad = (1 - K_\infty[M])(K_\infty[M])^{i-1}$ (Type I; $i \geq 2$)

$(7\text{-}13\text{-}II) \quad p_i = [M_i^*]/(\Sigma_i\,[M_i^*]) \;= (1 - K_\infty[M])(K_\infty[M])^{i-2}$ (Type II; $i \geq 1$)

$(7\text{-}13\text{-}III) \quad p_i = [IM_iJ]/(\Sigma_i\,[IM_iJ]) = (1 - K_\infty[M])(K_\infty[M])^{i-1}$ (Type III; $i \geq 1$)

These equations contain only the monomer concentrations in equilibrium, [M], the equilibrium constants K_∞, and the degrees of polymerization, $i = X_i$.

7.2.4 Incomplete Equilibria

The equations presented above assume *complete* equilibria, i.e., those in which *all* reacting molecules participate. Many "living" polymerizations proceed, however, to an "equilibrium" between active polymer chains and monomer molecules but, within reasonable times, not to an equilibrium between the active chains themselves. The equilibrium is thus not complete. The effect of incomplete equilibria between active chains manifests itself in the width of molecular weight distributions that are much smaller than demanded by Eq.(7-12). Instead of $\overline{X}_w/\overline{X}_n = 2 - (1/\overline{X}_n) \approx 2$ (Eq.(7-12)) for Type II, much lower polymolecularities are obtained, sometimes as low as $\overline{X}_w/\overline{X}_n = 1.02$.

Complete equilibria are attained by propagation/depropagation reactions of *active* chains *and* by exchange of segments of different polymer chains. In such equilibria, the number of participating molecules (active chains and monomer molecules) does not vary with time; the number-average degree of polymerization is a constant. Such complete equilibria may be achieved near the critical polymerization temperature (see Section 7.3).

At temperatures away from the critical polymerization temperature, polymerizing heterochains may still be in complete equilibrium because of the relative ease of exchange reactions (p. 193). For carbon chains, on the other hand, exchanges of chain segments can occur only by breaking of carbon-carbon bonds and subsequent recombination of the macroradicals, a process which requires much higher energies and has great potential for side reactions.

In the absence of exchange reactions between chain segments, complete equilibria can only be achieved by depropagations $P_i^* \to P_{i-1}^* + M$ of *active* chains with the rate R_{dp} = $-d[P_i^*]/dt = k_{dp}[P_i^*]$ and competing propagations $P_i^* + M \to P_{i+1}^*$ with the rate R_p = $-d[P_i^*]/dt = k_p[P_i^*][M]$. The variation of the weight-average degree of polymerization, $\overline{X}_w$, has been shown by a lengthy derivation to depend on the rate constant of depolymerization, k_{dp}, the weight-average and number-average degrees of polymerization, and the mole fractions of the monomer at time t, x_{mon}, and in equilibrium, $x_{mon,eq}$:

(7-14) $\quad$ d $\overline{X}_w/dt = k_{dp}(x_{mon} - x_{mon,eq})(\overline{X}_w - 1)/\overline{X}_n$

The establishment of equilibrium between active polymer chains and monomers is fairly fast; for example, ca. 5 seconds in the living anionic polymerization of styrene at 25°C, according to theoretical calculations. Since segment exchange reactions are completely absent, polymer-polymer equilibria are attained only after extraordinarily long time spans of up to 100 years, depending on the molar mass.

7.3 Critical Polymerization Temperatures

7.3.1 Fundamentals

Gibbs polymerization energies ΔG_p become zero at a critical temperature T_{crit}, i.e., Eq.(7-1) converts to $T = T_c = \Delta H_p/\Delta S_p$ if $\Delta G_p = 0$. In the most common case of negative polymerization enthalpies, ΔH_p, and negative polymerization entropies, ΔS_p, "no polymerization is possible" (but see below) above the critical temperature T_c since the depropagation rate $R_{dp} = k_{dp,i}[M_{i+1}]$ of the reaction $M_{i+1} \to M_i + M$ will be greater than the propagation rate $R_p = k_{p,i}[M_i][M]$ of the reaction $M_i + M \to M_{i+1}$.

In equilibrium, rates of propagation and depropagation equal each other, $R_p = R_{dp}$. Since the equilibrium constant K_i is defined as the ratio of these two rate constants, one obtains $K_i \equiv k_{p,i}/k_{dp,i} = [M_{i+1}]/([M_i][M])$.

The rate constants are independent of the degree of polymerization according to the principle of equal chemical reactivity (Sections 6.2.4, 13.5.1). Furthermore, the molar concentrations $[M_i]$ and $[M_{i+1}]$ are practically independent of the degree of polymerization at infinitely high degrees of polymerization ($i \to \infty$). Setting $K_i = K_\infty$, one obtains

(7-15) $\quad$ $K_\infty \equiv k_p/k_{dp} = 1/[M]$ $\quad$ or $\quad$ $K_\infty[M] = 1$ $\quad\quad$ (for $\overline{X}_n \to \infty$)

and Eq.(7-1) becomes after a rearrangement

(7-16) $\quad$ $-\ln[M] = (\Delta S_p/R) - (\Delta H_p/R)(1/T) = \ln K_\infty$

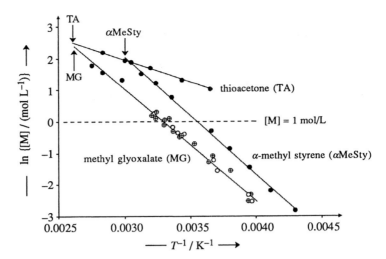

Fig. 7-2 Natural logarithm of equilibrium concentration, [M], as a function of the inverse temperature, $1/T$. Polymerizations of thioacetone CH_3–CS–CH_3 [5], α-methylstyrene CH_2=$C(CH_3)C_6H_5$ [6], and methyl glyoxalate CHO–$COOCH_3$ [7] in bulk (●) or in solution in $CHCl_3$ or CH_2Cl_2 by an anionic ($N(C_2H_5)_3$, ○) or a cationic (BF_3, ⊕) initiator. Arrows indicate the temperature T_c at which the concentration $[M]_o$ of the pure monomer is observed.

According to Eq.(7-16), a plot of the natural logarithm of the monomer concentration as a function of the inverse polymerization temperature should give a straight line (Fig. 7-2). The slope of this line contains the enthalpy of polymerization, and the intercept the entropy of polymerization. At a certain temperature $T_{c,o}$, the concentration $[M]_o$ of the neat monomer is obtained for polymerizations in *bulk*. This temperature is obviously a critical polymerization temperature at which polymer molecules are no longer present.

The situation is different for polymerizations *in solution*. At [M] = 1 mol/L, a value of ln {[M]/(mol L^{-1})} = 0 is obtained and thus $T = \Delta H_p/\Delta S$ according to Eq.(7-16) and ΔG_p = 0 according to Eq.(7-1). This temperature T_c is a critical polymerization temperature albeit one for a standard monomer concentration of $[M]^o$ = 1 mol/L. The entropy ΔS_p is similarly a standard polymerization entropy, ΔS_p^o, and the enthalpy ΔH_p a standard polymerization enthalpy, ΔH_p^o, both for a standard monomer concentration of $[M]^o$.

The concentration-independent **Dainton-Ivin critical polymerization temperature** is therefore obtained from Eq.(7-16) as

$$(7\text{-}17) \qquad T_c = \frac{\Delta H_p^{\,o}}{\Delta S_p^{\,o} + R \ln [M]}$$

Since ΔG_p^o = 0 at [M] = 1 mol/L, it is often assumed that no polymer is present at temperatures of T_c and higher (if both ΔH_p^o and ΔS_p^o are negative). This reasoning neglects, however, that the inherent relationship K_∞ = 1/[M] applies only to infinitely high degrees of polymerization whereas at $T \approx T_c$ very low molecular weights are present.

The presence of polymer molecules at $T = T_c$ is caused by **consecutive equilibria** between polymer molecules of consecutive degrees of polymerization which demand that

$[M_{i+1}] < [M_i]$. It follows that $K_i[M] < 1$ (compare, for example, Eq.(7-11) for Type I). It follows also for $T = T_c$ that (a) the polymer concentration is not zero and (b) the degree of polymerization is greater than 2 (Type I) or 1 (Types II and III).

The degree of polymerization at T_c can be calculated exactly using the equations listed in Table 7-1 if thermodynamic activities equal molar concentrations (activity coefficients of unity). One obtains for Types I-III after some arithmetic:

(7-18 I) $\quad \bar{X}_n^3 - 4\,\bar{X}_n^2 + \bar{X}_n(K_\infty/K_1)\{4 + K_3[M]_0\} - (K_\infty/K_1)\{2 - K_\infty[M]_0\} = 0$

(7-18 II) $\quad \bar{X}_n^3 - \bar{X}_n^2 + \{\bar{X}_n/K_1\}\{1 - K_\infty[M]_0\} - (1/K_1) = 0$

(7-18 III) $\quad \bar{X}_n^3[IJ]_0 + \bar{X}_n^2\{(1/K_3) - [M]_0 - [IJ]_0\}$
$\qquad\qquad + \bar{X}_n\{[M]_0(1 - (K_\infty/K_1) + (1/K_1) - (2/K_\infty)\} + (1/K_\infty) - (1/K_1) = 0$

Table 7-2 lists number-average degrees of polymerization below and near the critical polymerization temperature for a hypothetical Type I equilibrium polymerization (experimental data are unknown). The table also shows the critical polymerization temperatures, T_c, and the number-average degrees of polymerization, $\bar{X}_n$. The ceiling temperatures were calculated for $K_\infty[M] \equiv 1$ (Dainton-Ivin definition) or $K_\infty[M]_0 \equiv 1$ (Tobolsky-Eisenberg definition). The table shows that considerable concentrations of polymers (because of $[M] < [M]_0 = 6$ mol/kg) are present at the two different ceiling temperatures and that the number-average degree of polymerization at T_c is greater than 2 (Type I).

Type III differs from Types I and II because the number-average degree of polymerization at T_c depends not only on the initial monomer concentration $[M]_0$ but also on the initial initiator concentration $[IJ]_0$ and the equilibrium constant K_I of the initiation equilibrium. Assuming $[M]_0 = 1$ mol/kg and $[IJ]_0 = 0.1$ mol/kg, one obtains, for example, $\bar{X}_n \approx 3.2$ at $K_I = 10$ kg/mol and $\bar{X}_n \approx 5.6$ at $K_I = 0.1$ kg/mol. A lowering of the initiator concentration to $[IJ]_0 = 0.001$ kg/mol increases the number-average degree of polymerization to $\bar{X}_n \approx 5000$ at $K_I = 10$ kg/mol and $\bar{X}_n \approx 5010$ at $K_I = 0.1$ kg/mol!

Ceiling temperatures can therefore not be obtained accurately from $\bar{X}_n \to 1$ via $\bar{X}_n = f(T)$, from $[M] \to [M]_0$ via $[M] = f(T)$, or from $\phi_{pol} \to 0$ via $\phi_{pol} = f(T)$ (Fig. 7-3).

Table 7-2 Values of $K_\infty[M]_0$, $K_\infty[M]$, $[M]$ in equilibrium, and $\bar{X}_n$ for a hypothetical Type I equilibrium polymerization (all from Eq.(7-18-I)) for various temperatures, calculated for a system with temperature-independent values of $[M]_0 = 6$ mol/kg and $K_1 = 1$ kg/mol. K_∞ was calculated from Eq.(7-16) with $\Delta H_p° = -30$ kJ/mol and $\Delta S_p° = -100$ J/(K mol). Characteristic polymerization temperatures (here ceiling temperatures, see below) were calculated for $K_\infty[M] = 1$ according to Dainton-Ivin (D-I, [8]) and for $K_\infty[M]_0 = 1$ according to Tobolsky-Eisenberg (T-E, [1, 2]).

Assumptions		Values calculated from Eq.(7-18-I)			
T/K	$K_\infty/(\text{kg mol}^{-1})$	$K_\infty[M]_0$	$K_\infty[M]$	$[M]/(\text{mol kg}^{-1})$	$\bar{X}_n$
200	409.1	2454.6	0.998	0.002 44	1003	
260	6.362	38.17	0.935	0.147	16.9	
280	2.361	14.17	0.838	0.355	7.2	
300	1.000	6.00	0.667	0.667	4.0	T_c (D-I)
352.5	0.1667	1.00	0.212	1.276	2.27	T_c (T-E)
463.4	0.0144	0.0864	0.0196	1.362	2.02	

7.3.2 Ceiling and Floor Temperatures

Depending on the signs of polymerization enthalpy, ΔH_p^o, and polymerization entropy, ΔS_p^o, four different classes of equilibrium polymerizations can be distinguished.

Class C. In this most common case, both polymerization enthalpy and polymerization entropy are negative. The entropy term $-T\Delta S_p^o$ becomes less positive with increasing temperature until, at a **ceiling temperature** T_c (hence, Class C), it no longer compensates the negative polymerization enthalpy and the standard Gibbs polymerization energy ΔG_p^o becomes zero. At temperatures at and above T_c, no *high-molecular weight* polymers exist (Fig. 7-3) but oligomers do (Table 7-2). *Thermodynamic* ceiling temperatures have been observed for living polymerizations of many vinyl and heterocyclic monomers. An example is the equilibrium polymerization of neat liquid α-methyl styrene to a solution of poly(α-methyl styrene) in its own monomer ($T_c = 60°C$).

Class F. Equilibrium polymerizations with positive polymerization enthalpies and positive polymerization entropies are rare. For such systems, the entropy term $-T\Delta S_p^o$ becomes less negative with decreasing temperature until it just compensates the positive polymerization enthalpy at a lower critical polymerization temperature, aptly called the **floor temperature**. Below this temperature, no *high-molecular* weight polymers exist.

Floor temperatures are observed if cyclic monomers with tight macroconformations are polymerized to linear chains that form loose statistical coils. Examples are the polymerization of *cyclo*-octasulfur (Fig. 2-3), oxacycloheptane (= oxepane), and cyclic phosphonium esters with large substituents. The ring opening increases the number of rotatory degrees of freedom; the polymerization entropy is thus positive.

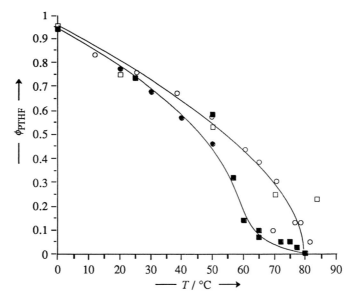

Fig. 7-3 Temperature dependence of the volume fraction ϕ_{PTHF} of poly(tetrahydrofuran) in the equilibrium polymerization of neat tetrahydrofuran with cationic initiators: O *p*-chlorophenyldiazonium hexafluorophosphate [9], □ phosphorus pentafluoride [10], ● [11] (unknown initiator; data taken from [12]), ■ triphenylmethyl hexachloroantimonate [13]. Solid lines serve to guide the eye. For unknown reasons, the data seem to fall on two curves (O and □ versus ● and ■).

Class A. Polymerizations should be *possible* at **all temperatures** (hence Class A) if the polymerization enthalpy is negative or zero but the polymerization entropy is positive. This case has been found for higher cyclic formals (see below, Fig. 7-5) An unconfirmed example is the polymerization of crystalline trioxane to crystalline poly(oxymethylene) ($\Delta H_p = -4.6$ kJ/mol, $\Delta S_p = 18$ J/(K mol)).

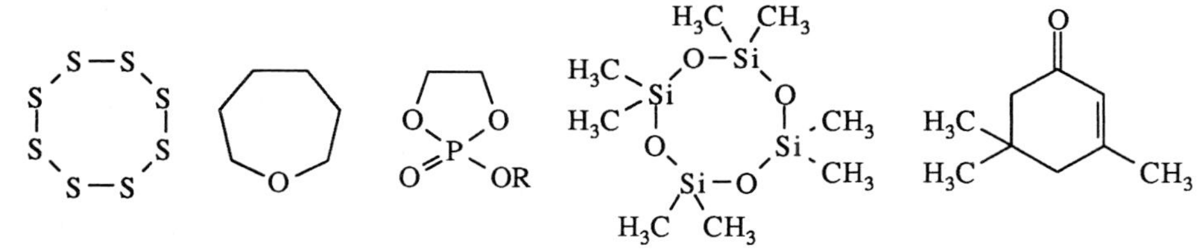

cyclo-octasulfur	oxepane	cyclic phosphonium ester	octamethyl cyclotetrasiloxane	isophorone

Class N. The reverse case, positive or zero polymerization enthalpy and negative polymerization entropy, indicates that **no polymerization** is possible at *all* temperatures. Such a case cannot be checked experimentally since a negative cannot be proven.

For example, a $\Delta G_p{}^\circ = +82.7$ kJ/mol has been calculated theoretically for the chain polymerization of acetone (I) to poly(oxy(dimethylmethylene)) (II) at $T = 298°C$ (Eq.(7-19)). The polymerization should thus be impossible at all temperatures but there are reports of its success with Mg atoms and other initiators at very low temperatures. The structure of these products is not very clear, however.

Theory predicts a $\Delta G_p{}^\circ = +10.1$ kJ/mol for the polymerization of acetone (I) to poly-(1-methyl vinylene) (III). Experimentally, however, acetone reacts at high temperatures with proton, Brønsted, or Lewis acids to phorone, $(CH_3)_2C=CH-CO-CH=C(CH_3)_2$, as main product and isophorone as a byproduct and (probably hyperbranched) oligomers with molecular weights between 600 and 1000. Alkali catalyzes a selfcondensation of 3 acetone molecules to isophorone and the condensation of acetone and isophorone.

(7-19)

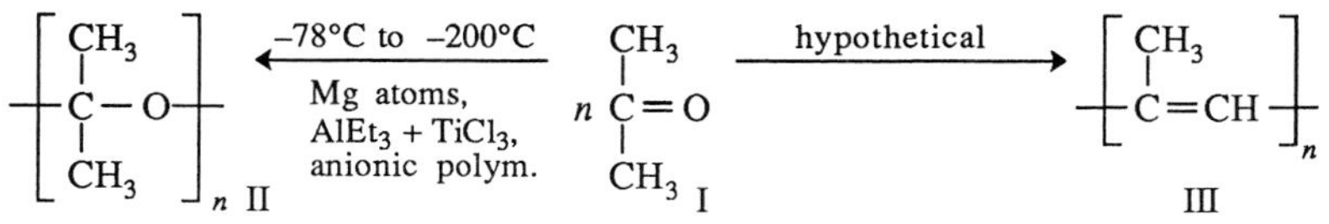

7.3.3 Effect of Pressure

The change of characteristic temperatures, T_c, with pressure p is given by the **Clausius-Clapeyron** equation, Eq.(7-20), and its integrated form, Eq.(7-21), (see textbooks of physical chemistry) where $V_M{}^m$ = molar volume of monomer, $V_u{}^m$ = molar volume of monomeric units in polymer molecules, and $\Delta H_p{}^\circ$ = standard enthalpy of polymerization:

(7-20) $dT_c/dp = T_c(V_u{}^m - V_M{}^m)/\Delta H_p{}^\circ$

(7-21) $\ln (T_c)_p = \ln (T_c)_{p\,=\,1\ bar} + [(V_u{}^m - V_M{}^m)/\Delta H_p{}^\circ]p$

The term $(V_u{}^m - V_M{}^m)$ is generally negative because the molar volume of monomeric units is usually smaller than that of monomer molecules. The polymerization enthalpy is also mostly negative (Classes C and A, p. 202-203). The natural logarithm of the ceiling temperature thus increases linearly with increasing pressure. An example is the increase of the ceiling temperature of the polymerization of neat α-methyl styrene to 143°C at 4860 bar from 60°C at 1 bar. An increase of pressure also causes an increase of melting temperatures, however, so that the polymerization may freeze in and no polymerization equilibrium can be established.

Similar to the ceiling temperature, a **ceiling pressure** can be defined as the pressure above which no polymerization takes place at a constant polymerization temperature. For 25°C, examples are ca. 0.2 bar for the polymerization of 0.1 mol/L chloral in pyridine, ca. 5 kbar for neat butyraldehyde, and more than 30 kbar for carbon disulfide.

7.3.4 Thermodynamic Non-Ideality

All discussions presented in the preceding sections relate to the relatively rare case in which thermodynamic activities are identical with molar concentrations. In most cases, however, activity coefficients deviate from unity and vary with the thermodynamic state of monomer and polymer. The various states are thus characterized by indices (Table 7-3). Three cases need to be especially considered:

1. Equilibrium polymerizations of liquid monomers to crystalline polymers are fairly rare. The ceiling temperature corresponds here to a phase transition temperature below which the monomer is *completely* polymerized. Examples are polymerizations of thioacetone ($T_c = 95$°C), sulfur trioxide ($T_c = 30.4$°C), and chloral ($T_c = 58$°C).

2. The polymerization liquid monomer $\rightleftarrows$ dissolved polymer is characterized by the presence of a considerable proportion of monomer below the ceiling temperature (Fig. 7-3). The calculation of the molar Gibbs energy of polymerization, $\Delta G_{lc}{}^m$, as the difference between Gibbs energies of monomer and polymer must therefore also consider the interaction monomer-polymer which can be expressed by the Flory-Huggins interaction parameter (Volume III), χ_{MP}, resulting in

$$(7\text{-}22) \qquad \Delta G_{lc}{}^m = \Delta G_M{}^m - \Delta G_P{}^m = RT(1 - \ln \phi_M + \chi_{MP}(\phi_P - \phi_M))$$

Table 7-3 Indices x for molar thermodynamic quantities ($\Delta G_{xx}{}^\circ$, $\Delta H_{xx}{}^\circ$, $\Delta S_{xx}{}^\circ$) at equilibrium polymerizations. U = monomeric unit.

Monomer		Polymer	Index x
Gas (1 atm)	$\rightarrow$	gas (1 atm)	gg
Gas (1 atm)	$\rightarrow$	condensed, amorphous (liquid or solid)	gc
Gas (1 atm)	$\rightarrow$	condensed, crystalline	gc'
Neat liquid	$\rightarrow$	condensed, amorphous (liquid or solid)	lc
Neat liquid	$\rightarrow$	condensed, crystalline	lc'
Neat liquid	$\rightarrow$	solution of polymer in monomer (1 mol U/L)	ls
Solution (1 mol/L)	$\rightarrow$	solution of polymer in monomer + solvent (1 mol U/L)	ss
Solution (1 mol/L)	$\rightarrow$	insoluble polymer (liquid or amorphous)	sc

Eq.(7-22) allows one to calculate the reduced Gibbs energy of polymerization, $\Delta G_{lc}{}^m/RT$, for various interaction parameters, χ_{MP}, and volume fractions, $\phi_M \equiv 1 - \phi_P$. These calculations show that $\Delta G_{lc}{}^m/RT$ does not vary much with χ_{MP} if it is in the usual range of $0.3 \leq \chi_{MP} \leq 0.5$ (Volume III). At a volume fraction of $\phi_M = 0.5$, it assumes a value of $\Delta G_{lc}{}^m/RT = 0.31$ that is independent of the interaction parameter χ_{MP}. At $\phi_M = 0.1$, the corresponding values are -1.06 for $\chi_{MP} = 0.3$ and -0.90 for $\chi_{MP} = 0.5$; at $\phi_M = 0.9$, the values are 0.65 at $\chi_{MP} = 0.3$ and 0.49 for $\chi_{MP} = 0.5$.

3. The polymerization of a dissolved monomer to a polymer in a solution of monomer and solvent requires two additional parameters for the interaction of the solvent (S) with the monomer ($\chi_{M,S}$) and with the polymer ($\chi_{P,S}$). These interactions give rise to less negative Gibbs energies of polymerization that differ from solvent to solvent and approximately parallel those of the neat monomer (Fig. 7-4). The Gibbs energy ΔG_{ss} of the polymerization of a dissolved monomer to a dissolved polymer is

(7-23) $\Delta G_{ss} = \Delta H_{ss} - T\Delta S_{ss}$

The reduced quantity $\Delta G_{ss}/RT$ thus depends on the inverse temperature:

(7-24) $\Delta G_{ss}/RT = - (\Delta S_{ss}/R) + (\Delta H_{ss}/R)(1/T)$

The standard state is characterized by $\Delta G_{ss}{}^\circ = \Delta H_{ss}{}^\circ - T\Delta S_{ss}{}^\circ = - RT \ln K$ (Eq.(7-1)) and, for high degrees of polymerization (see Eq.(7-16)), also by $\Delta G_{ss}{}^\circ = RT \ln [M]$. The standard Gibbs energy $\Delta G_{ss}{}^\circ$ becomes zero at the ceiling temperature and therefore:

(7-25) $T_c = \Delta H_{ss}/\Delta S_{ss} = \Delta H_{ss}{}^\circ/(\Delta S_{ss}{}^\circ + R \ln [M])$

The values of $\Delta H_{ss}{}^\circ$ and $\Delta S_{ss}{}^\circ$ are affected by the activity coefficients since the standard state relates to a monomer concentration of $[M] = 1$ mol/L.

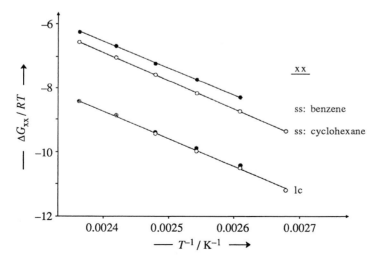

Fig. 7-4 Temperature dependence of reduced Gibbs polymerization energies, $\Delta G_{xx}/RT$, of the anionic polymerization of styrene in bulk (xx = lc), the thermodynamically bad solvent cyclohexane, and the good solvent benzene (both: xx = ss) (data of [14], calculated from Eqs.(7-23) and (7-26)).

Depending on the solvent and its interactions with monomer and polymer, respectively, various values of $\Delta G_{ss}/RT$ are obtained (Fig. 7-4). These interactions can be evaluated by the Flory-Huggins theory of polymer solutions (volume III) which predicts

(7-26) $\Delta G_{ss}/RT = 1 + \ln \phi_M + \chi_{MP}(\phi_P - \phi_M) + (\chi_{M,S} - \chi_{P,S}(V_M^m/V_S^m))\phi_S$

where χ_{MP} = interaction parameter monomer-polymer, χ_{MS} = interaction parameter monomer-solvent, χ_{PS} = interaction parameter polymer-solvent, V_S^m = molar volume of monomer, and V_S^m = molar volume of solvent. ΔG_{ss} can be converted into ΔG_{lc} (Eq.(7-22)) if the last term of Eq.(7-26) is known. In general, $\chi_{MS} > \chi_{PS}$ and $V_M^m \approx V_S^m$, so that the last term of Eq.(7-26) can be approximated by $\chi_{MS}\phi_S$. ΔG_{lc} then becomes practically independent of the solvent (line lc in Fig. 7-4).

7.4   Effect of Constitution

7.4.1   Polymerization Entropies

Entropies of polymerization can be evaluated from the temperature dependence of equilibrium concentrations of the monomer, heat capacities, pre-exponential factors of the Arrhenius equations for polymerization and depolymerization, or summations of empirically determined increments of the various groups of monomers and polymers.

Heat capacities can be used to calculate polymerization entropies because the ratio $\Delta s^o/c_p^o$ of specific entropy, Δs^o, and specific heat capacity, c_p^o, of polymers practically equals unity at a temperature of $T = 298$ K, regardless of the polymer constitution. This rule does not apply if monomer molecules associate in the vapor or polymers exhibit physical transformations in the temperature range between calorimetric and equilibrium measurement.

Polymerization entropies depend on both constitution and the state-of-matter of monomers and polymers. The change of standard entropies, ΔS_{gg}^o, during the polymerization of a gaseous monomer to a (hypothetical) gaseous polymer depends on the contributions of four entropy changes: translational entropy ΔS_{tr}^o, external rotational entropy ΔS_{er}^o, internal rotational entropy ΔS_{ir}^o, and vibrational entropy, ΔS_{vb}^o:

(7-27) $\Delta S_{gg}^o = \Delta S_{tr}^o + \Delta S_{er}^o + \Delta S_{ir}^o + \Delta S_{vb}^o$

The number of molecules in the system decreases during the polymerization. Translational and external rotational entropies must thus be negative. The linking of monomer molecules to polymer molecules, on the other hand, increases the degrees of freedom of vibration and internal rotation: vibrational entropies and internal rotational entropies must thus be positive.

Calculations show for the polymerizations of gaseous ethene, isobutene, and styrene to their (hypothetical) gaseous polymers that the loss of external rotational entropy is just compensated by the gain of internal rotational entropy and vibrational entropy:

(7-28) $- \Delta S_{er}^o \approx \Delta S_{ir}^o + \Delta S_{vb}^o$;   i.e., for olefins: $\Delta S_{gg}^o \approx \Delta S_{tr}^o$

The polymerization of gaseous monomers to amorphous polymers also involves the evaporation entropy, $\Delta S_V°$,

(7-29) $\qquad \Delta S_{gc}° = \Delta S_{gg}° - \Delta S_V° \qquad ; \qquad$ i.e., for olefins: $\Delta S_{gc}° \approx \Delta S_{tr}° - \Delta S_V°$

and the polymerization of gaseous monomers to crystalline polymers must include the melting entropy, $\Delta S_M°$:

(7-30) $\qquad \Delta S_{gc'}° = \Delta S_{gc}° - \Delta S_M° = \Delta S_{tr}° - \Delta S_V° - \Delta S_M°$

During polymerization, changes of evaporation entropies are always much larger than changes of external rotational entropies. The polymerization entropy of gaseous monomers must therefore always be more negative than that of liquid monomers. Because melt entropies are positive, polymerization entropies of polymerizations of fluid monomers to crystalline polymers are furthermore always more negative than those to amorphous polymers. Hence: crystallization fosters the polymerization. Conversely, polymerizations of crystalline monomers are thermodynamically less favored.

The polymerization entropy is thus most negative for the polymerization of a gaseous monomer to a crystalline polymer (g → c'). It becomes less negative in the following order: gc' < gc < gg < lc' < lc < c'c' (Table 7-4).

Table 7-4 Standard polymerization entropies for various states of matter at corresponding temperatures (e.g., ethene → poly(ethylene): gc' at 25°C, c'c' at –173°C, etc.). The standard state of gaseous monomers refers to a pressure of 1 atm. (→) indicates polymerizations to isotactic (it), cis-tactic (ct) and trans-tactic (tt) configurations. Values in ( ) obtained by semiempirical rules.

Monomer			$\Delta S_{xx}°/(\text{J K}^{-1} \text{ mol}^{-1})$ for xx =					
			gc'	gc	gg	lc'	lc	c'c'
Olefins								
Ethene		$CH_2=CH_2$	–174	–156	–142			(–66)
Propene	(→ it)	$CH_2=CHCH_3$	–205		(–167)	–136	–116	
1-Butene	(→ it)	$CH_2=CHC_2H_5$	–213	–190	(–166)	–141	–113	
1-Heptene	(→ it)	$CH_2=CHC_5H_{11}$			(–168)			
Styrene		$CH_2=CHC_6H_5$		–149	(–149)	–111	–104	
Vinylycyclohexane		$CH_2=CHC_6H_{11}$				–120	–85	
Aldehydes								
Formaldehyde		$O=CH_2$	–174		–124			
Acetaldehyde		$O=CHCH_3$			(–119)		–18	+6
Rings (in parentheses: number of ring atoms)								
Cyclopentene	(→ tt)	C_5H_8 (5)					–53	–18
Cycloheptene	(→ ct$_{30}$tt$_{70}$)	C_7H_{12} (7)				–70	–37	–6
Cyclooctene	(→ ct$_{52}$tt$_{48}$)	C_8H_{14} (8)				–63	–9	–10
Oxirane		C_2H_4O (3)	–173					
Oxetane		C_3H_6O (4)		–163			–68	
Tetrahydrofuran		C_4H_8O (5)	–177	–139		–100	–43	
1,3-Dioxolane		$C_3H_6O_2$ (5)	–205	–151		–100	–62	
1,3-Dioxepane		$C_5H_{10}O_2$ (7)	–163	–131		–77	–25	
Trioxane		$C_3H_6O_3$ (6)	–156		–64			–18
Tetraoxane		$C_4H_8O_4$ (8)			–51			+3

The polymerization entropy ΔS_{gg}° of gaseous 1-olefins to gaseous poly(1-olefin)s is essentially only due to the loss of translational entropy which is independent of the chemical structure. The value of ΔS_{gg}° is therefore practically the same for propene, 1-butene, and 1-heptene (Table 7-4).

Not only translational but also external rotational entropy is lost in the polymerization of liquid 1-olefins to condensed poly(1-olefin)s. Internal rotational and vibrational entropies remain approximately the same so that $\Delta S_{gg}^{\circ} < \Delta S_{lc}^{\circ} \approx \Delta S_{tr}^{\circ} + \Delta S_{er}^{\circ}$. Most olefins have the same linear momentums and moments of inertia so that losses of external rotational entropies and thus values of ΔS_{lc}° are about the same for all of them.

The rotation about bonds between ring atoms is very difficult for small *rings* that therefore have very negative polymerization entropies (Table 7-4). The rotation becomes easier with increasing ring size, leading to less negative polymerization entropies ΔS_{lc}° and, at larger oxygen-containing rings, even to positive ones (Fig. 7-5). Strong effects of monomer conformations (Pitzer strain, etc.) on polymerization entropies are observed for medium-sized rings.

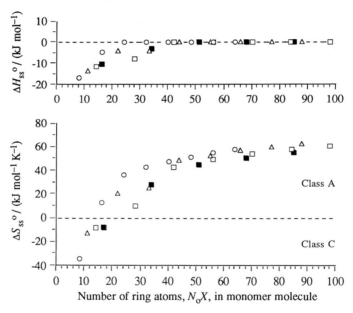

Fig. 7-5 Standard polymerization enthalpies, ΔH_{ss}°, and standard polymerization entropies, ΔS_{ss}°, of the polymerization of cyclic formals, *cyclo*-[OCH$_2$O(CH$_2$CH$_2$O)$_i$CH$_2$CH$_2$]$_N$ with i = 1-4 CH$_2$CH$_2$O units and X = 1-8 identical segments by BF$_3$–O(C$_2$H$_5$)$_2$ as catalyst in CH$_2$Cl$_2$ at temperatures of $-30°C$ to $+30°C$ (i = 1 (O) and 2 (△)), $-20°C$ to $+30°C$ (i = 3 (□)), and $-30°C$ to $+25.5°C$ (i = 4 (■)) [15]. 1,3,6-Trioxacyclooctane is also called 1,3,6-trioxocane. The simplest rings (all X = 1) are

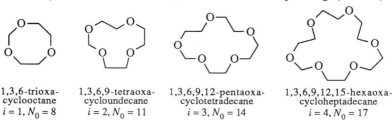

1,3,6-trioxa-cyclooctane	1,3,6,9-tetraoxa-cycloundecane	1,3,6,9,12-pentaoxa-cyclotetradecane	1,3,6,9,12,15-hexaoxa-cycloheptadecane
i = 1, N_0 = 8	i = 2, N_0 = 11	i = 3, N_0 = 14	i = 4, N_0 = 17

The polymerization of *strained oxygen-containing rings* releases more rotational entropy than the polymerization of the corresponding cycloalkanes (Table 7-5) since in linear chains the potential barriers for a rotation about $O-CH_2$ bonds are lower than those about CH_2-CH_2 bonds (Table 5-2). Gains in rotational entropies are however much smaller for larger oxacycles which may not only have less negative polymerization entropies but even positive ones (Fig. 7-5). The class of polymerization (p. 202) changes here from C (polymerization with ceiling temperature) at small ring sizes to A (polymerization at all temperatures) at larger ones.

The polymerization of cycloolefins leads to poly(alkylene)s with cis and/or trans units in the chain. An example is the ring-opening metathesis polymerization (ROMP, Section 9.4.4) of cyclopentene to *cis* or *trans*-poly(1-pentenylene), $+CH=CH-(CH_2)_3+_n$. The polymerization entropies $\Delta S_{lc}°$ are practically independent of the macroconformation of the resulting polymers (Table 7-5).

Table 7-5 Polymerization entropies of cyclic compounds, *cyclo*-$[-(CH_2)_{N-3}Z-]$ with N ring atoms.

| N | $\Delta S_{lc}°/(J\ K^{-1}\ mol^{-1})$ for Z = | | | |
	$CH_2CH_2CH_2$	CH_2CH_2O	OCH_2O	CH_2OCO
3	Cyclopropane −69	oxirane −78		
4	Cyclobutane −55	oxetane −68		β-propiolactone −54
5	Cyclopentane −43	tetrahydrofuran −41	1,3-dioxolane −61	γ-butyrolactone −30
6	Cyclohexane −11	tetrahydropyran −26	1,3-dioxane	δ-valerolactone −14
7	Cycloheptane −16	oxepane +12	1,3-dioxepane −32	ε-caprolactone −6
8	Cyclooctane −3	cyclooxaoctane −23	1,3-dioxocane −10	enanthlactone +1
9				capryllactone −5

7.4.2 Polymerization Enthalpies

Polymerization enthalpies are obtained experimentally by direct measurements of heats of polymerization, calculated from the temperature dependence of equilibrium concentrations of monomers, or estimated from group contributions. Data from these methods usually agree well.

Very many, but not all, polymerizations are exothermic. The heat of polymerization is sometimes considerable: it has been calculated that the adiabatic polymerization of ethene to 100 % poly(ethylene) would lead to a temperature increase of 1800 K! Industrial large-volume reactors therefore need efficient devices to quickly remove the heat from the polymerization systems.

The change of enthalpy, $\Delta H_{gg}°$, during the polymerization of a gaseous monomer to a (hypothetical) gaseous polymer is brought about by three factors: (1) the difference of molar bond energies $E_\sigma{}^m$ of the polymer (usually for σ bonds) and $E_\sigma{}^m$ of the monomer (often π bonds) whereby 1 bond in the monomer molecule generates N_b bonds in the polymer molecule; (2) the delocalization energy E_D; (3) the difference of strain energies of polymer, E_{sP}, and monomer, E_{sM}:

(7-31) $\Delta H_{gg}° = (N_b E_\sigma{}^m - E_\pi{}^m) + E_D + (E_{sP} - E_{sM})$

Table 7-6 Bond energies of multiple and single bonds and calculated contributions to the polymerization enthalpy (first term in parentheses on the right hand side of Eq.(7-31)). $N_b = 2$ for the polymerization of double bonds to single bonds and 3/2 for the conversion of triple bonds into double bonds.

Monomer bond		Polymer bond		Change upon polymerization
Type	$E_\pi{}^m/(kJ\ mol^{-1})$	Type	$E_\sigma{}^m/(kJ\ mol^{-1})$	$[N_b E_\sigma{}^m - E_\pi{}^m]/(kJ\ mol^{-1})$
C=C	−611	C–C	−346	−81
C=O	−737	C–O	−357	23
C=N	−615	C–N	−305	5
C≡N	−889	C=N	−615	−33.5
C=S	−540	C–S	−272	−4
S=O	−435	S–O	−232	−29

The difference of bond energies of double and single bonds is very negative for the polymerization of carbon-carbon double bonds (Table 7-6). The somewhat smaller difference for the polymerization of >S=O to –S–O– is about the same as that for the polymerization of the nitrile group –C≡N to >C=N–. The difference is positive for the polymerization of >C=O to >C–O–. This, in combination with a positive polymerization entropy, does not mean, however, that the polymerization of, e.g., $H_2C=O$, is thermodynamically impossible. The reason is that the data in Table 7-6 are averages of the bond energies of several representatives of the same class of compounds, for example, carbonyl compounds such as $H_2C=O$, $CH_3C(H)=O$, $(CH_3)_2C=O$, etc. The individual compounds, however, have different proportions of resonance and strain energies. The resulting polymerization enthalpies of individual compounds may thus vary widely depending on the constitution of monomer and polymer as well as the states of matter.

For the polymerization of gaseous monomers to condensed polymers, the evaporation enthalpy, and for that to crystalline polymers the melting entropy, has to be considered as well. Polymerization enthalpies vary therefore with the state of matter in the same way as polymerization entropies: they become less negative in the order gc' < gc < gg < lc' < lc < c'c' (Table 7-7). The differences between the polymerization enthalpies for the various states of matter are, however, not as large as those of polymerization entropies.

Polymerization enthalpies of 1-olefins are practically independent of their constitution since they do not differ much in their delocalization or strain energies (Table 7-7). For polymerizations of multiple bonds, delocalization energies are given by the difference in resonance energies which is determined experimentally as difference of heats of combustion of monomers and monomeric units. For example, styrene $CH_2=CH(C_6H_5)$ is resonance stabilized but poly(styrene) $-\!\!\!+\!CH_2–CH(C_6H_5)\!+\!\!\!-_n$ is not. The polymerization enthalpy of styrene is thus somewhat less negative than that of 1-butene where neither monomer nor polymer is stabilized by resonance.

Strain energies arise in double-bond polymerizations from conformational ("steric") effects. They are controlled by van der Waals radii and not atomic radii. Fluorine has, for example, a smaller van der Waals radius than hydrogen because the greater atomic mass of fluorine leads to smaller average vibration amplitudes. Hydrogen thus produces a larger steric effect than fluorine: the polymerization enthalpy of tetrafluoroethene is more negative than that of ethene. Other examples of steric effects are styrene versus α-methylstyrene and propene versus isobutene (Table 7-7).

Table 7-7 Effect of state of matter on the polymerization enthalpy of unsaturated compounds at corresponding temperatures (e.g., ethene → poly(ethylene): gc' at 25°C, c'c' at –173°C, etc.) The standard state of gaseous monomers refers to a pressure of 1 atm.

Monomer		$\Delta H_{xx}°/(kJ\ mol^{-1})$ for xx =					
		gc'	gc	gg	lc'	lc	c'c'
Olefins							
Tetrafluoroethene		–172		–155	–163	–155	–173
Ethene		–108	–102	–94		–92	
Propene	(→ it)	–104		–87		–84	
1-Butene	(→ it)	–108	–101	–87	–81	–77	
1-Hexene	(→ it)			–86		–83	–84
Styrene				–75		–71	
Vinyl cyclohexane					–106	–83	–88
Isobutene			–72			–48	
Halogenated olefins							
Vinyl chloride			–132			–84	
Vinylidene chloride		–108	–96		–73	–68	–67
Cycloolefins							
Cyclopentene	(→ cis)					–16	
Cyclopentene	(→ trans)					–18	
Aldehydes and ketones							
Formaldehyde		–61			–68	–31	
Acetaldehyde	(→ –CH(CH_3)–O–)				–65	–35	
Acetaldehyde	(→ –CH_2–CH(OH)–)					–64	
Chloral		–68	–51		–36	–20	
Acetone			–10	+12		+25	
Ring ethers							
Ethylene oxide		–140	–127	–104	–102	–95	
Oxetane			–82			–51	
Tetrahydrofuran			–43	–17	–34	–13	
Lactams							
γ-Butyrolactam					–12	–4	+5.5
δ-Valerolactam						–9	–4.5
ε-Caprolactam						–16	–12.5
Laurolactam (dodecanolactam)						–11	

Polymerization enthalpies may vary widely for the same monomer. The polymerization of liquid acetaldehyde CH_3CHO via the C=O double bond to condensed poly(acetaldehyde) $+OCH(CH_3)+_n$ has a polymerization enthalpy of $\Delta H_{1c}° = -35$ kJ/mol. Because of the keto-enol equilibrium, a small proportion of liquid acetaldehyde is present as tautomeric vinyl alcohol $CH_2=CHOH$. Metalorganic catalysts attack the enol and polymerize it to poly(vinyl alcohol), $+CH_2–CH(OH)+_n$. New enol is formed by reestablishment of the keto-enol equilibrium, etc. The polymerization enthalpy of this process is much more negative, –64 kJ/mol (enol polymerization) versus –35 kJ/mol (keto polymerization).

σ-Bonds are opened and newly formed in the polymerization of cyclic monomers. The difference of bond energies of monomer and polymer is therefore practically zero. Important here are, however, delocalization and strain energies. Some rings may even have zero or positive polymerization enthalpies (Fig. 7-5, Fig. 7-6).

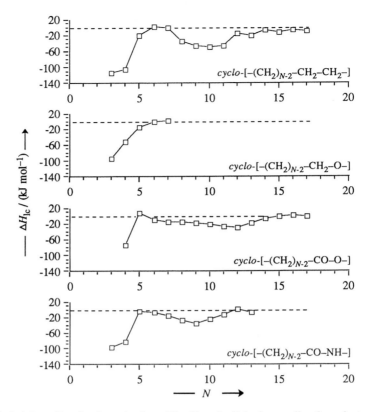

Fig. 7-6 Enthalpies ΔH_{lc} of polymerization of liquid cycloaliphatics, cyclic ethers, lactones, and lactams with various numbers of ring atoms to their condensed polymers.

In the series of unsubstituted cycloaliphatic rings, cyclopropane and cyclobutane have very negative polymerization enthalpies because considerable strain is released upon their thermodynamically favored polymerization to linear chains (hypothetical because of strong side reactions). This strain is caused by strong deviations of their C–C–C valence angles (60° in C_3H_6!)) from the typical angle $\angle_{C-C-C} = 111.5°$ (p. 144) of aliphatic chains. Cyclohexane, on the other hand, has a positive polymerization enthalpy since it has neither strain energy nor dislocation energy because of the easily established chair-boat equilibrium of its molecules. Cyclopentane C_5H_{10} and cycloheptane C_7H_{14} experience changes of dislocation energies upon polymerization since they release the Pitzer strain caused by the steric hindrance of their ecliptic hydrogen atoms. C_8H_{16} through $C_{11}H_{22}$ can neither form planar rings (valence angles) nor crown structures (not enough chain atoms): the transannular hindrance of hydrogen atoms increases the ring strain and $\Delta H_p°$ is again strongly negative. The same phenomenona are observed for unsubstituted heterocycles with one heterogroup per ring (Fig. 7-6): very negative polymerization enthalpies at very small ring sizes, slightly negative to slightly positive values at $N = 5$-6, then a decrease to a minimum at N = 9-12, followed by an increase and another maximum. The polymerization enthalpies presumably become zero or almost zero for rings with ca. 15 or more ring atoms (see Fig. 7-6).

Disubstituted rings polymerize less well than unsubstituted ones. In 1,4-disubstituted rings, both substituents are further apart in the monomer than in the polymer: the polymerization enthalpy is less negative than that for the unsubstituted compound. Polymers from 1,3-disubstituted monomers cannot assume all-trans conformations because of the strong steric hindrance of the two adjacent substituents: the polymerization enthalpy is considerably less negative compared to the unsubstituted rings.

7.4.3 Critical Thermodynamic Temperatures

Ceiling temperatures T_c and floor temperatures T_f of polymerizations in bulk are usually reported for $\Delta G_p = \Delta G_{lc'} = 0$ or $\Delta G_{lc} = 0$, i.e., for polymerizations of liquid monomers (index l) to amorphous (index c) or crystalline (index c') polymers. For polymerizations in solution, they refer in general to the polymerization of a 1 mol/L monomer solution (index s) to a dissolved polymer (index s), i.e., $\Delta G_{ss}^{\circ} = 0$.

Ceiling temperatures $T_c(lc')$ of polymerizations of neat monomers to crystalline polymers are always smaller than ceiling temperatures $T_c(ss)$ of polymerizations of dissolved monomers to dissolved polymers (Table 7-8). A monomer may thus be polymerized in bulk ($\Delta G_{lc} \leq 0$) even if it cannot be polymerized in dilute solution ($\Delta G_{ss}^{\circ} > 0$). The same reasoning applies to dilute versus concentrated solutions. At 25°C for example, tetrahydrofuran does not polymerize in 1 mol/L solution but it does in 5 mol/L. Polymerizations in solution thus require certain minimum concentrations of monomer (see also Section 7.5.1, Fig. 7-7).

Table 7-8 Ceiling temperatures T_c of polymerizations of neat gaseous (g) and liquid (l) monomers (lc': [M] = [M]$_o$) to crystalline polymers (c') and of dissolved monomers (s) to dissolved polymers (s) at standard conditions (ss: [M] = 1 mol/L). The last column indicates the solvent for "ss".

Monomer		gc'	lc'	sc'	lc	ls	ss	Solvent
					$T_c/°C$ for			
Ethene	$CH_2=CH_2$					367		
Butene ($\rightarrow$ it)	$CH_2=CH(CH_2CH_3)$					247		
1-Pentene	$CH_2=CH(CH_2CH_2CH_3)$					317		
Styrene	$CH_2=CH(C_6H_5)$		310				150	C_6H_6
Vinyl cyclohexane	$CH_2=CH(C_6H_{11})$					567		
α-Methyl styrene	$CH_2=C(CH_3)(C_6H_5)$		61				0	THF
							58	$C_6H_5CH_3$
Methyl methacrylate	$CH_2=C(CH_3)(COOCH_3)$	164	220		102	164	156	o-$C_6H_4Cl_2$
Formaldehyde	$O=CH_2$	-119	120	27				
Acetaldehyde	$O=CH(CH_3)$		-31				-39	
Propanal ($\rightarrow$ it)	$O=CH(C_2H_5)$				83	-35		
Methyl glyoxylate	$O=CH(COOCH_3)$		109				26	
Chloral	$O=CHCCl_3$	83	58	19			11	
Acetone	$O=C(CH_3)_2$	-220	<-80					
Thioacetone	$S=C(CH_3)_2$		95				-46	
Tetrahydrofuran	C_4H_8O	83	80				23	C_6H_6
1,3-Dioxolane	$C_3H_6O_2$	87	91				1	CH_2Cl_2
1,3-Dioxepane	$C_5H_{10}O_2$	149					78	C_6H_6
ε-Caprolactam	$C_6H_{11}ON$					223		

7.5 Ring-Expansion and Ring-Formation Polymerizations

The previous sections of Chapter 7 were concerned with the polymerization of non-cyclic and cyclic monomers to open (linear) chain polymers with corresponding mono-meric units. The special case of non-cyclic monomers polymerizing via internal back-biting to linear chains with incorporated small rings was discussed in Section 6.1.5.

This section will discuss two other special cases. **Ring-expansion polymerization** is the polymerization of cyclic monomers to larger rings; it often competes with the ring-opening polymerization of cyclic monomers to open linear chains. Another special case is the **ring-formation polymerization** of non-cyclic monomers to closed, cyclic oligo-mers and polymers.

Ring formations are not only controlled by thermodynamics but also by kinetics, which is also treated in this Section.

7.5.1 Thermodynamics

The thermodynamic equilibrium between chain or ring molecules M_i with i mono-meric units and cyclic molecules c-M_j with j monomeric units is described by

$$(7\text{-}32) \qquad M_i \rightleftarrows M_{i-j} + c\text{-}M_j \qquad\qquad ; \; K_{cycl} = [M_{i-j}][c\text{-}M_j]/[M_i]$$

The molar concentration $[M_i] = (n_R/V_{sln})p^{i-1}(1-p)$ of molecules M_i is calculated from the amounts n_R of all reactants per volume V_{sln} of the solution and the extent of reaction, p, (see Eq.(13-38)), and, by analogy, also for the molar concentration $[M_{i-j}]$ of M_{i-j} molecules. The equilibrium constant K_{cycl} of cyclization thus becomes

$$(7\text{-}33) \qquad K_{cycl} = \frac{[M_{i-j}][c-M_j]}{[M_i]} = \frac{p^{i-j-1}[c-M_j]}{p^{i-1}} = p^{-j}[c-M_j]$$

The probability term approaches $p^{-j} \rightarrow 1$ for high extents of reaction, $p \rightarrow 1$, and small ring sizes j. Values of K_{cycl} at $p = 0.999$ are, for example, 1.005 (for $j = 5$), 1.008 ($j = 8$), and 1.011 ($j = 11$). For $p = 0.99$, the corresponding numbers are 1.052 ($j = 5$), 1.084 ($j = 8$), and 1.117 ($j = 11$). Eq.(7-33) thus reduces to

$$(7\text{-}34) \qquad K_{cycl} \approx [c\text{-}M_j]$$

The equilibrium constant of cyclization to a ring molecule of size j thus approximate-ly equals its equilibrium concentration. The equilibrium concentration of units, u_j, in c-M_j equals $[u_j] = j[c\text{-}M_j]$.

The equilibrium concentration, $[u]$, of ring units in *all* rings depends on the initial monomer concentration (Fig. 7-7). At low concentrations, only rings are formed and the molar concentration $[u]$ of monomeric units in all rings equals the initial monomer con-centration $[M]_0$. At a cut-off concentration of $[M]_0$, the molar concentration $[u]$ of ring units in all rings reaches the value given by the equilibrium constant. Above this concen-tration, the concentration of ring units stays constant. The excess monomer concentra-tion is used for the formation of linear molecules.

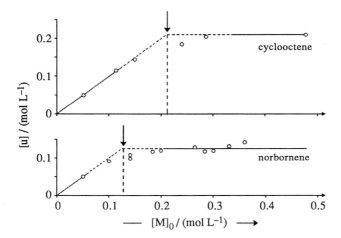

Fig. 7-7 Molar concentration, [u], of monomeric units in all cyclic oligomers as function of the initial molar monomer concentration, $[M]_0$. Polymerizations of cyclooctene and norbornene in chlorobenzene at room temperature. Arrow indicates the cut-off concentration of $[M]_0$. Data of [16a].

These equilibrium concentrations deliver rings with various sizes j. The concentration of a certain size j increases with the probability that the two ends can meet; for the reaction of the two ends of a chain, this is the end-to-end distance r_j of the chain (not the contour length). In the three-dimensional case, the probability density, $W(r_j)$, for such an event per volume to happen for one molecule is given by the degree of polymerization X, the number $N_{b,u}$ of chain bonds per monomeric unit, the length b of these bonds, and the so-called characteristic ratio C_N of the molecule (a measure of the flexibility of the chain, see Volume III):

$$(7\text{-}35) \qquad W(r_j)|_{r=0} = \left(\frac{3}{2\pi N_{b,u} b^2 C_N} \right)^{3/2} X^{-3/2}$$

At the same probability density, the concentration of ring molecules is lower, the greater the number of chain bonds, $N_a = N_{sym}X$, that can be opened per ring molecule with the degree of polymerization, X. The symmetry number N_{sym} equals 1, for example, for cyclic lactams, $c\text{-}[\text{NH–CO–(CH}_2)_i]_n$ since only the bond NH–CO can be opened. It is 2, however, for cyclic siloxanes, $c\text{-}[\text{O–SiR}_2]_n$, since both the bond O–SiR$_2$ and the bond SiR$_2$–O can be opened in a chain section ~O–SiR$_2$–O–SiR$_2$~. The equilibrium constant of cyclization, K_{cycl}, can then be calculated from Eq.(7-34) with the molar concentration of ring molecules, $[c\text{-}M_j] = W(r_j)/(N_A N_a)$, and the total number $N = N_{b,u}X$ of chain units per ring:

$$(7\text{-}36) \qquad K_{cycl} = \frac{N_{b,u}}{N_{sym} N_A} \left(\frac{3}{2\pi b^2 C_N} \right)^{3/2} N^{-5/2} \quad ; \quad \textbf{Jacobson-Stockmayer equation}$$

The theory thus predicts for unstrained rings in the unperturbed state that the equilibrium constant of cyclization, K_{cycl}, decreases with the 5/2 power of the number N of

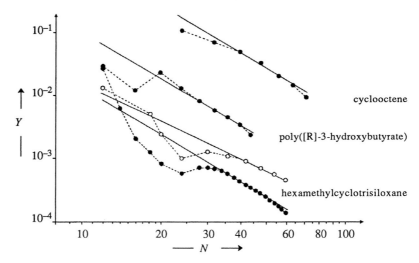

Fig. 7-8 Parameter Y (see below) as a function of the number N of ring atoms. ●---● equilibrium data, ○---○ kinetic data.

●—● Cyclooctene: equilibrium constant $Y = K_{cycl}/(\text{mol L}^{-1})$ for the polymerization in bulk at 25°C [16b]. The line with a slope of –5/2 was calculated with the known value of C_N.

●—● Poly([R]-3-hydroxybutyrate): equilibrium constant $Y = K_{dep}/(\text{mol L}^{-1})$ for depolymerization of a 5 wt% solution of polymer in boiling toluene [17]. The line with the slope –5/2 is empirical.

○—○ Hexamethylcyclotrisiloxane: inverse monomer concentration $Y = 1/[\text{c-M}_i]_t$ for the cationic polymerization of a 40 wt% monomer solution in heptane at 30°C to a monomer conversion of $u = 30$ % [18]. The indicated theoretical slope is –3/2.

●—● Hexamethylcyclotrisiloxane: equilibrium constant $Y = K_{cycl}/(\text{mol kg}^{-1})$ for the polymerization of a 40 wt% monomer solution in heptane at 30°C [18]. The line with the slope of –5/2 was calculated with Eq.(7-36) [19] and the known value of C_N.

chains. This prediction is realized for medium-sized homologs of cyclooctene, [R]-3-hydroxybutane, and hexamethylcyclotrisiloxane (Fig. 7-8). Deviations from Eq.(7-36) are found for small rings as well as for very large ones.

Small rings do not follow theory because they are strained. Large rings are not strained but are not necessarily in an unperturbed state since monomers and solvents are usually thermodynamically good solvents for polymers. For such monomers and solvents, scaling theories predict exponents of less than –5/2 for $K_{cycl} = f(N)$. In a first approximation, a slope of –(2 + α) is predicted where α is the exponent in the KMHS equation, $[\eta] = K_v M^\alpha$ (p. 98). An example is the polymerization of 0.28 mol/L cyclooctene in chlorobenzene at room temperature where an exponent of –2.68 was found for 100 % monomer conversion.

Chain molecules and medium-sized ring molecules do adopt their unperturbed dimensions, however. They thus show the theoretically predicted exponent –5/2, albeit only for a limited range (Fig. 7-8).

7.5.2   Kinetics

Ring molecules arise from chain molecules by intramolecular reactions of their end-groups A and D. First, a tight contact ~~A----D~~ is created by an equilibrium reaction.

In a rate process, the contact is then converted into a bond ~~A–D~~:

(7-37)

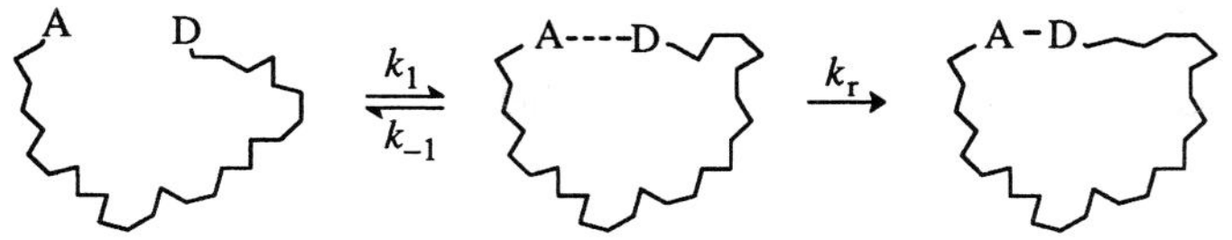

The endgroups A and D disappear with a rate of $-d[\sim A]/dt = k_1[\sim A] = k_1[\sim D] = -d[\sim D]/dt$ whereas rings are formed with a rate of $d[\text{c-M}_j]/dt = k_j[\sim A]$. If no strain energy is released upon ring formation (i.e., if the polymerization enthalpy equals zero, see Section 7.4.2), then the rate constant $k_{r,j}$ of the ring formation from a chain of size j is proportional to the probability $W(r_j) \sim X^{-3/2} \sim N^{-3/2}$ (Eq.(7-35)). The integration of $d[\text{c-M}_j]/dt = k_r[\sim A]$ with $k_r \sim W(r_j) \sim N^{-3/2}$ delivers

$$(7\text{-}38) \qquad [\text{c-M}_j]_t = A_t N^{-3/2}$$

where all proportionality constants are assembled to a time-dependent pre-exponential factor A_t. The concentration of rings is thus proportional to the $-3/2$ power of the number N of ring atoms (Fig. 7-8) in kinetically controlled ring formations from chain molecules whereas it is proportional to the $-5/2$ power in thermodynamically controlled ones.

Kinetically controlled ring closures may be conformation or diffusion controlled. If each encounter of A and D leads to a reaction with a small probability (small rate constant k_r), then the equilibrium $\sim A + D\sim \rightleftarrows \sim A\text{-}\text{-}\text{-}\text{-}D\sim$ is important and the ring formation is conformationally controlled. The rates of these cyclizations do not depend on the viscosity of the medium. Different solvents do lead to different solvent-solute interactions, however, and thus to different end-to-end distances $\sim A\text{-}\text{-}\text{-}\text{-}D\sim$ and probability densities $W(r_j)$.

If each encounter of chain ends results in a reaction, however, then $k_{1,D}/k_{-1,D} = const.$ and the reaction is diffusion controlled. In this case, the rate of ring formation is proportional to the temperature T and inversely proportional to the viscosity η_S of the medium since rate constants $k_{1,D}$ are proportional to the ratio T/η_S.

A-7 Appendix: Degrees of Polymerization

Equilibria I-III assume that equilibrium constants $K_2 = K_3 = ... = K_\infty$ do not depend on the degree of polymerization and the equilibrium constant K_1 does not depend on the initiator concentration. The relationships between the number-average degree of polymerization, the equilibrium constant, and the monomer concentration can be derived by assuming that the thermodynamic acitivities equal the amount concentrations. This assumption is justified if specific interactions are absent and the polymer remains dissolved in its monomer or solvent, respectively.

In Type III equilibria, molar equilibrium concentrations of various species are symbolized by [IJ] (initiator), [M] (monomer), $[P_1] \equiv [IMJ]$ ("polymer" with the degree of polymerization, $X = 1$), $[P_2] \equiv [IM_2J]$, etc. The equilibrium constants are then

$$IJ + M \rightleftarrows IMJ \qquad ; K_1 = \frac{[IM_1J]}{[IJ][M]} = \frac{[P_1]}{[IJ][M]}$$

$$IMJ + M \rightleftarrows IM_2J \qquad ; K_2 = \frac{[IM_2J]}{[IMJ][M]} = \frac{[P_2]}{[P_1][M]} = \frac{[P_2]}{K_1[IJ][M]^2}$$

$$IM_2J + M \rightleftarrows IM_3J \qquad ; K_3 = \frac{[IM_3J]}{[IM_2J][M]} = \frac{[P_3]}{[P_2][M]} = \frac{[P_3]}{K_1K_2[IJ][M]^3} \quad \text{etc.}$$

With $K_2 = K_3 = \ldots = K_\infty$, one obtains $[P_2] = K_1K_\infty[IJ][M]^2$, $[P_3] = K_1K_\infty^2[IJ][M]^3$, etc.

The amount concentration (molar concentration) $[P] = \Sigma_i \, [P_i]$ of all polymer types equals the sum of all molar concentrations of molecules with 1, 2, 3 ... monomeric units:

$$(A-7-1) \qquad \Sigma_i \, [P_i] = [P_1] + [P_2] + [P_3] + \ldots = [P_1]\{1 + K_\infty[M] + (K_\infty[M])^2 + \ldots\}$$

$$= \frac{[P_1]}{1 - K_\infty[M]}$$

This series of the type $1 + x + x^2 + \ldots$ can be expressed by $1/(1 - x)$ since $x \equiv K_n[M]$ must always be smaller than unity (Section 7.2.1).

The total molar concentration of all monomeric units in polymer molecules is

$$(A-7-2) \qquad \Sigma_i \, i[P_i] = [P_1] + 2\,[P_2] + 3\,[P_3] + \ldots = [P_1]\{1 + 2\,K_n[M] + 3\,(K_n[M])^2 + \ldots\}$$

Because $x = K_n[M] < 1$, this series $1 + 2\,x + 3\,x^2 + \ldots$ can be replaced by $1/(1 - x)^2$.

The number-average degree of polymerization of the polymer is given by the total molar concentration of all monomeric units in polymer molecules divided by the molar concentration of all polymer molecules:

$$(A-7-3) \qquad \overline{X}_n \equiv \frac{\Sigma_i \, i[P_i]}{\Sigma_i \, [P_i]} = \frac{1}{1 - K_\infty[M]}$$

With $K_2 = K_3 = K_4 = \ldots = K_\infty$, exactly the same relationship is obtained for Type II polymerization equilibria since formally the same species are present, i.e., monomer molecules M and "activated monomer molecules" $*M_1*$ with one monomeric unit, "activated dimers" $*M_2*$ with two monomeric units, etc. (Table 7-1). Type II equilibrium is also present if an initiator molecule IJ is first converted quantitatively into a polymer molecule $IMJ \equiv P_1$ which then reacts in equilibrium with monomer to $IM_2J \equiv P_2$, $IM_3J \equiv P_3$, etc.

Eq.(A-7-3) has to be replaced by Eq.(A-7-4), if $K_2 \gg K_3 = K_4 = \ldots = K_\infty$ (Type IIa):

$$(A-7-4) \qquad \overline{X}_n = (2 - K_\infty[M])/(1 - K_\infty[M])$$

For Type I equilibria with species $M_1, M_2, M_3, \ldots$, one obtains $\overline{X}_n = 1 + 1/(1 - K_\infty[M])$.

Literature to Chapter 7

7.2 POLYMER-MONOMER EQUILIBRIA
F.S.Dainton, K.J.Ivin, Some Thermodynamic and Kinetic Aspects of Addition Polymerization, Quarterly Revs. **12** (1958) 61
K.E.Waele, Addition Polymerisation at High Pressure, Quart.Rev. **16** (1962) 267
H.Sawada, Thermodynamics of Polymerization, Dekker, New York 1976
K.J.Ivin, Zur Thermodynamik von Additionspolymerisationsprozessen, Angew. Chem. **85** (1973) 533

7.3 CRITICAL POLYMERIZATION TEMPERATURES
A.L.Kovarskii, Ed., High-Pressure Chemistry and Physics of Polymers, CRC Press, Boca Raton (FL), 1993

7.5 RING-EXPANSION AND RING-FORMATION POLYMERIZATIONS
H.R.Allcock, Ring-Chain Equilibria, J.Macromol.Sci. [Revs.] C **4** (1970) 149
J.A.Semlyen, Ring-Chain Equilibria and the Conformation of Polymer Chains, Adv.Polym.Sci. **21** (1976) 41
J.A.Semlyen, B.R.Wood, Advances in Cyclic Polymer Chemistry, Polymer News **21** (1996) 335

References to Chapter 7

[1]   A.V.Tobolsky, A.Eisenberg, J.Am.Chem.Soc. **82** (1960) 289
[2]   A.V.Tobolsky, A.Eisenberg, J.Colloid Sci. **17** (1962) 49
[3]   H.-G.Elias, J.Semen, Makromol.Chem. **177** (1976) 3465, (a) Table 1, (b) Table 2
[4]   H.-J.Kress, W.Stix, W.Heitz, Makromol.Chem. **185** (1984) 173, Table 9 (experimental data for $\overline{X}_n$ and [M]; values of K_i have been averaged for the various degrees of polymerization i)
[5]   V.C.E.Burnop, K.G.Latham, Polymer **8** (1967) 589
[6]   H.W.McCormick, J.Polym.Sci. **25** (1957) 488, Table 1
[7]   J.P.Vairon, E.Muller, C.Bunel, Macromol.Symp. **85** (1994) 307, Table 1
[8]   F.S.Dainton, K.J.Ivin, Quart.Rev., Chem.Soc. **12** (1958) 61
[9]   M.P.Dreyfuss, P.Dreyfuss, J.Polym.Sci. [A-1] **4** (1966) 2179, Fig. 2
[10]   D.Sims, J.Chem.Soc. (1964) 864 (data taken from [12])
[11]   B.A.Rozenberg, O.M.Chekhuta, E.B.Ludwig, A.R.Gantmakher, S.S.Medvedev, Vysokomol.Soedineniya **6** (1964) 2030 (data taken from [12])
[12]   K.J.Ivin, L.Leonard, Polymer **6** (1965) 621, Fig. 1
[13]   C.E.H.Bawn, R.M.Bell, A.Ledwith, Polymer **6** (1965) 95, Fig. 1
[14]   D.J.Worsfold, S.Bywater, J.Polym.Sci. **26** (1957) 299, data of Fig. 1
[15]   Y.Yamashita, J.Mayumi, Y.Kawakami, K.Ito, Macromolecules **13** (1980) 1075, Tables V-VIII
[16]   L.Reif, H.Höcker, Macromolecules **17** (1984) 952, Figs. 3 and 4
[17]   M.Melchiors, H.Keul, H.Höcker, Macromolecules **29** (1996) 6442, Fig. 10
[18]   J.Chojnowski, M.Scibiorek, J.Kowalski, Makromol.Chem. **178** (1977) 1351, Fig. 3
[19]   H.Jacobson, W.H.Stockmayer, J.Chem.Phys. **18** (1950) 1600

8 Ionic Polymerization

8.1 Survey

8.1.1 Fundamentals

Ionic polymerizations are chain polymerizations that are initiated by electron transfer from ions of dissociating initiators, for example, anions $I^\ominus$ from an initiator $I^\ominus Q^\oplus$ (Section 8.3) or cations $I^\oplus$ from an initiator $I^\oplus Q^\ominus$ (Section 8.4). These initiators are usually added to the monomers; they may, however, also be present in the monomer as impurities. Spontaneous ionic polymerizations of pure monomers via macromonoions are unknown but there are catalyst-less polymerizations via macrozwitterions (Section 8.5).

The initiation reaction $I^\ominus + M \rightarrow IM^\ominus$ of an initiator anion $I^\ominus$ and a monomer molecule M delivers so-called **monomer anions** $IM^\ominus$ in which the initiator residue is covalently bonded to the monomeric unit and the monomeric unit now carries an anionic charge or a very polar group. Initiator cations likewise deliver **monomer cations** $IM^\oplus$. The terms "monomer anion" and "monomer cation" thus do not refer to $M^\ominus$ and $M^\oplus$ but to $IM^\ominus$ and $IM^\oplus$, respectively.

Monomer ions $IM^* = (^\ominus, ^\oplus)$ react with additional monomer molecules whereby new covalent bonds are formed between the incoming monomer molecules and the already present last monomeric unit. The reactive centers * are transferred to the chain ends:

(8-1) $\quad I(M_i)^\ominus \quad \rightarrow \quad I(M_{i+1})^\ominus \quad \rightarrow \quad I(M_{i+2})^\ominus \quad$ etc. $\quad$ (anionic)

(8-2) $\quad I(M_i)^\oplus \quad \rightarrow \quad I(M_{i+1})^\oplus \quad \rightarrow \quad I(M_{i+2})^\oplus \quad$ etc. $\quad$ (cationic)

Anionic chain polymerizations thus proceed via **macroanions** (Eq.(8-1)) and cationic chain polymerizations via **macrocations**, both with $i \geq 2$ (Eq.(8-2)). The initiator residues $I^\ominus$ or $I^\oplus$, respectively, become part of the resulting **macroions**: initiators are not catalysts! In special cases, polymerizations proceed via **macrozwitterions**, $^\ominus M(M_{i-2})M^\oplus$.

In the most simple cases, monomer molecules are only consumed by **initiation reactions**, $I^* + M \rightarrow IM^*$, and **propagation reactions**, Eqs.(8-1) and (8-2). Very often, however, additional elementary reactions are present. In **chain transfer reactions** between the growing polymer chain and another molecule, $I(M_i)^* + L \rightarrow I(M_i) + L^*$, the active center is transferred to another molecule L (monomer, solvent, polymer, etc.). The old chain is now dead and the newly formed ion L^* starts a new polymer chain.

Termination reactions lead to the complete destruction of active ions; no new active centers are formed. Such reactions may be external with impurities present in the polymerization system (bimolecular reaction) or internal within the macroion and its counterion (monomolecular reaction, see below).

In ionic polymerizations, both transfer and bimolecular termination involve a *chemical* transfer of atoms or groups (see the following sections). "Transfer" will be used hereafter only in the kinetic sense for the transfer of active centers (anions, cations, radicals).

Anionic and cationic polymerizations show many similarities, for example, the existence of ion pairs (Section 8.1.2) and the occurrence of so-called "living polymerizations" (Section 8.2). Anionic polymerizations are discussed in Section 8.3, cationic ones in Section 8.4, and zwitterion polymerizations in Section 8.5.

8.1.2 Ion Equilibria

Ionic polymerizations may involve various types of initiator ions, monomer ions, and macroions. Each of these three species may exist as free ions, loose ion pairs, contact (tight) ion pairs, solvated (loose) ion pairs, and various types of ion associates. In most cases, it is not known which species are present in which concentrations. In all reaction schemes that follow, *all* ionic species are thus symbolized by $\ominus$ or $\oplus$. These symbols do not necessarily symbolize *free* ions.

An **ion pair** consists of two oppositely charged ions. It is held together by Coulomb interactions without forming a covalent bond between the two species. Ion pairs were first postulated by Niels Bjerrum because some electrolyte solutions were less electrically conductive than expected. Ion pairs can be detected spectroscopically (see below).

Both positive and negative ions are in direct contact in **contact ion pairs (tight ion pairs)**. The bond between positive and negative ions is partially covalent. Sometimes it is necessary to distinguish between **externally non-solvated contact ion pairs** and **externally solvated contact ion pairs**.

In **loose ion pairs**, cations and anions are either connected by a solvent molecule (**solvent-shared ion pairs**) or separated by two or more solvent molecules (**solvent-separated ion pairs**). The heat of solvation is considerably larger than the electrostatic work that is needed to separate the ions; the process is thus exothermic.

Generally, rapid dynamic equilibria exist between the various ionic species. The position of the equilibrium depends on the type of solvent, the temperature, and the type of **counterions (gegenions; G: gegen = counter)** as well as the global and local concentrations of the various species. One can write schematically for anions $R^\ominus$ of initiators, monomers, or macromolecules:

$$
\begin{array}{ccccc}
\text{covalent} & \text{polarized} & \text{tight} & \text{loose} & \text{free} \\
\text{compound} & \text{molecule} & \text{ion pair} & \text{ion pair} & \text{ions} \\
\end{array}
$$

(8-3) $\qquad R{-}Q \rightleftharpoons \overset{\delta^-}{R}{-}\overset{\delta^+}{Q} \rightleftharpoons R^\ominus Q^\oplus \rightleftharpoons R^\ominus /\!/ Q^\oplus \rightleftharpoons R^\ominus + Q^\oplus$

$\qquad\qquad$ polarization $\longrightarrow$ ionization $\longrightarrow$ solvation $\longrightarrow$ dissociation

Free ions may furthermore associate with tight ion pairs to form **triple ions**, for example, $R^\ominus + {}^\oplus QR^\ominus \rightleftharpoons {}^\ominus RQ^\oplus R^\ominus$ or $2\,{}^\ominus RQ^\oplus \rightleftharpoons {}^\ominus RQ^\oplus R^\ominus + Q^\ominus$. Several ion pairs can furthermore combine into dimeric, and/or trimeric **ion associates** (Eq.(8-4)). Hexameric ion associates are generated by stacking two trimers on top of each other, etc.

$$(8\text{-}4) \qquad 6\,R^\ominus Q^\oplus \;\rightleftharpoons\; 3\;{}^\ominus R \underset{Q}{\overset{\overset{\oplus}{Q}}{\diagup\diagdown}} R^\ominus \;\rightleftharpoons\; 2 \;\text{[trimeric ring]}\; \text{etc.}$$

$$
\begin{array}{ccc}
\text{ion pair} & \text{dimeric} & \text{trimeric} \\
 & \text{ion associate} & \text{ion associate} \\
\end{array}
$$

The presence and proportion of the various ionic species can be detected, identified, and/or determined by conductivity measurements, spectroscopic investigations, and polymerization kinetics. Free ions and triple ions conduct electricity but ion pairs and ion associates do not since they are electroneutral species.

Some ions absorb ultraviolet or even visible light. Tetrahydrofuran solutions of naphthalene radical anions I are green, those of distyryl dianions III red, and those of poly-(isoprenyl) anions V yellow. Poly(styryl) anions IV show a strong charge delocalization according to NMR measurements. They thus absorb strongly in the near ultraviolet.

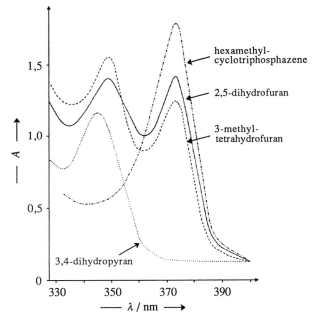

The same species may show one, two, or more absorption bands (Fig 8-1). An example is fluorenyl lithium with one band each in 3,4-dihydropyran or hexamethylcyclotriphosphazene but with two bands each in 3-methyl tetrahydrofuran or 2,5-dihydrofuran.

UV bands can be often correlated with the type of ions. Fluorenyl anions II in tetrahydrofuran show at $T = 25°C$ a band at ca. 355 nm, at $T \leq -55°C$ a band at 373 nm, and both bands in the temperature range 25°C to –55°C (not shown). The relative band heights are not affected by diluting the solutions, which would shift an equilibrium between loose ion pairs and free ions toward the latter. The relative band heights are also not influenced by added $NaB(C_6H_5)_4$ which dissociates easily in tetrahydrofuran. The addition of this salt would shift that equilibrium toward ion pairs.

Fig. 8-1 Absorption A (formerly called extinction or optical density) of fluorenyl lithium in different solvents at 25°C [1].

The solutions also do not conduct electricity; free ions, if any, can therefore be present in only very small proportions. The two UV bands of fluorenyl sodium in tetrahydrofuran must thus come from two different types of ion pairs, for example, tight and loose ion pairs. The equilibrium tight ion pairs ⇄ loose ion pairs is practically independent of the concentration of the fluorenyl salt since the solvent is present in large excess and only a small fraction of the solvent is used for the formation of the loose (solvated) ion pairs. The band at the higher wavelength (= lower energy) must come from the solvated ion pair since solvations occur preferably at lower temperatures (solvation: positive ΔH, negative ΔS). The band at the lower wavelength is thus caused by the tight ion pair.

Two bands, and thus two types of ion pairs, are not only found for fluorenyl sodium in tetrahydrofuran but also for fluorenyl lithium in 2,5-dihydrofuran or 3-methyltetrahydrofuran (Fig. 8-1). Only one band corresponding to a tight ion pair is found for fluorenyl lithium in 3,4-dihydropyran. Hexamethylcyclotriphosphazene as solvent also delivers only one band albeit for the solvated ion pair.

Tight ion pairs are not solvated. Their band positions are therefore little influenced by the solvent (Fig. 8-1). It is remarkable, however, that the band positions of the postulated loose (solvated) ion pairs also do not vary much with the solvent. In loose ion pairs, it is not the fluorenyl anion that is solvated but the non-absorbing lithium cation.

Electron spin resonance spectroscopy (ESR = EPR = electron paramagnetic resonance spectroscopy) delivers direct proof of the existence of ion pairs. Sodium naphthalene NpNa dissociates in hexamethylphosphoric triamide (HMTA) into free naphthalene radical ions $Np^{\ominus\bullet}$ and sodium counterions $Na^{\oplus}$. The ESR spectrum of NpNa shows 25 hyperfine lines that are caused by the interaction of the odd electron with the spins of the α and β protons (spin 1/2) of naphthalene. Each line of the spectrum is split into four lines if NpNa is dissolved in tetrahydrofuran or tetrahydropyran because the odd electron here interacts with the sodium nucleus (spin 3/2) indicating the presence of tight ion pairs $Np^{\ominus\bullet}Na^{\oplus}$.

Addition of tetraglyme (tetraethyleneglycol dimethyl ether), $CH_3(OCH_2CH_2)_4OCH_3$, decreases the coupling constant from 1.4 gauss to 0.4 gauss, indicating that the ions of the tight ion pair have moved further apart because of a solvation of the sodium cations. The ions of the tight ion pair are now present as loose (solvated) ion pairs $Np^{\ominus\bullet}//Na^{\oplus}$. The solvation lowers the proportion of covalency in the bond between $Np^{\ominus\bullet}$ and $Na^{\oplus}$ and the bond in loose ion pairs is thus only electrostatic.

8.2   Living Polymerization

8.2.1   Phenomena

In the simplest ionic polymerizations, initiator molecules IQ dissociate fast and completely into initiator ions and counterions, e.g., $IQ \rightarrow I^{\ominus} + Q^{\oplus}$. The initiation reaction I* + M → IM* (* = $\oplus$ if cationic, $\ominus$ if anionic) to monomer ions IM* is also fast and complete. The initial molar initiator concentration is thus given by $[IQ]_o = [I^*] = [IM^*]$.

If all monomer ions are completely converted into macroions $P^* \equiv IM_{i+1}^*$ ($i = 1, 2,$ ..., N), i.e., the monomer conversion equals $u = ([M]_o - [M])/[M]_o = 1$, then the molar concentration of macroions P* equals the molar concentration of monomer ions IM* or initiator ions I* or initiator molecules IQ, i.e., $[IQ]_o = [I^*] = [IM^*] = [P^*]$.

An initiator ion I* consumes only one monomer molecule to become a monomer ion IM* but a monomer ion adds many monomer molecules in the propagation reaction. The monomer consumption by initiation is thus negligible compared to the one by propagation. The gross polymerization rate R_{gross} is therefore practically identical with the propagation rate R_p that, in turn, equals the decrease of monomer concentration with time. It is also proportional to the molar concentrations of macroions and monomer molecules

$$(8-5) \qquad R_{gross} \approx R_p = - d[M]/dt = k_p[P^*][M] = k_p[IQ]_o[M]$$

where k_p = rate constant of propagation. The actual monomer concentration, $[M]$, decreases with time and so does the polymerization rate R_p (Eq.(8-5)) and the monomer conversion, $u = ([M]_o - [M])/[M]_o$, (Fig. 8-2). An almost 100 % monomer conversion is obtained if the polymerization equilbrium $IM_i^* + M \rightleftarrows IM_{i+1}^*$ is strongly shifted to the right, i.e., the depropagation reaction is negligible; the reaction is irreversible.

Macroions obtained at $u \approx 1$ are theoretically "**immortal**" if termination and transfer reactions are absent. Addition of new monomer causes the polymerization to "restart" and proceed until all monomer molecules are again used up (Fig. 8-2). The polymerization is "**living**". The polymerization rate and the maximum monomer conversion are lower, however, since the reaction mixture is diluted by the polymer from the first polymerization experiment.

The growth of a macroion is, however, often stopped by a termination or transfer reaction. Both reactions lead to "dead" macromolecules and a lowering of the degree of polymerization. Transfer reactions preserve kinetic chains, however, whereas termination reactions do not. In ionic polymerizations, these reactions are sometimes so slow that they are not noticeable during the experiment; such polymerizations are still said to be "living" (see Section 8.2.2).

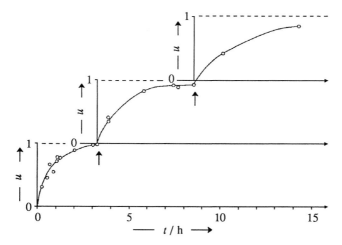

Fig. 8-2 Time dependence of monomer conversion u in the living cationic polymerization of 0.5 mol/L p-methyl styrene, $CH_2=CH(p\text{-}C_6H_4CH_3)$, in methylene choride at $-78°C$ by 0.002 mol/L acetylperchlorate, $CH_3CO^{\oplus}ClO_4^{\ominus}$ as initiator in the presence of 0.01 mol/L $(C_4H_9)_4NClO_4$ (data of [2]). Renewed monomer addition (↑) causes the polymerization to start again.

Table 8-1 Final monomer concentrations $[M]_\infty$ in polymerizations with spontaneous unimolecular termination and different ratios of rate constants of propagation, k_p, and termination, k_t; number-average degrees of polymerization, $\overline{X}_n$, at various ratios of rate constants of chain transfer to monomer, $k_{tr,m}$, and propagation, k_p; and times $t_{y\%}$ at which 5 % or 1 %, respectively, of macroions are deactivated by chain transfer or termination with rate constants k_v. $[M]_o = 1$ mol/L, $[I]_o = 10^{-3}$ mol/L.

Unimolecular termination		Bimolecular transfer to monomer		Termination or transfer		
$\dfrac{k_p / k_t}{\text{L mol}^{-1}}$	$\dfrac{[M]_\infty}{\text{mol L}^{-1}}$	$\dfrac{k_{tr,M}}{k_p}$	$\overline{X}_n$	$\dfrac{k_v}{\text{s}^{-1}}$	$t_{5\%}$	$t_{1\%}$
10	0.990	10^{-1}	9.9	10^{-1}	0.5 s	0.1 s
10^2	0.905	10^{-2}	90.9	10^{-2}	5.1 s	1.0 s
10^3	0.368	10^{-3}	500	10^{-3}	51.3 s	10.1 s
10^4	$4.54 \cdot 10^{-5}$	10^{-4}	909	10^{-4}	8.5 min	1.7 min
10^5	$3.72 \cdot 10^{-44}$	10^{-5}	990	10^{-5}	85.5 min	16.8 min
10^6	~ 0	10^{-6}	999	10^{-6}	14.2 h	2.8 h
∞	0	0	1000	0	∞	∞

For example, chains may be terminated by a spontaneous *unimolecular termination* such as $\sim CH_2CHR^{\oplus}[AsF_6]^{\ominus} \rightarrow \sim CH_2CHRF + AsF_5$. The rate of termination is given by $-d[P^*]/dt = k_t[P^*]$, and the rate of propagation by $-d[M]/dt = k_p[P^*][M]$ (Eq.(8-5)). Solving these equations for $[P^*]$ results in $d[M]/[M] = (k_p/k_t)d[P^*]$. Integration of this equation delivers $\ln ([M]_o/[M]_\infty) = (k_p/k_t)[IQ]_o$, if $[P^*] = [IQ]_o$. The residual monomer concentration *approaches* but never becomes zero. The monomer can be polymerized to 100 % conversion only if $k_p/k_t = 0$ (Table 8-1; $[M]_\infty = 0$).

The time $t_{y\%}$ for the disappearance of a fraction y of macroions can be calculated from the rate constant k_t of a unimolecular termination as $t_{y\%} = [\ln \{1/(1-y)\}]/k_t$ (Section 8.2.5). Calculations show that the "living character" of a polymerization is only maintained for sufficiently long times if the rate constant k_t is very small (Table 8-1). Such long **shelf lives** are sometimes needed for preparative work, for example, for the sequential synthesis of block copolymers (Section 16.2.2).

The transfer of the active center * to a monomer molecule from a growing polymer molecule, $\sim\sim P^* + M \rightarrow \sim\sim P + M^*$, is a *bimolecular transfer reaction* that lowers the degree of polymerization. The number-average degree of polymerization is calculated (Eq.(8-13)) to be $\overline{X}_n = [M]_0/\{[I]_0 + [M]_0(k_{tr,M}/k_p)\}$ if the monomer is polymerized completely and termination reactions are absent. The degree of polymerization of such polymerizations with transfer approaches that of truly living polymerizations only if the ratio $k_{tr,M}/k_p$ of transfer to propagation becomes very small (Table 8-1, $\overline{X}_n = 1000$). Termination and chain transfer may also be caused by external agents (impurities in monomer and solvent; atmospheric agents such as water, oxygen, or carbon dioxide).

8.2.2   Definitions

Polymer science borrows terms like "living" and "immortal" from biology but ignores other essential attributes of "life" such as metabolism, reproduction, response to stimulation, and evolution. It rather uses "living" for only one aspect of life: growth of an entity (i.e., a polymer molecule) which is curiously called "propagation" (L: *propages* = off-spring!). **Living polymerization** in polymer science refers to a chain polymerization with neither chain termination nor chain transfer.

If such a center remains active for an infinite time, then the polymerization would be truly **immortal** (this term has been used in the literature for a very special case); such a polymerization has also been called an **ideal living polymerization**. In many cases (but not all!), the rate of kinetic-chain initiation is much faster than the rate of propagation. This feature leads to narrow molecular weight distributions (Section 8.2.3).

These polymerizations remain "living" even if there are *noticeable* equilibria between growing chains and monomer molecules (this resembles periodic weight losses/gains of true living beings). *All* reactions between growing polymer chains and monomer molecules are equilibrium reactions $\sim M_N^* + M \rightleftarrows \sim M_{N+1}^*$. They are labelled **irreversible** if the equilibrium is shifted far to the right (note: these polymerization equilibria are *consecutive* equilibria involving active chains of different degrees of polymerization!).

To spin the somewhat inaccurate analogy to life further: the *irreversible* removal of an active center by unimolecular termination would be truly "suicidal"; an example is the reaction $\sim CH_2CHR^{\oplus}[AsF_6]^{\ominus} \rightarrow \sim CH_2CHRF + AsF_5$. The corresponding *reversible* reaction $\sim CH_2CHR^{\oplus}[AsF_6]^{\ominus} \rightleftarrows \sim CH_2CHRF + AsF_5$ amounts to a "sleep" or "dormancy".

The last reaction belongs to a group of polymerizations that has been called **quasi-living polymerization**, **pseudoliving polymerization**, or **quasiliving equilibrium**. The reactions of this group are characterized by equilibria between actively propagating ("living") species and their nonactive ("**dormant**") counterparts on one hand and the absence of any reactions that break the kinetic chain such as termination and transfer of active centers on the other. Again, a bad choice of words: a sleeping (dormant) being is very much alive and not "quasiliving" or "pseudoliving".

Dormancy of propagating species can be self-induced, caused by self-association with like molecules, or brought about by other agents. *Self-inflicted dormancy* is observed for certain species in cationic chain polymerizations such as $\sim CH_2-^{\oplus}CHR[AsF_6]^{\ominus} \rightleftarrows$ $\sim CH_2CHRF + AsF_5$ (see Section 8.4), cationic ring-opening polymerizations of the type $P-^{\oplus}M\bigcirc X^{\ominus} \rightleftarrows P-M-X$ (Section 8.4), or free-radical polymerizations $P^{\bullet} + R^{\bullet} \rightleftarrows P-R$ by stable radicals $R^{\bullet}$ or so-called iniferters (Chapter 10).

Dormant species in equilibrium with propagating ones occur if N growing macro-anions $P^{\ominus}$ with their countercations $Q^{\oplus}$ *self-associate* to N-mers according to $N P^{\ominus}Q^{\oplus} \rightleftarrows (P^{\ominus}Q^{\oplus})_N$ (Section 8.3.4). Dormancy may also be caused by other agents, for example, in transition-metal complex mediated polymerizations or ring-opening polymerizations (Section 9.4.4).

8.2.3 Degree of Polymerization

In ideal living polymerization with infinitely fast initiation compared to propagation ($k_i \gg k_p$), monomeric units present at a monomer conversion $u = ([M]_o - [M])/[M]_o$ are distributed evenly among all active centers *, i.e., with the molar concentration $[*] =$ $[IQ]_o$. An active polymer *molecule* may have f_c active centers, for example, $f_c = 1$ for the reaction $C_4H_9^{\ominus} + M \rightarrow C_4H_9M^{\ominus}$ or $f_c = 2$ for $2 \cdot M^{\ominus} \rightarrow {^{\ominus}}MM^{\ominus}$ (see also Eq.(8-23)).

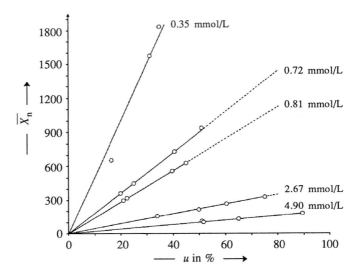

Fig. 8-3 Number-average degree of polymerization as a function of conversion u of styrene in tetrahydrofuran using sodium naphthalene as initiator ($f_u = 2$) at 25°C [3a]. Numbers indicate initial initiator concentrations. From top to bottom: $[M]_o/(mol\ L^{-1}) = 1.02, 0.79, 0.53, 0.58$, and 0.51. The authors did not determine number averages but viscosity averages and assumed $\overline{X}_n \approx \overline{X}_v$ because of the very narrow molecular weight distributions.

The number-average degree of polymerization of such living polymers

(8-6) $$\overline{X}_n = f_c u[M]_o/[IQ]_o = f_c([M]_o - [M])/[IQ]_o$$

increases linearly with increasing initiator concentration and monomer conversion (Fig. 8-3). It is $\overline{X}_n = f_o[M]_o/[IQ]_o$ at 100 % monomer conversion ($u = 1$, $[M] = 0$).

The *hypothetical* case of a truly ideal living polymerization should lead to an extremely narrow distribution of degrees of polymerization, provided all initiator anions and monomer anions are already present at the beginning of the polymerization ($k_i \gg k_p$) and are homogeneously distributed in the reactor vessel. All polymer chains should thus start growing simultaneously. Monomer molecules add to monomer anions or macroanions at random and independently of the preceding monomer additions. It results in a Poisson distribution of the degree of polymerization of the polymer molecules (Appendix A-8 and Section 3.3.3). A polymer with $\overline{X}_n = 500$ should thus have a weight/number polymolecularity index of $\overline{X}_w/\overline{X}_n = 1.002$.

However, the addition of a monomer molecule to a growing polymer molecule is not truly irreversible but an equilibrium reaction, $I–M_i^{\ominus} + M \rightleftarrows I–M_{i+1}^{\ominus}$. Such an equilibrium is obtained in few seconds of polymerization according to theoretical calculations. At any given time, growing polymer molecules possess neither an exactly equal number of monomeric units nor the one dictated by a Poisson process (p. 73). Hence, the molecular weight distribution rather becomes broader than a true Poisson distribution.

The distribution of the degrees of polymerization is also broadened by diffusion effects. Thus, on addition of the initiator to the monomer, the first polymer molecules are already formed before the last initiator molecules arrive.

A more homogeneous distribution of initiator molecules is obtained by **seed techni-ques**. Monomers and initiators are mixed conventionally. The solution of the resulting living polymers is then heated to temperatures *far above* the ceiling temperature, T_c (note: $\overline{X}_n > 1$ at T_c, see Section 7.3.2), until all macroanions have depolymerized to monomer anions. After a *very fast* cooling to the desired polymerization temperature, all polymer molecules start to grow *almost* simultaneously and deliver a much narrower molecular-weight distribution. If the ceiling temperature is very high, one can alter-natively let the total of the initiator react with a small proportion of the monomer and use the resulting oligomers as "seed" for the remaining monomer.

Sometimes slow initiation reactions are suspected of causing broader molar mass dis-tributions. To eliminate this effect, it was proposed to react all of the initiator with a small proportion of the monomer as above and then to add all of the bulk of the monomer. However, such a procedure does not narrow the molar mass distribution as can be seen from the following.

Monofunctional initiators I disappear according to $-d[I]/dt = k_i[I][M]$. In slow initia-tion reactions, the monomer is consumed according to $-d[M]/dt = k_i[I][M] + k_p[P^*][M]$. Since the concentration of active polymer molecules is given by $[P^*] = [I]_0 - [I]$, the monomer concentration changes with the initiator concentration according to

(8-7) $\qquad d[M]/d[I] = 1 + (k_p/k_i)([I]_0 - [I])/[I].$

The change of concentration, $d[M]/d[I]$, at constant initiator concentration is thus in-dependent of the initial monomer concentration, $[M]_0$: the molar mass distribution is *theoretically* independent of the way the monomer is added. Contrary to theory, how-ever, experiments do deliver narrower molecular weight distributions that may result either from decreased diffusion effects or from the removal of those impurities on the surface of the reaction vessel that may terminate the kinetic chain.

Integration of Eq.(8-7) delivers

(8-8) $\qquad [M] = [M]_0 + \{1 - (k_p/k_I)\}\{[I] - [I]_0\} + (k_p/k_i)[I]_0 \ln ([I]/[I]_0)$

which allows one to calculate the initiator concentration $[I]$ that remains after a complete monomer consumption ($[M] = 0$) in polymerizations with slow initiation reactions.

Most reasonably conducted living polymerizations result in polymers with $1,03 \leq \overline{X}_w/\overline{X}_n \leq 1,05$, usually said to be "practically molecularly uniform." In reality, such mo-lecular weight distributions are fairly broad, however. A polymer with $\overline{X}_n = 500$ and $\overline{X}_w/\overline{X}_n = 1.04$ possesses a number-standard deviation of $\sigma_n = 100$ (p. 71) which means that 15.87 % of the molecules have degrees of polymerization of $X < 400$ and 15.87 % of $X > 600$ if the distribution is assumed to be Gaussian or nearly so.

The kinetic control of living polymerizations prevents the macroions from reaching a complete thermodynamic equilibrium with their fellow macroions. Such an equilibrium can be obtained only by depolymerization/polymerization reactions at active chain ends if exchange reactions between chain segments are absent, for example, in vinyl polymer-izations. Such approaches to equilibrium require, however, up to 100 years according to theoretical calculations. There is indeed a broadening of molecular weight distributions with time in living vinyl polymerizations but this is not caused by an approach to equibrium but by kinetic chain transfer reactions.

Living polymerizations with exchange reactions between two chain segments or an end segment and a chain segment lead to the establishment of true equilibria at higher monomer conversions and/or temperatures. An example is the equilibration of polysiloxanes according to Eq.(7-3) or the polymerization of lactams with carboxylic acids as initiators at higher temperatures (Fig. 8-4). The polymerizations lead in equilibrium to a Schulz-Flory distribution of the degrees of polymerization, resulting in $\bar{X}_w/\bar{X}_n = 2$ at high degrees of polymerization (Appendix A-7).

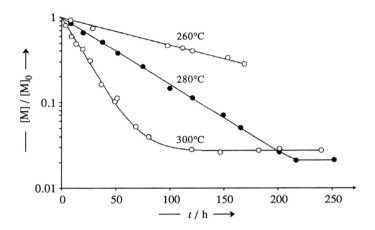

Fig. 8-4 Relative monomer proportion (logarithmic scale) as a function of time for the polymerization of neat laurolactam with 1 % lauric acid as initiator at various temperatures [4]. Polymerization equilibria are established after ca. 130 h at 300°C and ca. 220 h at 280°C.

8.2.4 Kinetics

Integration of the rate equation (8-5) delivers

(8-9) $\ln ([M]/[M]_0) = - k_p[I]_0 t$

setting IQ = I. For *irreversible* polymerizations, a plot of lg ([M]/[M]$_0$) = f(t) gives a straight line with a slope of $-k_p[I]_0/2.303$ (Fig. 8-4, 260°C). In sufficient time, many apparently irreversible polymerizations approach equilibrium (Fig. 8-4, 280°C, 300°C).

Reversible polymerizations are increasingly affected by back reactions and, at large times, equilibrium with [M]/[M]$_0$ = [M]$_\infty$/[M]$_0$ = *const.* is established ([M]$_\infty$ = equilibrium concentration of monomer) (Fig. 8-4, 300°C). Because of depolymerization, ~P$_i$* → ~P$_{i-1}$* + M, Eq.(8-5) has to be replaced by –d[M]/dt = $k_p[P^*][M] - k_{dp}[P^*]$. At high degrees of polymerization, one can set $k_{dp} \approx k_p[M]$ (Eq.(7-15)) and thus –d[M]/dt = $k_p[P^*]\{[M] - [M]_\infty\}$. Integration of this equation and rearrangement delivers Eq.(8-10) which allows straightening of the curves resulting from a plot of lg ([M]/[M]$_0$) = f(t):

(8-10) $\ln\left\{\dfrac{[M]-[M]_\infty}{[M]_0-[M]_\infty}\right\} = -k_p[P^*]t$

8.2.5 Polymerization with Transfer or Termination

Individual macroions can be destroyed by kinetic termination or transfer reactions. In both cases, the ion of the macroion is replaced by an atom or a group R of another molecule. The newly formed species is unable to start a new polymer chain in termination reactions but may do so in transfer reactions where it may be even as reactive as the disappearing macroion. Termination and transfer may be unimolecular or bimolecular reactions with other molecules (monomer, polymer, solvent, initiator, etc.).

Transfer to Monomer

In transfers to monomer, $P^* + M \rightarrow P + M^*$, the newly formed ion M^* is consumed in a new initiation reaction $M^* + M \rightarrow M_2^*$. The rate of this initiation reaction, $R_i = k_i[M^*][M]$, must thus equal the rate of the transfer reaction, $R_{tr,M} = k_{tr,M}[P^*][M]$. It follows that $[P^*]/[M^*] = k_i/k_{tr,M} \equiv C$. If each initiator molecule I starts a kinetic chain, the total concentration of all initiating molecules is $[I]_0 = [P^*] + [M^*] = [P^*](1 + C^{-1})$ and the rate of polymerization, R_p, becomes

(8-11) $R_p = -d[M]/dt = k_p[P^*][M] = k_p[I]_0(1 + C^{-1})^{-1}[M] = k_pC(1 + C)^{-1}[I]_0[M]$

(8-12) $\ln ([M]_0/[M]) = k_pC(1 + C)^{-1}[I]_0t = const [I]_0t ; \quad const = k_pC/(1 + C)$

The rate constant of the start reaction, k_i, must be larger than the rate constant of the transfer to monomer, $k_{tr,M}$, because no polymer molecules would be formed otherwise. The condition $C = k_i/k_{tr,M} > 1$ means, however, that $k_pC(1 + C)^{-1}$ in Eq.(8-12) must be larger than k_p in Eq.(8-9). It also follows that $\ln ([M]/[M]_0$ increases linearly with time even in polymerizations with transfer to monomer (Fig. 8-5, II) albeit less than in ideal living polymerizations (Fig. 8-5, I). Rate studies thus do not detect transfer to monomer.

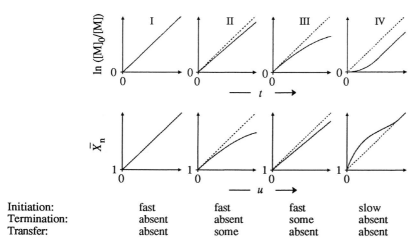

Initiation:	fast	fast	fast	slow
Termination:	absent	absent	some	absent
Transfer:	absent	some	absent	absent

Fig. 8-5 Ideal living polymerization (I and - - - in II-IV) and polymerizations with chain transfer (II), termination (III), and slow initiation reactions (IV). $u = ([M]_0 - [M])/[M]_0$.
Top: Time dependence of $\ln ([M]_0/[M])$.
Bottom: Number-average degree of polymerization as a function of relative monomer conversion, u.

Chain transfer to monomer can be detected from the time dependence of number-average degrees of polymerization. With each new ion M^*, a new dead polymer molecule P is also formed ($P^* + M \rightarrow P + M^*$), thus $[M^*] = [P]$. Molecules P appear with a rate of $d[P]/dt = k_{tr,M}[P^*][M]$ whereas monomer disappears with a rate of $-d[M]/dt = k_p[P^*][M]$. It follows that $d[P]/d[M] = -(k_{tr,M}/k_p)$. The integration of this equation from 0 to $[P]_\infty$ for polymer formation and from $[M]_0$ to $([M]_0 - [M])$ for the change in monomer concentration delivers the molar concentration of dead polymer molecules as $[P]_\infty = (k_{tr,M}/k_p)([M]_0 - [M])$. Since $[M^*] = [P]_\infty$, $[P^*] = [I]_0(1 + C^{-1})$, $C \equiv k_i/k_{tr,M}$ (p. 231), and $k_i \gg k_{tr,M}$, the number-average degree of polymerization becomes

$$(8\text{-}13) \qquad \overline{X}_n = \frac{[M]_0 - [M]}{[I]_0} = \frac{[M]_0 - [M]}{[M^*] + [P^*]} = \frac{[M]_0 - [M]}{[P]_\infty + [I]_0(1 + C^{-1})^{-1}}$$

$$(8\text{-}14) \qquad \frac{1}{\overline{X}_n} = \frac{k_{tr,M}}{k_p} + \left(\frac{k_{st}}{k_{st} + k_{tr,M}}\right)\left(\frac{[I]_0}{[M]_0 - [M]}\right) \approx \frac{k_{tr,M}}{k_p} + \frac{[I]_0}{[M]_0 - [M]}$$

Hence, number-average degrees of polymerization do not increase linearly with monomer conversion if chain transfer to monomer is present (Fig. 8-5, II). This dependence can be rectified by plotting $1/\overline{X}_n$ as a function of $[I]_0/([M]_0 - [M])$ which delivers a straight line with a slope of unity and an intercept of $k_{tr,M}/k_p$ (Fig. 8-6).

Termination

The rate of polymerization is $R_p = -d[M]/dt = k_p[P^*][M]$ for both bimolecular and unimolecular termination reactions. The concentration $[P^*]$ of macroions is, however, neither constant nor identical with the initial initiator concentration, $[I]_0$, as it is in ideal living polymerizations. Instead, it decreases with time, and the term $\ln([M]_0/[M])$ is therefore not a linear function of time (Fig. 8-5, III).

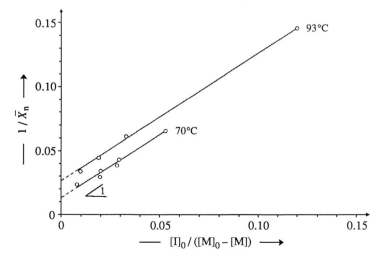

Fig. 8-6 Effect of chain transfer to monomer in the anionic polymerization of propylene oxide with CH_3ONa as initiator [5]. With permission of the Royal Society of Chemistry, London.

At the end of the polymerization, the concentration [P*] of macroions is smaller than the initial initiator concentration by a factor of f_∞, i.e., $[P*]_{t \to \infty} = [P]_\infty = f_\infty[I]_0$. Hence, at a monomer conversion of $u < 1$, a certain monomer concentration $[M]_\infty$ remains unpolymerized. The two types of termination reactions thus deliver the following expressions for the rates R_t of termination, the relative change of concentrations, $d[M]/d[P*]$, and the integrated rate equations:

	1st order termination	2nd order termination by polymer

$$(8\text{-}15) \qquad R_t = -d[P*]/dt = k_t[P*] \qquad\qquad R_t = -d[P*]/dt = k_t[P*]([M]_0 - [M])$$

$$(8\text{-}16) \qquad \frac{d[M]}{d[P*]} = \frac{k_p}{k_t}[M] \qquad\qquad \frac{d[M]}{d[P*]} = \frac{k_p}{k_t}\left(\frac{[M]}{[M]_0 - [M]}\right)$$

$$(8\text{-}17) \qquad \ln\left(\frac{[M]_0}{[M]_\infty}\right) = f_\infty \frac{k_p}{k_t}[I]_0 \qquad\qquad \ln\left(\frac{[M]_0}{[M]_\infty}\right) = \left(\frac{[M]_0 - [M]_\infty}{[M]_0}\right) + f_\infty \frac{k_p}{k_t}\frac{[I]_0}{[M]_0}$$

The number-average degree of polymerization is generally given by

$$(8\text{-}18) \qquad \overline{X}_n = \frac{\text{concentration of monomeric units}}{\text{concentration of polymer chains}} = \frac{[M]_0}{[P]_{tot}} u$$

where $[M]_0$ = initial monomer concentration, $[P]_{tot}$ = total concentration of living and dead polymer chains, and $u = ([M]_0 - [M])/[M]_0$ = fractional monomer conversion. Termination (spontaneous or by non-polymer molecules) will thus still deliver a linear dependence of $\overline{X}_n$ on u (Fig. 8-5, III). Any deviation from linearity points to the successive formation of additional chains (transfer, slow initiation) or to chain coupling (combination of two polymer molecules).

Active chain ends of macroions can be sealed to prevent undesired reactions such as chain transfer and depolymerization, for example, by addition of water or alcohols in anionic polymerizations ($RM_i^\ominus + H_2O \to RM_iH + HO^\ominus$). The newly formed hydroxyl anion is unable to start a new chain. Clever choice of sealing agents allows one to **func-tionalize** chains by introducing endgroups such as COOH or OH (Section 8.3.7).

Slow Initiation

In slow initiation reactions, polymer chains are started successively and not all at once as in ideal living polymerizations. The change of monomer concentration with time thus depends not only on the concentration [I] of initiator ions but also on the concentration $[P*] = [I]_0 - [I]$ of active polymer chains. The polymerization rate $-d[M]/dt = k_i[I][M] + k_p[P*][M] = -d\ln[M]/dt = (k_i - k_p)[I] + k_p[I]_0$ can be integrated to deliver

$$(8\text{-}19) \qquad \ln([M]_0/[M]) = k_p[I]_0 t + (k_i - k_p)\int_0^t [I]\,dt$$

Since $k_i \ll k_p$, $\ln([M]_0/[M])$ increases at first slowly before it becomes directly proportional to time (Fig. 8-5, IV). The degree of polymerization, on the other hand, is initially higher than it is later because fewer initiator ions are present at the early stage.

8.3   Anionic Polymerization

8.3.1   Monomers

Anionically polymerizing monomers contain electron accepting groups in ring molecules or in substituents of monomer molecules with multiple bonds. The latter comprise, for example, styrene $CH_2=CH(C_6H_5)$ and its phenyl-substituted derivatives, α-methylstyrene $CH_2=C(CH_3)C_6H_5$, acrylic compounds $CH_2=CHR$ (e.g., R = COOR', CN), methacrylic compounds $CH_2=C(CH_3)R$ (R = COOR', CN), vinylidene cyanide $CH_2=C(CN)_2$, 1,3-dienes $CH_2=CR–CH=CH_2$ (R = CH_3, Cl), certain aldehydes R"CHO, some ketones R"–CO–R, as well as isocyanates R–N=C=O. Oxiranes (ethylene oxide and derivatives), thiiranes, glycolides (e.g., lactide), N-carboxy anhydrides of α-amino acids, cyclosiloxanes, and some lactams and lactones polymerize by ring-opening. Examples are:

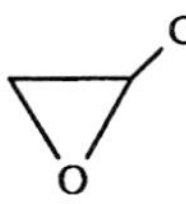

propylene oxide (methyl oxirane) 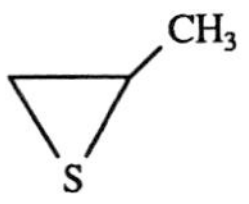   methyl thiirane (methyl episulfide) 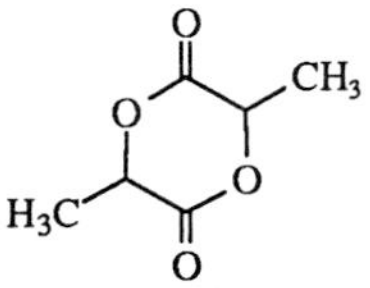   lactide 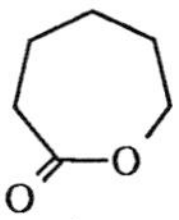 ε-caprolactone 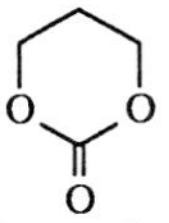   1,3-propylene carbonate

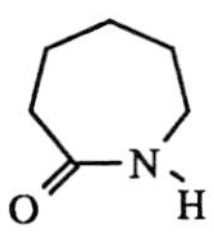

ε-caprolactam    laurolactam (dodecanolactam) 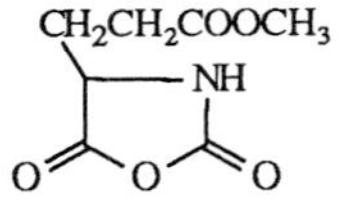 γ-methyl glutamate N-carboxy-anhydride (Leuchsanhydride)   hexamethyl cyclotrisiloxane

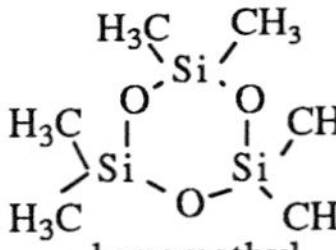

Some of these monomers polymerize only anionically. Others can also be polymerized by cationic initiators (Section 8.4), free radicals (Chapter 10), or coordination catalysts (Chapter 9) but the polymers from these processes often find other applications than the ones from anionic polymerizations. Table 8-2 lists important industrial anionic polymerizations and the main uses of the resulting polymers.

Despite the greater number of types of anionically polymerizable monomers, anionic homopolymerizations are used far less industrially than free-radical polymerizations (Chapter 10). Reasons include (1) higher costs of anionically polymerizable monomers, (2) required use of organic solvents (cost, health and environmental problems, expensive recovery), (3) expensive anionic initiators, and (4) the fact that high molecular weights are only obtained at high monomer conversions (see Fig. 6-1). Higher molecular weights can, of course, be obtained by lowering the initiator concentration but this would slow down the polymerization considerably; furthermore, the effect of impurities increases.

Many growing macroanions react fast with proton-donating compounds (e.g., water) and usually also with oxygen. Such chain-terminating compounds must be rigorously excluded which is more difficult to achieve in the laboratory than in industrial reactors since the former have a greater surface to volume ratio.

Anionic polymerizations, on the other hand, do allow much better stereocontrol of the polymerization than free-radical ones. Many anionic polymerizations can also be used for the synthesis of block copolymers (Chapter 16).

Table 8-2 Industrial anionic homopolymerizations. THF = tetrahydrofuran

Monomer	Monomeric unit	Applications
Butadiene	$-CH_2-CH=CH-CH_2-$ [1) $-CH_2-CH(CH=CH_2)-$	elastomers [2)
Alkyl cyanoacrylates	$-CH_2-C(CN)(COOR)-$	adhesives (crazy glue)
Formaldehyde	$-O-CH_2-$	engineering plastics
Ethylene oxide	$-O-CH_2-CH_2-$	thickeners
Propylene oxide	$-O-CH_2-CH(CH_3)-$	polyols
Glycolide	$-(O-CO-CH_2)_2-$	sutures
Lactide	$-(O-CO-CH(CH_3))_2-$	fibers
ε-Caprolactone	$-O-CO-(CH_2)_5-$	polymeric plasticizers
ε-Caprolactam	$-NH-CO-(CH_2)_5-$	thermoplastics
Laurolactam (dodecanolactam)	$-NH-CO-(CH_2)_{11}-$	fibers, sausage casings
Hexamethyl cyclotrisiloxane	$-O-Si(CH_3)_2-$	elastomers

[1) Cis and trans units. [2) Three industrial types of poly(butadiene)s by anionic polymerization: (1) with medium content of cis units (40 % cis, 50 % trans, 10 % vinyl) from the direct polymerization of butadiene in the C_4 cut of naphtha cracking by BuLi; (2) with medium content of vinyl units (35-55 %) by alkyl lithium + Lewis bases as randomizers in tetrahydrofuran (THF) or by alcoholates ROMt (Mt = Na, K, etc.); (3) liquid poly(butadiene)s by high concentrations of alkyl lithium in hydrocarbons (10-20 % vinyl), sodium alkyls in hydrocarbons (30-70 % vinyl), or sodium in hydrocarbons + a small proportion of THF (90 % vinyl). See also Volume II.

8.3.2 Initiators

Anionic polymerizations are initiated by bases or Lewis bases. Examples are alkali metals, alcoholates, metal ketyls, amines, phosphines, and Grignard compounds. The action of these and other initiators depends not only on the type of monomer but also on the solvent and the temperature. For example, tertiary amines initiate not only anionic polymerizations but also those via zwitterions (Section 8.5). Polymerizations in less polar systems may also lead to insertion polymerizations (Chapter 9). The nature of the initiator thus does not necessarily allow one to deduce the type of polymerization.

Anionic polymerization initiators may act in either of two ways: the initiating species may already be present in the added initiator or they may result from a reaction of the initiator with the monomer or a solvent.

Base Strength

An anionic initiator usually initiates the polymerization of an *unsaturated monomer* better, the smaller the ratio of the energy of the lowest unoccupied π-orbital to the basicity of the initiating anion. A monomer polymerizes more easily, the lower the pK_a value of the initiator. However, the ability to initiate depends not only on the basicity of the initiator. For example, styrene is polymerized in ammonia as solvent by sodium xanthenyl ($pK_a = 29$) but not by sodium fluorenyl ($pK_a = 31$). Additional factors include steric effects, resonance stabilization of initiator anions, and complexation of counterions by solvent and monomer molecules.

The ability to polymerize anionically decreases for olefin derivatives $CH_2=CHR$ in the following order of substituents R:

$$NO_2 > COR' > COOR'' \approx CN > C_6H_5 > CH=CH_2 \gg CH_3$$

Monomers with strong electron-accepting groups need only weak bases as initiators. The polymerization of 1-nitro-1-propene, $CH_2=C(CH_3)(NO_2)$ is thus initiated by the weak base $KHCO_3$. The nitrile group $-C\equiv N$ is a weaker electron acceptor than the nitro group $-NO_2$; acrylonitrile $CH_2=CH(CN)$ therefore requires a strong base such as sodium methylate, CH_3ONa, and acrylic esters $CH_2=CH(COOR'')$ an even stronger one such as sodium. Methyl methacrylate $CH_2=C(CH_3)(COOCH_3)$, on the other hand, is more difficult to polymerize anionically because of the electron-donating methyl group.

The effects are even stronger if the monomer contains two electron-accepting substituents. For example, the two nitrile groups of $CH_2=C(CN)_2$ allow the polymerization of vinylidene cyanide by the very weak bases water, alcohol, or ketone.

Metal Alkyls

Metal alkyls directly attack vinyl compounds nucleophilically. Preferred are lithium alkyls, RLi, because Li has the smallest radius and the highest electron affinity, electronegativity, and ionization potential. Only hydrocarbons and ethers can be used as solvents since halogenated compounds, esters, ketones, and alcohols react with RLi.

The alkyl group R of RLi becomes a covalently bonded polymer endgroup, for example, in the RLi-initiated polymerization of styrene:

(8-20) $(C_4H_9)^{\ominus} Li^{\oplus} \xrightarrow{+ C_8H_8} C_4H_9CH_2\overset{\ominus}{C}H \; Li^{\oplus} \xrightarrow{+ C_8H_8} C_4H_9CH_2\overset{}{C}HCH_2\overset{\ominus}{C}H \; Li^{\oplus}$

$\qquad\qquad\qquad\qquad\qquad\qquad\quad C_6H_5 \qquad\qquad\qquad\qquad C_6H_5 \;\; C_6H_5$ etc.

The negative charge of the macroanion is internally stabilized by the electron accepting group of the monomeric units. Macroanions may be also stabilized externally, however. Methyl methacrylate, for example, cannot be polymerized by C_6H_5Li in tetrahydrofuran because phenyl lithium reacts with ester groups. The polymerization succeeds at $-78°C$ if LiCl is added to C_6H_5Li. Even ethene can be polymerized anionically if alkyl lithium is combined with $(CH_3)_2NCH_2CH_2N(CH_3)_2$.

Lithium alkyls may lead to undesirable reactions. t-BuCl is, for example, a common impurity of s-BuLi. These two compounds react in the presence of dienes according to $C_2H_5-CH(CH_3)-Li + (CH_3)_3CCl \rightarrow C_2H_5-{}^{\bullet}CH-CH_3 + LiCl + (CH_3)_3C^{\bullet}$. The radicals then start a free-radical polymerization of dienes to randomly branched poly(diene)s.

In the polymerization of methyl thiirane in tetrahydrofuran at $-78°C$, ethyl lithium CH_3CH_2Li does not act directly as initiator. C_2H_5Li reacts first with methyl thiirane to form propene, $CH_2=CH(CH_3)$, and C_2H_5SLi, the latter being the true initiator.

Alcoholates

Alcoholate anions $RO^{\ominus}$ often add directly to monomer molecules. In the polymerization of ε-caprolactone, for example, the strongly nucleophilic alcoholates attack preferentially the carbonyl carbon atom. The newly formed alcoholates $RO–CO(CH_2)_5O^{\ominus}$ then initiate the polymerization of the lactone by an acyl cleavage of the latter.

The initiator is not always the genuine initiating species, however. The initiation of the polymerization of β-propiolactone (PL) by alcoholates, for example, leads to an unreactive 2-alkyl acrylic ester. The by-product KOH then initiates the polymerization which proceeds via carboxylate anions:

(8-21)

$$\underset{\substack{\downarrow + KOH}}{\overset{\substack{O \diagup R \\ \diagup}}{\square}} \xrightarrow{+ ROK} [RO–CO–CH_2–CHR–OK] \xrightarrow[-KOH]{} RO–CO–CH=CHR$$

$$[KO–CHR–CH_2–COOH] \longrightarrow HO–CHR–CH_2–COOK \xrightarrow[-H_2O]{} CHR=CH–COOK$$

$$\overset{\diagup + n\,PL}{} \qquad\qquad\qquad\qquad \downarrow + n\,PL$$

$$H[O–CHR–CH_2–CO]_{n+1}OK \qquad\qquad CHR=CH–CO[O–CHR–CH_2–CO]_n OK$$

True initiating species may also be generated by a reaction of the initiator with the solvent. Strong bases such as t-C_4H_9OK, for example, react with dimethylsulfoxide as solvent to give CH_3SOCH_2K whose anion is the true initiating species:

(8-22) $\qquad C_4H_9O^{\ominus} K^{\oplus} + (CH_3)_2SO \longrightarrow CH_3SOCH_2^{\ominus} K^{\oplus} + C_4H_9OH$

Alkali Metals

Solid alkali metals transfer an electron on initiation, e.g., from the sodium atom to the π orbital of a conjugated diene. The resulting diene radical anion then either (I) undergoes a second electron transfer with a sodium atom to a butadiene dianion or (II) couples with another diene radical anion to form a dibutadiene dianion:

(8-23)

$$\qquad\qquad\qquad\qquad\qquad\qquad\qquad\qquad \longrightarrow {}^{\oplus}Na\,{}^{\ominus}[CH_2–CH=CH–CH_2]^{\ominus}\,Na^{\oplus}$$

$$I \quad + Na^{\bullet}$$

$$[CH_2=CH–CH=CH_2]^{\bullet\ominus}\,Na^{\oplus}$$

$$II \quad + [CH_2=CH–CH=CH_2]^{\bullet\ominus}\,Na^{\oplus}$$

$$\qquad\qquad\qquad\qquad\qquad\qquad \longrightarrow {}^{\oplus}Na\,{}^{\ominus}[CH_2–CH=CH–CH_2–CH_2–CH=CH–CH_2]^{\ominus}\,Na^{\oplus}$$

Reaction I is much more probable than reaction II since the concentration of sodium atoms on the surface of the sodium metal is far greater than the concentration of radical anions. Furthermore, electron transfer is a relatively slow process.

Complexed metals also transfer electrons if monomers possess a sufficiently large electron affinity. In the reaction of naphthalene with sodium, for example, the transferred electron is placed into the lowest unoccupied π orbital of the naphthalene. The resulting naphthalene radical-anion $Np^{\ominus\bullet}$ (see Eq.(8-24)) stabilizes itself by interaction with solvating aprotic solvents such as tetrahydrofuran.

Addition of styrene monomer triggers an equilibrium reaction in which electrons are transferred to styrene molecules from naphthalene radical-anions (Eq.(8-24), left). Since

(8-24)

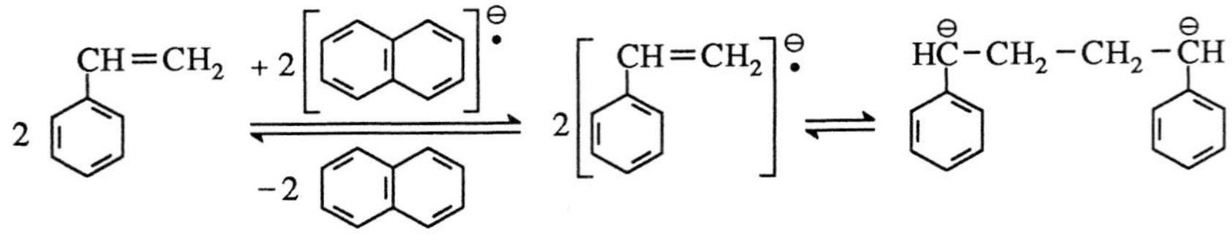

$$2 \; \substack{CH=CH_2} + 2 \left[\text{naphthalene} \right]^{\ominus \bullet} \underset{-2 \; \text{naphthalene}}{\rightleftharpoons} 2 \left[\substack{CH=CH_2} \right]^{\ominus \bullet} \rightleftharpoons \overset{\ominus}{HC}-CH_2-CH_2-\overset{\ominus}{CH}$$

naphthalene radical-anions $Np^{\ominus \bullet}$ are more strongly resonance stabilized than styrene radical-anions $S^{\ominus \bullet}$ the equilibrium $S + Np^{\ominus \bullet} \rightleftharpoons S^{\ominus \bullet} + Np$ should be strongly shifted to the left. The styrene radical anions are however fairly unstable and dimerize to distyryl dianions (Eq.(8-24) right). Since styrene radical-anions are removed from the equilibrium, new electron transfer takes place until practically all naphthalene radical anions are exhausted and only styrene dianions remain.

The dianions start the polymerization of styrene. The growing poly(styryl) dianions are, however, not infinitely stable, even if chain-terminating impurities are absent. After a few hours at room temperature, a unimolecular termination reaction and a subsequent bimolecular reaction leads to products that are no longer capable of polymerizing:

(8-25) $\sim\sim CH_2-\overset{\ominus}{\underset{C_6H_5}{CH}} Na^{\oplus} \longrightarrow \sim\sim CH=\underset{C_6H_5}{CH} + NaH$

$\sim\sim CH=\underset{C_6H_5}{CH} + Na^{\oplus} \overset{\ominus}{\underset{C_6H_5}{CH}}-CH_2\sim\sim \longrightarrow \sim\sim CH_2-\underset{C_6H_5}{CH_2} + Na^{\oplus} \underset{C_6H_5}{CH} \doteq CH \doteq \underset{C_6H_5}{CH} \doteq CH_2\sim\sim$

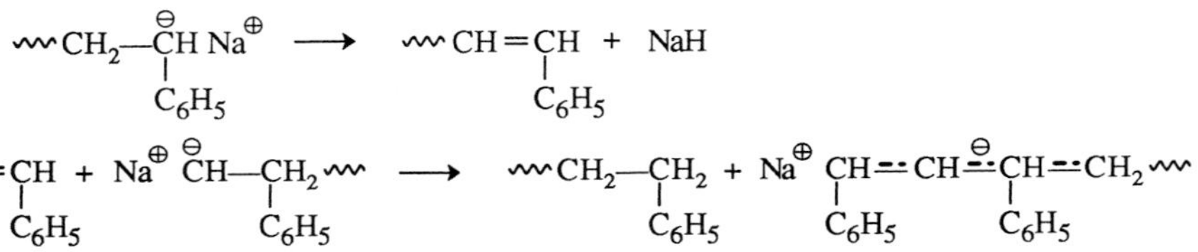

The polymerization of heterocycles by sodium naphthalene does not always occur by electron transfer. Ethylene oxide EO (oxirane), for example, adds directly to the naphthalene radical-anion $Np^{\ominus \bullet}$:

(8-26)

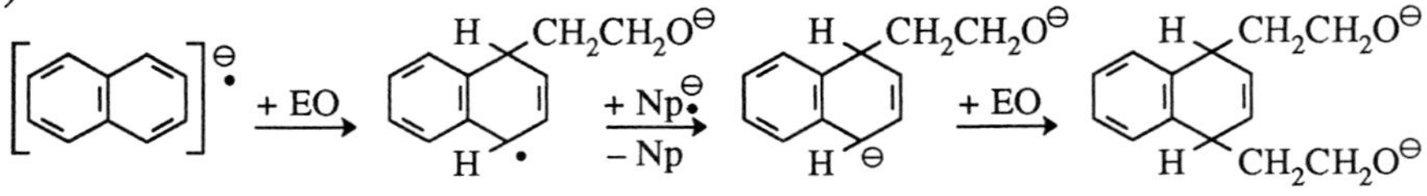

$$\left[\text{naphthalene} \right]^{\ominus \bullet} \xrightarrow{+ EO} \text{...} \xrightarrow[- Np]{+ Np^{\ominus \bullet}} \text{...} \xrightarrow{+ EO} \text{...}$$

8.3.3 Ion-Pair Equilibria

The rate constant k_p of propagation from Eq.(8-9) is independent of the initial initiator concentration $[I]_0$ if the polymerization proceeds via only one type of propagating species, e.g., free ions *or* loose ion pairs *or* tight ion pairs. A variation of k_p with $[I]_0$ or P^* points to the presence of more than one propagating species.

An example is the polymerization of styrene by sodium naphthalene in tetrahydrofuran. The polymerization is so fast that the concentration $[P^*]$ of growing polymer molecules is usually measured with the help of a flow tube. Separately prepared monomer and initiator solutions are mixed by a mixing jet. The mixture flows turbulently through a flow tube in order to prevent a distribution of growth times of macroanions as would happen in a laminar flow. Macroanions are then quickly killed by injecting an effective terminating agent. This **short-stop** technique allows one to calculate the effective

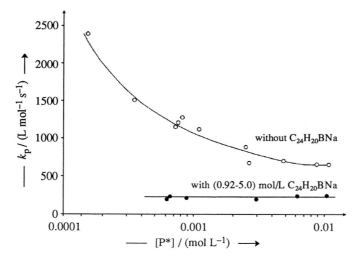

Fig. 8-7 Average propagation rate constants, k_p, of the polymerization of styrene by sodium naphthalene without and with addition of sodium tetraphenylborate at 25°C in tetrahydrofuran as a function of the concentration [P*] of active chain ends [3b]. O Polymerization of (0.25-1.02) mol/L styrene. ● Polymerization of 0.5 mol/L styrene with added sodium tetraphenylborate (Kalignost®).

polymerization time from the volume of the flow tube and the volume that flowed during the experiment. The apparent rate constant k_p of propagation is obtained from the monomer conversion and the total concentration [P*] of active chains.

The average propagation rate constant decreases with increasing concentration of active chain ends (Fig. 8-7). Since free ions $P^\ominus$ are more reactive than ion pairs $P^\ominus Na^\oplus$, this decrease points to an increasing proportion of ion pairs with increasing total concentration of active chains, [P*]. The association to ion pairs also increases with increasing concentration of compounds with the same counterion such as sodium tetraphenylborate, $C_{24}H_{20}BNa$. The propagation rate constant k_p becomes constant at high concentrations of $C_{24}H_{20}BNa$ (Fig. 8-7) since here practically only ion pairs are present.

The observed propagation rate constant is thus an average of the rate constants $k_{(\pm)}$ (ion pairs) and $k_{(-)}$ (free ions). Therefore, Eq.(8-5) needs to be modified to

(8-27) $R_p = - d[M]/dt = k_p[P^*][M] = k_{(\pm)}[P^\ominus Na^\oplus][M] + k_{(-)}[P^\ominus][M]$

(8-28) $k_p = \{k_{(\pm)}[P^\ominus Na^\oplus] + k_{(-)}[P^\ominus]\}/[P^*]$

The **Ostwald dilution law** allows one to eliminate the concentration of ion pairs and free ions. Since $[Na^\oplus] = [P^\ominus] = \alpha[P^*]$ and $[P^\ominus Na^\oplus] = (1 - \alpha)[P^*]$, resp., where α = degree of dissociation of ion pairs, the equilibrium constant K_d of dissociation becomes

(8-29) $K_d = \dfrac{[P^\ominus][Na^\oplus]}{[P^\ominus Na^\oplus]} = \dfrac{\alpha^2[P^*]}{1 - \alpha}$

$1 - \alpha \approx 1$ for small α. Since $K_d \approx \alpha^2[P^*]$ and also $[P^\ominus Na^\oplus] \approx [P^*]$, Eq.(8-28) becomes

(8-30) $k_p = k_{(\pm)} + k_{(-)}K_d^{1/2}[P^*]^{-1/2}$ (measurements without added salt)

Table 8-3 Rate constants $k_{(\pm)}$ of the polymerization of styrene via ion pairs at 25°C.

Monomer	Solvent	$k_{(\pm)}/(\text{L mol}^{-1}\text{ s}^{-1})$ for the counterion				
		$Li^\oplus$	$Na^\oplus$	$K^\oplus$	$Rb^\oplus$	$Cs^\oplus$
α-Methyl styrene	1,4-dioxane	-	0.016	0.098	0.062	-
	tetrahydropyran	2.6	0.047	0.25	0.35	0.26
Styrene	1,4-dioxane	0.9	4.0	20	22	25
	tetrahydropyran	-	14	60	80	-
	tetrahydrofuran	160	80	70	60	22
	dimethoxyethane	-	3600	-	-	150
2-Vinyl pyridine	tetrahydropyran	-	4500	-	-	-

The dissociation constant K_d of ion pairs is difficult to obtain with sufficient accuracy from conductivity measurments but can be obtained from the reaction kinetics. The degree of dissociation, $\alpha = [P^\ominus]/[P^*]$, equals the fraction of free macroanions $P^\ominus$ in the total concentration $[P^*]$ of active species. One obtains with Eq.(8-29) and $1 - \alpha \approx 1$:

(8-31) $k_p = k_{(\pm)} + k_{(-)}[P^\ominus][P^*]^{-1} = k_{(\pm)} + k_{(-)}K_d[Na^\oplus]^{-1}$ (with added salt)

The product $k_{(-)}K_d$ is obtained from the slope of a plot of $k_p = f(Na^\oplus]^{-1}$ and the various rate and equilibrium constants from combining data of Eqs.(8-30) and (8-31). The $k_{(\pm)}$ value follows from the intercept.

Rate constants $k_{(\pm)}$ of propagation by ion pairs vary strongly with solvent and the type of the counterion (Table 8-3, Fig. 8-8) whereas the rate constants $k_{(-)}$ of propagation by free macroanions are independent of solvent and counterion. Free macroanions have much greater propagation rate constants than their corresponding ion pairs, for example, $k_{(-)} = 65\ 000$ L mol $^{-1}$ s^{-1} for free poly(styryl) anions whereas $0.9 \leq k_{(\pm)}/(\text{L mol}^{-1}\text{ s}^{-1})$ ≤ 3600 for poly(styryl) ion pairs in various solvents and with different counterions.

Ion pairs however contribute a lot to the polymerization rates. For example, the dissociation constant of poly(styrene) anion pairs in tetrahydrofuran is only $K_d \approx 10^{-7}$ mol/L at 25°C (Table 8-4). At $[P^*] = 0.001$ mol/L, the fraction of free ions is therefore $\alpha = (10^{-7}/10^{-3})^{1/2} = 0.01$, i.e., only 1 %!

Table 8-4 Equilibrium constants K_d of dissociation of ion pairs into free ions in the anionic polymerization of different monomers with various counterions.

Monomer	Solvent	$T/°C$	$K_d/(\text{mol L}^{-1})$ for the counterion			
			$Li^\oplus$	$Na^\oplus$	$K^\oplus$	$Cs^\oplus$
Styrene	dimethoxyethane	25		$1.4 \cdot 10^{-6}$		$0.9 \cdot 10^{-7}$
	tetrahydrofuran	25	$1.9 \cdot 10^{-7}$	$1.5 \cdot 10^{-7}$	$0.7 \cdot 10^{-7}$	$4.7 \cdot 10^{-9}$
	oxepane	30		$7.0 \cdot 10^{-12}$		
Ethylene oxide	dimethylsulfoxide	25		$3.0 \cdot 10^{-4}$	$4.7 \cdot 10^{-2}$	$9.4 \cdot 10^{-2}$
	tetrahydrofuran	20		–	$1.8 \cdot 10^{-10}$	$2.7 \cdot 10^{-10}$
Propylene sulfide	tetrahydrofuran	20		$3.0 \cdot 10^{-9}$	$5.4 \cdot 10^{-9}$	
2-Vinyl pyridine	tetrahydrofuran	20		$8.3 \cdot 10^{-10}$	$2.5 \cdot 10^{-9}$	$1.1 \cdot 10^{-9}$

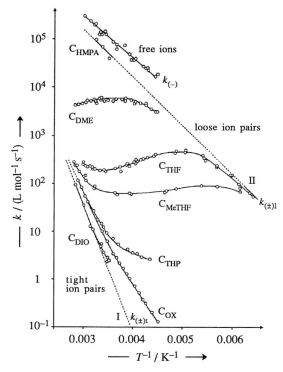

Fig. 8-8 Arrhenius plot of the rate constants k of propagation of poly(styryl) macroanions via free ions ($k_{(-)}$), loose ion pairs ($k_{(\pm)l}$), and tight ion pairs ($k_{(\pm)t}$) in dimethyl ethylene glycol (C_{DME}), 1,4-dioxane (C_{DIO}), hexamethyl cyclotriphosphamide (C_{HMPA}), 3-methyltetrahydrofuran (C_{MeTHF}), oxepane (C_{OX}), tetrahydrofuran (C_{THF}), and tetrahydropyran (C_{THP}) [6]. With the permission of the American Chemical Society, Washington, DC.

The temperature dependence of propagation rate constants of ion pairs, $k_{(\pm)}$, does not follow the Arrhenius equation. $k_{(\pm)}$ must therefore be an average of at least two types of ion pairs (Fig. 8-8) which is also indicated by the transition from one Arrhenius-type behavior (C_{DIO}) to another (C_{HMPA}). These limiting lines are not affected by the type of solvent. Line II does not differ much from the one for free ions and it must therefore indicate the behavior of loose ion pairs, $\sim P_i^{\ominus}/S/Na^{\oplus}$. Consequently, line I must then be assigned to tight ion pairs, $\sim P_i^{\ominus}Na^{\oplus}$, since these are certainly less reactive than loose ion pairs as indicated by their smaller propagation rate constants.

Tight ion pairs, loose ion pairs, and free ions are in thermodynamic equilibria with equilibrium constants K_{tl} for the equilibrium between tight and loose ion pairs and K_{lf} for the equilibrium between loose ion pairs and free ions:

(8-32)

$$\sim\!\!\sim P_i^{\ominus}Na^{\oplus} + j\,S \overset{K_{tl}}{\rightleftharpoons} \sim\!\!\sim P_i^{\ominus}/S_j/Na^{\oplus} \overset{K_{lf}}{\rightleftharpoons} \sim\!\!\sim P_i^{\ominus} + S_j\,Na^{\oplus}$$

$$+\,M\!\downarrow\! k_{(\pm)t} \qquad\qquad +\,M\!\downarrow\! k_{(\pm)l} \qquad\qquad +\,M\!\downarrow\! k_{(-)}$$

$$\sim\!\!\sim P_{i+1}^{\ominus}Na^{\oplus} \overset{K_{tl}}{\rightleftharpoons} \sim\!\!\sim P_{i+1}^{\ominus}/S_j/Na^{\oplus} \overset{K_{lf}}{\rightleftharpoons} \sim\!\!\sim P_{i+1}^{\ominus} + S_j\,Na^{\oplus}$$

The equilibrium constant $K_{lf} = [P_i^\ominus[Na^\oplus]/[P_i^\ominus/S_jNa^\oplus])$ of the non-conducting loose ion pairs into free ions can be obtained from conductivity measurements which deliver the experimental equilibrium constant $K_{exp} = [P_i^\ominus])[Na^\oplus]/([P^\ominus Na^\oplus] + [P_i^\ominus/S_jNa^\oplus])$ with regard to both loose and tight ion pairs, using the relationship $K_{exp} = K_{lf}K_{tl}/(1 + K_{tl})$. Poly(styrene) macroanions at 25°C showed, for example, K_{tl} values of 0.14 (dimethoxy ethane), 0.002 (tetrahydrofuran), and 0.0001 (tetrahydropyran).

The propagation rate constants k_{tl} of tight ion pairs and k_{lf} of loose ion pairs can be obtained from the equilibrium constant K_{tl} and the experimental average rate constant k_{exp} of propagation by ion pairs:

(8-33) $k_{exp} = k_{tl} + k_{lf}K_{tl}/(1 + K_{tl})$

The non-Arrhenius behavior of k_{exp} (Fig. 8-8) is thus caused by the equilibrium constant K_{tl} and the temperature dependence of the two rate constants, k_{tl} and k_{lf}.

8.3.4   Ion Associates

According to Eq.(8-9), polymerization rates of anionic polymerizations with fast initiation reactions should be directly proportional to the initial concentration of the initiator, $[I]_0$. However, a proportionality to $[I]_0^\alpha$ with $\alpha < 1$ is usually observed for the polymerization of apolar monomers such as styrene and isoprene (Table 8-5). This phenomenon is interpreted as an effect of association of initiator molecules.

Table 8-5 Association numbers N of initiators and inverse orders α of the initiator concentration for the addition of RLi to 1,1-diphenyl ether (DPE) and the polymerizations of styrene, butadiene, and isoprene in various solvents at 25°C [7a].

n-BuLi = $CH_3CH_2CH_2CH_2Li$ s-BuLi = $CH_3CH_2CH(CH_3)Li$ PBuLi = poly(butadienyl) lithium
i-BuLi = $(CH_3)_2CHCH_2Li$ t-BuLi = $(CH_3)_3CLi$ PILi = poly(isoprenyl) lithium
                                                                                           PSLi = poly(styryl) lithium.

Initiator	Solvent	N	$1/\alpha$ for DPE	$1/\alpha$ for the polymerization of		
				Styrene	Isoprene	Butadiene
n-BuLi	Benzene	6.25 ± 0.06	5.6	3 - 6	-	-
s-BuLi	Benzene	4.13 ± 0.05	-	4	4	1 - 3
t-BuLi	Benzene	4	3.6	-	-	-
PBuLi	Benzene	3.7 - 2	-	-	-	-
PILi	Benzene	2	-	-	-	-
PSLi	Benzene	2	-	3 - 6	-	-
n-BuLi	Cyclohexane	6.17 ± 0.12	-	1 - 2	1	1 - 2
s-BuLi	Cyclohexane	4.12 ± 0.09	-	0.7 - 1	0.9 - 1.5	1.0
i-BuLi	Cyclohexane	4.00 ± 0.08	-	1.0	1.0	-
t-BuLi	Cyclohexane	4.00 ± 0.05	-	1.0	1.4 - 5	1.0
PSLi	Cyclohexane	2	-	-	-	-
n-BuLi	hexane	6.25 ± 0.06	5.6	-	-	1
n-BuLi	diethyl ether	4	3.3	-	-	-
n-BuLi	tetrahydrofuran	2 - 3	2 - 2.5	-	-	-

Lithium alkyls RLi associate, for example, in apolar liquids to N-mers with equilibrium constants $K = [(RLi)_N]/[RLi]^N$ ($N = 2, 3, 4, 6$, etc.):

(8-34) $6 \, RLi \rightleftarrows 3 \, (RLi)_2 \rightleftarrows 2 \, (RLi)_3 \rightleftarrows (RLi)_6$

Association numbers N decrease with increasing polarity of the solvent and/or monomer and with increasing steric hindrance near the lithium atom (Table 8-5).

The association lowers the concentration of the non-associated alkyl lithium to $[RLi] = K^{-1/N}[(RLi)_N]^{1/N}$. The initiation rate by free alkyl lithium molecules is then given by

(8-35) $R_i = - \, d[M]/dt = k_i[RLi][M] = k_i K^{-1/N}[(RLi)_N]^{1/N}[M]$

The rate of polymerization is by analogy

(8-36) $R_p = - \, d[M]/dt = k_p[\sim P_iLi][M] = k_{p(\pm)}K^{-1/N}[(\sim P_iLi)_N]^{1/N}[M]$

where the rate constant k_p is replaced by the rate constant $k_{p(\pm)}$ for the propagation via ion pairs. Propagations by free macroanions need not be considered for these apolar systems. If only one type of associated species is present (e.g., with $N = 6$), then $[(\sim P_iLi)_N] = [\sim P_iLi]/N$ and thus also $R_p = k_{p(\pm)}(KN)^{-1/N}[\sim P_iLi]^{1/N}[M]$. The order of the initiator concentration should adopt $\alpha = 1/N$ in this borderline case (Table 8-5).

There may exist however multiple equilibria between species with different degrees of association. At very low initiator concentrations, association should be absent ($N = 1$) and the polymerization rate should be proportional to $[RLi]_0$. At very high initiator concentrations, almost all initiator should be associated to the highest degree of association, N, and the polymerization rate should thus be proportional to $[RLi]_0^{1/N}$ (Fig. 8-9). Intermediate exponents $1/N$ should be observed for medium initiator concentrations.

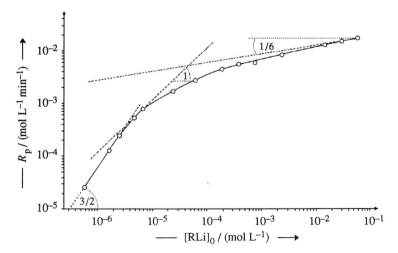

Fig. 8-9 Change of the initial polymerization rate R_p of the polymerization of 2.57 mol/L isoprene $CH_2=C(CH_3)-CH=CH_2$ in heptane at 20°C with the initial concentration $[RLi]_0$ of lithium isoprenyl as initiator. Numbers indicate the order $1/N$ with respect to the molar concentration of lithium isoprenyl $RLi = RCH_2C(CH_3)=CHCH_2CH_2C(CH_3)=CHCH_2Li$ [8]. The exponent 3/2 (if true) at very small values of $[RLi]_0$ is unexplained.

If only a small range of initiator concentrations is investigated (usually 1-2 decades) and not an extremely broad one as in Fig. 8-9 (almost 6 decades!), then any exponents α between 1 and 1/6 may be observed because of the consecutive equilibria between, e.g., RLi, $(RLi)_2$, $(RLi)_3$, and $(RLi)_6$. Even smaller values of α, and thus higher values of N, are observed for very high initiator concentrations if the initiator molecules associate in a chain-like manner. An example is the association of allyl lithium, $CH_2=CH-CH_2-Li$, in diethyl ether with $N = 2$ at $[I]_0 = 0.1$ mol/L but $N > 12$ at $[I]_0 = 1.6$ mol/L. Another example is the polymer $C_4H_9\text{+}CH_2CH(C_6H_5)\text{+}_n CH_2-CH=CH-CH_2Li$ that forms not only dimers but also cylindrical associates with very large association numbers N.

Differences between $1/\alpha$ from polymerization rates and N from association of the initiator in the absence of monomer may also arise for other reasons. Initiators RLi and growing polymers $\sim M_N Li$ may associate differently because of their different structure. There may also be an effect of different molecular weights, the possibility of cross-associations between RLi and $\sim M_N Li$, etc.

8.3.5 Propagation

Vinyl and Diene Polymerization

In classic anionic polymerizations, propagation steps comprise the addition of a new monomer molecule to the anionic end of a growing polymer chain. Since electrophilic monomers are nucleophilically attacked, an attack on the β-carbon atom of vinyl and acrylic compounds $^\beta CH_2 = {}^\alpha CHR$ can be expected. The new anion is formed at the α-carbon atom which carries the electron-accepting substituent R (C_6H_5, CN, COOR', etc.). The polymerization of acrylamide, $CH_2=CH(CONH_2)$ by weak bases thus leads to growing macroanions $\sim CH_2 - {}^\ominus CH(CONH_2)$.

Strong bases such as $(CH_3)_3 CO^\ominus K^\oplus$ do not polymerize acrylamide to poly(acrylamide), $\text{+}CH_2-CH(CONH_2\text{+}_n$ but to poly(β-alanine), $\text{+}CO-CH_2-CH_2-NH\text{+}_n$ (i.e., the initiator removes a proton from the amide group). The resulting acrylamide anion, and not an initiator fragment, is the true initiating species. Since $\sim CH_2-CH_2-CO-^\ominus NH$ is a stronger base than $-CH_2-^\ominus CH(CONH_2)$, the polymerization proceeds via the former:

(8-37)
$$CH_2=CH-CONH_2 \; \underset{+\,H^\oplus}{\overset{-\,H^\oplus}{\rightleftarrows}} \; CH_2=CH-CONH^\ominus \; \xrightarrow{+\,CH_2=CH-CONH_2}$$

$$CH_2=CH-CO-NH-CH_2-\overset{\ominus}{C}H-CONH_2 \rightarrow CH_2=CH-CO-NH-CH_2-CH_2-CONH^\ominus$$

Lactam Polymerization

Alkoxides as initiators also remove a proton from lactams, L, similar to Eq.(8-37). The resulting lactam anion $L^\ominus$(I) adds a monomer molecule to give an acyllactam anion (II). Transfer of II to a lactam molecule L regenerates a monomer anion $L^\ominus$ and delivers the ω-aminoacyl lactam (III), for example, in the ε-caprolactam polymerization:

(8-38)

The monomer anion $L^\ominus$ (I) can attack either a lactam molecule L or an ω-aminoacyl lactam molecule III. The latter reaction is, however, much faster than the former since the lactam ring of III is additionally activated by a second CO group:

(8-39)

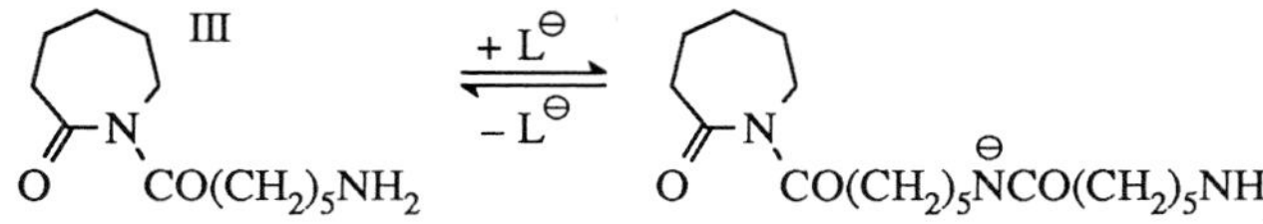

For this reason, lactam polymerizations by strong bases can be accelerated if "activators" such as acyl lactams are added. Such acyl lactams may also be formed *in situ* by adding acetic anhydride or ketenes.

Three cases can be distinguished for the propagation reaction: (A) via macroanions similarly to Eq.(8-21), (B) via monomer anions with activation, and (C) via monomer anions without activation. The initiating base usually reacts very fast with the monomer in Case A but very fast with the acyl lactam in Case B. All initiating species are therefore formed at the beginning of propagation. The polymerization rate is thus given by the ratio of the molar concentrations of monomer and base. The number-average degree of polymerization is controlled by the ratio of monomer to initiator molecules if termination and transfer reactions are absent. Such living polymerizations are therefore also called **stoichiometric polymerizations**.

The situtation is different for polymerizations via monomer anions without additional activators. The rate-determining step in Case C here is the slow formation of the activator, ω-aminoacyl lactam, which, once formed, starts the propagation in a subsequent fast reaction. More ω-aminoacyl lactam molecules are formed during the polymerization. Only if all base molecules have reacted with monomer molecules before all of the latter are consumed by polymerization is the degree of polymerization controlled by the ratio of monomer and base concentrations. This is usually not the case; so no direct correlation exists between the degree of polymerization and the monomer/base ratio.

Certain substituted lactams R–L present special cases. *N*-(1-Chloroacetyl)lactams $ClCH_2CO-(L)$ with 6-8 methylene groups per lactam ring L split off HCl on polymerization and lead to polymers with lactam rings as substituents.

p-Aminobenzoyl lactams I with large rings ($n > 9$) react to 90-100 % by ring-opening to polyamides II with alternating amide units. Smaller rings with $5 \le n \le 7$ polymerize to 70-80 % to poly(*p*-benzamide) III with elimination of lactam rings IV (Eq.(8-39)):

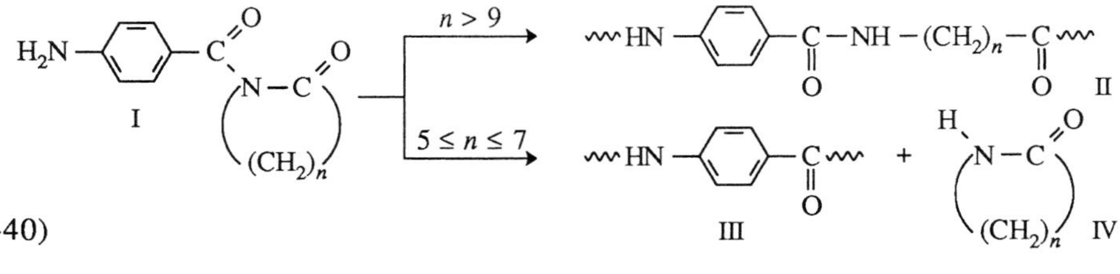

(8-40)

Polymerization of α-Amino Acid *N*-Carboxy Anhydrides

Primary amines attack *N*-carboxy anhydrides of α-amino acids (Leuchs anhydrides, NCAs) nucleophilically at the C^5 atom and generate carbamate anions II. Secondary amines react similarly if *N*-substituted Leuchs anhydrides are polymerized. Tertiary amines however cause a deprotonation of NCA to an NCA anion III (Eq.(8-41)):

(8-41)

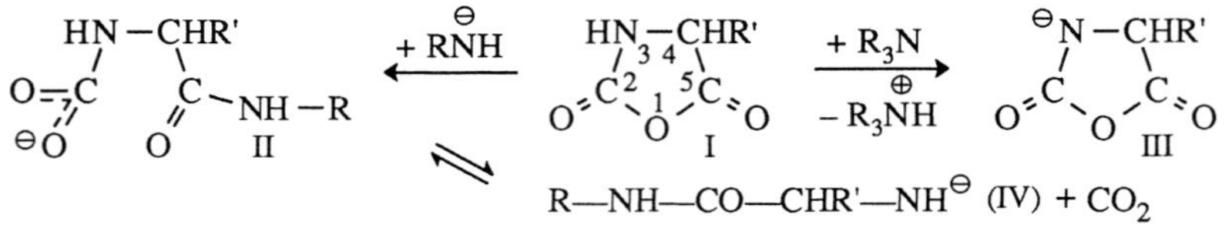

$$ R—NH—CO—CHR'—NH^{\ominus} \text{ (IV)} + CO_2 $$

An equilibrium similar to that between II and IV exists for amine initiators RNH_2 after decarboxylation of II or if CO_2 flows constantly through the reacting mixture. In the latter case, the monocarbamate, $H_2N(CH_2)_6NHCOOH$, of the initiator hexamethylene diamine could be isolated but not the dicarbamate. The equilibrium is shifted to the amine-anion if the liberated CO_2 is constantly removed by passing N_2 through the mixture.

NCA molecules can be attacked by carbamate anions $R–NH–CO–CHR'–NH–COO^{\ominus}$ (II) and amine-anions $R–NH–CO–CHR'–NH^{\ominus}$;

(8-42)

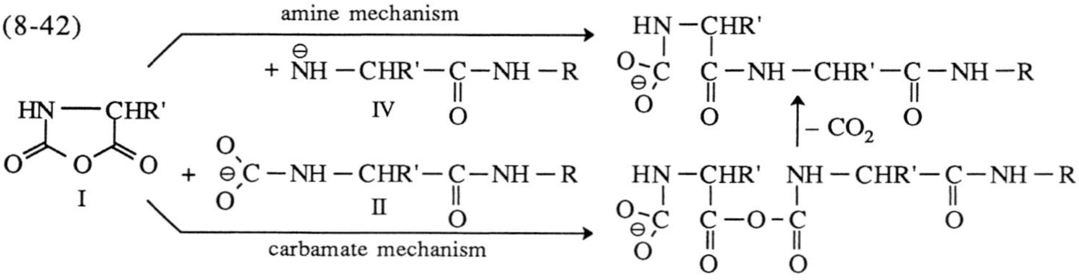

The newly formed carbamate ions can also be decarboxylated. The amine mechanism dominates if the liberated CO_2 is constantly removed by a nitrogen stream, leading to faster polymerizations than by the carbamate mechanism in the presence of CO_2.

NCA polymerizations in CO_2 atmosphere do not show an induction period (Fig. 8-10) because the primary amine initiators are already present as carbamates. The induction period 1 is short if the same polymerization is conducted in N_2 atmosphere; it ends after ca. 1 hour at lg $([M]_0/[M]) \approx 0.046$ which corresponds to a number-average degree of polymerization of $\overline{X}_n = ([M]_0 - [M])/[I]_0 \approx 5$ (Eq.(8-6)).

The induction periods are followed by relatively long periods 2 where lg $([M]_0/[M])$ increases linearly with time and polymers start to precipitate. The periods 2 are replaced by periods 3 with even faster polymerizations if $\overline{X}_n$ surpasses 17 (DL-NCA) and 28 (L-NCA), respectively.

The "2-period" polymerization is caused by the onset of association at the transition 1→2 and subsequent aggregations. It is not completely clear which additional role is played by β-pleated sheet (period 2) and α-helix structures (period 3).

The polymerization of the L-monomer by the chiral initiator L-α-methylbenzylamine is twice as fast as the polymerization of the racemic D,L-monomer. The initiator reacts with only one of the two enantiomers: the polymerization is thus enantioasymmetric (Section 6.4.3). The polymerization by the non-chiral initiator 1,6-hexamethylene diamine should thus be (almost completely) enantiosymmetric and should lead to a 1:1

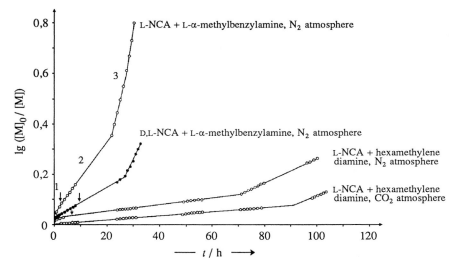

Fig. 8-10 Polymerization of 0.095 mol/L L-leucine Leuchs anhydride (O) and 0.095 mol/L D,L-leucine Leuchs anhydride (●), resp., both by 0.0019 mol/L L-α-methylbenzylamine in nitrogen atmosphere and 0.1 mol/L L-leucine Leuchs anhydride by 0.0005 mol/L 1,6-hexamethylene diamine in nitrogen or CO_2 atmosphere. All polymerizations in *p*-dioxane at 25°C [9]. Arrows indicate the beginning of precipitation.

mixture of the (almost completely) isotactic L and D polymers. Degradation studies of the polymer from the D,L monomer with the stereospecifically acting enzyme carboxypeptidase A delivered proportions of L units that correspond to 94-98 % isotactic diads (control: 98 % in poly(L-leucine), 0 % in poly(D-leucine)).

Very nucleophilic primary amines such as *n*-alkyl amines and benzylamine lead to $k_i > k_p$; the number-average degree of polymerization can be calculated from Eq.(8-6). Because of termination reactions and the limited solubility of the polymers in the solvents, such polymerizations are only "living" up to $\overline{X}_n \approx 100$. However, the highest degree of polymerization obtained was only 42 (at lg $[M]_0/[M] = 0.8$).

Far higher degrees of polymerizations of up to 5000 are achieved with tertiary amines as initiators. For such polymerizations, the existence of the NCA anion III has been proven (Eq.(8-41)). It is thus assumed that initiation proceeds according to reaction 8-43a to a carbamate anion V.

(8-43)

If the subsequent propagation reaction follows an **activated monomer mechanism**, Eq.(8-42b), both *N*-acyl-NCA dimers (V) and polymers must be more electrophilic than the NCA monomers (I) which implies $k_p > k_i$. Similarly to lactam polymerizations, Eq.(8-40), added N-acyl-NCA accelerates polymerizations.

In this scenario, the slow initiation reaction between the NCA anion $M^{\ominus}$ (III) and the monomer M (I) in Eq.(8-43) is the rate-determining step of polymer formation, $d[P]/dt = k_i[M^{\ominus}][M]$. The low initiator concentration means, however, that practically all the monomer is only consumed by the propagation reaction polymer P (VI) + NCA anion (III), i.e., according to $-d[M]/dt = k_p[P][M^{\ominus}]$. Solving these equations for $[M^{\ominus}]$ leads to $-k_i[M]d[M] = k_p[P]d[P]$. Integration with initial conditions $[M] = [M]_0$ and $[P]_0$ and the final conditions $[M] = 0$ and $[P] = [P]_\infty$ delivers the maximum number-average degree of polymerization as $\bar{X}_n = [M]_0/[P]_\infty = (k_p/k_i)^{1/2}$. The observed values of $30 < k_p/k_i < 90$ correspond to $5 < \bar{X}_n < 10$. Contrary to these kinetic assumptions, however, far higher degrees of polymerization have been measured, which is probably caused by the topo-logical peculiarities (see Fig. 8-10 for the polymerization kinetics).

8.3.6 Stereocontrol

The stereospecificity of anionic polymerizations is mainly controlled by the type of monomer and initiator (Table 8-6) and the proportion of the various types of propaga-ting macroanions (free ions, loose ion pairs, tight ion pairs, type and extent of solvation). The latter, in turn, depend on the solvent, the temperature, added complexing agents, the concentration of the initiator, and even the extent of monomer conversion. The degree of stereocontrol increases in general in the order free ions < loose ion pairs < tight ion pairs < ion associates. It is especially prominent in polymerizations by coordination cata-lysts (Chapter 9).

With increasing initiator concentration at constant monomer concentration, the equili-brium free ions $\rightleftarrows$ ion pairs is shifted toward ion pairs and maybe even further to ion associates. Free ions, loose ion pairs, tight ion pairs, and the various types of ion associ-ates (p. 242) control propagation steps differently. Tacticities thus depend on the initial initiator concentration, $[I]_0 \approx [P^{\ominus}]$, and therefore on the concentration of macroanions, $[P^{\ominus}]$ (Fig. 8-11). Since $[P^{\ominus}] \sim 1/\bar{X}_n$, tacticities will vary with the number-average degree of polymerization at otherwise constant polymerization conditions.

Table 8-6 Mole fractions of isotactic (x_{ii}), heterotactic (x_{is+si}), and syndiotactic (x_{ss}) triads of poly-(methyl methacrylate)s from anionic polymerization by alkyl lithiums and cationic polymerization by butyl magnesium bromides [7b]. TMEDA = tetramethylethylene diamine.

Initiator	Solvent	$T/°C$	x_{ii}	x_{is+si}	x_{ss}
$C_5H_{11}C(C_6H_5)_2Li$	toluene	− 78	0.87	0.10	0.03
n-C_4H_9Li	toluene	− 78	0.68	0.19	0.13
n-C_4H_9Li	pyridine	− 60	0.07	0.33	0.60
t-C_4H_9Li	toluene	0	0.02	0.26	0.72
t-C_4H_9MgBr	toluene	− 78	0.967	0.03	0.003
i-C_4H_9MgBr	toluene	− 78	0.925	0.054	0.021
n-C_4H_9MgBr	toluene	− 78	0.11	0.153	0.737
n-C_4H_9MgBr + TMEDA	tetrahydrofuran	− 78	0.002	0.096	0.902

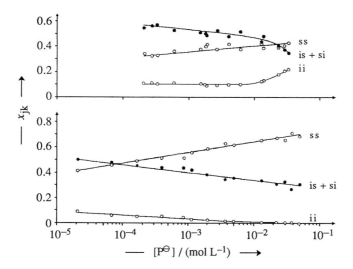

Fig. 8-11 Change of mole fractions x of isotactic (ii), heterotactic (is + si), and syndiotactic triads with the concentration $[P^{\ominus}]$ of macroanions of α-methyl styrene in tetrahydrofuran.

Top: sodium naphthalene as initiator at $-78°C$ and a solvent/monomer ratio of V/V = 10:1 [10].

Bottom: lithium butyl or lithium naphthalene as initiator at $-75°C$ and an initial monomer concentration of 0.75 mol/L [11].

Increasing concentrations of macroanions lead to increasing concentrations of tighter ion species and thus to enhanced stereocontrol: the proportion of heterotactic triads decreases with increasing concentration of macroanions (Fig. 8-11) and decreasing degree of polymerization. Lithium as a counterion leads to higher syndiotacticities than sodium.

The data of Fig. 8-11 show that the usually reported tacticities (such as in Table 8-6) are just snapshots for certain concentrations of initiator and monomer. For example, the fraction of syndiotactic triads varies by about a factor of 2 (Fig. 8-11, bottom). In general, anionic polymerizations of polar monomers in polar solvents are not very stereospecific but those of apolar monomers in apolar solvents very much so.

Ionic polymerizations can furthermore proceed by one-way or two-way mechanisms (Section 6.3.2). The active centers of growing macroanions of acrylates, methacrylates, vinyl aryls, and dienes exist as planar delocalized systems. Since the barrier for a rotation around the $^{\gamma}C-^{\beta}C$ bond is fairly high, active centers exist in both E and Z types (p. 127) which, in turn, leads to 2-way mechanisms. The proportion of tactic triads at the growing chain ends determines the proportion of E and Z types. They can be determined by NMR spectroscopy if suitable ^{13}C-labeled terminating agents are used.

(8-44)

Table 8-7 Mole fractions of configurational triads from the anionic polymerization of 2-vinyl pyridine [12]. jk = ii, is, si, or ss.

Counter-ion	Solvent	x_{jk} in chains			x_{jk} in endgroups			
		ii	is + si	ss	ii	is	si	ss
Mg	tetrahydrofuran	0.37	0.56	0.07	0.30	0.25	0.30	0.15
Li	tetrahydrofuran	0.44	0.44	0.12	0.55	0.01	0.43	0.01
Li	toluene	0.47	0.39	0.14	0.42	0.18	0.23	0.17
Mg	toluene	0.77	0.18	0.05	0.50	0.21	0.19	0.10
Mg	benzene	0.90	0.08	0.02	0.50	0.17	0.16	0.16

An example is poly(2-vinyl pyridine) where the tacticities of chains and endgroups do not agree (Table 8-7). In this system, the metal ion is probably associated with the unpaired electron pair of the nitrogen which makes this anionic polymerization more like an insertion polymerization with an enantiomorphic catalyst. Eq.(6-66) for 1-way mechanisms thus does not apply. It may be also possible that the formation of polymer molecules changes the polarity of the solution and thus the relative proportion of E and Z types and active species, respectively.

E types are especially prominent in polymerizations with lithium counterions in apolar solvents since lithium compounds have a higher degree of covalency than the other alkali compounds and the E type is preferred over the Z type for steric reasons. In kinetically controlled diene polymerizations, E types (trans centers) are frozen in if a new cis-1,4 unit is formed (Table 8-8).

Slow propagation steps or fast relaxations to the thermodynamically favored Z type (cis centers) lead to more trans-1,4 units. It is the opposite in polar solvents: cis is thermodynamically more stable and trans is kinetically preferred.

Table 8-8 Stereocontrol in the polymerization of 1,4-dienes, $CH_2=CR–CH=CH_2$.

Substituent R in diene	Initiator	Solvent	$T/°C$	Percentage in polymer			
				1,4-cis	1,4-trans	3,4	1,2
H	Li	-	50	35	52	-	13
H	Na	-	50	10	25	-	65
H	K	-	50	15	40	-	45
H	Rb	-	60	7	31	-	62
H	Cs	-	60	6	35	-	59
H	Li	1,4-dioxane	15	0	13	-	87
CH3	Li	heptane	25	94	0	6	0
CH3	Na	heptane	25	0	43	51	6
CH3	Na	diethyl ether	25	------- 19 -------		75	6
CH3	C4H9Li	heptane	40	70	22	7	0
C2H5	C4H9Li	heptane	40	78	14	8	0
C3H7	C4H9Li	heptane	40	91	4	5	0
C4H9	C4H9Li	heptane	40	62	35	3	0
C4H9	C4H9Li	diethyl ether	25	-------- 56 -------		44	0

8.3.7 Termination Reactions

Living anionic polymerizations stop at the monomer conversion that is dictated by the equilibrium macroanion $\rightleftarrows$ monomer, i.e., at nearly 100 % monomer conversion in the ideal case. The resulting macroanions, however, may act as initiators for the polymerization of other monomers, which will lead to block copolymers (p. 594). They may also react with non-monomers to give **functionalized polymers** (consisting of macromolecules with reactive endgroups) or non-functionalized polymers (consisting of "capped" macromolecules with non-reactive endgroups).

System-imminent termination reactions are fairly rare in anionic polymerizations. They do occur in tetrahydrofuran with sodium counterions in an **elimination** reaction:

$$(8\text{-}45) \quad \underset{I}{\sim\!\!\sim\!\!\overset{|}{\underset{C_6H_5}{C}}H\!-\!CH_2\!-\!\overset{\ominus}{\underset{C_6H_5}{C}}H\ \overset{\oplus}{Na}} \longrightarrow \underset{II}{\sim\!\!\sim\!\!\overset{|}{\underset{C_6H_5}{C}}H\!-\!CH\!=\!\overset{|}{\underset{C_6H_5}{C}}H} + NaH$$

$$I + II \longrightarrow \sim\!\!\sim\!\!\overset{|}{\underset{C_6H_5}{C}}H\!-\!CH_2\!-\!\overset{|}{\underset{C_6H_5}{C}}H_2 + \sim\!\!\sim\!\!\overset{\ominus}{\underset{C_6H_5}{C}}\!-\!CH\!=\!\overset{|}{\underset{C_6H_5}{C}}H\ \overset{\oplus}{Na}$$

The color of the poly(styrene) solution changes here from cherry-red to purple. Termination by elimination is also observed for butyl lithium polymerized poly(butadiene) where LiH is split off from the $\sim CH_2\text{–}CH_2\text{–}CH=CH\text{–}CH_2^{\ominus}$ $Li^{\oplus}$ ends. The double bonds of the resulting end units $\sim CH_2\text{–}CH=CH\text{–}CH=CH_2$ add monomer molecules and lead to branched polymers. Such elimination reactions are unimolecular. For initial conditions with $t = 0$ and $[P^{\ominus}] = [IQ]_0$, they follow the relationship $\ln([M]_0/[M]) = (k_p/k_t)[I]_0$ (Section 8.2.1).

Unintended cappings are bimolecular reactions that result from traces of impurities such as water, alcohols, and ammonia, which should be rigorously excluded. These compounds transfer protons according to $\sim P_i^{\ominus} + RH \rightarrow \sim P_iH + R^{\ominus}$. Water (R = OH) always leads to a termination whereas alcohols (R = R'O) and ammonia (R = NH$_2$) cause both termination and kinetic chain transfer (Section 8.3.8).

Deliberate capping reactions of polymer molecules with functional endgroups are **functionalizations**. For example, macroanions $P_i^{\ominus}$ react with carbon dioxide to polymers $P_iCOO^{\ominus}$ with carboxyl endgroups whereas ethylene oxide delivers $P_iCH_2CH_2O^{\ominus}$ that is converted to $P_iCH_2CH_2OH$ on further reaction with water.

Such cappings are not always complete. The reaction of poly(styryl) lithium (I) with CO_2 transforms I to the lithium carboxylate (II) with 60 % yield whereas 40 % of I reacts further via III to the ketone (IV) (25 %) and the tertiary alcoholate (V) (15 %):

$$(8\text{-}46)$$

$$\underset{I}{\sim\!\!\sim\!(sty)_m Li} \qquad \underset{II}{\sim\!\!\sim\!(sty)_m COOLi} \qquad \underset{III}{\overset{(sty)_m}{\underset{(sty)_n}{>}}C\overset{OLi}{\underset{OLi}{<}}} \qquad \underset{IV}{\overset{(sty)_m}{\underset{(sty)_n}{>}}C\!=\!O} \qquad \underset{V}{\overset{(sty)_m}{\underset{(sty)_p}{>}}(sty)_n\!-\!C\!-\!O\!-\!Li}$$

Capping reactions allow transformations of macroanions into macrocations and vice versa. Such reactions permit the synthesis of block copolymers if one monomer can only be polymerized anionically and the other one only cationically.

8.3.8 Transfer Reactions

Chain transfer may occur to all participants in anionic polymerizations, i.e., to monomers, polymers, solvents, initiators, or counterions. All true transfer reactions terminate
the growth of individual polymer chains but do not terminate that of *kinetic* chains since,
by definition, the newly formed anions start new polymer chains.

Transfer to monomer is especially prominent if the monomer molecules contain
groups that are easily attacked by macroanions. Functional groups such as OH and NH_2
must therefore always be protected because they would undergo proton transfer. The
CH_3 of the ester group of methyl methacrylate monomer and polymer is another sensitive groups.

Transfer reactions to monomers are often not very prominent and therefore not easily
detected analytically. Polymerization rate studies do not deliver hints since kinetics leads
to the equation $\ln([M]/[M]_0 = k_p[C(1 + C)^{-1}][I]_0 t$ that differs from the equation for a
propagation without transfer to monomer, $\ln([M]/[M]_0 = k_p[I]_0 t$, only by another constant (Fig. 8-5). Transfer to monomer can be deduced, however, from the dependence of
the number-average degree of polymerization on the initiator concentration (Eq.(8-14)
versus Eq.(8-6)). As a rule, transfer to monomer leads to bimodal molecular weight distributions.

Transfers to regulators are used industrially to control the degree of polymerization.
The polymerization of 1,3-butadiene, $CH_2=CH-CH=CH_2$ in hydrocarbons by lithium
butyl as initiator, for example, is regulated by the addition of 1,2-butadiene (methyl al-
lene), $CH_2=C=CH-CH_3$. The ratios k_p/k_{tr} of the rate constants of propagation and transfer are here 1250 (50°C), 670 (70°C), and 500 (90°C). At a ratio of $[1,2\text{-Bd}]/[\text{BuLi}] = 5$
and a monomer conversion of 95 %, the transfer to 1,2-Bd is ca. 3 % at $k_p/k_{tr} = 500$ and
ca. 14 % at $k_p/k_{tr} = 100$ for polymerizations in batch reactors. It is about 6 times higher
in continuous stirred tank reactors because of the change in diffusion patterns.

8.4 Cationic Polymerization

8.4.1 Overview

Cationic polymerizations are initiated by the reaction of electrophilic initiators $E^{\oplus}$
with electron-donating monomers M. The resulting monomer cations $EM^{\oplus}$ add further
monomer molecules if the nucleophilic groups are present in a part of the monomer
molecule that can participate directly in the propagation step. Cationically polymerizable
monomers are thus (1) olefins $CH_2=CHR'$ with electron-rich substituents R', (2) compounds $R_2C=Z$ with heteroatom-containing groups Z, and (3) heterocyclic rings.

Cationically polymerizable *electron-rich olefin derivatives* include π-donors such as
olefins, dienes, and vinyl aromatics as well as $(\pi + n)$-donors such as *N*-substituted vinyl
amines and vinyl ethers:

$CH_2=CRR'$	$CH_2=CR'$	$CH_2=CR$	$CH_2=CH$	$CH_2=CH$
	$CR=CH_2$	Ar	NRR'	OR
olefins	dienes	vinyl aromatics	vinyl amines	vinyl ethers

The propagating group is always the most nucleophilic part of the monomer molecules. In acrylonitrile, it is the nitrile group and not the vinyl group. Because of the resulting resonance stabilization, acrylonitrile cannot be polymerized cationically:

$$(8\text{-}47) \qquad CH_2=CH-C\equiv N \xrightarrow{+R^\oplus} \left[CH_2=CH-C\overset{\oplus}{\equiv}N-R \longleftrightarrow CH_2=CH-C\overset{\oplus}{=}N-R \right]$$

Carbon-carbon double bonds in cyclic structures also polymerize. Examples are benzofuran (coumarone), indene, cyclopentadiene, dicyclopentadiene, α-pinene, and β-pinene. α-pinene isomerizes first to D,L-limonene, the real monomer. β-pinene, however, polymerizes spontaneously (initiator-free) to a phantom polymer.

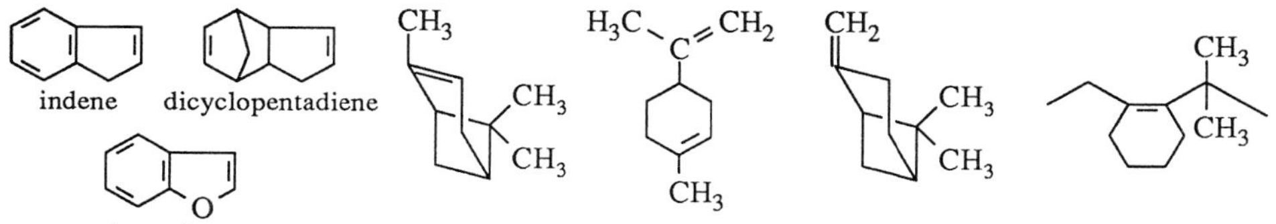

indene   dicyclopentadiene α-pinene   D,L-limonene β-pinene   poly(β-pinene)

benzofuran

Cationically polymerizable monomers with *heteronuclear multiple bonds* include aldehydes, RHC=O, certain ketones, RR'C=O, thioketones, RR'C=S, and diazoalkanes, $RR'C=N^\ominus=N^\oplus \leftrightarrow RR'C^\ominus-N^\oplus\equiv N$.

Cationically polymerizable *heterocyclic rings* are all n-donors. They include cyclic ethers, acetals, sulfides, imines, esters (lactones), and amides (lactams), for example:

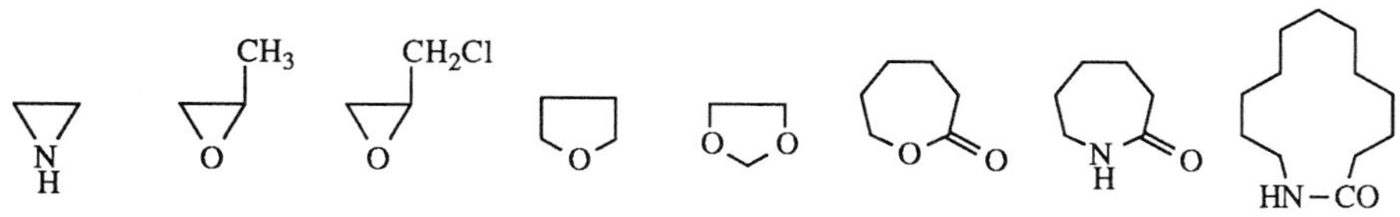

ethylene imine | propylene oxide | epichlorohydrin | tetrahydrofuran | 1,3-dioxolane | ε-caprolactone | ε-caprolactam | laurolactam

Many more monomer *types* can be subjected to cationic polymerizations than to anionic or free-radical ones yet far less monomer *species* are cationically polymerized in industry (Table 8-9). The main reason is the great reactivity (and thus the high instability) of many macrocations which leads to a great number of different termination and transfer reactions. Furthermore, practically all cationic polymerizations require expensive and difficult to reprocess organic solvents.

8.4.2 Initiators

Cationic polymerizations can be initiated via one-electron or two-electron mechanisms. One-electron mechanisms are homolytic; they comprise charge transfers or direct oxidation of radicals. Much more common are heterolytic two-electron mechanisms that are caused by three types of initiators: Brønsted acids, Lewis acids, and carbenium salts.

All polymerizations require non-nucleophilic gegenions since nucleophilic gegenions would combine with growing macrocations and terminate the chains. Anions are easily

Table 8-9 Industrial cationic homopolymerizations. [a] Less than 3 % isoprene as comonomer; [b] aliphatics from the C_5-C_6 fraction of the steam cracking of crude oil (volume II), aromatics from the C_8-C_{10} fraction; [c] + 6 % allyl glycidyl ether; [d] e.g., glycidyl ether based on bisphenol A, cycloaliphatic epoxides, etc.; $AlCl_3$ and BF_3 are usually used with water as "cocatalyst" (see p. 256).

Monomer	Initiator	Polymer application
Isobutene [a]	$AlCl_3$	elastomers, adhesives, viscosity improvers
Indene-coumarone resin	$AlCl_3$, BF_3	sealants, coatings
Petroleum resins [b]	$AlCl_3$, BF_3	adhesives, printing inks, rubber additives
Alkyl vinyl ethers	BF_3	adhesives, plasticizers, textile additives
Epichlorohydrin	$AlEt_3/H_2O$/acetyl acetone	elastomers
Propylene oxide [c]	$AlEt_3/H_2O$/acetyl acetone	elastomers
Epoxides [d]	$BF_3 \cdot C_2H_5NH_2$	coatings, adhesives, casting resins
Tetrahydrofuran	$HClO_4$, H_2SO_4, HSO_3F+SO_3	soft segments for elastomers
Ethyleneimine	proton and Lewis acids	paper additives, flocculants
Cyclosiloxanes	strong acids	oils, elastomers, resins

solvated, however. Since solvation changes the nucleophilicity, only a few solvents are suitable. Common solvents are benzene, nitrobenzene, and methylene chloride.

Initiating cations and propagating macrocations may be carbon-based (**carbocations**), silicon-based (**silicon cations**), etc. They may be either enium ions or onium ions.

Enium ions are classic cations with electrophilic centers. The simplest **carbenium cation** is the methyl cation $H_3C^\oplus$, the simplest silicon cation the **silicenium cation** $R_3Si^\oplus$. In low-molecular weight model compounds, they can be identified by UV spectroscopy. In polymerizations, their concentration is usually too low to be detected and their presence is therefore deduced from kinetics, plausible mechanisms, electrical conductivity, product analysis, and, most often, by analogy.

Onium ions are non-classical cations that are formed by the addition of a positively charged group to a saturated atom. An example is **carbonium ions** that are four-fold or five-fold coordinated, non-classical carbocations. The simplest carbonium cation is $[CH_5]^\oplus$. True carbonium ions are rare; most so-called carbonium ions are in reality carbenium ions.

Onium ions of heteroatoms include oxonium ions $R_3O^\oplus$, immonium ions $R_2N^\oplus=CR_2$, aryl diazonium ions $ArN_2^\oplus$, sulfonium ions $R_3S^\oplus$, etc. The formal addition of hetero compounds to carbenium ions delivers a class of more complex onium ions, for example, carboxonium ions from the reaction $H_3C^\oplus + OR_2 \rightarrow H_3C{-}^\oplus OR_2$. Such onium ions are much more stable than carbonium ions because they are somewhat stabilized by the strongly nucleophilic ligands. They can often be detected by NMR spectroscopy.

All macrocations are thermodynamically and kinetically unstable. Their attempt to stabilize themselves by addition of nucleophilic species leads not only to often very fast propagation reactions but also to fast transfer and termination reactions, especially with water. Cationic polymerizations are, however, not sensitive toward oxygen.

Initiation by Salts

Certain salts dissociate directly into initiating cations and counterions in the presence of solvents that are commonly used for cationic polymerizations. These cations can

Table 8-10 Dissociation constants K of some organic salts in different solvents. Dissociation enthalpies and entropies refer to the temperature range $-45°C < T < 0°C$.

Cation	Anion	Solvent	$\dfrac{T}{°C}$	$\dfrac{10^4\,K}{\text{mol L}^{-1}}$	$\dfrac{\Delta H°}{\text{kJ mol}^{-1}}$	$\dfrac{\Delta S°}{\text{J mol}^{-1}\,\text{K}^{-1}}$
$(C_6H_5)_3C^{\oplus}$	$[SbCl_6]^{\ominus}$	CH_2Cl_2	0	3.1	-8.4	-97
	$[SbCl_6]^{\ominus}$	CH_2Cl_2	25	1.4	-8.4	-97
	$[SbF_6]^{\ominus}$	CH_2Cl_2	25	1.7		
	$[ClO_4]^{\ominus}$	CH_2Cl_2	25	2.5		
$C_7H_7^{\oplus}$	$[ClO_4]^{\ominus}$	CH_2Cl_2	0	0.3	-5.0	-105
	$[SbCl_6]^{\ominus}$	CH_2Cl_2	0	0.3	-10.0	-126
$(C_2H_5)_4N^{\oplus}$	$[SbCl_6]^{\ominus}$	CH_2Cl_2	0	0.84	-2.3	-90
$(C_2H_5)_3O^{\oplus}$	$[BF_4]^{\ominus}$	CH_2Cl_2	0	0.04		
$(C_2H_5)_3S^{\oplus}$	$[BF_4]^{\ominus}$	CH_2Cl_2	20	0.44		
	$[BF_4]^{\ominus}$	$C_6H_5NO_2$	20	165		

sometimes be identified by spectroscopy. The dissociation is promoted by complexation of the gegenion. For example, the dissociation of trityl chloride, $(C_6H_5)_3CCl$, into the triphenyl carbenium ion $(C_6H_5)_3C^{\oplus}$ and the chloride counterion $Cl^{\ominus}$ is advanced by complexation of $Cl^{\ominus}$ by $SbCl_5$:

(8-48) $(C_6H_5)_3CCl + SbCl_5 \rightarrow (C_6H_5)_3C^{\oplus}[SbCl_6]^{\ominus}$

The dissociation of such salts depends little on the nature of the complexed counterion, somewhat more on the cation, and strongly on the solvent (Table 8-10). Dissociation constants are small: ion pairs practically do not exist, only free ions.

Initiator cations often add directly to monomer molecules. Examples are polymerizations of tetrahydrofuran by acetyl perchlorate $[CH_3CO]^{\oplus}[ClO_4]^{\ominus}$, of epoxides or p-methoxy styrene by trityl hexachloroantimonate $(C_6H_5)C^{\oplus}[SbCl_6]^{\ominus}$, and of vinyl ethers or N-vinyl carbazole by tropylium hexachloroantimonate $C_7H_7^{\oplus}[SbCl_6]^{\ominus}$.

Monomer cations are sometimes not formed directly from initiator cations and monomer molecules but only after a reaction of the initiator salts with the monomer molecules. In the polymerization of tetrahydrofuran by trityl carbenium ions $(C_6H_5)_3C^{\oplus}$ or dications $(C_6H_5)_2C^{\oplus}-CH_2CH_2-^{\oplus}C(C_6H_5)_2$, the same molecular weight was found for the same molar ratio monomer/initiator. Since the concentration of cations from the dications is twice as large as that of the monocations, it was assumed that the two types of cations first dehydrate tetrahydrofuran (Eq.(8-49)). The resulting oxonium ions then start the polymerization.

(8-49)

Other cationic polymerizations get started by electron transfer to the monomer, for example, $(C_6H_5)_3C^{\oplus} + CH_2=CH(OCH_2R) \rightarrow (C_6H_5)_3C^{\bullet} + {}^{\bullet}[CH_2=CH(OCH_2R)]^{\oplus}$. Very few radical cations dimerize, however, to the corresponding dications.

Initiation by Brønsted and Lewis Acids

Perchloric acid $HClO_4$ and other *Brønsted acids* do not conduct electricity in oxygen-free solvents. Under these conditions, they are "covalent" compounds that do not dissociate into protons $H^{\oplus}$ and perchlorate anions $ClO_4^{\ominus}$. The protons of Brønsted acids thus do not initiate cationic polymerizations.

Cationically polymerizable monomers such as isobutene are, however, Brønsted bases. The initiation by Brønsted acids is therefore an acid-base reaction:

$$(8\text{-}50) \qquad HClO_4 + CH_2=C(CH_3)_2 \rightarrow (CH_3)_3C^{\oplus}[ClO_4]^{\ominus}$$

This reaction lets one understand why cationic polymerizations are not always initiated by proton acids. The cation should not combine reversibly with the anion, in Eq.(8-50), for example, not to an ester $(CH_3)_2CHCH_2ClO_4$, which can be prevented by a suitable stabilization (see below).

Another example is the polymerization of styrene by trifluoroacetic acid. Addition of the initiator to styrene delivers only a small amount of poly(styrene) of low molecular weight. Addition of styrene to the initiator, however, produces high molecular-weight poly(styrene) in high yield, because the trifluoroacetate ions are stabilized by a surplus of acid in the latter case but not in the former.

In general, *Lewis acids* such as BF_3, $AlCl_3$, $TlCl_4$, etc., do not start polymerizations directly. The true initiators are rather complexes, either self-complexes of Lewis acids themselves or complexes of Lewis acids and so-called co-initiators.

Solutions of some Lewis acids do not conduct electricity since the acids self-ionize and form non-conducting ion pairs (salts). Examples are $2\ AlCl_3 \rightleftarrows [AlCl_2]^{\oplus}[AlCl_4]^{\ominus}$, $RAlCl_2 \rightleftarrows [RAlCl]^{\oplus}[RAlCl_3]^{\ominus}$, $2\ TiCl_4 \rightleftarrows [TiCl_3]^{\oplus}[TiCl_5]^{\ominus}$, $2\ I_2 \rightleftarrows I^{\oplus}[I_3]^{\ominus}$, and $2\ PF_5 \rightleftarrows [PF_4]^{\oplus}[PF_6]^{\ominus}$. No self-association seems to take place in CH_2Cl_2.

Other Lewis acids do not self-ionize. They need a "co-catalyst" or "co-initiator" such as water, trichloroacetic acid, alkyl halogenides, ethers, methylene chloride, or the monomer. These co-catalysts form dissociable compounds with Lewis acids, for example:

$$(8\text{-}51) \qquad BF_3 \quad + H_2O \qquad \rightleftarrows H^{\oplus}[BF_3OH]^{\ominus}$$

$$(8\text{-}52) \qquad BF_3 \quad + (C_2H_5)_2O \rightleftarrows [C_2H_5]^{\oplus}[BF_3OC_2H_5]^{\ominus}$$

$$(8\text{-}53) \qquad AlCl_3 + CH_2Cl_2 \qquad \rightleftarrows [CH_2Cl]^{\oplus}[AlCl_4]^{\ominus}$$

The cations $H^{\oplus}$, $C_2H_5^{\oplus}$, $CH_2Cl^{\oplus}$, etc., are the true initiators of polymerizations. The term "co-catalyst" is thus a misnomer since co-catalysts neither catalyze nor co-catalyze anything. They are rather fodder for the formation of initiating species. But it is also wrong to call them "initiators" because it is the cation of the salt that really initiates.

Some polymerizations by Lewis acids are extremely sensitive toward the presence of water and other "co-catalysts." Only small proportions are needed of co-catalysts (Friedel-Crafts reactions do not proceed in absolutely dry solutions!). Larger proportions may, however, "poison" the polymerization since the polymerization rate passes through a maximum with increasing concentration of H_2O. It is not clear whether this poisoning happens before the initiation step, for example, by reactions such as $H^{\oplus}[BF_3OH]^{\ominus}$ + $H_2O \rightleftarrows [H_3O]^{\oplus}[BF_3OH]^{\ominus}$ or later, for example, by $\sim^{\oplus}CR_2\ X^{\ominus} + H_2O \rightarrow \sim CR_2OH + HX$.

8.4.3   Propagation

In the simplest case, cationic polymerizations proceed by repeated addition of mono-mer molecules to the propagating centers which are either carbocations (in general, carb-enium ions) or to onium ions of a hetero atom (hereafter simply called "onium ion" since *carb*onium ions are very rare (p. 254)).

In *carbocation* polymerizations, a center with a high positive charge density attacks the slightly negative β-atom of a molecule of, e.g., an olefin, creating a dipole moment in the transition state which must be nearly linear. The activation energy is thus low.

Growing *onium ions* are strongly solvated; their charge density is thus lower than that of carbocations. The transition state must be non-linear since the monomer molecules possess strong dipoles and approach the onium ion with their slightly negative hetero atom. The reaction must thus require a high activation energy.

The growing macrocation is stabilized by delocalization of its positive charge. Ex-amples are the macrocations of isobut(yl)ene, I, vinyl methyl ether, II, acetaldehyde, III, tetrahydrofuran, IV, dimethyl thiirane, V, and nitrosobenzene, VI. The electrophilicity decreases in the order $C^\oplus > O^\oplus > S^\oplus > N^\oplus$.

$$
\begin{array}{llllll}
\text{www}CH_2-\overset{\overset{\displaystyle CH_3}{|}}{\underset{\underset{\displaystyle CH_3}{|}}{C}}{}^\oplus &
\text{www}CH_2-\overset{\overset{\displaystyle H}{|}}{\underset{\underset{\displaystyle OCH_3}{|}}{C}}{}^\oplus &
\text{www}O-\overset{\overset{\displaystyle H}{|}}{\underset{\underset{\displaystyle CH_3}{|}}{C}}{}^\oplus &
\text{www}\overset{\oplus}{O}\!\!\square &
\text{www}\overset{\overset{\displaystyle H\ H}{|\ \ |}}{\underset{\underset{\displaystyle CH_3CH_3}{|\ \ \ |}}{C-C}}-S^\oplus &
\text{www}O-\overset{\oplus}{N} \\
\text{I} & \text{II} & \text{III} & \text{IV} & \text{V} & \text{VI}\quad C_6H_5
\end{array}
$$

In the polymerization of unsaturated monomers, the driving force is the exothermic transformation of double bonds into single bonds whereas in cyclic compounds, it is the decrease of ring strain. Isomerizations are also important in some cases (Section 8.4.4).

The conventionally reported rate constants k_p are averages of the rate constants of propagation of free ions, various types of ion pairs, and ion associates. They extend over the extremely large range $10^9 \geq k_p/(\text{L mol}^{-1}\text{ s}^{-1}) \geq 10^{-5}$ (Table 8-11). Very large differences in k_p are observed for the polymerization of the same monomer in different solvents and with different initiators, especially between carbocations and onium ions.

Table 8-11 Average rate constants of cationic polymerizations.

Monomer	Solvent	$\dfrac{T}{°C}$	Initiator	$\dfrac{k_p}{\text{L mol}^{-1}\text{ s}^{-1}}$
Cyclopentadiene	-	− 78	γ radiation	600 000 000
Isobutene	-	0	γ radiation	200 000 000
Isobutene	CH_2Cl_2	0	$[C_7H_7]^\oplus[SbCl_6]^\ominus$	70 000 000
Styrene	-	15	γ radiation	4 000 000
Styrene	$ClCH_2CH_2Cl$	30	I_2	4 000
Styrene	$ClCH_2CH_2Cl$	25	$HClO_4$	20
Styrene	CCl_4	20	$HClO_4$	0.001
N-Vinyl carbazole	CH_2Cl_2	− 25	$[C_7H_7]^\oplus[SbCl_6]^\ominus$	200 000
i-Butyl vinyl ether	CH_2Cl_2	− 25	$[C_7H_7]^\oplus[SbCl_6]^\ominus$	2 000
1,3-Dioxepane	CH_2Cl_2	0	$HClO_4$	3 000
1,3-Dioxolane	CH_2Cl_2	20	$[(C_2H_5)_3O]^\oplus[BF_4]^\ominus$	0.000 05

These differences are not only caused by different ion equilibria but also by ion fragmentations and intramolecular isomerizations. Ion equilibria have been observed in polymerizations of olefin derivatives but not in those of cyclic compounds, probably because the positive charges are better screened in the latter. It is remarkable that the rate constants of propagation of olefin derivatives such as styrene and *p*-methoxy styrene differ only by a factor 20 for propagations by free cations and cation pairs whereas much higher differences (up to a factor of 10^5) were found for anionic polymerizations.

The stereocontrol of cationic polymerizations depends strongly on the chemical nature of the system monomer-initiator-solvent (Table 8-12). Some systems deliver predominantly isotactic but others highly syndiotactic polymers. Possible reasons are complexations of counterions with polar chain ends and the presence of ion associates.

Table 8-12 Triad tacticities of polymers from cationic (C) and anionic (A) polymerizations.

Monomer	Solvent	Initiator	$T/°C$	x_{ii}	x_{is+si}	x_{ss}
Ethyl vinyl ether	toluene	C $BF_3 \cdot OEt_2$	-78	0.56	0.36	0.08
	CH_2Cl_2	C $BF_3 \cdot OEt_2$	-78	0.38	0.50	0.12
α-Methyl styrene	CH_2Cl_2	C BF_3	-78	0.00	0.11	0.89
Methyl-α-chlorostyrene	heptane	A EtMgBr	30	0.22	0.37	0.41
Methyl methacrylate	toluene	A C_6H_5MgBr	30	0.99	0.01	0

8.4.4 Phantom Polymerization

Some polymerizations are accompanied by isomerizations of the supposed monomer to the real monomer before or during the propagation steps. These polymerizations are sometimes called **phantom** or **exotic polymerizations** since the structure of their monomeric units cannot be deduced from the structure of the original monomer.

In these polymerizations, primary macrocations rearrange themselves to more stable structures if the structural conditions are right and the primary macrocations have relatively long life times. The isomerizations proceed by intramolecular transfer reactions, either by bond isomerization or by "material transport."

In the polymerization of β-pinene, for example, double bonds are shifted

(8-54)

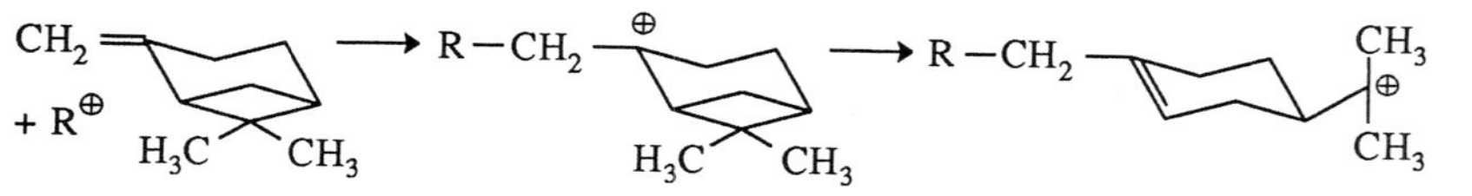

whereas in the cationic polymerization of norbornene, bonds are isomerized:

(8-55)

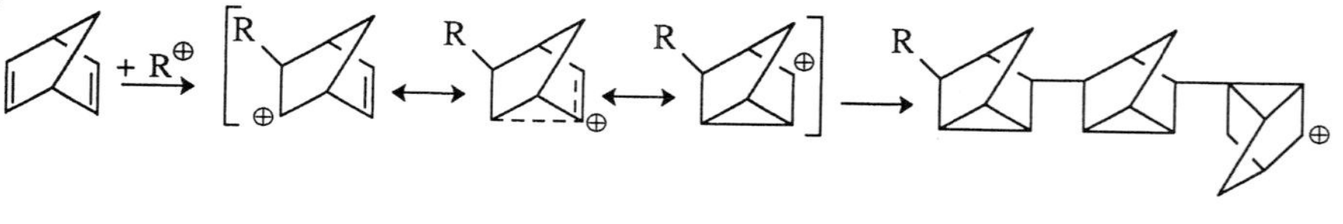

A hydride shift occurs in the polymerization of 3-methyl-1-butene (and by analogy in that of 4-methyl-1-pentene and 4-methyl-1-hexene)

(8-56)

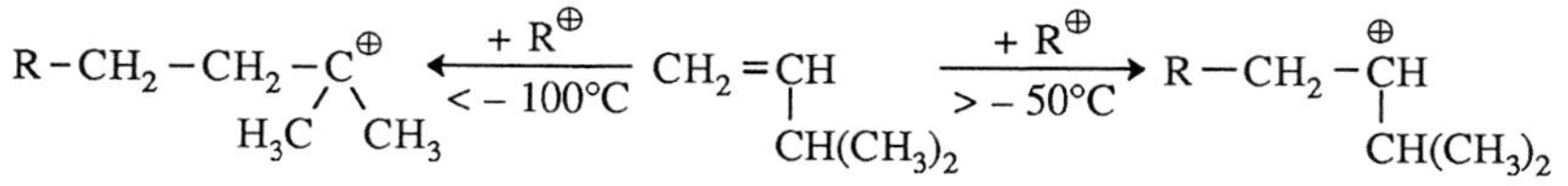

$$R-CH_2-CH_2-C^{\oplus} \underset{\substack{<-100°C}}{\overset{+\,R^{\oplus}}{\longleftarrow}} CH_2=CH \underset{\substack{>-50°C}}{\overset{+\,R^{\oplus}}{\longrightarrow}} R-CH_2-CH^{\oplus}$$

with H_3C, CH_3 below the left cation; $CH(CH_3)_2$ below the central and right structures.

whereas a methide shift to ~CH$_2$–CH(CH$_3$)-C(CH$_3$)$_2$~ happens in the polymerization of 3,3-dimethyl-1-butene, $CH_2=CH-C(CH_3)_3$.

Rings open in **isomerization polymerizations**, including those in side groups:

(8-57 a-f)

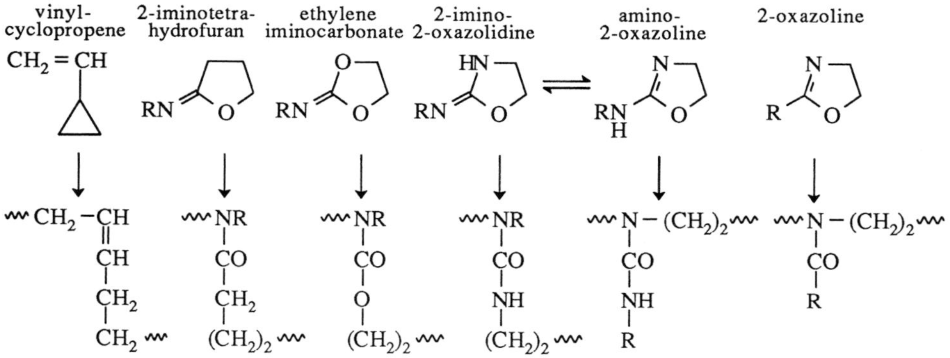

| vinyl-cyclopropene | 2-iminotetra-hydrofuran | ethylene iminocarbonate | 2-imino-2-oxazolidine | amino-2-oxazoline | 2-oxazoline |

$CH_2=CH$ (cyclopropane ring)

RN—O (furan ring) ; RN—O ; RN—O (with HN); RN—O (with N, and H) $\rightleftharpoons$; R—O (with N)

↓ ↓ ↓ ↓ ↓ ↓

~CH$_2$–CH	~NR	~NR	~NR	~N–(CH$_2$)$_2$~	~N–(CH$_2$)$_2$~
CH	CO	CO	CO	CO	CO
CH$_2$	CH$_2$	O	NH	NH	R
CH$_2$~	(CH$_2$)$_2$~	(CH$_2$)$_2$~	(CH$_2$)$_2$~	R	

8.4.5 Termination and Transfer

Overview

Cationic polymerizations are distinguished by many termination and transfer reactions. Water sometimes serves as "cocatalyst" (Eq.(8-51)) but also acts as a terminating agent if it is an undesired impurity:

(8-58)

$$\sim CR_2\,Q^{\ominus} \overset{+\,H_2O}{\longrightarrow} \sim CR_2-\overset{\oplus}{O}H_2\,Q^{\ominus} \underset{-H_2O}{\overset{+\,H_2O}{\rightleftharpoons}} \sim CR_2-OH + H_3\overset{\oplus}{O}\,Q^{\ominus}$$

The *monomer* itself can be a terminator as in the polymerization of propene, Eq.(8-59), or a transfer agent as in the polymerization of isobutene, Eq.(8-60):

(8-59)

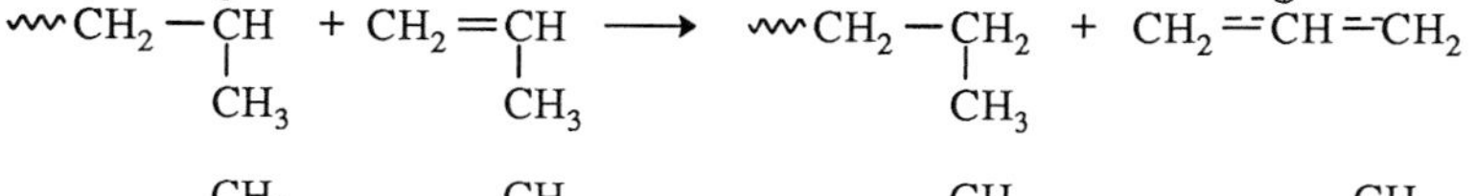

$$\sim CH_2-\overset{\oplus}{CH} + CH_2=CH \longrightarrow \sim CH_2-CH_2 + CH_2=CH=CH_2$$

with CH_3 below each of the first three structures.

(8-60)

$$\sim CH_2-\overset{\oplus}{C} + CH_2=C \longrightarrow \sim CH_2-C + H-CH_2-\overset{\oplus}{C}$$

with CH_3 above and below the cation, CH_3 above and below the second, CH_2 above and CH_3 below the third, CH_3 above and below the fourth.

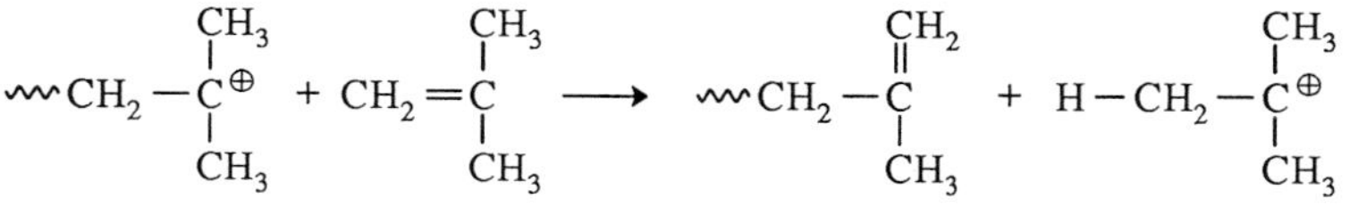

A hydride shift takes place from the monomer to the macrocation in the polymerization of propene, Eq.(8-59). The resulting allyl structure of the monomer is stabilized by resonance and is unable to initiate a polymerization: the monomer commits suicide. In the isobutene polymerization, Eq.(8-60), on the other hand, the hydrogen is transferred to the monomer from the macrocation. The new monomer cation starts a new polymer chain because it practically has the same structure as the macrocation.

Solvents may also act as transfer agents. An example is the polymerization of isobutene by $AlCl_3/HCl$ in $^{13}CH_3Cl$ where the resulting $^{13}CH_3$ cation (Eq.(8-61)) initiates new chains and is thus found as endgroup of the polymer molecules.

$$
(8\text{-}61) \qquad \text{ww} CH_2 - \overset{\overset{\displaystyle CH_3}{|}}{\underset{\underset{\displaystyle CH_3}{|}}{C^{\oplus}}} + \, ^{13}CH_3Cl \longrightarrow \text{ww} CH_2 - \overset{\overset{\displaystyle CH_3}{|}}{\underset{\underset{\displaystyle CH_3}{|}}{C}} - Cl \; + \; ^{13}\overset{\oplus}{C}H_3
$$

Transfer to *initiator* is utilized in the **inifer method** where the initiator both *ini*tiates and trans*fers*. Such initiators may be salts of the type $R[BCl_4]$ that are formed from BCl_3 and an excess of RCl. The growth of the macrocations is terminated here by its own counterion (Eq.(8-62)). The newly formed BCl_3 is again complexed by the surplus RCl whereupon it initiates a new chain. The method is especially suitable for isobutene since termination and transfer reactions of isobutene macrocations are practically absent at temperatures below 0°C.

$$
(8\text{-}62) \qquad \text{ww} CH_2 - \overset{\overset{\displaystyle CH_3}{|}}{\underset{\underset{\displaystyle CH_3}{|}}{C^{\oplus}}} [BCl_4]^{\ominus} \longrightarrow \text{ww} CH_2 - \overset{\overset{\displaystyle CH_3}{|}}{\underset{\underset{\displaystyle CH_3}{|}}{C}} - Cl \; + \; BCl_3
$$

Fragmentations

Counterions may be complex anions such as $[AsF_6]^{\ominus}$, $[BF_4]^{\ominus}$, $[PF_6]^{\ominus}$, $[SbF_6]^{\ominus}$, etc., or uncomplexed anions such as $ClO_4^{\ominus}$, $CF_3SO_3^{\ominus}$, $FSO_3^{\ominus}$, etc.

Complex counterions fragment relatively easily. $[BF_4]^{\ominus}$ and $[SbCl_6]^{\ominus}$ are only stable up to ca. 30°C whereas $[PF_6]^{\ominus}$ and $[SbF_6]^{\ominus}$ are stable up to 80°C. Complex counterions dissociate markedly above these temperatures. The new anions, such as $F^{\ominus}$ from the reaction $[BF_4]^{\ominus} \rightleftarrows BF_3 + F^{\ominus}$, may then become involved in a variety of side reactions, including those with initiator cations.

Uncomplexed anions do not fragment but may form covalent bonds with the propagating macrocations. Such reactions are especially prominent in cationic polymerizations that are initiated by **super acids**. These acids are, by definition, more acidic than 100 % sulfuric acid. An example is trifluoromethane sulfonic acid, CF_3SO_3H (triflic acid).

Esters of triflic acid initiate cationic polymerizations in which propagating macrocations are in equilibrium with macroesters, e.g., polymerization of tetrahydrofuran:

$$
(8\text{-}63) \qquad CF_3SO_3R \; + \; O\!\!\left\langle\!\!\begin{array}{c}\\\\\end{array}\!\!\right\rangle \; \rightleftharpoons \; R\!-\!\overset{\oplus}{O}\!\!\left\langle\!\!\begin{array}{c}\\\\\end{array}\!\!\right\rangle [CF_3SO_3]^{\ominus} \; \rightleftharpoons \; RO(CH_2)_4SO_3CF_3
$$

Polar solvents here lead to a high proportion of macrocations (in CH_3NO_2: 92 % macroions, 8 % macroesters) whereas apolar solvents deliver more macroesters (in CCl_4:

4 % macrocations, 96 % macroesters). The polymer growth takes place almost exclusively via macrocations since the propagation rate constants of macrocations are 50-2000 times greater than those of macroesters. The degrees of polymerization of polymers from macrocations are therefore much greater than those from macroesters (10^3-10^6 versus 10-10^3). Correspondingly, molecular weight distributions are bimodal (see below). The polymerization via macrocations is affected by traces of water whereas relatively large amounts of water do not influence the formation of macroesters.

Initiations by less strong acids, HA, such as perchloric acid, $HClO_4$, probably involve dimeric acids. The dimeric acid $(HA)_2$ reacts here with the monomer M, e.g., styrene, to $HM^{\oplus}[HA_2]^{\ominus}$. The monomer cation $HM^{\oplus}$ adds monomer molecules and forms macrocations $HM_i^{\oplus}$ that are in equilibrium, $HM_i^{\oplus}[HA_2]^{\ominus} \rightleftarrows HM_iA + HA$. This unimolecular decomposition is a transfer reaction since HA dimerizes spontaneously to $(HA)_2$ and $(HA)_2$ reacts very fast with the monomer to $HM^{\oplus}[HA_2]^{\ominus}$, an initiating species. The concentration $[P^{\oplus}]$ of this species is obtainable from the propagation rate, $-d[M]/dt = k_p[P^{\oplus}][M]$.

The total concentration $[P^{\oplus}]_{tot}$ of all propagating species is given by $[P^{\oplus}]$ and, in infinitely fast initiations, by $[I]_0$, the initial concentration of the initiator. Introduction of the rate of formation of $HM_i^{\oplus}$, $d[P^{\oplus}]/dt = k_{tr}[HM_i^{\oplus}]$, delivers $[P^{\oplus}]_{tot}$:

$$(8\text{-}64) \qquad [P^{\oplus}]_{tot} = [I]_0 + [HM_i^{\oplus}] = [I]_0 + k_{tr}\int_0^t [P^{\oplus}]dt$$

$$= [I]_0 + (k_{tr}/k_p)\int_{[M]_0}^{[M]} -d[M]/[M] = [I]_0 + (k_{tr}/k_p)\ln([M]_0/[M])$$

Inserting Eq.(8-64) into Eq.(8-6) and setting $[I]_0 = [P^{\oplus}]_{tot}$ and $f_c = 1$ delivers an expression for the number-average degree of polymerization:

$$(8\text{-}65) \qquad \overline{X}_n = \frac{[M]_0 - [M]}{[P^{\oplus}]_{tot}} = \frac{[M]_0 - [M]}{[I]_0 + (k_{tr}/k_p)\ln([M]_0/[M])}$$

The number-average degree of polymerization is thus a complicated function of the monomer conversion, $u = ([M]_0 - [M])/[M]_0$, (Fig. 8-12). The second term of the denominator dominates at higher temperatures but is negligible at lower ones.

8.4.6 Living Cationic Polymerization

The nucleophilic stabilization of unstable carbenium ions leads to living polymerizations, including those with equilibria. Such stabilizations are obtained (I) by adding certain counterions B in order to reduce the cationic charge and (II) by lowering the acidity of β-protons through addition of weak Lewis bases Y:

$$(8\text{-}66)$$

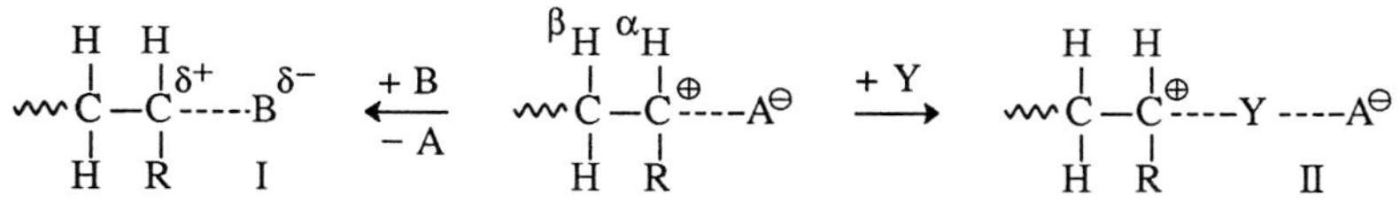

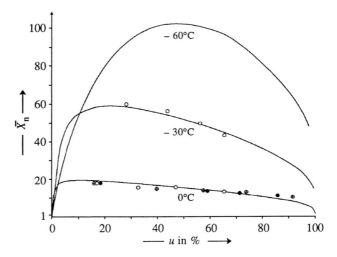

Fig. 8-12 Influence of chain transfer by unimolecular decomposition of the initiator on the dependence of the number-average degree of polymerization on monomer conversion, $u = ([M]_0 - [M])/[M]_0$. Polymerization of styrene by $HClO_4$ in 1,2-dichloroethane [13].

o   Experimental data for $[I]_0 = 0.0018$ mol/L and $k_{tr}/k_p = 0.0058$ mol/L at –30°C.
o,●,⊕   Experimental data for $0.9 < [I]_0/(\text{mol L}^{-1}) < 4.1$ and $k_{tr}/k_p = 0.020$ mol/L at 0°C.

——   Calculated with Eq.(8-64) and the following data for $[I]_0$ and k_{tr}/k_p:
$T = -60°C$, $[M]_0 = 0.435$ mol/L, $[I]_0 = 6.0 \cdot 10^{-4}$ mol/L, $k_{tr}/k_p = 0.001$ mol/L
$T = -30°C$, $[M]_0 = 0.435$ mol/L, $[I]_0 = 1.8 \cdot 10^{-4}$ mol/L, $k_{tr}/k_p = 0.0058$ mol/L
$T = 0°C$, $[M]_0 = 0.435$ mol/L, $[I]_0 = 2.0 \cdot 10^{-2}$ mol/L, $k_{tr}/k_p = 0.020$ mol/L

An example of method I is the polymerization of isobutyl vinyl ether in toluene by hydrogen iodide, HI. At –40°C, only adducts are formed. Addition of the electrophilic I_2 converts HI into $H^\oplus [I_3]^\ominus$ which initiates a living polymerization.

An example of method II is the polymerization of styrene by "wet" $TiCl_4$ plus triethylamine in the mixed solvent dichloromethane/hexane (Fig. 8-13). At small monomer conversions, u, number-average degrees of polymerization increase almost linearly with monomer conversion. The polymolecularities are initially large and become smaller later (Fig. 8-13, top).

Overall molar mass distributions of such polymerizations may be broad as evidenced by polymolecularities (Table 8-13, last column; see also Fig. 8-13).

Table 8-13 Number-average degrees of polymerization, $\bar{X}_n$, and polymolecularities, $\bar{X}_w/\bar{X}_n$, of the total polymer and its fractions I and II (from the bimodal distribution) of the cationic polymerization of 1 mol/L styrene in CH_2Cl_2 at –20°C by 0.01 mol/L 1-phenyl ethyl chloride as initiator and 0.1 mol/L $SnCl_4$ as activator [14].

Monomer conversion	I	$\bar{X}_n$ II	Total polymer	I	$\bar{X}_w/\bar{X}_n$ II	Total polymer
10 %	3.2	2900	12	1.15	2.20	400
50 %	12.2	2300	51	1.10	2.10	70
90 %	22.8	1330	91	1.05	2.78	30

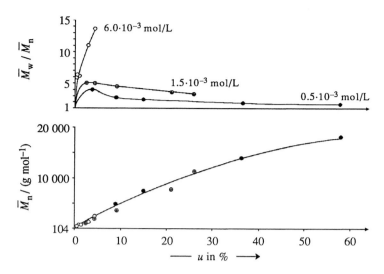

Fig. 8-13 Number-average degrees of polymerizations, $\overline{M}_n$, and polymolecularities, $\overline{M}_w/\overline{M}_n$, as a function of the monomer conversion, u, in the polymerization of 0.655 mol/L styrene in hexane-dichloromethane ($V/V = 40/60$) at –60°C by 0.041 mol/L "wet" TiCl$_4$ and different concentrations of triethylamine [15].

Such broad molar mass distributions are caused either by an uneven distribution of the reacting species or by differently propagating macrocations (free ions, ion pairs, etc.). The first case may deliver unimodal or bimodal distribution curves, depending on the polymerization system. The second case always leads to broad bimodal distributions; the distribution of each of the fractions may be narrow, however (Table 8-13).

8.5 Zwitterion Polymerization

Several ionic polymerizatins involve **zwitterions**, i.e., molecules that carry both positive and negative charges (Fig. 8-14) (German: *Zwitter* = mongrel, hybrid). Ion pairs from zwitterions may be intramolecular (molecules very flexible) or intermolecular (molecules very rigid). Intermolecular zwitterion pairs resemble polybetaines.

Polymerizations *via* zwitterions are different from polymerizations *by* zwitterions. In polymerizations via zwitterions, the zwitterion is generated by the addition of a monomer molecule to an initiator molecule in a Michael addition-type reaction.

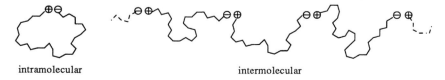

intramolecular   intermolecular

Fig. 8-14 Intramolecular and intermolecular ion pairs.

Examples of polymerizations via zwitterions are the reaction of tertiary amines and β-lactones

(8-67) R_3N + [β-lactone] $\longrightarrow$ $R_3\overset{\oplus}{N}-CH_2-CH_2-COO^{\ominus}$

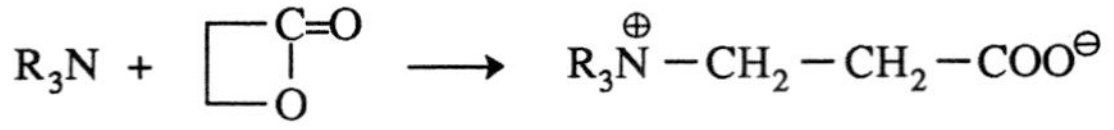

and the addition of sulfur trioxide to thiiranes:

(8-68) O_3S + $+S$[thiirane] $\longrightarrow$ $^{\ominus}O_3S-\overset{\oplus}{S}$[thiirane] $\xrightarrow{+S\,[thiirane]}$ $^{\ominus}O_3S-S-CH_2-CH_2-\overset{\oplus}{S}$[thiirane]

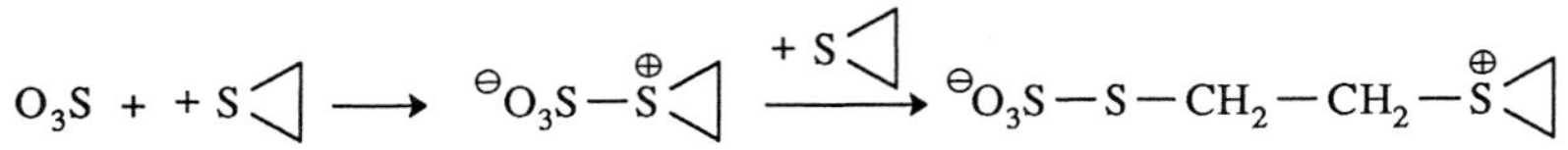

Another example is the formation of ylides, i.e., zwitterions in which a carbanion is directly bound to a positively charged heteroatom. An example is the reaction of olefin derivatives with tertiary amines:

(8-69) R_3N + $CH_2=\underset{R''}{\overset{R'}{C}}$ $\longrightarrow$ $\left[\underset{\text{zwitterion}}{R_3\overset{\oplus}{N}-CH_2-\underset{R''}{\overset{R'}{\underset{|}{\overset{|}{C}}}}{}^{\ominus}} \quad\longleftrightarrow\quad \underset{\text{ylide}}{R_3\overset{\oplus}{N}-\overset{\ominus}{C}H-\underset{R''}{\overset{R'}{\underset{|}{\overset{|}{C}}}H}}\right]$

These zwitterions then initiate regular polymerizations: anionic ones as in Eqs.(8-67) and (8-68) and cationic ones as in Eq.(8-69). The participation of zwitterions in such polymerizations was proven for Eq.(8-69) by the nitrogen content of the polymer, IR and NMR measurements, the positive charge during esterification, and the electrophoretic mobility of the esterified products.

The zwitterion is sometimes only the *apparent* initiator. In the polymerization of acrylonitrile $CH_2=CHCN$ by triarylphosphines, R_3P, a zwitterion $R_3P^{\oplus}–CH_2–^{\ominus}CHCN$ is formed. It does not initiate the polymerization but accepts a proton from another acrylonitrile molecule:

(8-70)

$R_3\overset{\oplus}{P}-CH_2-\overset{\ominus}{C}HCN + CH_2=CHCN \longrightarrow R_3\overset{\oplus}{P}-CH_2-CH_2CN + CH_2=\overset{\ominus}{C}CN$

The resulting acrylonitrile anion then starts an *anionic* polymerization. All phosphorus is in the counterion, none was found in the polymer. An analogous polymerization is that of nitroethene, $CH_2=CHNO_2$.

Polymerizations of preformed zwitterions are very rare. An example is the polymerization of arylcyclosulfonium ions by charge saturation which converts a water-soluble monomer into water-insoluble (unfortunately toxic) surface coatings:

(8-71)   [cyclic S] + [C₆H₄]-OH $\xrightarrow[-\,HCl]{+\,Cl_2}$ [cyclic S]$\overset{\oplus}{}$-[C₆H₄]-OH $Cl^{\ominus}$

$\xrightarrow[-\,CH_3OH\,;\,-\,NaCl]{+\,CH_3ONa}$ [cyclic S]$\overset{\oplus}{}$-[C₆H₄]-O$^{\ominus}$ $\longrightarrow$ $\sim\!\!\sim(CH_2)_4-S$-[C₆H₄]-O$\sim\!\!\sim$

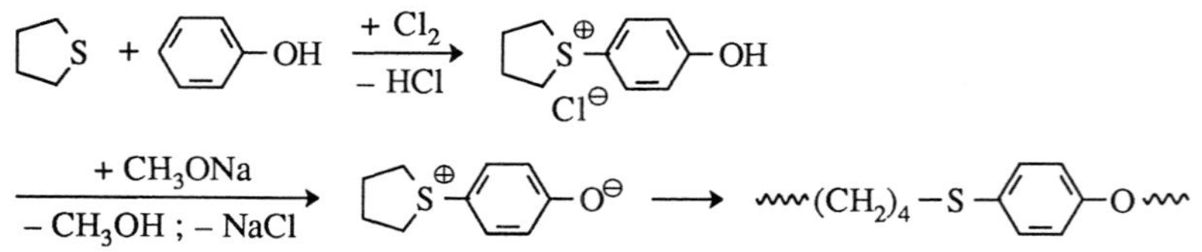

A-8 Appendix: Derivation of the Poisson Distribution

The derivation assumes that miscibility and diffusion problems are absent: all polymer chains are initiated simultaneously. At infinitely small times, all initiator molecules are thus present as monomer anions P_1^*. Since the monomer anion adds a monomer molecule, the concentration of P_1^* decreases with time according to

(A-8-1) $\quad - d[P_1^*]/dt = k_p[P_1^*][M]$

where k_p is the rate constant of propagation which is assumed to be independent of the degree of polymerization. i-Mers are formed from monomer molecules with a rate of

(A-8-2) $\quad d[P_i^*]/dt = k_p[P_{i-1}^*][M] - k_p[P_i^*][M] \quad ; \quad i = 1, 2, 3 \ldots$

The decrease in monomer concentration depends on the concentration of all growing monomer anions ($i = 1$) and macroanions ($i = 2, 3 \ldots$):

(A-8-3) $\quad - d[M]/dt = k_p[M] \, \Sigma_i \, [P_i^*] = k_p[M][C^*] \quad ; \quad 1 < i < \infty$

At the time t_0, the total concentration $[C^*]$ of active sites equals the concentration $[P_1^*]$ of monomer anions. Terminations are assumed to be absent; $[C^*]$ thus remains constant throughout the polymerization. Eq.(A 8-3) can thus be integrated to give

(A 8-4) $\qquad \displaystyle\int_{[M]_0-[C^*]}^{[M]} -d[M] = \int_0^t k_p[C^*][M]dt = [C^*]\int_0^t dv$

$$\int_0^t dv = k_p \int_0^t [M]dt = \frac{[M]_0 - [C^*] - [M]}{[C^*]} = \frac{[M]_0 - [M]}{[C^*]} - 1 = \bar{X}_n - 1 \equiv v$$

The newly introduced quantity $v \equiv \bar{X}_n - 1$ is the kinetic chain length of the system; it indicates how many monomer molecules have been added to one monomer anion.
Introducing $dv = k_p[M]dt$ into Eqs.(A-8-1) and (A-8-2) results in

(A-8-5) $\quad d[P_1^*] = - [P_1^*]dv$

(A-8-6) $\quad d[P_2^*] = [P_1^*]dv - [P_2^*]dv$

(A-8-7) $\quad d[P_i^*] = [P_{i-1}^*]dv - [P_i^*]dv$

The integration of Eq.(A-8-5) from $t = 0$ to $t = t$ delivers $\ln [P_1^*] = - v + const$. At zero time, v is also zero and the molar concentration of growing ions is $[C^*]$, resulting in $\ln [P_1^*] = - v + \ln [C^*]$ and also in

(A-8-8) $\quad [P_1^*] = [C^*] \exp (- v)$

Introduction of Eq.(8-8) into Eq.(8-6) delivers

(A-8-9) $\quad d[P_2^*] = [C^*] \exp(- v)dv - [P_2^*]dv$

This equation is integrated by the method of the integrating factor. Multiplication of the result by e^v leads to

(A-8-10) $e^v d[P_2^*] + e^v [P_2^*]dv = [C^*]dv$

The second term on the left side of Eq.(A-8-10) is replaced by $[P_2^*]d(e^v)$ because $d(e^v)/dv = e^v$. The resulting complete differential (Eq.(A-8-11)) can be integrated:

(A-8-11) $e^v d[P_2^*] + [P_2^*]d(e^v) = [C^*]dv$

(A-8-12) $e^v [P_2^*] = [C^*]v + const$

The integration constant is zero since $[P_2^*] = 0$ at $t = 0$ ($v = 0$). Eq.(A-8-12) thus becomes

(A-8-13) $[P_2^*] = e^{-v} [C^*]v$

The concentration of trimers is obtained in a similar way:

(A-8-14) $d[P_3^*] = [P_2^*]dv - [P_3^*]dv = e^{-v}[C^*]vdv - [P_3^*]dv$

(A-8-15) $[P_3^*] = e^{-v}[C^*]v^2/2!$

The general expression for an i-mer is therefore

(A-8-16) $[P_i^*] = \dfrac{[C^*]v^{i-1}\exp(-v)}{(i-1)!}$

The mole fraction $x_i = [P_i^*]/[C^*]$ of i-meric anions in all anions equals the mole fraction of i-mers in all molecules. Since $i \equiv X_i$, one obtains a Poisson distribution (cf. Section 3.3.3):

(A-8-17) $x_i = \dfrac{v^{i-1}\exp(-v)}{(i-1)!} = \dfrac{v^{X_i-1}\exp(-v)}{(X_i-1)!}$

v steps are necessary to generate $(v + 1)$ monomeric units in a polymer molecule, hence $X_i = v + 1$ and also $\overline{X}_n = v + 1$. From the definition of the weight fraction and the result for the mole fraction, from Eq.(A-8-17), one gets

(A-8-18) $w_i = \dfrac{x_i X_i}{\overline{X}_n} = \dfrac{x_i i}{v+1} = \dfrac{i v^{X_i-1}\exp(-v)}{(i-1)!\,(v+1)}$

The weight-average degree of polymerization is obtained from X_i and Eq.(A-8-18):

(A-8-19) $\overline{X}_w = \Sigma_i w_i X_i = \Sigma_i \dfrac{i^2\, v^{i-1}\exp(-v)}{(i-1)!(v+1)} = \dfrac{v^2+3v+1}{v+1} = 1 + \overline{X}_n - \dfrac{1}{\overline{X}_n}$

The polymolecularity is thus $\overline{X}_w/\overline{X}_n = (\overline{X}_n^2 + \overline{X}_n - 1)/\overline{X}_n^2$. At a number-average degree of polymerization of $\overline{X}_n = 1000$, it is just $\overline{X}_w/\overline{X}_n \approx 1.001$.

Historical Notes to Chapter 8

Archdeacon Watson, Chemical Essays, 1st edition, Cambridge, UK; 3rd edition, R.Moncrieffe, London; 5 volumes, T.+R.Merrill, London 1781-1789
  First cationic polymerization with sulfuric acid as initiator.

N.Bjerrum, Kon.Danske Videsk. Selsk. **7** (1926) 9
  Postulates ion pairs because some electrolyte solutions are less conducting than expected.

K.Ziegler, O.Schäfer, Ann.Chem. **479** (1930) 150
  Note that the anionic polymerization of dienes proceeds without termination.

M.Szwarc, Nature **178** (1956) 1168
  Develops the concept of living polymerizations. Experimental confirmation by
M.Szwarc, M.Levy, R.Milkovich, J.Am.Chem.Soc. **78** (1956) 2656
  Polymerization of styren by sodium naphthalene. Further development of the concept by
S.N.Khanna, M.Levy, M.Szwarc, Trans.Faraday Soc. **58** (1962) 747

Literature to Chapter 8

8.1 GENERAL OVERVIEWS
M.Szwarc, Ed., Ions and Ion Pairs in Organic Reactions, Wiley-Interscience, New York 1972-1974 (2 vols.)
R.Foster, Molecular Complexes, Crane-Russak, New York, 1973-1974 (2 vols.)
C.H.Bamford, C.F.H.Tipper, Eds., Comprehensive Chemical Kinetics, vol. 15, Non-Radical Polymerizations, Elsevier, Amsterdam 1976
M.D.Glasse, Spontaneous Termination in Living Polymers, Progr.Polym.Sci. **9** (1983) 133
B.L.Erusamlimskij, Mechanisms of Ionic Polymerization. Current Problems, Plenum, New York 1986
G.Allen, J.C.Bevington, Eds., Comprehensive Polymer Science, vol. 3, G.C.Eastmond, A.Ledwith, S.Russo, P.Sigwalt, Eds., Chain Polymerization I, Pergamon Press, Oxford 1989
M.Szwarc, Ionic Polymerization Fundamentals, Hanser, Munich 1996
G.Litvinenko, A.H.E.Müller, General Kinetic Analysis and Comparison of Molecular Weight Distributions for Various Mechanisms of Activity Exchange in Living Polymerizations, Macromolecules **30** (1997) 1253

8.1.a RING-OPENING POLYMERIZATION (anionic and cationic)
N.Spassky, Ring-Opening Polymerisation (Rapra Review Report 85), RAPRA Technol. Ltd., Shawbury, UK
D.J.Brunelle, Ring-Opening Polymerization, Hanser, Munich 1992

8.3 ANIONIC POLYMERIZATION
M.Szwarc, Carbanions, Living Polymers and Electron Transfer Processes, Interscience, New York 1968
M.Imoto, T.Nakaya, Polymerization by Carbenoids, Carbenes, and Nitrenes, J.Macromol.Sci. C **6** (1972) 1
D.H.Richards, Anionic Polymerization, Dev.Polym. **1** (1979) 1
A.F.Halasa, D.N.Schulz, D.P.Tate, V.D.Mochel, Organolithium Catalysis of Olefin and Diene Polymerization, Adv. Organometallic Chem. **18** (1980) 55
M.Morton, Anionic Polymerization: Principles and Practice, Academic Press, New York 1983
M.Szwarc, M. van Beylen, Ionic Polymerizations and Living Polymers, Chapman and Hall, New York 1993
H.L.Hsieh, R.P.Quirk, Anionic Polymerization. Principles and Practical Applications, Dekker, New York 1996

8.4 CATIONIC POLYMERIZATION

G.A.Olah, P.R. von Schleyer, Eds., Carbonium Ions, Interscience, New York 1968 (4 vols.)

P.H.Plesch, The Propagation Rate-Constants in Cationic Polymerizations, Adv.Polym.Sci. **8** (1971) 137

M.Perst, Oxonium Ions in Organic Chemistry, Academic Press, New York 1971

G.A.Olah, Ed., Friedel-Crafts Chemistry, Interscience, New York 1973

J.P.Kennedy, Cationic Polymerization of Olefins: A Critical Inventory, Wiley, New York 1975

A.Ledwith, D.C.Sherrington, Stable Organic Cationic Salts: Ion Pair Equilibria and Use of Cationic Polymerizations, Adv.Polym.Sci. **19** (1975) 1

J.P.Kennedy, P.D.Trivedi, Cationic Olefinic Polymerization Using Alkyl Halide Alkylaluminum Initiator Systems, Adv.Polym.Sci. **28** (1978) 83, 113

A.Gandini, H.Cheradame, Cationic Polymerization. Initiation Processes with Alkenyl Monomers, Adv.Polym.Sci. **34/35** (1980) 1

S.Penczek, P.Kubisa, K.Matyjaszewski, Cationic Ring-Opening Polymerization of Heterocyclic Monomers. I. Mechanisms, Adv.Polym.Sci. **37** (1980) 1

J.P.Kennedy, E.Marechal, Chemistry of Initiation in Carbocationic Polymerization, J.Polym.Sci.-Macromol.Revs. **16** (1981) 123

J.P.Kennedy, E.Maréchal, Carbocationic Polymerization, Wiley, New York 1982

E.J.Goethals, Ed., Cationic Polymerization and Related Processes, Academic Press, New York 1984

S.Penczek, P.Kubisa, K.Matyjaszewski, Cationic Ring-Opening Polymerization. 2. Synthetic Applications, Adv.Polym.Sci. **68/69** (1985) 1

J.P.Kennedy, B.Iván, Designed Polymers by Carbocationic Macromolecular Engineering: Theory and Practice, Hanser, Munich 1992

P.Sigwalt, Some Still Unsolved Problems in Carbocationic Polymerization, Macromol.Symp. **132** (1998) 127

K.Matyjaszewski, Ed., Cationic Polymerizations. Mechanisms, Synthesis, and Applications, Dekker, Monticello (NY) 1999

8.5 ZWITTERION POLYMERIZATION

H.Zweifel, Th.Völker, Polymerisation via Zwitterionen, Chimia [Aarau] **26** (1972) 345

D.S.Johnston, Macrozwitterion Polymerization, Adv.Polym.Sci. **42** (1982) 51

I.J.McEwen, Alternating Copolymers from Macrozwitterions, Progr.Polym.Sci. **10** (1984) 317

References to Chapter 8

[1]   J.Smid, in M.Szwarc, Ed., Ions and Ion Pairs in Organic Reactions, Wiley, New York, vol. I (1972), p. 85, Fig. 5

[2]   A.Tanizaki, M.Sawamoto, T.Higashimura, J.Polym.Sci.–Polym.Chem.Ed. **24** (1986) 87, Fig. 7

[3]   H.Hostalka, G.V.Schulz, Z.physik.Chem. NF **45** (1965) 286; (a) Table 2 and Fig. 2; (b) Fig. 4, Tables 2 and 3

[4]   H.-G.Elias, A.Fritz, Makromol.Chem. **114** (1968) 31, Fig. 3

[5]   G.Gee, W.C.E.Higginson, K.J.Taylor, M.W.Trenholme, J.Chem.Soc. (1961) 4298, Fig. 2

[6]   G.V.Schulz, Chem.Technol. **3/4** (1973) 222, Fig. 8

[7]   H.L.Hsieh, R.P.Quirk, Anionic Polymerization. Principles and Practical Applications, Dekker, New York 1996, (a) Tables 1.2-1.5 and 6.1, (b) Tables 23.10 and 23.11

[8]   H.Sinn, C.Lundberg, T.Onsager, Makromol.Chem. **70** (1964) 222, Fig. 12

[9]   M.M.B.El-Sabbah, H.-G.Elias, J.Macromol.Sci.-Chem. **A 16** (1981) 579, Fig. 1; Makromol.Chem. **182** (1981) 1617, Fig. 1

[10] H.-G.Elias, V.S.Kamat, Makromol.Chem. **117** (1968) 61, Table 1

[11] K.-F.Elgert, E.Seiler, Makromol.Chem. **145** (1971) 95, Table 1

[12] S.S.Huang, A.H.Soum, T.R.Hogen-Esch, J.Polym.Sci.-Letters **21** (1983) 559, Table I

[13] D.C.Pepper, Macromol.Chem.Phys. **196** (1995) 963, experimental data in Fig. 10

[14] K.Matyjaszewski, R.Szymanski, M.Teodorescu, Macromolecules **27** (1994) 7565

[15] I.Majoros, A.Nagy, J.P.Kennedy, Adv.Polym.Sci. **112** (1994) 1, Table 20

9 Coordination Polymerization

9.1 Survey

9.1.1 Introduction

Monomer molecules are *added* to growing active centers in ionic (Chapter 8) and free-radical (Chapter 10) *chain polymerizations*. These centers (anions, cations, radicals) and the covalently bound initiator fragments from which they originated are at opposite ends of the growing chains. An example of such control by active *chain ends* is the anionic polymerization of styrene, $CH_2=CHC_6H_5$, by butyl lithium, $C_4H_9{}^{\ominus}Li^{\oplus}$:

(9-1) $C_4H_9(CH_2-CHPh)_n{}^{\ominus} + CH_2=CHPh \longrightarrow C_4H_9(CH_2-CHPh)_n-CH_2-CHPh^{\ominus}$

In **coordination polymerizations**, monomer molecules are first coordinated with a polymer-initiator complex ("polymer-catalyst complex") containing a transition metal before they are *inserted* into the bond between the initiator moiety and the polymer chain, hence the also used term **insertion polymerization**.

An insertion is a "bifunctional reaction" of the monomer molecule. It obviously allows much better stereocontrol of the propagation step than the "monofunctional" monomer reaction in common chain polymerizations. An example is the isospecific polymerization of propene $CH_2=CHR$ (with R = CH_3) by the initiator complex of $TiCl_4$ + AlR'_3 (symbolized by [Ti]):

(9-2) $\sim(CHR-CH_2)_n-[Ti] + CH_2=CHR \longrightarrow \sim(CHR-CH_2)_n-CHR-CH_2-[Ti]$

The term "coordination polymerization" refers to the propagation reaction of the coordinated monomer molecule, not to the coordination reaction of the monomer itself. For example, ethene coordinates with the initiator silver nitrate but the propagation proceeds by free radicals that are not coordinated by silver nitrate. "Insertion polymerization" is thus a less ambiguous term than "coordination polymerization."

It is also customary to distinguish between anionic and cationic coordination polymerizations. Again, these terms are somewhat misleading since coordination polymerizations do not proceed via free ions or loose ion pairs (solvated ion pairs). They rather involve electrophilic or nucleophilic species.

The initiating species is usually called the "catalyst". It is however not split off from the reacting site after *each* reaction step as true catalysts are (p. 163) but rather stays at that site for many propagation steps like an initiator. Chain transfer sees to it that not every polymer chain carries an initiator moiety, hence the resemblence to "catalysis".

The "bifunctional" nature of the insertion step makes it much more sensitive to the chemical nature of the participants than the classical ionic and free-radical polymerizations: only a few initiators are suitable for a certain monomer. There are, however, many different types of monomer-initiator combinations such as Ziegler polymerizations of olefins and dienes by transition metal catalysts (Section 9.2), so-called metallocene polymerizations (Section 9.3), metathesis polymerizations by Group 8 compounds (Section 9.4), group-transfer polymerizations (Section 9.5), and enzymatic polymerizations (Section 14.4.6).

9.1.2   Historical Development

Coordination polymerizations owe their discovery and development to both industrial necessities and chance discoveries. Ethene was considered a non-polymerizable compound until researchers at ICI in England tried in 1933 to compress ethene in newly developed Dutch high-pressure vessels. To their surprise, the pressure dropped strongly and a white, solid material was formed, obviously from the polymerization of ethene by traces of oxygen. The 1937 ICI patent started the development of high-pressure (up to p = 280 MPa $\approx$ 2800 atm), free-radical polymerization processes for ethene, at that time a coal-based chemical. These processes deliver highly branched poly(ethylene)s with low densities (LDPE = PE-LD).

LDPE served in World War II as a dielectric for radar installations; its usefulness for civilian purposes was recognized after the war ended. The economic upswing led to a strong demand for gasoline which was obtained by fractional distillation of crude oil and later also by cracking of petroleum. These processes also delivered vast amounts of ethene. Based on their knowledge of catalysts for the cracking process and propelled by the pending expiration of the ICI patents (1956), U.S. petroleum companies tried therefore to develop catalysts for the polymerization of ethene: partially reduced MoO_3 (Standard Oil of Indiana) and partially reduced Cr_2O_3 (Phillips Petroleum), both supported on Al_2O_3 (1954). The Phillips low-pressure polymerization of ethene ($p < 5$ MPa) is still one of the industrial processes for high-density poly(ethylene) (Table 9-1).

At about the same time, Karl Ziegler at the German Max-Planck Institute for Coal Research worked on the synthesis of higher aluminum alkyls by repeated addition of olefins to aluminum hydride in autoclaves ("Aufbaureaktion" = build-up reaction). One day, the reaction of tripropyl aluminum with ethene did not result in higher aluminum alkyls but in triethyl aluminum, propene, and 1-butene. This stray reaction was caused by a trace of metallic nickel from a previous experiment with nickel compounds that resided in a microcrack. Further experiments with many combinations of aluminum compounds and transition metal compounds led to the first generation of so-called **Ziegler catalysts**, composed of diethyl aluminum chloride and titanium tetrachloride (Table 9-1). Guilio Natta and coworkers found a short time later that Ziegler catalysts also polymerize propene and other α-olefins to highly stereoregular, crystallizing polymers. However, the composition-of-matter patent for crystallizing poly(propylene) is held by Phillips Petroleum (J.P.Hogan, R.L. Banks) while process patents have been awarded to Phillips Petroleum, Standard Oil, DuPont, and Natta and coworkers.

Most polymerizations by these catalysts proceed in heterogeneous phases; only the small proportion of transition metal compounds on the surface of the catalyst particles is polymerization active. The catalysts are furthermore **multi-site catalysts**, i.e., the transition metal atoms reside in centers of various activities which leads to polymers with high polymolecularities of $4 \leq \overline{M}_w/\overline{M}_n \leq 10$ (Section 9.2).

The next major developmental step was the discovery of **single-site catalysts** (SSCs), i.e., catalysts with only one defined center of catalytic activity (Section 9.3). SSCs greatly enhance polymerization rates in combination with **methyl alumoxanes = methyl aluminoxanes** (**MAO**s) which owe their discovery to the observation that rigorously dried Ziegler catalysts (and also Friedel-Crafts compounds) do not work. They are produced by addition of fairly large proportions of water to trimethyl aluminum (Section 9.3.1).

Table 9-1 Industrial polymerizations by transition-metal catalysts of high-density poly(ethylene)s (HDPE), linear low-density poly(ethylene)s (LLDPE), ethylene-propylene-diene rubbers (EPDM), isotactic poly(propylene)s (PP), poly(4-methyl-1-pentene)s (P4MP), *cis*-1,4-poly(butadiene) rubbers (BR), *cis*-1,4-poly(isoprene) rubbers (IR), and polymers of cyclopentene (PCP), dicyclopentadiene (PDCP), or norbornene (PNB). ED = electron donor, copm = copolymerization.

Polymer	Monomer	Catalysts	Comments
HDPE	ethene	partially reduced Cr_2O_3 on SiO_2	suspension in isobutane (Phillips)
	ethene	$TiCl_3 + (C_2H_5)_2AlCl + ED$	2nd generation (Montedison)
	ethene	$TiCl_4 + (C_2H_5)_3Al + MgCl_2 + ED$	4th generation (Spheripol®)
LLDPE	ethene + 1-butene	silylchromate + $Al(OC_2H_5)_3 + H_2$	gas phase (Unipol® process)
	ethene + 1-butene	$TiCl_4 + VOCl_3 + Al(C_2H_5)_3$	solution in cyclohexane (DuPont)
	ethene + 1-octene	$TiCl_4 + Mg/Al$	Isopar® process (Dow)
	ethene + 1-olefins	metallocene + methyl alumoxane	Insite® process (Dow)
EPDM	ethene + propene	$VCl_4 + R_2AlCl$	many other initiators
PP	propene	$TiCl_3 + R_2AlCl + MgCl_2 + ED$	similar to ethene (see above)
P4MP	4-methyl-1-pentene	heterogeneous Ziegler catalysts	also copm with 1-olefins (C_6-
C_{16})			
BR	1,3-butadiene	$TiI_4 + R_3Al$	
		$Co(OOCR_2)_2 + R_3Al_2Cl_3$	
IR	isoprene	$TiCl_3 + R_3Al$	
PCP	cyclopentene	$WCl_6 + R_3Al + C_2H_5OH$	
PDCP	dicyclopentadiene	$WCl_6 + WOCl_4 + (C_2H_5)_2AlCl$	
PNB	norbornene	$WCl_6 + R_3Al + I_2$	

The synthesis of chiral metallocenes with C_2 symmetry then enabled polymerizations to isotactic and syndiotactic polymers by a great variety of transition metal catalysts.

9.2 Multi-Site Polymerizations

9.2.1 Ziegler Catalysts

Ziegler-Natta polymerizations (**ZN polymerizations**) are chain polymerizations that are initiated and propagated by **Ziegler catalysts**. According to the basic German Ziegler patent, these catalysts result from the combination of metal compounds of elements of "Nebengruppen IV-VIII" of the Periodic System of Elements with hydrides or alkyl or aryl compounds of metals of "Hauptgruppen I-III". The old German designations of "Hauptgruppe" (main group) and "Nebengruppe" (subgroup) do not always match the old Anglo-Saxon designations of elements as "main group" vs. "subgroup", "group a" vs. "group b", or "group A" vs. "group B", nor do they totally coincide with the group designations of the new Periodic Table (Table 9-2) which caused some confusion in the scientific and patent literature. Elements of the new Groups 4-10 are now called **transition elements**. One distinguishes **early transition elements** (Groups 4-5 + lanthanides) from **late transition elements** (Groups 6 ff.). In the older literature, "transition elements" sometimes included elements of groups 11 and 12 or groups 11-13 but not lanthanides (in Table 9-2, these deviating classifications are not shown).

Table 9-2 Designations of groups in the Periodic Table of Elements.
  *   plus Lanthanides: Ce, Pr, Nd, Pm, Sm, Eu, Gd, Tb, Dy, Ho, Er, Tm, Yb, Lu (formerly Cp).
  **   plus Actinides: Th, Pa, U, Np, Pu, Am, Cm, Bk, Cf, Es, Fm, Md, No, Lr.

Elements	Catalyst component	Group Designations		
		Old German (Ziegler)	Old Anglo-Saxon	Modern System
Li, Na, K, Rb, Cs, Fr	yes	Hauptgruppe I	Group 1a or IA	Group 1
Be, Mg, Ca, Sr, Ba, Ra	yes	Hauptgruppe II	Group 2a or IIA	Group 2
B, Al	yes	Hauptgruppe III	Group 3a or IIIB	Group 3
Sc, Y, La *, Ac **	yes	Hauptgruppe III	Group 3b or IIIA	Group 3
Ga, In, Tl, 113	no	Nebengruppe III	Group 3a or IIIB	Group 13
Ti, Zr, Hf, Rf	yes	Nebengruppe IV	Group 4b or IVA	Group 4
V, Nb, Ta, Ha	yes	Nebengruppe V	Group 5b or VA	Group 5
Cr, Mo, W, 106	yes	Nebengruppe VI	Group 6b or VIA	Group 6
Mn, Tc, Re, 107	yes	Nebengruppe VII	Group 7b or VIIA	Group 7
Fe, Ru, Os, 108	yes	Nebengruppe VIII	Group 8 or VIIIA	Group 8
Co, Rh, Ir, 109	yes	Nebengruppe VIII	Group 8 or VIIIA	Group 9
Ni, Pd, Pt, 110	yes	Nebengruppe VIII	Group 8 or VIIIA	Group 10

The classic (patent) definition of a Ziegler catalyst comprises many millions of combinations with very different efficiencies (see below). Yet it is both too narrow and too wide since Ziegler-Natta polymerizations are not only initiated by metal compounds of "Hauptgruppen I-III" (in combination with compounds of elements of "Nebengruppen IV-VIII") but also by organometallic compounds of tin and lead, both Group 14 elements (see Periodic Table of Elements, Fig. 2-1). On the other hand, not all possible combinations are effective.

Ziegler catalysts may also not necessarily initiate Ziegler-Natta polymerizations (coordination or insertion polymerizations) but occasionally also classic (anionic, cationic, free-radical) chain polymerizations. It is furthermore useful to separate polymerizations by insoluble Ziegler catalysts (multi-site polymerizations, Section 9.2) from polymerizations by soluble organometallic catalysts (mostly single-site polymerizations, Section 9.3). Even polymerizations with soluble catalysts may proceed heterogenously, however, since the newly formed polymers are usually insoluble in the reaction medium.

9.2.2 Monomers

Monomers for Ziegler-Natta polymerizations comprise in general aliphatic and cyclo-aliphatic olefins and dienes, in special cases also vinyl and acrylic compounds. Not every catalyst system is suitable and efficient for each monomer. All Ziegler catalysts that polymerize 1-olefins are also catalysts for ethene but the reverse is not true. Organometallics of Group 4-6 elements initiate the polymerization of 1-olefins and dienes but transition metal compounds of Group 8-10 elements are only effective for dienes but not for 1-olefins.

Ethene is polymerized by Ziegler catalysts to poly(ethylene)s with far fewer branches than poly(ethylene)s from free-radical polymerizations. Fewer branches lead to more efficient packing of chains in crystal lattices and hence to higher crystallinities, densities, and melting temperatures. High-density poly(ethylene)s from ZN polymerizations

(HDPE, PE-HD) are therefore distinguished from low-density poly(ethylene)s (LDPE, PE-LD) from free-radical polymerizations.

Propene and *styrene* have been polymerized by various catalysts to isotactic, syndiotactic, and atactic polymers. Other *1-olefins* deliver only isotactic or atactic polymers.

Ziegler catalysts often isomerize *2-olefins* before the polymerization. In fast isomerizations, the newly formed isomer is rapidly removed from the monomer-isomer mixture by polymerization. The resulting polymer is then exclusively composed of units of the isomerization product. An example is 4-methyl-2-pentene, $CH_3-CH=CH-CH(CH_3)_2$, which is polymerized to poly(4-methyl-1-pentene), $\{CH_2-CH(CH_2-CH(CH_3)_2)\}_{\overline{n}}$, by $TiCl_3$-$CrCl_3$ (see also 2-butene in Section 2.4.1).

Cycloolefins polymerize either by opening of the double bond and preservation of the ring or by ring-opening and preservation of the double bond. In ZN polymerizations, the polymerization via the double bond is more likely, the more electronegative the transition metal (Cr, V, Ni, Rh). Organometallics with electropositive transition metals catalyze ring-opening polymerizations.

1,3-Dienes deliver predominantly 1,4-cis, 1,4-trans, or 1,2-st polymers, depending on the Ziegler catalyst. All polymers usually contain small proportions of 3,4-structures.

Ziegler catalysts can also accommodate *copolymerizations*. Examples are elastomeric compounds from ethene + propene (EPM) or ethene + propene + non-conjugated dienes (EPDM).

9.2.3 Processes

Ziegler-Natta polymerizations can be conducted in various states of matter. *Ethene* is polymerized industrially in the gas phase or in slurries of Ziegler catalysts in organic solvents. Gas phase polymerizations are the least expensive since no investments or capital outlays are necessary for the purchase, storage, and recovery of solvents. The copolymerization of ethene with several percent of 1-butene is also a gas-phase process whereas that of ethene with several percent of 1-hexene or 1-octene proceeds in hydrocarbon solution; both types of copolymerization lead to so-called "linear low-density poly(ethylene)s" (LLDPE, PE-LLD). The polymerization of ethene by partially hydrogenated chromium oxide proceeds on silica, aluminum silicates, or aluminum oxide as carriers (Phillips process).

Propene and higher 1-olefins are usually polymerized by slurries of the catalyst since the volume-time yield is higher and less solvent has to be recovered. *Dienes*, on the other hand, are mostly polymerized in solution.

Neither the structure of the catalyst nor the concentration of the active species is often known with certainty. Catalyst efficiencies are therefore not described by rate constants but by

catalyst yield	m_P/m_{cat} or m_P/m_T
catalyst productivity	$m_P/(n_T t)$
catalyst activity	$m_P/(n_T [M] t)$, $m_P/(n_Y c_M t)$, or $m_P/(n_T p_M t)$

where m_P = mass of polymer P, m_{cat} = mass of catalyst, m_T = mass of transition metal T, n_T = amount of transition metal (in moles), and t = time. Monomer concentrations are

Table 9-3 Catalyst yields, Y_{cat}, of Ziegler catalysts in the polymerization of ethene (En), propene (Pr), and butadiene (Bu). Cp = cyclopentadienyl, Et = ethyl, Mt = metal, P = polymer.

Monomer and catalyst	Polymerization Conditions			Y_{cat} in	Tacti-
	State of matter	$T/°C$	p/bar	g P/g Mt	city
En $CrO_3 + SiO_2$	suspension	95	31	10 000	-
En silyl chromate + Al(OEt)$_3$	gas phase	95	20	9 000	-
En $TiCl_4 + R_3Al + MgCl_2$	suspension	60	0.25	50 000	-
En $Ti[CH_2C(CH_3)_3]_4$	solution	80	2.1	19	-
En $Ti[CH_2C(CH_3)_3]_4$	on SiO_2	50	2.1	48	-
En $Zr[CH_2C(CH_3)_2C_6H_5]_4 + H_2$	on Al_2O_3	150	2.8	2 800	-
En Cp_2ZrR_2 + methyl alumoxane	solution	90	8	3 000 000	-
Pr $TiCl_4 + AlR_3$	suspension	60	1	5 000	it
Pr $TiCl_4 + AlR_3 + MgCl_2$	suspension	60	1	14 500	it
Pr $Ti[CH_2C(CH_3)_3]_4$	on SiO_2	50	2.8	3	
Pr $Zr[CH_2C(CH_3)_2C_6H_5]_4$	on Al_2O_3	60	8	400	
Pr Cp_2ZrR_2 + methyl alumoxane	solution	5	8	2 000	at
Pr $VCl_4 + Et_2AlCl$-anisole	solution	−78			st
Bu $TiI_4 + AlEt_3$	suspension				1,4-cis
Bu $CoCl_2 + Et_2AlCl$ + pyridine	solution				1,4-cis
Bu V(acetylacetonate)$_3$ + AlEt$_3$	solution ?	25			1,2-st

characterized by amount-of-substance concentrations ("molar concentrations") [M], mass concentrations c_M, or pressures p_M. Catalyst yields, and also catalyst productivities and activities, vary widely as shown in Table 9-3 for several monomers, catalysts, and polymerization conditions.

9.2.4 Catalyst Structure

Differences in catalyst yields, productivities, and activities may have different origins. Catalyst components and thus catalyst complexes may be chemically different. The same components may furthermore give several complexes with different activities. The resulting complexes may also be soluble or insoluble. In the latter case, active centers may either be inefficient because they are buried in the catalyst particle or they may be especially efficient because of their interaction with the crystal lattice of the insoluble component or the carrier as it is generally known for heterogeneous catalyses. In the polymerization of ethene by Phillips catalysts, the catalyst yield increased, for example, with increasing mass fraction of chromium whereas the yield of poly(ethylene) per mass of chromium decreased (Fig. 9-1).

Depending on proportion, kind of mixing, temperature, time, etc., the same catalyst components can thus exhibit different efficiencies. Mixing of $TiCl_4$ and (i-$C_4H_9)_3$Al at a temperature of −78°C results in a soluble dark red complex that is efficient for the polymerization of ethene but not for that of propene. Mixing at −25°C delivers an insoluble black-brown complex that does not dissolve at lower temperatures. This complex consists of a mixture of (i-C_4H_9)$TiCl_3$ and (i-$C_4H_9)_4Al_2Cl_2$; it readily polymerizes propene or butadiene.

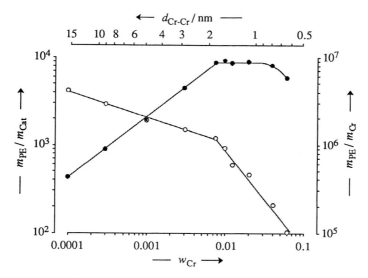

Fig. 9-1 Catalyst yields of the chromium-SiO$_2$ catalyst of Phillips Petroleum with respect to the mass m_{Cat} of the catalyst (●) or the mass m_{Cr} of chromium (O) as a function of the mass fraction w_{Cr} of chromium in the catalyst after 80 min during the polymerization of ethene at 10°C and 38 bar [1]. d_{Cr-Cr} = calculated average distance between chromium atoms. With permission by Academic Press.

Not all Ziegler catalysts initiate insertion polymerizations. Anionic polymerizations arise if the metal alkyl can initiate polymerizations itself; an example is the polymerization of isoprene by TiCl$_4$/C$_4$H$_9$Li. Cationic polymerizations are triggered if monomers with electron donor groups, e.g., vinyl ethers, are polymerized by catalysts with electron-accepting components, e.g., TiCl$_4$.

The actual structure of the active species in classic Ziegler catalysts is not easy to determine since (a) many different structures exist side by side that (b) cannot be investigated separately because of their insolubility. Each complex with a coordination gap and an uneven electron distribution is a potential Ziegler catalyst, i.e., compounds with

    (1) two different metal atoms, e.g., Ti and Al;
    (2) two identical metal atoms with different valency, e.g., Ti(II) and Ti(III);
    (3) two identical metal atoms with the same valency but different ligands, e.g., RTiCl$_2$ and TiCl$_3$.

Proposed structures of hetereogeneous Ziegler catalysts based on Ti and Al include the structures I, II, and III (Fig. 9-2). I is a bimetallic complex for a bimetallic mechanism (see below), II a bimetallic complex for a monometallic mechanism, and III a monometallic complex for a monometallic mechanism.

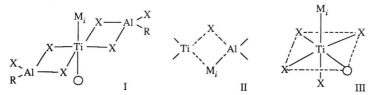

Fig. 9-2 Proposed structures of active species in heterogeneous Ziegler catalysts. M_i = polymer chain, R = ligand (e.g., C$_2$H$_5$), X = "anion", O = unoccupied ligand site.

The polymerization activity is controlled by the stability of the Mt–X bond. A very stable Mt–X bond cannot be activated if a monomer molecule coordinates with the unoccupied ligand site. An unstable Mt–X bond, on the other hand, dissociates under the conditions of polymerization. The ligands X must thus be balanced carefully with respect to their electron-donating properties in order to have the right degree of destabilization of the Mt–X bond. In heterogeneous Ziegler catalysts, this degree can be adjusted to some extent by the stabilizing effect of the crystal field.

9.2.5 Active Centers

The mechanism of Ziegler-Natta polymerizations has been the object of many experiments, hypotheses, and theories. It is certainly not a free-radical polymerization since hydrogen and deuterium act as transfer agents (Eq. (9-4)). The polymer is radioactive if the Ziegler catalyst contains $(^{14}C_2H_5)_3Al$; the polymer chain must therefore propagate at the metal-carbon bond (Et = ethyl):

(9-3) $Mt-^{14}Et + n\ CH_2=CH_2 \rightarrow Mt-(CH_2-CH_2)_n-^{14}Et$

Addition of deuterium delivers deuterated polymers, and that of ROT tritiated ones:

(9-4) $Mt-(CH_2-CH_2)_n-^{14}Et + D_2 \quad \rightarrow \quad Mt-D \quad + \ D-(CH_2-CH_2)_n-^{14}Et$

(9-5) $Mt-(CH_2-CH_2)_n-^{14}Et + ROT \rightarrow \quad Mt-OR + \ T-(CH_2-CH_2)_n-^{14}Et$

Aluminum alkyls in Ziegler catalysts are reducing agents that reduce $TiCl_4$ to compounds with lower valencies:

(9-6)

$$
TiCl_4 \underset{- EtAlCl_2}{\overset{+ Et_3Al}{\left\langle \begin{matrix} - Et_2AlCl \\ + Et_2AlCl \end{matrix} \right\rangle}} \rightarrow EtTiCl_3 \xrightarrow[- Et_2AlCl]{+ Et_3Al} Et_2TiCl_2 \xrightarrow{- Et^{\bullet}} EtTiCl_2
$$

$EtTiCl_3 \downarrow {-Et^{\bullet}}$

$$TiCl_3$$

These exchange reactions are so slow that such Ziegler catalysts contain at any given moment very different compounds in various proportions; for example, the compounds $TiCl_4$, $EtTiCl_3$, Et_2TiCl_2, $TiCl_3$, and $EtTiCl_2$ (Eq.(9-6)). These compounds complex with the aluminum compounds Et_3Al, Et_2AlCl, and $EtAlCl_2$ and/or other titanium compounds (Fig. 9-2). Activities of heterogeneous Ziegler catalysts thus vary with time.

The ethyl radicals of Eq.(9-6) dimerize to butane. To some extent, they may also initiate free-radical polymerizations of ethene. Free-radical polymerizations have also been observed in the polymerization of vinyl chloride by $Ti(OC_4H_9)_4 + (C_2H_5)_2AlCl$ in the presence of some CCl_4. Here, an inactive Ti(III) complex of unknown structure is formed first. The reaction of this complex with CCl_4 generates a Ti(IV)Cl compound and a $^{\bullet}CCl_3$ radical that initiates a free-radical polymerization.

Because of the exchange reaction of alkyl groups, one cannot decide whether the polymerization proceeds at the Group 1-3 metal atoms (e.g., at Al–C) or at the Group 4-10 metal atoms (e.g., at Ti–C). There are, however, some observations which, if taken together, speak for a polymerization at the bond Group 4-10 metal-carbon.

1. Ethene and 1-olefins can be polymerized to high-molecular weight polymers by transition metal halides without added metal alkyls. The poly(1-olefin)s are isotactic but polymerization rates are lower than those with metal alkyls. Such catalysts include $TiCl_2$, $TiCl_3$, $TiCl_3 + Et_3N$, CH_3TiCl_3, $Zr(CH_2C_6H_5)_4$, and $Cp_2Ti(C_6H_5)_2 + TiCl_4$. Other catalysts such as $Ti + I_2$, $Zr + ZrCl_4$, or $(C_6H_6)_2Cr$ polymerize ethene but not 1-olefins.

2. Polymerization rates increase by a factor between 10 and 10 000 if metal alkyls are added to these catalysts. The addition generates either more monometallic positions at the transition metal halide or points toward a bimetallic mechanism. Since the catalysts are active up to 100 hours, one can assume that one chain is formed per active site. If the polymerization is terminated by tritiated isopropanol, polymers from the polymerization by true Ziegler catalysts (transition metal compound + metal alkyls) contain 10^3-10^4 times more tritium than those from polymerizations without metal alkyls.

3. Organic chlorides react with metal-carbon bonds, for example, according to $ZnEt_2 + t\text{-}BuCl \rightarrow EtZnCl + t\text{-}BuEt$, and terminate the chain. Both true Ziegler catalysts and transition metal halides without metal alkyls show the same ranking with respect to efficiencies. which is improbable if both groups of catalysts act differently.

4. The fraction of propylene units in polymers from the copolymerization of ethene and propene by catalysts from $Al(i\text{-}Bu)_3$ + halides or oxyhalides of transition metals increases in the order $HfCl_4 < ZrCl_4 < VOCl_3 < VOCl_4$. The copolymerization by VCl_4 and different metal alkyls such as $Al(i\text{-}Bu)_3$, $Zn(C_6H_5)_2$, $Zn(i\text{-}Bu)_2$, and CH_3TiCl_3, on the other hand, always delivers copolymers with the same fraction of propylene units. The polymer chains must therefore grow at the transition metal-carbon bond.

9.2.6 Industrial Catalyst Systems for Ethene and Propene

Industry uses Ziegler catalysts practically only for heterophase homo and copolymerizations of olefins (Table 9-1). The transition metal compounds are almost exclusively $TiCl_3$ (catalyst generations 1 and 2) and the less expensive $TiCl_4$ (catalyst generations 3 and 4) (Table 9-4). The added $(C_2H_5)_3Al$ reduces $TiCl_4$ to $TiCl_3$ which resides in the system as crystalline aggregates. The resulting $(C_2H_5)_2AlCl$ forms a soluble dimer (similar to I on p. 279, with Cl instead of CH_2–CH_3).

$TiCl_3$ has four crystal modifications. It is hexagonally packed in the α modification and cubically in the γ form. The δ modification is intermediate between α and γ. The brown β modification has a chain-like (polymeric) structure with three chlorine bridges between two Ti atoms each; it delivers only low isotacticities. High isotacticities are obtained from the α, γ, and δ modifications.

The active complexes in all four types of catalysts thus consist of solid $TiCl_3$ particles whose surfaces are dotted with aluminum compounds. According to electron microscopy, the polymerization starts at the edges and sides of the aggregates but not on the base plane. The polymer then grows along the spiral steps of the catalyst particles.

The *first generation* of industrial catalysts for the polymerization of ethene and propene used the system $\beta\text{-}TiCl_3 + (C_2H_5)_2AlCl$. Residual $TiCl_3$ was removed by reaction with alcohols. The resulting TiO_2 remained in the polymer.

These first-generation catalysts had relatively low catalyst yields (Table 9-4). The isotacticity was ca. 90 %. They also delivered large amounts of so-called atactic fractions

Table 9-4 Catalysts for the bulk polymerization of propene (PP) at 70°C and 7 bar (4 h), catalyst yields, isotacticities (= % insoluble PP in boiling xylene), and necessary work-up by removal of catalyst residues (C), "atactic" fractions (A), and pelletizing (P) [2]. ED = electron donor.

Gene-ration	Initial components				Catalyst yield g PP/g cat.	g PP/g Ti	Isotacticity in wt%	Work-up C	A	P
1	$TiCl_3$ + Et_2AlCl + $AlCl_3$				1 000	4 000	90	+	+	+
2	$TiCl_3$ + Et_2AlCl			+ ED	4 000	16 000	95	+	–	+
3	$TiCl_4$ + Et_3Al	+ $MgCl_2$	+ ED		5 000	300 000	92	–	+	+
4	$TiCl_4$ + Et_3Al	+ $MgCl_2$	+ ED		15 000	600 000	98	–	–	–

which were really highly branched low-molecular weight polymers with many head-to-head connections. These "at-PP" were deleterious to the properties of it-PP. They were thus extracted and then either burned or buried. Later, a market developed for at-PP as adhesives, coatings, etc.. The newer generations of catalyst do not deliver at-PP as by-product; it is now produced separately.

Electron donors such as ethers, esters, amines, etc. were added in the *second generation catalyst*. Small proportions of these Lewis bases lower the polymerization rate, probably by reducing the number of active centers. Larger proportions increase the rate, however, most likely by destroying crystal aggregates, thus producing larger catalyst surfaces and more active sites.

Catalysts of the *third generation* use $MgCl_2$ or Mg(OH)Cl as carriers. $TiCl_4$ is, for example, ground in a ball mill with a surplus of $MgCl_2$ in the presence of the electron donor ethyl benzoate. This process produces very small $MgCl_2$ crystallites that do not agglomerate because ethyl benzoate is adsorbed on their surfaces.

$TiCl_4$ is reduced to γ-$TiCl_3$ by Et_3Al (see above). γ-$TiCl_3$ can complex with the co-ordinatively unsaturated surface of $MgCl_2$ since both have the same lattice constants, for example, according to

$(C_2H_5)_3Al$, $C_6H_5COOC_2H_5$, and $TiCl_4$ are usually used in the ratio 300:100:1. Since the rate constants of polymerizations by first, second, and third generation catalysts do not vary much, the high activity of the third generation catalyst is probably not caused by a different structure of the active centers but by a large increase of the number of active centers. Much less catalyst is therefore needed which in turn reduces the necessity to remove catalyst residues from the polymer.

The *fourth generation* of catalysts uses, for example, dried silicic acids as carriers on which non-crystalline δ-$MgCl_2$ is deposited by reaction of magnesium alkyls with chlorinating agents such as $SiCl_4$ or $SnCl_4$. The particles are subsequently treated with $TiCl_4$ and electron donors. The resulting polymerization-active particles act as "micro-reactors" that produce spherical pellets of highly isotactic poly(propylene) that does not need a work-up. These poly(propylene) particles are only slightly crystalline. Less energy is thus required for their melting in extruders, injection-molding machines, etc. The resulting parts are, however, highly crystalline since the high isotacticity leads to very rapid crystallizations on cooling.

9.2.7   Mechanisms

The growth at the transition metal-carbon bond may proceed via monometallic or bimetallic mechanisms. In **monometallic mechanisms**, only the transition metal participates in the propagation step; in **bimetallic mechanisms**, both transition metal and metal halide do. It thus does not matter for monometallic *mechanisms* whether the active *complex* is monometallic or bimetallic.

Bimetallic mechanisms were proposed as early as 1958 and 1960. They were later supported by the discovery of dimeric aluminum alkyls such as I and various titanium-aluminum complexes such as II-IV. For example, the reaction of dicyclopentadienyl titanium dichloride Cp_2TiCl_2 and triethyl aluminum Et_3Al with ethene in heptane at temperatures below 70°C generates ethane and forms a dark blue solution from which blue crystals appear on cooling. X-ray and molecular-weight studies indicate a compound II with electron-deficient bonds. II polymerizes ethene but III and IV do not.

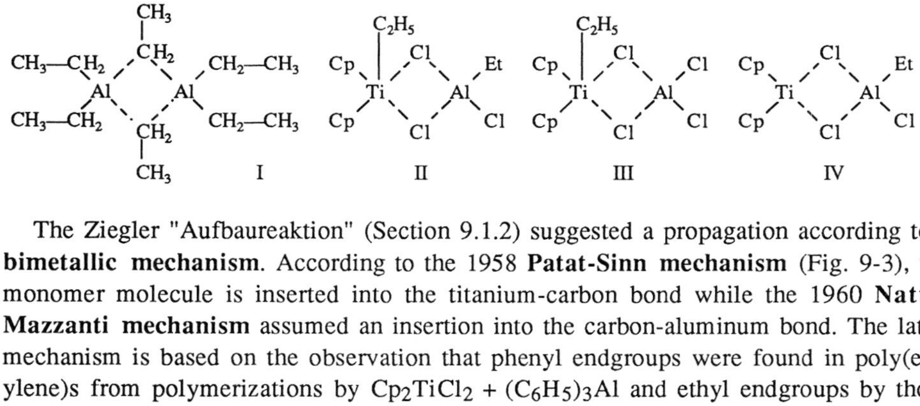

The Ziegler "Aufbaureaktion" (Section 9.1.2) suggested a propagation according to a **bimetallic mechanism**. According to the 1958 **Patat-Sinn mechanism** (Fig. 9-3), the monomer molecule is inserted into the titanium-carbon bond while the 1960 **Natta-Mazzanti mechanism** assumed an insertion into the carbon-aluminum bond. The latter mechanism is based on the observation that phenyl endgroups were found in poly(ethylene)s from polymerizations by $Cp_2TiCl_2 + (C_6H_5)_3Al$ and ethyl endgroups by those from $Cp_2Ti(C_6H_5)_2 + (C_2H_5)_3Al$ catalysts. This led to the conclusion that the polymerization proceeds at the Al atom by insertion of ethene into the C–Al bond. However, if Cp_2TiCl_2 is used with a mixture of $(C_6H_5)_3Al + (C_2H_5)_3Al$, only ethyl endgroups are found. Also, exchange reactions are observed if $Cp_2Ti(C_6H_5)_2$ and $(C_2H_5)_2AlCl$ are mixed. These findings point toward a polymerization by insertion of the monomer into the C–Ti bond, i.e., by the Patat-Sinn mechanism.

The Patat-Sinn and Natta-Mazzanti mechanisms both apply to molecularly soluble catalysts. Ziegler catalysts are, however, in general partially crystalline materials. In order to maintain electroneutrality, crystals must therefore possess on their surface unoccupied ligand sites. The **Cossee-Arlman** monometallic mechanism thus postulates an octahedral structure of the catalytically active titanium complex. According to this mechanism (Fig. 9-4), the π bond of ethene (and also of 1-olefins) approaches the unoccupied ligand site O of the transition metal (X). The resulting coordination (XI) destabilizes the bond Ti–R

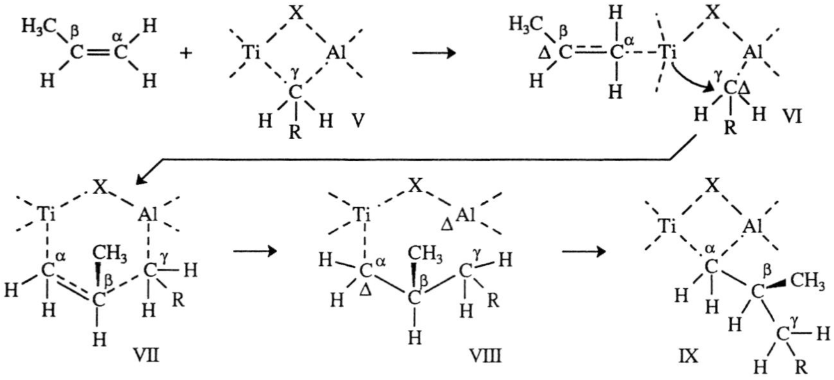

Fig. 9-3 Polymerization of propene according to the bimetallic Patat-Sinn mechanism with insertion into the Ti–C bond. The π-electron system of the 1-olefin interacts first with the p or d orbitals of the transition metal (V) which generates a new electron-deficient bond (VI). Residual valences (indicated in VI by Δ) remain at $^\beta$C and $^\gamma$C. Since the double bond is only partially abolished and the 2p-3d overlap at the $^\alpha$C–Ti bond is planar, no free rotation is possible around these bonds. The relatively rigid $^\beta$C–$^\gamma$C–Ti system rotates around the Ti–X bond (VI) and the residual valences at $^\beta$C and $^\gamma$C are saturated (VII). The subsequent hybridization at the $^\beta$C and $^\gamma$C atoms opens the $^\gamma$C–Al bond (VIII). The newly formed residual valences at $^\alpha$C and Al saturate and the original complex is restored (IX) but the polymer chain is extended by one monomeric unit.

The similar Natta-Mazzanti mechanism assumes that the monomer first forms a π bond with the Ti atom. The simultaneous opening of the Ti–$^\gamma$C bond delivers ~Ti$^{\delta+}$ and $^{\delta-}$CH$_2$R–Al~. The coordinated monomer molecule is polarized by the slightly positive Ti atom and inserts into the C–Al bond.

between the transition metal (e.g., Ti) and the alkyl group R (XII) as was shown by quantum mechanical calculations and magnetic measurements on nonpolymerizing olefins. The destabilization activates the alkyl group which can subsequently react with the double bond of the coordinated monomer molecule: the olefin molecule is inserted between the transition metal and the alkyl group or the polymer group (XIII).

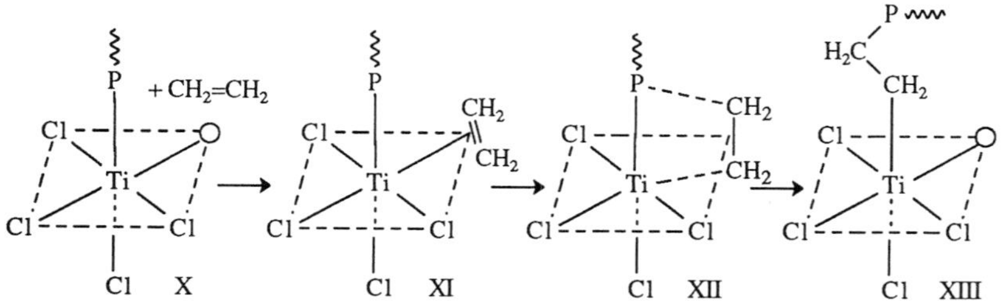

Fig. 9-4 First step in the polymerization of ethene according to the monometallic Cossee-Arlman mechanism. In subsequent propagation steps, ~P is the poly(ethylene) chain –(CH$_2$–CH$_2$)$_i$R.

9.2.8   Stereocontrol in Polymerizations of 1-Olefins

1-Olefins and dienes are prochiral compounds (Fig. 4-4). Their insertion into the bond between the transition metal and polymer chain is controlled by four factors: (1) origin of the stereocontrol, (2) approach of the monomer molecule to the unoccupied ligand site, (3) geometry of insertion, and (4) type of interconnection.

Stereocontrol

The stereospecificity of the insertion step is controlled by the last unit(s) of the growing polymer chain (chain-end model; usually found in soluble catalysts) or by the structure of the catalytic site (enantiomorphic-site model; usually found in insoluble catalysts) (Section 6.3.3). The chain-end model accommodates the whole range of J-ad tacticities, for example, diad tacticities in the range $0 \le x_i \le 1$.

In the most simple case of *enantiomorphic-site* control, the connection of a new monomeric unit is independent of that of the preceding unit. The mole fraction of the syndiotactic diads never exceeds $x_s = 1/2$ (Fig. 9-5).

The simple enantiomorphic-catalyst mechanism assumes a catalyst with two mirror-image (D,L) types of catalyst sites (Section 6.3.3). The addition of the prochiral monomer molecules should be independent of the type of the last incorporated unit, i.e., it should obey Bernoullian statistics with respect to conditional probabilities ($p_{L/D} = p_{D/D} = p_D$). One obtains for the relationships between the mole fractions of triads and diads

(9-7) $x_{ii} = -(1/2) + (3/2)\, x_i = 1 - (3/2)\, x_s$

(9-8) $x_{ss} = (1/2)\, x_s$

In Fig. 9-5, the slight deviation from the theoretical lines might be caused by an effect of the already incorporated unit, i.e., $p_{L/D} \ne p_{D/D}$.

In predominantly isotactic polymers, configurational errors should be of the type ···iiiissiiii··· if the polymerization is enantiomorphic-site controlled and ···iiiisiiii··· if it is chain-end controlled. Such stereo errors are difficult to detect in the highly isotactic pentane-*insoluble* fractions of poly(propylene)s. The pentane-*soluble* fractions of such polymers indicate considerable syndiotactic pentads ssss and especially syndiotactic sequences such as heptads ssisis with (at least) two nonconsecutive errors (Fig. 4-17). This has been taken as an indication of a chain-end control. However, it may also be that the soluble and insoluble fractions result from two different mechanisms.

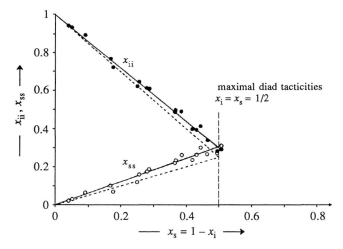

Fig. 9-5 Mole fraction of isotactic and syndiotactic triads as a function of the mole fraction of syndiotactic diads [3]. Polymerization of propene by insoluble catalysts ($C_6H_5CH_2)_4$Ti-BF$_3$, ZrCl$_4$-($C_2H_5)_3$Al, or MtCl$_3$-($C_2H_5)_2$AlCl (Mt = Ti, V, Cr). - - - - Prediction of Eqs.(9-7) and (9-8).

Approach to the Catalyst

1-Olefins are prochiral compounds with two enantiotopic sites, *re* and *si* (Fig. 4-4). Monomer molecules can thus approach the catalyst always from the same side (*...re-re-re...* or *...si-si-si...*) or from alternating sites (*...re-si-re-si...*).

Geometry of Insertion

A prochiral monomer molecule can be inserted in either a cis or trans position (Fig. 9-6). Unfortunately, this can only be detected for unsymmetrically substituted monomers. The polymerization of *cis*-1-alkyl-2-alkoxy ethene R-CH=CH(OR') and *cis*-1-chloro-2-alkoxy ethene Cl–CH=CH(OR') by Ziegler catalysts leads to erythrodiisotactic polymers (eit) whereas the corresponding trans monomers deliver threodiisotactic ones (tit) (see p. 126 for eit and tit). The double bond must therefore be opened in the cis position.

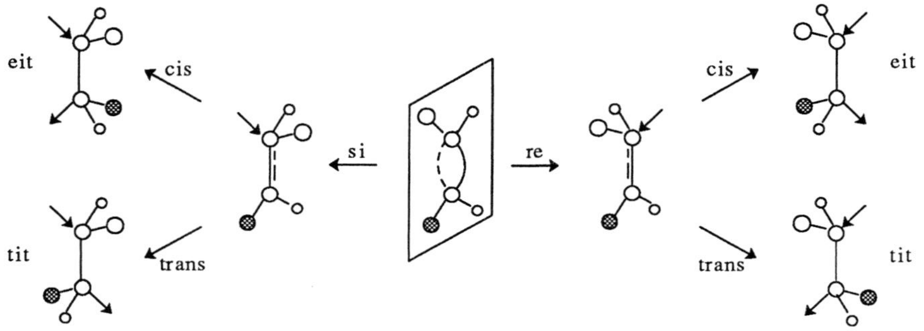

Fig. 9-6 Geometry of the insertion of unsymmetrically substituted 1-olefins.

Correspondingly, one also finds for differently deuterated propenes that high isotacticities with respect to D-substituents at β-carbon atoms result only if the D-substituent is in the cis position to the CD_3 group. The proportion of it-diads with respect to the CD_3 groups are however independent of the type of substitution:

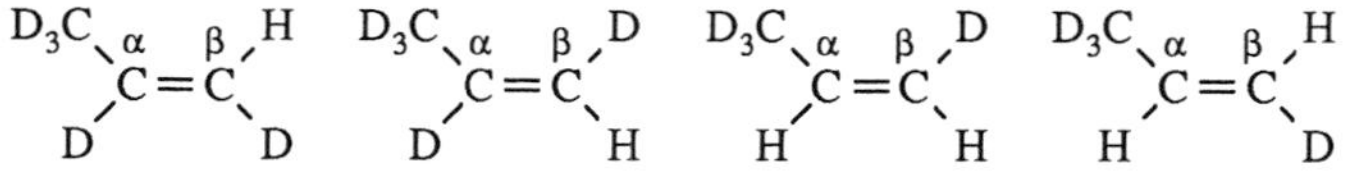

	D_3C α β H / C=C / D ... D	D_3C α β D / C=C / D ... H	D_3C α β D / C=C / H ... H	D_3C α β H / C=C / H ... D
$x_i (D_\alpha)=$	< 0.02	> 0.98	> 0.98	< 0.05
$x_i (CD_3) =$	> 0.98	> 0.98	> 0.95	> 0.95

Bonding

On insertion into the metal-carbon bond, 1-olefins can bond to the polymer chain with either the α-carbon atom (C^1) or the β-carbon (C^2). Both types can be distinguished by reaction of the monomer with surplus metal alkyl and analysis of the resulting compounds after killing the active centers by reaction with methanol (Eq.(9-5)). According to these experiments, the isospecific polymerization of propene by $TiCl_4$–AlR_3 occurs with certainty by a 1,2-insertion whereas the syndiospecific polymerization of propene by VCl_4–$(C_2H_5)_2AlCl$–anisole very probably takes place by a 2,1-insertion:

(9-9)
$$L_pMtH + n\,CH_2{=}\overset{\overset{\displaystyle CH_3}{|}}{CH} \xrightarrow[\alpha]{1,2} L_pMt{-}(\overset{2}{CH_2}{-}\overset{\overset{\displaystyle CH_3}{1\,|}}{CH})_n{-}CH_2{-}CH_2{-}CH_3$$

(9-10)
$$L_pMtH + n\,\overset{\overset{\displaystyle CH_3}{|}}{CH}{=}CH_2 \xrightarrow[\beta]{2,1} L_pMt{-}(\overset{\overset{\displaystyle CH_3}{1\,|}}{CH}{-}\overset{2}{CH_2})_n{-}\overset{\overset{\displaystyle CH_3}{}}{\underset{\underset{\displaystyle CH_3}{}}{CH}}$$

Both insertions lead to head-to-tail structures in the chain. However, the structure of the endgroups depends on the ligands L and R if catalysts L_pMtR are used. It was found for the polymerization of propene by L_pMtH that n-propyl endgroups result from 1,2-insertion (α-insertion) (Eq.(9-9)) but i-propyl endgroups result from 2,1-insertion (β-insertion) (Eq.(9-10)).

The catalyst system $TiCl_4 + (C_2H_5)_3Al$ first exchanges ligands and gives $(C_2H_5)TiCl_3 + (C_2H_5)_2AlCl$. A 1,2-insertion delivers here $-CH_2{-}CH(CH_3){-}CH_2{-}CH(CH_3){-}C_2H_5$ as endgroup whereas a 2,1-insertion results in $-CH(CH_3){-}CH_2{-}C(CH_3){=}CH_2 + C_2H_6$.

All propagation reactions thus proceed via *chiral centers* which differentiate between the two sides (image and mirror-image) of the prochiral monomer molecules. Isospecific catalyst systems such as δ-$TiCl_3$-$(CH_3)_2AlCl$ polymerize [S]-3-methyl-1-pentene to isotactic polymers with [S]-monomeric units. The [R,S] monomer, on the other hand, delivers a mixture of the two isotactic enantiomers that contain either only [R] units *or* only [S] units. The two enantiomers can be separated chromatographically on chiral columns into fractions of opposite optical rotation. Although the polymerization of, e.g., propene proceeds via chiral centers, the resulting isotactic polymers are *not* optically active (Section 4.2.2).

Control of Molecular Weight

Homogeneous Ziegler catalysts usually deliver polymers with relatively narrow molecular weight distributions. In a few cases, both termination and transfer reactions are absent, and the molar-mass ratio $\overline{M}_w/\overline{M}_n$ approaches that of Poisson distributions (Section 3.3.3). Strong transfer reactions, however, result approximately in Schulz-Flory distributions with $\overline{M}_w/\overline{M}_n \approx 2$.

Heterogeneous polymerizations are industrially often controlled by hydrogen (Eq.(9-4)). The molecular-weight distributions can here be described in general by Wesslau distributions (Section 3.3.3). This type of distribution is probably due to the inhomogeneous distribution of active sites on the catalyst surface. The resulting broad molecular-weight distributions ($\overline{M}_w/\overline{M}_n \approx 4$-10) are industrially often narrowed by thermal chain degradations, usually with added peroxides. Even broader distributions result from polymerizations in reactor cascades.

9.2.9 Other Ziegler-Natta Polymerizations

Classical Ziegler catalysts also allow the copolymerization of ethene with 4-10 % of 1-butene, 1-hexene, or 1-octene to so-called linear low-density poly(ethylene)s (LLDPE) (Table 9-1). Larger proportions of higher 1-olefins can be copolymerized by metallocene-MAO catalysts (Section 9.3). Copolymers of ethene, propene, and a nonconjugated diene are usually obtained by vanadium-containing catalysts (Table 9-1).

The polymerization of dienes leads to polymers that are high in cis structures if catalysts are used similar to the ones for the polymerization to isotactic poly(1-olefin)s.
Butadiene is polymerized to butadiene rubbers by titanium catalysts. Such catalysts consist, for example, of TiI_4-R_3Al or $TiCl_4$-R_2AlI. They deliver polydienes with 94 % cis-
1,4-, 2 % trans-1,4-, and 4 % vinyl-1,2 structures. Higher cis contents are obtained with
systems containing cobalt or nickel (97 % cis-1,4; 2 % trans-1,4; 1 % vinyl-1,2) or urani-
um or neodymium (99 % cis-1,4, < 1 % trans-1,4, < 0.5 % vinyl-1,2). Isoprene delivers
an isoprene rubber on polymerization with $TiCl_4$–R_3Al (> 96 % cis-1,4).

The molar ratio of transition metal to aluminum should not exceed unity for Ziegler-
Natta polymerizations of dienes to cis structures. Polymerizations to trans structures with
Mt/Al > 1 probably follow bimetallic mechanisms. In industry, dienes are also polymer-
ized anionically by organolithium compounds.

Very many other transition metal catalysts have been reported. However, none have
gained any importance. Some of these polymerize monomers to rather peculiar structures. For example, the system nickel(0) compound + bis(trimethylsilylamino)-bis(trimethylsilylimino)phosphorane polymerizes 1-olefins to 2,ω-type structures such as

(9-11)

9.2.10 Kinetics of Living Polymerizations

Ziegler-Natta polymerizations by preformed, stable, soluble Ziegler catalysts proceed
without termination and chain transfer. They are living; their polymerization rates $R_p =$
$d[P]/dt = k_p[C^*][M]$ are first order with respect to the molar concentrations $[M]$ of
monomer and $[C^*]$ of active centers. The concentration of active centers equals the concentration of transition metal-carbon bonds. It can be determined by tagged initiation or
termination reactions (Eqs.(9-3)-(9-5)).

In heterogeneous Ziegler-Natta polymerizations, however, active centers are neither
preformed nor stable or soluble. They are rather formed very slowly, often only after
addition of the monomer. Such polymerizations show "induction" periods that decrease
with increasing temperature because of the increasing rate of formation of active centers
(Fig. 9-7).

The formation of active centers is a true chemical reaction. In a physical adsorption
of metal alkyls on the surface of the transition metal compound, the concentration of
active centers should be proportional to the catalyst surface that is covered by the metal
alkyl. However, this was not found to be the case.

The total surface is relatively small according to measurements of the nitrogen adsorption (Brunauer-Emmet-Teller method). It is only partially covered by metal alkyls.

Newer catalyst systems often show no induction periods (Fig. 9-8). The concentration
of active centers is not constant, though, but rather must be determined as a function of
time or monomer conversion. This time dependence can have many reasons, including
formation or disappearance of active centers by dynamic exchange reactions, blocking
of active centers by precipitated polymer, breaking-up of catalyst particles and
formation of new surfaces, loss of active centers by termination reactions, etc.

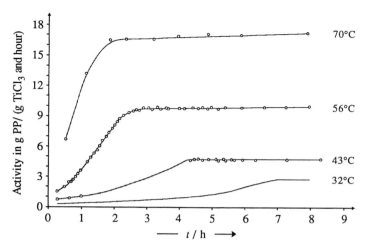

Fig. 9-7 Catalyst activity in the polymerization of propene by TiCl$_3$ + (C$_2$H$_5$)$_3$Al [4]. By permission of Academic Press.

The same catalyst components may thus lead to very different types of polymerization. In the polymerization of ethene by TiCl$_4$ + ((CH$_3$)$_2$CH–CH$_2$)$_3$Al + MgCl$_2$, the catalyst productivity increases first rapidly (formation of more active centers) and then more slowly with time until it finally becomes constant (active centers remain constant; Fig. 9-8, top). In the polymerization of propene, however, the catalyst productivity also increases very rapidly because of the formation of more active centers but then passes through a sharp maximum and subsequently decreases first rapidly and then more slowly because more and more active centers are destroyed (Fig. 9-8, bottom). These two polymerization types are distinguished as **acceleration type** and **decay type**.

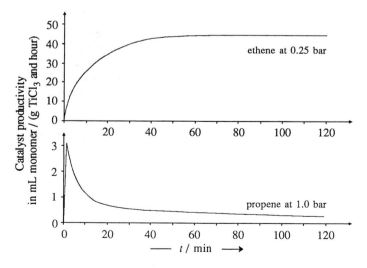

Fig. 9-8 Time dependence of the catalyst productivity as a measure of the polymerization rate in the polymerization of ethene and propene in heptane by the catalyst system TiCl$_4$ + (*i*-Bu)$_3$Al + MgCl$_2$ at 60°C and constant pressure [5].

The monomer conversion $u = ([M]_0 - [M])/[M]_0$ thus depends on many factors. The monomer concentration may stay constant by maintaining a constant pressure (gaseous monomers) or continuous addition of monomer (liquid monomers). The change of monomer conversion with time is given by $du/dt = k_p[C^*](1 - u)$ if the concentrations $[M]$ of monomer and $[C^*]$ of active centers both stay constant.

The monomer conversion may also depend on the rate of catalyst decomposition which may be of the first or second order with rate constants k_d. The time dependence is then given by Eqs.(9-12)-(9-14) where Y is a monomer function that adopts the value $Y = u = ([M]_0 - [M])/[M]_0$ if $[M] = const$ and $Y = -\ln([M]/[M]_0) = -\ln(1 - u)$ if the monomer concentration decreases with time by a first order reaction:

(9-12) $Y = k_p[C^*]_0 t$ no catalyst decomposition

(9-13) $Y = (k_p/k_d)[C^*]_0\{1 - \exp(-k_d t)\}$ 1st order catalyst decomposition

(9-14) $Y = (k_p/k_d)\ln\{1 + k_d[C^*]t\}$ 2nd order catalyst decomposition

The polymerization rate is no longer first order with respect to monomer if the molecules coordinate with the transition metal before the propagation step. Similar to enzymatic reactions (Chapter 14), an equilibrium reaction is followed here by the rate-determining insertion of the monomer molecule:

(9-15) $\text{wM}_iC^* + M \underset{k_{-1}}{\overset{k_1}{\rightleftharpoons}} \text{wM}_iC^*/M \overset{k_p}{\longrightarrow} \text{wM}_{i+1}C^*$

The total concentration $[C^*]_0 = \Sigma_i [\sim M_iC^*] + \Sigma_i [\sim M_iC^*/M]$ of all active centers C^* is very small and the concentration of all monomer-catalyst complexes even smaller. Concentrations $[\sim M_iC^*/M] \equiv [C^*/M]$ can then not be measured directly. This type of reaction can be described by the **Michaelis-Menten formalism** of enzymatic reactions (see Chapter 14). Replacing k_2 of Eq.(14-20) by k_p and $[E]_0$ by $[C^*]$ and setting $K_m = (k_{-1} + k_p)/k_1$, one obtains for the propagation rate

(9-16) $R_p = \left(\dfrac{k_p}{K_m + [M]}\right)[C^*][M] = \dfrac{k_1 k_p[C^*][M]}{k_{-1} + k_p + k_1[M]} = k_p[C^*][M]_{eff}$

According to Michaelis-Menten, the concentration of the monomer-catalyst complex is given by $[C^*/M] = [C^*][M]/([M] + K_m)$ where $K_m = (k_{-1} + k_p)/k_1$ is the Michaelis-Menten constant. The effective monomer concentration $[M]_{eff} = [M]/([M] + K_m) = f_M$ equals the fraction f_M of the catalyst surface that is covered with monomer molecules.

In the case of weakly coordinated monomer molecules, one has $k_1[M] \ll k_{-1}$. The polymerization rate $R_p = k[C^*][M]_{eff} \approx \{kk_1/(k_{-1} + k_p)\}[C^*][M]$ is then first order with respect to monomer. It cannot be distinguished from that of the polymerization of an uncoordinated monomer where $R_p = k_p[C^*][M]$, except for the meaning of the proportionality constant. Such a linear relationship between R_p and $[C^*][M]$ was, for example, found for the polymerization of 1-methyl-1-pentene by $VCl_3 + R_3Al$.

The concentration $[C^*]$ of active centers is usually not evaluated. One rather knows only the molar concentrations $[T]$ of transition metal compounds and $[A]$ of metal alkyls. $[C^*]$ can be calculated, however, from $[C^*] = K_A[A][T] = f_T[T]$ if the active centers are formed in an equilibrium reaction with an equilibrium constant K_A. The quantity $f_T = K_A[A]$ can be interpreted as the effective usage of T in an equilibrium reaction.

The rate of formation of the T-A complex is proportional to $(1 - f_A)$ if only part of T complexes with A. The dissociation rate of the complex is proportional to f_T. Both rates are equal in the steady state which results in $k_A[A](1 - f_T) = k_{-A}f_T$. Introduction of the equilibrium constant $K_A = k_A/k_{-A}$ then leads to the **Langmuir-Hinshelwood isotherm**:

(9-17) $\qquad f_T = \dfrac{K_A[A]}{1 + K_A[A]}$

The polymerization rate is thus given by

(9-18) $\qquad R_p = k_p[C^*][M] = k_p f_T[T][M] = \dfrac{k_p[M]K_A[A][T]}{1 + K_A[A]}$

This equation applies to non-associated metal alkyls. Lower metal alkyls exist as dimers in solution but coordinate in monomeric form. Because of the dimerization equilibrium, the concentration $[A]$ in Eq.(9-18) has to be replaced by $([A_2]/K_D)^{1/2}$ where $K_D = [A_2]/[A]^2$ is the equilibrium constant of dimerization.

9.2.11 Kinetics of Nonliving Polymerizations

True termination reactions are rare for Ziegler-Natta polymerizations at room temperature. However, polymer chains are terminated at higher temperatures (ca. 100°C) by a β-elimination of transition metal hydrides:

(9-19)

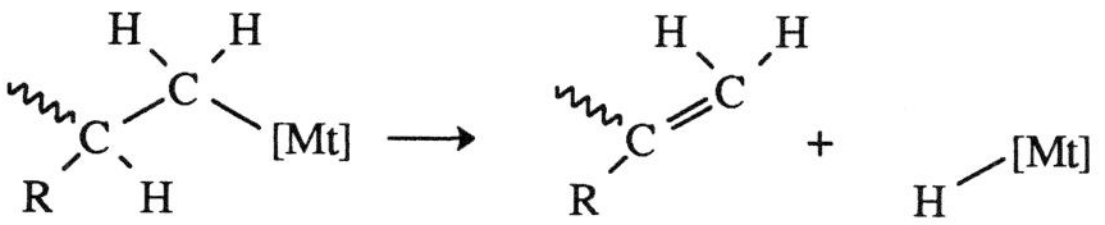

The total number of active centers in the system (and thus the kinetic chain) remains constant, however, since the metal hydride [Mt]–H is realkylated by the monomer $CH_2=CHR$. The polymerization rate remains constant but the number-average degree of polymerization decreases by this reaction.

However, Ziegler-Natta polymerizations are prone to transfer reactions to the monomer and the metal alkyls:

(9-20) $\quad \text{\textasciitilde CHR}-CH_2-Mt + CHR{=}CH_2 \xrightarrow{k_{tr,M}} \text{\textasciitilde CHR}{=}CH_2 + R-CH_2-CH_2-Mt$

(9-21) $\quad \text{\textasciitilde CHR}-CH_2-Mt + R'_3Al \xrightarrow{k_{tr,A}} \text{\textasciitilde CHR}-CH_2-AlR'_2 + R'-Mt$

The newly formed transition metal compounds ~~CHR–CH₂–Mt and R'–Mt are potential active centers C^*_{pot} that convert to true active centers C^* in an initiation step with the rate $R_i = k_i[C^*]_{pot}[M]_{eff}$ where $[M]_{eff}$ is the effective monomer concentration.

The total concentration $[C^*]_{tot} = [C^*]_{pot} + [C^*]$ of all transition metal–carbon bonds, i.e., all active centers, remains constant. In the steady state, the rate of initiation must equal the sum of the rates of the two transfer reactions (9-20) and (9-21)

(9-22) $R_i = R_{tr,M} + R_{tr,A}$

(9-22a) $k_i[C^*]_{pot}[M]_{eff} = k_i([C^*]_{tot} - [C^*])[M]_{eff} = k_{tr,M}[C^*][M]_{eff} + k_{tr,A}[C^*][A]_{eff}$

where $[A]_{eff} = f_A[A]$ is the effective concentration of A, i.e., the fraction f_A of the catalyst surface that is covered by the metal alkyl.

Insertion of $[C^*]$ from Eq.(9-22a) into the standard rate equation for R_p delivers

(9-23) $R_p = k_p[C^*][M]_{eff} = \dfrac{k_i k_p[M]_{eff}^2[C^*]_{tot}}{(k_{tr,i} + k_{tr,M})[M]_{eff} + k_{tr,A}[A]_{eff}}$

On the catalyst surface, the concentrations of metal alkyl and monomer are fractions f_A and f_M, resp., of their total concentrations, i.e., $f_A = [A]_{eff}/[A]$ and $f_M = [M]_{eff}/[M]$. Introduction of these quantities into Eq.(9-23) delivers a linear relationship between $[M][C^*]_{tot}/R_p$ and $1/[M]$ if $[A] = const.$ (Fig. 9-9):

(9-24) $\dfrac{[M][C^*]_{tot}}{R_p} = \dfrac{k_i + k_{tr,M}}{k_i k_p f_M} + \dfrac{k_{tr,A} f_A[A]}{k_i k_p f_M^2} \cdot \dfrac{1}{[M]}$

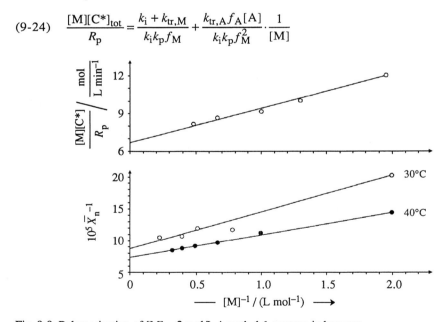

Fig. 9-9 Polymerization of $[M] = 2$ mol/L 4-methyl-1-pentene in benzene.
Top: $[M][C^*]_{tot}/R_p$ as a function of $1/[M]$ at 30°C according to Eq.(9-24);
          $[C^*]_{tot} = 17.8$ mmol VCl$_3$/L, $[A]_0 = 35.6$ mmol (i-Bu)$_3$Al/L $= const.$ [6].
Bottom: $1/\overline{X}_n$ as a function of $1/[M]$ according to Eq.(9-24) with Al/V $= 2:1$ mol/mol;
          $[C^*]_{tot} = 18.8$ mmol VCl$_3$/L (30°C) and 8.8 mmol VCl$_3$/L (40°C) [7].

The number-average degree of polymerization of such a polymerization is given by the ratio of the propagation rate to the concentration of all active centers plus the sum of all rates that terminate a certain polymer chain by transfer or start a new one:

(9-25) $\bar{X}_n = \dfrac{\int_0^t R_p dt}{[C^*]_{tot} + \int_0^t R_{tr,M} dt + \int_0^t R_{tr,A} dt} = \dfrac{k_p [M]_{eff} t}{1 + (k_{tr,M} [M]_{eff} + k_{tr,A} [A]_{eff}) t}$

(9-26) $\dfrac{1}{\bar{X}_n} = \dfrac{k_{tr,M}}{k_p} + \dfrac{1}{k_p f_M [M] t} + \dfrac{k_{tr,A} f_A}{k_p f_M} \cdot \dfrac{[A]}{[M]}$

The second term on the right side of Eq.(9-26) becomes very small after a sufficiently long time and the number-average degree of polymerization is independent of time.

According to Eq.(9-26), the inverse of the degree of polymerization increases with increasing concentration [A] of the metal alkyl A at constant monomer concentration [M]. At constant [A], it is inversely proportional to the monomer concentration (Fig. 9-9). The two intercepts allow the calculation of the rate constant $k_{tr,M}$ for the transfer to monomer if the propagation rate constant k_p is known. For example, the following values have been obtained for the polymerization of 4-methyl-1-pentene at 30°C: $k_p = 53$ s^{-1} and $k_{tr,M} = 0.0043$ s^{-1}. The ratio $k_p/k_{tr,M} \approx 12\ 300$ indicates that 12 300 propagation steps are followed by a transfer to monomer.

9.3 Single-Site Polymerizations

9.3.1 Aluminoxanes

Soluble Ziegler catalysts became known a few years after Ziegler-Natta polymerizations were discovered. Polymerizations by these catalysts are living. They also allow the preparation of block copolymers, for example, of propene and methyl methacrylate at 25°C by vanadium acetylacetonate + $(C_2H_5)_2AlCl$ as catalyst. The catalyst productivities were low, however, and the polymerizations often led only to low molecular weights or low tacticities. Industrially, soluble Ziegler catalysts became interesting only for the polymerization of dienes and the copolymerization of some 1-olefins.

The break-through, however, came with the discovery of so-called metallocene catalysts (Section 9.3.2) and alum(in)oxanes as co-components. It was known for many years that aluminum alkyls (such as Friedel-Crafts catalysts) are destroyed by water but it was not realized that *totally dry* aluminum alkyls are completely inactive while "wet" aluminum alkyls are very active.

For example, the addition of some water to triethyl or trimethyl aluminum produces oligomeric **aluminoxanes** with molecular weights of ca. 1000-1500 which may be linear, self-associates, or cyclic compounds (Fig. 9-10).

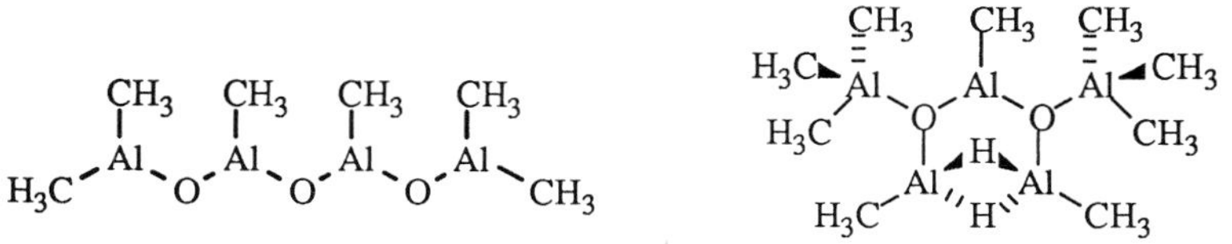

Fig. 9-10 Possible structures of methyl alumoxanes.

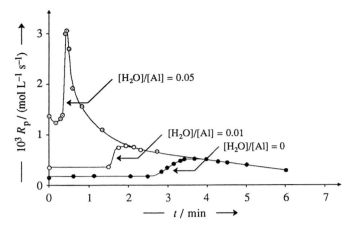

Fig. 9-11 Effect of water on the time dependence of the polymerization rate of ethene in toluene by a soluble catalyst from 3 mmol/L $Cp_2Ti(C_2H_5)Cl$ and 6 mmol/L $C_2H_5AlCl_2$ at 10°C [8]. The ethene concentration was kept constant at 52 mmol/L. Cp = Cyclopentadienyl group.

Controlled addition of a little water to alkyl aluminum compounds before their combination with titanium compounds increases the activity of such catalysts dramatically in the early stages of polymerization (Fig. 9-11).

9.3.2 Metallocene Catalysts

Metallocene catalysts usually consist of transition metal catalysts and aluminoxanes. The transition metal Mt is bound to one (I) or two cyclopentadiene units (II) which may be single or part of a larger ring system such as indene (III, IV) or fluorene (V, VI) (Fig. 9-12). The rings may be substituted (not shown) or not. They may be bridged by one single transition metal group (II-IV), or by an additional bridge that does contain a transition metal group (V, VI).

In combination with aluminoxanes, metallocenes are very efficient soluble catalysts for ethene polymerizations (Table 9-5). Their catalyst activities often surpass those of insoluble Ziegler catalysts. In general, zirconium catalysts are more effective than titanium or hafnium catalysts and chlorine containing metallocenes more effective than chlorine-free ones.

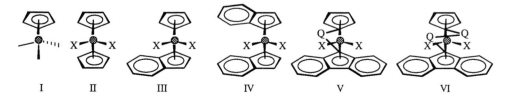

Fig. 9-12 Different types of metallocene catalysts. The transition metal ⊛ is mostly an early transition metal element such as Ti, Zr, or Hf (Group 4). The substituents X are often Cl but may also be phenyl groups. The bridge group Q< can be a methylene group $-CH_2-$, a dimethylmethylene group $-C(CH_3)_2-$, a dimethylsilylene group $-Si(CH_3)_2-$, etc.

Table 9-5 Catalyst activities A_C and propagation rate constants k_p for the polymerization of ethene E in toluene by metallocene and Ziegler catalysts at constant pressure of ethene of 8 bar [9, 10].
n_T = amount of transition metal T, n_{AL} = amount of aluminum compound AL, $\overline{M}_v$ = viscosity average of molecular weight, A_C = catalyst activity in $m_P/(m_{Mt} \, t \, p)$ (m_P, m_{Mt} in g; t in h; p in bar).

Catalyst T	$\dfrac{n_T}{\mu\text{mol}}$	AL	$\dfrac{n_{AL}}{\mu\text{mol}}$	$\dfrac{T}{°C}$	$\dfrac{A_C}{g_E/(g_T \text{ h bar})}$	$\dfrac{k_p}{\text{L mol}^{-1}\text{s}^{-1}}$	$\dfrac{\overline{M}_v}{\text{g mol}^{-1}}$
Metallocene catalysts							
Cp_2ZrCl_2	0.030	MAO	5 000	70	1 000 000		
$Cp_2Zr(CH_3)_2$	0.033	MAO	2 200	90	3 100 000	105 000	150 000
	0.0005	MAO	5 000	70	440 000		~ 1 000 000
	0.10	MAO	5 000	70	700 000		
	0.10	MAO	5 000	20	9 600		
$Cp_2Ti(CH_3)_2$	3.0	MAO	5 000	20	500		
Cp_2TiCl_2	3.0	MAO	5 000	20	90 000	1 100	
Cp_2TiCl_2		Me_2AlCl		45		39	
Ziegler catalysts							
$TiCl_4$ + RMgCl		Et_3Al		70		13 000	
$TiCl_3$		Et_2AlCl		50		540	2 000 000
$TiCl_4$		$(iBu)_3Al$		30		33	1 000 000
$TiCl_4$ + $(EtO)_2Mg$		$(iBu)_3Al$		30		2 000	10 000 000
				60		12 000	

The catalyst productivity increases with increasing temperature (larger propagation constants) whereas the molecular weights decrease (more β-eliminations, see below).

A catalyst activity of $3.1 \cdot 10^6$ g PE/(g Zr h bar) (Table 9-5, Line 2) produces $m_{PE} = 24.8 \cdot 10^6$ g poly(ethylene) per gram zirconium ($m_{Zr} = 1$ g) in 1 hour at 8 bar. With $M_{Zr} \approx 91.22$ g/mol and $M_E \approx 28.05$ g/mol, this leads to $N_E = (m_{PE}/m_{Zr})(M_{Zr}/M_E) \approx 80.7 \cdot 10^6$ inserted ethene molecules per zirconium atom and hour. Since the observed viscosity-average molar mass of $\overline{M}_v = 150\,000$ g/mol corresponds to a number-average molar mass of $\overline{M}_n \approx 100\,000$ g/mol ($\overline{X}_n \approx 3600$) for this type of molar mass distribution, it follows that $N_{PE} = t_M / \overline{X}_n \approx 22\,400$ poly(ethylene) molecules are generated per zirconium atom and hour. A poly(ethylene) molecule is thus formed within $t_M = 3600$ s/22 400 $\approx$ 0.16 s. Only $t_E = t_M/\overline{X}_n = 0.16$ s/3600 $\approx 4.5 \cdot 10^{-5}$ s are needed for the insertion of one ethene molecule into the active complex. This time t_E is called the **turn-over time**.

Methyl aluminoxanes (MAO) are more effective than ethyl aluminoxanes. They are usually employed in great excess, for example, up to a molar ratio of 10^7:1 (mol Al/mol Zr). Industry often uses molar ratios of 10^4:1.

The abstraction of a methyl group from metallocenes by alumoxanes according to

(9-27) $Cp_2Zr(CH_3)_2 + MAO \rightarrow [Cp_2ZrCH_3]^{\oplus} [CH_3MAO]^{\ominus}$

generates the catalytically active species which is a contact ion pair with a coordinatively unsaturated zirconium atom in the cation. The successive insertion of monomeric units into the bond between the transition metal and the already present polymer chain results in a polymer cation $[Cp_2Zr-CH_2-CH_2-CH_2CH_2(CH_2CH_2)_{n-2}CH_3]^{\oplus}$.

Intramolecular β-elimination regenerates the original complex:

(9-28) $[Cp_2Zr-CH_2-CH_2-CH_2CH_2[CH_2CH_2]_{n-2}CH_3]^{\oplus}$

$\quad\quad\quad\quad \longrightarrow [Cp_2ZrCH_3]^{\oplus} + CH_2{=}CHCH_2[CH_2CH_2]_{n-2}CH_3$

that starts a new polymer chain. The generated 1-olefin acts as a comonomer. The resulting polymer chains thus have long-chain branches:

$$[Cp_2Zr[CH_2CH_2]_{n-2}CH_3]^{\oplus} + CH_2{=}CHCH_2[CH_2CH_2]_{n-2}CH_3$$

(9-29)
$$\longrightarrow [Cp_2Zr{-}CH_2{-}\underset{\underset{CH_2(CH_2CH_2)_{n-2}CH_3}{|}}{CH}{-}[CH_2CH_2]_{n-2}CH_3]^{\oplus}$$

The Cossee-Arlman mechanism for polyinsertions by Ziegler catalysts (Fig. 9-4) involves the reaction sequences A-B-C-D-A or D-E-F-A-D (Fig. 9-13) in which the monomer molecules approach the active site always in the same manner, i.e., ...*si-si-si*... or ...*re-re-re*... (Fig. 9-6). This mechanism is also found for some single-site polymerizations with soluble metallocene-MAO catalysts. It obviously requires catalytic complexes in which two active sites are *homotopic*. Examples are the C_2-symmetric (one symmetry axis) Group 4 *rac-ansa*-metallocenes with two equal active sites 1 (VII in Fig. 9-14).

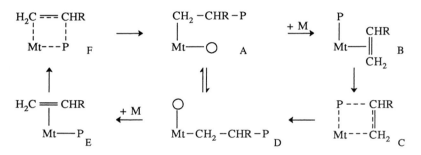

Fig. 9-13 Polymerization of 1-olefins by the Cossee-Arlman mechanism (see text) [11]. ○ indicates an unoccupied ligand site. P symbolizes polymer chains with different numbers of monomeric units in A and D (P = N units) vs. B and E (P = (N+ 1) units).

Catalysts with C_1-symmetry (no symmetry operation possible, see Section 4.1.3), on the other hand, have two *diastereotopic* active sites with different reactivities 1 and 2 (examples are the ansa compounds VIII-X in Fig. 9-14). The monomer insertion follows the sequence A-B-C-D-E-F (Fig. 9-13). The successive approaches of monomer molecules alternate according to ...*si-re-si-re*... and the monomer molecules are inserted according to M–MtQ$_2$–P$_i$ → P$_{i+1}$–MtQ$_2$–M → M–MtQ$_2$–P$_{i+2}$.

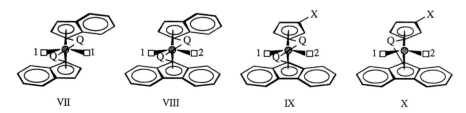

VII   VIII   IX   X

Fig. 9-14 Examples of catalyst complexes of *ansa*-metallocenes with C_2 symmetry (VII) and C_1 symmetry (VIII-X). X may be a bulky substituent such as –C(CH$_3$)$_3$ or –Si(CH$_3$)$_3$. Q is a bridge group such as >CR$_2$ in bridges –Q– (X) or –Q–Q– (VII-IX). Ansa compounds are bridged compounds consisting of at least two rings where one ring acts as "handle" (G: *ansa* = handle) for the other one(s).

XI XII XIII XIV

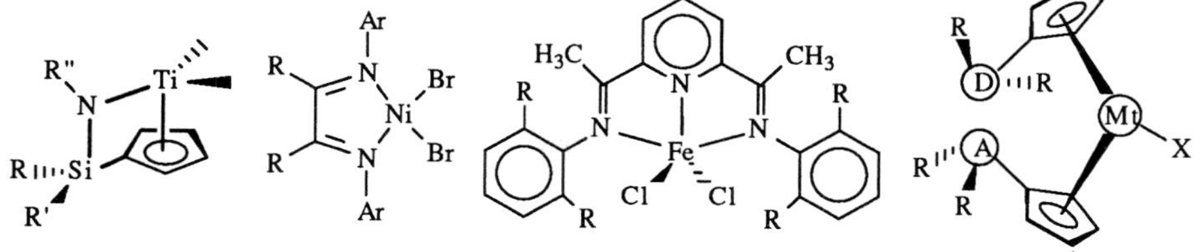

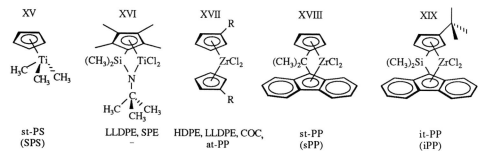

Fig. 9-15 Some newer metallocene compounds. R, R', R" = aliphatic groups (often bulky), Ar = aromatic groups, X = halogen, D = electron donor, A = electron acceptor, Mt = metal.

A great number of different metallocene compounds can obviously be designed as precursors for single-site catalysts (Fig. 9-15). Examples are mono-cyclopentadienyl compounds with restrained geometry (XI), late transition metal (Groups 8-10) compounds (XII, XIII), and so-called **donor-acceptor metallocenes (D/A-metallocenes)** (XIV). The latter molecules resemble a "mouth" where the two cyclopentadiene rings act as upper and lower jaws. A strong electron transfer from the donor to the acceptor group leads to a strong interaction between the two cyclopentadiene rings. The resulting strong bending of the angle Cp–Mt–Cp opens the "mouth" and lets large monomer molecules access the active metal center. Conversely, a weak D–A interaction allows only small monomer molecules to enter.

9.3.3 Polymerizability

Simple SSCs (see p. 272) based on Cp_2MtR_2 or Cp_2MtCl_2 plus MAO are excellent catalysts for the polymerization of ethene (Table 9-5) but polymerize propene only to "atactic" polymers (Fig. 9-16). Subsequently developed SSCs with open (XVII) or sterically less hindered structures allowed the copolymerization of ethene with bulkier comonomers such as 1-butene, 1-hexene, or 1-octene to linear low-density poly(ethylene)s (LLDPEs) by XVI and XVII, with styrene to ethene-styrene copolymers (SPE) by XVI, and with cycloalkenes to cycloolefin-copolymers (COC) by XVII. Stereospecific polymerizations require more rigid SSCs with two bridges between the cyclopentadienyls such as XVIII for the polymerization of propene to isotactic polymers and XIX to syndiotactic ones. "Atactic" poly(propylene)s are produced by the open structure XVII.

XV XVI XVII XVIII XIX

st-PS (SPS) | LLDPE, SPE – | HDPE, LLDPE, COC, at-PP | st-PP (sPP) | it-PP (iPP)

Fig. 9-16 Some metallocene-MAO catalysts for homo and copolymerizations (see text).

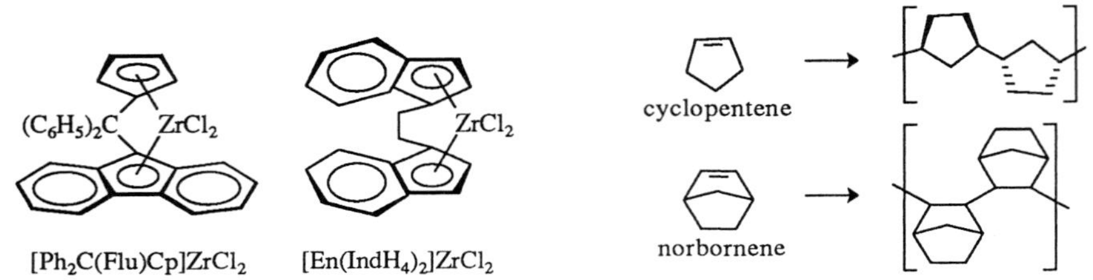

$(C_6H_5)_2C$ $ZrCl_2$ $ZrCl_2$ cyclopentene

$[Ph_2C(Flu)Cp]ZrCl_2$ $[En(IndH_4)_2]ZrCl_2$ norbornene

Fig. 9-17 Polymerization of cyclopentene and norbornene (right) by the metallocenes shown left. Cp = Cyclopentadiene, En = ethylene ($-CH_2-CH_2-$), Flu = fluorene, Ind = indene, Ph = phenyl.

Similar SSCs such as $Ph_2C(Flu)Cp]ZrCl_2$-MAO or $[En(IndH_4)_2]ZrCl_2$-MAO allow the polymerization of cycloolefins with preservation of ring structures (Fig. 9-17). Cyclopentene rings are connected via the 1,3-positions, all other cycloolefines such as norbornene via 1,2 positions. The double bond of the monomer molecule is opened in the cis position (Fig. 9-6). C_2-symmetric metallocenes lead to erythrodiisotactic structures, C_s-symmetric ones to erythrodisyndiotactic ones (Fig. 4-13). The high melting temperatures of poly(cyclopentene) (395°C) and poly(norbornene) (> 500°C), however, prevent processing. Industrial products are thus copolymers of cycloalkenes and ethene.

1,5-Hexadiene is inserted in the 1,2-position and undergoes a cyclopolymerization. Depending on the catalyst, eit, tit, est, or tst structures of poly(methylene-1,3-cyclopentylene) are obtained:

(9-30)

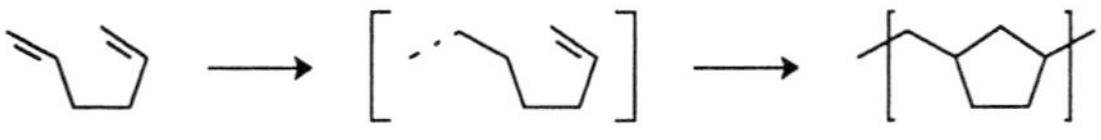

Coating of filler particles with MAO and then metallocenes delivers carrier-bound catalysts that can replace the previously used heterogeneous Ziegler catalysts in large industrial plants (drop-in technology).

9.3.4   Stereocontrol and Regiocontrol

In *regiospecific* polymerizations, 1-olefins are inserted into the catalyst-polymer bond exclusively in tail-to-head position by 1,2-insertions (Eq.(9-9)) and exclusively in head-to-tail position by 2,1-insertions (Eq.(9-10)). Both head-to-head and tail-to-tail structures are observed in less regiospecific polymerizations.

The *stereoregularity* of poly(1-olefin)s is largely controlled by the symmetry of the metallocene component L_2MtX_2 (Fig. 9-18) of the catalyst and its cation $[L_2MtMP]^{\oplus}$ where L, X = ligands, M = inserted monomer molecule, and P = polymer chain.

Types I and II have mirror symmetries around the horizontal axis X–Mt–X; their ligands L are either in the same environment (Type I) or in different environments (Type II). No mirror symmetries exist around the horizontal axis X–Mt–X in Types III-V. In Type III, the two ligands are in the same environment (homotopy). The ligands exist in Type IV as vertical mirror-mirror image (enantiotopy) but are sterically different in Type V (diastereotopy).

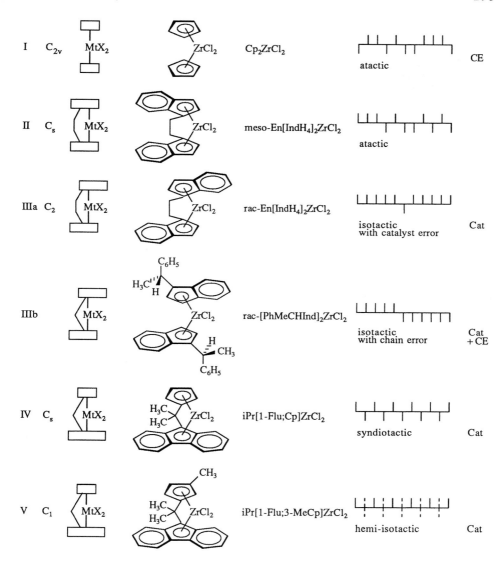

Fig. 9-18 Stereoregularities obtained by different types of transition metal components L_2MtX_2 of single-site catalysts [11]. Columns from left to right: Types I-V; symmetry groups C_{2v}, C_s, C_2, or C_1 (Table 4-3, Fig. 4-2); schematic metallocene structures; examples of chemical metallocene structures; structural formulas of these metallocenes; schematic representations of stereostructures of poly-(propylene)s; stereocontrol by enantiomorphic catalyst (Cat) or chain end (CE).
Cp = cyclopentadiene, En = ethylene (–CH₂–CH₂–), Flu = fluorene, Ind = indene, iPr = isopropyl, Me = methyl, Mt = transition metal, X = ligand (often Cl).

The symmetry properties control the stereospecificity of the insertion step. For alternating monomer insertions (Fig. 9-13), Types I (C_{2v} symmetry) and II (symmetric C_s (mirror plane)) deliver at room temperature only atactic poly(propylene)s. Type III (C_2 symmetry (one symmetry axis)) lead to isotactic poly(propylene)s and Type IV (unsymmetric C_2) to syndiotactic ones.

The stereoregularity produced by Type V catalysts is not predictable. For example, hemi-isotactic poly(propylene)s were obtained with iPr[1-Flu;3-MeCp]ZrCl$_2$ as catalyst (Fig. 9-18, V). CH$_3$CH[Me$_4$Cp;Ind]ZrCl$_2$, on the other hand, delivered poly(propylene)s with isotactic and atactic blocks.

The resulting degree of stereoregularity seems to depend on the relative time required for the rearragements of ligands L and X of the metallocene versus the turn-over time of the monomer, i.e., on the relative conformational rigidity of the metallocene. In principle, metallocenes become less rigid with increasing temperature and the stereoregularity decreases (Fig. 9-19).

Fig. 9-19 Effect of polymerization temperature on the stereospecificity of the polymerization of propene by Cp$_2$Ti(C$_6$H$_5$)$_2$ + MAO.

Stereoblocks are created by different mistakes during the insertion. As mentioned before, chain-end control leads mainly to mistakes ...iiii-s-iiii... (Fig. 9-18, IIIb) whereas enantiomorphic catalysts should predominantly deliver ...iiii-s-s-iiii... (Fig. 9-18, IIIa).

For the same metallocene, the type and extent of stereocontrol is also given by the type of transition metal and the polymerizing monomer (Fig. 9-20). Systematic investigations are rare, however.

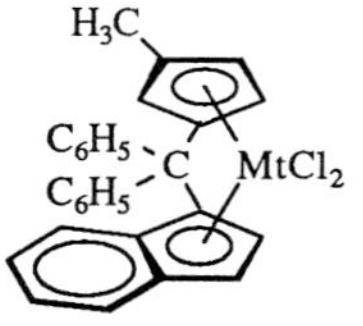

Mt	Poly(propylene)	Poly(styrene)
Ti	$x_{ii} = 0.57$	$x_{ssss} = 1.0$
Zr	$x_{ii} = 0.19$	atactic
Hf	$x_{ii} = 0.52$	atactic

Fig. 9-20 Effect of the type of transition metal in metallocenes (C$_6$H$_5$)$_2$C[Ind;3-MeCp]MtCl$_2$ on the triad and pentad tacticities of poly(propylene) and poly(styrene).

9.3.5 Polymerization Kinetics

Generalizations about the kinetics of polymerizations by single-site catalysts are difficult because these polymerizations can be affected by many factors. Some of the possible effects were touched upon in previous sections. There is certainly a difference in the action of SSCs from early transition metals (Lanthanides of Group 3: Sm, Yb; Group 4: Ti, Zr, Hf; Group 5: V, Nb) and electron-rich late transition metals (Group 6 ff.): the Mt-C bonds of the former are partly ionic (and thus intolerant of air, moisture, and heteroatoms) whereas the latter are generally covalent. Late transition metal complexes are also prone to β-hydrogen elimination and to reductive elimination. Other factors to be considered are the structure and instability of methyl aluminoxanes, enantiomorphic site control versus chain-end control, 2-way mechanisms versus 1-way mechanisms, etc.

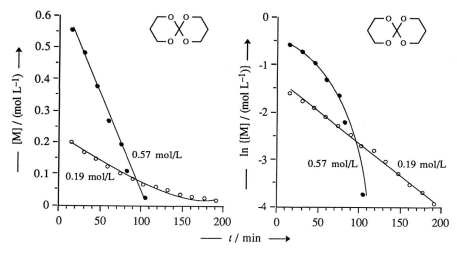

Fig. 9-21 Time dependence of monomer concentration (left) and the natural logarithm of monomer concentration (right) as a function of time in the polymerization of 1,5,7,11-tetraoxaspiro[5.5]undecane in deutero-benzene by [CpZrCH$_3$][CH$_3$B(C$_6$F$_5$)$_3$] [12]. [M]$_0$: 0.57 mol/L (●; 55°C) and 0.19 mol/L (O; 65°C). Left: zeroth-order plot of [M] = f(t); right: first-order plot of ln [M] = f(t).

For example, the reaction kinetics of the polymerization of a tetraoxospiro compound changed from zero order with respect to monomer at high monomer concentrations to first order at lower ones (Fig. 9-21). This effect was explained by a solvation of the catalyst complex by the monomer.

The existence of non-linear Arrhenius plots (Fig. 9-22) may be an indication of the presence of two active species with different reactivities that are in slow equilibrium with each other. Such polymerizations lead to bimodal molar mass distributions.

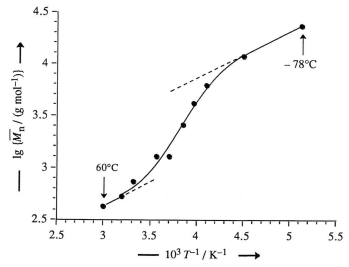

Fig. 9-22 Dependence of the logarithm of the number-average molar mass on the inverse of kelvin temperature. Data for the polymerization of 1-hexene by Cp$_2$ZrCl$_2$–MAO in toluene [13].

9.4 Metatheses

9.4.1 Introduction

Metatheses are exchange and disproportionation reactions of carbon–carbon multiple bonds of olefins, cycloolefins, dienes, and alkynes (G: *meta* = involving change; *tithenai* = to place). They are mediated by late transition-metal compounds based on tungsten, molybdenum, or ruthenium (Section 9.4.2). These reactions proceed via carbenes (Section 9.4.3 ff.).

Metatheses may involve intramolecular rearrangements, or addition reactions between two like molecules (homo-metatheses) or two unlike molecules (cross-metatheses) to either low-molecular weight compounds or polymers (Fig. 9-23). Metatheses of simple acyclic olefins deliver only low-molecular weight compounds (Section 9.4.3). Metatheses of cycloolefins at sufficiently high monomer concentrations result in linear polymers plus a small proportion of cyclic oligomers (Section 9.4.4). The metathesis polymerization of acyclic dienes and certain olefin derivatives delivers polymers (Section 9.4.5).

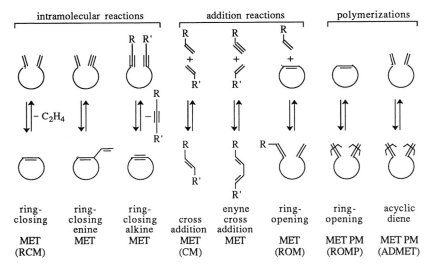

Fig. 9-23 Types of metatheses (MET) and metathesis polymerizations (MET PM): ring-closing metathesis (RCM), cross-addition metathesis (CM), ring-opening metathesis (ROM), ring-opening metathesis polymerization (ROMP), and acyclic metathesis polymerization (ADMET).

9.4.2 Catalysts

Early (first generation) metathesis catalysts comprised three types:

a. Preformed metal carbene complexes such as $(C_6H_5)_2C=W(CO)_5$ which act as metathesis catalysts only after losing a ligand thermally, photochemically, or by reaction with another chemical compound.

b. Metal carbenes formed by reaction of the transition metal with the monomer. This group comprises systems such as $Re_2O_7–Al_2O_3$ and $MoO_3–Al_2O_3$.

c. The most effective metal carbenes result from the reaction of a transition metal compound with the alkyl or allyl group of a second catalyst component which leads to a carbon ligand at the transition metal and then to the carbene, for example:

$$\text{(9-31)} \qquad WCl_6 \xrightarrow[-\,(CH_3)_3SnCl]{+\,(CH_3)_4Sn} CH_3WCl_5 \xrightarrow[-\,HCl]{} CH_2{=}WCl_4$$

Examples are WCl_6–$EtAlCl_2$–EtOH (1:4:1), WCl_6–Me_4Sn (1:5), $TiCl_4$–$LiAlR_4$ (1:1), and $[(C_6H_5)_3P]_2(NO)_2MoCl_2$–$EtAlCl_2$ (1:5).

Metathesis catalysts of the first generation were highly active but could be used only for metatheses of fairly stable compounds such as olefins, cycloolefins, and dienes. Newer catalysts such as the Schrock wolfram and ruthenium catalysts and the Grubbs ruthenium carbene complexes (Fig. 9-24) tolerate more sensitive functional groups. The Grubbs catalysts can also be used in the presence of additives, stabilizers, fillers, etc. For these reasons and because of their commercial availability, they have found wide applications in low-molecular weight organic chemistry. In macromolecular chemistry, they serve for ring-opening metathesis polymerizations (Section 9.4.4) and acyclic diene metathesis polymerizations (Section 9.4.5). The newest cyclic Grubbs carbene catalysts (Fig. 9-24) also allow the synthesis of very large rings, e.g., cyclooctene to cyclic poly(octenamer), $\nparallel CH(CH_2)_6CH \nparallel_n$, with molecular weights in the millions.

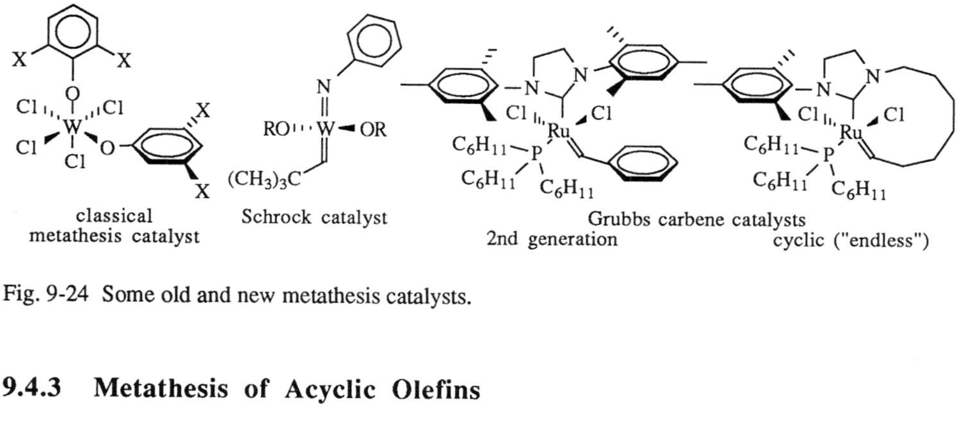

classical metathesis catalyst — Schrock catalyst — Grubbs carbene catalysts: 2nd generation, cyclic ("endless")

Fig. 9-24 Some old and new metathesis catalysts.

9.4.3 Metathesis of Acyclic Olefins

The metathesis of simple acyclic olefins proceeds formally by the exchange of parts of the molecules. For example, a mixture of 2-butene, 2-pentene, and 3-hexene is formed in the molar ratio of 1:2:1 within minutes after 2-pentene is treated by the catalyst WCl_6-$C_2H_5AlCl_2$-C_2H_5OH:

$$\text{(9-32)} \qquad 2\begin{array}{c} CH_3{-}CH \\ \| \\ C_2H_5{-}CH \end{array} + 2\begin{array}{c} CH{-}CH_3 \\ \| \\ CH{-}C_2H_5 \end{array} \rightleftharpoons \begin{array}{c} CH_3{-}CH{=}CH{-}CH_3 \\ + \\ C_2H_5{-}CH{=}CH{-}C_2H_5 \end{array} + 2\begin{array}{c} CH_3{-}CH \\ \| \\ CH{-}C_2H_5 \end{array}$$

The composition of the resulting mixture corresponds to the statistical expectation for an equilibrium reaction in which molecular fragments are "exchanged" around the double bonds. The enthalpy of the reaction is zero since similar bonds are exchanged. The reaction is propelled only by the increase in entropy.

Such metatheses are enthalpically controlled, however, if acyclic olefins are sterically hindered. An example is the metathesis of styrene to ethene, styrene, and stilbene in the molar ratio 2.5:94.0:3.5:

(9-33)

$$C_6H_5{-}\underset{\underset{CH_2}{\|}}{CH} \; + \; \underset{\underset{CH_2}{\|}}{CH}{-}C_6H_5 \quad \rightleftharpoons \quad C_6H_5{-}CH{=}CH{-}C_6H_5 \; + \; C_6H_5{-}\underset{\underset{CH_2}{\|}}{CH}$$
$$+$$
$$CH_2{=}CH_2$$

Metatheses do not proceed spontaneously but only in the presence of metallocenes of the type R'HC=[Mt] where [Mt] is the complexed atom of a transition metal (see below). Monomers and catalysts form unstable intermediates (carbenes) which then dissociate:

(9-34)

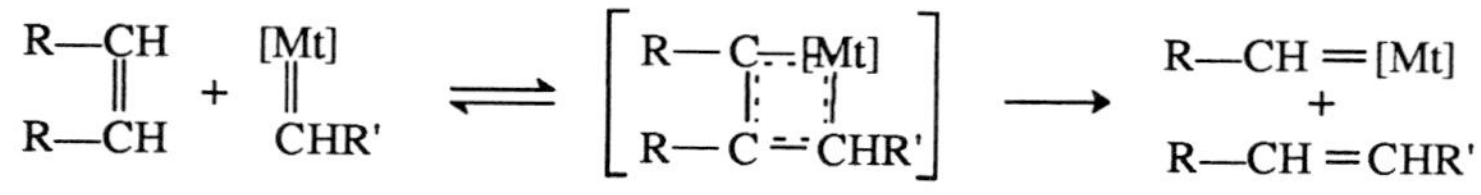

$$\underset{\underset{R-CH}{\|}}{R{-}CH} \; + \; \underset{\underset{CHR'}{\|}}{[Mt]} \quad \rightleftharpoons \quad \left[\begin{matrix} R{-}C\cdots[Mt] \\ \vdots \quad \vdots \\ R{-}C\cdots CHR' \end{matrix}\right] \quad \longrightarrow \quad \begin{matrix} R{-}CH{=}[Mt] \\ + \\ R{-}CH{=}CHR' \end{matrix}$$

9.4.4 Ring-Opening Polymerization of Cycloolefins

Products

Cycloolefins polymerize cationically and by metallocene catalysts with preservation of rings (Eq.(9-35)). Metathesis catalysts, however, lead to a **ring-*opening* *metathesis* polymerization (ROMP)** (Fig. 9-23). An example is the polymerization of norbornene (= bicyclo[2.2.1]-2-heptene) where one ring is opened and the other one preserved:

(9-35)   Monomer $\longrightarrow$ monomeric units by

| norbornene | cationic polymerization | metallocene polymerization | ring-opening metathesis polymerization |

The metathesis polymerization of simple cycloolefins delivers both linear polymers and cyclic oligomers. The *inital* product distribution is *kinetically* controlled. If the initial monomer concentration is *lower* than the critical one (Fig. 7-7), first cyclic dimers, cyclic trimers, etc., are formed and finally the whole series of oligo(cycloolefin)s (Fig. 9-25). Above a critical monomer concentration, polymers are formed immediately. The proportion of cyclic oligomers remains small and their equilibrium concentrations are independent of the initial monomer concentration. At $[M]_0 > [M]_{0,crit}$, the polymer yield increases therefore with increasing monomer concentration.

In the final stage, a defined distribution of monomer, linear polymers, and cyclic oligomers is obtained. This product distribution is also observed if the same metathesis catalyst is added to linear polymers with narrow molecular-weight distributions. In the *final* stage of the polymerization, thermodynamic equilibria exist between monomer and polymer molecules on one hand and linear polymers and cyclic oligomers on the other. Similarly to living chain polymerizations, these equilibria are often *incomplete* since exchange equilibria are either absent or slow to establish (Section 7.2.4).

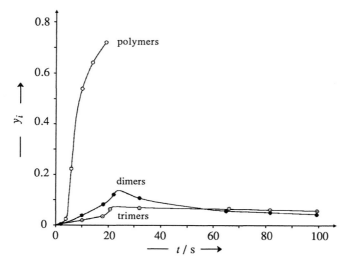

Fig. 9-25 Time dependence of the relative yield y_i of dimers ($i = 2$), trimers ($i = 3$), and polymers ($i = N \geq 4$) during the ring-opening polymerization of 0.025 mol/L cyclopentene in chlorobenzene by the catalyst system WCl$_6$–C$_2$H$_5$AlCl$_2$–C$_2$H$_5$OH at 0°C [14]. Molar concentrations of WCl$_6$: $2.5 \cdot 10^{-4}$ mol/L (polymers) and $2.5 \cdot 10^{-3}$ mol/L (oligomers).

The ROMP of simple cycloolefins preserves the carbon-carbon double bonds. Whether cis or trans polymers are formed depends on the types and molar ratio of catalyst and cocatalyst as well as the temperature (Table 9-6). This catalyst dependence shows that the polymer configuration is controlled by kinetics and not by thermodynamics. It seems that the (mostly) cis-cycloolefins deliver cis polymers that are subsequently isomerized to trans by certain catalysts. Indeed, cis-olefins are more easily metathesized than trans-olefins.

Table 9-6 ROMP catalysts for simple cycloolefins.

Monomer	cis-Polymer by		trans-Polymer by	
Cyclobutene	TiCl$_4$–(C$_2$H$_5$)$_3$Al–heptane	(– 50°C)	RuCl$_3$–C$_2$H$_5$OH	(+ 20°C)
Cyclopentene	WCl$_6$–(C$_2$H$_5$)$_4$Sn–(C$_2$H$_5$)$_2$O	(– 30°C)	WCl$_6$–(C$_2$H$_5$)$_4$Sn–(C$_2$H$_5$)$_2$O	(0°C)
Cyclooctene	(C$_6$H$_5$)$_2$C=W(CO)$_5$		WCl$_6$–(C$_2$H$_5$AlCl$_2$–C$_2$H$_5$OH	

Polymerizability

Ring-opening metathesis polymerizations may be used for many monomers that cannot be polymerized by other means. *Cyclobutenes* are easily polymerized because of their large ring strains (Fig. 9-26). For example, cyclobutene (I) is polymerized by many metathesis catalysts to *cis*-1,4-poly(butadiene) BR, and 1-methylcyclobutene (II) to *cis*-1,4-poly(isoprene) IR (both rubbers). The ROMP of (III) also opens the ring but the polymer P$_{III}$ converts to poly(acetylene), $\{CH=CH\}_n$ by losing 1,2-di(trifluoromethyl)benzene, C$_6$H$_4$(CF$_3$)$_2$. Monomer IV delivers a polymer P$_{IV}$ that crosslinks.

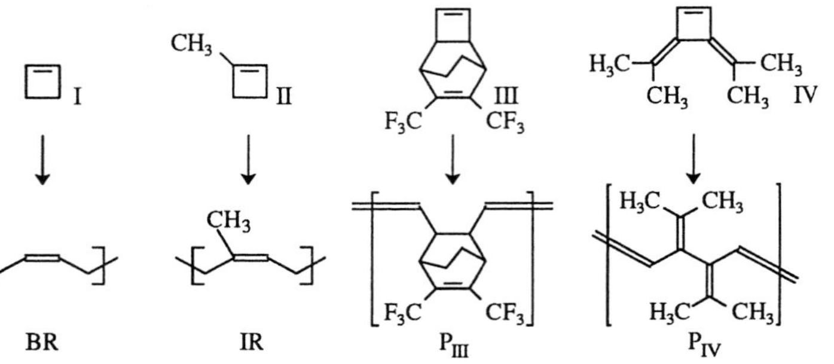

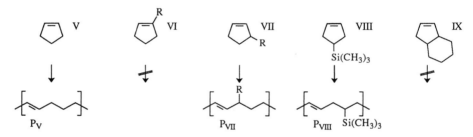

Fig. 9-26 ROMP of some cyclobutenes.

Cyclopentenes also have strained rings but their polymerizability depends strongly on the type, number, and position of their substituents. Cyclopentene is a byproduct of petroleum refining and thus an industrially important monomer (Volume II). Its ROMP leads to the so-called *trans*-poly(pentenamer) P_V, an all-purpose rubber. The 2-methyl and 2-ethoxy cyclopentenes VI do not polymerize but the 3- and 4-substituted compounds VII and VIII do (Fig. 9-27).

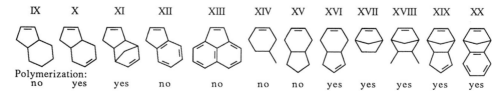

Fig. 9-27 ROMP of some cyclopentenes.

No polymerization is observed for these cyclopentene derivatives when the C_3-C_4 bond is part of a cyclohexane ring (hexahydroindene IX), a benzene ring (indene XII), or, with C_3–C_4–C_5, part of a naphthalene system (acenaphthylene XIII) (Fig. 9-28). Polymerizations are possible, however, if the joined 6-membered ring is unsaturated (X, XI) but these compounds are also cyclohexene derivatives (X = XVI; XI = XIX).

Neither cyclohexene nor 4-substituted cyclohexenes (XIV) nor the cyclohexene with a C_4-C_5 annellated (L: *anellus* = small ring) cyclopentane ring can be polymerized. Polymerizable are norbornene (XVII) and its 5,6-disubstituted (XVIII) and 5,6-annellated compounds (XIX, XX).

IX	X	XI	XII	XIII	XIV	XV	XVI	XVII	XVIII	XIX	XX

Polymerization:

no	yes	yes	no	no	no	no	yes	yes	yes	yes	yes

Fig. 9-28 ROMP of some ring-substituted cyclopentenes and cyclohexenes.

Norbornene (XVII) is industrially polymerized to a thermoplastic poly(1,3-cyclopentylene vinylene) (Eq.(9-35)) that becomes a crosslinkable rubber on plasticization by mineral oil. Dicyclopentadiene (XIX) is processed by reaction injection molding (Volume IV) to crosslinked molding compounds XXI (Eq.(9-36)).

(9-36)

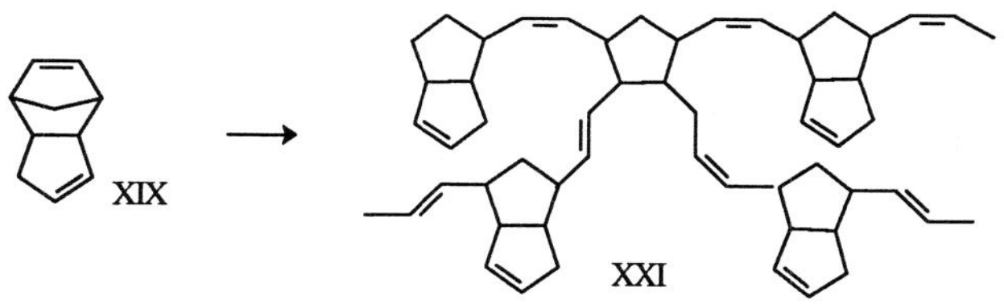

Neither cyclohexene nor cycloheptene polymerize. Cyclooctene is industrially polymerized to poly(octenamer)s, $+CH=CH-(CH_2)_6+_n$; the cis and trans polymers are both elastomeric. Cyclooctatetraene leads to poly(acetylene), $+CH=CH+_n$.

9.4.5 Acyclic Diene Metathesis

Like acyclic olefins (Section 9.4.3), conjugated dienes do not undergo metathesis polymerizations. 1,4-Pentadiene, the simplest non-conjugated diene, delivers mainly octa-1,4,7-triene, $CH_2=CH+CH_2-CH=CH+_nCH_2-CH=CH_2$ ($n = 1$), and, in smaller proportions, higher oligomers ($n \geq 2$).

However, nonconjugated dienes with double bonds that are separated by at least *two* methylene groups can be polymerized by typical metathesis catalysts in so-called *acyclic diene metathesis polymerizations* (ADMET). 1,5-Hexadiene, 1,7-octadiene, and 1,9-decadiene polymerize to 1,4-poly(butadiene), 1,6-poly(hexadiene), and 1,8-poly(octadiene), resp., and 5-methyldeca-1,5,9-triene to poly(butadiene-*alt*-isoprene) (Eq.(9-37)), all with release of ethene:

(9-37)

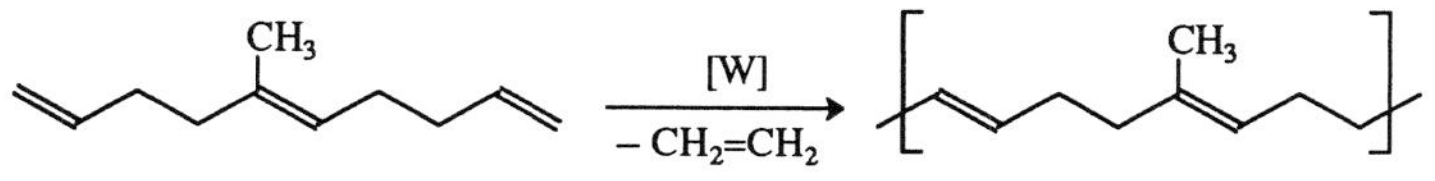

Based on the release of a low-molecular weight compound, the absence of high-molecular weight polymers at lower monomer conversions, and a ratio of $\overline{M}_w / \overline{M}_n \approx 2$ at higher ones, it was concluded that these reactions are polycondensations. However, these criteria apply also to polyeliminations (condensative chain polymerizations). Indeed, the reported $\overline{M}_n \approx 57\,000$ g/mol for a poly(octenylene) $[C_8H_{14}]_n$ would correspond to $\overline{X}_n \approx 518$ and thus to a monomer conversion of $p = 0.998$, an unlikely value for a simple equilibrium polycondensation (see Chapter 13). Equilibrium polycondensations and living polyeliminations can be distinguished by their $\overline{X}_n = f(p)$ functions but these were not reported.

Similar eliminations are observed if carbon-carbon double bonds are conjugated with carbonyl groups. Benzylidene acetophenone, $C_6H_5-CH=CH-CO-C_6H_5$, releases benzaldehyde, C_6H_5CHO, and polymerizes to poly(1-phenyl vinylene) = poly(phenyl acetylene), $+C(C_6H_5)=CH+_n$. Mesityloxide delivers poly(1-methylvinylene):

(9-38)

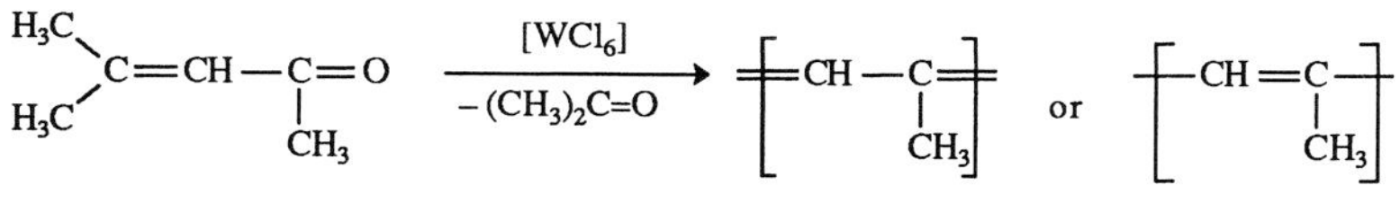

9.5 Group-Transfer Polymerizations

In group-transfer polymerizations (GTP), the active group of an initiator molecule is transferred to a monomer molecule (or vice versa) under the action of a nucleophilic or an electrophilic catalyst. The classic GTP of (meth)acrylates uses a silyl ketene acetal initiator (I) and, as nucleophile $Nu^{\ominus}$, a carbonyl-activating catalyst such as $[HF_2]^{\ominus}$, $CN^{\ominus}$, $N_3^{\ominus}$, or $[(CH_3)_3SiF_2]^{\ominus}$. The mechanism is still not very clear. The low rate constants correspond to a propagation by ion pairs. The small activation energies as well as the formation of predominantly syndiotactic diads rule out a concerted mechanism. Rather a two-step mechanism is favored in which the rate-determining step R of the addition of the activated chain end to the vinyl group of the monomer is followed by the transfer T of the silyl group to the newly formed enol (Eq.(9-39)):

(9-39)

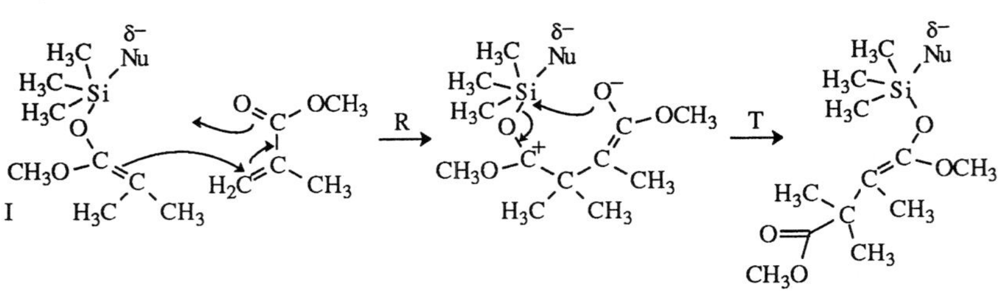

The polymerization rate $R_p = -d[M]/dt = const \cdot [C]_0 [I]_0^{\beta} [M]$ has been found to be first order with respect to both the initial catalyst concentration $[C]_0$ and the actual monomer concentration $[M]$. However, the exponent β of the initial initiator concentration $[I]_0$ varies between 0 and 1 which can be explained by an equilibrium reaction $C + I \rightarrow I^*$ between the nucleophilic catalyst C and the initiator I to an activated initiator I^* that is followed by the rate-determining step R (Eq.(9-39)). Monomer M, initiator I, and activated initiator I^* are very similar so that $I \approx M \equiv P_1$ and thus $I^* \approx P^*$. The equilibrium constant $K = [I^*]/[C][I]$ thus becomes $K = [P^*]/[C][P]$ where $[P]$ = concentration of the uncoordinated polymer.

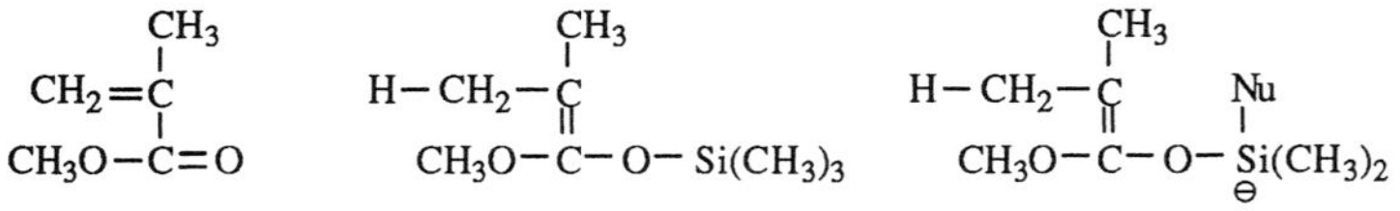

The polymerization rate of such a reaction (see Eq.(9-16)) is given by

$$(9-40) \qquad R_p = k_p \left(\frac{K[I]_0}{1 + K[I]_0} \right) [C]_0 [M]$$

A strong coordination favors activated initiators and polymers ($K[I]_0 \gg 1$). Eq.(9-40) becomes $R_p = k_p [C]_0 [M]$ and also $\beta = 0$ and $const = k_p$. A weak coordination leads to $K[I]_0 \ll 1$ and thus to $R_p = k_p K[I]_0 [C]_0 [M]$, i.e., $\beta = 1$ and $const = k_p K[I]_0$.

Propagation rate constants from kinetic data correspond to those from polymerizations via ion pairs. They are far lower than those found for polymerizations via free anions. Pre-exponential factors also correspond to those for polymerizations via ion pairs.

Group-transfer polymerizations of (meth)acrylates can also be initiated and catalyzed by transition metal compounds, for example, neutral zirconocene enolate II as initiator and the conjugated zirconocene III as catalyst (Cp = cyclopentadienyl; Me = methyl, M_n, M_{n+1} = polymers):

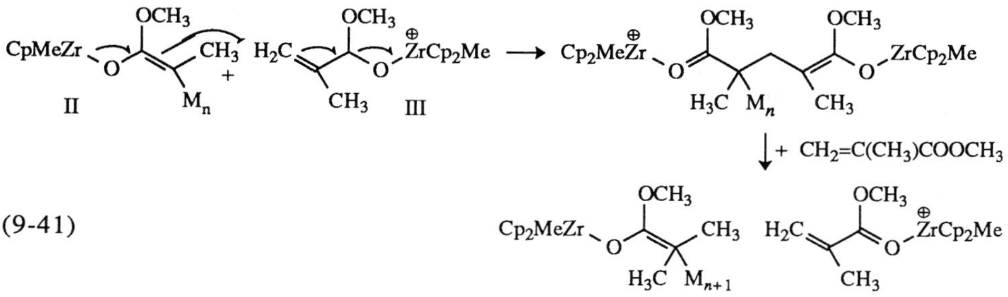

(9-41)

$$\downarrow + CH_2=C(CH_3)COOCH_3$$

Historical Notes to Chapter 9

Transition-Metal Catalysts

A.Clark, J.P.Hogan, W.Lancing, Ind.Eng.Chem. **48** (1956) 1152
   Development of chromium oxide catalysts for the polymerization of ethene at low pressures by Phillips Petroleum in 1953 and beyond.

K.Ziegler, E.Holzkamp, H.Breil, H.Martin, Angew.Chem. **67** (1955) 426, 541
   Addition of olefins to aluminium hydride in the presence of nickel and other trace catalysts does not lead to the desired higher aluminum alkyls (Aufbaureaktion) but to the low-pressure polymerization of ethene. Umbrella patent for many catalyst systems. For the historical development, see also K.Ziegler, in Les Prix Nobel en 1963, The Nobel Foundation 1964.

G.Natta, Acc.Naz.Lincei Mem. **4/2** (1955) 61; G.Natta, P.Corradini, Acc.Naz.Lincei Mem. **4/2** (1955) 73; G.Natta, J.Polym.Sci. **16** (1955) 143; G.Natta, P.Pino, P.Corradini, F.Danusso, E.Mantica, G.Mazzanti, G.Moraglio, J.Am.Chem.Soc. **77** (1955) 1708
   Ziegler catalysts polymerize propene and other monomers to stereoregular polymers.

Mechanism of Coordination Polymerization

B.Eistert, Tautomerie und Mesomerie, Enke, Stuttgart 1938, p. 108
   Proposes insertion of dienes into metal-carbon bonds.

F.Patat, H.Sinn, Angew. Chem. **70** (1958) 496
   Bimetallic insertion mechanism with insertion into the Ti-C bond.
G.Natta, G.Mazzanti, Tetrahedron **8** (1960) 86
   Bimetallic insertion mechanism with insertion into the Al-C bond.

P.Cossee, Tetrahedron Lett. **17** (1960) 12; -, Proc. 6th Int.Congr.Coordination Chem., Macmillan, New York 1961, p. 241;
E.J.Arlman, Proc. 3rd Int.Congr.Catal. **2** (1964) 957; E.J.Arlman, P.Cossee, J.Catal. **3** (1964) 99; E.J.Arlman, J.Catal. **5** (1966) 178;
P.Cossee, in A.D.Ketley, Ed., The Stereochemistry of Macromolecules, Dekker, New York 1967, vol. 1
   Monometallic mechanism.

R.A.Shelden, T.Fueno, T.Tsunetsugu, J.Furukawa, J.Polym.Sci. (Letters) **3** (1965) 23
   Model of enantiomorphic catalyst.

Metallocene Catalysts

G.Natta, P.Pino, G.Mazzanti, U.Giannini, J.Am.Chem.Soc. **79** (1957) 2895 (Cp$_2$TiCl$_2$/Me$_2$AlCl)
D.S.Breslow, N.R.Newburg, J.Am.Chem.Soc. **79** (1957) 5072 (Cp$_2$TiCl$_2$/Et$_2$AlCl)
  First homogeneous catalyst systems for the polymerization of ethene.

K.H.Reichert, K.R.Meyer, Makromol.Chem. **169** (1973) 163 (Cp$_2$TiEtCl/EtAlCl$_2$)
W.P.Long, D.S.Breslow, Liebigs Ann.Chem. (1975) 463 (Cp$_2$TiCl$_2$/Me$_2$AlCl)
A.Andresen, H.G.Cordes, J.Herwig, W.Kaminsky, A.Merck, R.Mottweiler, J.Pein, H.Sinn, H.J.Voll-
  mer, Angew.Chem. **88** (1976) 689 (Cp$_2$TiCl$_2$/Me$_2$AlCl)
J.Cihlář, J.Mejzlík, O.Hamřík, Makromol.Chem. **179** (1978) 2553 (Cp$_2$TiEtCl/EtAlCl$_2$)
  Enhancement of catalyst acticity of metallocene catalysts by addition of water:

W.Kaminsky, H.Sinn, Liebigs Ann.Chem. (1975) 424 (Cp$_2$TiCl$_2$/Al(C$_2$H$_5$)$_3$)
  First highly active homogeneous catalysts for olefin polymerization.

H.Sinn, W.Kaminsky, Adv.Organomet.Chem. **18** (1980) 99; H.Sinn, W.Kaminsky, H.J.Vollmer,
R.Woldt, Angew.Chem. **92** (1980) 396; -, Angew.Chem.Internat.Ed.Engl. **19** (1980) 390
  Efficacy of methyl alumoxanes (MAO) (from Me$_3$Al + H$_2$O = 1:2) in combination with transition
metal compounds (zirconocenes) (single-site metallocene catalysts). Poly(ethylene) with narrow molecular weight distribution. Poly(propylene) only atactic because zirconocenes were flexible.

F.R.W.P.Wild, L.Zsolanai, G.Huttner, H.H.Brintziger, J.Organomet.Chem. **232** (1982) 233
  Synthesis of sterically rigid chiral metallocenes with C$_2$ symmetry (*ansa*-bis(indenyl) titanocene).

J.A.Ewen, J.Am.Chem.Soc. **106** (1984) 6355
W.Kaminsky, K.Külper, H.H.Brintziger, F.R.W.P.Wild, Angew.Chem.Int.Ed.Engl. **24** (1985) 507
  Chiral metallocene catalysts + MAO lead to isotactic poly(propylene)s.

J.A.Ewen, R.L.Jones, A.Razavi, J.P.Ferrara, J.Am.Chem.Soc. **110** (1988) 6255
  Chiral metallocene catalysts produce syndiotactic poly(propylene)s.

Metatheses

K.Ziegler, H.G.Gellert, W.Pfohl, Angew.Chem. **67** (1955) 424; K.Ziegler, E.Holzkamp, H.Breil,
H.Martin, Angew.Chem. **67** (1955) 541
  Discovery of disproportionation of simple olefins by Ni catalysts.

H.C.Eleuterio, U.S. Patent 3,074,913 (1957)
  First ring-opening metathesis polymerization.

E.F.Peters, B.L.Evering, Standard Oil of Indiana, U.S. Patent 2,963,447 (1960)
  Disproportionation of propene to ethene and butenes by MoO$_3$/alumina + (*i*-C$_4$H$_9$)$_3$Al.

R.L.Banks, G.C.Bailey, Phillips Petroleum, Ind.Eng.Chem.Prod.Res.Dev. **3** (1964) 170
  Disproportionation of propene to ethene and butenes by Mo(CO)$_6$/alumina.

G.Natta, G.Dall'Asta, G.Mazzanti, Angew.Chem.Int.Ed.Engl. **3** (1964) 723
  Discovery of ring-opening polymerization of cyclopentene by WCl$_6$ + (C$_2$H$_5$)$_3$Al.

E.O.Fischer, Angew.Chem.Int.Ed.Engl. **3** (1964) 580
  New type of metal-carbon bond: carbene (CO)W$_5$=C(CH$_3$)(OCH$_3$).

N.Calderon, E.A.Ofstead, W.A.Judy, J.Polym.Sci. [A] **5** (1967) 2209
  Proposal of the term "olefin metathesis."

J.-L.Hérisson, Y.Chauvin, Makromol.Chem. **141** (1971) 161
  Propose that olefin metatheses proceed via metal carbenes.

M.Lindmark-Hamburg, K.B.Wagener, Macromolecules **20** (1987) 2949
  Discovery of acyclic diene metathesis (ADMET).

Literature to Chapter 9

9.1 SURVEY

P.Pino, R.Mülhaupt, Stereospecific Polymerization of Propylene: An Outlook 25 Years after Its Discovery, Angew.Chem.Int.Ed.Engl. **19** (1980) 857

F.M.McMillan, The Chain Straighteners: Fruitful Innovation. The Discovery of Linear and Stereoregular Polymers, MacMillan Press, London 1981

R.B.Seymour, T.Cheng, Eds., History of Polyolefins, Reidel Publ., Hingham (MA) 1985

H.Martin, Polymere und Patente. Karl Ziegler, das Team 1953-1998, Wiley-VCH, Weinheim 2001 (history of the Ziegler invention by one of Ziegler's coworkers (in German))

9.2 POLYMERIZATIONS BY MULTI-SITE CATALYSTS

T.Keii, Kinetics of Ziegler-Natta Polymerization, Kodansha, Tokyo 1972

J.Boor, Jr., Ziegler-Natta Catalysts and Polymerizations, Academic Press, New York 1979

P.D.Gavens, M.Bottrill, J.W.Kellend, J.McMeeking, Ziegler-Natta Catalysis, in G.Wilkinson, Ed., Comprehensive Organometallic Chemistry, Pergamon, New York 1982

R.P.Quirk, Ed., Transition Metal Catalyzed Polymerizations. Alkenes and Dienes, Harwood Academic Publ., Chur (GR), Switzerland, 1983 (2 vols.)

Y.V.Kissin, Isospecific Polymerization of Olefins, Springer, Berlin 1986

W.Kaminsky, H.Sinn, Eds., Transition Metals and Organometallics as Catalysts for Olefin Polymerizations, Springer, Berlin 1988

G.Allen, J.C.Bevington, Ed., Comprehensive Polymer Science, vol. 4, G.C.Eastmond, A.Ledwith, S.Russo, P.Sigwalt, Eds., Chain Polymerization II, Pergamon Press, Oxford 1989

I.Tritto, U.Giannini, Ed., Synthetic, Structural and Industrial Aspects of Stereospecific Polymerization (Stepol '94), Macromol.Symp. **89** (1995)

K.Soga, M.Terano, Ed., Catalyst Design for Tailor-Made Polyolefins, Kodansha, Tokyo 1994

G.Fink, R.Mülhaupt, H.H.Brintziger, Eds., Ziegler Catalysts. Recent Scientific Innovations and Technological Improvements, Springer, Berlin 1995

W.Kaminsky, M.Arndt, Metallocenes for Polymer Catalysis, Adv.Polym.Sci. **127** (1997) 143

K.Soga, T.Shiono, Ziegler-Natta Catalysts for Olefin Polymerizations, Progr.Polym.Sci. **22** (1997) 1503 (includes metallocene catalysts)

E.P.Moore, Jr., The Rebirth of Polypropylene: Supported Catalysts, Hanser Gardner, Cincinnati 1998

V.Draguten, R.Streck, Catalytic Polymerization of Cycloolefins: Ionic, Ziegler-Natta, and Ring-opening Metathesis Polymerization, Elsevier Science, Amsterdam 2001

W.Kuran, Principles of Coordination Polymerization. Heterogeneous and Homogeneous Catalysis in Polymer Chemistry–Polymerization of Hydrocarbon, Heterocyclic, and Unsaturated Monomers, Wiley, Hoboken (NJ) 2002

B.Rieger, L.Saunders Baugh, S.Kacker, S.Striegler, Eds., Late Transition Metal Polymerization Catalysis. Wiley-VCH, Weinheim 2003

K.Osaka, D.Takeuchi, Coordination Polymerization of Dienes, Allenes, and Methylenecycloalkanes, Adv.Polym.Sci. **194** (2004) 171

M.Suginome, Y.Ito, Transition Metal-Mediated Polymerization of Isocyanides, Adv.Polym.Sci. **171** (2004) 77

9.3 POLYMERIZATIONS BY SINGLE-SITE CATALYSTS (see also literature to Section 9.2)

H.Sinn, W.Kaminsky, Eds., Alumoxanes, Macromol.Symp. **97** (1995)

W.Kaminsky, New Polymers by Metallocene Catalysis (Feature Article), Macromol.Chem.Phys. **197** (1996) 3907

W.Kaminsky, M.Arndt, Metallocenes for Polymer Catalysis, Adv.Polym.Sci. **127** (1997) 143

A.Togni, R.L.Halterman, Metallocenes: Synthesis, Reactivity, Applications, Wiley-VCH, Weinheim 1998 (2 vols.)

G.M.Benedict, B.L.Goodall, Metallocene-Catalyzed Polymers, Plastics Design Library, Norwich (NY) 1998

J.Scheirs, W.Kaminsky, Eds., Metallocene-Based Polyolefins, Wiley, New York 1999 (2 vols.)

9.4 METATHESES

V.Dragutan, A.T.Balaban, M.Dimonie, Olefin Metathesis and Ring-Opening Polymerization of Cyclo-Olefins, Editura Academiei, Bukarest 1985; Wiley, London, 2nd Ed. 1985

R.P.Quirk, Ed., Transition Metal Catalyzed Polymerizations: Ziegler-Natta and Metathesis Polymerizations, Cambridge University Press, New York 1988

D.S.Breslow, Metathesis Polymerization, Prog.Polym.Sci. **18**/6 (1993) 1141

A.J.Amass, Metathesis Polymerization: Chemistry, in G.Allen, J.C.Bevington, Eds., Comprehensive Polymer Science **4** (1989) 109

W.J.Feast, Metathesis Polymerization: Applications, in G.Allen, J.C.Bevington, Eds., Comprehensive Polymer Science **4** (1989) 135

K.J.Ivin, J.C.Mol, Eds.,Olefin Metathesis and Metathesis Polymerization, Academic Press, San Diego (CA) 1997

Y.Imamoglu, Metathesis Polymerization of Olefins and Polymerization of Alkynes, Kluwer, Dordrecht 1998

L.S.Boffa, An Overview of Transition Metal-mediated Polymerizations: Catalysts for the 21st Century, in C.D.Craver, C.E.Carraher, Jr., Applied Polymer Science–21st Century, Elsevier, Amsterdam 2000

R.H.Grubbs, Ed., Handbook of Metathesis, Wiley-VCH, Weinheim, 3 vols. 2003

M.R.Buchmeiser, Ed., Metathesis Polymerization, Adv.Polym.Sci. **176** (2005)ß

9.5 GROUP-TRANSFER POLYMERIZATIONS

R.A.Haggard, S.N.Lewis, Methacrylate Oligomers Via Alkoxide-Initiated Polymerizations, Prog.Org.Coatings **12** (1984) 1

O.W.Webster, Group Transfer Polymerization: Mechanism and Comparison with Other Methods of Controlled Polymerization of Acrylic Monomers, Adv.Polym.Sci. **167** (2004) 1

References to Chapter 9

[1] J.P.Hogan, in B.E.Leach, Ed., Applied Industrial Catalysis, Academic Press, New York 1983, Fig. 7

[2] P.Galli, Macromol.Symp. **89** (1995) 13, Table II

[3] H.-G.Elias, T.Ogawa, Makromol.Chem., Rapid Commun. **2** (1981) 247, Fig. 3; data of C.Wolfsgruber, G.Zannoni, E.Rigamonte, A.Zambelli, Makromol.Chem. **175** (1975) 2765

[4] G.Natta, I.Pasquon, Adv.Catal. **11** (1959) 1, Fig. 1

[5] P.Pino, B.Rotzinger, Makromol.Chem.Suppl. **7** (1984) 41, Fig. 4

[6] I.D.McKenzie, P.J.T.Tait, D.R.Burfield, Polymer **13** (1972) 307

[7] I.D.McKenzie, P.J.T.Tait, Polymer **13** (1972) 510, Figs. 5 and 6

[8] K.H.Reichert, K.R.Meyer, Makromol.Chem. **169** (1973) 163, Fig. 1

[9] K.H.Reichert, Chem.-Ing.Techn. **49** (1977) 626, Tables 1 and 2

[10] W.Kaminsky, M.Miri, H.Sinn, R.Woldt, Makromol.Chem., Rapid Commun. **4** (1983) 417, Tables 1-3

[11] M.Farina, Trends Res.Polym. **2** (1994) 80; Macromol.Symp. **89** (1995) 489

[12] Z.Ariffin, D.Wang, Z.Wu, Macromol.Rapid Commun. **19** (1998) 601, data of Figs. 1 and 2

[13] H.Frauenrath, H.Keul, H.Höcker, Macromol.Rapid Commun. **19** (1998) 391, Fig. 2

[14] H.Höcker, H.Reimann, L.Reif, K.Riebel, J.Mol.Catal. **8** (1980) 191, Figs. 1 and 3

10 Free-Radical Polymerization

10.1 Introduction

10.1.1 Overview

Free-radical chain polymerizations are initiated by small free radicals and propagated by macroradicals, i.e., chemical compounds with one unpaired electron. Historically, they were distinguished as the "free radicals" from the "bound radicals" that are today called substituents or ligands (L: *radix* = root, i.e., something fundamental).

In very rare cases, initiating radicals are generated by the monomer molecules themselves (Section 10.2.1). In general, though, they are formed by the thermal, electrochemical, or photochemical dissociation of added initiators (Section 10.2.2), usually pairwise according to $I_2 \rightarrow 2\ I^\bullet$.

In the initiation reaction, initiator radicals react with a monomer molecule M to form so-called **monomer radicals** $I-M^\bullet$. In the subsequent propagation reaction, monomer radicals add $i - 1$ additional monomer molecules. In the resulting **macroradicals** $I-M_i^\bullet$, active centers are separated by many monomeric units $-M-$ from initiator groups $I-$. In contrast to ionic polymerizations, the active center-monomer interaction is not affected by other initiator fragments. The propagation steps are only influenced by the macroradical and the monomer molecules and, to a minor extent, by the solvent. Macroradicals grow until they are terminated by reaction with another macroradical or initiator radical (Section 10.3.2) or transfer their radical to another compound (Section 10.4). Terminations restrict the concentration of active radicals to ca. 10^{-8} mol/L which is far lower than that of active ions in ionic polymerizations (ca. 10^{-3} mol/L).

In transfer reactions, the free electron of the macroradical $P_i^\bullet$ is exchanged for a group Q of another molecule RQ, resulting in a dead polymer molecule P_iQ and a new radical $R^\bullet$ that starts a new polymer molecule and thus preserves the kinetic chain. RQ may be a monomer, polymer, solvent, initiator, regulator, etc.

True living free-radical polymerizations may be achieved in principle by using biradicals if disproportionations of polymer molecules, intramolecular ring closure reactions, and reactions with initiator monoradicals or transfer agents can be avoided. So-called **controlled polymerizations** utilize exchange reactions between the growing polymer radicals and a low molecular weight radical that helps to maintain the quasi-living nature of these free-radical polymerizations (Section 10.5).

10.1.2 Polymerizable Monomers

Free-radical polymerizations can be subdivided into two groups. Couplings of biradicals to macrobiradicals and related reactions are polyadditions and not chain polymerizations. Examples are the formations of poly(*p*-xylylene) (Eq.(7-8)) and poly(*p*-phenylene sulfide) (Sections 13.1.5). In free-radical chain polymerizations, on the other hand, an initiator generates monoradicals (Section 10.2) that add monomer molecules until the growing chains are terminated (Section 10.3) or the radical is transferred to another molecule (Section 10.4).

The attack of the monomer molecule by a radical causes the former to open a bond, either a double bond in unsaturated monomers or a single bond in cyclic molecules. Polymerizations of monomers with carbon-carbon double bonds are economically very important. Free-radical ring opening polymerizations are rather academic curiosities.

Free-Radical Polymerization of Cyclic Monomers

The opening of *strain-free* saturated carbon rings requires activation energies of ca. 250 kJ/mol. Since the abstraction of a hydrogen atom needs only 40-80 kJ/mol, radicals will attack cycloaliphatics such as cyclohexane rather unspecifically. The reaction leads to a mixture of differently branched low-molecular weight hydrocarbons. Organic heterocycles are also not polymerized by free radicals. However, *cyclo*-octasulfur polymerizes, to unbranched *catena*-poly(sulfur) and other sulfur rings (Fig. 2-3).

Highly strained cycloaliphatics polymerize by ring-opening, for example, the cyclopropane ring of 1-bicyclo[1.1]butanitrile (= 1-cyanobicyclo[1.1]butane) (I). The more stable cyclobutane ring stays intact. Methylene heterocyclopentanes (II-IV) often polymerize via the carbon-carbon double bond (II, III). Ring opening occurs if more stable units are generated, for example, esters from III or amides from IV.

(10-1)

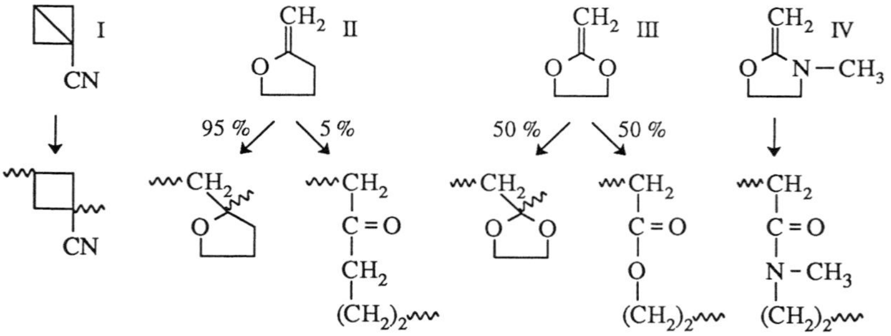

For the same reason, the vinylidene groups of unsaturated spiro-orthoesters remain intact during polymerization:

(10-2) CH_2=$\langle$... $\rangle$=CH_2 $\longrightarrow$ $\text{~~}CH_2-\overset{\text{CH}_2}{\underset{\parallel}{C}}-CH_2-O-\overset{O}{\underset{\parallel}{C}}-O-CH_2-\overset{\text{CH}_2}{\underset{\parallel}{C}}-CH_2-O\text{~~}$

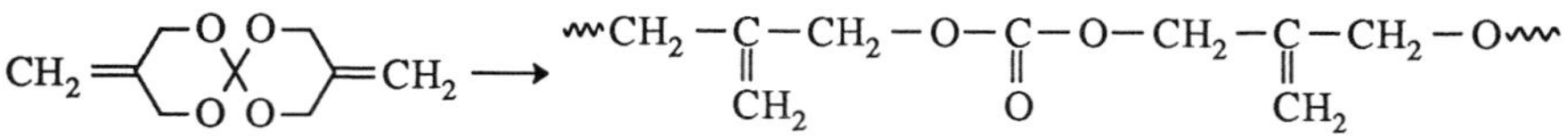

Vinyl cyclopropanes polymerize by ring-opening and a shift of the double bond (R, R' = H, Cl, CN, $COOC_2H_5$ or R = CN, R' = $COOC_2H_5$):

(10-3) CH_2=$CH-\overset{H}{\underset{}{C}}-CH_2$ $\longrightarrow$ $-CH_2-CH=CH-CH_2-CRR'-$

80 % 1,5-structures if R = R' = H

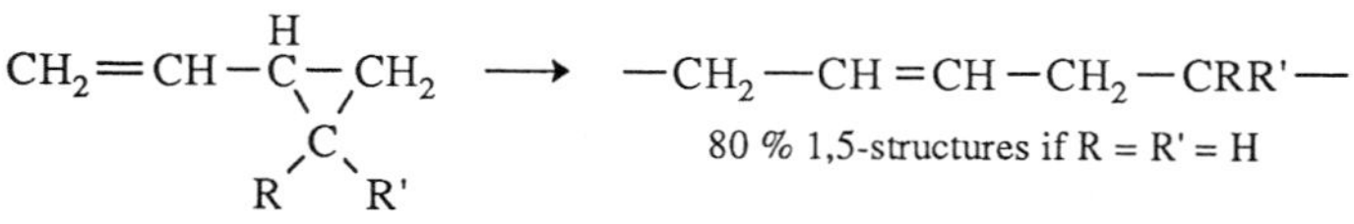

An "isomerizing" polymerization is observed for substituted vinyl oxiranes:

(10-4) CH_2=$CH-\overset{H}{\underset{}{C}}\overset{H}{\underset{}{C}}\langle_{C_6H_5}$ $\longrightarrow$ $\text{~~}CH_2-CH=CH-O-\underset{C_6H_5}{CH}\text{~~}$

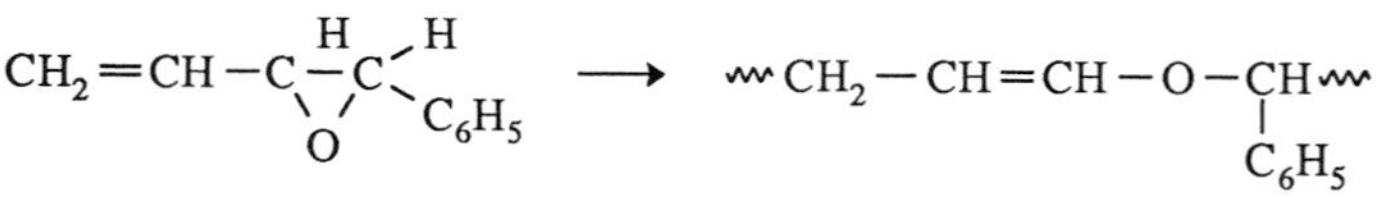

In very special cases, the shift of double bonds leads to an elimination of fragments. The free-radical polymerization of the 4-vinyl compound of 2,2-diphenyl-4-methylene-1,3-dioxolane (I) in chlorobenzene solution delivers crosslinked polymers. In *N,N*-dimethylformamide as solvent, however, an **elimination polymerization** produces a mixture of two polymers I and II:

(10-5)

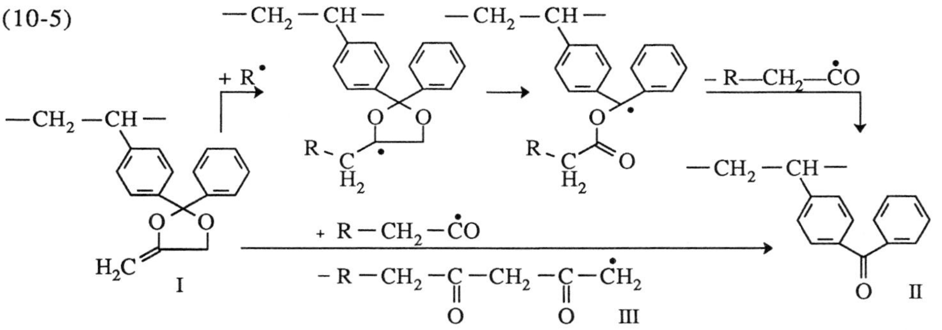

$$-CH_2-CH- \qquad -CH_2-CH-$$

$$+ R^{\bullet}$$

$$-CH_2-CH-$$

$$R-\underset{H_2}{\overset{\bullet}{C}} \qquad R-\underset{H_2}{\overset{}{C}}=O$$

$$+ R-CH_2-\overset{\bullet}{C}O$$

$$-R-CH_2-\overset{}{\underset{O}{C}}-CH_2-\overset{}{\underset{O}{C}}-\overset{\bullet}{CH_2} \quad III$$

$$- R-CH_2-\overset{\bullet}{C}O$$

$$-CH_2-CH- \quad II$$

$$H_2C \quad I$$

Free-Radical Polymerization of Double Bonds

In principle, all double bonds can be polymerized (Table 7-6). However, the average changes of enthalpies are only slightly negative for >C=S and even slightly positive for >C=N– and >C=O, so that free-radical polymerizations of such double bonds are only possible in very special cases. An example is trifluoroacetaldehyde that polymerizes free-radically via the >C=O bond opening because the electron-withdrawing CF_3 group stabilizes the resulting macroradicals $R-[CH(CF_3)-O]_{n-1}-CH(CF_3)-O^{\bullet}$.

Free-radical polymerizations are therefore usually restricted to carbon-carbon double bonds. Important are the polymerizations of **vinyl** compounds, $CH_2=CHR$ (R = Cl, F, OOCR', C_6H_5), **vinylidene** compounds, $CH_2=CR_2$ (R = Cl, F, CN), **acrylic** compounds, $CH_2=CHR$ (R = CN, COOH, COOR'), **methacrylic** monomers, $CH_2=C(CH_3)R$ (R = COOH, COOR', CN), **allyl** compounds, $CH_2=CH–CH_2R$ (R = OH, OR', OOCR'), the corresponding **divinyl**, **diacryl**, and **diallyl** monomers, $CH_2=CH–Z–CH=CH_2$, and **1,3-dienes**, $CH_2=CR–CH=CH_2$ (R = H, CH_3, Cl). From these monomers, more than 50 % of *all* synthetic polymers are produced (Table 10-1).

The free-radical polymerization of 1-olefins $CH_2=CHR$ (R = CH_3, C_2H_5, etc.) leads only to branched oligomers. Monounsaturated cycloolefins are not polymerized by free radicals. Derivatives of 1,3-cyclohexadiene can be polymerized, however, albeit by shifting the remaining double bond similarly to the polymerization of 1,3-dienes.

Advantages and Disadvantages of Free-Radical Polymerizations

Free-radical polymerizations have technically many advantages. Rigorous cleaning-up procedures are not necessary since the effect of impurities, if any, can be industrially overcome by adjusting the polymerization system. In general, oxygen has to be removed, however, since its free-radical nature allows it act as both initiator and inhibitor.

Furthermore, free-radical polymerizations, except those of ethene, are pressure-less and can be performed at convenient temperatures of 20-80°C. They can be conducted in

Table 10-1 Industrial free-radical homopolymerizations in the gas phase (G) (with precipitation of the polymers), in bulk (B), in suspension (S), in emulsion (E), in solution (L), or by precipitation (P). Main application as adhesive (A), coating (C), elastomer (E), fiber (F), thermoplastic (T), thermoset (S), or various others (V). + Mainly, (+) less often.

| Monomer | Polymerization in | | | | | | Appli‑ |
	G	B	S	E	L	P	cation
Polymerization to linear or slightly branched polymers							
Ethene		+	+	(+)	(+)	(+)	T,F
Styrene		+	+	(+)	(+)		T
Methyl methacrylate		+	+	+	+		T
Vinyl chloride	+	(+)	(+)	+	(+)		T
Vinyl acetate		(+)	(+)	+	(+)		C,A
N-Vinyl pyrrolidone		+			+		V
Vinyl fluoride		+					T,C
p-Methyl styrene			+	+			T
Vinylidene fluoride			+	+			T,C
Trifluorochloroethene			+				T
Tetrafluoroethene			+				T,C
Acrylic esters				+			A
Chloroprene				+			E
Acrylamide					+		V
Acrylic acid					+		V
Acrylonitrile						+	F
Crosslinking polymerizations							
Diallyl compounds			+				S

bulk, or in aqueous solutions, suspensions, or emulsions (Section 10.6), often by both discontinuous and continuous processes (Volume II). Polymerizations in organic solvents are possible but are generally avoided because such solvents are expensive, toxic, and/or flammable, and also difficult and expensive to recover and work up. Free-radical initiators are usually not very expensive and often relatively easy to store or to handle.

Free-radical polymerizations are relatively unspecific in their stereocontrol. Most polymers by free-radical polymerizations, except those to polydienes, are not predominantly stereoregular. Free-radical polymerizations also allow copolymerizations with many different comonomers, including some that cannot be homopolymerized (Chapter 12). Disadvantageous for some purposes is the statistical variation of segment lengths that is not easy to modify.

10.2   Initiation and Start

10.2.1   Thermal and Spontaneous Polymerizations

Thermal polymerizations are true **spontaneous** or **self-initiated polymerizations** that proceed in the dark. They differ from those that are initiated by light or spurious impurities, including oxygen from the air and trace compounds on the surface of reactor vessels. The latter ones are also erroneously called thermal, spontaneous, or self-initiated.

True thermal polymerizations to high-molecular weight polymers have been observed for styrene and some of its derivatives as well as for 2-vinyl pyridine, 2-vinyl furan, 2-vinyl thiophene, acenaphthalene, and methyl methacrylate, and possibly also for *p*-vinyl phenol. Vinyl mesitylene, 9-vinyl anthracene, and methyl acrylate do not polymerize thermally. Some of the thermal polymerizations are free-radical processes since they are stopped by free-radical inhibitors, but some are not.

Best known is the thermal polymerization of styrene, C_8H_8, which requires a high activation energy: a monomer conversion of 50 % is obtained after 400 days at 29°C, 253 min at 127°C, and 16 min at 167°C. Two styrene molecules first form a Diels-Alder product (II) with an axial phenyl group (plus a considerably less reactive product with an equatorial phenyl group). The Diels-Alder product then reacts with styrene to form the radicals III and IV:

(10-6)

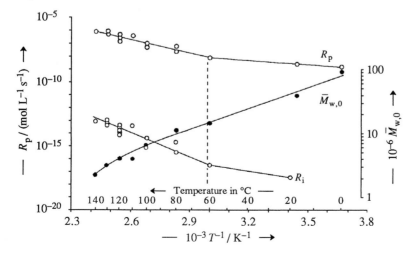

Radicals III and IV initiate the polymerization. The polymerization proceeds at small monomer conversions to very high molecular weights up to ca. $80 \cdot 10^6$ since the total concentration of radicals (III, IV, macroradicals) is too small for substantial termination or transfer reactions. Molecular weights are smaller at higher monomer conversions because of increasing chain transfer to the Diels-Alder product II.

High molecular weights have also been observed in the thermal polymerization of methyl methacrylate, MMA (Fig. 10-1). Polymerization rates R_p are small (50 % monomer conversion in 5.3 years at 29.8°C). Rates R_p and initiation rates R_i show different temperature dependencies below and above 60°C. There must therefore be two different influences on polymerization rates: a true thermal one at $T > 60$°C and an almost temperature-independent ("athermal") one at $T < 60$°C.

Fig. 10-1 Temperature dependence of observed polymerization rates R_p, calculated initiation rates R_i = $k_i[I^\bullet][M] = k_{t(pp)}[P^\bullet]^2$, and experimental weight-average molecular weights, $\overline{M}_w$, in the thermal polymerization of methyl methacrylate to low monomer conversions [1].

The athermal polymerization must be caused by natural radiation (cosmic rays, radon, radiation from ^{40}K in rocks, etc.). From radiation polymerization (Chapter 11), it is known that the initiation rate $R_{i\cdot rad} = \beta \dot{D}$ of MMA is directly proportional to the absorbed dose rate $\dot{D}$ of radiation. Using the β value from these radiation experiments and the absorbed dose rate of natural radiation of $\dot{D} = 2.38 \cdot 10^{-11}$ J kg^{-1} s^{-1} at 50° northern latitude (sea level) where the thermal polymerizations of MMA were performed, one arrives at a calculated thermal initiation rate of $R_{i,calc} \approx (3\pm 1) \cdot 10^{-17}$ mol L^{-1} s^{-1} which agrees well with the experimental value of $R_i = 3.7 \cdot 10^{-17}$ mol L^{-1} s^{-1} for 0°C.

10.2.2 Thermal Initiators

Styrene is the only monomer that is industrially polymerized by thermal polymerization at temperatures between 100°C and 220°C. The resulting non-crystallizing, atactic polymers are glass clear and therefore called "crystal poly(styrene)".

Practically all other free-radical polymerizations employ initiators that deliver initiating radicals. The required energy for the homolysis of initiator molecules can be put into the system thermally, chemically, electrochemically, or photochemically. The lower the dissociation energy, the more stable are the radicals. Very stable free radicals do not initiate polymerizations, however. An example is the triphenylmethyl radical $(C_6H_5)_3C^\bullet$.

The most commonly used free-radical initiators comprise hydroperoxides such as cumene hydroperoxide (CHP), dialkyl peroxides such as dicumene peroxide (Dicup), diperoxy ketals such as 1,1-di(*t*-butylperoxy)cyclohexane, peracid esters such as *t*-amyl peroxyacetate, diacyl peroxides such as dibenzoyl peroxide (BPO), peroxy dicarbonates such as diisopropyl peroxydicarbonate (IPP), and methyl ethyl ketone peroxide (MEKP). Other initiators are dipotassium persulfate $K_2S_2O_8$, *N,N*-azo*bis*isobutyronitrile (AIBN), dibenzyl derivatives (DBD), and benzpinacols (DBD with R' = OH).

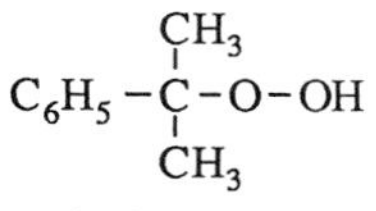

cumene hydroperoxide (CHP)	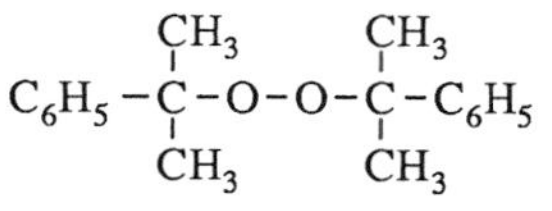 dicumene peroxide (Dicup)	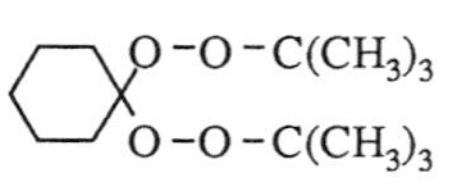 1,1-di(*t*-butylperoxy)cyclohexane
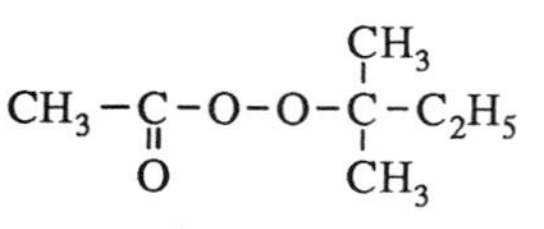 *t*-amyl peroxyacetate	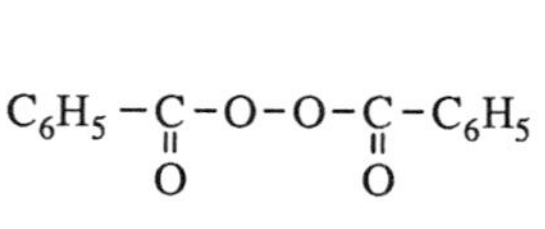 dibenzoyl peroxide (BPO)	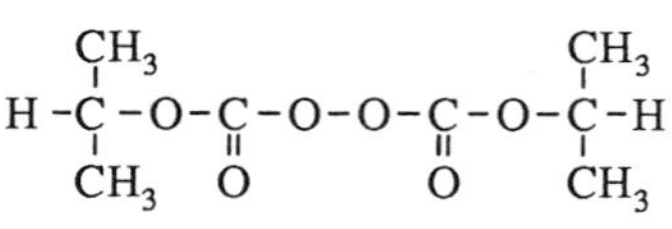 diisopropyl peroxydicarbonate (IPP)
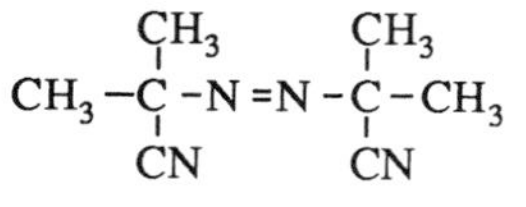 *N,N*-azo*bis*isobutyronitrile (AIBN)	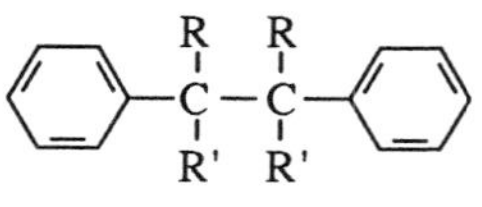 dibenzyl derivatives (DBD)	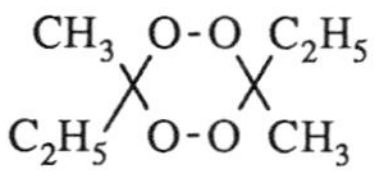 methylethylketone peroxide (MEKP)

10.2.3 Initiator Decomposition and Radical Yield

Chemical Reactions

Industry uses a great number of very different free-radical initiators (Section 4.2.3 in Volume II) whereas academic research employs mainly AIBN and BPO, in part because of tradition and in part because of their supposedly simple reactions. But this is not necessarily true.

Azo compounds R–N=N–R' and tetrazenes R_2N–N=N–NR'$_2$ decompose irreversibly because of the formation of the highly stable nitrogen molecule N_2. *N,N*-Azo*bis*isobutyronitrile I dissociates into two isobutyronitrile radicals (II) that may start the polymerization. However, since these radicals reside in a "cage" of monomer and/or solvent molecules shortly after decomposition of I, they may also combine to III or disproportionate to IV + V. Radicals II may also attack VI (mesomeric to II) and form VII. The compounds V (= methacrylonitrile) and VII, as well as I itself, can also act as comonomers.

(10-7)

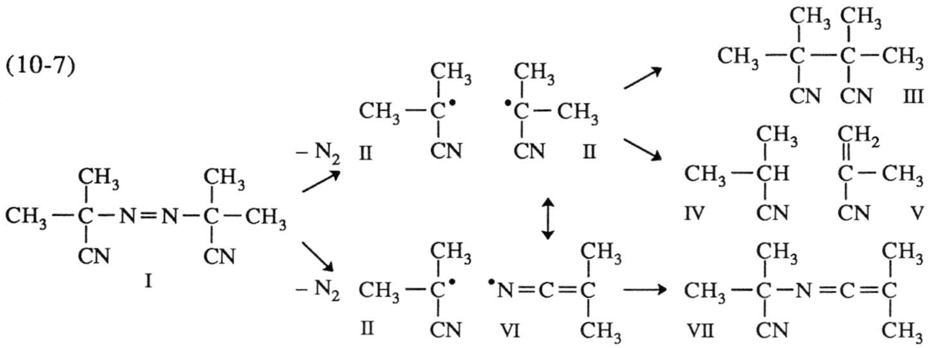

Dibenzoyl peroxide ("benzoyl peroxide") dissociates first irreversibly into two dibenzoyloxy radicals, $C_6H_5COO^\bullet$, and, in part, further to phenyl radicals and CO_2. The dibenzoyl radicals start the polymerization by adding to the β-carbon or α-carbon (less frequently) of monomers such as styrene, $^\beta CH_2{=}^\alpha CHC_6H_5$. They may also, albeit rarely, undergo reactions such as an aromatic substitution to $CH_2{=}CH(^\bullet C_6H_3OOCC_6H_5)$ (reversible) or a hydrogen abstraction from the α-methyl group of methyl methacrylate, delivering $CH_2{=}C(^\bullet CH_2)(COOCH_3)$ and C_6H_5COOH (see also Eq.(10-16) ff.).

Some initiators dissociate reversibly, for example, dibenzyl derivatives at elevated temperatures. They are used for the hot curing of unsaturated polyester resins.

Persulfate anions dissociate in water or alkaline solutions into two radical anions (used in emulsion polymerization). No radicals, however, are formed in acidic environments:

(10-8) $(S_2O_8)^{2-}$ → $2\ ^\bullet(SO_4)^-$ (alkaline to neutral)

(10-9) $(S_2O_8)^{2-} + 3\ H^+$ → $H_2SO_4 + HSO_4^-$ (acidic)

Initiator Efficiency

The **initiator efficiency** f_{eff} (**radical yield**) is always smaller than unity since only a part of the primary radicals initiates the polymerization, Thermal decomposition of *one* initiator molecule I_2 generates *two* initiator radicals $I^\bullet$ so that the rate of formation R_I of initiator radicals is $R_I = d[I^\bullet]/dt = 2\ k_d[I_2]$. The initiator radicals then react in the initiation (or start) reaction with monomer molecules M to form monomer radicals $IM^\bullet$. The rate of formation of these monomer radicals is $R_{st} = d[IM^\bullet]/dt = k_{st}[I^\bullet][M]$. The overall initiation rate R_i is, however, controlled by the much slower formation of initiator radicals, so that $R_i = R_I = 2\ k_d[I_2]$ if *all* initiator radicals react with monomer molecules. Since some of the initiator radicals are wasted (for example, as III, IV, V, and VII from II in Eq.(10-7)), an efficiency factor $f_{eff} \leq 1$ is introduced:

(10-10) $R_i = 2 f_{eff} k_d [I_2]$

The rate of reaction of two initiator radicals II to III or IV and V, respectively, is proportional to one-half of the concentration $[I^\bullet]$ of initiator radicals:

(10-11) $R_{re} = (1/2)\, k_{re} [I^\bullet]$

The theoretical radical yield, f_{theor}, is calculated from the ratio of the rate R_i of formation of all initiator radicals to the sum of the rates of all reactions of initiator radicals:

(10-12) $f_{theor} = \dfrac{R_i}{R_i + 2 R_{re}} = \dfrac{1}{1 + \{k_{re} / (k_i [M])\}}$

The often found values of $k_i \approx 100$ L mol^{-1} s^{-1} and $k_{re} \approx 10^9$ s^{-1} predict theoretical radical yields of only $f_{theor} \approx 10^{-6}$ for bulk polymerizations ($[M] \approx 10$ mol/L). Experimental radical yields are far higher, however; for example, $f_{eff} \approx 0.7$ for AIBN in styrene and various solvents at 50°C. The experimental values are far larger than theory because initiator radicals react preferentially with the monomer molecules of their cage instead of recombining. Radical yields may therefore approach unity if the initiator radicals cannot recombine for steric reasons; for example, the radicals from 1,2-diacetoxy-1,1-diphenyl azoethane. In general, radical yields decrease with monomer conversion (see Fig. 10-13 in Section 10.3.9).

Radical yields are often determined from the numbers of decomposed initiator molecules and resulting polymer endgroups by using radioactively or otherwise tagged initiators. These methods fail, however, if the initiating radicals are secondary radicals and not the primary ones. For example, the polymerization by dipotassium persulfate, $K_2S_2O_8$, does not lead to polymers with sulfate endgroups. The primary sulfate anion radical, $^\bullet SO_4^\ominus$, probably transfers an electron to the surrounding water, thus generating anions $HSO_4^\ominus$ and initiating hydroxy radicals $HO^\bullet$.

Radicals may also be transferred from primary radicals to other molecules. For example, benzpinacol radicals $(C_6H_5)_2C^\bullet(OH)$ react with monomerss $CH_2{=}CHR$ to form benzophenone $(C_6H_5)_2CO$ and radicals $CH_3-^\bullet CHR$ that initiate the polymerization.

Endgroups from primary and secondary radicals may, however, not always come from initiation reactions. They may also be generated if macroradicals are terminated by initiator or monomer radicals.

10.2.4 Rate of Decomposition

Initiators I_2 decompose monomolecularly with a rate of decomposition, R_d:

(10-13) $R_d = -\, d[I_2]/dt = k_d [I_2]$; $[I_2] = [I_2]_0 \exp(-k_d t)$

The time t_x for the thermal decomposition of x % of an initiator is calculated by integrating Eq.(10-13) between $[I_2]_0$ at $t = 0$ and $[I_2] = [I_2]_0/x$ at $t = t_x$ which delivers $t_x = \{\ln([I_2]/[I_2]_0)\}/k_d$.

Table 10-2 Activation energies $\Delta E_d^{\ddagger}$, half-lives t_{50}, and half-life temperatures T_{50} of the decomposition of some free-radical initiators. Abbreviations of initiator names, see Section 10.2.2.

Initiator	Solvent	$\Delta E_d^{\ddagger}$ in kJ/mol	t_{50}/h at 40°C	70°C	110°C	T_{50}/°C in 1 h
AIBN	dibutyl phthalate	122.2	303	5.0	0.057	
	benzene	125.5	354	6.1	0.076	79
	styrene	127.6	414	5.7	0.054	
BPO	acetone	111.3	443	10.6	0.180	
	benzene	128.9	2 130	23.7	0.177	92
	styrene	132.8	3 525	29.2	0.231	
	poly(styrene)	146.9	11 730	84.6	0.392	
MEKP	ethyl acetate			217		
IPP	dibutyl phthalate	115.0	21	0.32	0.0044	
Dicup	benzene	170	3 000 000	11 200	27	128
CHP	benzene	100	4 000 000	60 000	760	
$K_2S_2O_8$	0.1 mol NaOH/L H_2O	140	1 850	8.3		
	water			0.0117	0.0038	

For scientific investigations, a time t_5 for a 5 % decomposition is relevant since it signals a practically constant initiator concentration which in turn eases the mathematical treatment of polymerization kinetics. Industry prefers the **half-life** $t_{50} = 0.693/k_d$ (Table 10-2), i.e., the time for a 50 % decomposition, or the **half-life temperature** T_{50}, i.e., the temperature at which 50 % of the initiator is decomposed after a specified time, for example, 10 hours, 1 hour, 1 minute, etc.

Activation energies of initiation, and thus rates of decomposition, often depend on the solvent and/or monomer. The effects are relatively small for AIBN but can be dramatic for BPO (Table 10-2). After 60 min at 79.8°C, BPO decomposes 13 % in CCl_4, 16 % in C_6H_6, 51 % in C_6H_{12}, and 82 % in 1,4-dioxane. Ten minutes in i-propanol is sufficient to decompose 95 % BPO; in amines, it explodes. Small proportions of, e.g., dimethylaniline are therefore used in industry as **promotors** of free-radical polymerizations.

The reason for these variations is the solvent-induced decomposition of the initiators which is much faster than the thermal decomposition. The reaction of BPO with dimethylaniline generates first radical cations and then free radicals:

(10-14) $(CH_3)_2NC_6H_5 + (C_6H_5COO)_2 \rightarrow [(CH_3)_2N^{\bullet}C_6H_5]^{\oplus} C_6H_5COO^{\ominus} + C_6H_5COO^{\bullet}$
$[(CH_3)_2N^{\bullet}C_6H_5]^{\oplus} C_6H_5COO^{\ominus} \rightarrow (CH_3)_2NC_6H_4{}^{\bullet} + C_6H_5COOH$

The decomposition of BPO by dibutylether generates α-butoxybutyl radicals:

(10-15) $(C_4H_9)_2O + C_6H_5COO^{\bullet} \rightarrow C_3H_7-{}^{\bullet}CH-OC_4H_9 + C_6H_5COOH$
$C_3H_7-{}^{\bullet}CH-OC_4H_9 + (C_6H_5COO)_2 \rightarrow C_3H_7-CH(OC_4H_9)-(OOCC_6H_5) + C_6H_5COO^{\bullet}$

10.2.5   Initiation Reactions

The thermal decomposition of initiator molecules generates many more initiating radicals per unit of time than a thermal polymerization (Section 10.2.1). Since the increase in production of free radicals increases the rate of polymerization, thermal initiators were formerly also called **accelerators**.

Initiation reactions are exothermic. The generated radicals are excited by additional vibrations and must be therefore "hot". However, the excess vibration energy is much lower than that of the "hot species" of radiation chemistry. The collision of the "hot" initiator radicals with monomer molecules furthermore transfers the energy from the former to the latter in about 10^{-9} s. The excess energy is however disbursed by collisions with other entities long before the initiator radicals add to monomer molecules, a much slower process of 10^{-3} s. "Hot radicals" thus play no role in initiation reactions.

The decomposition of thermal initiators creates primary radicals and sometimes also secondary radicals which may undergo a variety of initiation reactions, for example, in the styrene polymerization by BPO:

(10-16)

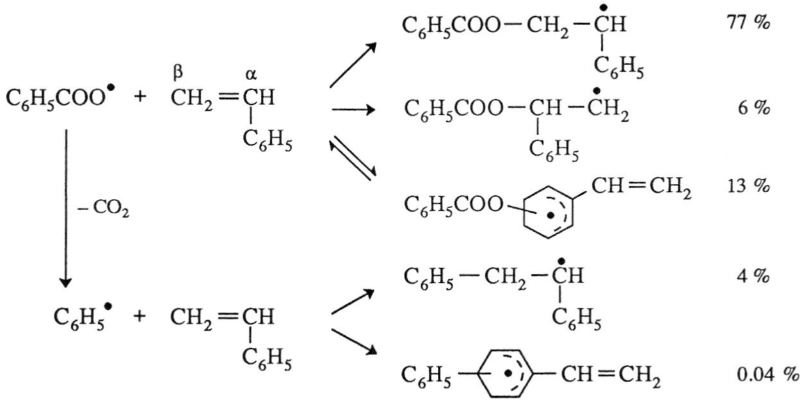

$$C_6H_5COO\overset{\bullet}{} + \underset{C_6H_5}{\overset{\beta}{CH_2}}\overset{\alpha}{=CH}$$

$$C_6H_5COO-CH_2-\overset{\bullet}{CH} \qquad 77\ \% \\ C_6H_5$$

$$C_6H_5COO-\underset{C_6H_5}{\overset{|}{CH}}-\overset{\bullet}{CH_2} \qquad 6\ \%$$

$$C_6H_5COO\overset{CH=CH_2}{\underset{(\overset{\bullet}{})}{}} \qquad 13\ \%$$

$-CO_2$

$$C_6H_5\overset{\bullet}{} + \underset{C_6H_5}{\overset{|}{CH_2}}=CH$$

$$C_6H_5-CH_2-\overset{\bullet}{CH} \qquad 4\ \% \\ C_6H_5$$

$$C_6H_5\overset{(\overset{\bullet}{})}{}-CH=CH_2 \qquad 0.04\ \%$$

The various radicals react at various rates; the rate of the initiation reaction is therefore not a simple function of the initiator concentration. A faster dissociation of an initiator thus does not necessarily mean a faster polymerization. For example, dibenzoyl peroxide dissociates 1000 times faster than cyclohexyl peroxide but causes only 5 times as high a rate of polymerization of styrene.

The various initiator radicals not only generate different endgroups (Table 10-3) but may also cause radical transfers that lead to branching and crosslinking of polymers.

Table 10-3 Reactions of benzoyloxy radicals with vinyl and acrylic monomers (for styrene, see also Eq.(10-16)). The resulting macroradicals were trapped by spin markers (free radicals at nitrogen atoms in substituted cycloazaalkanes). Data reported by [2].

Monomer	Percent probability of attack by radicals (rad.)			
	Primary rad. at β	Primary rad. at α	Secondary rad. at β	Other radicals
Styrene	77	6	4	13
Vinyl acetate	65	20	15	0
Methyl methacrylate	66	5	28	1
Methyl acrylate	32	6	62	0
Acrylonitrile	30	1	69	0

Branching and crosslinking is caused by transfer reactions of initiator radicals to polymer molecules (Section 10.4.5). The polymerization of *p*-vinyl benzyl methyl ether, $CH_2=CH(p\text{-}C_6H_4\text{-}CH_2\text{-}O\text{-}CH_3)$, is an example. AIBN produces linear polymers at low monomer conversions but slightly branched polymers at higher ones. BPO causes crosslinking at high monomer conversions and diacetyl, $CH_3CO\text{-}COCH_3$, even at low ones.

Some initiator-monomer combinations even produce no polymers at all. Vinyl mercaptals, $CH_2=CH\text{-}S\text{-}CH_2\text{-}S\text{-}R$, are polymerized by AIBN but not by BPO because the mercaptal group induces the decomposition of BPO to benzoic acid, C_6H_5COOH, and an unstable ester, $CH_2=CH\text{-}S\text{-}CH(OOCC_6H_5)\text{-}S\text{-}R$. This reaction consumes all initiator.

Phenol group containing vinyl monomers such as *p*-vinyl phenol cannot be polymerized if the initiator forms oxy radicals; an example is BPO (see also Section 10.4.7). The polymerization of *p*-vinyl phenol is possible with AIBN.

10.2.6 Redox Initiators

Redox initiators generate initiating radicals by reaction of a reducing agent with an oxidizing agent. Such reactions require only small thermal activation energies. Polymerizations thus proceed at much lower temperatures than the ones that are initiated by a thermal decomposition of an initiator such as peroxy compounds. Five types of redox systems can be distinguished:

1. Systems of peroxides + amines (see Eq.(10-14)) are relatively insensitive toward oxygen. They are often used in industry for polymerizations in bulk, especially for crosslinking reactions.

2. Systems of a hydroperoxide, especially dihydrogen peroxide, H_2O_2, and a reducing metal ion, for example, Fe^{2+}:

$$(10\text{-}17) \quad HOOH + Mt^{n+} \quad \rightarrow HO^\bullet + HO^\ominus + Mt^{(n+1)+}$$

$$(10\text{-}18) \quad HOOH + Mt^{(n+1)+} \rightarrow HOO^\bullet + H \quad + Mt^{n+}$$

Such systems are much more sensitive toward oxygen. They produce hydroxy radicals $HO^\bullet$ that are so small that they can also attack the α-carbon atoms of vinyl and acrylic monomers. The extent of this anomalous start reaction depends on the pH which points towards participation of complexes of hydroxy radicals and metal ions.

The metal ion cannot be reduced according to Eq.(10-18) if peroxides ROOH are used instead of dihydrogen peroxide HOOH. However, the metal ion can be regenerated if reducing agents are added, for example, glucose that is oxidized to glucuronic acid.

3. The metal in transition metal carbonyls is in the valence state zero. Such carbonyls generate alkyl radicals $R^\bullet$ from organic halogenides RX:

$$(10\text{-}19) \quad Mt^0 + RX \rightarrow MtX + R^\bullet$$

4. Boron alkyls react with oxygen and form alkyl radicals (Eq.(10-20)). This reaction requires such low activation energies that it is possible to initiate free-radical polymerizations at $-100°C$. The resulting alkyl radicals are very little resonance stabilized and therefore very reactive.

$$(10\text{-}20) \quad R_3B + O_2 \rightarrow R_2BOOR; \quad R_2BOOR + 2\,R_3B \rightarrow R_2BOBR_2 + R_2BOR + 2\,R^\bullet$$

5. Systems 1-4 generate single radicals; the radical efficiency is always unity. Certain other redox systems, hoever, produce pairs of radicals which lead to cage effects and thus lower radical yields. An example is the system potassium persulfate-mercaptan:

(10-21) $K_2S_2O_8 + RSH \rightarrow RS^\bullet + KSO_4^\bullet + KHSO_4$

Certain salts of transition metals seem to polymerize certain monomers via redox reactions. For example, $TiCl_3$ or $PdCl_2$ polymerize styrene but not methyl methacrylate; it is just the opposite for $RhCl_3$.

Slow redox reactions generate only a few radicals per time unit and the polymerization is slow. If the redox reaction is much faster than the initiation reaction, however, then most initiator radicals are used up before they can react with monomer molecules. Redox systems are thus regulated by additives. For example, the reactivity of heavy metal ions changes by complexation with citrates. Redox systems are also very sensitive toward the medium and the concentration of reactants since some components can decompose initiators. Industrial redox systems are thus often very complex.

Some redox systems use oxygen as an oxidant. Oxygen can however also react directly with some monomers and form hydroperoxides that decompose to initiator radicals. Oxygen itself is also a biradical and can thus react with monomer radicals and macroradicals. The newly formed peroxy and peroxide radicals are often very sluggish to react; oxygen thus inhibits polymerizations. After all the oxygen has been used up, the peroxy radicals decompose into radicals that initiate polymerizations. Peroxide radicals may also react to form aldehydes which are strong transfer agents. Depending on the conditions, oxygen may thus either accelerate or decelerate free-radical polymerizations.

10.2.7 Photo Initiators

Polymerization-initiating radicals may also be produced photochemically. For example, the azo group of $(CH_3)_2(CN)C–N=N–C(CN)(CH_3)_2$ absorbs light at 350 nm. Nitrogen is split off and the resulting isobutyronitrile radicals $(CH_3)_2(CN)C^\bullet$ initiate the polymerization (Eq.(10-7)). Light also induces the formation of radicals from certain aliphatic ketones (see Chapter 11). Such photochemical generations of radicals do not need thermal activation energies. Polymerizations by such systems have therefore low activation energies (Section 10.3.4) and proceed at low temperatures.

10.2.8 Electrolytic Polymerizations

Electrolytic (electrochemical) polymerizations are electrolyses in the presence of a monomer. The electrolysis of salts of fatty acids generates alkyl radicals according to

(10-22) $R–CH_2–CH_2–COO^\ominus – e^\ominus \rightarrow R–CH_2–CH_2–COO^\bullet \rightarrow R–CH_2–CH_2^\bullet + CO_2$

The alkyl radicals initiate the polymerization. In the absence of monomers, alkyl radicals will combine, disproportionate, or react with acyloxy radicals to form esters.

Such polymerizations may also proceed anionically or cationically. The anionic discharge of acetate ions in the homogeneous phase leads to free-radical polymerizations of styrene and acrylonitrile. The anionic discharge of perchlorate or boron tetrafluoride ions, on the other hand, results in cationic polymerizations of styrene, *N*-vinyl carbazole, and *i*-butyl vinyl ether. The polymerization of acrylonitrile by the cathodic discharge of tetraalkyl ammonium salts is also anionic.

In such polymerizations, electrodes are often covered with layers of polymers. Such an effect is unwanted for the preparation of polymers *per se* but can be utilized to form protective layers on metals, depending on the right combination of monomer and electrode material. Steel is useful for acrylonitrile or vinyl acetate but zinc, lead, or tin are used for *p*-xylylene or diacetone acrylamide, $CH_3COCH_2-C(CH_3)_2-NH-CO-CH=CH_2$, all in dilute sulfuric acid as electrolyte.

10.3 Propagation and Termination

10.3.1 Polymerizability, Regiocontrol, and Stereocontrol

In free-radical chain polymerizations, an initiator radical, a monomer radical, or a macroradical attacks an atom of the double bond of a monomer molecule. In α-olefins $^{\beta}CH_2=^{\alpha}CHR$, it is preferentially the β-carbon atom (Table 10-3). In general:

(10-23)

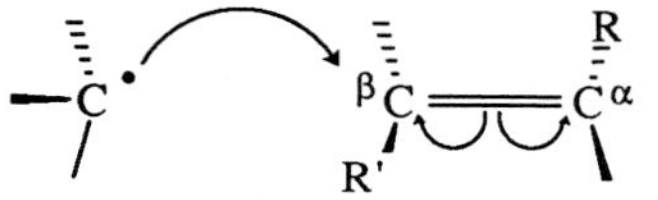

The addition of a radical opens the π-bond and generates a σ-bond. Theoretical calculations favor an unsymmetric transition step in which unequal distances exist between the attacking radical and the α and β carbon atoms of the double bond (note that α and β have opposite meanings in organic and macromolecular chemistry!). There is therefore a strong steric effect of the β-substituent R' and a very small one of the α-substituent R. Both substituents exert polar effects.

The **polymerizability** of *monosubstituted ethene derivatives*, $CH_2=CHR$ is higher, the stronger the resonance stabilization of the new radical, i.e., the better the substituent R of $\sim CH_2-^{\bullet}CHR$ is conjugated with the unpaired electron. Relative to the methyl radical $H_3C^{\bullet}$, resonance stabilizations are 17 kJ/mol for $CH_3CH_2^{\bullet}$, 50 kJ/mol for $(CH_3)_3C^{\bullet}$ or $Cl_3C^{\bullet}$, and 104 kJ/mol for $C_6H_5CH_2^{\bullet}$ or $CH_2=CH-CH_2^{\bullet}$. The resonance stabilization of radicals of monomers $CH_2=CHR$ decreases for various R in the order

$$C_6H_5 > CH=CH_2 > COCH_3 > CN > COOR' > Cl > CH_2R' > OOCR' > OR'$$

The decrease corresponds approximately to the decrease in polymerizability by free radicals. Exceptions are allyl monomers with R = CH_2R' and vinyl ethers with R = OR". Monoallyl compounds deliver only branched oligomers because of a strong radical transfer to monomers (Section 10.4.3). Vinyl ethers do not polymerize free-radically, probably for kinetic reasons.

Styrene with R = C_6H_5 is thus more easy to polymerize than vinyl acetate with R = $OOCCH_3$. In general, the more easily activated monomer molecules give the more stable radicals; poly(vinyl acetate) radicals are approximately 1000 times more reactive than poly(styrene) radicals. The reactivity can be effected by complexation of the substituents, for example, by $ZnCl_2$ in monomers with nitrile or carboxy groups.

1,1-Disubstituted monomers $CH_2=CRR'$ are usually more reactive than monosubstituted ones since their macroradicals are more strongly resonance stabilized due to two substitutents. *1,2-Disubstituted* and *1,1,3-trisubstituted* monomers are mostly less reactive since they are less resonance stabilized and their attack by radicals is sterically hindered.

1,3-Dienes $CH_2=CR–CH=CH_2$ can polymerize in two ways. Each double bond either reacts "independently" of the other to 1,2-units $–CH_2–CR(CH=CH_2)–$ or in a "concerted" manner to 1,4-units $–CH_2–CR=CH–CH_2–$ with double bonds in the 2,3-position. 1,3-Cyclohexadienes polymerize similarly with a shift of the double bond. The remaining double bond polymerizes at higher monomer conversions: the polymers crosslink.

The two carbon-carbon double bonds of *divinyl* and *diallyl* compounds are sufficiently separated from each other so that they react "independently." However, as tetrafunctional compounds, they become crosslinked at fairly small monomer conversions.

Large steric effects and resonance stabilizations of macroradicals exert strong **regiocontrols**. Poly(vinyl acetate), $+CH_2–CH(OOCCH_3)+_n$ has only 1-2 % head-to-head units. The considerably smaller fluorine substituents lead to much higher proportions of such connections: 6-10 % in $+CH_2–CHF+_n$ and 10-12 % in $+CH_2–CF_2+_n$.

Carbon radicals at growing chain ends have either a planar structure or that of a rapidly oscillating pyramid. The stereocontrol is thus exerted by the prochiral side of the threefold-substituted carbon atom. Experiments with deuterated monomer molecules showed that the corresponding carbon atom of the monomer does not contribute to the stereocontrol. The monomer addition to the macroradical is therefore sterically controlled by the interaction of the monomer with at least the penultimate monomeric unit of the growing chain, i.e., the last configurational diad. Free-radical polymerizations thus follow in general first-order Markov statistics since here the formation of a diad depends on the preceding diad ($p_{s/s} \neq p_{i/s}$, $p_{i/i} \neq p_{s/i}$) (Table 10-4).

Free-radical polymerization usually leads to more syndiotactic than isotactic diads. An exception is monomers with strong neighboring group effects, such as vinyl formate, $CH_2=CH(CHO)$. The syndiotacticity increases with decreasing temperature which is assumed to indicate that the stereocontrol in free-radical polymerizations is due to steric effects. However, the stereocontrol must also be influenced by electronic factors.

Table 10-4 Mole fractions x_s of syndiotactic diads and conditional probabilities p of the formation of isotactic ($p_{i/i}$, $p_{s/i}$) and syndiotactic ($p_{s/s}$, $p_{i/s}$) diads in free-radical polymerizations of neat ethene derivatives $CH_2=CRR'$. $p_{i/i} + p_{i/s} \equiv 1$, $p_{s/s} + p_{s/i} \equiv 1$. * In benzene.

R	R'	$T/°C$	x_s	$p_{s/s}$	$p_{i/s}$	$p_{i/i}$	$p_{s/i}$
H	HCOO	40	0.48	0.54	0.42	0.58	0.46
H	CH_3COO	30	0.54	0.63	0.41	0.59	0.37
H	Cl	25	0.56	0.60	0.52	0.48	0.40
H	C_6H_5	100	0.74	0.74	0.77	0.23	0.26
CH_3	$COOCH_3$	60	0.74	0.80	0.67	0.33	0.20
CH_3	$COO(CH_2)_4H$	60	0.72	0.82	0.62	0.38	0.18
CH_3	$COOC(CH_3)_3$	60	0.72	0.72	0.71	0.29	0.28
CH_3	$COOC(C_6H_5)_3$ *	60	0.44	0.55	0.32	0.68	0.45

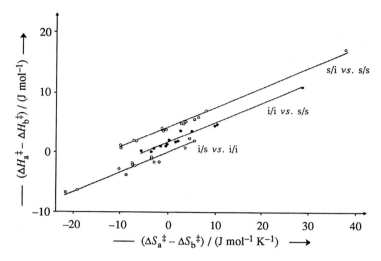

Fig. 10-2 Difference of activation enthalpies as a function of the difference of activation entropies of two different monomer additions a and b (a, b = i/i, i/s, s/i, s/s). Free-radical polymerization of 9.1 mol% methyl methacrylate in 14 different solvents. Only three of the six possible combinations are independent of each other. Data from [3].

Conditional probabilities and mole fractions allow the calculation of ratios of propagation rate constants for the various types of addition and thus also differences of activation enthalpies and entropies (Section 6.3.5). A plot of differences of activation enthalpies as a function of differences of activation entropies

$$(10\text{-}24) \qquad \Delta H_a^\ddagger - \Delta H_b^\ddagger = \Delta\Delta H_0^\ddagger - T_0(\Delta S_a^\ddagger - \Delta S_b^\ddagger)$$

delivers straight lines, i.e., a **compensation effect** (Fig. 10-2). The slope T_0 of the plot must have the physical unit of temperature. This **compensation temperature** is independent of the type of addition steps. It indicates the temperature at which the polymerization delivers the same tacticity regardless of the solvent. Such temperatures are 330 K ≈ 57°C for methyl methacrylate and 350 K ≈ 77°C for vinyl trifluoroacetate.

The compensation enthalpies $\Delta\Delta H_0^\ddagger$ of the cross steps s/i vs. i/s are always large and positive for methacrylates but very small or negative for vinyl compounds (Table 10-5). The sign should always be the same, however, if only steric effects are present. One can thus conclude that both steric and electronic effects control the stereoregularity.

Table 10-5 Compensation temperatures and compensation enthalpies for s/i vs. i/s.

Methacrylates	T_0/K	$\Delta\Delta H_0^\ddagger$/(J mol^{-1})	Vinyls	T_0/K	$\Delta\Delta H_0^\ddagger$/(J mol^{-1})
glycidyl	345	3800	acetate	1100	−2050
hexyl	303	3260	trifluoroacetate	340	167
butyl	236	3580	chloroacetate	300	419
propyl	248	3300	formate	330	− 290
methyl	362	1470	chloride	305	−220

10.3.2   Termination

The concentration of new initiator radicals $I^\bullet$ decreases with time since (a) the initiator concentration decreases and (b) initiator radicals are converted to monomer radicals $IM^\bullet$. The concentration of $IM^\bullet$ increases from 0 at $t = 0$ and then passes through a maximum since more and more monomer radicals are converted to macroradicals $IM_i^\bullet \equiv P^\bullet$. The total concentration of all radicals ($I^\bullet + IM^\bullet + P^\bullet$) increases with time until it becomes so large that radicals react with other radicals and a **steady state** is attained in which the radical production just equals the radical consumption (Section 10.3.3).

The reaction between two radicals to inactive molecules is called **termination**. Since macroradicals dominate, termination usually happens between two macroradicals, either by **combination** (for unclear reasons sometimes called **recombination**) or by **disproportionation**, usually with transfer of a hydrogen atom:

(10-25)   ~CH$_2$–$^\bullet$CHR + $^\bullet$CHR–CH$_2$~ → ~CH$_2$–CHR–CHR–CH$_2$~   combination

(10-26)   ~CH$_2$–$^\bullet$CHR + $^\bullet$CHR–CH$_2$~ → ~CH$_2$–CH$_2$–R + CHR=CH~ disproportionation

These two termination reactions are also called **terminations by mutual deactivation**. They share the same rate equation

(10-27) $R_{t(pp)} = -\,d[P^\bullet]/dt = 2\,k_{t(pp)}[P^\bullet]^2$ (U.S. convention)

The proportionality constant is written $2\,k_{t(pp)}$ because two radicals are destroyed. In the German (D) and British (UK) literature, the factor 2 is included in the rate constant. The termination rate constants are therefore related by $k_{t(pp)}$ (US) = (1/2) $k_{t(pp)}$ (D, UK). The rates of polymer formation are correspondingly $d[P]/dt = (k_{tc} + 2\,k_{td})[P^\bullet]^2$ (US) and $d[P]/dt = ((1/2)\,k_{tc} + k_{td})[P^\bullet]^2$ (D, UK) where the indices tc refer to *t*ermination by *c*ombination and td to that by *d*isproportionation.

Disproportionations generate at random two dead polymer molecules from two propagating macroradicals. The molecular weight distribution of polymers from free-radical polymerizations with 100 % termination by disproportionation is therefore identical with that of stoichiometric AB or AA+BB polycondensations at very high monomer conversions (Section 13.2): the molar mass distribution is of the Schulz-Flory type and the polymolecularity index approaches $\overline{M}_w/\overline{M}_n = 2$ (Appendix A-10).

A termination by combination produces one dead polymer molecule from two macroradicals. Random reactions lead to Schulz-Zimm type distributions (Section 3.3.3) and $\overline{M}_w/\overline{M}_n$ approaches 3/2. At 60°C, 100 % combination was found for acrylonitrile, 77 % for styrene, 21 % for methyl methacrylate, and 0 % for vinyl acetate. Combination decreases with increasing temperature for styrene but increases for methyl methacrylate.

At higher initiator concentrations, a termination by initiator radicals is also observed:

(10-28)   ~CH$_2$–$^\bullet$CHR + $I^\bullet$ → ~CH$_2$–CHR–I

In polymerizations of allyl compounds, termination is by the monomer:

(10-29)   ~CH$_2^\bullet$CHCH$_2$R + CH$_2$=CHCH$_2$R $\longrightarrow$ ~CH$_2$CH$_2$CH$_2$R + [CH$_2$⋯CH⋯CHR]$^\bullet$

The resulting allyl monomer radical is so strongly stabilized by resonance that resonance energy would have to be released on monomer addition. Allyl radicals thus do not initiate polymer chains (p. 347).

10.3.3 Steady State

At not too high monomer conversions, the continuous generation of radicals by initi- ator decomposition and removal by termination leads to a **steady state (stationary state)** with a constant radical concentration. Steady states of the first kind with constant *total* radical concentrations are distinguished from steady states of the second kind with constant *individual* radical concentrations of initiator, monomer, or polymer radicals.

Steady states of the 1st kind (**Bodenstein approximation**) are established after rela- tively short times. The rate of formation of macroradicals is given by the difference of the rate of the initiation reaction and the rate of termination by mutual deactivation:

(10-30) $d[P^\bullet]/dt = R_i - R_{t(pp)} = R_i - 2\,k_{t(pp)}[P^\bullet]^2$

$d[P^\bullet]/dt = 0$ in the steady state and the concentration of macroradicals is then

(10-31) $[P^\bullet] = [P^\bullet]_{stat} = [R_i/(2\,k_{t(pp)})]^{1/2}$

Exprimentally observed are rates of $R_i = (10^{-6}-10^{-8})$ mol L^{-1} s^{-1} and termination constants of 2 $k_{t(pp)} = (10^7-10^8)$ L mol^{-1} s^{-1}. The steady-state radical concentration is therefore $[P^\bullet]_{stat} \approx (10^{-7}-10^{-8})$ mol/L.

The time required to reach the steady state is obtained by integration of Eq.(10-30):

(10-32) $\displaystyle\int_0^{[P\bullet]_t} \frac{d[P^\bullet]}{R_i - 2\,k_{t(pp)}[P^\bullet]^2} = \int_0^t dt$

(10-33) $\left[\dfrac{1}{(2\,R_i k_{t(pp)})^{1/2}} \tanh^{-1}\left(\dfrac{(2\,k_{t(pp)})^2[P^\bullet]}{(R_i)^{1/2}} \right) \right]_0^{[P\bullet]_t} = t$

(10-34) $[P^\bullet] = \{R_i/(2\,k_{t(pp)})\}^{1/2} \tanh\,[(2\,R_i k_{t(pp)})^{1/2}t]$

The combination of Eqs.(10-34) and (10-30) delivers for the steady state

(10-35) $[P^\bullet]/[P^\bullet]_{stat} = \tanh\,\{(2\,R_i k_{t(pp)})^{1/2}t\}$

The true steady-state concentration is practically identical with the actual one if the above ratio reaches 99.5 %, thus

(10-36) $\tanh\,\{(2\,R_i k_{t(pp)})^{1/2}t\} \geq 0.995$

According to the tables of hyperbolic functions, Eq.(10-36) is fulfilled for the condi- tion $(2\,R_i k_{t(pp)})^{1/2}t \geq 3$. For the polymerization of neat styrene at $T = 50°C$, one obtains with Eq.(10-36), Eq.(10-10), $[AIBN]_0 = 0.005$ mol/L, $f_{eff} = 0.7$, $k_d = 2 \cdot 10^{-6}$ s^{-1}, and $k_{t(pp)} = 5 \cdot 10^7$ L mol^{-1} s^{-1} a value of $[4\,f_{eff}k_d[AIBN]_0 k_{t(pp)}]^{1/2}t = [1.4\ s^{-2}]^{1/2}t \geq 3$. The steady state is thus reached after $t \approx 2.54$ s. The relative monomer conversion u at this time can be calculated from $u = ([M]_0 - [M])/[M]_0$, $([M]_0 - [M])/\Delta t = k_p[P^\bullet][M]_0$, $\Delta t = t$ = 2.54 s, $[P^\bullet] = [P^\bullet]_{stat} = 1 \cdot 10^{-7}$ mol/L, and the experimental value of the propagation constant of $k_p = 215$ L mol $^{-1}$ s^{-1} as $u \approx 0.0055$ %.

10.3.4   Ideal Kinetics

The rate R_p of an irreversible free-radical propagation reaction is given by the molar concentrations of macroradicals $R^\bullet$ and monomer molecules M and the propagation rate constant k_p:

(10-37) $R_p = k_p[P^\bullet][M]$

Kinetics usually assumes that monomer is consumed only in the propagation reaction and not by any other elementary reaction such as initiation, termination, and transfer. This condition is always fulfilled if the number-average degree of polymerization is greater than 100 since then the other monomer consumption must be less than 1 %. In this case, the gross rate of polymerization equals approximately the propagation rate, i.e.,

(10-38) $R_{gross} = -\,d[M]/dt \approx R_p = k_p[P^\bullet][M]$

The gross rate of isothermal linear free-radical polymerization shows a very peculiar behavior (Fig. 10-3). It is non-stationary at very small monomer conversions and reaches the steady state at $u = ([M]_0 - [M])/[M]_0 = (0.001\text{-}0.1)$ %. In this state, polymerization rates are independent of monomer conversions since the radical concentration is constant and the small monomer consumption is negligible ($[M] \approx [M]_0$). After a few percent of monomer conversion, polymerization rates drop because of $[M] < [M]_0$. The polymerization rates then increase first dramatically at higher monomer conversions ("**gel effect**") and finally drop very drastically ("**glass effect**") (Section 10.3.9).

The polymerization stops at a monomer conversion of less than 100 % for both thermodynamic and kinetic reasons (see Section 10.3.8). Residual monomer is often difficult to remove (Section 6.5.2). In industry, residual monomers are usually burned (if volatile) or extracted by water (if non-volatile), both because of their toxic, carcinogenic, and/or environmentally deleterious properties. The remaining monomer concentrations are usually in the range of 1 ppm to 1 ppb ($10^{-4}\text{--}10^{-7}$ %).

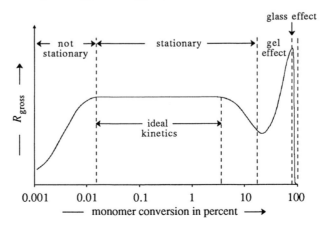

Fig. 10-3 Gross polymerization rate as a function of the *logarithm* of the monomer conversion (schematic) [4]. Ideal polymerization kinetics applies only to a fairly small range of monomer conversion.

The steady state can be described by simple kinetic expressions. For the derivation of these equations, one assumes:

(a) All reactions are irreversible.
(b) The principle of equal chemical reactivity is applicable (rates are independent of the molecular weight of growing macroradicals).
(c) There are only three elemenatary reactions: initiation, propagation, and termination.
(d) Termination is only by mutual deactivation of macroradicals.
(e) The initiator concentration remains practically constant ($[I_2] \approx [I_2]_0$).
(f) The concentration of initiator radicals does not change with time. All initiator radicals cals are generated by initiator decomposition and consumed by the initiation reaction:

(10-39) $d[I^\bullet]/dt = R_d - R_{st} = 2 f_{eff} k_d [I_2] - k_{st}[I^\bullet][M] = 0$

(g) The concentration of macroradicals is also stationary. Macroradicals are generated by the initiation reaction and consumed by mutual deactivation:

(10-40) $d[P^\bullet]/dt = R_{st} - R_{t(pp)} = k_{st}[I^\bullet][M] - 2 k_{t(pp)}[P^\bullet]^2 = 0$

The polymerization rate is obtained from Eqs.(10-38)-(10-40) and condition (e), using $u = ([M]_0 - [M])/[M]_0$ for the relative monomer conversion:

(10-41) $R_p = - d[M]/dt = k_p \{f_{eff} k_d/k_{t(pp)}\}^{1/2}[I_2]^{1/2}[M]$

$$= k_p \{f_{eff} k_d/k_{t(pp)}\}^{1/2}[I_2]_0^{1/2}[M]_0(1 - u)$$

The polymerization rate is directly proportional to the monomer concentration $[M]$. At very small monomer conversions u, R_p is practically constant. It does decrease linearly with u over greater conversion intervals as long as conditions (a)-(g) apply (see also Fig. 10-11, p. 337). The polymerization rate R_p increases only with the square root of the initiator concentration (see also Fig. 10-4, next page).

The activation energy $\Delta E_p^{\ddagger}$ of the polymerization is given by the sum of weighted activation energies of elementary reactions, each of which is assumed to follow the Arrhenius equation $k = A^{\ddagger} \exp(- \Delta E^{\ddagger}/RT)$. Hence, from Eq.(10-41) assuming that f_{eff} does not depend on temperature:

(10-42) $\Delta E_{gross}^{\ddagger} = \Delta E_p^{\ddagger} + (1/2) \Delta E_d^{\ddagger} - (1/2) \Delta E_{t(pp)}^{\ddagger}$

An example is the polymerization of neat styrene ($\Delta E_p^{\ddagger} = 32.5$ kJ/mol and $\Delta E_{t(pp)}^{\ddagger} = 2$ kJ/mol) by dibenzoyl peroxide as initiator ($\Delta E_d^{\ddagger} = 133.9$ kJ/mol) in the temperature range $-12 \le T/°C \le +93$. The overall activation energy of $\Delta E_{gross}^{\ddagger} \approx 108.5$ kJ/mol indicates a strong increase of the polymerization rate with increasing temperature (see also Section 10.4.3).

After introduction of condition (e), integration of Eq.(10-41) delivers an expression for the time dependence of the natural logarithm of the inverse relative monomer concentration, $[M]_0/[M]$, which allows a linearization of experimental results:

(10-43) $\ln ([M]_0/[M]) = k_p \{f_{eff} k_d/k_{t(pp)}\}^{1/2}[I_2]_0^{1/2}t$

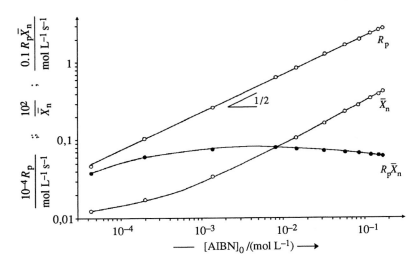

Fig. 10-4 Dependence of polymerization rates R_p, inverse number-average degrees of polymerization, $1/\overline{X}_n$, and products $R_p\overline{X}_n$ on the initial concentration of the initiator *N,N*-azobisisobutyronitrile, [AIBN]$_0$, for the polymerization of neat styrene at 60°C at monomer conversions of 1-3 % [5].

10.3.5 Kinetic Chain Length

Polymerization kinetics does not deliver *a priori* the number-average degree of poly-merization, $\overline{X}_n$, but the **kinetic chain length**, ν. This quantity indicates the number of monomer molecules that are polymerized by one initiator radical before the resulting macroradical is converted to a dead molecule by a termination reaction. It is thus given by the ratio of the propagation rate to the sum of the rates of all termination reactions:

$$(10\text{-}44)\qquad \nu \equiv R_p / \Sigma_t\, R_t$$

A termination by disproportionation (Eq.(10-26)) creates two dead chains from two macroradicals ($\nu = \overline{X}_n$). A termination by combination, on the other hand, unites two macroradicals to one dead polymer molecule ($2\,\nu = \overline{X}_n$).

The kinetic chain length of polymerization, ν_{pp}, with termination by mutual deactiva-tion of macroradicals is calculated from the rates R_p of polymerization (Eq.(10-41)) and $R_{t(pp)}$ of termination which equals the rate R_i of the start reaction, Eq.(10-10):

$$(10\text{-}45)\qquad \nu_{pp} = \frac{R_p}{R_i} = \frac{k_p(f_{eff}k_d / k_{t(pp)})^{1/2}[I_2]^{1/2}[M]}{2\,f_{eff}k_d[I_2]} = \frac{k_p}{2(f_{eff}k_dk_{t(pp)})^{1/2}} \cdot \frac{[M]}{[I_2]^{1/2}}$$

Multiplication of both sides by R_p and introduction of Eq.(10-41) delivers

$$(10\text{-}46)\qquad \nu_{pp}R_p = \{k_p^2/(2\,k_{t(pp)})\}[M]$$

The product $\nu_{pp}R_p$ of kinetic chain length and polymerization rate should thus be a constant that does not depend on the initiator but only on the two rate constants k_p and

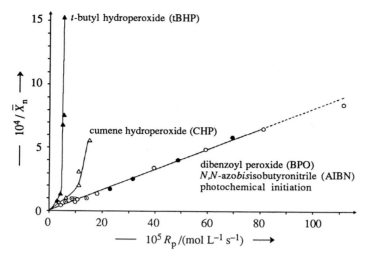

Fig. 10-5 Inverse number-average degree of polymerization as a function of polymerization rate for the polymerization of methyl methacrylate at 60°C to monomer conversions of 1-15 % by initiator concentrations of 0.0004-0.16 mol/L [6]. O BPO, ● AIBN, ⊙ photochemically, △ CHP, ▲ tBHP.

$k_{t(pp)}$ and the monomer concentration [M] (assumed to be constant in ideal kinetics, see above). For an exclusive termination by disproportionation, $v_{pp}R_p$ can be replaced by $\overline{X}_n R_p$ and for an exclusive termination by combination by $\overline{X}_n R_p/2$. Both quantities should be independent of the initiator concentration which is indeed approximately found for the usual initiator concentrations (Fig. 10-4). Eq.(10-46) shows that an increase in the polymerization rate leads to a decrease in the degree of polymerization.

According to Eq.(10-46), $1/v_{pp}$ (and also $1/\overline{X}_n$ if $k_{t(pp)} = const$) should be a linear function of R_p. Such a behavior was found for the polymerization of methyl methacrylate by AIBN or BPO but not by cumene hydroperoxide or t-butyl hydroperoxide (Fig. 10-5). The latter behavior indicates a transfer to the initiator (Section 10.4.5).

Based on Eq.(10-45), the expression for the activation energy for kinetic chain length (and thus also for the degree of polymerization) in polymerizations with mutual deactivation of macroradicals, assuming $f_{eff} \neq f(T)$, is given by

$$(10\text{-}47) \qquad \Delta E_v^{\ddagger} = \Delta E_p^{\ddagger} - (1/2)\,\Delta E_d^{\ddagger} - (1/2)\,\Delta E_{t(pp)}^{\ddagger}$$

For the neat polymerization of styrene by BPO (Section 10.3.4), a value of $\Delta E_v^{\ddagger} = -37$ kJ/mol is obtained. The kinetic chain length (and thus the degree of polymerization) decreases with increasing temperature.

Initiator radicals and macroradicals are formed successively. Macroradicals can furthermore be deactivated by other macroradicals and by initiator radicals. Radicals can also be transferred to other compounds, such as monomers, solvents or additives. Polymers by free-radical polymerization thus have a distribution of degrees of polymerization (see Appendix A-10). In the case of termination by disproportionation, the distribution function of the degrees of polymerization is identical with that obtained by polycondensation to high monomer conversions (Section 13.2.5) since polymer chains are formed at random in both types of polymerization (Section 10.3.2).

10.3.6 Determination of Rate Constants

Rate studies of free-radical polymerizations do not deliver propagation constants k_p *per se* but always combinations of k_p with other constants, for example, $k_p(f_{eff}k_d/k_{t(pp)})^{1/2}$ (Eq.(10-41)). The constants k_p, $f_{eff}k_d$, and $k_{t(pp)}$ can be determined as follows.

The *propagation rate constant* k_p is obtainable from Eq.(10-38) if the concentration of macroradicals, $[P^\bullet]$, can be measured directly by electron spin resonance (ESR). However, this concentration is often too small for ESR measurements (Section 10.3.3).

The *product of initiator efficiency and decomposition rate constant*, $f_{eff}k_d$, can be calculated from the rate $-d[IHB]/dt$ by which an added inhibitor IHB is consumed as a function of the initiator concentration, $[I_2]$:

$$(10\text{-}48) \qquad -\,d[IHB]/dt = f_{eff}k_d[I_2]$$

The *rate constant of mutual deactivation*, $k_{t(pp)}$, cannot be determined separately. It can be calculated from the ratio $k_p/k_{t(pp)}^{1/2}$ (Eq.(10-41)) and the ratio $k_p/k_{t(pp)}$ that is obtained from the **average lifetime** τ^* of a polymer chain. The latter quantity is defined as the ratio of the concentration of macroradicals and the rate of termination by mutual deactivation of two macroradicals. τ^* is obtained from Eqs.(10-37) and (10-27):

$$(10\text{-}49) \qquad \tau^* = \frac{[P^\bullet]}{R_{t(pp)}} = \frac{1}{2\,k_{t(pp)}[P^\bullet]} = \left(\frac{k_p}{k_{t(pp)}}\right)\frac{[M]}{2\,R_p}$$

The average lifetime τ^* is available from various experiments. In principle, it can be obtained from the time that is needed to reach the steady state. This time is, however, on the order of seconds (Section 10.3.3) and therefore difficult to determine.

In photochemically initiated polymerizations, the radical production may also be confined to small volumes of the reactor vessel by constantly sending light through regularly spaced narrow slots. The average lifetime of radicals is then calculated from the diffusion rate of the polymer molecules. In a variant of this method, the light shines intermittently through the whole volume. Both variants are combined in the rotating-sector method. A newer method is the pulsed-laser polymerization (see below).

Rotating-Sector Method

The rotating-sector method measures rates of photochemically initiated polymerizations by a rotating sector (Fig. 10-6) that controls the ratio r of dark and light periods. Focusing the light beam provides sharp transitions between dark and light periods.

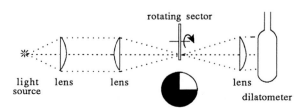

Fig. 10-6 Rotating-sector method (here: $r = 3{:}1$). For viewing, the sector is rotated by 90°.

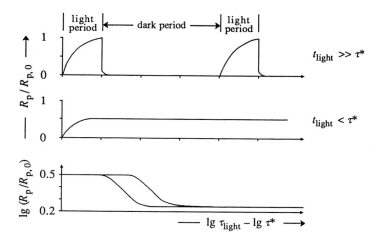

Fig. 10-7 Schematic representation of the relative rate of polymerization, $R_p/R_{p,0}$, as a function of time (top and center) and the difference of logarithms of the length t_{light} of the light period and the lifetime τ^* of radicals. The ratio of dark and light periods is $r = 3:1$.

The polymerization rate is proportional to the square root of the intensity I_{abs} of the absorbed light on steady illumination ($R_{p,0} = K I_{abs}^{1/2}$). The steady state is interrupted by the rotating sector. In the light period, radicals are generated and the polymerization rate approaches the steady rate ($R_p \rightarrow R_{p,0}$), provided the illumination time is much greater than the average lifetime τ^* of radicals. Radicals are terminated in the dark period and the polymerization rate rapidly becomes zero (Fig. 10-7, top). Since the illumination time is inversely proportional to the rotating velocity of the sector, the polymerization rate becomes $R_p = R_{p,0}/(r + 1)$.

At high rotating speeds, illumination times become smaller than the lifetimes of radicals. The polymerization rate never reaches that of the steady state, never is interrupted, and never becomes zero (Fig. 10-7, center). For $r = 3$, the total illumination is only $1/(r + 1) = 1/4$ of that by steady illumination. The polymerization rate is thus

$$(10\text{-}50) \qquad R_p = K(I_{abs}/4)^{1/2} = R_{p,0}/2 = R_{p,0}/(r + 1)^{1/2}$$

and the polymerization rate drops to one-half of that of the steady state. Experimentally, one observes a dependence of the relative polymerization rate on the average illumination time (Fig. 10-7, bottom).

Experimental curves can be compared with theoretical ones. Expansion of the term in square brackets in the equation for steady-state conditions, Eq.(10-35), by $k_{t(pp)}/k_{t(pp)}$ leads to

$$(10\text{-}51) \qquad [P^\bullet]_t/[P^\bullet]_{stat} = \tanh\{(2R_i/k_{t(pp)})^{1/2}k_{t(pp)}t\}$$

Introduction of $[P^\bullet] = [R_i/(2\,k_{t(pp)})]^{1/2}$ (Eq.(10-31)), and the lifetime of a macroradical, $\tau^* = 1/(2\,k_{t(pp)}[P^\bullet])$ (Eq.10-49)), delivers for the end of the light period at time t

$$(10\text{-}52) \qquad [P^\bullet]/[P^\bullet]_{stat} = \tanh(t/\tau^*) = R_p/R_{p,0} \qquad\qquad \text{(light period)}$$

Radicals disappear in the dark period by mutual deactivation but no new ones are formed. The rate of disappearance is thus $-d[P^\bullet]/dt' = 2\,k_{t(pp)}[P^\bullet]^2$ (Eq.(10-27)). Integration of this equation for the radical concentration $[P^\bullet]_1$ at the beginning of the dark period at time t and $[P^\bullet]_2$ for the end of the dark period at time t' delivers with $t' = rt$

$$(10\text{-}53)\qquad \frac{[P^\bullet]_{stat}}{[P^\bullet]_1} - \frac{[P^\bullet]_{stat}}{[P^\bullet]_2} = 2\,k_{t(pp)}[P^\bullet]_{stat}\,t' = \frac{t\,r}{\tau^*} \qquad \text{(dark period)}$$

The equations for the light and dark periods can be combined. A corresponding plot of $R_p/R_{p,stat} = f(\lg t_{light} - \lg \tau^*)$ shows a shift of the experimental curve relative to the theoretical one which allows one to calulate the life time τ^* and thus the rate constant k_p.

An initial polymerization rate of $R_p = 5\cdot10^{-5}$ mol L^{-1} s^{-1} and a number-average degree of polymerization of $\overline{X}_n = 1670$ was observed for the polymerization of styrene in bulk ($[M]_0 = 8.5$ mol/L) at 60°C by 0.005 mol AIBN/L. Experimental rate constants were $k_p = 285$ L mol^{-1} s^{-1} and $k_{t(pp)} = 1.2\cdot10^8$ L mol^{-1} s^{-1}. The average lifetime of a macroradical is thus $\tau^* = 0.40$ s and the time t for the addition of one monomeric unit, $t = (0.40/1670) = 0.24$ ms.

Fig. 10-8 compares propagation rate constants of methyl methacrylate from rotating sector experiments ($\circ$) with those by electron-spin resonance ($\square$), various other methods ($\triangle$), and pulsating laser polymerization ($\blacksquare$, see below). The latter method clearly emerged as the method of choice.

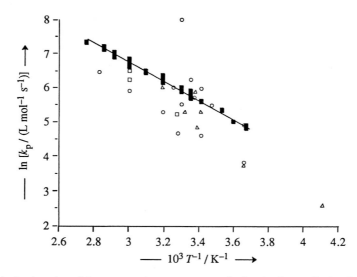

Fig. 10-8 Arrhenius plot of the propagation rate constant k_p for the free-radical polymerization of neat methyl methacrylate at 1 bar. Data from pulsed-laser polymerization ($\blacksquare$), rotating-sector experiments ($\circ$), ESR ($\square$), and various others ($\triangle$). Data from [7] ($\blacksquare$) and [8] ($\circ$, $\square$, $\triangle$).

Pulsating Laser Polymerization

Polymerization by a pulsating laser (**PLP method, Olaj method**), followed by the measurement of the molar mass distribution (MMD) of the specimen, allows one to determine propagation rate constants by a single experiment. The method neglects chain transfer to monomer.

The PLP-MMD method was introduced as "laser-flash polymerization" and also became known as "pulsed-initiation polymerization". In this method, a mixture of monomer and photoinitiator is subjected to periodic laser flashes. The generated initiator radicals add monomer molecules at random so that each flash produces polymer radicals with a certain molar mass distribution. After the flash, the concentration of these radicals declines because of the mutual deactivation of radicals. At the end of the dark period between two flashes, the remaining macroradicals will have a degree of polymerization of

(10-54) $X_1 = k_p[M]\Delta t_o$; Δt_o = time period between flashes

The next flash produces a large concentration of new radicals which terminate with high probability those macroradicals that survived the preceding dark period. The dead polymer molecules will thus have a degree of polymerization of X_1.

The rate constant k_p is calculated from Eq.(10-54) using an X_1 that is taken from the first maximum of the first derivative of the differential mass distribution of the logarithm of relative molecular masses (Fig. 10-9). Such distributions are directly obtained by size-exclusion chromatography (Section 3.4.2). The correlation of this maximum and X_1 is empirical. It is, however, supported by model calculations, simulations, and comparison with k_p values that were obtained by other methods. An internal control of PLP-MMD data is provided by the second maximum $M_{r,2} = M_{r,u}X_2$ which allows one to calculate k_p from $X_2 = 2 \ k_p[M]\Delta t_o$ where $M_{r,u}$ is the relative molar mass of the monomeric unit.

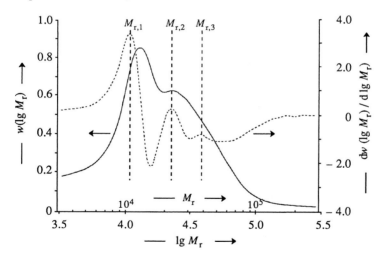

Fig. 10-9 Differential mass distribution —— of decadic logarithms of relative molar masses M_r (left scale) of a poly(styrene) by PLP at 30°C and 1 bar and its first derivative (- - - -, right scale) [9].

10.3.7   Propagation Rate Constants

Propagation rate constants $k_p/(L \ mol^{-1} \ s^{-1})$ of vinyl and acrylic monomers range between 10 and 10^5, while rate constants $k_{t(pp)}/(L \ mol^{-1} \ s^{-1})$ for termination by mutual deactivation are between 10^6 and 10^{11} (Table 10-6). However, although termination constants by far exceed propagation constants, free-radical polymerizations still proceed

Table 10-6 Rate constants k_p of propagation and $k_{t(pp)}$ of termination by mutual deactivation of two macroradicals at 25°C as well as the corresponding activation energies. * By PLP-MMD.

Monomer	Solvent	$\dfrac{k_p}{\text{L mol}^{-1}\text{ s}^{-1}}$	$\dfrac{k_{t(pp)}}{\text{L mol}^{-1}\text{ s}^{-1}}$	$\dfrac{\Delta E_p^{\ddagger}}{\text{kJ mol}^{-1}}$	$\dfrac{\Delta E_{t(pp)}^{\ddagger}}{\text{kJ mol}^{-1}}$
Styrene	-	86 *	6 000 000	32.5 *	2
Methyl methacrylate	-	325 *	30 000 000	22.4 *	6
	benzene	260	21 000 000		
	benzonitrile	330	17 000 000		
Vinyl acetate	-	1 000	200 000 000	29	21
Vinyl chloride	-	3 100	2 300 000 000		
Butyl acrylate	-	15 200 *	180 000 000	17.4 *	
Acrylonitrile	-	90	9 000 000		
	DMF	380	48 000 000		
	water	21 000	2 800 000 000		
Acrylamide	water	18 000	15 000 000		
Methacrylamide	water	800	17 000 000		

because the usual usual monomer concentrations of $[M]/(\text{mol L}^{-1}) = 10^{-1}\text{-}10$ are much larger than the macroradical concentrations of $[P^{\bullet}]/(\text{mol L}^{-1}) = 10^{-7}\text{-}10^{-9}$. Propagation rates $R_p = k_p[P^{\bullet}][M]$ are thus much greater than termination rates $R_{t(pp)} = 2\,k_{t(pp)}[P^{\bullet}]^2$.

The propagation rate constants of non-resonance stabilized monomers (e.g., vinyl acetate, vinyl chloride) are greater than those of resonance stabilized ones (e.g., styrene, methyl methacrylate) (Table 10-6). The larger the substituents, the smaller the rate constants of propagation and termination. 1,1-Disubstituted monomers $CH_2=CRR'$ always polymerize much more slowly than the corresponding monosubstituted ones (examples: acrylamide-methacrylamide, vinyl chloride-vinylidene chloride). 1,2-Disubstituted monomers (e.g., dialkyl fumarates) polymerize only if they carry large substitutents.

It is generally believed that rate constants are little or not at all influenced by solvents. This may be true for apolar solvents (Table 10-6) but these monomers are usually polymerized near their compensation temperatures (p. 179, 323) where solvent effects are not expected (methyl methacrylate: $T_0 = 57°C$). Solvent effects are very pronounced for polar monomers such as acrylonitrile (Table 10-6). Another example is the ionizable monomer 1,2-dimethyl-5-vinylpyridiniummethylsulfate (IV) with a propagation rate constant $k_p/(\text{L mol}^{-1}\text{ s}^{-1})$ of 5000 in acetic acid but only 300 in $CH_3OH:H_2O$ (1:1).

Polymers by free-radical polymerizations are usually atactic to slightly syndiotactic (p. 322), even if the substitutent is as large as in I. Monomer II shows, however, $x_i = 0.6$ and monomer III even $x_i = 0.98$, probably due to the helical chain structure of their polymers that is caused by the steric hindrance between the bulky substituents.

Activation energies of propagation steps do not vary much with the constitution of monomers (Table 10-6). Differences in propagation rate constants are therefore mainly caused by differences in the pre-exponential factor of the Arrhenius equation.

Activation energies of termination vary with the type of termination. In the combination of two macroradicals, no mass is transferred and the activation energy is small. It is not zero, however, since the combination of two radicals causes a spin reversal. The low activation energy of termination of styrene macroradicals is caused predominantly by combination (77 % at 60°C). Macroradicals of vinyl acetate, on the other hand, terminate mainly by disproportionation (100 % at 60°C). The activation energy of termination by disproportionation is high (Table 10-6) because mass is transferred.

10.3.8 Dead-End Polymerization

Free-radical polymerizations proceed only to a monomer conversion of $u_{\infty,1} < u_{compl}$ = 100 % because all initiator molecules are decomposed before all monomer molecules are polymerized. New addition of initiator rekindles the polymerization and the polymerization now proceeds to a new maximum conversion of $u_{\infty,2} < u_{compl}$. The maximum monomer conversions $u_{\infty,1}$, $u_{\infty,2}$, etc., are not thermodynamically controlled.

The polymerization thus comes to a dead end which is calculated as follows. Combination of equations for initiation by thermal decomposition of the initiator (Eq.(10-13)) and propagation (Eq.(10-41)) delivers after separation of the variables

$$(10\text{-}55) \qquad -\,d \ln [M] = k_p (f_{eff} k_d / k_{t(pp)})^{1/2} [I_2]_0^{1/2} \{\exp(-k_d t)\}^{1/2} dt$$

and, for $[M] = [M]_0(1 - u)$ on integration from $[M]_0$ at $t = 0$ to $[M]$ at $t = t$,

$$(10\text{-}56) \qquad -\ln(1 - u) = k_p \left(\frac{f_{eff} k_d}{k_{t(pp)}} \right)^{1/2} [I_2]_0^{1/2} \left\{ \frac{1 - \exp(-k_d t/2)}{k_d/2} \right\}$$

At infinitely long times, $u \rightarrow u_{\infty}$, and $t \rightarrow \infty$, Eq.(10-56) becomes

$$(10\text{-}57) \qquad -\ln(1 - u_{\infty}) = 2\,k_p \left(\frac{f_{eff}}{k_d k_{t(pp)}} \right)^{1/2} [I_2]_0^{1/2}$$

Example: Polymerization of styrene in bulk with $[I_2]_0 = 0.0582$ mol AIBN/L (1 wt% AIBN) and $f_{eff} = 1/2$, $k_p = 285$ L mol^{-1} s^{-1}, $k_d = 1.35 \cdot 10^{-5}$ s^{-1}, and $k_{t(pp)} = 1.2 \cdot 10^8$ L mol^{-1} s^{-1}. The maximum monomer conversion is therefore 91.1 % ($u_{\infty} = 0.911$). Doubling the initiator concentration increases the maximum monomer conversion of 96.7 % but reduces the number-average molecular weight by a factor of $2^{1/2}$ (Eq.(10-46)).

In principle, dead-end polymerizations allow the determination of rate constants k_d of initiator decomposition. Dividing Eq.(10-56) by Eq.(10-57) delivers

$$(10\text{-}58) \qquad [\ln(1 - u)] / [\ln(1 - u_{\infty})] = 1 - \exp(-k_d t/2)$$

$$(10\text{-}59) \qquad -\ln\left[1 - \frac{\ln(1-u)}{\ln(1-u_{\infty})} \right] = \frac{k_d}{2} t$$

A plot of the left side of Eq.(10-59) against time should thus deliver a straight line with a slope of $k_d/2$ which is indeed observed. However, decomposition constants k_d by this method are usually lower than the ones observed directly (Section 10.2.4) because the high initiator concentration leads to transfer and termination reactions by initiator radicals (Section 10.4.5). For example, the dead-end polymerization of isoprene delivers $k_d' = 0.44 \cdot 10^{-5}$ s^{-1} whereas the true decomposition rate constant is $k_d = 1.35 \cdot 10^{-5}$ s^{-1}.

10.3.9 Gel and Glass Effect

Ideal polymerization kinetics applies to small monomer conversions. But even then the relative monomer conversion, $u = ([M]_0 - [M])/[M]_0$, is not a linear function of time as one can see from Eq.(10-41), $u = k_p\{f_{eff}k_d/k_{t(pp)}\}^{1/2}[I_2]^{1/2} ([M]/[M]_0)t$. The relative monomer conversion rather increases less than proportionally with time since the monomer concentration [M] is getting smaller and smaller. An example is the polymerization of a 10 % solution of methyl methacrylate (Fig. 10-10).

Ideal polymerization kinetics also predicts that the curves $u = f(t)$ are independent of the absolute monomer concentration, provided, of course, that types and relative proportions of elementary reactions do not change. However, Fig. (10-10) shows that the relative change of u with time is greater for a 40 % monomer solution than a 10 % one although the relative shape of the curves is similar. A 60 % monomer solution shows after some time a drastic increase in the relative monomer conversion. This effect occurs at shorter times and is more pronounced, the higher the initial monomer concentration.

The self-acceleration is called **gel effect, Trommsdorff effect, Trommsdorff-Norrish effect**, or **Norrish-Smith effect**. It is characterized by an initial decrease of the polymerization rate R_p as a function of monomer conversion, followed by an increase and then a steep decrease of R_p (Fig. 10-11). The number-average degree of polymerization stays constant at first and then increases sharply (Fig. 10-11). The radical concentration is also constant at first, then increases, and finally becomes constant again (Fig. 10-12).

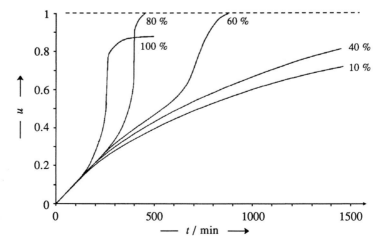

Fig. 10-10 Self-acceleration during the polymerization of methyl methacrylate in bulk and in benzene solutions of different concentrations at 50°C by 10 g/L dibenzoyl peroxide as initiator [10].

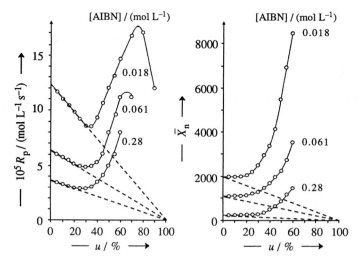

Fig. 10-11 Polymerization rate, R_p, and number-average degree of polymerization, $\overline{X}_n$, as function of relative monomer conversion, u, for the styrene polymerization in bulk by AIBN at 50°C [11].
- - - - Predictions of ideal polymerization kinetics.

These effects are observed for isothermal polymerizations; they are thus not caused by heat build-up. They are stronger, the more viscous the polymerization system; i.e., they increase at higher monomer conversions, with lower initiator concentrations, higher monomer concentrations, addition of inert polymers, and in thermodynamically bad solvents (at *high* polymer concentrations, viscosities are *higher* in bad solvents than in good ones (see Volume III) whereas at low polymer concentrations, they are lower). The influence of viscosity points toward a diffusion effect.

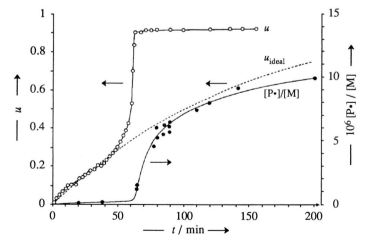

Fig. 10-12 Time dependence of relative monomer conversion, u, and reduced macroradical concentration, $[P^\bullet]/[M]$, during the polymerization of neat methyl methacrylate ($[M]_0 = 9.25$ mol/L) at 60°C by an initiator concentration of $[AIBN]_0 = 0.1$ mol/L [12]. - - - - Ideal polymerization kinetics (no gel effect). At the onset of the gel effect, the concentration $[P^\bullet]$ of macroradicals increases strongly. It runs through a maximum at $t \approx 500$ min and then decreases (not shown).

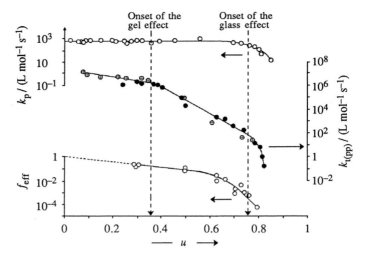

Fig. 10-13 Propagation rate constants k_p, termination rate constants $k_{t(pp)}$, and initiator efficiencies f_{eff} as a function of relative monomer conversion u. Polymerization of neat methyl methacrylate at 60°C by photolysis of $[I_2]_0 = 0.109$ mol/L dimethyl-2,2'-azodiisobutyrate (O), at 60°C by thermal decomposition of $[I_2]_0 = 0.337$ mol/L dicumene peroxide (⊕), and at 26°C by thermal decomposition of $[I_2]_0 = 0.335$ mol/L dibenzoyl peroxide (●) [13]. Broken lines indicate onsets of gel and glass effects.

The strong increase of concentration of macroradicals with time (Fig. 10-12) may be suspected to come from an increased production of initiator radicals (by heat build-up or induced decomposition), but the initiator efficiency decreases (not increases) with increasing relative monomer conversion (Fig. 10-13) because of the cage effect (Section 10.2.3). The mutual deactivation of two macroradicals, on the other hand, must be hindered at small u since the termination rate constant $k_{t(pp)}$ decreases even at small monomer conversions (Fig. 10-13). $k_{t(pp)}$ decreases more strongly with u after the onset of the gel effect and even more dramatically after the onset of the **glass effect** caused by the freezing-in of motions of polymer segments.

The termination rate constant must therefore be diffusion controlled since both the gel and the glass effect depend strongly on the viscosity. Since $k_{t(pp)} = f(u)$ at even the smallest monomer conversions (Fig. 10-13), the self-acceleration of the polymerization cannot come from a sudden *onset* of diffusion control but must be due to a *change* of that control. Two main hypotheses have been advanced to explain this behavior: chain entanglement (p. 149) and decrease of free volume (unoccupied regions of ca. atomic diameter, Volume III).

Entanglements may be acting because both gel effects and entanglements increase with increasing polymer concentration and molecular weight. The addition of inert, low-molecular weight polymer to the polymerizing system should reduce entanglements, but here the gel effect sets in *earlier* and at *lower* monomer conversions. A decrease of free volume during polymerization is appealing since large intermolecular distances between monomer molecules are replaced by short intramolecular bond lengths between chain atoms (Section 6.5.1) which should hinder local motions of chain segments. However, it was found that gel effects set in earlier in thermodynamically bad solvents than in good ones although both had the same free volume. Hence, it seems that it is not either-or but that both effects may act simultaneously.

The strong decrease in termination rate constants with increasing monomer conversion (Fig. 10-13) leads to a strong increase in kinetic chain lengths (Eq.(10-45)) and thus to a large increase in degrees of polymerization (Fig. 10-11). The dead polymer molecules stemming from the deactivation of the now much larger macroradicals create another type of molecular weight distribution. The total molecular weight distribution becomes bimodal and much broader.

At small initial initiator concentrations, polymerization rates often pass through a maximum (Fig. 10-11) because the polymer-monomer-(solvent) mixture solidifies to an amorphous "glassy" mass. This **glass effect** causes a freezing of the movements of chains, chain segments, and monomer molecules. No longer can the latter diffuse easily to macroradicals. The propagation rate constant k_p decreases (Fig. 10-13) and thus also the polymerization rate (Fig. 10-11). Finally, the polymerization stops without ever attaining a monomer conversion of 100 % (Fig. 10-10 and 10-12).

10.3.10 Frontal Polymerizations

The gel effect is utilized in so-called **isothermal frontal polymerizations (IFP, interfacial gel polymerizations)** in which a neat monomer containing a free-radical initiator and an inhibitor ("liquid") is layered over a seed polymer (Fig. 10-14, left). The inhibitor delays the start of polymerization so that the liquid has time to diffuse into the polymer which is transformed into a gel. The polymerization proceeds fast in the gel because of the gel effect but is much slower in the liquid so that a propagating front is formed. The resulting polymer has a gradient refractive index (GRIN). It is used for magnification or focalizing, for example, in gradient contact lenses.

In **non-isothermal frontal polymerizations**, an initiator-containing, high-boiling monomer is heated from above by an external heat source (Fig. 10-14, right). The exothermic polymerization leads to a very narrow propagating front that travels fast downward with velocities up to 20 cm per minute. The method can be used to synthesize interpenetrating networks, polymer blends, and *functional gradient materials* (**FGMs**).

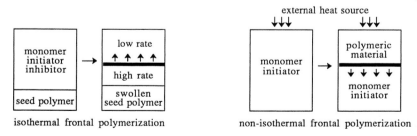

Fig. 10-14 Isothermal and non-isothermal frontal polymerizations. ━━ Advancing front.

10.3.11 Non-Ideal Kinetics

Ideal polymerization kinetics assumes a steady state and predicts the polymerization rate to depend on the first power of monomer concentration and on the square root of initiator concentration (Eq. (10-41)). These proportionalities do not apply to non-steady state conditions such as the gel and glass effects. Additional deviations from ideal conditions stem from a number of other effects. They are usually all described globally by a power expression:

(10-60) $R_p = - d[M]/dt = k[I_2]^p[M]^q$

Effect of Monomer Concentration

Polymerization rates are directly proportional to monomer concentrations only as long as the rate of formation of initiator radicals does not depend on the monomer concentration. However, monomers may induce the decomposition of initiators. If this dominates the thermal decomposition, the decomposition rate will be $R_d = - d[I_2]/dt = k_d[I_2][M]$ instead of $R_d = k_d[I_2]$ (Eq.(10-13)). The polymerization rate will then be proportional to $[M]^{3/2}$ (p = 3/2 in Eq.(10-60)) and not to $[M]$ (cf. the derivation of Eq.(10-41)). k_d may also depend on the solvent (if any) and its concentration.

In non-ideal systems, the exponent q of the monomer concentration in Eq.(10-60) may by smaller or larger than 1 which may be caused by a termination of macroradicals by primary (initiator) radicals, a diffusion-controlled mutual deactivation of two macroradicals, or by an induced decomposition of the initiator.

The mutual deactivation of macroradicals is even diffusion controlled at small monomer conversions (Fig. 10-13) and at very low degrees of polymerization (Fig. 10-15). Since diffusion coefficients are inversely proportional to viscosities (Volume III), termination rate constants decrease with increasing viscosities (Fig. 10-15). A correction for the effect of viscosity η of the polymerizing system on the polymerization rate can be made by assuming $k = k_0\eta^{-b}$ for the proportionality constant of Eq.(10-60). This correction is not sufficient for solution polymerizations and an additional correction must by made for preferential solvation since the concentration of monomer molecules near propagating polymer radicals is different from the average concentration (Table 10-7).

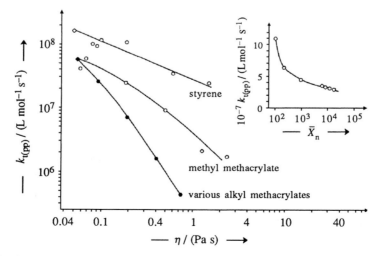

Fig. 10-15 Viscosity dependence of rate constants $k_{t(pp)}$ of styrene at 20°C [14], methyl methacrylate at 20°C [14], and alkyl methacrylates at 25°C [15]. Insert: dependence of $k_{t(pp)}$ on the number-average degrees of polymerization during the polymerization of methyl methacrylate in bulk at 25°C [14].

Table 10-7 Exponents q of monomer concentrations in Eq.(10-60) for polymerizations of alkyl methacrylates, $CH_2=C(CH_3)(COOR)$, in benzene at 30°C [16]. b = exponent in $k_{t(pp)} = k_{(t,pp),0}\eta^{-b}$; q = exponents in Eq.(10-60) before corrections (q_{exp}) and after correction for viscosity (q_v) and viscosity and preferential solvation (q_{vs}).

Substituent R	b	q_{exp}	q_v	q_{vs}
Methyl	1.0	1.0		
Butyl	0.8	1.45	1.3	0.96
i-Butyl	0.35			
Dodecyl	0	1.8		1.01
Hexadecyl	0	1.9		

Effect of Initiator Concentration

Similar considerations apply to the dependence of polymerization rate on initiator concentration. Exponents p > 1/2 in Eq.(10-60), as found for initiation by dibenzoyl peroxide, may be caused by a monomer-induced initiator decomposition that can be described by

(10-61) $- d[I_2]/dt = k_d[I_2] + k_{ind}[I_2]^{\beta}$

Exponents β may vary between 1/2 and 2, depending on the mechanism of the induced decomposition. If $\beta = 2$ and if the induced decomposition is much faster than the thermal one, then $k_{ind}[I_2]^2 \gg k_d[I_2]$ and

(10-62) $R_p = k_p(f_{eff}k_{ind}/k_{t(pp)})^{1/2}[M][I_2] = const\ [M][I_2]$

The polymerization rate no longer depends on the square root of the initiator concentration. Note that Eq.(10-62) is formally identical with the kinetic equation for the rate of propagation of a living polymerization.

Effect of Monomer Conversion

Eq.(10-45) for ideal polymerization kinetics predicts that the kinetic chain length v_{pp} is proportional to the monomer concentration [M] and inversely proportional to the square root of initiator concentration. The kinetic chain length is proportional to the number-average degree of polymerization ($v_{pp} = q\bar{X}_n$) if the type or proportion of mutual deactivation does not change with time ($q = 1$ for disproportionation; $q = 1/2$ for combination) (Section 10.3.5). $\bar{X}_n$ should thus stay constant as long as $[M]/[I_2]^{1/2}$ remains constant (Eq.(10-45)). Since $[M] = [M]_0(1 - u)$, $\bar{X}_n$ should decrease linearly with the relative monomer conversion u if $[I_2] \approx [I_2]_0$ which applies to Fig. 10-11 because of the very long half-time of decomposition of AIBN at 50°C (Table 10-2).

Experimentally, number-average degrees of polymerization increase first slightly and then sharply with increasing monomer conversion (Fig. 10-13). A part of this effect is caused by the decrease of the termination constants with conversion (p. 340). Another part stems from a radical transfer to polymer molecules which generates branched polymers. The tendency toward such reactions increases with the number of transferable entities per monomeric unit (Section 10.4). Increase of branching lowers the probability of entanglement; gel effects become less pronounced. Hexyl, dodecyl, and hexadecyl methacrylates thus show no gel effect whereas methyl methacrylate has a strong one.

Kinetics, degrees of polymerization, and polymer constitutions may also be changed by a number of other reactions. Certain polymerization conditions may, for example, lead, in a so-called **popcorn polymerization**, to cauliflower-like polymers that are sometimes, but not always, crosslinked.

10.3.12   Crosslinking Polymerizations

Monomer *molecules* with functionalities of three and higher produce crosslinked polymers at higher monomer conversions. Such polymerizations require two or more bifunctional vinyl, vinylidene, acryl, methacryl, allyl etc. *groups* per monomer molecule. Crosslinking may also be caused by radical transfer to polymers.

The course of the polymerization depends on whether the polymerizable groups are part of a resonance system (such as 1,3-dienes or divinylbenzenes) or not (such as glycol dimethacrylate or diallyl compounds). In the first case, the polymerization of the first group changes the reactivity of the second one so that these polymerizations can be viewed as intramolecular copolymerizations of groups with different reactivities. In the second case, both groups have the same reactivity if the reaction of the second group is not diffusion controlled.

At low monomer conversions of N-functional monomer molecules ($N = 4, 6, 8$, etc.), probabilities are small for an incorporation of the second, third ... group into the same polymer chain; these polymer molecules are not at all or only slightly branched. The polymerizable functional groups in these polymer molecules are side-groups that can be attacked by radicals. The probability of the attack is higher, the greater the mass-average degree of polymerization since it relates to the number of monomeric units whereas the number-average relates to the number of polymer molecules. The change of the weight-average degree of polymerization with the relative monomer conversions is thus

$$(10\text{-}63) \qquad d\,\overline{M}_w/du = const\cdot\overline{M}_w \qquad ; \qquad \ln \overline{M}_w = \ln \overline{M}_{w,0} + const\cdot u$$

The natural logarithm of the weight-average degree of polymerization should thus increase linearly with relative monomer conversion (Fig. 10-16) which is observed for both inherently multifunctional monomer molecules (e.g., diallyl phthalate) and mixtures of bifunctional and multifunctional ones (e.g., styrene + divinyl benzene). At these relatively small monomer conversions, polymers consist of branched polymer molecules with many polymerizable groups and an ever-increasing number of radicals per molecule. The combination of radicals of two different molecules and the polymerization of incorporated functional groups by polymer radicals leads to crosslinking at higher monomer conversions. The polymer "gels."

The **gel point** is defined as that relative conversion p_{gel} of *functional groups* at which the weight-average molecular weight of the polymer molecules or the viscosity of the polymerizing system tends toward infinity (see Sections 13.9.1 and 13.9.7). The relative

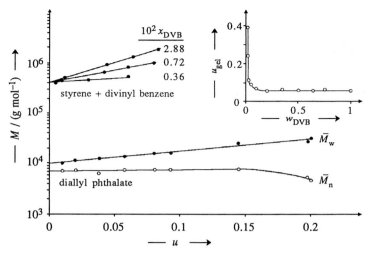

Fig. 10-16 Weight-average (●) and number-average (○) molecular weight as a function of the relative monomer conversion u. Polymerization of neat styrene with various mole fractions x_{DVB} of 1,4-divinyl benzene (DVB) [17] and of neat diallyl phthalate at 60°C [18]. Insert: Monomer conversion u_{gel} at the gel point as a function of the initial mass fraction w_{DVB} during the polymerization of neat styrene with divinyl benzene by 1 wt% BPO at 89.7°C [19].

molecule conversion at the gel point, u_{gel}, and the extent of reaction, p_{gel}, of functional groups are related to the fraction γ_{gel} of unsaturated groups by $p_{gel} \equiv (1 - \gamma_{gel})u_{gel}$. The gel point u_{gel} decreases at first strongly with increasing concentration of added multifunctional monomer molecules but then becomes constant (Fig. 10-16, insert).

A fraction γ_{cycl} of functional side groups is wasted by intramolecular ring closures. The network formation is thus controlled only by the fraction $p_{eff} = (1 - \gamma_{cycl})p$. At the gel point, this fraction becomes $p_{eff,gel} = (1 - \gamma_{cycl,gel})p_{gel}$.

The greater the molecular weight $\overline{M}_{w,0}$ of primary polymer molecules (those formed at very small monomer conversions), the more functional side groups are present and the stronger is the increase of the molecular weight $\overline{M}_w$ with increasing conversion. This behavior can be described by $1 - (\overline{M}_{w,0}/\overline{M}_w) = p_{eff}/(p_{eff})_{gel}$ because of the limiting conditions $\overline{M}_w = \overline{M}_{w,0}$ at $p_{eff} = 0$ and $\overline{M}_{w,0}/\overline{M}_w = 0$ at $p_{eff} = (p_{eff})_{gel}$. Rearrangement of this equation and introduction of expressions for p_{eff} and $p_{eff,gel}$ results in

$$(10\text{-}64) \qquad \overline{M}_w = \overline{M}_{w,0}\left[1 - \frac{(1-\gamma_{cycl})p}{(1-\gamma_{cycl,gel})p_{gel}}\right]^{-1}$$

Equation (10-64) reduces to

$$(10\text{-}65) \qquad \overline{M}_w \approx \overline{M}_{w,0}\left[1 - (p/p_{gel})\right]^{-1}$$

if the fraction of intramolecular ring formations is small or constant over the total range of conversion from zero to the gel point. A plot of $\overline{M}_w$ vs. lg $[1 - (p/p_{gel})]$ should thus lead to a straight line with a slope of -1 which was indeed observed for the polymerization of diallyl phthalate to small monomer conversions (up to ca. 50 % of the group conversion needed for gelling) (Fig. 10-17).

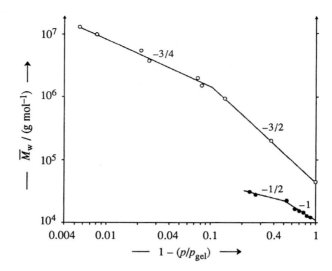

Fig. 10-17 Weight-average molecular weight as a function of the reduced group conversion for the polymerization of diallyl phthalate (●) [18] and the copolymerization of 83.3 wt% methyl methacrylate and 14.61 wt% ethylene dimethacrylate by 0.02 wt% AIBN and addition of 2.04 wt% 1-dodecanethiol (○) [20]. Group conversions increase from right to left. Numbers indicate slopes.

At group conversions greater than ca. 50 %, the slope increases to $-1/2$ from -1. A similar increase by a factor of 2 was also observed for the crosslinking copolymerization of methyl methacrylate and ethylene dimethacrylate albeit for an initial slope of $-3/2$ instead of -1 (Fig. 10-17).

These experimental findings, however, do not agree with theoretical predictions. Experimental gel points are usually always higher than theoretical ones: a factor of 7 for diallyl phthalate and factors up to 100 for other multifunctional monomers.

The failure of theory has several reasons. Theoretical predictions for the whole range of monomer conversions are difficult because the size of primary polymer molecules before the gel point changes with increasing monomer conversion. In copolymerizations, monomer compositions are also changing. Theories thus usually apply only for the gel point. They deviate from experiment here because the thermodynamically excluded volume (Section 5.3.2 of Volume III) effects intermolecular crosslinking between growing primary polymer radicals.

10.4   Chain Transfer

10.4.1   Introduction

"Transfer" has two different meanings: either the transfer of atoms or groups to another molecule (atom or group transfer) or the transfer of the kinetic chain from one active molecule to another non-active molecule (chain transfer). For example, disproportionation reactions of two macroradicals involve the chemical transfer of an atom (sometimes a group). On the other hand, chain transfer in free-radical polymerization comprises the exchange of an electron of the macroradical with an atom X of another non-active molecule AX according to $\sim CH_2-{}^{\bullet}CHR + AX \rightarrow \sim CH_2-CHRX + A^{\bullet}$, resulting in a dead macromolecule and a new, *active* radical $A^{\bullet}$ that initiates the polymerization of monomer molecules. Such transfers can occur to all molecules of a polymerizing system: polymers, monomers, initiators, solvents, etc. Transfer of a lone electron to another molecule that does not become active is kinetically not a transfer but a termination reaction (Eq.(10-29) and Section 10.4.3, last paragraph).

Transfer reactions by deliberately added compounds such as solvents, inhibitors, etc. are usually not called "transfers" but are given special names (Table 10-8). Since actions depend on the relative reactivities of the reaction partners, one and the same compound can act as telogen, retardant, or inhibitor.

Table 10-8 Kinetic classification of chemical transfers.

Reactivity of the new radical	Name of the reaction	
	without additive	with additive
equally reactive	transfer	telomerization
less reactive	retardation	retardation
inactive	degradative chain transfer	inhibition

10.4.2 Mayo Equation

The transfer of an electron from a macroradical to a molecule of a monomer, initiator, solvent, or additive terminates the macromolecule and creates a radical that starts a new macromolecule (note that an electron transfer to a polymer chain leads to branching but not to a radical that starts a new macromolecule!). The number-average degree of polymerization of all macromolecules in the system is given by the ratio of the rate R_p of the propagation reaction to the rates of all reactions that terminate a macroradical, i.e., by termination (index t) caused by mutual deactivation of two macroradicals (pp) and/or by transfer (tr) to monomer molecules (m) or other molecules (ax):

$$(10\text{-}66) \qquad \overline{X}_n = \frac{R_p}{R_{t(pp)} + R_{tr(m)} + R_{tr(ax)}}$$

In the steady state, the rate $R_{t(pp)}$ of termination by mutual deactivation of two macroradicals equals the rate R_i of the initiation reaction. The rate of transfer to monomer is given by $R_{tr(m)} = k_{tr(m)}[P^{\bullet}][M]$ and that to an additive AX by $R_{tr(ax)} = k_{tr(ax)}[P^{\bullet}][AX]$. The ratio R_p/R_i equals the degree of polymerization, $\overline{X}_{n,0}$, if transfer to monomer or additives is absent. Since $R_p = k_p[P^{\bullet}][M]$, Eq.(10-66) transforms into the **Mayo equation**

$$(10\text{-}67) \qquad \frac{1}{\overline{X}_n} - \frac{1}{\overline{X}_{n,0}} = \frac{k_{tr(m)}}{k_p} + \frac{k_{tr(ax)}}{k_p}\frac{[AX]}{[M]} = C_m + C_{ax}\frac{[AX]}{[M]}$$

where C_m and C_{ax} are **transfer coefficients** of monomer and additive.

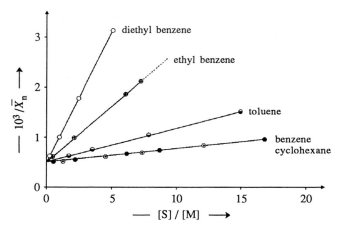

Fig. 10-18 Mayo plot [21]. Dependence of inverse number-average degrees of polymerization on dilution ratios [S]/[M] of solvents S (= AX) to styrene M at 100°C.

A plot of $1/\overline{X}_n$ against the dilution ratio [AX]/[M] delivers $(\overline{X}_n^{-1} + C_m)$ as intercept and C_{ax} as slope (Fig. 10-18). For the data of Fig. 10-18, $\overline{X}_{n,0} \approx 2000$ and $C_m = 6 \cdot 10^{-5}$ so that C_m can be neglected.

Alternatively, transfer coefficients may also be determined directly from the consumption of the transfer agent:

(10-68) $d[AX]/d[M] = C_{ax}[AX]/[M]$

The transfer coefficient of the monomer, $C_m \equiv k_{tr(m)}/k_p$, can be obtained similarly from the initiator concentration $[I_2]$ and measurements of the number-average degree of polymerization, $\overline{X}_n$, and the polymerization rate, R_p, by using Eq.(10-66) with $R_{t(pp)} = R_{st}$, Eq.(10-10), $R_{tr(ax)} = 0$, $R_{tr(m)} = k_{tr(m)}[P^\bullet][M]$, and Eq.(10-38):

(10-69) $\dfrac{1}{\overline{X}_n} = C_m + 2 f_{eff} k_d \dfrac{[I_2]}{R_p}$

Eqs (10-67) and (10-68) can only be used if the additive AX does not affect the initiation reaction, the rate constants depend neither on the size of macroradicals nor on the concentration of AX, and newly formed radicals do not terminate the chain.

Because these restrictions are often not noticed, many reported transfer coefficients are distorted. For example, the radical transfer to dihydrogen sulfide generates radicals $HS^\bullet$ that initiate the polymerization of styrene, $CH_2{=}CHR$. However, the resulting dimer radicals $HS{-}CH_2{-}CHR{-}CH_2{-}{}^\bullet CHR$ undergo an intramolecular **secondary transfer**

(10-70)

$$
\begin{array}{ccc}
\overset{\displaystyle H_2}{\underset{\diagup\diagdown}{C{-}\overset{\bullet}{C}HR}} & & \overset{\displaystyle H_2}{\underset{\diagup}{C{-}CH_2R}} \\
HCR\quad\ H & \longrightarrow & HCR \\
\underset{\displaystyle H_2}{C{-}S} & & \underset{\displaystyle H_2}{C{-}S^\bullet}
\end{array}
$$

that leads to polymers of the type $H{-}(CHR{-}CH_2)_2{-}S{-}(CH_2{-}CHR)_{i\sim}$.

Secondary transfers are common to all multifunctional transfer agents. Their reported transfer coefficients do not relate to the primary transfer agent but to the secondary products. For this reason, and also because of the high consumption of transfer agents with large $C_{tr(ax)}$, transfer coefficients should always be determined at very low monomer conversions. A transfer coefficient of 6000 means, for example, that 55 % of the agent has already reacted at a monomer conversion of only 0.1 % if monomer and transfer agents were used in equal initial concentrations.

Transfer reactions (except those to polymers) generate new types of endgroups in polymer molecules. The proportion of such groups from the transfer to monomer and other molecules AX decreases with increasing monomer concentration.

10.4.3 Transfer of Radicals to Monomer Molecules

Transfer coefficients of vinyl and acrylic monomers are usually in the range $10^{-4} \geq C_m \geq 10^{-5}$ (Table 10-9): 10 000 to 100 000 propagation steps are thus followed by a transfer step. The newly formed monomer radicals then initiate new polymer chains.

The transfer of an H atom from an allylic monomer to a macroradical does not lead to kinetic chain transfer, however, but to chain termination (p. 324). Monoallyl compounds $CH_2=CH-CH_2R$ (including propene) polymerize only sluggishly and only to low molecular weights because a resonance-stabilized monomer radical is formed:

$$(10\text{-}71) \quad \text{\textasciitilde}CH_2-\overset{\bullet}{C}H + CH_2{=}CH \longrightarrow \text{\textasciitilde}CH_2-CH_2 + \left[\begin{array}{c} CH_2{=}CH-\overset{\bullet}{C}HR \\ \updownarrow \\ {}^{\bullet}CH_2-CH{=}CHR \end{array} \right]$$
$$\quad\quad\quad\quad CH_2R \quad\quad CH_2R \quad\quad\quad CH_2R$$

The resonance stabilization *per se* is not the cause of the kinetic chain termination, however. Poly(styryl) radicals, for example, are resonance stabilized but still furnish poly(styrene)s with molecular weights in the millions because the addition of a monomer molecule to a monostyryl radical leads to a macroradical with the same resonance stabilization. The addition of an allyl monomer to an allyl monomer radical, on the other hand, results in a more energy-rich dimer radical.

The energy increase from monomer radical to dimer radical cannot be the only reason for the difficult polymerization of monoallyl compounds. Another reason seems to be that allyl monomers $CH_2=CH-CH_2R$ (R = Cl, OH, OOCCH$_3$, etc.) do not possess strong electron donor or acceptor groups. The charged monoallyl ammonium chloride, $[CH_2{=}CH-CH_2NH_3]^{\oplus}Cl^{\ominus}$, can be polymerized to high-molecular weight polymers by the ionic initiator, $^{\ominus}Cl^{\oplus}[H_3N{=}C(NH_2)-C(CH_3)_2-N{=}N-C(CH_3)_2-C(NH_2){=}NH_3]^{\oplus}Cl^{\ominus}$.

The polymerization of uncharged monoallyl monomers thus proceeds by a strong lowering of degrees of polymerization, often to not more than $\overline{X}_n = 15\text{-}25$. The reaction is thus called **degradative chain transfer**. Macroradicals are here generated with a rate of R_i and terminated by chain transfer to the monomer with a rate of $R_{t(pm)}$. In the steady state, the change $d[P^{\bullet}]/dt$ of radical concentration with time is zero: $R_{st} - R_{t(pm)} = k_{st}[I^{\bullet}][M] - k_{t(pm)}[P^{\bullet}][M] = 0$. Replacing $R_{st} = k_{st}[I^{\bullet}][M]$ by $R_i = 2 f_{eff}k_d[I_2]$ (Eq.(10-10)), solving for $[P^{\bullet}]$, and introducing Eq.(10-38) results in

$$(10\text{-}72) \quad R_p = -\,d[M]/dt = (2 f_{eff}k_d k_p/k_{t(pm)})[I_2]$$

The initial polymerization rate thus does not depend on the monomer concentration but it is directly proportional to the initiator concentration. The activation energy

$$(10\text{-}73) \qquad \Delta E_{gross}^{\ddagger} = \Delta E_d^{\ddagger} + \Delta E_p^{\ddagger} - \Delta E_{t(pm)}^{\ddagger}$$

and thus the temperature dependence of polymerization rates, is far greater than that of common free-radical polymerizations. Typical activation energies are $\Delta E_d^{\ddagger} = 130$ kJ/mol, $\Delta E_p^{\ddagger} = 25$ kJ/mol, and $\Delta E_{t(pm)}^{\ddagger} = 25$ kJ/mol, resulting in $\Delta E_{gross}^{\ddagger} = 130$ kJ/mol.

The kinetic chain length of monoallyl polymerizations

$$(10\text{-}74) \qquad v_{pm} = R_p/R_{t(pm)} = k_p[P^{\bullet}][M]/(k_{t(pm)}[P^{\bullet}][M]) = k_p/k_{t(pm)}$$

is independent of monomer and initiator concentrations. For the example given above, no temperature dependence of kinetic chain lengths (and thus also of degrees of polymerization) is expected since $\Delta E_p^{\ddagger} = \Delta E_{t(pm)}^{\ddagger}$ and thus $\Delta E_X^{\ddagger} = \Delta E_p^{\ddagger} - \Delta E_{t(pm)}^{\ddagger} = 0$.

Number-average degrees of polymerization are small, usually 15-25. Termination by the monomer must therefore consume a considerable fraction of macroradicals. In order to maintain a steady state, however, many initiator radicals must be produced per unit time. These initiator radicals can also terminate macroradicals. The initiator is thus often completely consumed before all monomer has polymerized, for example, at 52.5 % in the polymerization of diallyl phthalate by 1 wt% dibenzoyl peroxide. Allyl polymerizations are thus much more complicated than the simple scheme presented above.

10.4.4   Transfer from Macroradicals to Polymer Molecules

Polymers have greater transfer coefficients than their monomers (Table 10-9). These coefficients sometimes depend on the degree of polymerization (e.g., poly(methyl methacrylate)) but sometimes not (e.g., poly(styrene)). It seems that center and terminal groups behave differently to radical transfer in poly(methyl methacrylates)s.

Chain transfers to polymer molecules generate polymer radicals that may add monomer molecules. The kinetic chain length does not change but the polymer becomes branched and the average degree of polymerization increases. Because of the greater polymer concentration, degrees of branching increase with increasing monomer conversion. *Intramolecular* chain transfer, such as in poly(ethylene)s, generates short chain branches by a back-biting reaction

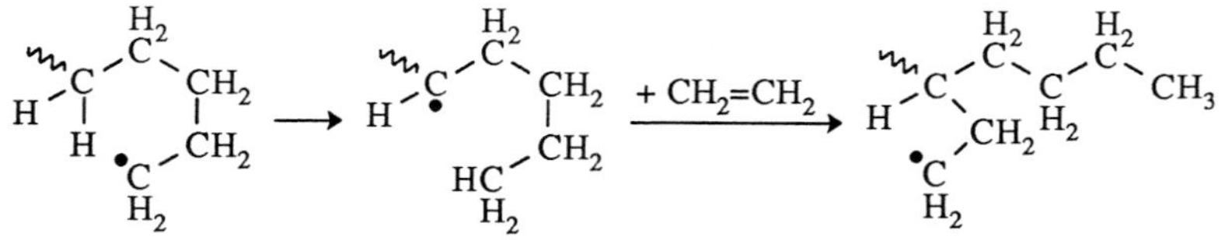

whereas *intermolecular* transfers lead to long-chain branching as in poly(vinyl acetate):

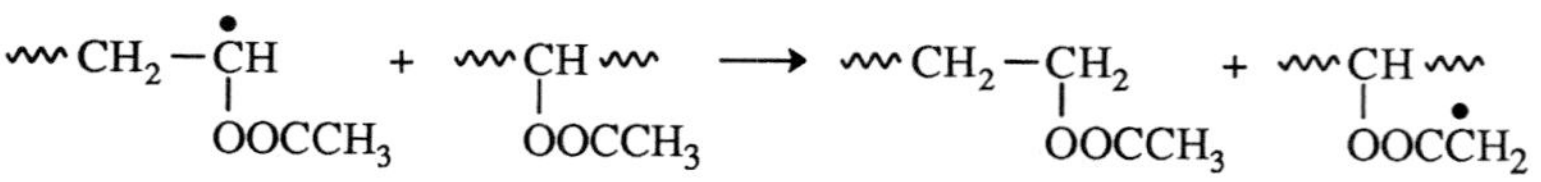

Table 10-9 Transfer coefficients C_m and C_{ax} of some compounds at 60°C [22, 23]. * Chloranil.

Transferring compounds	$10^6 C$ for the polymerization of		
	Styrene	Acrylonitrile	Vinyl acetate
Monomers (C_m)	72	26	102
Polymers ($C_{ax} = C_p$)			
of the above monomers	860	350	340
Solvents ($C_{ax} = C_s$)			
Benzene	2	250	250
Cyclohexane	3	210	210
Ethyl benzene	70	3 600	5 500
Carbon tetrachloride	12 000	110	1 100 000
Carbon tetrabromide	2 500 000	50 000	74 000 000
Initiators ($C_{ax} = C_i$)			
N,N-Azobisisobutyronitrile	10	0	55 000
Dibenzoyl peroxide	68 000		90 000
Bis(2,4-dichlorobenzoyl) peroxide	2 900 000		
Inhibitors ($C_{ax} = C_h$)			
Nitrobenzene	326 000		11 200 000
p-Dinitrobenzene			55 000 000
p-Benzoquinone	230 000 000	130 000 000	54 000 000
Hydroquinone	360 000 000		70 000 000
2,3,5,6-Tetrachloro-1,4-benzoquinone *	950 000 000		
2,3,5,6-Tetramethyl-1,4-benzoquinone (80°C)	67 000 000		9 500 000 000
Regulators ($C_{ax} = C_r$)			
Butyl mercaptan (butanethiol)	22 000 000		48 000 000
1- Dodecyl mercaptan (1-dodecanethiol	15 000 000	730 000	

10.4.5 Transfer from Initiator Radicals to Polymer Molecules

Branched polymers may not only be produced by chain transfer from macroradicals to polymer molecules but also from initiator radicals. For example, *bis*(2,4-dichloro-benzoyl) peroxide (C_i = 2.9 for poly(styrene), Table 10-9) and *t*-butyl hydroperoxide (C_i = 0.1 [I_2] for poly(methyl methacrylate), Fig. 10-5) generate very strong transfers that reduce kinetic chain lengths and thus degrees of polymerization (Fig. 10-5). The product $v_{pp}R_p$ of kinetic chain length v_{pp} and polymerization rate (Eq.(10-46)) is no longer constant.

Since $v_{pp}R_p = \overline{X}_n R_p \approx const$ for AIBN initiated polymerizations (see Fig. 10-4), it may be concluded that this initiator is not a transfer agent. However, the rate $R_{tr(i)}$ of transfer from initiator to the rate R_{st} of the start reaction was found be 0.35 for the polymerization of styrene by AIBN:

$$(10\text{-}77) \qquad \frac{R_{tr(i)}}{R_{st}} = \frac{k_{tr(i)}[I^\bullet][P]}{k_{st}[I^\bullet][M]} = 0.35 \frac{[P]}{[M]}$$

The ratio $R_{tr(i)}/R_{st}$ becomes $0.35 \cdot 2.5/97.5 \approx 0.01$ at a monomer conversion of 5 % (i.e., a time average of 2.5 %) and $0.35 \cdot 25/75 \approx 0.12$ at a conversion of 50 %. At higher monomer conversions, a large number of polymer chains are started at branching points!

10.4.6 Transfer from Macroradicals to Solvents and Regulators

Transfer constants of *solvents* vary widely, for example, from $2 \cdot 10^{-6}$ for benzene in styrene polymerizations to 74 for carbon tetrabromide in vinyl acetate (Table 10-9). They increase with the number of transferable atoms per solvent molecule, the strength of severable bonds, and the resonance stabilization of the new polymer radical. Halogens are usually, but not always, more easily transferred than hydrogen.

The high transfer coefficients of some solvents are utilized in **telomerization**. The reaction of ethene with CCl_4 in the presence of rather large proportions of free-radical initiators $R_I–R_I$ generates various radicals, and compounds, by transfer and propagation

$$R_I–CH_2–CH_2^{\bullet} \quad R_I–CH_2–CH_2–Cl \quad R_I–CH_2–CH_2–CCl_3 \quad R_i–(CH_2–CH_2)_2–Cl$$
$$R_I^{\bullet} \quad R_I–Cl \quad R_I–CCl_3 \quad Cl–CH_2–CH_2–CCl_3 \quad Cl–(CH_2–CH_2)_2–CCl_3$$
$$CCl_3–CH_2–CH_2–CCl_3 \quad CCl_3–(CH_2–CH_2)_2–CCl_3 \quad \text{etc.}$$

that can be separated by distillation, etc., and utilized as, e.g., intermediates. The transfer coefficients vary with the degree of polymerization. They become constant at $X = 3\text{-}6$.

Regulators are chemical compounds with high transfer coeffcients that are utilized purposely to lower the degree of polymerization. Examples are thiols (Table 10-9). Polymers with high degrees of polymerization are not always desirable since such polymers are difficult to process because of their high viscosity or may crosslink in the polymerization reactor. The alternative, lowering the degree of polymerization by a higher initiator concentration, is not very practical for polymerizations of neat monomers since it would lead to higher polymerization rates, a build-up of heat, and, finally, a reactor run-away that may lead to an explosion.

The new radicals resulting from transfer to solvents or regulators may participate in termination reactions, thus lowering the degree of polymerization, the concentration of growing radicals, and also the polymerization rate. For this reason, degradative chain transfer causes the polymerization rate to decrease more strongly than would be expected from the dilution considerations alone.

10.4.7 Inhibitors and Retardants

Practically no industrial monomers are "pure" substances but contain **stabilizers** that inhibit undesired polymerizations during monomer synthesis and storage. These stabilizers are usually not removed before polymerization. Their action is rather annulled by careful overdosages of initiators.

Quinones and nitro compounds are the most often used **inhibitors**. Both types of compounds act very differently (Fig. 10-19).

Benzoquinone produces in the thermal polymerization of styrene a pronounced inhibition period with no polymerization that is followed by a polymerization with the same rate as that without inhibitor. Benzoquinone is thus a true **inhibitor**.

Nitrobenzene, on the other hand, shows no inhibition period. The polymerization starts immediately albeit with a lower rate. Nitrobenzene is thus a **retarder** but not an inhibitor. Nitrosobenzene, on the other hand, is both an inhibitor and a retarder.

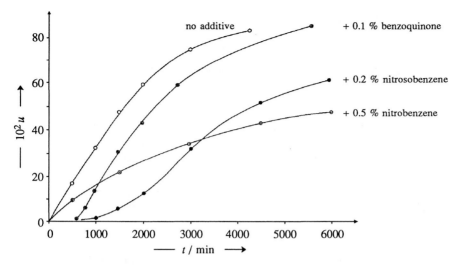

Fig. 10-19 Influence of additives on the thermal polymerization of styrene at 100°C [24].

In polymerizations with benzoquinone as inhibitor, all initiator and polymer radicals react immediately with benzoquinone to form polymerization-inactive radicals, either by radical transfer or by two types of addition, for example:

(10-78)

Radicals generated from benzoquinone and other quinones with low redox potentials do not initiate polymerizations. Quinones with a very high redox potential, on the other hand, do not act as inhibitors but as comonomers, for example, of styrene. An example is 2,5,7,10-tetrachlorodiphenolquinone. However, this copolymerization does not seem to follow the usual copolymerization mechanism (Chapter 12) since this quinone does not copolymerize with acrylonitrile or vinyl acetate.

Industry often uses hydroquinone as a **stabilizer**. This compound is oxidized by oxygen to benzoquinone; hence it removes oxygen traces from the polymerization system before the oxygen can form very reactive hydroperoxides (Volume II). Benzoquinone then acts as an inhibitor while hydroquinone itself is neither an inhibitor nor a retarder.

Nitro compounds, such as nitrobenzene $C_6H_5NO_2$, are extraordinary radical catchers (Table 10-9). They add macroradicals $P^\bullet$ without chain transfer (Eq.(10-79)). The resulting nitro radicals then react with another radical. A polymer molecule POP is thus formed and nitrosobenzene C_6H_5NO is released. One molecule of nitrobenzene consumes 2 macroradicals, and one molecule of 1,3,5-trinitrobenzene 6 of them.

(10-79)

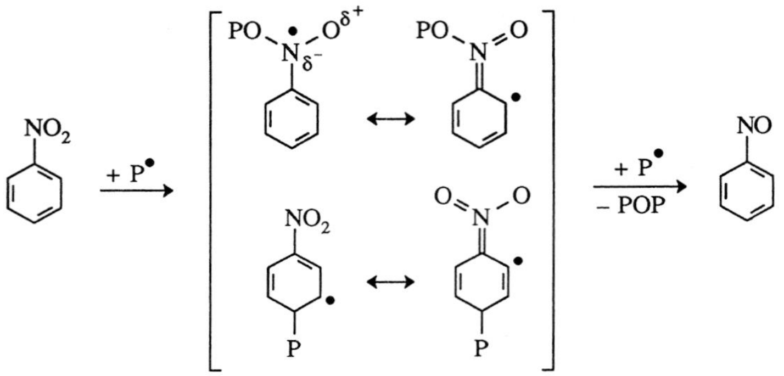

Hence, nitrosobenzene behaves first as an inhibitor but later as a retarder (Fig. 10-19). Such double roles are played by many molecules. For example, tetraphenyl hydrazine inhibits the polymerization of methyl methacrylate but induces the radical decomposition of dibenzoyl peroxide.

1,1-Diphenyl-2-picryl hydrazide (DPPH), $(C_6H_5)_2N-^\bullet N-(C_6H_2)-2,4,6-(NO_2)_3$ is a stable free radical and, with its three nitro groups, an excellent inhibitor. It prevents the polymerization of styrene or vinyl acetate in concentrations as low as ca. 10^{-4} mol/L.

10.5   Non-Ideal Living Polymerizations

Conventional free-radical polymerizations deliver polymers with polymolecularity indices that cannot be lower than $\overline{X}_w/\overline{X}_n = 1.5$ (termination by disproportionation of two macroradicals) and may approach 4 or more. These broad molecular weight distributions are caused by reactions that stop the growth of polymer chains, i.e., by terminations and/or kinetic transfers, and compete with the propagation reaction. The differential number distribution of conventional free-radical polymerizations is thus controlled by kinetic constants and not, as in living anionic polymerizations, by the initial ratios of monomer and initiator concentration.

Any attempt to narrow molecular weight distributions of polymers by free-radical polymerization below $\overline{X}_w/\overline{X}_n = 1.5$ must therefore eliminate chain termination and reduce chain transfer reactions. No termination is possible in **poly(re)combinations of biradicals** such as the polymerization of p-cyclophane to poly(p-xylylene) (Eq.(7-8)). The mechanism of this reaction is, however, not that of a chain polymerization but that of a polyaddition. The same is true for poly(re)combinations where biradicals are produced by transfer reactions of initiator radicals. An example is the formation (and subsequent polycombination) of biradicals from p-diisopropyl benzene (Eq.(10-80)). However, radicals also attack other groups so that branched and finally also crosslinked polymers are produced.

(10-80)

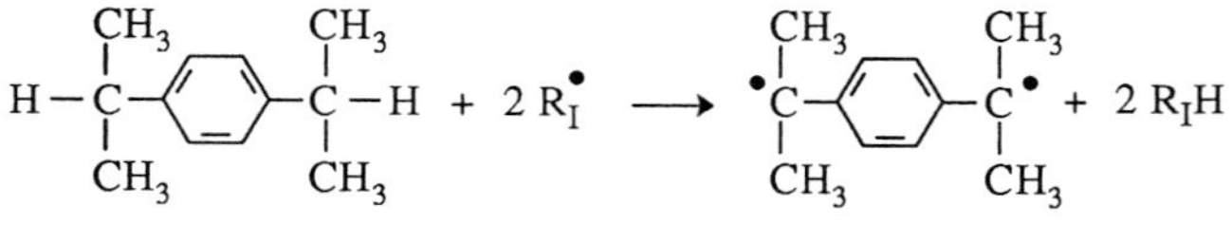

Recent attempts to utilize free-radical polymerization for the synthesis of polymers with narrow molecular weight distributions have thus concentrated on reversible terminations by suitable agents via equilibria between active and dormant chain ends, usually at elevated temperatures. These polymerizations have been called **quasi-living, pseudoliving, controlled**, or **mediated**. A generally accepted term does not exist. Since the polymerizations share certain features with (ideal) living polymerizations, such as linear time dependencies of monomer conversions and number-average degrees of polymerization, they will be summarily called non-ideal living polymerizations.

10.5.1 Polymerizations with Iniferters

Polymerizations with free-radical **iniferters** R_I–R_I use *ini*tiators that also act as trans*fer* agents and *ter*minators, thus producing polymer molecules R_I–M_i–R_I with two initiator endgroups –R_I. The endgroups should be reversibly cleavable so that the resulting macroradicals R_I–$M_{i-1}M^\bullet$ or macrobiradicals $^\bullet M$–$M_{i-2}M^\bullet$ can add further monomer molecules before they are terminated again by initiator radicals. During chain extension by monomers, the temporarily "free" initiator radicals $R_I^\bullet$ should not initiate the polymerization of a new chain.

An example is the polymerization of styrene by the iniferter tetraethylthiuram disulfide, $(C_2H_5)_2N$–$C(=S)$–S–S–$C(=S)$–$N(C_2H_5)_2$, which delivers polymer molecules with the constitution $(C_2H_5)_2N$–$C(=S)$–S–$[CH_2$–$CH(C_6H_5)]_n$–S–$C(=S)$–$N(C_2H_5)_2$. The molecular weight of these polymers did not change when they were heated in benzene solution in the dark. It increased strongly, however, if they were irradiated by a light source, indicating a cleavage of endgroups and subsequent recombination of the monoradicals R_I–$M_i^\bullet$ + $^\bullet M_j$–R_I to R_I–M_i–M_j–R_I. Photolysis in the presence of ethyl acrylate E produced block copolymers R_I–E_k–M_i–E_l–R_I.

Polymerizations with iniferters are fairly slow and deliver polymers with broad molecular weight distributions.

10.5.2 Stable Free-Radical Polymerizations

"Stable free-radical polymerizations" (**SFRP**) are polymerizations that are mediated by nitroxide radicals, borinate radicals, or cobalt complexes. These "persistent" radicals $Y^\bullet$ should reversibly form dormant compounds PY with the transient macroradicals according to $P^\bullet + {}^\bullet Y \rightleftarrows PY$ but should not undergo self-termination. Most often used is the nitroxyl radical TEMPO = 2,2,6,6-*tetra*methyl-1-*pi*peridinyl-1-*oxy* that combines reversibly with macroradicals before the latter undergo irreversible mutual deactivation by combination or disproportionation of two macroradicals:

(10-81)

$$\text{≈}CH_2-\overset{\bullet}{\underset{C_6H_5}{C}}H \; + \; {}^\bullet O-N\overset{H_3C \quad CH_3}{\underset{H_3C \quad CH_3}{\bigvee}} \quad \underset{k_{-L}}{\overset{k_L}{\rightleftarrows}} \quad \text{≈}CH_2-\underset{C_6H_5}{C}H-O-N\overset{H_3C \quad CH_3}{\underset{H_3C \quad CH_3}{\bigvee}}$$

TEMPO

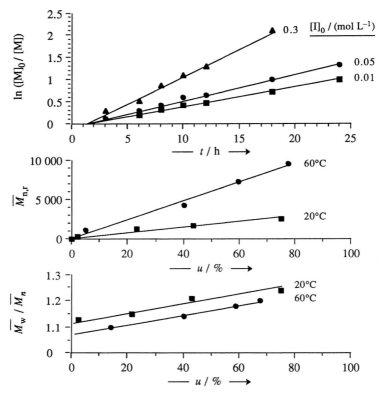

Fig. 10-20 First order time-conversion plot of 2.5 mol/L vinyl acetate at 60°C with various concentrations $[I]_0$ of Al(*i*Bu)$_3$·bpy·TEMPO (1:1:2) (top). Conversion dependence of number-average degrees of polymerization (center) and polymolecularity indices (bottom), both with $[I]_0$ = 0.05 mol/L [25].

Nitroxide-mediated polymerizations with dibenzoyl peroxide or similar radical delivering initiators are presently restricted to styrene and its derivatives. TEMPO-mediated polymerizations of vinyl acetate have also been performed with Al(*i*Bu)$_3$·2,2'-bipyridyl (Fig. 10-20), an initiator that delivers persistent radicals (Fig. 10-21).

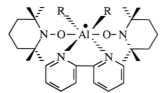

Fig. 10-21 Suggested structure of the complex of Al(*i*Bu)$_3$, TEMPO, and 2,2'-bipyridyl. Methyl groups of the nitroxyl molecule are not shown; R = CH(CH$_3$)$_2$. The complex becomes an active radical after loss of a radical R•.

The polymerization shows an initial time lag (due to formation of an initiator complex?) and is then first order with respect to the monomer (Fig. 10-20, top). The linearity of the plot indicates a constant initiator concentration. Number-average molecular weights (Fig. 10-20, center) and polymolecularity indices (Fig. 10-20, bottom) both increase linearly with relative monomer conversion, the latter being small and far below the theoretical lower limit of $\overline{M}_w/\overline{M}_n$ = 1.5 for conventional free-radical polymerizations.

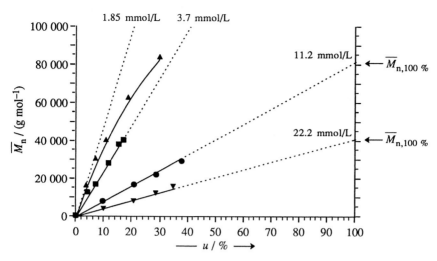

Fig. 10-22 Dependence of number-average molar mass on relative monomer conversion in the poly-merization of neat styrene at 90°C by various concentrations of the dinitroxide TAT [26a].
TAT = [(CH$_3$)$_3$C]$_2$N–O–CH(CH$_3$)–(1,4-C$_6$H$_4$)–CH(CH$_3$)–O–N[C(CH$_3$)$_3$]$_2$.

Polymerizations of styrene with the dinitroxide TAT to end-capped polymers T–P–T do not require an extra thermal or photo initiator. At fairly low relative monomer con-versions u, number-average molar masses of polymers P increase linearly with u but de-viations from linearity are seen at higher initiator concentrations (Fig. 10-22).

The linearity of ln {[M]$_0$/[M]} = f(t) (Fig. 10-23) indicates a first-order polymeriza-tion rate with respect to the monomer concentration and a time-independent concentra-tion of the active species T–P$^{\bullet}$. The plot is also independent of the initiator concentra-tion (see Fig. 10-20). This behavior has been interpreted as a consequence of small equilibrium constants of the dissociation equilibrium T–P–T $\rightleftarrows$ T–P$^{\bullet}$ + T and an association equilibrium (T–P–T)$_N$ $\rightleftarrows$ N (T–P–T) with high association numbers N.

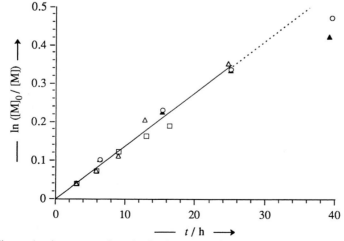

Fig. 10-23 First-order time-conversion plot for the polymerization of neat styrene by TAT [26b].

10.5.3 Atom-Transfer Radical Polymerization

Atom-transfer radical polymerization (**ATRP**) employs transition metal compounds T as mediators for the transfer of halogen atoms X between active and dormant chain ends. The method is suitable for the homopolymerization of styrenes, (meth)acrylates, and acrylonitrile but not for less reactive monomers such as olefins and vinyl acetate.

An example is the polymerization of methyl acrylate M by CuCl as transition metal T, complexed by 2,2'-bipyridine (B), and 1-phenylethyl chloride R–X as initiator. In the early stage of the reaction, the initiator is completely consumed by the formation of oligomeric molecules R~M_n–X which leads to a time lag in the polymerization (Fig. 10-24, bottom). The reaction of the oligomers with the complexed transition metal compound to oligoradicals (and later macroradicals) proceeds by oxidation of T:

$$(10\text{-}82) \quad R\text{~~}M_n\text{–}X + T^{i+}/B \; \rightleftharpoons \; R\text{~~}M_n^{\bullet} + X\text{–}T^{(i+1)+}/B$$

The resulting transient macroradicals add only a few monomer molecules before they become dormant by the back reaction. Because there are so few monomer molecules added, polymolecularities are low and become smaller with increasing molecular weight (Fig. 10-24, top; see Poisson distribution, p. 73). During the active time of these radicals, chain termination and transfer should be small, thus guaranteeing high molar masses and degrees of polymerization that are given by the ratio of monomer conversion to initiator concentration, $\overline{X}_n = ([M]_0 - [M])/[I]_0$ (Fig. 20-24, top). The polymerization rate is first-order with respect to monomer, indicating a constant concentration of active species.

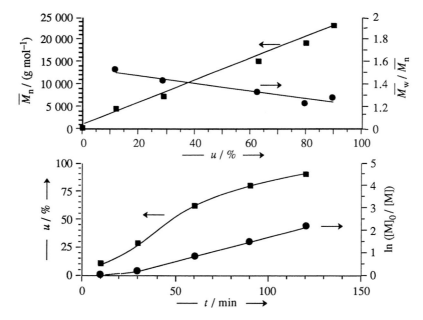

Fig. 10-24 Polymerization of neat (11 mol/L) methyl acrylate at 130°C by 0.1 mol/L CuCl, 0.3 mol/L 2,2'-bipyridine, and 0.1 mol/L 1-phenylethyl chloride. Bottom: Time dependence of relative monomer conversion u and ln ($[M]_0/[M]$) [27a]. Top: Number-average molar mass and polymolecularity index as a function of relative monomer conversion [27b].

10.5.4 Reversible Addition-Fragmentation Chain Transfer

Reversible addition-fragmentation chain transfer (**RAFT**) is a free-radical polymerization with conventional peroxide or azo initiators and addition of dithioesters such as R–S–C(=S)–Q where R is an alkyl or aryl residue and Q another group such as CH_3, $CH_2C_6H_5$, $C(CH_3)_2CN$, etc. Polymerizations of styrene, acrylic acid, methyl methacrylate, vinyl benzoate, etc., have been performed in bulk, solution, emulsion, or suspension; there does not seem to be a limitation with respect to solvents or reaction temperatures.

Number-average molar masses depend linearly on relative monomer conversions up to very high values (Fig. 10-25). Polymolecularity indices decrease with increasing monomer conversion, and block copolymerization is possible.

The key to successful RAFT polymerization is the extremly high transfer coefficient of dithioesters, most likely $C_{tr} = k_{tr}/k_p > 100$ (note the transfer coefficients of regulators in Table 10-9: $C_{tr} = 0.73$-48). RAFT appears to be more general, simple, and cost effective than ATRP but seems to be slower.

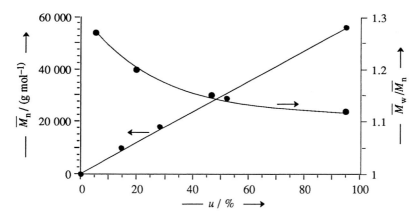

Fig. 10-25 Number-average molar mass and polymolecularity index as a function of conversion during polymerization of 7.01 mol/L methyl methacrylate in benzene at 70°C by 0.0061 mol/L AIBN and 0.0111 mol/L cumyl dithiobenzoate [28].

10.6 Industrial Polymerizations

10.6.1 Overview

Industrial polymerizations are discussed in greater detail in Volume II. This section therefore provides only a short survey with special emphasis on differences to laboratory procedures.

Free-radical polymerizations can be performed in many states of matter: in the gas phase, in bulk (melt), solution, emulsion, suspension, by polymer precipitation, and even in the solid state. The various processes not only call for different procedures but they also affect polymer properties (molar mass, molar mass distribution, constitution, configuration, etc.).

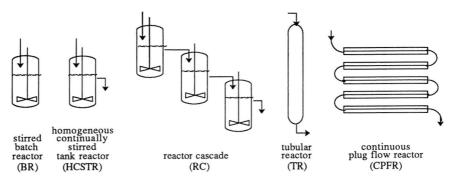

stirred batch reactor (BR)   homogeneous continually stirred tank reactor (HCSTR)   reactor cascade (RC)   tubular reactor (TR)   continuous plug flow reactor (CPFR)

Fig. 10-26 Some simple polymerization reactors.

The choice of the polymerization process and the type of polymerization reactor for a given monomer depends on the economy and reaction engineering of the process as well as on the desired properties of the resulting polymer (cf. Table 10-1). Reactors come in various shapes and sizes (Fig. 10-26). Discontinuous reactors with volumes up to 200 m^3 are used for free-radical polymerizations in bulk, solution, suspension, and emulsion, continuous reactors for solution, precipitation, and emulsion polymerization, and cascades also for emulsion polymerizations (Table 10-10; see Volume II).

The technologically most important factor is viscosity which controls the diffusion of the different species as well as the heat dissipation. Since heats of polymerization are usually high and heat conductivities of polymers are low, strong coupling effects exist between chemical reactions and the transport of matter and heat. This leads not only to large differences between the macrokinetics in industrial reactors and the microkinetics in laboratory experiments but also to variations in polymers with respect to branching, tacticities, molecular weights, and molecular weight distributions.

Table 10-10 Applications of polymerization reactors [29]: BR = stirred batch reactor, RC = reactor cascade, HCSTR = homogeneous continually stirred tank reactor, CPFR = continuous plug flow reactor, FBR = fluidized-bed reactor, TR = tubular reactor [28]. G = Gas, L = liquid, S = solid.
EPS = expandable poly(styrene), EVAC = ethylene-vinyl acetate copolymer, HDPE = high-density poly(ethylene), HIPS = high-impact poly(styrene), LDPE = low-density poly(ethylene), PA = polyamide, PB = poly(1-butene), PE = poly(ethylene), PMMA = poly(methyl methacrylate), PP = it-poly-(propylene), PS = poly(styrene), PTFE = poly(tetrafluoroethylene), PVC = poly(vinyl chloride), SBR = styrene-butadiene rubber. * By non-radical ring-opening polymerization.

Reactor	Medium	Polymerization in			
		Gas phase	Melt or Solution	Suspension	Emulsion
BR	L	PVC	LDPE, HDPE, PS, PB, PMMA, PC	PVC, PTFE, PMMA, EPS	PVC, PTFE, SBR, many copolymers
HCSTR	L		LDPE, PS, PMMA		PVC
RC	L		PE, PP, HIPS	LDPE	SBR
FBR	G/S	HDPE, LDPE, PP			
TR	L		HIPS, PA *		
CPFR	L		LDPE, EVAC, PA *		

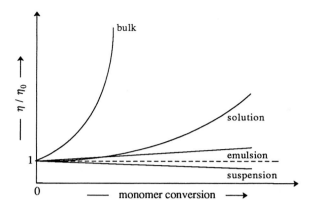

Fig. 10-27 Variation of relative viscosity during polymerization in various media (schematic).

During polymerization, polymer concentrations increase and usually also molecular weights and molecular weight distributions (exception: molecular weights of conventional free-radical polymerizations up to medium monomer conversions). Since viscosities not only increase with concentration and the weight-average of molecular weight but in melts to the 3.4 power of the latter at higher molecular weights ($\eta = K_\eta \overline{M}_w^{3.4}$ (Volume III)), very strong viscosity increases are observed for polymerizations in bulk (Fig. 10-27): up to 6 decades during the neat polymerization of styrene between monomer conversions of 0 % and 80 %. Viscosities change only very little in emulsion and suspension polymerizations.

Most of the obtained polymers are linear or slightly branched. They are used as thermoplastics, fibers, elastomers, adhesives, thickeners, etc. Crosslinking free-radical homopolymerizations are rare; the most important example is diallyl compounds. Thermosetting resins are usually prepared in batch reactors as soluble prepolymers that are then crosslinked (cured, hardened) during application.

10.6.2 Bulk Polymerization

Polymerizations of monomer melts are called **bulk polymerizations** or **polymerizations of neat monomers**. In the older literature, one can sometimes find the term "block polymerization" because the polymer became a solid block and had to be removed from the reactor by miner's means. "Block polymerization" now refers to the synthesis of polymer molecules with large segments of different monomeric units (p. 607).

Only monomers, polymers, and initiators are present in bulk polymerizations. The resulting polymers are thus relatively pure. Industrial bulk polymerizations sometimes do contain 5-15 % solvents, however, either as processing aids or as transfer agents.

Some styrene is polymerized thermally without added initiator to an amorphous crystal-clear poly(styrene) ("crystal poly(styrene)"). Most styrene, however, is polymerized by high-temperature initiators such as 1,2-dimethyl-1,2-diethyl-1,2-diphenylethane or vinyl silanetriacetate $CH_2=CH-Si(OOCCH_3)_3$. Most other bulk polymerizations prefer peroxydicarbonates.

Bulk polymerizations create large heats of polymerization per unit volume which are difficult to remove rapidly from large reactors. Additional heat build-ups are caused by gel effects. Local overheating in the reactor may lead to multi-modal molecular weight distributions, branching, degradation, and/or discoloration of polymers, and even to re-actor explosions. Industrial bulk polymerizations are therefore often terminated at 40-60 percent monomer conversion. Residual monomer is distilled off. An alternative method uses polymerizations in two steps: first in large batch reactors, then in thin layers.

10.6.3   Solution Polymerization

Inert solvents act as diluents. They lower the monomer concentration and thus the polymerization rate (Eq.(10-41)). They furthermore decrease radical transfer to poly-mer: degrees of branching are smaller and molecular weight distributions more narrow. However, chain-transfer-causing solvents may increase the polymerization rate.

Diluents also reduce gel effects and the energy input for stirring. They furthermore allow faster removal of heats of polymerization. However, they are expensive, costly to recover and work up, and damaging to one's health and the environment.

Some polymers are insoluble in their own monomers and precipitate during the polymerization. In these **precipition polymerizations**, macroradicals remain occluded in the polymer particles. Similarly to the gel effect, diffusion of macroradicals and their terminations by mutual deactivation are severely hindered so that molecular weights in-crease sharply. Polymerization rates remain high, however, since monomer molecules can still diffuse relatively easily to the growing macroradicals. The polymers precipitate as particles or even as powders. Unlike polymers from bulk polymerizations, they re–quire neither grinding nor pelletizing.

10.6.4   Polymerization in the Gas Phase

Free-radical polymerizations are rarely performed with gaseous monomers (Tables 10-1, 10-10); if at all, then by photo-initiation. The polymers precipitate as particles that contain only one macroradical because of the high dilution. New monomer molecules are delivered to macroradicals via the gas phase. The resulting polymers are very pure.

10.6.5   Suspension Polymerization

In suspension polymerizations, water-insoluble monomers are dispersed in aqueous media as small droplets of 0.0001 cm to 1 cm diameter by "suspending" agents such as ionic detergents, poly(vinyl alcohol), barium sulfate, etc. The polymerization is a "water-cooled bulk polymerization" in droplets acting as "microreactors." Heat of polymeriza-tion is rapidly removed by the water.

The polymerization is initiated by "oil-soluble" (i.e., soluble in the monomer) initia-tors. The monomer droplets are converted into pearls; the process is therefore also called **pearl** or **bead polymerization**. "Reverse bead polymerizations" polymerize aqueous monomer droplets in organic solvents. They are known as **dispersion polymerizations**.

The terms "suspension polymerization" and "dispersion polymerization" differ from the established use of the words "suspension" and "dispersion". In colloid chemistry, *all* systems comprised of a continuous and a discontinuous phase are called **dispersions**. Such dispersions may consist of a solid or liquid phase in a gas (**aerosol**), a liquid in another liquid (**emulsion**), or solid particles in a liquid (**suspension**). "Suspension polymerizations" always start with an *emulsion* of water-insoluble monomer molecules in water and "dispersion polymerizations" always start with an *emulsion* of water-soluble monomers in organic solvents. Depending on the polymerization temperature T_P and the glass temperature T_G of the resulting polymer, "suspension polymerizations" deliver either suspensions (if $T_G > T_P$; example: poly(styrene)) or emulsions ($T_G < T_P$; example: styrene-butadiene rubbers).

Advantages of suspension polymerizations comprise easy control of polymerization and the pearly shape of polymer particles; no pelletizing is required. Disadvantages are the need for the removal and clean-up of water and the residual suspending agent in the pearls that lowers certain material properties (haziness, accelerated aging).

10.6.6 Emulsion Polymerization

Emulsion polymerizations are free-radical polymerizations of water-insoluble monomers in water by water-soluble initiators ($K_2S_2O_8$, redox initiators) in the presence of surfactants. Surfactant molecules S are amphiphilic and associate reversibly in water to micelles that contain $N = 15\text{-}100$ surfactant molecules, depending on the constitution of S. The concentration of micelles is regulated by the association equilibrium $N\,S \rightleftarrows S_N$ with an equilibrium constant of $K_N = [S_N]/[S]^N$. The concentration of S_N is very low at low concentrations of S but increases dramatically above the so-called critical micelle concentration (Volume III). These surfactant micelles have diameters of 4-10 nm.

The interior of micelles is hydrophobic and can take up to ca. 100 monomer molecules, which increases slightly with micelle diameter (Fig. 10-28). Another proportion of the monomer is dispersed in large monomer droplets of ca. 1000 nm diameter that are stabilized by surfactant molecules.

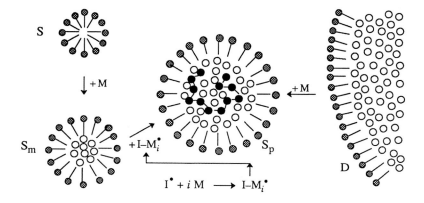

Fig. 10-28 Surfactant molecules ⊗— form micelles S that take up monomer molecules M and swell to become S_m. Water-soluble initiators dissociate in water into initiator radicals I• that add some of the molecularly water-dissolved monomer molecules and become oligomer radicals I–M_i•. The amphiphilic oligomer radicals enter the micelle where they add monomer molecules and become macroradicals ●–●–●–●. The growth of macroradicals in the polymerizing micelle S_P is sustained by additional monomer molecules M that diffuse from large monomer droplets D (shown as a partial cross-section).

The water-soluble initiator molecules dissociate in water into two initiator radicals $I^\bullet$ that add dissolved monomer molecules and become oligomer radicals $I-M_i^\bullet$. These radicals enter the monomer-loaded micelle and start the polymerization. An entry of oligomer radicals into monomer droplets is not likely since a typical polymerization system contains per liter ca. 10^{20} monomer-loaded micelles of ca. 8 nm diameter and also ca. 10^{12} monomer droplets of ca. 5000 nm diameter. The surface of the micelles is thus much larger than that of the monomer droplets (20 000 m²/L versus 80 m²/L).

Each micelle contains only one oligomer radical since an entering second radical would immediately terminate the growing chain because of the very small micelle volume. Since each micelle contains only ca. 100 monomer molecules, the polymerization would stop soon if there were no diffusion of monomer molecules from the monomer droplets into the polymerizing micelle. The micelles swell, add further monomer molecules and surfactant molecules, and finally become **latex particles** with diameters of 500-5000 nm.

In principle, all monomer molecules in and arriving at latex particles are polymerized by 1 oligomer radical. At a monomer density of ca. 1 g/cm³, a latex particle of 1000 nm diameter contains $5.24 \cdot 10^{-13}$ g monomer. The polymer molecule in that latex particle would thus have a molar mass of $M = 5.24 \cdot 10^{-13}$ g $\times 6.02 \cdot 10^{23}$ mol$^{-1} \approx 3.5 \cdot 10^{11}$ g/mol. These "theoretical" molar masses are far higher than the experimental ones which usually reach several millions. The growing macroradical in the latex particle must therefore be terminated by an entering second oligomer radical. It is also possible (and experimentally confirmed) that radicals can leave micelles and latex particles.

Emulsion polymerizations have many industrial advantages: easy removal of heat of polymerization, high polymerization rates by redox initiators, "self-pelletization" of products, and easy removal of residual monomer by, e.g., a steam purge. The main disadvantage is the difficult removal of last traces of surfactants that may affect product properties (clarity, chemical aging, etc.).

A-10   Appendix: Schulz-Flory Distribution

Kinetically controlled free-radical chain polymerizations with termination by disproportionation of macroradicals and thermodynamically controlled equilibrium polycondensations differ in both the participation of reactants (addition of monomer molecules to growing macroradicals versus random reaction of functional groups of monomer and polymer molecules) and the absence or presence of leaving groups. However, both types of polymerizations lead to the same type of molecular weight distribution (Schulz-Flory distribution, SF distribution) because macromolecules are formed at random.

The SF distribution is derived in this section from the kinetics of chain polymerizations. It may also be obtained from probability considerations, similar to those for polycondensations and polyadditions (Chapter 13).

This derivation assumes ideal polymerization kinetics with photo-initiated formation of monomer radicals and chain termination by disproportionation. Two macroradicals are thus converted into two dead polymer molecules.

In the initiation reaction, light quanta of intensity I_{hv} generate monomer radicals $M^\bullet$

≡ $P_1^\bullet$ with a rate of $R_{st} = k_I[M]I_{hv}$. These monomer radicals disappear by either prop-agation or termination. The net rate of formation of monomer radicals is thus

(A-10-1) $d[P_1^\bullet]/dt = k_I[M]I_{hv} - k_p[P_1^\bullet][M] - k_{t(pp)}[P_1^\bullet][P^\bullet]$

where $[P_1^\bullet] + [P_2^\bullet]... + [P_i^\bullet] = \Sigma_j [P_j^\bullet] \equiv [P^\bullet]$. In the steady state, one has $d[P_1^\bullet]/dt = 0$ and $k_I[M]I_{hv} = k_{t(pp)}[P^\bullet]^2$ and thus:

(A-10-2) $(k_I[M]I_{hv})/(k_{t(pp)}[M]) = [P_1^\bullet](1 + \beta)$

$$\beta \equiv (k_{t(pp)}[P^\bullet])/(k_p[M]) = (k_I k_{t(pp)}I_{hv})^{1/2}/(k_p[M]^{1/2})$$

Since the concentrations of $P^\bullet$ and M are constant at small monomer conversions, β is also a constant.

Dimer, trimer, ... , polymer radicals ($i \geq 2$) are formed with a rate of

(A-10-3) $d[P_i^\bullet]/dt = k_p[P_{i-1}^\bullet][M] - k_p[P_i^\bullet][M] - k_{t(pp)}[P_i^\bullet][P_1^\bullet]$

$$- k_{t(pp)}[P_i^\bullet][P_2^\bullet] ...... - k_{t(pp)}[P_i^\bullet][P_i^\bullet]$$

By analogy to the formation of monomer radicals, one obtains $[P_{i-1}^\bullet] = [P_i^\bullet](1 + \beta)$. Successive introduction of expressions for all values of $i > 1$ and Eq.(A-10-2) delivers

(A-10-4) $[P_i^\bullet] = \{(k_I[M]I_{hv})/(k_{t(pp)}[M])\}(1 + \beta)^{-i}$

Dead polymer molecules are generated with a rate of $d[P_i]/dt = k_{t(pp)}[P_i^\bullet][P^\bullet]$. Inte-gration leads to $[P_i] = k_{t(pp)}[P_i^\bullet][P^\bullet]t$. Introduction of Eqs.(A-10-4) and (A-10-2 b) delivers $[P_i^\bullet] = k_I[M]I_{hv}\beta(1 + \beta)^{-i}t$ for the concentration of macroradicals.

Per unit time, a total of $k_I[M]I_{hv}t$ radicals are generated that deliver the same number of dead polymer molecules. With the condition $\Sigma_i \beta(1+\beta)^{-i} = 1$, the sum of the concen-trations of all polymer molecules becomes

(A-10-5) $\Sigma_i [P_i] = \Sigma_i k_I[M]I_{hv}\beta(1 + \beta)^{-i}t = k_I[M]I_{hv}t$

Because of $\beta << 1$, the mole fraction x_i of i-mers with a degree of polymerization of X_i is given by

(A-10-6) $x_i = [P_i]/\Sigma_i [P_i] = \beta(1 + \beta)^{-i} = \beta[\exp(\beta)]^{-1}$

The number-average degree of polymerization of polymer molecules is calculated by replacing the sum by an integral. After integration from zero to infinity, one obtains for $i \equiv X_i$:

(A-10-7) $\overline{X}_n = \dfrac{\Sigma_i x_i X_i}{\Sigma_i x_i} = \dfrac{\Sigma_i (1+\beta)^{-i} X_i}{\Sigma_i (1+\beta)^{-i}} = \dfrac{\int X_i[\exp(-\beta X_i)]dX_i}{\int [\exp(-\beta X_i)]dX_i} = \dfrac{\beta^{-2}}{\beta^{-1}} = \dfrac{1}{\beta}$

Similar calculations give $\overline{X}_w = 2/\beta$ for the weight-average degree of polymerization, leading to the weight-number polymolecularity index of $\overline{X}_w/\overline{X}_n = 2$.

Historical Notes to Chapter 10

Mechanism of Chain Polymerizations

M.Gomberg, Chem.Ber. **33** (1900) 3150
   Discovery of stable free trityl radicals $(C_6H_5)_3C^\bullet$ (I) from the dissociation of hexaphenyl ethane. It was found by Lamberg in 1968 that hexaphenyl ethane is not $(C_6H_5)_3C–C(C_6H_5)_3$, but the product II:

M.Bodenstein, S.C.Lind, Z.phys.Chem. **57** (1907) 168
   Discovery of chain reactions of low-molecular weight molecules $(H_2 + Br_2 \rightarrow 2\ HBr)$.

H.Stobbe, G.N.Posnjak, Liebigs Ann.Chem. **371** (1909) 259
G.Stobbe, Liebigs Ann.Chem. **409** (1915) 1
   The polymerization of styrene (note: without initiators!) is due to the formation of polymerization "seeds" (original German: "Polymerisationskeime"; G: *Keim* = germ, seed, bud, shoot, sprout, depending on context). These seeds were said to be formed by light and heat; they "accelerate" the polymerization.

J.A.Christiansen, Kgl.Danske Videnskab.Selskab., Mat.-fys.Medd. **1** (1919) 14
K.F.Herzfeld, Ann.Phys. **59** (1919) 635
M.Polanyi, Z.Elektrochem. **26** (1920) 50
   Established chain mechanism of the reaction $H_2 + Br_2 \rightarrow 2\ HBr$

H.Staudinger, E.Urech, Helv.Chim.Acta **12** (1929) 1107
   Polymerization consists of a successive addition of monomer molecules to an "active seed" of undisclosed nature (original German: "Keim"; see above).

W.Chalmers, Can.J.Res. **1** (1930) 32; -, **7** (1932) 113; -, J.Am.Chem.Soc. **56** (1934) 912
H.W.Starkweather, G.B.Taylor, J.Am.Chem.Soc. **52** (1930) 4708
H.Staudinger, H.W.Kohlschütter, Ber.Dtsch.Chem.Ges. **64** (1931) 2093
   Vinyl polymerizations are kinetic chain reactions.

W.Chalmers, Can.J.Res. **7** (1932) 113, 464, 472; -, J.Am.Chem.Soc. **56** (1934) 918
   Established of mechanism of vinyl polymerization (still valid).

W.Chalmers, Can.J.Chem. **1** (1930) 32; -, J.Am.Chem.Soc. **56** (1934) 912
H.Staudinger, W.Frost, Ber.Dtsch.Chem.Ges. **68** (1935) 2351
H.W.Melville, Proc.Roy.Soc. [London] A **163** (1937) 511
   The "active seed" is a radical.

G.V.Schulz, E.Husemann, Z.Phys.Chem. B **39** (1938) 246
   Radical polymerizations involve activated complexes between monomer molecules and thermal initiators (not true).

P.J.Flory, J.Am.Chem.Soc. **63** (1941) 2798
   Comprehensive scheme of free-radical polymerizations: first-order decomposition of initiator, fast propagation reaction, fast bimolecular termination of growing chains.

C.C.Price, R.W.Kell, E.Kred, J.Am.Chem.Soc. **63** (1941) 2708; -, **64** (1942) 1103
W.Kern, H.Kämmerer, J.Prakt.Chem. **161** (1942) 81, 289
P.D.Bartlett, S.D.Cohen, J.Am.Chem.Soc. **65** (1943) 543
   Tagging of initiator molecules showed that initiator fragments are incorporated into chains as endgroups.

Non-Ideal Living Free-Radical Polymerizations

T.Otsu, M.Yoshida, Makromol.Chem.-Rapid Comm. **3** (1982) 127
   Iniferter method of polymerization.

D.H.Solomon, E.Rizzardo, P.Cacioli, U.S. Patent 4,581,429 (27 March 1985)
   Stable free radicals form adducts with monomers that initiate oligomerization.

M.K.Georges, P.R.N.Veregrin, P.M.Kazmeier, G.K.Hamer, Macromolecules **26** (1993) 2987
   Development of stable free-radical (nitroxide mediated) polymerization (SFRP).

J.S.Wang, K.Matyjaszewski, J.Am.Chem.Soc. **117** (1995) 5614
   Development of atom transfer radical polymerization (ATRP).

J.Chiefari, Y.K.Chong, F.Ercole, J.Krstina, J.Jeffery, T.P.T.Lee, R.T.A.Mayadunne, G.F.Meijs,
C.L.Moad, G.Moad, E.Rizzardo, S.H.Thang, Macromolecules **31** (1998) 5559
   Development of living free-radical polymerization by reversible addition-fragmentation chain
   transfer (RAFT).

Literature to Chapter 10

10.1 INTRODUCTION: General Overviews
L.Küchler, Polymerisationskinetik, Springer, Heidelberg 1951
J.C.Bevington, Radical Polymerization, Academic Press, London 1961
C.H.Bamford, C.F.H.Tipper, Eds., Free-Radical Polymerization (= Comprehensive Chemical
   Kinetics **14 A**), Elsevier, Amsterdam 1976
R.N.Haward, Ed., Developments in Polymerization, Vol. 2, Free Radical, Condensation,
   Transition Metal, and Template Polymerization, Appl.Sci.Publ., London 1979
G.Allen, J.C.Bevington, Eds., Comprehensive Polymer Science. Vols. 3 and 4: G.C.Eastmond,
   A.Ledwith, S.Russo, P.Sigwalt, Eds., Chain Polymerization, Parts I and II, Pergamon Press,
   Oxford 1989
G.Moad, D.H.Solomon, The Chemistry of Free Radical Polymerization, Pergamon Press,
   Oxford 1995
M.K.Mishra, Y.Yagci, Eds., Handbook of Radical Vinyl Polymerizations, Dekker, New York 1998
K.Matyjaszewski, T.P.Davis, Eds., Handbook of Radical Polymerization, Wiley, New York 2002

10.2 INITIATION AND START
C.Walling, Free Radicals in Solution, Wiley, New York 1957
D.Swern, Ed., Organic Peroxides, Wiley, New York, 3 Vols. (1970-1972)
G.Parravano, Electrochemical Polymerization, in M.M.Bazier, Ed., Organic Electrochemistry,
   Dekker, New York 1973
P.L.Nayak, S.Leuka, Redox Polymerization Initiated by Metal Ions, J.Macromol.Sci.-
   Rev.Macromol.Chem. **C 19** (1980) 83
G.Silvestri, S.Gambino, G.Filardo, Electrochemical Production of Initiators for Polymerization
   Processes, Adv.Polym.Sci. **38** (1981) 27
G.S.Misra, U.D.N.Bajpai, Redox Polymerization, Prog.Polym.Sci. **8** (1982) 61
S.Patai, Ed., The Chemistry of Peroxides, Wiley, New York 1984
N.N.Dass, Charge-Transfer Initiation and Termination, Prog.Polym.Sci. **10** (1984) 51
H.J.Hageman, Photoinitiators for Free Radical Polymerization, Prog.Organic Coatings **13/2**
   (1985) 123
Cr.I.Simionescu, E.Comanita, M.Pastravanu, S.Dumitriu, Progress in the Field of Bi- and
   Polyfunctional Free-Radical Polymerization Initiators, Prog.Polym.Sci. **12** (1986) 1
W.Ando, Ed., Organic Peroxides, Wiley, Chichester 1992
J.E.Leffler, An Introduction to Free Radicals, Wiley, Somerset, NJ, 1993
W.Yuan, J.O.Iroh, Electropolymerization of Olefinic Monomers, Trends Polym. Sci. **1** (1993) 388

10.3 PROPAGATION AND TERMINATION

D.J.T.Hill, J.J.O'Donnell, P.W.O'Sullivan, The Role of Donor-Acceptor Complexes in Polymerization, Prog.Polym.Sci **8** (1982) 215

O.F.Olaj, I.Bitai, F.Hinkelmann, The Laser-flash initiated Polymerization as a Tool of Evaluating (Individual) Kinetic Constants of Free-radical Polymerization, 2. The Direct Determination of the Rate Constant of Chain Propagation, Makromol.Chem. **188** (1987) 1689; see also O.F.Olaj, I.Bitai, Angew.Makromol.Chem. **155** (1987) 177

M.Kamachi, ESR Studies of Radical Polymerization, Adv.Polym.Sci. **82** (1987) 207

I.Mita, K.Horie, Diffusion-Controlled Reactions in Polymer Systems, J.Macromol.Sci.-Revs.Macromol.Chem.Phys. **C 27** (1987) 91

M.Buback, Free-Radical Kinetics to High Conversions, Makromol.Chem. **191** (1990) 1575

G.A.O'Neil, J.M.Torkelson, Recent Advances in the Understanding of the Gel Effect in Free-Radical Polymerization, Trends Polym.Sci. **5** (1997) 349

M.L.Coote, M.D.Zammit, T.P.Davis, Determination of Free-Radical Rate Coefficients Using Pulsed-Laser Polymerization, Trends Polym.Sci. **4** (1996) 189

R.P.Washington, O.Steinbock, Frontal Free-Radical Polymerization: Applications to Materials Synthesis, Polym. News **28** (2003) 303

10.4 CHAIN TRANSFER

G.Henrici-Olivé, S.Olivé, Kettenübertragungen bei der radikalischen Polymerisation, Fortschr.Hochpolym.Forschg. **2** (1960/61) 496

C.M.Starks, Free Radical Telomerization, Academic Press, New York 1974

E.J.Goethals, Ed., Telechelic Polymers: Synthesis and Applications, CRC Press, Boca Raton, FL 1988

B.Boutevin, Telechelic Oligomers by Radical Reactions, Adv.Polym.Sci. **94** (1990) 69

A.V.Ambade, A.Kumar, Controlling the Degree of Branching in Vinyl Polymerization, Prog.Polym.Sci. **25/8** (2000) 1141

10.5 NON-IDEAL LIVING POLYMERIZATIONS (see also Historical Notes)

J.Barton, E.Borsig, Complexes in Free-radical Polymerization, Elsevier, New York 1988

D.Colombani, Chain-Growth Control in Free-radical Polymerization, Prog.Polym.Sci. **22** (1997) 1649

T.Otsu, A.Matsumoto, Controlled Synthesis of Polymers Using the Iniferter Technique: Devlopments   in Living Radical Polymerization, Adv.Polym.Sci. **136** (1998) 75

H.S.Bisht, A.K.Chatterjee, Living Free-Radical Polymerization, J.Macromol.Sci.-Polymer Reviews **C 41** (2001) 139

A.Goto, T.Fukud, Kinetics of Living Radical Polymerization, Progr.Polym.Sci. **29** (2004) 329

10.6 INDUSTRIAL POLYMERIZATIONS

F.A.Bovey, I.M.Kolthoff, A.J.Medalia, E.J.Mehan, Emulsion Polymerization, Interscience, New York 1955

D.C.Blackley, Emulsion Polymerization: Theory and Practice, Halsted, New York 1975

K.E.J.Barrett, Ed., Dispersion Polymerization in Organic Media, Wiley, New York 1975

I.Piirma, Ed., Emulsion Polymerization, Academic Press, New York 1982

R.G.Gilbert, D.H.Napper, The Direct Determination of Kinetic Parameters in Emulsion Polymerization Systems, J.Macromol.Sci.–Revs.Macromol.Chem.Phys. **C 23** (1983) 127

Y.Ogo, Polymerizations at High Pressure, J.Macromol.Sci.–Revs.Macromol.Chem.Phys. **C 24** (1984) 1

R.Buscall, T.Corner, J.F.Stageman, Polymer Colloids, Elsevier, New York 1985

D.Hunkeler, F.Candau, C.Pichot, A.E.Hamielec, T.Y.Xie, J.Barton, V.Vaskova, J.Guillot, M.V.Dimonie, K.H.Reichert, Heterophase Polymerizations: A Physical and Kinetic Comparison and Categorization, Adv.Polym.Sci. **112** (1994) 115

J.Barton, I.Capek, Radical Polymerization in Disperse Systems, Ellis Horwood, Hemel Hempstead, UK 1994

Q.Wang, S.Fu, T.Yu, Emulsion Polymerization, Prog.Polym.Sci. **19** (1994) 703

R.G.Gilbert, Emulsion Polymerization. A Mechanistic Approach, Academic Press, San Diego (CA), 1995

P.A.Lovell, M.S.El-Aasser, Eds., Emulsion Polymerization and Emulsion Polymers, Wiley, New York 1997

References

[1] M.Stickler, G.Meyerhoff, Makromol.Chem. **179** (1978) 2729, Tables 1 and 3
[2] G.Moad, E.Rizzardo, D.H.Solomon, Makromol.Chem.-Rapid Commun. **3** (1982) 533, Table 1
[3] P.Göldi, H.-G.Elias, Makromol.Chem. **153** (1972) 277, Tables 7 and 8
[4] K.F.O'Driscoll, Pure Appl.Chem. **53** (1981) 617, modified Fig. 1
[5] W.A.Pryor, J.H.Coco, Macromolecules **3** (1970) 500, different tables
[6] B.Baysal, A.V.Tobolsky, J.Polym.Sci. **8** (1952) 529, data of Table II (see also Fig. 3)
[7] S.Beurmann, M.Buback, T.P.Davis, R.G.Gilbert, R.S.Hutchinson, O.F.Olaj, G.T.Russell, J.Schweer, A.M. van Herk, Macromol.Chem.Phys. **198** (1997) 1545, Table 1
[8] M.Kamachi, B.Yamada, in J.Brandrup, E.H.Immergut, E.A.Grulke, Eds., Polymer Handbook, Wiley, New York, 4th ed. (1999), p. II/77
[9] M.Buback, R.G.Gilbert, R.A.Hutchinson, B.Klumperman, F.-D.Kuchta, B.G.Manders, K.F.O'Driscoll, G.T.Russell, J.Schweer, Macromol.Chem.Phys. **196** (1995) 3267, Fig. 1
[10] G.V.Schulz, G.Harborth, Makromol.Chem. **1** (1948) 106, Fig. 1
[11] G.Henrici-Olivé, S.Olivé, Kunststoffe-Plastics **5** (1958) 315, Fig. 1 (modified)
[12] T.G.Carswell, D.J.T.Hill, D.I.Londero, J.H.O'Donnell, R.J.Polmery, C.L.Winzor, Polymer **33** (1992) 137; data of Figs. 2 and 3
[13] J.Shen, Y.Tian, G.Wang, M.Yang, Makromol.Chem. **192** (1991) 2669, Figs. 3, 5, 9
[14] G.V.Schulz, Chem.Techn. **3/4** (1973) 221, data of Figs. 2a and 2b
[15] H.K.Mahabady, K.F.O'Driscoll, J.Polym.Sci.-Polym.Letts. **14** (1976) 351; see also [14]
[16] K.F.O'Driscoll, Pure Appl.Chem. **53** (1981) 617, Table 1
[17] B.Soper, R.N.Haward, E.F.T.White, J.Polym.Sci. [A-1] **10** (1972) 2545, Fig. 4
[18] K.Ito, Y.Murase, Y.Yamashita, J.Polym.Sci.-Chem.Ed. **13** (1975) 87, Table I
[19] B.T.Storey, J.Polym.Sci. **A 3** (1965) 265, Table IV
[20] R.S.Whitney, W.Burchard, Makromol.Chem. **181** (1980) 869, Table 3
[21] F.R.Mayo, J.Am.Chem.Soc. **65** (1943) 2324; data of G.V.Schulz, A.Dinglinger, E.Husemann, Z.Phys.Chem. **B 43** (1939) 385, Fig. 3, Table 6
[22] A.Ueda, S.Nagai, in J.Brandrup, E.H.Immergut, E.A.Grulke, Polymer Handbook, Wiley, New York, 4th Ed. (1999), p. II/97
[23] L.Küchler, Polymerisationskinetik, Springer, Berlin 1951, Tables 15-18
[24] G.V.Schulz, Ber.Dtsch.Chem.Ges. **80** (1947) 232, Fig. 1
[25] D.Mardare, K.Matyjaszewski, Macromolecules **27** (1994) 645, data of Figs. 1, 6, 7
[26] S.O.Hammouch, J.-M. Catala, Macromol.Rapid Commun. **17** (1996) 149, data of Figs. 1 (a) and 3 (b)
[27] J.S.Wang, K.Matyjaszewski, J.Am.Chem.Soc. **117** (1995) 5614, (a) Fig. 1, (b) replotted data of Fig. 2
[28] J.Chiefari, Y.K.Chong, F.Ercole, J.Krstina, J.Jeffery, T.P.T.Lee, R.T.A.Mayadunne, G.F.Meijs, C.L.Moad, G.Moad, E.Rizzardo, S.H.Thang, Macromolecules **31** (1998) 5559 Fig. 2
[29] H.Gerstenberg, in K.Winnacker, L.Küchler, Eds., Chemische Technologie, Hanser, Munich, 4th Ed. (1982), p. 343

11 Polymerization by Radiation or in Ordered States

11.1 Survey

Polymerizations are most often performed in bulk and solution and only occasionally in the gaseous state. In fluid states (gases and liquids), molecules are disordered and have to diffuse to each other in order to attain the short distances that are required for initiation, propagation, transfer, and termination steps. The required high mobility of entities is provided by the low viscosity of the system, at least initially.

In solids, molecules and chain segments are far less mobile but their reactive sites may be so near to each other that reactions are possible. Because of such spatial arrangements, some monomers polymerize even only in ordered states but not in fluid phases. An example is $[H_3C(CH_2)_{17}O(p-C_6H_4)CH=CHCOOCH_2CH_2]_2N^\oplus(CH_3)_2Br^\ominus$ that polymerizes as vesicle or in monolayers but only dimerizes in solution.

Propagation reactions are usually initiated by radiation since diffusion of initiators into solids is difficult or impossible. Polymerizations initiated or propagated by electromagnetic radiation are called **radiation-activated polymerizations**, regardless of the state of matter. They are subdivided into **radiation-initiated polymerizations** and **radiation polymerizations**. In radiation-initiated polymerizations, the radiation initiates, but does not propagate, the polymerization. In radiation polymerizations, on the other hand, radiation causes every propagation step.

Radiation-activated polymerizations are further subdivided according to the energy of radiation (Table 11-1). High-energy radiation comprises β-rays, γ-rays and slow neutrons. Low-energy radiation, caused by visible or ultraviolet light, leads to **photo-activated polymerizations** with the subclasses of **photo-initiated polymerizations** and **photopolymerizations**, similar to the subclasses of radiation-activated polymerizations.

Table 11-1 Wavelengths λ_0 and average energies E of radiation.

Radiation	λ_0/nm	E/MJ	Polymer Applications
γ-Rays	0.001	117 000	polymerization
Electrons (150 kV)	0.008	15 000	hardening of thermosets
Ions (100 kV)	0.013	8 000	etching of resists
Electrons (20 kV)	0.060	2 000	resists
X-Rays	0.1 - 1	1 000	resists
Ultraviolet light	100 - 300	0.8	curing of resins
Ultraviolet light	300 - 400	0.4	hardening of printing inks
Laser	300 - 600	0.3	holography
Visible light	400 - 500	0.25	photoresists
Laser	10 600	0.013	evaporation of plastics

All radiation-initiated polymerizations, whether high-energy or photochemical, proceed like regular chain polymerizations, except for the initiation reaction. However, the type of propagation cannot always be predicted for radiation-initiated polymerizations. Radiation polymerizations, on the other hand, often do not proceed according to ionic or free-radical mechanisms.

Radiation may not only initiate and/or propagate polymerizations. It may also modify existing polymers. Such polymers are called **photo-active polymers** or simply **photopolymers**. A subgroup of photopolymers are **photo-crosslinkable polymers**. In photolithography, photo-active polymers are known as **photo resists**.

11.2 Photochemical Polymerization

11.2.1 Excited States

A molecule M is usually in the **ground state** S_0, which is the state with the lowest electronic energy. In this state, π electrons reside in a singlet state (L: *singulus* = single, sole). This is always the case if the spins of two similar particles such as electrons compensate each other: paired electrons have antiparallel spins.

Collisions with other molecules provide the molecule with thermal vibration energy of ca. 0.4-40 kJ/mol but the ground state will not be vacated (indicated in Fig. 11-1 by a band of parallel lines). Upon irradiation by visible or ultraviolet light, a molecule M will completely absorb in $t = 1/v \approx 10^{-15}$ s one light quantum hv (= photon) with the energy $E = hv$ according to $M + hv \rightarrow {}^1M^*$ where h = Planck constant and v = frequency. At $\lambda = 300$ nm, the absorbed *molar* energy is $E_m = hvN_A \approx 400$ kJ/mol.

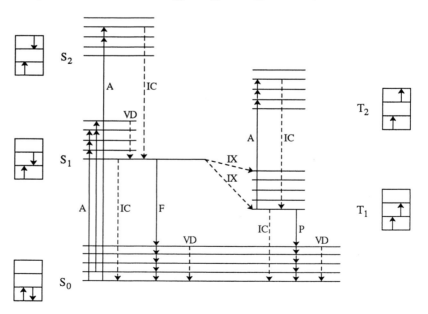

Fig. 11-1 Energy diagram (Jablonsky diagram) for simple organic molecules.
    Left and right: Relative energy levels and spin directions of electrons. S_0 = Singlet ground state, S_1 = first excited singlet state, S_2 = second excited singlet state, T_1 = first excited triplet state, T_2 = second excited triplet state.
    Vertical lines: Transitions with radiation (solid lines) or without radiation (broken lines). A = Absorption to excited singlet states, F = fluorescence, IC = internal conversion, IX = intersystem crossing, P = phosphorescence, VD = deactivation of vibrations.

This high (and thermally not achievable) energy lifts π electrons from the highest occupied molecular orbital of the ground state (HOMO) into the lowest unoccupied molecular orbital of the excited **singlet** state (LUMO). The electron thus undergoes a so-called $\pi \to \pi^*$ transition. The electron spins are no longer paired but still antiparallel in the excited S_1 state. Each of the excited states can be deactivated in 10^{-11}–10^{-13} s by molecule collisions but the excited state is not vacated.

Electron spins are also no longer paired in the excited S_1 state but still antiparallel (the same is true for the higher excited singlet states S_2, S_3, ...). However, these higher states are transformed rapidly by internal conversions into the first excited state S_1. They thus play no role compared to the other, slower photochemical processes.

The first excited state S_1 returns to the ground state S_0 by loosing excitation energy through various photophysical processes (Fig. 11-2):

- radiation-less, either according to $^1M^* \to M$ by losing heat, or according to $^1M^* \to {}^3M^*$ in 10^{-7}-10^{-8} s into the excited triplet state T_1 (intersystem crossing);
- by radiation in ca. 10^{-8} s according to $^1M^* \to M + h\nu$ with fluorescence;
- by energy transfer to a quencher.

In **triplet** states, electrons are unpaired with parallel spins. Because repulsive forces exist between parallel spins, triplet states must be of lower energy than the corresponding singlet states.

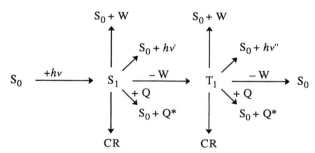

Fig. 11-2 Transitions from ground state, S_0, first excited singlet state, S_1, and first excited triplet state, T_1. CR = chemical reaction, Q = quencher, Q* = excited quencher, W = heat.

Triplet states cannot be formed directly from the ground state by absorption of a photon (**forbidden transition**). The first excited triplet state is rather obtained from the lowest excited singlet state by a radiation-less transfer (**intersystem crossing**, sometimes called **intercombination**).

Higher excited triplet states are exclusively generated from the lowest triplet state by addition of new photons and not from singlet states. Triplet states are deactivated similarly to singlet states. However, radiation is not transferred by **fluorescence** but, in more than 10^{-4} s, by **phosphorescence**.

Absorption of photons of sufficiently high energy is the necessary condition for photochemical reactions. It is followed by the so-called primary photochemical reaction of electronically excited states and then by the secondary or dark reaction of the chemical species that was generated by the primary photochemical reaction.

Molecules in excited states deliver energy to other molecules by direct transfer or by formation of excimers, exciplexes, or excitons. In **direct energy transfers**, an excited donor molecule D* transfers the electronic excitation energy to an acceptor A and returns to its ground state. The acceptor molecule becomes excited:

(11-1) $D* + A \rightarrow D + A*$

The process is also called **sensibilization** for $D \neq A$ and **energy migration** for $D = A$.

Excimers MM* are dimers of identical molecules that are stable if electronically excited but unstable in the ground state. They show fluorescence but not phosphorescence.

Exciplexes are complexes of two different species of which one is in the ground state and the other in the excited state. Exciplexes are only stable in this excited state. They are generated by an excitation of a primary charge-transfer complex (D,A) between a donor D and an acceptor A

(11-2) $D + A \; \rightleftharpoons \; (D,A) \; \xrightarrow{h\nu} \; (D^{+},A^{-})^*$

or by reaction of an excited donating molecule with an acceptor molecule in the ground state (or vice versa):

(11-3) $D \xrightarrow{h\nu} D* \xrightarrow{+A} (D*,A) \longrightarrow (D^{+},A^{-})* \longleftarrow (D,A*) \xleftarrow{+D} A* \xleftarrow{h\nu} A$

In **excitons** (pairs of electrons and holes), excited states are delocalized over a large group of highly ordered molecules, for example, in crystals. The diffusion length of excitons here may be ca. 50 nm.

11.2.2   Photoinitiation

In the simplest case, photoinitiation occurs on the monomer molecules themselves. Upon radiation, the monomer molecule becomes excited and dissociates into two radicals that initiate the polymerization. Such photoinitiations are, however, very rare.

Much more common are initiations by **photoinitiators**. Examples are *N,N*-azo*bis*isobutyronitrile and other azo compounds that decompose into two isobutyronitrile radicals, both thermally and by the action of light (Eq.(10-7)). Such **photofragmentations** are also shown by some peroxides and disulfides, and for example, benzoin derivatives:

(11-4) $C_6H_5-CO-C(OR)_2-C_6H_5 \; \xrightarrow[-\,C_6H_5\dot{C}O]{h\nu} \; {}^{\bullet}C(OR)_2-C_6H_5 \longrightarrow ROOC-C_6H_5 + R^{\bullet}$

Some free-radical polymerizations are triggered by **photosensitizers**. For example, UV light causes benzophenone, $C_6H_5-CO-C_6H_5$, to assume its excited triplet state which allows the transfer of energy to monomer molecules. **Photoreductions** occur in the presence of amines or sulfur compounds, possibly via exciplexes as intermediates:

(11-5) $Ar_2CO* \,(T_1) + R_2N-CH_2R' \longrightarrow Ar_2\dot{C}-O^{\ominus} \,/\, R_2\overset{\bullet\;\bullet}{N}{}^{\oplus}-CH_2R'$

$Ar_2\dot{C}-OH + R_2N-\dot{C}HR' \qquad Ar_2CO \,(S_0) + R_2N-CH_2R'$

Benzophenone is also photoreduced in the presence of hydrogen transfer agents:

$$(11\text{-}6) \quad (CH_3)_2CHOH \xrightarrow[\text{– Ar}_2\overset{\bullet}{C}\text{—OH}]{\text{+ Ar}_2CO^* \ (T_1)} (CH_3)_2\overset{\bullet}{C}OH \xrightarrow[\text{– Ar}_2\overset{\bullet}{C}\text{—OH}]{\text{+ Ar}_2CO} (CH_3)_2CO$$

The free-radical photopolymerization of styrene, isoprene, and vinyl acetate is accelerated by traces of oxygen. According to ultraviolet spectra, this acceleration is caused by an intermediate formation of charge-transfer complexes.

Applications of *anionic* photoinitiators are not known. *Cationic* photoinitiators are of interest for solvent-free surface coatings (paints, printing inks, adhesives) because they initiate cationic polymerizations that are insensitive toward oxygen. The top layers of these coatings are thus cured better than they are by free-radical polymerization.

Cationic photoinitiators should be completely stable in the dark, even if monomer is present. Only upon irradiation should they initiate the polymerization with high quantum yields. The process should not produce byproducts that may retard the polymerization and impair product properties. Some photoinitiators are shown in Table 11-2.

Most effective are diaryldiazonium salts, $[Ar_2I]^{\oplus}[MtX_i]^{\ominus}$, with weakly nucleophilic counterions $[MtX_i]^{\ominus}$ such as $BF_4^{\ominus}$, $SbF_6^{\ominus}$, $AsF_6^{\ominus}$, and $PF_6^{\ominus}$. Strongly nucleophilic counterions would lead to charge neutralization. At wavelengths below 200-300 nm, such cationic photoinitiators react with compounds RH and form Brønstedt acids $HMtX_i$:

$$(11\text{-}7) \quad [Ar_2I]^{\oplus}[MtX_i]^{\ominus} + RH + h\nu \rightarrow ArI + Ar^{\bullet} + R^{\bullet} + HMtX_i$$

The radicals $Ar^{\bullet}$ and $R^{\bullet}$ combine; the acids $HMtX_i$ initiate cationic polymerizations.

Charge transfer from a monomer to a non-polymerizable organic acceptor or donor or an inorganic salt, respectively, can generate complexes that are excitable by light. Polymerization may then involve free-radical or cationic steps.

N-Vinyl carbazole (VCz), for example, forms a complex with $NaAuCl_3 \cdot 2\ H_2O$ but it is unknown whether Au(III) or Au(I) is responsible for the electron transfer. The polymerization of VCz by this complex is strongly retarded by NH_3 but only slightly by O_2. It is thus presumably a cationic polymerization.

Charge transfer between a polymerizable donor and a polymerizable acceptor may also lead to a charge-transfer complex that may be excited by irradiation with light. This type of phototoinitiation either starts the homopolymerization of one of the two monomers or copolymerization of both of them (Chapter 12).

Table 11-2 Cationic photoinitiators $R^{\oplus}A^{\ominus}$ and their absorption maxima λ_{max}, absorption coefficients ε_{max}, singlet energies E_S, triplet energies E_T, and quantum yields Φ.

$R^{\oplus}$	$\dfrac{\lambda_{max}}{nm}$	$\dfrac{\varepsilon_{max}}{L\ cm^{-1}\ mol^{-1}}$	$\dfrac{E_S}{kJ\ mol^{-1}}$	$\dfrac{E_T}{kJ\ mol^{-1}}$	Φ
$(C_6H_5)_2I^{\oplus}$	270	4 300	385	310	0.95
4-Morpholino-2,5-C_4H_9-$C_6H_5N_2^{\oplus}$	407	30 000			0.52
$(C_6H_5)_3S^{\oplus}$	298	10 000	398	314	0.19
$(C_6H_5)_3Se^{\oplus}$	275	2 100			0.15
$(C_6H_5)COCH_2Pyr^{\oplus}$	278	2 340			0.0017

11.2.3   Photopolymerization

In photopolymerizations, each propagation step is activated photochemically. The steps may involve reactive ground states, excited singlets, or triplets.

Reactive ground states from a photochemical reaction participate in the photoreduction of aromatic diketones to high-molecular weight poly(benzpinacol)s. The reducing agent *i*-propanol is oxidized to acetone:

$$
(11\text{-}8)\quad
\underset{\substack{|\\C_6H_5}}{\overset{\substack{O\\\parallel}}{C}}\!-\!Ar\!-\!\underset{\substack{|\\C_6H_5}}{\overset{\substack{O\\\parallel}}{C}}
\xrightarrow[-(CH_3)_2CO]{+\,(CH_3)_2CHOH,\ h\nu}
\www\underset{\substack{|\\C_6H_5}}{\overset{\substack{OH\\|}}{C}}\!-\!Ar\!-\!\underset{\substack{|\\C_6H_5}}{\overset{\substack{OH\\|}}{C}}\www
$$

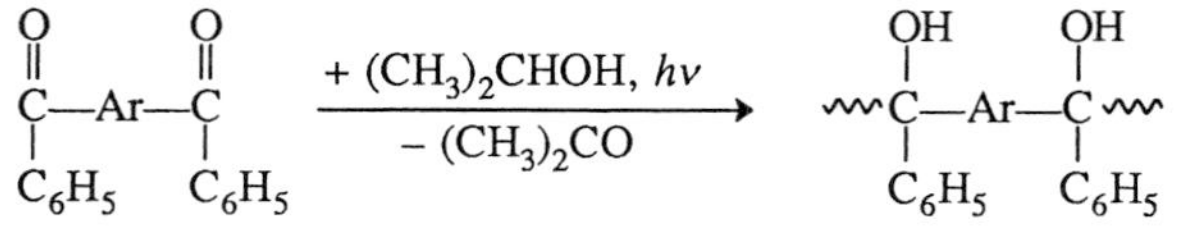

The photopolymerization of certain anthracene derivatives with bridging groups Z (e.g., $-COO(CH_2)_nOOC-$ or $-CH_2-OOC(CH_2)_nCOO-CH_2-$) proceeds via *singlet states*. This polymerization is a $(4\pi + 4\pi)$ cycloaddition:

(11-9)

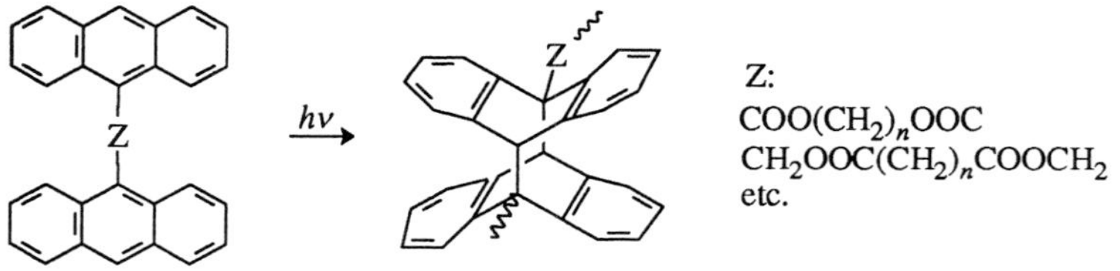

Z:
COO(CH$_2$)$_n$OOC
CH$_2$OOC(CH$_2$)$_n$COOCH$_2$
etc.

The 4-center photopolymerization of distyryl pyrazines is a $(2\pi + 2\pi)$-cycloaddition via *triplet states*. It proceeds with reasonable speed only in the solid state, e.g., in a suspension of monomer crystals:

(11-10)

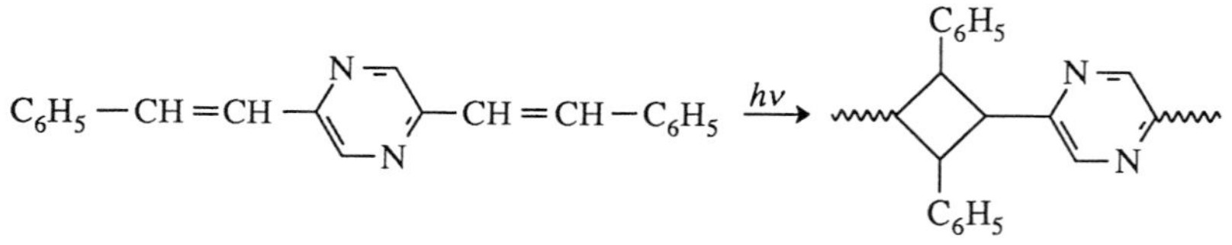

An example of an industrial free-radical photopolymerization is the crosslinking addition of the thiol groups of polythiols to the double bonds of the endgroups of polyenes, $\sim SH + CH_2{=}CH\sim \rightarrow \sim S{-}CH_2{-}CH_2\sim$. The resulting polymers are used as adhesion promoters or for the packaging of electronic parts.

The initiation rate, $R_{st} = N_{rad}N_p I_{abs}$, of photopolymerizations is controlled by the amount I_{abs} of light quanta that are absorbed per unit volume and time, the number N_p of polymer chains that are started per absorbed photon, and the number N_{rad} of radicals that are generated per initiator molecule (2 or 1). In the steady state, $R_{st} = R_{t(pp)}$, and for a mutual deactivation of macroradicals, $R_{t(pp)} = k_{t(pp)}[P^\bullet]^2$, one obtains for the rate of polymerization (see also Section 10.3.4)

$$
(11\text{-}11)\quad R_p = k_p(N_{rad}/k_{t(pp)})^{1/2}[M](N_p I_{abs})^{1/2}
$$

11.2.4 Photoactive Polymers

Some polymers are intrinsically photoactive. On radiation, they change their chemical structures which, in turn, may lead to some undesirable properties. An example is the **photo-Fries rearrangement** of polycarbonates that converts some carbonate groups into keto groups. The polymer becomes yellowish and, because of the changed packing of polymer chains, also brittle:

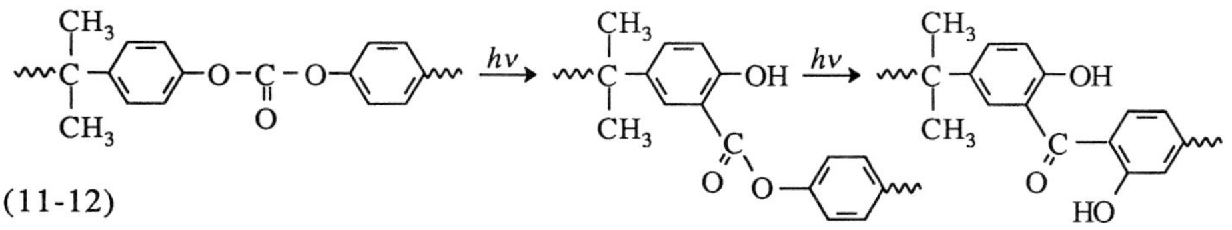

(11-12)

Phtotochemical reactions are utilized in the crosslinking of thin layers of polymers, for example, for lacquers, adhesives, or photoresists (microlithography and electronics). Crosslinkable groups may be present in the chain itself or as substituents. An example of the former is the chalcon group, $-(p-C_6H_4)-CH=CH-CO-(p-C_6H_4)-$, that undergoes a 4-center polymerization via the carbon-carbon double bond (Eq.(11-9)). Oligomeric epoxides with chalcon groups are used as photoresists or as solder resists.

Photosensitive substituents comprise azides $-N_3$, carbazides $-CON_3$, sulfonazides $-SO_2N_3$, diazonium salts $-N_2X$, and certain cycloaliphatic diazoketones. Industry also uses dimethylmaleimide groups that form cycloaddition products on irradiation by ultraviolet light in the presence of a sensitizers. These structures crosslink polymer chains:

(11-13)

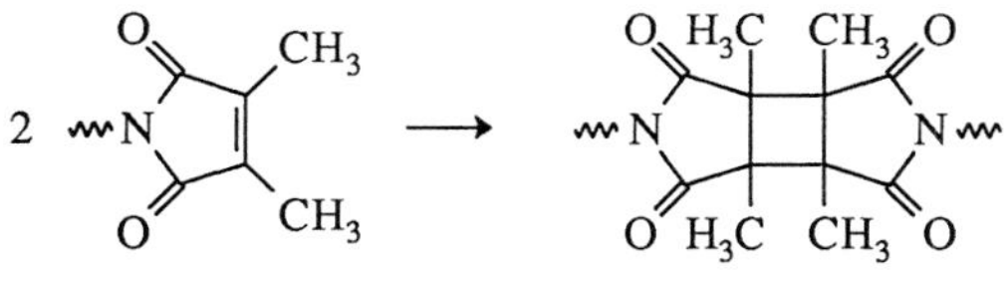

11.3 Radiation Polymerization

11.3.1 Radiation and Initiation

By passing through matter, electromagnetic radiations lose intensity by the photo-electric effect (ca. 9.6 fJ), Compton scattering (9.6-4000 fJ), and/or the generation of electron-positron pairs (ca. 160 fJ).

For high energy radiation, Compton scattering is the most important factor. A ^{60}Co source emits γ-rays that are photons of such high energy that they non-selectively displace electrons from their orbitals (primary radiation effect). In this process, photons lose part of their energy. However, photons and knocked-out electrons often retain enough energy to displace additional electrons in secondary processes. Local ionizations

are thus created near the position of the primary interaction and additional ionized products along the path of the travelling photons and their daughter products until all energy is exhausted. The resulting cations and electrons initiate ionic polymerizations.

Excited electrons with insufficient energy for the production of additional ionizations
are called **thermal electrons**. On return to the ground state, they emit low-energy photons that may initiate the polymerization of suitable monomers. The thermal electrons
finally combine with previously formed cations.

Ions may combine to form new, excited molecules. They may also react with other
molecules and start a cationic polymerization. Such polymerizations are, however, observed only for ultrapure monomers in solvents with high relative permittivity at low
temperatures. Impurities should be less than 10^{-5} %.

At not too high energy doses, the initiation rate $R_{st} = N_{rad}N_p I_{abs}$ of such polymerizations is directly proportional to the absorbed energy I_{abs} and independent of temperature (Eq.(11-11)). The absorbed energy depends to a first approximation on the total
number of electrons in the system. In dilute monomer solutions, the initiation is therefore dominated by the solvent and not by the monomer.

A 0.1 mol/L solution of ethyl vinyl ether in benzene corresponds to a solution of 11.1 mol/L
benzene. Ethyl vinyl ether possesses 33 electrons per molecule and benzene 42. The solvent benzene
thus absorbs $100 \cdot (42 \cdot 11.1/(33 \cdot 0.1 + 42 \cdot 11.1)) = 99.3$ % of the total electrons of the system. It thus
provides most of the initiating ions (e.g., $C_6H_5^\oplus$) and radicals (e.g., $C_6H_5^\bullet$).

Initiation reactions are thus not selective and it makes no sense to calculate a quantum
yield. The energy yield of such reactions is instead described by a $G(R^\bullet)$ value (**G
value**) that indicates the number of radicals that are created per 100 electron volts of absorbed energy (1 eV $= 1.6021 \cdot 10^{-19}$ J).

Resonance-stabilized molecules distribute the absorbed energy over the whole molecule. Since this leads to fewer broken bonds per molecule than in non-resonance stabilized molecules, G values of the former are lower than those of the latter. Examples are
styrene and poly(styrene) (Table 11-3). Vinyl chloride (VCM) and poly(vinyl chloride)
(PVC), on the other hand, generate chlorine radicals on irradation. Since these radicals
start chain reactions, both VCM and PVC therefore have high G values. Ethene with its
double bonds absorbs more energy than the fully saturated poly(ethylene) before radicals are formed. The G value of poly(ethylene) is thus greater than that of ethene.

The formation of ion pairs requires G $\approx$ 3-4. However, most of the initially formed
ion pairs immediately neutralize their charges. Only a few "initiating" ions survive, for
example, only (0.1/3)-(0.1/4) (= 3.3-2.5 %) in hydrocarbons (G $\approx$ 0.1) while in water (G
= 2.5), the surviving fraction is much higher since (2.5/3)-(2.5/4) (= 83-63 %).

Table 11-3 G values of some solvents, monomers, and polymers.

Substrate	G	Substrate	$G_{monomer}$	$G_{monomeric\ unit}$
Benzene	0.66	Styrene	0.66	1.5 - 3.0
Toluene	2.4			
Hexane	5.8	Ethene	4.0	6.0 - 8.0
		Isobutene	3.9	
Methyl methacrylate	6.1	Methyl acrylate	6.3	6 or 12
		Vinyl acetate	9.6	6 or 12
		Vinyl chloride	10	10 - 15

11.3.2   Polymerization

The polymerization rate $R_p = k_p[M][C^*] = (k_p/k_{t(pp)})^{1/2}[M]R_i^{1/2}$ of chain polymerizations is generally given by the rate constants k_p of propagation and $k_{t(pp)}$ of termination by mutual deactivation, the concentrations [M] of monomer and [C*] of active species, and the rate R_i of formation of the active species (see Eq.(11-11)).

The rate R_i of formation of initiating ions is smaller than that of radicals by a factor of 10-100 whereas the rate constant $k_{t(pp)}$ is about 100 times greater. The concentration of propagating ions is thus 32-100 times smaller than that of propagating radicals. Most radiation-initiated polymerizations are therefore free-radical polymerizations.

The rate equation shown above can be written as $\ln (R_p/[C^*]) = \ln k_p + K \ln [M]$ where $K \equiv 1$. In the radiation-initiated cationic polymerization of ethyl vinyl ether, the required linearity of $\ln (R_p/[C^*])$ on $\ln [M]$ was found for polymerizations in benzene, diethyl ether, and diglyme for the whole range of monomer concentrations and in methylene chloride up to high monomer concentrations. $K = 1$ was observed for CH_2Cl_2 but $K = 3/2$ for other solvents.

Radiation-initiated polymerizations of gaseous, liquid, and dissolved monomers require large capital investments. They are therefore only used for specialties such as PMMA from the polymerization of methyl methacrylate in bulk (ca. 2000 t/a) or very high-molecular weight poly(acrylamide). Smaller processes include the polymerization of acrylate- or methacrylate-impregnated wood, the grafting of surfaces of fibers and fabrics by, e.g., *N*-methylol acrylamide, the crosslinking of styrene-fumarate resins in the hardening of lacquers, the vulcanization of rubbers, and the crosslinking of poly(ethylene) for cable sheatings.

The irradiation of polymers generates radicals on chains or by homolytic chain scission. The radicals may either combine to produce branched and/or crosslinked polymers or disproportionate and form degraded polymers. In general, G values for degradation are larger than those for crosslinking. Degradation thus dominates crosslinking.

Mechanical properties depend on the degrees of degradation and crosslinking. They do not vary much upon small extents of degradation but change quite strongly for large ones. A little crosslinking, on the other hand, creates dramatic property changes but these do not change much for high degrees of crosslinking (see Volume III). Crosslinking by radiation thus leads to a maximum of properties as function of the radiation dose.

11.4   Plasma Polymerization

Plasma polymerizations are conversions of low-molecular weight compounds into high-molecular weight polymers that are performed in gas plasmas. **Plasma-initiated polymerizations** are conventional polymerizations of monomers that are initiated by one or more unconventional species; they are also called **plasma-induced polymerizations**. **Plasma-propagated polymerizations** of conventional monomers, on the other hand, involve unusual structures and mechanisms (see below).

Plasma polymerizations operate solvent-free in vacuum. They may be continuous or discontinuous and are mainly used for the coating of surfaces.

11.4.1   Plasmas

A gas plasma is a partially ionized gas consisting of ions, electrons, and neutral species (G: *plassein* = to mold). It is often produced by glow discharges but may also be created by flames, electric discharges, electron beams, lasers, or nuclear fusion. Gas plasmas by arcs or plasma jets are also called **equilibrium plasmas** since the temperature T_{el} of the resulting electrons equals the temperature T_{gas} of the gas. These temperatures may be several thousand kelvins. They are thus unsuitable for plasma polymerizations.

Plasmas from molecules, molecular fragments, atoms, and electrons that are produced by glow discharges are **non-equilibrium plasmas** of considerably lower temperatures ($T_{el}/T_{gas} \approx 10\text{-}100$). The gaseous *atoms* and *molecules* of these plasmas have kinetic energies that correspond to those of species at relatively low temperatures of several hundred degrees Celsius. The free *electrons* of these plasmas possess far higher kinetic energies, however: they correspond to temperatures of more than 10 000°C.

Plasmas by glow discharge have average electron energies of $(2\text{-}20)\cdot10^{-19}$ J and electron concentrations of $(10^9\text{-}10^{12})$ electrons/cm^3. The free electrons pick up energy from the applied electric fields. If the resulting "hot" electrons collide with neutral gas molecules, energy is transferred to the latter which leads to the breakage of covalent bonds. The resulting molecular fragments become precursors of plasma polymers.

11.4.2   Monomers and Polymerizations

Plasma polymerizations are usually conducted in vacua of (0.13-1300) Pa (equals ca. (10^{-3}-10) torr). The "monomers" are therefore very dilute. For example, gaseous styrene at 133 Pa (= 1 torr) is ca. $6\cdot10^6$ times less concentrated than liquid styrene.

Practically all organic compounds that can be used as educts ("monomers") for plasma polymerizations have vapor pressures of less than 1 mbar at room temperature. Special functional groups are not required. Examples include ethane, tetrafluoromethane, or mixtures of CO, H_2, and N_2. Chemical compounds with olefinic carbon-carbon double bonds, aromatic groups, amino groups, nitrile groups, or silicon are easier to polymerize than aliphatics, cycloaliphatics, or compounds with hydroxy, ether, carbonyl, or chlorine groups. Conventional monomers for plasma polymerizations include ethene, propene, butadiene, styrene, trioxane, and hexamethyl cyclotrisiloxane.

The constitutions of educts ("monomers") and resulting products ("polymers") are not stoichiometrically related. For example, the plasma polymerization of ethane, C_2H_6, leads to a polymer $(C_2H_3)_n$, i.e., to a net loss of hydrogen. The polymerization of ethene, C_2H_4, in the presence of air generates a polymer $(C_2H_{2.6}O_{0.4})_n$.

The number N_H of lost hydrogen atoms per unit depends linearly on the number $N_{H,0}$ of hydrogen atoms per educt. The loss is greatest for aliphatics and smallest for highly unsaturated educts:

(11-14) $N_H =\ \ \ 0.74\ +\ \ 0.12\ N_{H,0}$;   aliphatics

(11-15) $N_H =\ \ \ 0.21\ +\ \ 0.10\ N_{H,0}$;   cycloaliphatics, olefins

(11-16) $N_H = -\ 0.20\ +\ \ 0.10\ N_{H,0}$;   cycloolefins, diolefins, alkynes

(11-17) $N_H =\ \ \ 0\ \ \ \ +\ \ 0.02\ N_{H,0}$;   benzene, benzene derivatives

Plasma polymerizations lead mostly to crosslinked films but sometimes to oils of highly branched oligomers. Plasma "polymerization" is thus a misnomer since it is not a molecular polymerization of monomers but a reaction of molecule fragments. Plasma polymers of ethane or ethene contain not only aliphatic units but also unsaturated and aromatic ones.

The rate of plasma polymerization depends on the chemical structure of educts, the gap between electrodes, the frequency and intensity of the glow discharge, the type of reactor, the pressure, and the gas velocity. Reactive species dominate at low flow rates and the polymerization rate is controlled by the rate of the educt feed-in. At high flow rates, plenty of educt is offered per unit time and the polymerization rate depends on the residence time. With increasing flow rate, the polymerization rate thus passes through a maximum.

11.5 Solid-State Polymerization

11.5.1 Survey

Some monomers can be polymerized not only in bulk, in solution, or in the gaseous state but also in the solid state. There are even polymerizations that proceed exclusively in the crystalline state. Some of these polymerizations are astonishingly fast. Others deliver polymers with constitutions that differ from those obtained in fluid phases. For example, the polycondensation of salts of p-halogen thiophenols

$$(11\text{-}18) \qquad \text{Hal–}(p\text{-}C_6H_4)\text{–SMt} \quad \longrightarrow \quad \sim(p\text{-}C_6H_4)\text{–S}\sim \ + \ \text{MtHal}$$

with Hal = F, Cl, or Br, and Mt = Li, Na, or K delivers crosslinked poly(p-thiophenylene)s ("poly(phenylene sulfide)s") in melt reactions but linear high-molecular weight polymers if monomer crystallites are polymerized.

Most solid-state polymerizations are academic curiosities but some are industrially important, for example, after-polycondensations of melt-polymerized polyamides as a means to increase molecular weights. The polymerizations are usually performed at temperatures T_G below the glass temperature $T_{G,\infty}$ of polymers of infinitely high molecular weights in order to prevent polymer degradation. T_G is, however, surpassed at fairly low monomer conversions and thus at low molecular weights (see Chapter 13).

Some monomers polymerize "spontaneously", for example, crystals of p,p'-divinyldiphenyl at room temperature. In most cases, however, solid-state polymerizations must be externally initiated. The methods of choice are photochemical or radiation initiations since initiators diffuse into crystals either with difficulty or not at all.

Chemical reactions in crystals, including polymerizations of monomer crystals, are called **topochemical reactions** (G: *topos* = a place). These reactions are either controlled by the reactants or by the resulting products. In homogeneous topochemical reactions, structure and symmetry of products are regulated by the packing of educts in the lattice. In heterogeneous topochemical reactions, lattice defects initiate the reaction which is guided by the order of monomer molecules near these defects.

Topotacticity is the geometric control of a topochemical reaction. A topotactical reaction is thus always topochemical but a topochemical reaction is not necessarily topotactic.

11.5.2 Initiation Reactions

There are many indications that topochemical reactions begin at defects if crystals are irradiated by high-energy beams: the reaction starts where the crystal surface has been scratched. The starting points are distributed irregularly.

Monomers and polymers usually differ in their densities. The polymerization of monomer crystals thus leads to stresses that cause new lattice defects which, in turn, initiate polymerizations. Electron micrographs of such systems show craters that are later surrounded by satellite craters.

It is not easy to determine whether an initiation by radiation leads to a free-radical or an ionic polymerization. Electron spin resonance often shows signals, but that does not mean that these radicals did indeed initiate the polymerization. Therefore, one relies in many cases on "chemical experience": if a monomer polymerizes only cationically in solution, it should not polymerize free-radically in crystals and vice versa.

The action of inhibitors is also not a sure proof for or against a free-radical mechanism. For example, addition of 5 % benzoquinone, an inhibitor of free-radical polymerization (p. 350 ff.), lowers the polymerization rate of crystalline acrylonitrile by 50 % but the same effect is also caused by addition of 5 % toluene, an inactive compound. Proof of free-radical polymerization by the action of added inhibitors can only be assumed if inhibitor and monomer are isomorphous, concentrations of lattice defects stay constant, and inhibitors are present in high concentration. The same is true for copolymerizations as criteria of mechanisms (Chapter 12).

Other criteria are also not necessarily proof of a certain mechanism. Activation energies of topochemical polymerizations are often unusually low (see below). However, this may be caused by the crystal structure since irradiation does require an activation energy for the initiation reaction. The same reasoning applies to the solid-state polymerization of monomers that polymerize only ionically in fluid phases.

Polymerizations by high-energy radiation usually proceed free-radically (Section 11.3.1). According to ESR measurements, the initiation reaction seems to be a disproportionation of monomer molecules, for example:

$$(11\text{-}19) \quad 2 \ CH_2{=}CHR \quad \longrightarrow \quad CH_3{-}C^\bullet HR \ + \ CH_2{=}C^\bullet R$$

11.5.3 Propagation, Termination, and Transfer

The polymerization of a crystalline monomer to a crystalline polymer corresponds to a phase transition. A thermodynamically possible polymerization should thus lead to 100 % polymerization (see Section 7.3.4). Maximum monomer conversions of less than 100 % therefore point toward kinetic effects (Table 11-4).

Table 11-4 Maximum monomer conversions u_{max} during the polymerization of crystalline monomers at various temperatures. T_M = melting temperature of monomer; T_p = polymerization temperature.

Monomer	$T_M/°C$	$T_p/°C$	$u_{max}/\%$	
Acrylamide	85	27	100	
Formaldehyde	- 92	- 196	45	
		- 131	23	
Acrylonitrile	- 82	- 196	4	without after-polymerization
		- 196	11	with after-polymerization
		- 90	22	

For example, polymerizations are kinetically inhibited if already formed macromolecules block the path of growing polymer chains. The polymerization stops if all monomer is used up in this region even though macroradicals are preserved.

Solid-state polymerizations can be divided into two classes. One group of monomers polymerize just below the melting temperature of the monomer (Fig. 11-3) with high activation energies, for example, 26.4 kJ/mol for β-propiolactone, 40.2 kJ/mol for hexamethyl cyclotrisiloxane, 77 kJ/mol for trioxane, and 96 kJ/mol for acrylamide. Only a fraction of these overall activation energies, however, is needed for the propagation step (Table 10-6). The overall activation energy must therefore be increased by the mobility of monomer molecules in the crystal. Since the polymerization rates are also high (about 20 times higher below T_M than above), high preexponential factors must accompany the high activation energies. The preexponential factors are high because the steric factor is considerably reduced by the orientation of the monomer molecules in the crystal.

The second group of monomers polymerize far below the melting temperature albeit very slowly. The polymerization rates of these monomers do not depend on temperature; the thermal activation energy is zero. This group of monomers may include formaldehyde although it has an activation energy of 11.7 kJ/mol.

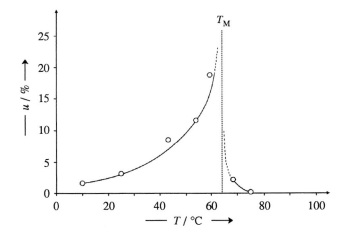

Fig. 11-3 Temperature dependence of conversion u of crystalline hexamethylcyclotrisiloxane by γ-rays. Exposure: 590 C/kg (= $2.3 \cdot 10^6$ roentgen); dose rate $5 \cdot 10^{-5}$ C/(kg s)) [1].

Many monomers continue to polymerize after the irradition stops. Such **after-polymerizations** may continue for weeks. They are probably caused by trapped radicals or
ions (see previous page). The contribution of after-polymerizations to the total polymerization is difficult to determine. One way is to heat the samples with different rates, determine the monomer conversion, and extrapolate the data to a heating time of zero.

Solid-state polymerizations may also show chain transfers. For example, as little as
2 % propionamide may lower the molecular weight of the isomorphous acrylamide but
ca. 50 % is needed to affect the polymerization rate. The non-isomorphous system
acrylamide-acetamide shows no chain transfer. True termination reactions are unknown.

11.5.4   Morphology

The radiation-induced polymerization of crystalline monomers usually delivers atactic, non-crystalline polymers. The monomer orientation in the crystal does not effect the
propagation because of the difference of densities of monomer and polymer.

Topotactic polymerizations can be expected if monomers and their polymers have
similar crystal lattices. Example are diacetylenes, $R-CH_2-C\equiv C-C\equiv C-CH_2-R$, that are arranged in crystal lattices in such a manner that the conjugated triple bonds form the
rungs of a ladder (Fig. 11-4). The substituents R constitute the side-rails; they are held
together by hydrogen bonds between suitable substituents R– such as $C_6H_5-NH-COO-$,
$HOCH_2-$, or $HOOC(CH_2)_8-$.

The polymerization of the colorless diacetylene crystals can be triggered by heat,
ultraviolet light, X-rays, or mechanical stress. During propagation, groups are sheared;
densities of monomers and polymers thus do not differ much.

The structure of monomeric units is $=C(CH_2R)-C\equiv C-C(CH_2R)=$ according to Raman
spectroscopy. The polymers are deeply colored because of the alternating double and
triple bonds. The color change on polymerization can be used as an indicator: at elevated temperatures, colorless diacetylenes in printing inks of food labels become colored
polymers and indicate that the allowed storage temperature was exceeded.

Acetylenes also polymerize topotactically to cis or trans polymers, depending on initiation (Volume II). On suitable doping, completely topotactic poly(acetylene)s have electrical conductivities that approach those of metals.

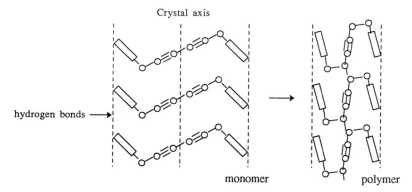

Fig. 11-4 Crystal structures of 1,6-di(*N*-carbazolyl)-2,4-hexadiyne and its polymers.

The conformation of topochemical produced polymers often depends strongly on the crystal structure of their monomers. For example, the polymerization of tetraoxane, *cycl*-$(OCH_2)_4$, leads to poly(oxymethylene)s, $+OCH_2+_n$ in helical conformations whereas that of trioxane, *cycl*-$(OCH_2)_3$, delivers these polymers in all-trans conformations.

A high orientation of polymer chains is mainly due to the crystallizability of polymer molecules and less to the orientation of molecules in monomer crystals. In all cases in which a polymerization is possible below the melting temperature of monomers, one also observes a certain mobility of monomer molecules in their crystal lattices. However, high mobilities of monomer molecules reduce the ability of polymer molecules to orient themselves in lattices. A high orientation of polymer chains must therefore be due to the crystallizability of polymer molecules.

11.6 Polymerization in Other Ordered States

The polymerization of monomer crystals (Section 11.5) is a topochemical reaction in a three-dimensionally ordered state. However, topochemical reactions may also proceed in two-dimensionally and one-dimensionally ordered states, both with and without carriers. In water, amphiphilic monomer molecules arrange themselves spontaneously in polymerizable micelles or vesicles (Section 11.6.1). In matrix polymerizations, monomer molecules are one-dimensionally bound to carrier molecules by physical bonds and then polymerized (Section 11.6.2). In Langmuir troughs, amphiphilic molecules may also be spread two-dimensionally in monomolecular, bimolecular, etc., layers that are subsequently polymerized (Section 11.6.3. Monomers may also orient themselves in lyotropic or thermotropic mesophases (Volume III) and subsequently be polymerized (see also Section 16.3.2 ff.).

Polymerizations of monomer molecules in clathrates do not seem to offer special advantages with respect to the polymerizability of monomer molecules and to the chemical and physical structure of the resulting polymers. Such clathrates are formed by monomer molecules in cyclodextrins (p. 629 and Volume II), thiourea, or perhydrotriphenylene. For example, copolymerizations in clathrates lead to the same polymer composition as copolymerizations of "free" monomer molecules.

11.6.1 Polymerization of Micelles and Vesicles

Amphiphilic monomer molecules consisting of a hydrophilic head group and *one* polymerizable hydrophobic tail group associate spontaneously to form **micelles** in aqueous solutions (Fig. 11-5, see also Fig. 10-28) (L: *mica* = grain). Micelles are spherical at low concentrations and often rod-like at higher ones. At still higher concentrations, they usually form mesophases (Volume III).

Monomer molecules with one hydrophilic head group and *two* polymerizable hydrophobic tails and similar compounds aggregate on sonication of their aqueous solutions to **vesicles** and not to micelles (L: *vesicula*, diminutive of *vesica* = bladder, blister). Vesicles are spherical particles with far greater diameters than spherical micelles (Fig. 11-5). They consist of an aqueous interior that is surrounded by a bimolecular "membrane" of

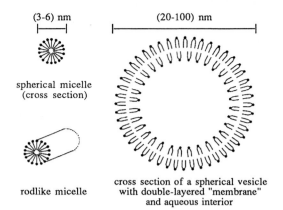

spherical micelle
(cross section)

rodlike micelle

cross section of a spherical vesicle
with double-layered "membrane"
and aqueous interior

Fig. 11-5 Associates of amphiphilic monomer molecules in aqueous solution: Micelles from molecules with one hydrophilic head group (●) and one hydrophobic tail (——), and vesicles with one hydrophilic headgroup and two hydrophobic tails (⊂). Polymerizable groups are usually in the hydrophobic parts.

ordered monomer molecules. The membranes form fairly stable cages for enclosed molecules or particles such as active substances. In contrast to micelles, vesicles are not in equilibrium with their non-associated monomer molecules (= "unimers"). They are rather metastable entities with lifetimes of days or weeks.

Depending on the arrangement of hydrophilic head groups and polymerizable hydrophobic tails, four simple types of vesicle-forming monomer molecules and resulting polymerized vesicles can be distinguished (Fig. 11-6). On polymerization, the shape of the vesicle is preserved.

Monomers (examples)   Polymer types

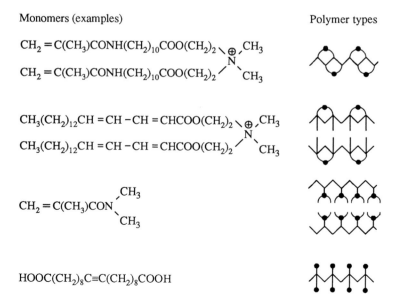

Fig. 11-6 Examples of monomers that form vesicles (left) and their polymers (right; schematic). Hydrophilic groups are indicated by ● [2].

Polymerized vesicles have a much larger lifetime than unpolymerized ones. Such polymerized vesicles and other "nanoparticles" serve as carriers for pharmaceuticals that can be injected and thus transported to the desired site in the body without passing through the stomach and the digestive tract.

Micelles of polymerizable monomers, on the other hand, are in dynamic equilibrium with their unimers. Polymerizations *of* such micelles are obviously different from that of emulsion polymerizations, which are the polymerizations of solubilized monomer molecules *in* micelles of molecules that are not monomers themselves.

Micelles of monomer molecules should be polymerizable if the lifetimes of micelles and growing polymer chains are comparable. The average residence time of a monomer molecule in such a micelle is ca. 10^{-6} s and the average lifetime of a micelle itself about 10^{-2} s. In bulk or solution, a propagation step requires 10^{-3}-10^{-4} seconds. In a micelle, however, polymerizable hydrophobic groups are much more tightly packed and oriented than the same groups in bulk. The resulting higher preexponential factors and smaller entropy losses should thus lead to much faster propagation steps in micelles and should approach the average residence time of a monomer molecule in the micelle. Since a micelle consists of 10-100 monomer molecules, the time requird to polymerize a micelle should be much smaller than the average lifetime of a micelle.

Styrene trimethylene glucoside, $CH_2=CH(p\text{-}C_6H_4(CH_2)_3\text{-glucose}$, for example, forms micelles in water that solubilize the "oil-soluble" free-radical initiator 2-(phenylazothio)-naphthalene (ATE). The degree of polymerization of the polymerized micelles (P) was found to equal the degree of association of the micelles before the polymerizaation. The polymolecularity index was low ($\overline{X}_w/\overline{X}_n = 1.04$) and did not change with monomer conversion (Fig. 11-7). All molecules of a micelle thus became one single polymer molecule. Bigger polymer molecules were observed only at larger monomer conversions (tail to the left side of P in the polymerization by ATE, Fig. 11-7, right).

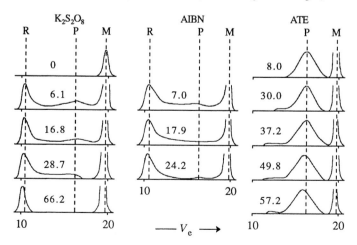

Fig. 11-7 Molecular weight distributions (by SEC, V_e in 5 mL) during the polymerization of dilute aqueous solutions of styrene trimethylene glucoside by the free-radical initiators ATE, AIBN, or $K_2S_2O_8$ at 50°C [3]. Curves were standardized. Numbers indicate monomer conversion in percent.
M = monomer, P = low-molecular weight polymer (polymerized micelles with $\overline{X}_n = 24.3$ (see above), R = high-molecular weight polymer ($\overline{X}_n \approx 1050$). Degree of association before polymerization: 23.7.

The oil-soluble initiator *N,N-azobis*isobutyronitrile (AIBN) and the water-soluble initiator dipotassium sulfur peroxide ($K_2S_2O_8$) give relatively few polymerized micelles (P in Fig. 11-7). They rather lead to higher molecular weights (R) that become more uniform with increasing monomer conversion (R peak: $\overline{X}_w/\overline{X}_n \approx 1.09$ for $u = 66.2$ % by $K_2S_2O_8$). ATE seems to polymerize almost exclusively spherical micelles whereas AIBN and $K_2S_2O_8$ attack predominantly larger, rodlike micelles. In the latter cases, the proportion of polymerized smaller micelles disappears at high monomer conversions.

11.6.2 Matrix Polymerization

In matrix polymerizations, monomers M are attached to macromolecular matrices (templates T) and polymerized. Subsequently, the bonds between template units T and monomer units m are severed:

(11-20)   ⌇⌇T—T—T⌇⌇ ⟶ ⌇⌇T—T—T⌇⌇ ⟶ ⌇⌇T—T—T⌇⌇ ⟶ ⌇⌇T—T—T⌇⌇
                               │ │ │   │ │ │   +
                              M M M   ⌇⌇m–m–m⌇⌇   ⌇⌇m–m–m⌇⌇

In these structure-controlled syntheses, molecular weights and chemical structures (monomer sequences) of the resulting polymers ~~m–m–m~~ are governed by morphological patterns. Examples are the biosyntheses of nucleic acids and enzymes (Chapter 14 and Volume II) that lead to completely uniform compounds with respect to monomeric sequences and molecular weights.

Matrix polymerizations to synthetic polymers were only partially successful. As a rule, neither chemical nor physical bonds of monomers to templates guarantee constitutionally and configurationally uniform polymers. In the free-radical polymerization of methyl methacrylate in the presence of isotactic poly(methyl methacrylate) as template, the predominantly formed partially syndiotactic polymer chains combine with it-PMMA chains. The resulting stereocomplex is the driving force for an increased syndiospecific polymerization if the growing chains surpass a certain minimum length: 10-20 for it-PMMA and ca. 600 for st-PMMA.

11.6.3 Carrier-Supported Polymerization

Monomers can form monomolecular and multimolecular monomer layers on solid or liquid surfaces, for example, by the Langmuir-Blodgett method (Fig. 11-8). In this method, a dilute solution of an amphiphilic monomer in an organic solvent ($\rho < \rho_{water}$) is placed on the water surface of a Pockels-Langmuir trough (I, left). The hydrophilic heads ● of the monomer molecules face the water surface (I, right); the molecules themselves are distributed at random or in clusters. The molecules are pushed together by a movable barrier (II, left) to form a tightly packed surface film (II, right) in which all hydrophobic tails are now parallel. By pulling out the barrier (III, left), a surface film of oriented monomer molecules is deposited on the barrier (III, right). A renewed dipping of the barrier (IV, left) into the trough generates a second layer (IV, right) in which the hydrophobic tails of the first and second layer face each other.

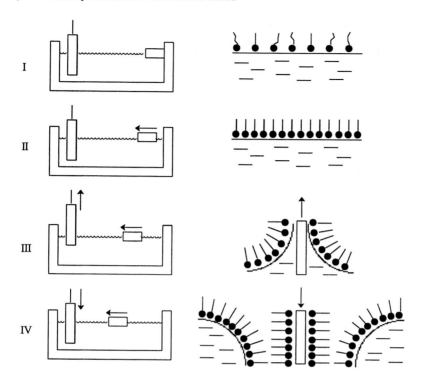

Fig. 11-8 Formation of mono and double layers on water by the Langmuir-Blodgett method.

Polymerizable groups in hydrophobic tails of suitable monomers can be polymerized by irradiation, leading to oriented monolayers, bilayers, etc., of polymers. The polymerization depends strongly on the surface pressure. It is faster in islands of monolayers than in continuous surface layers, probably because islands can much more easily expand laterally on polymerization than continuous layers. The chemical structure, the morphology, and the properties of polymerized monolayers depend strongly on the structure and properties of the monolayers (see Volume III).

Literature to Chapter 11

11.1 SURVEY
G.Allen, J.C.Bevington, Eds., Comprehensive Polymer Science; Vol. 3, G.C.Eastmond, A.Ledwith, S.Russo, P.Sigwalt, Eds., Chain Polymerization I, Pergamon Press, Oxford 1989

11.2 PHOTOCHEMICAL POLYMERIZATIONS
D.R.Arnold, N.C.Baird, J.R.Bolton, J.C.D.Brand, P.W.M.Jacobs, P.de Mayo, W.R.Ware, Photochemistry - An Introduction, Academic Press, New York 1974
M.Gordon, W.R.Ware, Eds., The Exciplex, Academic Press, New York 1975
B.Rånby, J.F.Rabek, Eds., Singlet Oxygen, Wiley, New York 1978
N.J.Turro, Modern Molecular Photochemistry, Benjamin-Cummings, Menlo Park (CA), 1978

J.F.McKellar, N.S.Allen, Photochemistry of Man-Made Polymers, Appl.Sci.Publ., London 1979
C.G.Roffey, Photopolymerization of Surface Coatings, Wiley, Chichester 1982
A.Ledwith, Photochemical Cross-Linking in Polymer-Based Systems, Dev.Polym. **3** (1982) 55
W.L.Dilling, Polymerization of Unsaturated Compounds by Photocycloaddition Reactions,
    Chem.Revs. **83** (1983) 1
J.V.Crivello, Photoinitiated Cationic Polymerization, Ann.Rev.Mater.Sci. **13** (1983) 173
H.Baumann, H.-J.Timpe, H.Böttcher, Initiatorsysteme für kationische Polymerisationen, Z.Chem.
    23 (1983) 394
N.S.Allen, W.Schnabel, Eds., Photochemistry and Photophysics in Polymers, Elsevier, New York
    1984
J.Guillet, Polymer Photophysics and Photochemistry, Cambridge Univ.Press, Cambridge, UK, 1984
D.Phillips, Ed., Polymer Photophysics, Luminescence, Energy Migration and Molecular Motion in
    Synthetic Polymers, Chapman and Hall, New York 1985
J.F.Rabek, Mechanisms of Photophysical Processes and Photochemical Reactions in Polymers:
    Theory and Applications, Wiley, Chichester 1987
A.Reiser, Photoreactive Polymers. The Science and Technology of Resists, Wiley, Chichester 1989
H.Böttcher, J.Bendig, M.A.Fox, G.Hopf, H.-J.Timpe, Technical Applications of Photochemistry,
    Deutscher Verlag für Grundstoffindustrie, Leipzig 1991
G.J.Kavernos, Ed., Fundamentals of Photoinduced Electron Transfer, VCH, Weinheim 1993
J.-P.Fouassier, Photoinitiation, Photopolymerization, and Photocuring, Hanser, Munich 1996

11.3 RADIATION POLYMERIZATION
A.Charlesby, Atomic Radiation and Polymers, Pergamon Press, Oxford 1960
A.Chapiro, Radiation Chemistry of Polymeric Systems, Interscience, New York 1962
M.Dole, Ed., The Radiation Chemistry of Macromolecules, Academic Press, New York, 2 vols. 1972
J.E.Wilson, Radiation Chemistry of Monomers, Polymers, and Plastics, Dekker, New York 1974
F.A.Makhlis, Radiation Physics and Chemistry of Polymers, Halsted, New York 1975
D.R.Randall, Ed., Radiation Curing of Polymers, CRC Press, Boca Raton (FL) 1987
J.G.Drobny, Radiation Technology for Polymers, CRC Press, Boca Raton (FL) 2003

11.4 PLASMA POLYMERIZATION
M.Shen, Plasma Chemistry of Polymers, Dekker, New York 1976
A.T.Bell, The Mechanism and Kinetics of Plasma Polymerization, in S.Veprek, M.Venugopalan,
    Eds., Plasma Chemistry III (= Topics Current Chem. **94**), Springer, Berlin 1980
H.V.Boenig, Plasma Science and Technology, Cornell Univ.Press, Ithaca (NY) 1982; Hanser,
    Munich 1982
H.Yasuda, Plasma Polymerization, Academic Press, Orlando (FL) 1985
H.Biederman, Y.Osada, Plasma Polymerization Processes, Elsevier, Amsterdam 1993
R.d'Agostino, Plasma Deposition, Treatment, and Etching of Polymers, Academic Press,
    San Diego 1990
N.Inagaki, Plasma Surface Modification and Plasma Polymerization, Technomic, Lancaster (PA), 1996

11.5 SOLID-STATE POLYMERIZATION
G.C.Eastwood, Solid State Polymerization, Prog.Polym.Sci. **2** (1970) 1
M.Nishii, K.Hayasaki, Solid State Polymerization, Ann.Rev.Mater.Sci. **5** (1975) 135
G.Wegner, Solid-State Polymerization Mechanisms, Pure Appl.Chem. **49** (1977) 443
R.H.Baughman, K.C.Yee, Solid-State Polymerization of Linear and Cyclic Acetylenes,
    Macromol.Revs. **13** (1978) 219
M.Hasegawa, Four-Center Photopolymerization in the Crystalline State, Adv.Polym.Sci. **42** (1982) 1
M.Hasegawa, Photopolymerization of Diolefin Crystals, Chem.Revs. **83** (1983) 507

11.6 POLYMERIZATIONS IN OTHER ORIENTED STATES
H.-G.Elias, Ed., Polymerization of Organized Systems, Gordon and Breach, New York 1977
E.M.Barall II, J.F.Johnson, A Review of the Status of Polymerization in Thermotropic Liquid
    Crystal Media and Liquid Crystalline Media, J.Macromol.Sci.-Revs.Macromol.Chem. C **17**
    (1979) 137
K.Takemoto, M.Miyata, Polymerization of Vinyl and Diene Monomers in Canal Complexes,
    J.Macromol.Sci.-Revs.Macromol.Chem. C **18** (1980) 83

J.Fendler, Polymerization in Organized Surfactant Assemblies, in K.Mittal, B.Lindman, Eds., Surfactants Solution (Proc.Int.Symp. (4th) **3** (1982) 1947), Plenum, New York 1984

H.Bader, K.Dorn, B.Hupfer, H.Ringsdorf, Polymeric Monolayers and Liposomes as Models for Biomembranes, Adv.Polym.Sci. **64** (1985) 1

P.Girot, P.Couvreur, Ed., Polymeric Nanoparticles and Microspheres, CRC Press, Boca Raton (FL) 1986

S.L.Regen, Polymerized Liposomes, in M.J.Ostro, Ed., Liposomes: From Biophysics to Therapeutics, Dekker, New York 1987

H.Ringsdorf, B.Schlarb, J.Venzmer, Molecular Architecture and Function of Polymeric Oriented Systems, Angew.Chem. **100** (1988) 117; Angew.Chem.Int.Ed.Engl. **27** (1988) 113

C.M.Paleos, Ed., Polymerization in Organized Media, Gordon and Breach, New York 1992

D.F.O'Brien, Polymerization of Supramolecular Assemblies, Trends Polym.Sci. **2** (1994) 183

K.Nagai, Radical Polymerization and Potential Applications of Surface-Active Molecules, Trends Polym.Sci. **4** (1996) 122

References to Chapter 11

[1]   E.J.Lawton, W.T.Grubb, J.S.Balwit, J.Polym.Sci. **19** (1956) 455, Fig. 1
[2]   H.Ringsdorf, B.Schlarb, J.Venzner, Angew.Chem. **100** (1988) 117; Angew.Chem.Int.Ed. Engl. **27** (1988) 113
[3]   K.Nagai, H.-G.Elias, Makromol.Chem. **188** (1987) 1095, Fig. 10

12   Copolymerization

12.1   Overview

Copolymers are defined by IUPAC as polymers that are "derived from more than one species of monomer". A copolymerization is therefore a "polymerization in which a co-polymer is formed". An example is the joint free-radical bipolymerization of methyl methacrylate, $CH_2=C(CH_3)COOCH_3$, and styrene, $CH_2=CH(C_6H_5)$, to a copolymer with statistically distributed methyl methacrylate and styrene units, $-CH_2-C(CH_3)COOCH_3-$ and $-CH_2-CH(C_6H_5)-$. In the older literature, such copolymerizations have been called **heteropolymerizations** or **interpolymerizations**.

Similarly, ionic copolymerizations of formaldehyde, HCHO, and ethylene oxide (oxi-rane), cyclo-$(CH_2)_2O$, usually deliver copolymers with randomly arranged formalde-hyde and ethylene oxide units, $-O-CH_2-$ and $-O-CH_2-CH_2-$, but a practically alterna-ting copolymer, $+O-CH_2-O-CH_2-CH_2+_n$, can also be obtained under certain copolymerization conditions. A "copolymer" of alternating $-O-CH_2-$ and $-O-CH_2-CH_2-$ units is obtained from the homopolymerization of 1,3-dioxolane:

(12-1)

$$\text{homo-polymerization} \qquad \text{co-polymerization}$$

The joint polymerization of two species of monomers is not called a "copolymeriza-tion" if the resulting alternating copolymer can be considered a homopolymer from an **"implicit monomer"**. An example is the polycondensation of stoichiometric amounts of adipic acid (**A**), $HOOC(CH_2)_4COOH$, and hexamethylene diamine (**H**), $H_2N(CH_2)_6NH_2$, to poly(hexamethylene adipamide), $HO[OC(CH_2)_4CONH(CH_2)_6NH]_nH$ (polyamide 66, nylon 6.6). This joint polymerization can be viewed as the homopolymerization of an implicit monomer, the so-called **nylon salt** (**AH salt**), $HOOC(CH_2)_4CONH(CH_2)_6NH_2 = [^{\ominus}OOC(CH_2)_4COO^{\ominus}][^{\oplus}NH_3(CH_2)_6NH_3^{\oplus}]$.

A polymer obtained by the *complete* transformation of another polymer (polymer analog reaction) is usually considered to be the homopolymer of a hypothetical monomer. An example is poly(vinyl alcohol), $+CH_2-CH(OH)+_n$, from the complete transesterification or saponification of poly(vinyl acetate), $+CH_2-CH(OOCCH_3)+_n$. An incomplete transformation delivers an irregular polymer with vinyl alcohol and vinyl acetate units similar to a true copolymer. Such polymers are therefore often called **pseudo copolymers**.

Pseudo copolymers also result if a monomer isomerizes before or during the homo-polymerization. Certain polymerization catalysts partially isomerize ethylene oxide to acetaldehyde; the resulting polymer contains both $-O-CH_2-CH_2-$ and $-CO-CH(CH_3)-$ units. Poly(butadiene)s are often pseudo copolymers because the monomer 1,3-buta-diene, $CH_2=CH-CH=CH_2$, may lead to cis and trans 1,4-units, $-CH_2-CH=CH-CH_2-$, and isotactic and syndiotactic 1,2-units, $-CH_2-CH(CH=CH_2)-$.

Copolymers are often synthesized in order to attain or improve certain properties; im-portant synthetic copolymers are shown in Table 12-1. In nature, all nucleic acids and proteins and many polysaccharides are copolymers. Other biocopolymers comprise, for example, block copolymers of α-amino acids and sugar units (Chapter 14).

Table 12-1 Important industrial copolymerizations in emulsion (E), in solution (S), in bulk (B), or by precipitation (P); concentration of comonomer B in parentheses. [a] In *t*-butanol; [b] in acetone, 1,4-dioxane, or hexane; [c] in water. Copolymerization parameters r_a and r_b indicate relative reactivities of monomers (Section 12.2.2). These parameters are independent of initiator and solvent in free-radical copolymerizations but not in ionic and coordination polymerizations. * After sulfonation.

Monomer A	Comonomer B		r_a	r_b	Type	Application
Free-radical polymerization to linear or slightly branched polymers						
Ethene	vinyl acetate	(10 %)	0.88	1.03	B	shrink films
Ethene	vinyl acetate	(10-35 %)	0.88	1.03	B	molding compounds
Ethene	vinyl acetate	(35-40 %)	0.88	1.03	P[a]	sheets
Ethene	vinyl acetate	(> 60 %)	0.88	1.03	E	rubbers
Ethene	methacrylic acid	(< 10 %)			B	coatings
Ethene	chlorotrifluoroethene		0.25	0.0025		molding compounds
Butadiene	styrene		1.44	0.84	E	SBR rubbers
Butadiene	acrylonitrile	(37 %)	0.36	0.04	E	NBR rubbers
Vinyl chloride	vinyl acetate	(3-20 %)	1.7	0.23	S[b]	floor coverings
Vinyl chloride	propene	(3-10 %)			B	molding compounds
Vinylidene chloride	acrylonitrile (or vinyl chloride)		0.32	0.92		packaging films
Acrylic ester	acrylonitrile	(5-15 %)	0.72	1.2		rubbers
Acrylonitrile	vinyl acetate	(4 %)			P[c]	acrylic fibers
Acrylonitrile	styrene		0.04	0.41		molding compounds
Acrylonitrile	styrene + butadiene					molding compounds
Tetrafluoroethene	propene		1.0	0.06	E	molding compounds
Methacrylic acid	methacrylonitrile		0.60	1.64		rigid foams
Free-radical copolymerization to crosslinked polymers						
Glycol methacrylate	glycol dimethacrylate	(2-4 %)				contact lenses
Styrene	unsaturated polyesters				B	thermosets
Styrene	divinylbenzene					ion exchangers *
Anionic copolymerization						
Styrene	butadiene				S	rubbers
Cationic copolymerization						
Isobutene	isoprene	(4 %)			S	butyl rubber
Trioxane	ethylene oxide				S	molding compounds
Ethylene oxide	propylene oxide				S	thickeners
Propylene oxide	non-conjugated dienes				S	rubbers
Coumarone	indene					bonding cement
Ziegler–Natta copolymerization (w/o block copolymers)						
Ethene	propene + non-conjugated dienes					rubbers
Ethene	1-butene, 1-hexene, 1-octene, or 4-methyl-1-pentene					molding compounds

12.2   Copolymerization Theory

12.2.1   Introduction

Copolymerizations are performed with initial mixtures of two or more monomers. According to combinatorics, N_{mon} different monomeric species can be combined in groups of $i \le N_{mon}$ species that will lead to N_{comb} different types of copolymers, each of which can furthermore vary with respect to composition and distribution of units:

$$(12\text{-}2) \qquad N_{comb} = \frac{N_{mon}(N_{mon}-1)\ldots\ldots(N_{mon}-i+1)}{i!}$$

In terpolymerizations ($i = 3$), $N_{mon} = 7$ different monomers lead to $N_{comb} = 35$ possible terpolymers whereas $N_{mon} = 100$ monomers deliver $N_{comb} = 161\ 700$ terpolymers. Depending on the monomeric species, the molar ratio of the monomers, and the type of polymerization, the constitutional make-up of each of these terpolymers may vary from alternating monomer units via statistical terpolymers to block copolymers. Very few different monomers can therefore deliver many different copolymers with correspondingly widely varying constitutions and properties.

Most copolymerizations are performed by free-radical mechanisms since (a) many monomers polymerize free-radically (Chapter 10), (b) most of these monomers are relatively inexpensive, and (c) the sequence statistics of monomeric units can be more easily manipulated in free-radical polymerizations than in any other type of polymerization.

Monomer compositions and sequence statistics of bipolymers are usually described by the so-called **terminal model** (Section 12.2.2). This model assumes that the addition of monomer molecules A and B to the growing chain ends is only controlled by the last monomeric unit ~a* or ~b*. The terminal model thus corresponds to first-order Markov statistics. The **penultimate model** assumes participation of penultimate units and thus four different active species, ~aa*, ~ab*, ~ba*, and ~bb* (2nd order Markov trials) (Section 12.3.3). Other non-terminal models consider additional reactions such as de-propagation, monomer partitioning, complex formation, etc.

Both the terminal and the penultimate model assume irreversible propagation steps; they thus exclude equilibrium reactions (no depolymerization). Both models also presume a steady state. The polymerization should furthermore be macroscopically and microscopically homogeneous: no heterogeneous phases (no precipitation, no emulsions, no suspensions) and neither monomer partitioning, complex formation nor preferential solvation. The local concentrations at the loci of the reaction should thus be identical with the global concentrations in the fluid phase.

With few exceptions, the terminal model describes reasonably well the instantaneous (incremental) bipolymer compositions as well as the distribution of diads, triads, etc. It is less successful, however, for the evaluation of rates and rate constants.

12.2.2 Terminal Model

In bipolymerizations, irreversible reactions of the two active chain ends ~a* and ~b* with monomers A and B lead to four different rates R_{jJ} and four rate constants k_{jJ} (j = a, b; J = A, B). For simplification, indices of rates and rate constants are written without a slash (aA instead of a/A, etc.):

(12-3)	~a* + A	$\rightarrow$	~a–a*	;	$R_{aA} = k_{aA}[a^*][A]$	(homopropagation)
(12-4)	~a* + B	$\rightarrow$	~a–b*	;	$R_{aB} = k_{aB}[a^*][B]$	(cross-propagation)
(12-5)	~b* + A	$\rightarrow$	~b–a*	;	$R_{bA} = k_{bA}[b^*][A]$	(cross-propagation)
(12-6)	~b* + B	$\rightarrow$	~b–b*	;	$R_{bB} = k_{bB}[b^*][B]$	(homopropagation)

In terpolymerizations, nine rates and nine rate constants must be considered, etc.

The ratio of rate constants of homopropagation to cross-propagation is called **copolymerization parameter (copolymerization ratio, copolymerization coefficient, monomer reactivity ratio)**:

(12-7) $r_a \equiv k_{aA}/k_{aB}$; $r_b \equiv k_{bB}/k_{bA}$

Five distinct cases can be distinguished for r_i $(= r_a$ or $r_b)$:

$r_i = 0$ The rate constants of both homopropagations are zero. Active centers add exclusively monomer molecules of the other type.

$r_i < 1$ The other type of monomer molecule is added preferentially.

$r_i = 1$ Both monomer types are added with equal probability if [A] = [B].

$r_i > 1$ The own monomer is added preferentially but not exclusively.

$r_i = \infty$ Only homopolymerization are present, no copolymerization occurs.

Copolymerization parameters thus measure the probability that a monomer species adds to an active site. Since both types of active sites compete for a monomer, the *product $r_a r_b$* must be considered, too (Section 12.3.2).

At high degrees of polymerization, neither initiation nor termination nor transfer reactions consume sizable fractions of monomers. Monomers are only consumed by the two homopropagations, Eqs.(12-3) and (12-6), and the two crosspropagations, Eqs.(12-4) and (12-5). The relative monomer consumption is thus

(12-8) $\dfrac{-d[A]/dt}{-d[B]/dt} = \dfrac{R_{aA} + R_{bA}}{R_{bB} + R_{aB}} = \left(\dfrac{k_{bA} + k_{aA}([a^*]/[b^*])}{k_{bB} + k_{aB}([a^*]/[b^*])} \right) \cdot \dfrac{[A]}{[B]} = \dfrac{d[A]}{d[B]}$

In cross-propagations, ~a* + B → ~ab* and ~b* + A → ~ba*, active units ~a* are replaced by active units ~b* and vice versa. In the steady state, the concentrations of active species thus do not change with time and the rates R_{aB} and R_{bA} must be equal to each other. Replacement of R_{bA} by R_{aB} in Eq.(12-8), division by R_{AB}, and introduction of Eqs.(12-3)-(12-7) leads to the **(differential) copolymerization equation**:

(12-9) $\dfrac{d[A]}{d[B]} = \dfrac{1 + r_a([A]/[B])}{1 + r_b([B]/[A])}$ **(Mayo-Lewis equation)**

The copolymerization equation can also be derived by probability considerations. The conditional probability $p_{a/A}$ for the formation of a constitutional diad of two a-units is given by the ratio of the addition rate of A monomer molecules at ~a* sites to the sum of the rates of all possible reactions at ~a*. $p_{a/A}$ depends only on the copolymerization parameter r_a and the ratio [B]/[A] of monomer concentrations but not on r_b:

(12-10) $p_{a/A} = \dfrac{R_{aA}}{R_{aA} + R_{aB}} = \dfrac{k_{aA}[a^*][A]}{k_{aA}[a^*][A] + k_{aB}[a^*][B]} = \dfrac{r_a}{r_a + ([B]/[A])}$

For the three other diads one obtains with $p_{a/A} + p_{a/B} \equiv 1$ and $p_{b/B} + p_{b/A} \equiv 1$:

(12-11) $p_{a/B} = \dfrac{[B]/[A]}{r_a + ([B]/[A])}$; $p_{b/A} = \dfrac{[A]/[B]}{r_b + ([A]/[B])}$; $p_{b/B} = \dfrac{r_b}{r_b + ([A]/[B])}$

Introduction of $p_{a/B}$ and $p_{b/A}$ into the condition $x_a p_{a/B} = x_b p_{b/A}$ (Eq.(6-36)) delivers

(12-12) $\quad \dfrac{x_a}{x_b} = \dfrac{1 + r_a([A]/[B])}{1 + r_b([B]/[A])}\quad$ (valid for a very small conversion interval ($\Delta u \to 0$)!)

Comparison of Eqs.(12-9) and (12-12) shows:

- Because $d[A]/d[B] = x_a/x_b$ for $\Delta u \to 0$, for example, from $u = 0$ % to $u = 1$ % or from $u = 49$ % to $u = 50$ %, the composition of copolymers refers to the *incremental (instantaneous)* composition and not to the integral one (such as $u = 0$ % to $u = 50$ %).
- The state is stationary with respect to the relative proportion of active centers but not with respect to the total concentrations of active species.
- The Mayo-Lewis equation does not contain a time parameter and therefore says nothing about copolymerization kinetics.

12.2.3 Types of Copolymerization

In bipolymerizations, one of the two monomers is usually consumed faster than the other. The instantaneous monomer compositions, and therefore also the instantaneous bipolymer compositions, thus change with increasing monomer conversion. This drift can be avoided if the more reactive monomer is continually fed to the reactor according to its consumption.

Conversion-independent copolymer compositions are also obtained by **azeotropic copolymerization** which may either lead to alternating or statistical azeotropic copolymers (Table 12-2). In these copolymerizations, the relative change of monomer concentrations equals the incremental (instantaneous) ratio of monomer concentrations, $d[A]/d[B] \equiv [A]/[B]$. Introduction of this expression into Eq.(12-9) delivers

(12-13) $\quad [A]/[B] = (1 - r_b)/(1 - r_a) = x_A/x_B$

Azeotropic copolymerizations are therefore characterized by $r_a < 1$ and $r_b < 1$. The formally possible case of $r_a > 1$ *and* $r_b > 1$ has never been observed experimentally.

Table 12-2 Special cases of bipolymerizations of monomers A and B

Name of bipolymerization	Product of parameters, $r_a r_b$	Copolymerization parameters azeotropic		non-azeotropic	
		r_a	r_b	r_a	r_b
Double-alternating azeotropic	0	0	0		
Simple alternating azeotropic	0	0	$0 < r_b < 1$		
Statistically azeotropic	< 1	$0 < r_a < 1$	$0 < r_b < 1$		
Ideal azeotropic	1	1	1		
Ideal non-azeotropic				$0 < r_a = 1/r_b < 1$	$1 < r_b = 1/r_a < \infty$
Statistically non-azeotropic	$\gtrless 1$			$0 < r_a \neq 1/r_b < 1$	$1 < r_b \neq 1/r_a < \infty$
Block-forming azeotropic	> 1	$1 < r_a < \infty$	$1 < r_b < \infty$		
Blend-forming azeotropic	∞	∞	∞		

In **double-alternating azeotropic** bipolymerizations with $r_a = r_b = 0$, each active center ~a* adds only monomer molecules B and vice versa resulting in ~(a–b)$_n$~, $x_a/x_b = 1$, and thus $x_a = x_b = 1/2$. The composition of initially formed copolymer molecules is thus independent of the monomer composition, $x_{A,0} = \{[A]/([A] + [B])\}_0$, (horizontal line for $x_a = f(x_{A,0})$ in Fig. 12-1). The polymerization stops if the minority monomer is totally consumed. Copolymer molecules do not contain homodiads in this ideal case ($x_{aa} = x_{bb} = 0$) .

In **simple alternating azeotropic** bipolymerizations ($r_a = 0, 0 < r_b < 1$), on the other hand, only one azeotropic monomer ratio [A]/[B] exists because only one of the co-polymerization parameters equals zero. If, for example, $r_b = 0$, then Eq.(12-9) becomes

(12-14) $d[A]/d[B] = 1 + r_a([A]/[B])$ (= x_A/x_B for $\Delta u \to 0$)

A great excess of B molecules thus leads to 1:1 copolymer molecules. The polymerization stops if the minority monomer A is completely polymerized. If $[A]_0/[B]_0 > 1$, however, copolymer molecules with $x_a > 0.5$ are formed, depending on monomer ratios and reactivities. Because of $0 < r_a < 1$, a-units will form longer sequences whereas b-units are always present as monads. An example is the copolymerization of maleic anhydride ($r_a \approx 0$) and styrene ($r_b \approx 0.03$) to small conversions (not shown in Fig. 12-1).

In **statistically azeotropic** copolymerizations, both copolymerization parameters are greater than zero but smaller than unity. Both types of monomeric units are added at random. The curves in copolymerization diagrams have points of inflection (Fig. 12-1). Such copolymerizations are common for comonomers with opposite polarities.

Ideal azeotropic copolymerizations would have $r_a = r_b = 1$ and Eq.(12-12) reduces to $x_a = x_A$ (45° line in Fig. 12-1). Apparently, such monomer pairs do not exist. Equally fictive seem to be the two azeotropic cases with $r_a > 1$ and $r_b > 1$ since monomer pairs with $r_a > 1$ *and* $r_a \neq r_b > 1$ have never been observed with certainty.

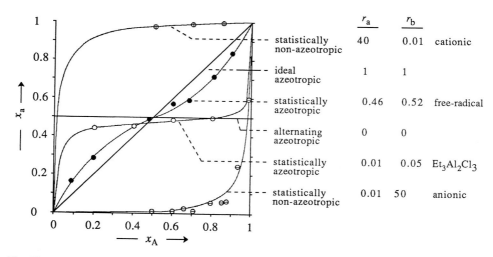

Fig. 12-1 Copolymerization diagram: Mole fractions x_a of styrenic units as a function of the mole fraction x_A of styrene monomer in cationic (⊕), free-radical (●), coordination (O), and anionic (⊖) co-polymerizations of styrene A and methyl methacrylate B [1].

Statistical non-azeotropic copolymerizations with $r_a < 1$ and $r_b > 1$ are most common. Again, there are two sub-classes. In the **ideal** case, $k_{aA}/k_{aB} = k_{bA}/k_{bB}$ and $r_a = 1/r_b$: the *relative* addition of monomers A and B is independent of the structure of the growing chain end. Eq.(12-9) reduces to

(12-15) $d[A]/d[B] = r_a[A]/[B] \ (= x_A/x_B \ \text{for } \Delta u \to 0)$

The molar ratio of added monomeric units always differs from the previous one by a factor r_a, i.e., for a small conversion interval of 1 %. The function $x_a = f(x_A)$ never intersects the 45° line of ideal azeotropic copolymerizations where $r_a = r_b = 1$ (Fig. 12-1). In the corresponding **nonideal** case, one also has $0 < r_a < 1$ and $1 < r_b < \infty$ but in addition $r_a \neq 1/r_b$ which leads to a statistical distribution of a-units and a blockiness of b-sequences.

All types of copolymerizations can sometimes be observed for a single pair of monomers, for example, for the bipolymerization of styrene and methyl methacrylate by different initiators (Fig. 12-1). The free-radical copolymerization is statistically azeotropic whereas both the cationic and the anionic polymerizations are statistically non-azeotropic. The cationic bipolymer has long sequences of styrene units ($r_s > r_{mma}$) and the anionic one long sequences of methyl methacrylate units ($r_{mma} > r_s$). Almost alternating bipolymers result from the initiation by $(C_2H_5)_3Al_2Cl_3$ in the presence of light and traces of oxygen.

The extent of alternation is difficult to determine from copolymerization diagrams since the curves for $r_a = r_b = 0.01$ and $r_a = r_b = 0.001$ become almost identical in the range $0.1 < x_A < 0.9$ (Fig. 12-2). A distinction between both sets of parameters is only possible at very small or very large mole fractions of A. However, even then, copolymer constitutions may differ considerably. In an alternating bipolymer $(a–b)_{1000}$ with a degree of polymerization of 2000, copolymerization parameters of $r_a = r_b = 0.001$ lead on average to one wrong -a–a- diad whereas $r_a = r_b = 0.01$ will produce about ten such diads per copolymer molecule.

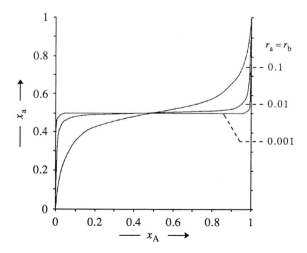

Fig. 12-2 Calculated copolymerization diagrams for almost alternating bipolymers.

12.2.4   Sequence Statistics

Truly alternating bipolymers consist of molecules with the sequences ...ababab... All other copolymerizations lead not only to copolymer molecules with homosequences of different lengths but also to variations in the average composition per polymer molecule unless the copolymerization proceeded under azeotropic conditions.

The heterogeneity of these copolymers results from

1. statistical variations in the composition of individual bipolymer chains;
2. conversion-dependent polymer compositions from non-azeotropic polymerizations;
3. more than one type of active center in certain types of polymerizations;
4. processing conditions (type of reactor, feeding of monomer, etc.).

The probability of the formation of homosequences of i equal monomeric units can be calculated from conditional probabilities, Eqs.(12-10) and (12-11). A sequence of i units of type a (a-block) requires addition of $(i - 1)$ monomer molecules A to a chain end ~b–a*. This probability is $(p_{a/A})^{i-1}$ (see also Section 13.2.5). The sequence is terminated by a b-unit, requiring a probability of $p_{a/B} = 1 - p_{a/A}$. The product of these two probabilities is the probability $(P_{a\text{-block}})_i$ that an a-block that consists of i a-units:

$$(12\text{-}16) \qquad (P_{a\text{-block}})_i = (p_{a/A})^{i-1}(1 - p_{a/A}) = (x_{a\text{-block}})_i$$

The probability of formation of an a-block, $(P_{a\text{-seq}})_i$, of length i must also be the mole fraction $(x_{a\text{-block}})_i$ of a-blocks with respect to all a-blocks of various lengths. By analogy, one can write $(P_{b\text{-block}})_i = (p_{b/B})^{i-1}(1 - p_{b/B}) = (x_{b\text{-block}})_i$.

Mole fractions $(x_{a\text{-block}})_i$ of a-blocks depend only on the conditional probability $p_{a/A}$. In the terminal model, they are controlled by the copolymerization parameter r_a and the *ratio* of monomer concentrations but not by r_b and the absolute monomer concentrations, (Eq.(12-9)). Since the monomer concentrations of Eq.(12-9) are *incremental* (instantaneous) concentrations, mole fractions $(x_{a\text{-block}})_i$ are differential fractions for conversion intervals, Δu. These fractions are only identical with measured fractions if $\Delta u \to 0$ since measured fractions are integral quantities for the range $\Delta u = 0$ to $\Delta u = u$.

Bipolymers from copolymerizations with equimolar initial monomer concentrations ($[A]_0 = [B]_0$), small monomer conversions ($\Delta u \to 0$), and small copolymerization parameters ($r_a \to 0$) contain a high percentage of a-monads (Table 12-3), resulting in a tendency to alternation. Large copolymerization parameters r_a and $[A]_0 >> [B]_0$ lead to a more flat distribution of sequence lengths and longer a-blocks.

According to the general definition of number averages, Eq.(3-23), the number-average **sequence length (block length)** of homosequences of a-units (= a-blocks) is

$$(12\text{-}17) \qquad \overline{X}_{a\text{-block,n}} \equiv \Sigma_i \, (x_{a\text{-block}})_i (X_{a\text{-block}})_i$$

Block lengths $(X_{a\text{-block}})_i$ of molecules can only be whole numbers. The mole fractions of the different homosequences are given by the conditional probabilities, i.e., by $(x_{a\text{-block}})_1 = p_{a/B}$ for the monoblock ~b$\underline{a}$b~, by $(x_{a\text{-block}})_2 = p_{a/A}p_{a/B}$ for the diblock ~b$\underline{aa}$b~, etc. The **number-average degree of block lengths (= number-average degree of sequence lengths)**, $\overline{X}_{a\text{-block,n}}$, of a-homosequences is therefore obtained from

(12-18) $\overline{X}_{\text{a-block},n} = p_{a/B} + 2\,p_{a/A}\,p_{a/B} + 3\,(p_{a/A})^2 p_{a/B} + ...$

$= p_{a/B}(1 + 2\,p_{a/A} + 3\,(p_{a/A})^2 + ...) = p_{a/B}(1 - p_{a/A})^{-2}$

that was converted to a closed expression because $p_{a/A} \leq 1$. Introduction of $1 - p_{a/A} \equiv p_{a/B}$ and Eq.(12-11) results in

(12-19) $\overline{X}_{\text{a-block},n} = p_{a/B}(1 - p_{a/A})^{-2} = 1/(1 - p_{a/A}) = 1/p_{a/B} = 1 + r_a([A]/[B])$

By analogy, the number-average degree of block lengths of b-blocks is

(12-20) $\overline{X}_{\text{b-block},n} = 1/(1 - p_{b/B}) = 1/p_{b/A} = 1 + r_b([B]/[A])$

Number-average degrees of sequence lengths are usually not very large (Table 12-3). They adopt values of 10 and more only for copolymers from copolymerizations with large ratios of monomer concentrations and/or large copolymerization parameters.

The ratio of block lengths of a-homosequences and b-homosequences equals the molar ratio of the corresponding monomeric units, i.e., $\overline{X}_{\text{a-block},n}/\overline{X}_{\text{b-block},n} = x_a/x_b$. Introduction of the **run number** $\overline{R}_n = 200/(\overline{X}_{\text{a-block},n} + \overline{X}_{\text{b-block},n})$ of the bipolymer (cf. the block character in Section 6.3.3) leads to

(12-21) $\overline{X}_{\text{a-block},n} = 200\, x_a/\overline{R}_n$

The mole fraction of a-units in these sequences, $(x_a)_i$, is calculated from the mole fraction, $(x_{\text{a-block}})_i$, and the number-average of block length, $(X_{\text{a-block}})_i$:

(12-22) $(x_a)_i = (x_{\text{a-block}})_i(X_{\text{a-block}})_i/\overline{X}_{\text{a-block},n}$

For small copolymerization parameters (e.g., $r_a = 0.1$ in Table 12-3), mole fractions of a-units in a-homosequences of length i, $(x_a)_i$, decrease monotonously with increasing sequence length, i. $(x_a)_i$ increases, however, with i for large copolymerization parameters, (e.g., $r_a = 10$), and then passes through a maximum (not shown in Table 12-3). The maximum is always at $(X_{\text{a-block}})_{i,\max} = \overline{X}_{\text{a-block},n} - (1/2)$; it is independent of copolymerization parameters and ratios of monomer concentrations. An example of such a maximum is $(X_{\text{a-block}})_{i,\max} = 10.5$ for $r_a = 10$ and $[A]/[B] = 1$.

Table 12-3 Calculated mole fractions of a-homosequences, $(x_{\text{a-block}})_i$, and a-units, $(x_a)_i$, in homosequences of length i for various values of r_a and ratios of initial monomer concentrations, $[A]_o/[B]_o$.

i		$(x_{\text{a-block}})_i$				$(x_a)_i$		
$[A]_o/[B]_o$ →		1	1	1	10	1	1	1
r_a →		0.1	1	10	10	0.1	1	10
1		90.91	50.00	9.09	9.90	82.64	25.00	0.83
2		8.26	25.00	8.26	9.80	15.03	25.00	1.50
3		0.75	12.50	7.51	9.71	2.05	18.75	2.05
4		0.07	6.75	6.83	9.61	0.25	12.50	2.48
etc.								
$\overline{X}_{\text{a-block},n}$		1.1	2.0	11	101	1.1	2.0	11

12.2.5 Determination of Copolymerization Parameters

From Mole Fractions

Copolymerization parameters can be calculated from the relationships between the mole fractions of the various i-ads, x_a, x_b, x_{aa}, $(x_{ab} + x_{ba})$, x_{bb}, x_{aaa}, etc. These mole fractions are sometimes obtainable from NMR spectroscopy. Knowledge of the mole fraction x_{a-a} of a–a bonds and the mole fraction x_a of a-units in bipolymers allows one to calculate the run number $\bar{R}_n$ from

$$(12\text{-}23) \qquad \bar{R}_n = 200(x_a - x_{a-a})$$

and thus the block length $\bar{X}_{a\text{-block},n}$ from Eq.(12-21), the conditional probability from Eq.(12-19), and also r_a from [A]/[B]. The mole fraction of a-units in the center of a<u>a</u>a triads or in the center of b<u>a</u>b triads is available from

$$(12\text{-}24) \qquad x_{a\underline{a}a} = \frac{\left(x_a - (\bar{R}_n/200)\right)^2}{x_a^2} \qquad ; \qquad x_{b\underline{a}b} = \frac{\bar{R}_n^2}{(200\,x_a)^2}$$

From Integral Copolymerization Equations

The Mayo-Lewis equation, Eq.(12-9), describes the *incrementally* formed copolymer composition as a function of the *instantaneous* ratio of monomer concentrations; i.e., for infinitesimally small intervals of conversion. These small concentration changes are difficult to determine at higher conversion intervals (e.g., from 50 % to 51 %). One therefore usually restricts the measurements to small conversions near zero conversion (e.g., from 0 % to 2 %) or measures copolymer compositions at various small monomer conversions Δu and extrapolates the data to $\Delta u \to 0$. The latter procedure is admissible for steady states and small differences in copolymerization parameters but not for living polymerizations since it includes in this case an extrapolation to low-molecular weights and the preferred initiation mechanism.

A correct computation of copolymerization parameters from measurements at larger conversion intervals requires the use of one of the following two **integrated copolymerization equations**

$$(12\text{-}25) \qquad \frac{[b]}{[b]_0} = \left(\frac{[B]_0[A]}{[A]_0[B]}\right)^{\frac{r_b}{1-r_b}} \left(\frac{1 - r_b + (r_a - 1)([A]/[B])}{1 - r_b + (r_a - 1)([B]/[A])}\right)^{\frac{1-r_a r_b}{(1-r_a)(1-r_b)}}$$

$$(12\text{-}26) \qquad r_b = \frac{\lg\dfrac{[B]_0}{[B]} - Z^{-1}\lg\dfrac{(1-Z)[A]/[B]}{(1-Z)[A]_0/[B]_0}}{\lg\dfrac{[A]_0}{[A]} + \lg\dfrac{(1-Z)[A]/[B]}{(1-Z)[A]_0/[B]_0}} \qquad ; \qquad Z = \frac{1-r_a}{1-r_b}$$

where $[A]_0$, $[A]$, $[B]_0$, and $[B]$ are the mole fractions of monomers A and B at times 0 and t and $[b]_0$ and $[b]$ the concentrations of b-units at infinitesimally small monomer conversions and at time t.

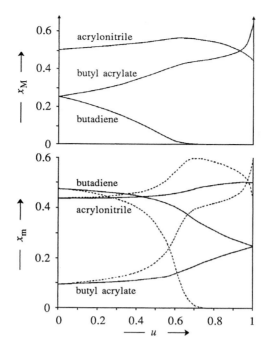

Fig. 12-3 Dependence of mole fractions x_M of acrylonitrile (AN), butyl acrylate (BA), and 1,3-buta-
diene (BU) in the monomer mixture and mole fractions x_m of corresponding monomeric units an, ba,
and bu as a function of total monomer conversion, u, in the terpolymerization of a mixture of $[AN]_0$:
$[BA]_0$: $[BU]_0 = 0.5 : 0.25 : 0.25$ at 60°C. Copolymerization parameters: $r_{anbu} = 0.7$, $r_{anba} = 12.0$,
$r_{buba} = 9.9$, $r_{buan} = 3.5$, $r_{baan} = 8.9$, $r_{babu} = 0.8$.
   Top: Composition of monomer mixture.
   Bottom: Integral (——) and differential (incremental) (····) copolymer composition.

At higher monomer conversions in non-azeotropic copolymerizations, integral copolymer compositions differ considerably from differential ones (Fig. 12-3). Copolymers become very non-uniform with respect to composition. In the example of Fig. 12-
3, butadiene is almost completely consumed at a total monomer conversion of ca. 70 %.
At monomer conversions of more than 75 %, newly formed copolymer molecules no
longer incorporate butadiene units. The total copolymer does of course contain such
units because they were incorporated at smaller monomer conversions.

Another integral equation is especially useful for free-radical copolymerizations. The
rate of bipolymerizations is given by $-d[A]/dt = k_{aA}[a^{\bullet}][A] + k_{bA}[b^{\bullet}][A]$. For steady-
state conditions, this can be integrated to $[A] = [A]_0 \exp \{-(k_{aA}[a^{\bullet}] + k_{bA}[b^{\bullet}])t\}$. For
rate constants of $k_p \approx 10^2$ L mol^{-1} s^{-1} (cf. Table 10-6) and radical concentrations of
$[R^{\bullet}] \approx 10^{-8}$ mol/L (cf. Section 10.3.3), the term in parentheses becomes ca. 10^{-4} min^{-1}.
At polymerization times of $t = 100$ min, one thus obtains $[A] = [A]_0 \exp \{-0.01\}$. The
exponential term $\exp \{-(k_{aA}[a^{\bullet}] + k_{bA}[b^{\bullet}])t\}$ can therefore be linearized to $1 - (k_{aA}[a^{\bullet}]$
$+ k_{bA}[b^{\bullet}])t$. The time dependences of monomer concentrations are therefore

$$(12\text{-}27) \qquad \frac{[A]_0 - [A]}{[A]_0} = (k_{aA}[a^{\bullet}] + k_{bA}[b^{\bullet}])t \; ; \quad \frac{[B]_0 - [B]}{[B]_0} = (k_{bB}[b^{\bullet}] + k_{aB}[a^{\bullet}])t$$

Division of the first equation by the second and introduction of the steady-state condition $k_{aB}[a^\bullet][B] = k_{bA}[b^\bullet][A]$ and the copolymerization parameters $r_a = k_{aA}/k_{aB}$ and $r_b = k_{bB}/k_{bA}$ leads to another **integrated copolymerization equation**:

$$(12\text{-}28) \qquad \frac{[A]_0 - [A]}{[B]_0 - [B]} = \frac{[A]_0([B] + r_a[A])}{[B]_0([A] + r_b[B])}$$

Introduction of the mole fractions $x_A = [A]/([A] + [B]) = 1 - x_B$ of monomers A and B and $x_a = ([A]_0 - [A])/\{([A]_0 - [A]) + ([B]_0 - [B])\} = 1 - x_b$ of monomeric units "a" and "b" (all cumulative to the total monomer conversion u) delivers a linear equation with intercept r_b and slope $-r_a$:

$$(12\text{-}29) \qquad \left\{ \frac{x_{A,0} x_B x_b - x_{B,0} x_A x_a}{x_{B,0} x_B x_a} \right\} = r_b - r_a \left\{ \frac{x_{A,0} x_A x_b}{x_{B,0} x_B x_a} \right\}$$

From Differential Copolymerization Equations

For small initial monomer conversions ($x_A \approx x_{A,0}$ and $x_B \approx x_{B,0}$), Eq.(12-29) reduces to the often used **Fineman-Ross equation**:

$$(12\text{-}30) \qquad \frac{\{(x_a/x_b) - 1\}\{x_{A,0}/x_{B,0}\}}{(x_a/x_b)} = -r_b + r_a \frac{\{x_{A,0}/x_{B,0}\}^2}{(x_a/x_b)}$$

This equation can also be obtained from Eq.(12-12) by setting $[A]/[B] = x_A/x_B \approx x_{A,0}/x_{B,0}$. Eq.(12-30) can be written $G = -r_b + r_a F$ and plotted correspondingly (Fig. 12-4). An alternative Fineman-Ross equation is $G/F = r_a - r_b F^{-1}$.

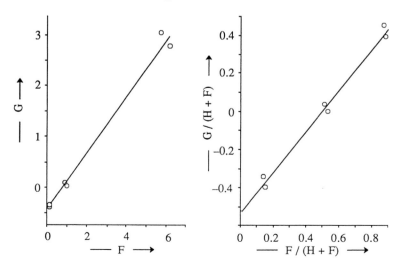

Fig. 12-4 Determination of copolymerization parameters for the free-radical copolymerization of styrene and methyl methacrylate at 60°C. Left: Fineman-Ross method (Eq.(12-30)); right: Kelen-Tüdös method (Eq.(12-31)).

The **Kelen-Tüdös** method is also often used. This method introduces another variable $H = (F_{min}/F_{max})$ that is calculated from the smallest and the largest experimental values of F. Plots of $G/(H + F)$ versus $F/(H + F)$ are said to allow the determination of copolymerization parameters from larger conversion intervals than the Fineman-Ross method:

$$(12-31) \qquad \frac{G}{H+F} = -\frac{r_b}{H} + \left(r_a + \frac{r_B}{H}\right)\left(\frac{F}{H+F}\right)$$

From Large Excess of One Comonomer

The Mayo-Lewis Equation, Eq.(12-9), reduces to $d[A]/d[B] = r_A([A]/[B])$ if $r_A[A]/[B]) \gg 1$ and $r_B[A]/[B]) \ll 1$, i.e., if the bipolymerization is performed with a large excess of one of the comonomers. The integrated equation

$$(12-32) \qquad \lg ([A]_t/[A]_0) = r_1 \lg ([B]_t/[B]_0)$$

is valid up to high monomer conversions provided that $[A] \gg [B]$ at all times. A corresponding **Jaacks plot** is shown in Fig. 12-5.

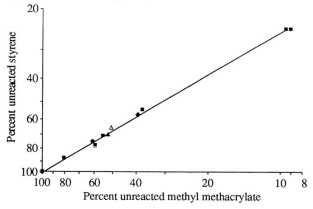

Fig. 12-5 Jaacks plot for the copolymerization of methyl methacrylate (MMA) with a large excess of styrene (S) in benzene at 60°C [2]. Ratios [S]/[MMA]: 193 (●), 50 (■), 19.8 (◆), and 10.6 before (△) and after (▲) correction for the small block length (see [2]).

12.2.6 Copolymerization with Depolymerization

A monomer will not polymerize at temperatures far above its thermodynamic ceiling temperature (Section 7.3.2). However, it may still copolymerize if the rate of the cross-propagation is greater than the rate of the depolymerization reaction

$$(12-33) \qquad k_{aB}[\sim a^*][B] > k_{-aB}[\sim b^*]$$

even if $k_{aA}[A] < k_{-aA}$ and $k_{bB}[B] < k_{-bB}$ for the two homopolymerizations. For example, tetrahydrofuran does not polymerize at temperatures above 80°C (Fig. 7-3) but it can be copolymerized with 3,3-dimethyloxetane at these temperatures.

The simplest case of a bipolymerization with depolymerization involves two irreversible cross-propagations, ~a* + B and ~b* + A, and one irreversible homopropagation, ~a* + A. However, i-diads ~abb*, ~abbb*, etc., may depolymerize reversibly to ~ab*. According to a lengthy derivation (not shown), Eq.(12-9) must be modified in this case by an additional term that includes a parameter σ and the equilibrium constant K of the polymerization-depolymerization equilibrium of monomer B:

(12-34) $$\frac{d[A]}{d[B]} = \frac{1 + r_a([A]/[B])}{1 + r_b([B]/[A])(1 - \sigma K^{-1}[B]^{-1})}$$

(12-35) $$\sigma = \frac{[A] + r_b([A] + K^{-1})}{2 r_b K^{-1}} - \left\{ \left(\frac{[A] + r_b([B] + K^{-1})}{2 r_b K^{-1}} \right)^2 - K[B] \right\}^{1/2}$$

Eq.(12-34) becomes more complicated if b-diads ~abb* do not depolymerize to ~ab* but b-triads ~abbb*, b-tetrads ~abbbb*, etc., depolymerize to b-triads ~abb*:

(12-36) $$\frac{d[A]}{d[B]} = \frac{1 + r_a([A]/[B])}{1 + r_b([B]/[A])\left(1 - \dfrac{\sigma r_b K^{-1}[A]^{-1}}{1 + r_b([B]/[A])} \right)}$$

Independent equilibrium polymerizations deliver the unknown equilibrium constant K. For $[B]/[A] \ll 1$, Eqs.(12-34) and (12-36) reduce to Eq.(12-14) from which one obtains the unknown copolymerization parameter r_a. The only adoptable parameter is r_b.

Contrary to the Mayo-Lewis Eq.(12-9), Eqs.(12-34) and (12-36) contain monomer concentrations in addition to the *ratio* of both monomer concentrations. The copolymer composition is thus also controlled by the total monomer concentration (Fig. 12-6). The experimental data of Fig. 12-6 indicate that in copolymers with methyl methacrylate, depolymerization of α-methyl styrene sequences includes diads but in copolymers with acrylonitrile only triads and higher sequences.

12.2.7   Living Copolymerization

In living copolymerizations with fast initiation reactions, the sum of the molar concentrations of active centers equals the initial initiator concentration, $[\sim a^*] + [\sim b^*] = [I]_0$. The total concentration of active species is furthermore constant, $d([\sim a^*] + [\sim b^*])/dt = 0$. However, the individual concentrations of the two active species ~a* and ~b* change with time and monomer conversion, respectively, because of cross-propagation steps. The molar ratio $[\sim a^*]/[\sim b^*]$ thus varies with the monomer conversion.

More simple are living copolymerizations without cross-propagations ($R_{aB} = R_{bA} = 0$). Here, the ratio $[\sim a^*]/[\sim b^*]$ does not depend on time and Eq.(12-8) converts to

(12-37) $$\frac{d[A]}{d[B]} = \left(\frac{k_{aA}[a^*]}{k_{bB}[b^*]} \right) \frac{[A]}{[B]} = const \cdot \frac{[A]}{[B]}$$

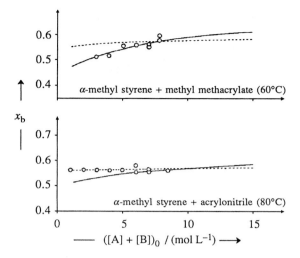

Fig. 12-6 Mole fractions x_b of monomeric units b of the reversibly polymerizing α-methyl styrene (B) as a function of the total initial monomer concentration, $([A] + [B])_0$, in free-radical bipolymerizations with methyl methacrylate or acrylonitrile. Lines: theoretical calculations for reversibly depolymerizing (——) and non-depolymerizing b-diads (- - -) [3].

Symmetric homopropagations, such as in the copolymerization of antipodes, are furthermore charcterized by $k_{aA} = k_{bB}$. In this case, the proportionality constant of Eq.(12-37) is just the ratio of concentrations of the two active species, $const = [a^*]/[b^*]$. In the formally identical equation for non-azeotropic copolymerizations, Eq.(12-15), it is the ratio of two rate constants, $const = r_a = k_{aA}/k_{aB}$. Such living copolymerizations without any cross-propagation lead to mixtures of homopolymers.

12.3 Free-Radical Copolymerization

12.3.1 Effect of Monomer Constitution

Free-radical bipolymerizations usually obey the Mayo-Lewis copolymerization equation, Eq.(12-9). The copolymerization parameters from this equation can be directly interpreted as the ratio of two rate constants and thus as relative reactivities. In general, they are most affected by resonance effects, less by polarities, and least by steric effects.

A resonance-stabilized radical will thus add preferably the monomer molecule that would lead to a new resonance-stabilized radical. An example of the dominance of resonance effects is the copolymerization of styrene S with vinyl esters or vinyl ethers V where $r_S = k_{SS}/k_{SV} \gg 1$ and $r_V = k_{VV}/k_{VS} \ll 1$ (Table 12-4).

The monomer addition to radicals is dominated by polarity effects if both monomers are resonance stabilized (example: styrene + acrylic esters) or not resonance stabilized (example: vinyl chloride + vinyl esters). The polymer radical adds preferably the monomer with the opposite polarity. The same is true for non-resonance-stabilized monomers of opposite polarity such as vinyl chloride and vinyl benzoate.

Table 12-4 Copolymerization parameters of free-radical copolymerizations in bulk at 60°C. R = Resonance-stabilized monomer, O = non-resonance-stabilized monomer, A = electron-accepting substituent, D = electron-donating substituent.

Monomer		Resonance stab.		Polarity		Parameters		
A	B	A	B	A	B	r_a	r_b	$r_a r_b$
Styrene	methyl acrylate	R	R	D	A	0.78	0.18	0.14
Styrene	butyl acrylate	R	R	D	A	0.79	0.17	0.13
Styrene	dodecyl acrylate	R	R	D	A	0.75	0.34	0.26
Styrene	methyl methacrylate	R	R	D	A	0.52	0.46	0.24
Styrene	butyl methacrylate	R	R	D	A	0.55	0.42	0.23
Styrene	dodecyl methacrylate	R	R	D	A	0.59	0.46	0.27
Styrene	acrylonitrile	R	R	D	A	0.41	0.04	0.016
Vinyl benzoate	vinyl chloride	O	O	D	A	0.28	0.72	0.20
Vinyl acetate	vinyl chloride	O	O	D	A	0.23	1.7	0.39
Styrene	vinyl chloride	R	O	D	A	17	0.02	0.34
Styrene	vinyl acetate	R	O	D	D	55	0.01	0.55
Styrene	methyl vinyl ether	R	O	D	D	100	0.01	1
Styrene	octyl vinyl ether	R	O	D	D	65	0	≈ 0
Styrene	dodecyl vinyl ether	R	O	D	D	27	0	≈ 0

Copolymerization parameters of monomers $CH_2=CRR'$ are usually only influenced by groups immediately adjacent to the ethylene moiety. The copolymerization parameters of methacrylates thus do not depend on the size of the ester substitutent (Table 12-4). However, an influence of the size of the substituent was observed for alkyl vinyl ethers + styrene (Table 12-4) and also for alkyl acrylates + acrylonitrile and alkyl methacrylates + vinyl acetate. Here the copolymerization parameter of the alkyl-containing monomer does not vary but that of the comonomer does. This effect is possibly caused by a penultimate effect or a variation in the association of the comonomer.

Steric effects are often subdued by those of resonance and/or polarity. 1,2-Disubstituted ethene monomers, CHR=CHR, copolymerize with monomers of similar polarity to statistical copolymers, e.g., dimethyl fumarate ($r_a = 0.47$) with vinyl chloride ($r_b = 0.12$). Monomers with strongly different polarities may even form charge-transfer complexes (CT complexes) that lead to alternating copolymers. An example is the copolymerization of maleic anhydride ($r_a = 5 \cdot 10^{-8}$) and *trans*-stilbene ($r_b = 0.32$) where the polar interaction in the transition state overcomes the steric hindrance. The additional stabilization of triple-substituted olefins, CHR=CR'R'', in the transition state leads to easy alternating copolymerizations with comonomers of opposite polarity.

Deviations from the Mayo-Lewis equation are often assumed to be caused by penultimate effects, especially for very polar monomers. Verification of such effects requires precise experiments at very small monomer ratios and a modification of the Mayo-Lewis equation (Section 12.3.3). However, these effects can often be equally well explained, or even better, by the formation of CT complexes (Section 12.5.3).

Solvents usually do not influence copolymerization parameters. They do effect them, however, if the molar ratio of comonomer molecules near the growing polymer radical differs from the average concentration of these molecules in the monomer mixture. This **bootstrap effect** seems to be prevalent for self-associating monomer molecules (e.g., alcohols) but is not caused by strong polarity *per se* (Table 12-5).

Table 12-5 Effect of solvents on copolymerization parameters of styrene S and methyl methacrylate M at 50°C. ε_r = Relative permittivity ("dielectric constant") of solvents at 20°C.

Solvent	ε_r	r_s	r_m	Solvent	ε_r	r_s	r_m
- (bulk)	≈ 3	0.52	0.46	Dichloromethane	9.08	0.50	0.63
Benzene	2.28	0.57	0.46	Acetone	20.7	0.49	0.50
1,4-Dioxane	2.22	0.56	0.53	Ethanol	24.3	0.41	0.41
N,N-Dimethylformamide	36.7	0.55	0.58	Benzyl alcohol	13.1	0.44	0.39
Acetonitrile	38.8	0.55	0.64	Phenol	9.8	0.35	0.35

Strong solvent effects can be expected for tautomeric monomers (G: *to auton meros* = to equal parts). According to NMR spectroscopy, acryloyl acetone (AA) is not present as ketone ($x_K = 0$), $CH_2=CH-CO-CH_2-CO-CH_3$, but completely ($x_E =1$) as a mixture of its two enols, $CH_2=CH-C(OH)=CH-CO-CH_3$ and $CH_2=CH-CO-CH=C(OH)-CH_3$. Ethyl acryloyl acetone (EAA) is a mixture of the ketone, $CH_2=C(C_2H_5)-CO-CH_2-CO-CH_3$, and the enol, $CH_2=C(C_2H_5)-CO-CH=C(OH)-CH_3$, depending on the solvent (Table 12-6). Copolymerization parameters therefore vary strongly with the solvent in these **tautomer polymerizations (isomerization polymerizations)**.

Table 12-6 Copolymerization parameters of styrene (S) with acryloyl acetone (AA) or ethyl acryloyl acetone (EAA). x_K = Mole fraction of keto-EAA in the various solvents [4].

Solvent	x_K of EAA	EAA + Styrene		AA + Styrene	
		r_{eaa}	r_s	r_{aa}	r_s
Benzene	0.224	0.89	0.15	1.38	0.0049
Carbon tetrachloride	0.24	2.68	0.16		
Ethyl acetate	0.336	1.48	0.27	1.91	0.171
Acetone	0.479			2.46	0.121
N,N-Dimethylformamide	0.606	0.76	0.18	3.19	0.044
Acetonitrile	0.759	0.57	0.22	2.69	0.240
Dimethyl sulfoxide	0.770	0.66	0.18	4.11	0.028

12.3.2 Cross-Termination Factor

In most cases, the simple terminal model describes reasonably well the *relative* kinetics of copolymerizations (i.e., the copolymerization parameters) and thus the incremental composition and sequence statistics of the resulting copolymer molecules. The model often fails, however, for *absolute* kinetics, i.e., for copolymerization rates.

The initial polymerization rates R_p of mixtures of chemically similar comonomers should follow the simple rule of mixtures. The rates should depend linearly on the initial monomer composition as it is found for the copolymerization of styrene and d_8-styrene (Fig. 12-7). Copolymerization rates of methyl methacrylate and d_8-methyl methacrylate, on the other hand, pass through a minimum. Such minima are quite common in copolymerizations (but see the maximum of catalyst productivity in Fig. 12-14). Copolymerizations are therefore slower than the homopolymerizations of their monomers.

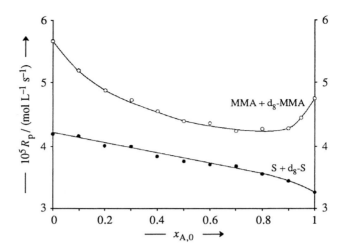

Fig. 12-7 Polymerization rate R_p of the free-radical copolymerization of styrene (S) with deutero-sty-rene (d_8-S) and methyl methacrylate (MMA) with deutero-methyl methacrylate (d_8-MMA) at 60°C as a function of the initial mole fraction, $x_{A,0}$, of non-deuterated monomers [5].

The terminal model ascribes the non-linear dependence of copolymerization rates on monomer composition to high **cross-termination factors**, $\Phi \equiv k_{t,ab}/(k_{t,aa}k_{t,bb})^{1/2}$. Values of $\Phi < 1$ indicate preferential terminations by like radicals while those of $\Phi > 1$ indicate those by unlike ones. Cross-termination factors are assumed to be independent of mono-mer concentrations because such a behavior is known from the reactions of gases.

The copolymerization rate for this model is given by

(12-38)

$$R_p = \left(\frac{r_a[A]^2 + 2[A][B] + r_b[B]^2}{\left\{ (k_{t,aa}/k_{aa}^2)r_a^2[A]^2 + 2\,\Phi(k_{t,aa}^{1/2}k_{t,bb}^{1/2}/k_{aa}k_{bb})r_ar_b[A][B] + (k_{t,bb}/k_{bb}^2)r_b^2[B]^2 \right\}^{1/2}} \right) R_{st}^{1/2}$$

Φ is calculated from the experimentally accessible polymerization rate (R_p), mono-mer concentrations ([A], [B]), relative rate constants ($r_a = k_{aA}/k_{aB}$, $r_b = k_{bB}/k_{bA}$), absolute homopropagation constants (k_{aA}, k_{bB}), termination constants ($k_{t,aa}$, $k_{t,bb}$), and the initia-tion rate ($R_i = 2\,fk_d[I_2]$) (Eq.(10-10)). However, such Φ values are high and also concen-tration dependent.

An example is the copolymerization of styrene (S) and methyl methacrylate (M) at 40°C where Φ values of 9 (at $x_S = 0.13$), 17 (at $x_S = 0.48$), and 23 (at $x_S = 0.85$) were found. The dependence of the *average* termination constants k_t on the monomer frac-tion x_S passes accordingly through a maximum (Fig. 12-8) where k_t was calculated from $k_t = k_{t,aa}(x_{a\bullet})^2 + 2\,k_{t,ab}x_{a\bullet}x_{b\bullet} + k_{t,bb}(x_{b\bullet})^2$ and mole fractions $x_{a\bullet} = [{\sim}a^\bullet]/([{\sim}a^\bullet] + [{\sim}b^\bullet])$ $= 1 - x_{b\bullet}$ of the two polymer radicals ${\sim}a^\bullet$ and ${\sim}b^\bullet$.

Some copolymerizations even show unrealistic cross-termination factors of up to $\Phi = 1000$. Since termination constants are controlled *physically* by diffusion (Section 10.3.9), a *chemical* interpretation of Φ values (preferred termination by unlike radicals) is physically meaningless.

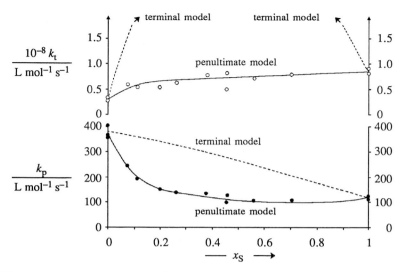

Fig. 12-8 Average rate constants of propagation, k_p, and termination, k_t, of the neat copolymerization of styrene and methyl methacrylate to $u < 3\%$ at 40°C [6]. For the terminal model, the maximum of $k_t \approx 4 \cdot 10^8$ L mol^{-1} s^{-1} at $x_S = 1/2$ is not shown.

12.3.3 Penultimate Model

The penultimate model describes the kinetics of copolymerization by effects of penultimate units of macroradicals on *propagation*; anomalous cross-termination factors (Section 12.3.2) are assumed to be absent. This approach requires 8 types of propagation steps (instead of 4 for the terminal model, Section 12.2.2) and 6 (instead of 2) penultimate copolymerization parameters:

$$
\begin{aligned}
(12\text{-}39) \quad r_{aa} &= k_{aaA}/k_{aaB} & r_{bb} &= k_{bbB}/k_{bbA} \\
r_{ba} &= k_{baA}/k_{baB} & r_{ab} &= k_{abB}/k_{abA} \\
s_A &= k_{baA}/k_{aaA} & s_B &= k_{abB}/k_{bbB}
\end{aligned}
$$

Four of these parameters describe the addition of different monomer molecules, A and B, to the same radical ($r_{aa} + r_{ba}$ and $r_{bb} + r_{ab}$). The other two parameters represent the addition of the same monomer molecule to different diads (s_A, s_B).

The overall propagation rate constant

$$
(12\text{-}40) \quad k_p = \frac{\bar{r}_{aa}x_A^2 + 2\,x_A x_B + \bar{r}_{bb}x_B^2}{(r_{ba}x_A/\bar{k}_{aA}) + (r_{ab}x_B/\bar{k}_{bB})}
$$

includes weighted copolymerization parameters and propagation rate constants:

$$
(12\text{-}41) \quad \bar{r}_{aa} = \frac{r_{ba}\,(r_{aa}x_A + x_B)}{r_{ba}x_A + x_B} \quad ; \quad \bar{r}_{bb} = \frac{r_{ab}\,(r_{bb}x_B + x_A)}{r_{ab}x_B + x_A}
$$

$$
(12\text{-}42) \quad \bar{k}_{aA} = \frac{k_{aaA}\,(r_{aa}x_A + x_B)}{r_{aa}x_A + (x_B/s_A)} \quad ; \quad \bar{k}_{bB} = \frac{k_{bbB}\,(r_{bb}x_B + x_A)}{r_{bb}x_B + (x_A/s_B)}
$$

Table 12-7 Terminal copolymerization parameters $r_a = k_{aA}/k_{aB}$ and $r_b = k_{bB}/k_{bA}$, penultimate co-polymerization parameters $s_A = k_{baA}/k_{aaA}$ and $s_B = k_{abB}/k_{bbB}$, termination parameters $\tau_a = (k_{t,ba}/k_{t,aa})^{1/2}$ and $\tau_b = (k_{t,ab}/k_{t,bb})^{1/2}$, and cross-termination factors $\Phi = \tau_a\tau_b$ of neat free-radical co-polymerizations of monomers A and B at 40°C (* 30°C, ** 25°C, *** at 50°C in benzene) [7].

Monomer A	Monomer B	r_a	r_b	s_A	s_B	τ_a	τ_b	Φ
Styrene	ethyl acrylate	0.78	0.18	0.27		0.48		
Styrene	diethyl fumarate	0.21	0.021	0.037	0.30	0.043	0.81	0.35
Styrene	acrylonitrile			0.74	1.65	0.57		
Styrene*	p-methoxy styrene*	1.12	0.82	1.00	1.00			
Styrene	methyl methacrylate	0.52	0.46	0.30	0.47	0.89	1.5	1.34
Styrene**	lauryl methacrylate**	0.57	0.45	0.59	0.33			
Butyl methacrylate**	methyl methacrylate**	1.27	0.79	1.2	1.2	1	1.5	1.5
Vinyl acetate	methyl methacrylate	0.014	27.8		0.40		0.7	
Acrylonitrile	methyl methacrylate	0.18	1.4	0.015	1.04	0.14	0.93	0.13
Dibutyl itaconate***	N-cyclohexyl maleimide***	0.34	0.21	1.4	0.09	1.1	0.3	0.33

In the terminal model, the weighted copolymerization parameters, Eqs.(12-41), reduce to $\bar{r}_{aa} = r_a$ (since $r_{ba} = r_{aa} = r_a$) and to $\bar{r}_{bb} = r_b$ (since $r_{ab} = r_{bb} = r_b$). The parameter s_A becomes $s_A = k_{baA}/k_{aaA} = k_{aA}/k_{aA} = 1$ so that the weighted propagation rate constant reduces to $\bar{k}_{aA} = k_{aA}$. The same reasonings apply to s_B and $\bar{k}_{bB}$.

The penultimate copolymerization parameters, s_A and s_B, usually deviate from unity (Table 12-7), indicating the presence of penultimate effects. In most cases, the monomer prefers to add to its homo-diad ($s_A < 1$, $s_B < 1$). Terminations are also influenced by penultimate units ($\tau_a \neq 1$, $\tau_b \neq 1$). Cross-propagation factors do not exceed unity very much; they do not attain the high values generated by the terminal model.

12.3.4   Q,e Scheme

Copolymerization parameters are relative reactivities that are specific for each mono-mer A–monomer B–temperature–initiator combination and have to be determined for each particular system. Therefore, many attempts heve been made to characterize the co-polymerization behavior by a set of monomer-specific and system-independent para-meters. The most successful of these is the **Q,e scheme** based on the terminal model.

For free-radical copolymerizations, monomer pairs can be arranged in a series ac-cording to the product r_ar_b of their copolymerization parameters (Table 12-8). In this series, monomers with electron donating groups are to the left and those with electron accepting groups to the right. In each column, the product r_ar_b falls to ca. 0 from ca. 1. In each row, r_ar_b increases to larger values (right) from smaller ones (left).

The product r_ar_b obviously mirrors the effects of polarity and resonance stabilization. Both quantities also affect the activation energy $\Delta E_{aB}^{\ddagger}$ in the Arrhenius equation $k_{aB} = A_{aB}^{\ddagger} \exp(-\Delta E_{aB}^{\ddagger}/RT)$. At constant temperature, the product $\Delta E_{aB}^{\ddagger}/RT$ can thus be re-placed by the contributions of the polymer radical ($p_a^{\bullet}$) and the monomer molecule (q_B) to resonance and the contributions of the polymer radical ($e_a^{\bullet}$) and the monomer molecule (e_B) to electrostatic interactions :

$$(12\text{-}43) \qquad k_{aB} = A_{aB}^{\ddagger} \exp(-p_a^{\bullet} + q_B + e_a^{\bullet}e_B)$$

Table 12-8 Products $r_a r_b$ of copolymerization parameters of some monomers in free-radical copolymerizations. For example, the copolymerization of butadien and vinyl acetate yields $r_a r_b = 0.31$.

Butadiene							
0.77	Styrene						
-	0.52	Vinyl acetate					
0.35	0.50	0.60	Vinyl chloride				
0.14	0.23	0.96	0.63	Methyl methacrylate			
0.09	0.22	0.66	0.80	0.58	Vinylidene chloride		
0.018	0.028	0.33	0.15	0.18	0.53	Acrylonitrile	
-	0.0003	0.0004	0.035	0.059	-	0	Maleic anhydride

In monomers $CH_2=CRR'$, growing macroradicals always attack methylene groups. The pre-exponential factor $A_{aB}^{\ddagger}$ is therefore assumed to be independent of the monomer. The terms $\exp(p_a^\bullet)$ and $\exp(q_B)$ are furthermore united with the corresponding pre-exponential factors in two new parameters, P_a and Q_B,

$$(12\text{-}44) \qquad k_{aB} = P_a Q_B \exp(-e_a^\bullet e_B)$$

or, for equal charges of radical and monomer $(e_a^\bullet = e_A, e_b^\bullet = e_B)$,

$$(12\text{-}45) \qquad k_{aB} = P_a Q_B \exp(-e_A e_B)$$

Analogous equations can be written for the three other propagation rate constants, k_{aA}, k_{bA}, and k_{bB}, so that the copolymerization parameters can be expressed by

$$(12\text{-}46a) \qquad r_a = k_{aA}/k_{aB} = (Q_A/Q_B) \exp(-e_A(e_A - e_B))$$
$$(12\text{-}46b) \qquad r_b = k_{bB}/k_{bA} = (Q_B/Q_A) \exp(-e_B(e_B - e_A))$$
$$(12\text{-}46c) \qquad r_a r_b = \exp(-(e_A - e_B)^2) \leq 1$$

Hence, each monomer is assigned a Q-value (resonance term) and an e-value (polarity term) which depend on the type of polymerization (free-radical, cationic, etc.) and the temperature. It is not clear, however, whether the Q-values and/or e-values are also influenced by penultimate effects.

In order to calculate the e_A value of a monomer A from Eq.(12-46c) and subsequently Q_A from Eq.(12-46a), the corresponding values of a reference monomer B must be known. Styrene was chosen as a reference because it copolymerizes free-radically with many monomers. Its Q value was set as unity because styrene was assumed to have the largest resonance stabilization of all polymer radicals (Table 12-9). Non-resonance-stabilized polymer radicals should have $Q = 0$. Experimentally, Q values between 0.0001 (tetrachloroethene) and 16 (vinyl chloromethylketone) are known.

The polarity term of styrene was set as $e = -0.800$. Monomers with $e < -0.8$ should be more nucleophilic and those with $e > -0.8$ more electrophilic than the poly(styryl) radical. Experimental values range from -8.53 (vinyl o-cresyl ether) to $+3.70$ (N-butyl maleimide).

The Q,e scheme allows one to estimate unknown copolymerization parameters and to judge the copolymerizability of monomers. As a rule: (1) monomers with very different Q values do not copolymerize and (2) at $Q_A \approx Q_B$, values of $e_A \approx e_B$ lead to ideal-azeotropic copolymerizations whereas widely different e_A and e_B give alternating ones.

Table 12-9 Q and e values of some monomers in free-radical copolymerizations.

Resonance stabilization	Q	e	No resonance stabilization	Q	e
m–Divinyl benzene	3.35	−1.77	Methyl vinyl ether	0.037	− 1.28
o–Divinyl benzene	1.64	− 1.31	Vinyl palmitate	0.025	− 0.82
p–Methyl styrene	1.27	− 0.98	Octyl vinyl ether	0.061	− 0.79
Styrene (reference)	1.000	− 0.80	Propene	0.002	− 0.78
o–Methyl styrene	0.90	− 0.78	Vinyl bromide	0.047	− 0.25
p–Chlorostyrene	1.03	− 0.32	Vinyl acetate	0.026	− 0.22
Methyl methacrylate	0.74	0.40	Ethene	0.015	− 0.20
Dodecyl methacrylate	0.70	0.35	Vinyl chloride	0.044	0.20
Acrylonitrile	0.60	1.20	Tetrafluoroethene	0.049	1.22
Maleic anhydride	0.23	2.25	Carbon monoxide	1.000	3.76

However, copolymerizabilities cannot always be predicted by the Q,e scheme, especially, if one monomer does not homopolymerize. For example, propene homopolymerizes free-radically only to highly branched oligomers. Its copolymerization with vinyl acetate or other comonomers with $e < 1.2$ delivers copolymers with molecular weights that decrease strongly with increasing proportion of propylene units. Strongly electrophilic monomers with $e > 1.2$, such as tetrafluorethylene ($e = 1.22$; $Q = 0.049$) or maleic anhydride ($e = 2.25$; $Q = 0.23$), copolymerize free-radically with propene to alternating copolymers with molecular weights of up to 200 000.

The Q,e scheme has nevertheless been quite successful in the interpretation of copolymerizabilities. It also predicts that $r_a r_b$ must always be smaller than unity (Table 12-4). However, its theoretical foundation is questionable since it (a) does not consider penultimate effects and (b) assumes electrostatic interactions between polymer radicals and monomer molecules. The latter assumption may be responsible for the finding that Q and e seem to be correlated for resonance-stabilized systems (Fig. 12-9). On the other hand, the Q values of non-resonance-stabilized monomers fluctuate about an average of 0.04 ± 0.025.

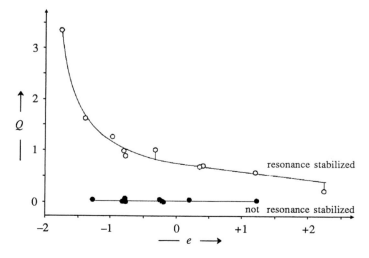

Fig. 12-9 Q and e values of various monomers (data of Table 12-9).

12.3.5 Terpolymerization

The terpolymerization of three monomers, A, B, and C is controlled by nine propagation rate constants that are combined to six copolymerization parameters:

$$(12\text{-}47) \quad r_{AB} = k_{aA}/k_{aB} ; \quad r_{BA} = k_{bB}/k_{bA} ; \quad r_{CA} = k_{cC}/k_{cA}$$
$$r_{AC} = k_{aA}/k_{aC} ; \quad r_{BC} = k_{bB}/k_{bC} ; \quad r_{CB} = k_{cC}/k_{cB}$$

A monomer can be consumed by three different propagation reactions, for example,

$$(12\text{-}48) \quad R_A = - d[A]/dt = k_{aA}[\sim\!a^*][A] + k_{bA}[\sim\!b^*][A] + k_{cA}[\sim\!c^*][A]$$
$$= R_{aA} \qquad\qquad + R_{bA} \qquad\quad + R_{cA}$$

Introduction of the steady-state conditions

$$(12\text{-}49) \quad \begin{aligned} R_{aB} + R_{aC} &= R_{bA} + R_{cA} \\ R_{bA} + R_{bC} &= R_{aB} + R_{cB} \\ R_{cA} + R_{cB} &= R_{aC} + R_{bC} \end{aligned}$$

leads to the terpolymerization equation of the terminal model

$$(12\text{-}50) \quad d[A] : d[B] : d[C] = Y_A : Y_B : Y_C$$

with

$$Y_A = [A] \cdot \left(\frac{[A]}{r_{CA}r_{BA}} + \frac{[B]}{r_{BA}r_{CB}} + \frac{[C]}{r_{CA}r_{BC}} \right) \cdot \left([A] + \frac{[B]}{r_{AB}} + \frac{[C]}{r_{AC}} \right)$$

$$Y_B = [B] \cdot \left(\frac{[A]}{r_{CA}r_{AB}} + \frac{[B]}{r_{AB}r_{CB}} + \frac{[C]}{r_{CB}r_{AC}} \right) \cdot \left([B] + \frac{[A]}{r_{BA}} + \frac{[C]}{r_{BC}} \right)$$

$$Y_C = [C] \cdot \left(\frac{[A]}{r_{AC}r_{BA}} + \frac{[B]}{r_{BC}r_{AB}} + \frac{[C]}{r_{AC}r_{BC}} \right) \cdot \left([C] + \frac{[A]}{r_{CA}} + \frac{[B]}{r_{CB}} \right)$$

In principle, the six copolymerization parameters can be determined by terpolymerization experiments. In most cases, they are taken from bipolymerizations, however. Unknown copolymerization parameters can be estimated as follows.

The Q,e scheme describes the rate constant for the addition of monomer molecules B to polymer radicals ~a* by

$$(12\text{-}51) \quad k_{aB} = P_a Q_B \exp(- e_A e_B)$$

and correspondingly for the other rate constants. Introduction of these equations into the definitions for the copolymerization parameters results in

$$(12\text{-}52) \quad r_{AB} r_{BC} r_{CA} = r_{AC} r_{CB} r_{BA}$$

and, because of the definition of conditional probabilities, Eq.(12-10), also in

$$(12\text{-}53) \quad P_{a/B} P_{b/C} P_{c/A} = P_{a/C} P_{c/B} P_{b/A}$$

Table 12-10 Averages of products of binary copolymerization parameters for the free-radical terpolymerization of the conjugated monomers methyl acrylate, methyl methacrylate, acrylonitrile, and styrene, and the non-conjugated monomers vinyl acetate, vinyl chloride, and vinylidene chloride at 60°C. All 21 copolymerization parameters are known.

Monomers in combination conjugated non-conjugated	Combinations	$r_{AB}r_{BC}r_{CA}$	$r_{AC}r_{CB}r_{BA}$	$\dfrac{r_{AB}r_{BC}r_{CA}}{r_{Ac}r_{CB}r_{BA}}$
3	0	4	0.073 ± 0.055	0.165 ± 0.169	0.442
2	1	18	0.372 ± 0.362	0.298 ± 0.315	1.248
1	2	12	0.359 ± 0.235	0.649 ± 0.553	0.553
0	3	1	0.362	0.317	1.142

A test of Eq.(12-53) indicates no significant differences between conjugated and non-conjugated monomers (Table 12-10) within the rather high limits of error.

Terpolymerizations may be azeotropic. Calculations for 653 monomer indicated 754 possible terpolymerizations with 36 ternary azeotropes, 598 quaterpolymerizations with 2 quaternary azeotropes, and 330 quinterpolymerizations with 1 quinternary azeotrope. Terpolymerizations, quaterpolymerizations, etc., may also have partial azeotropes where the mole fraction of only one type of monomeric unit equals its mole fraction in the monomer mixture.

12.4   Ionic and Coordination Copolymerization

12.4.1 Copolymerization Parameters

Free-radical copolymerizations usually lead to statistical copolymers while alternating copolymerizations and block copolymerizations are comparativebly rare. For ionic copolymerizations, it is however just the opposite (Fig. 12-1 and Table 12-11). Copolymerizations can thus diagnose whether a certain initiator leads to anionic, free-radical, cationic, etc., polymerizations. For example, the copolymerization of methyl methacrylate and acrylonitrile is initiated free-radically by boron alkyls ($r_{mma} = 1.24$; $r_{an} = 0.11$) but anionically by lithium alkyls ($r_{mma} = 0.34$; $r_{an} = 6.7$).

Table 12-11 Copolymerization parameters by anionic (an), free-radical (rad), and cationic (cat) mechanisms, assuming terminal models.

Monomers A	B	r_a anionic	radical	cationic	r_b anionic	radical	cationic
Styrene	vinyl acetate	0.01	55	8.3	0.1	0.01	0.015
Styrene	methyl methacrylate	0.12	0.52	10.5	6.4	0.46	0.1
Acrylonitrile	methyl methacrylate	7.9	0.18	-	0.25	1.35	-
Styrene	chloroprene	-	0.005	15.6	-	6.3	0.24

Such assignments are not without pitfalls, however. The anionic copolymerization of styrene and isoprene by lithium butyl leads to $r_s = 0.8$ and $r_i = 1.0$ in triethyl amine as solvent, similar to the values of $r_s = 0.4$ and $r_i = 2.0$ that were found for free-radical polymerizations. However, the same anionic initiator produced copolymerization parameters of $r_s = 0.046$ and $r_i = 16.6$ in cyclohexane and $r_s = 40$ and $r_i = 2.0$ in tetrahydrofuran. These effects are surely due to solvent-dependent types and proportions of free ions, ion pairs, and ion-pair complexes (Chapter 8). Polar solvents produce relatively more free ions than non-polar ones. For the same initiator, copolymerization parameters of ionic copolymerizations thus vary systematically with the polarity of the solvent as measured by its relative permittivity (Table 12-12).

Table 12-12 Influence of solvent on copolymerization parameters from the terminal model. Cp = cyclopentadienyl, Et = ethyl, iBu = isobutyl, MAO = methyl aluminoxane, Me = methyl, Ph = phenyl. ε_r = Relative permittivity of solvent. x unknown proportion.

Initiator	Solvent	T/°C	ε_r	r_a	r_b	$r_a r_b$
Cationic copolymerization of isobutene (A) and p-chlorostyrene (B)						
AlBr$_3$	benzene	25	2.28	1.14	0.99	1.13
	1,2-dichloroethylene	25	10	2.80	0.89	2.49
	nitrobenzene	25	36	14.9	0.53	7.90
	nitromethane	25	38	22.2	0.73	16.2
SnCl$_4$	benzene	25	2.28	12.2	2.8	34.2
	nitrobenzene	25	36	8.6	1.25	10.8
Anionic copolymerization of styrene (A) and butadiene (B)						
LiC$_4$H$_9$	cyclohexane	25	2.02	15.5	0.04	0.62
	benzene	25	2.28	10.8	0.04	0.43
	- (in bulk)	25	2.43	11.2	0.04	0.49
	diethyl ether	25	4.3	1.7	0.4	0.68
	tetrahydrofuran	25	7.6	0.3	4.0	1.2
Homogeneous Ziegler-Natta copolymerization of ethene (A) and propene (B)						
VOCl$_3$ + x iBu$_2$AlCl	alkanes	30		16.1	0.052	0.87
VOCl$_2$OEt + x iBu$_2$AlCl	alkanes	30		16.8	0.055	0.93
VOCl(OEt)$_2$ + x iBu$_2$AlCl	alkanes	30		18.9	0.069	1.06
VO(OEt)$_3$ + x iBu$_2$AlCl	alkanes	30		15.0	0.070	1.04
*Heterogeneous Ziegler-Natta copolymerization of ethene (A) and propene (B) or * styrene (B)*						
TiCl$_3$ + x (CH$_3$)$_3$Al	-	70		15.7	0.11	1.73
TiCl$_3$ + x Et$_3$Al	-	70		9.0	0.10	0.90
TiCl$_3$ + x Et$_3$Al *	-	70		81	0.012	0.97
Metallocene copolymerization of ethene (A) and propene (B)						
(Me$_5$Cp)$_2$ZrCl$_2$ + x MAO		50		250	0.002	0.50
Cp$_2$ZrCl$_2$ + x MAO		50		48	0.015	0.72
Cp$_2$TiPh$_2$ + x MAO		50		19.5	0.015	0.29
Metathesis copolymerization of cyclopentene (A) and norbornene (B)						
WCl$_6$ + 4 EtAlCl$_2$				0.32	13	4.16
WCl$_6$ + 10 EtAlCl$_2$				0.09	20	1.80
WCl$_6$ + (1/2) Bu$_4$Sn				0.27	12	3.24
WCl$_6$ + Ph$_4$Sn				0.55	2.6	1.43

In most ionic and coordinative copolymerizations, one copolymerization parameter is greater than unity and the other is smaller. Therfore, they cannot be azeotropic.

In contrast to free-radical copolymerizations, polarity is more important than resonance in ionic copolymerizations. Since cations and anions have opposite polarities, an $r_a > r_b$ in cationic polymerizations leads to $r_a < r_b$ in anionic ones and vice versa.

The product of copolymerization parameters seems always to adopt values of $r_a r_b \leq 1$ within limits of error for anionic, homogeneous Ziegler-Natta, and metallocene copolymerizations. Cationic and metathesis copolymerization have values of $r_a r_b > 1$.

Higher temperatures favor free ions over ion pairs. Copolymerization parameters should thus be very temperature dependent although this is not always found to be the case as they may increase or decrease with increasing temperature.

12.4.2   Anionic Copolymerization

Anionic polymerizations and copolymerizations are often living (Section 8.3). They are characterized by copolymerization parameters $r_a \ll 1$, $r_b \gg 1$, and $r_a r_b < 1$.

Butyl lithium-initiated anionic copolymerizations of butadiene and styrene in cyclohexane start slowly (Fig. 12-10, see also Fig. 12-12) but speed up dramatically some time after all butadiene has been consumed. This "critical point" corresponds to higher *total* monomer conversions than the initial proportion of butadiene, indicating an early polymerization of styrene. Styryl anions $\sim s^{\ominus}/Li^{\oplus}$ are rapidly converted to $\sim sb^{\ominus}/Li^{\oplus}$ with a very large rate constant k_{sB} (see Table 12-13). However, k_{sB} must also be larger than k_{bS}, and although k_{sS} is greater than k_{bB}, initially most butadiene is polymerized and only a little styrene, followed by a fast polymerization of mainly styrene. As a consequence, this copolymerization delivers gradient copolymers and not block copolymers.

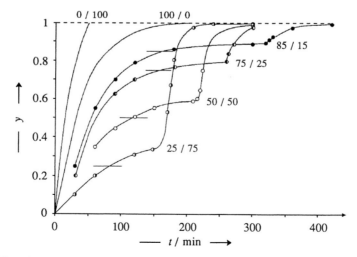

Fig. 12-10 Time dependence of the polymer yield, y, in the butyl lithium-initiated anionic copolymerization of butadiene and styrene for indicated initial monomer ratios of butadiene/styrene in cyclohexane at 50°C. Horizontal lines indicate the total consumption of butadiene for an independent and successive polymerization of the two monomers [8].

With kind permission by the Rubber Divison of the American Chemical Society.

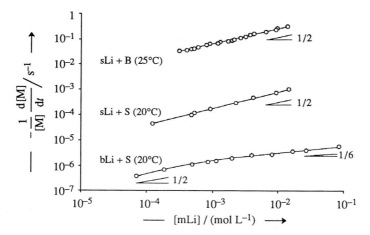

Fig. 12-11 Reduced polymerization rates, $-(1/[M])(d[M]/dt)$, of styrene S and 1,3-butadiene B as a function of the mole concentration of initiators styryl lithium (mLi = sLi) or butadienyl lithium (mLi = bLi) in toluene [9a].

A quantitative description of copolymerization rates requires detailed knowledge of all elementary reactions. Initiation reactions can be neglected for initiators that resemble propagating macroanions, for example, styryl lithium $C_4H_9-CH_2-CH(C_6H_5)Li$ for the styrene polymerization and butadienyl lithium $C_4H_9-CH_2-CH=CH-CH_2Li$ for that of butadiene. Homopolymerizations can be followed by UV spectroscopy (styrene) and dilatometry (butadiene), cross-propagations by the stopped-flow method.

The power α of the initiator concentration in the rate equation $-d[M]/dt = k_p[M][I]^\alpha$ indicates association of the initiators (see Eq.(8-35)): a dimerization of styryl lithium ($\alpha = 0.64 = 1/N \approx 2$) and a shift from dimerization to hexamerization for butadienyl lithium ($\alpha = 1/2$ at low [bLi] $\rightarrow \alpha = 1/6$ at high [bLi] (Fig. 12-11).

Copolymerization parameters from direct and indirect measurements agree well for slow reactions (bS, bB) (Table 12-13). But for fast reactions (sS, sB), a much too high copolymerization parameter r_a is obtained from relative kinetic data because the curves do not differ much in copolymerization diagrams (see similar situation in Fig. 12-2).

Table 12-13 Rate constants of anionic homo and copolymerizations of styrene S with various monomers B in different solvents from direct kinetic data (K) or indirectly from copolymerization parameters (C). Bu = 1,3-butadiene, MeS = α-methyl styrene, VPy = 2-vinyl pyridine.

Mono- mer B	Solvent	$T/°C$	k_p/(L mol^{-1} s^{-1}) for				Parameter			
			sS	sB	bS	bB	r_a	r_b	$r_a r_b$	
Bu	Toluene	20	0.45	110	0.0066	0.084	0.0041	12.7	0.052	K
Bu	Toluene	25					0.1	12.5	1.25	C
Bu	Benzene	30					0.05	15	0.75	C
Bu	Heptane	30					0	7	0	C
Bu	C$_6$H$_{12}$	40					0.04	2.6	0.104	K
MeS	THF	25	950		27	1200	2.5	35	0.0021	0.074
VPy	THF	25	950	100 000	<1	4500	0.0095	<4500	<43	C

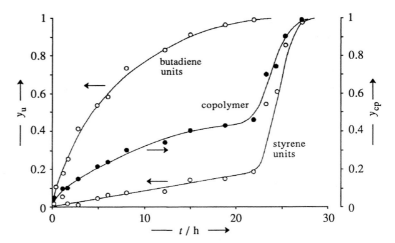

Fig. 12-12 Time dependence of yields y_u of butadiene or styrene units in copolymers [9b] and yields y_{cp} of copolymer molecules. —— Calculated with r_s and r_b from Table 12-13.

Time-conversion curves from kinetic data, as calculated with the terminal model, agree very well with experimental data (Fig. 12-12). The terminal model is applicable because cross-propagations can be neglected.

The calculated propagation rate constants k_{aA}, k_{aB}, k_{bA}, and k_{bB} are averages for the different ionic species (free ions, ion pairs, etc.) since they vary systematically with the total initiator concentration. If self-associations are present, shifts in self-association equilibria that cause deviations from the ideal values of $N = 1/\alpha$ will result.

A simple function $d[A]/d[B] = f([A]/[B])$ exists for the special case when two monomers A and B have widely differing polarities. If macroanions $\sim a^{\ominus}$ add monomer molecules B only slowly and macroanions $\sim b^{\ominus}$ add no monomer molecules A at all, then all a-units come from the initiation step and the immediate propagation steps. The rates of formation of macroions are then

$$(12\text{-}54) \quad d[a^{\ominus}]/dt = k_{IA}[I^{\ominus}][A] - k_{aB}[a^{\ominus}][B] \approx k_{IA}[I^{\ominus}][A]$$
$$d[b^{\ominus}]/dt = k_{IB}[I^{\ominus}][B]$$

At time t, the relative consumption of macroanions is

$$(12\text{-}55) \quad d[a^{\ominus}]/d[b^{\ominus}] = (k_{IA}/k_{IB})([A]/[B]) = [a^{\ominus}]/[b^{\ominus}]$$

The monomer consumption by initiation reactions is negligible for sufficiently high degrees of polymerization. A small monomer consumption by cross-propagation means

$$(12\text{-}56) \quad -d[A]/dt = k_{aA}[a^{\ominus}][A] + k_{bA}[b^{\ominus}][B] \approx k_{aA}[a^{\ominus}][A]$$
$$-d[B]/dt = k_{bB}[b^{\ominus}][B] + k_{aB}[a^{\ominus}][B] \approx k_{bB}[b^{\ominus}][B]$$

From Eqs.(12-55) and (12-56), the relative monomer consumption is obtained as

$$(12\text{-}57) \quad d[A]/d[B] = (k_{IA}/k_{IB})(k_{aA}/k_{bB})([A]/[B])^2 = const \cdot ([A]/[B])^2$$

Comparison of this equation with Eq.(12-15) shows that the exponent of the ratio of monomer concentrations may vary between 1 (Eq.(12-15)) and 2 (Eq.12-57)) which is indeed found for many ionic copolymerizations. An exponent of 2 indicates the formation of polymer blends which needs to be proven by fractionation, chromatography, etc.

For the preceding equilibria between macroanions and monomer molecules, the rate-determining step is the reaction of the intermediate, e.g., for the terminal model

$$(12\text{-}58) \quad \sim\!a^{\oplus} + A \ \rightleftarrows \ \sim\!a^{\oplus}\!/A \ \xrightarrow{\ k_{aA}\ } \ \sim\!aa^{\oplus}$$

and correspondingly for the other three equilibria and reactions as well. The conventional copolymerization parameters are here not the relative reactivities of the active centers but the concentrations and rate constants of the intermediate "compounds".

12.4.3 Coordination Copolymerization

Coordination copolymerizations in homogeneous phase are usually living polymerizations; their kinetics can be treated similarly to those in Section 12.4.2. Heterogeneous coordination polymerizations are much more complex since new active centers are constantly formed and destroyed. Instead of polymerization rates, one therefore often studies the change of catalyst activities or catalyst productivities (see p. 273) with time. The vast majority of these studies are performed in industry and never published.

The time and concentration dependencies of rates and catalyst activities and productivities are complex. They also differ widely for various monomers and catalyst systems (see also Fig. 9-8). For example, the catalyst productivity of a single-site catalyst as a function of the comonomer fraction x_B may either run through a maximum or drop drastically at low valus of x_B before it decreases linearly with x_B (Fig. 12-13). Note that the data for comonomer concentrations are obtained at different polymerization times.

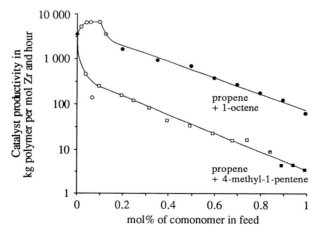

Fig. 12-13 Catalyst productivity in the copolymerization of propene with 1-octene or 4-methyl-1-pentene by Et[Ind]$_2$HfCl$_2$/MAO in toluene at 30°C [10]. ○ 30 min, ◎ 60 min, ● 90 min, □ 240 min, ◪ 480 min, ■ 600 min.

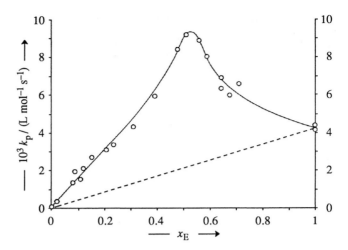

Fig. 12-14 Rate constant k_p as a function of the mole fraction x_E of ethene in the liquid phase for the copolymerization of ethene and propene by a metallocene-MAO catalyst at 37°C [11].

Rate constants may also show a curious behavior. For example, the gross propagation rate constant was found to pass through a maximum with increasing mole fraction of ethene in the copolymerization of ethene and propene by a metallocene-MAO catalyst (Fig. 12-14). Such behavior has never been observed for any other copolymerization.

Copolymerizations by single-site catalysts (+ MAO) may deliver random or even alternating copolymers depending on the metallocene structure. The situation is far from clear as the following examples show. The catalyst of Fig. 12-15 delivered alternating copolymers of ethene and propene if the substituent R of the cyclopentadienyl moiety was methyl but random ones if CH_3 was replaced by the bulky $C(CH_3)_3$.

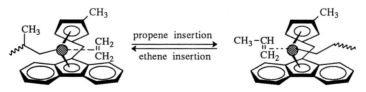

Fig. 12-15 Formation of alternating copolymers of ethene and propene.

Alternating copolymers of ethene and 1-octene were also obtained with $Me_2Si(2$-Me-1-Ind$)_2ZrCl_2$, a double-bridged sandwich compound with two rather large indenyl moieties. The related compounds meso-$Me_2Si(1$-Me-2-PhInd$)(2$-PhInd$)ZrCl_2$ (II-meso) and meso-$Me_2Si(2$-PhInd$)_2ZrCl_2$ (III-meso) delivered random copolymers of ethene and propene, however (Table 12-14). These double-bridged compounds yield the same copolymerization parameters r_e (ethene) and r_p (propene) as the single-bridged compound I; their products $r_e r_p$ are consistently greater than unity. All three catalysts (I, II-meso, III-meso) gave copolymers with propylene pentads of low isotacticity.

In contrast, the corresponding racemic compounds, II-rac and III-rac, have lower copolymerization parameters than the corresponding meso compounds, products $r_e r_p < 1$, and considerable isotacticities (Table 12-14).

Table 12-14 Copolymerization parameters of ethene (E) and propene (P) by the single-site catalysts I-III + MAO at 19°C [12] where O = Si(CH₃)₂, ● = ZrCl₂, Me = methyl, and PhInd = phenylindene. x_E/x_P = molar ratio of ethene and propene in feed; x_e = mole% ethylene units (e) in copolymer; r_e, r_p = copolymerization parameters of ethene (e) and propene (p), x_{iiii} = mole fraction of isotactic propylene pentads.

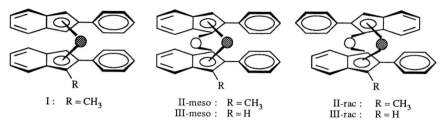

I : R = CH₃
I = (1-Me-2-PhInd)(2-PhInd)ZrCl₂

II-meso : R = CH₃
III-meso : R = H
II = Me₂Si(1-Me-2-PhInd)(2-PhInd)ZrCl₂

II-rac : R = CH₃
III-rac : R = H
III = Me₂Si(2-PhInd)(2-PhInd)ZrCl₂

Catalyst	CH₃	Bridge	x_E/x_P	x_e	r_e	r_p	$r_e r_p$	x_{iiii}
I	yes	no	0.08	34-41	6.4	0.18	1.15	0.14
II-meso	yes	yes	0.055	21	6.07	0.19	1.15	≈ 0.58
III-meso	no	yes	0.055	21	6.20	0.18	1.12	0.06
II-rac	yes	yes	0.058	25	4.36	0.15	0.65	≈ 0.58
III-rac	no	yes	0.084	27	3.12	0.21	0.67	0.87

12.5   Spontaneous Copolymerization

All free-radical, ionic, and coordination copolymerizations described above are initiated by added initiators. However, there are "spontaneous", initiator-free copolymerizations of comonomers with very different polarities. These copolymerizations may be subdivided into redox, zwitterion, and charge-transfer copolymerizations. There are also "regulated" copolymerizations, i.e., free-radical initiated (or also ionic) copolymerizations that are controlled (but not initiated) by added Lewis acids.

12.5.1   Redox Copolymerization

Redox copolymerizations are copolymerizations in which both monomers change their valence states. An example is the reaction of a five-membered phosphite with pyruvic acid to an alternating copolymer where the phosphite is oxidized to phosphoric acid triester units and the acid is reduced to propanoic acid units:

(12-59)

Six-membered phosphites undergo **deoxy copolymerizations** where only a part of the cyclic phosphite is oxidized and the copolymer thus contains more propanoic acid units than phosphate units.

The triethylamine-catalyzed reducing copolymerization of sulfur trioxide and dithiols leads to copolymers that contain three different types of units in statistical sequence:

(12-60) $SO_3 + HS-(CH_2)_2-SH \rightarrow$

$\sim[S-(CH_2)_2-S]$-*co*-$[S-(CH_2)_2-S-S]$-*co*-$[S-(CH_2)_2-S-SO_2]\sim$

12.5.2   Charge-Transfer Copolymerization

Much more common is the formation of charge-transfer complexes (CT complexes) that may transform into zwitterions $^\oplus DA^\ominus$ (Section 12.5.3; G: *zwitter* = mongrel, hermaphrodite, from *zwi* = twice, indicating a "hermaphroditic diion"). They may be incited by visible or ultraviolet light and then converted in polar solvents into diradicals or radical anions via CT singlets, CT triplets, or excited Franck-Condon states:

(12-61) $D + A \rightleftharpoons [D,A \longleftrightarrow D^\oplus,A^\ominus]$

$$\nearrow \quad ^\oplus D-A^\ominus \quad \nearrow \quad {}^\bullet D-A^\bullet$$
$$\searrow \quad [D^\oplus A^\ominus] \longrightarrow {}^\oplus D//A^\ominus \longrightarrow {}^\oplus D_{solv} + A^\ominus_{solv}$$

For example, if D is a vinyl monomer, $CH_2=CHR_d$ with an electron-donating group R_d and A is a vinyl monomer with an electron-accepting group R_a, then a tetramethylene moiety is formed in the transition state. Depending on the sustituents R_d and R_a, this moiety may then lead to biradicals or zwitterions:

(12-62)

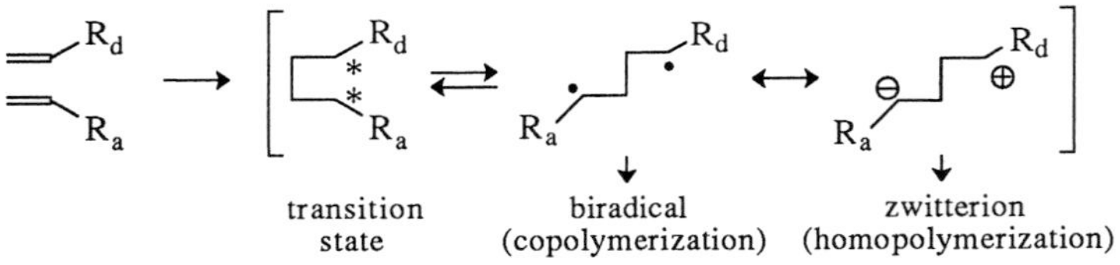

transition   biradical   zwitterion
state   (copolymerization) (homopolymerization)

Soft donors such as phenyl and vinyl groups stabilize diradicals whereas hard donors such as nitrogen and oxygen containing groups lead to zwitterions. From the electron-accepting groups, diester, cyanester, anhydride, and chlorine groups prefer to form diradicals whereas two cyano groups as substituents deliver zwitterions.

12.5.3   Zwitterion Polymerization

The joint polymerization of a vinyl monomer V_d with strong electron-donating groups (e.g., $CH_2=C(NR_2)_2$), with a vinyl monomer V_a with strong electron-accepting groups (e.g., $CH_2=C(CN)_2$) leads to zwitterions, $^\oplus V_d-V_a^\ominus$, that *homo*polymerize. Such zwitterions are also generated spontaneously from two suitable cyclic monomers or from a cyclic monomer and an olefin derivative.

Table 12-15 Comonomers D and A for zwitterion polymerizations to monomeric units -d- and -a-.

Comonomer D	Unit -d-	Comonomer A	Unit -a-
$CH_2=C(CN)(CN)$	$-CH_2-C(CN)(CN)-$	$CH_2=C(NR_2)(NR_2)$	$-CH_2-C(NR_2)(NR_2)-$
$CH=N$, Ar, Ar'	$-CH-N-$, Ar, Ar'	$CH_2=CH$, $O=COCH_2CH_2OH$	$-CH_2CH_2COCH_2CH_2O-$
 azetidine $R_2C{\cdots}N-R$	$-CH_2CR_2CH_2N-$, R	$CH_2=CH$, $COOH$	$-CH_2CH_2CO-$
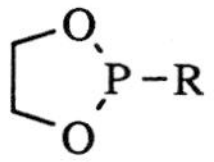 cyclic O_2P-R	$-CH_2CH_2OP(=O)-$, R	$CH_2=CH$, $CONH_2$	$-CH_2CH_2CO-NH$
lactam (O, NR)	$-CH_2CH_2CH_2NC(=O)-$, R	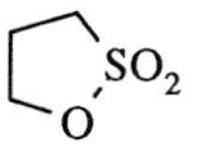 sultone SO_2, O	$-CH_2CH_2CH_2OS(=O)(=O)-$
oxazoline $N{=}$, O, R	$-CH_2CH_2N-$, $R{-}C(=O)$	β-lactone (oxetanone)	$-CH_2CH_2CO-$

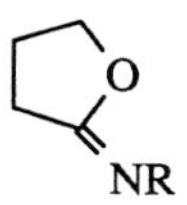

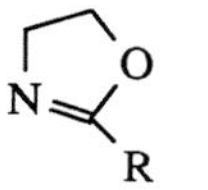

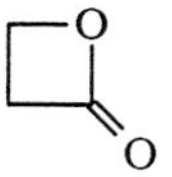

Suitable nucleophilic monomers are cyclic imino ethers and phosphites, azetidines, or Schiff bases in combination with electrophilic monomers such as lactones, sultones (cyclic esters of $HO–Z–SO_2OH$), cyclic anhydrides, and acrylic compounds (Table 12-15).

The constitution of the resulting monomeric units often deviates from those of conventional copolymerizations. For example, acrylamide, $CH_2=CH(CONH_2)$, does not copolymerize to units $–CH_2–CH(CONH_2)–$ (as in free-radical polymerizations) nor to units $–CH_2–CH_2–CO–NH–$ (as in anionic polymerizations with strong bases) but to iminoether structures, $–CH_2–CH_2–C(=NH)–O–$.

Some monomeric zwitterions, and even some polymeric ones, may neutralize their charges by cyclization, especially at elevated temperatures. The resulting macro-rings do not polymerize by ring-opening under these conditions. Such cyclizations are prevalent in dilute solutions and in solvents of low polarity.

The copolymerization of 2-methyl-2-oxazoline with acrylic acid leads first to a zwitterion I with β-propionic acid units, $–CH_2–CH_2–COO^{\ominus}$ (Eq.(12-63)). At 150°C, zwitterion I homopolymerizes spontaneously to the alternating poly(imino ester) II. This polymer is of low molecular weight because I may also dissociate into the two ions III and IV that act as terminating agents since they may add to the ends of the poly(zwitterion)s:

$$(12\text{-}63)$$

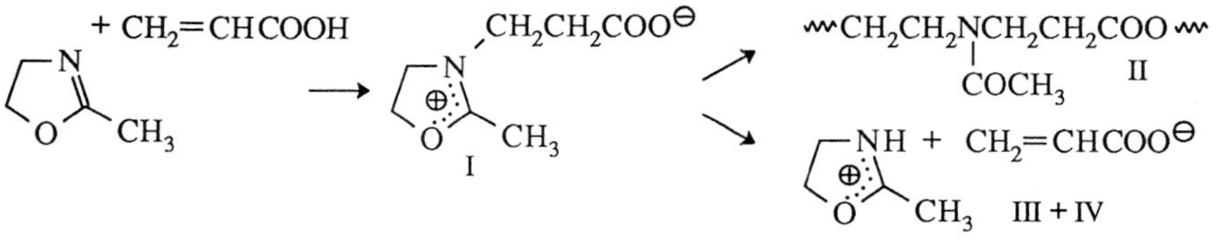

$$+ CH_2=CHCOOH \longrightarrow I \;(CH_2CH_2COO^{\ominus})$$

$$\nearrow \;\sim\!\!\!\sim CH_2CH_2NCH_2CH_2COO\!\sim\!\!\!\sim \quad (COCH_3) \quad II$$

$$\searrow \; -NH + CH_2=CHCOO^{\ominus} \quad (CH_3) \quad III + IV$$

In the ideal case, such copolymerizations proceed exclusively via zwitterions. Since zwitterions are the true monomers, the "copolymerization" is really a homopolymerization. The alternating "co"polymers may either be generated by successive additions of monomeric zwitterions to the cationic or anionic chain ends of macrozwitterions, i.e., by a living polymerization,

$$(12\text{-}64) \quad {}^{\oplus}D(AD)_nA^{\ominus} + {}^{\oplus}DA^{\ominus} \rightarrow {}^{\oplus}D(AD)_{n+1}A^{\ominus}$$

or by a combination of dimeric, oligomeric, ... polymeric zwitterions, i.e., a poly-addition:

$$(12\text{-}65) \quad {}^{\oplus}D(AD)_nA^{\ominus} + {}^{\oplus}D(AD)_mA^{\ominus} \rightarrow {}^{\oplus}D(AD)_{n+m+1}A^{\ominus} \quad ; \quad m \geq 1, n \geq 1$$

If dipole-ion reactions of monomers with zwitterions becomes comparable to the dipole-dipole interactions of the two monomers, the homopolymerization of the zwitterion competes with the copolymerization of the (macro)zwitterions with one or both of the original monomers:

$$(12\text{-}66) \quad {}^{\oplus}D_2(AD)_nA^{\ominus} \xleftarrow{+D} {}^{\oplus}DA^{\ominus} \xrightarrow{+A} {}^{\oplus}D(AD)_nA_2^{\ominus}$$

For example, at $T < 80°C$, acrylic acid reacts with 1,3,3-trimethyl azetidine to form an alternating copolymer. At higher temperatures, longer sequences of acrylic acid units are formed because, at $T > 150°C$, acrylic acid can homopolymerize to poly(β-propionic acid), $\{\text{O--CH}_2\text{--CH}_2\text{--CO}\}_n$. The attempted copolymerization of β-propiolactone with 1,3,3-trimethylazetidine delivered only poly(β-propiolactone) and not the copolymer, probably because a copolymerization would generate the less reactive ammonium ion.

Polyadditions and living-chain polymerizations differ characteristically in the dependence of the number-average degree of polymerization, $\overline{X}_n$, on conversion, u (Fig. 6-1). $\overline{X}_n$ increases linearly with u in living chain polymerizations but with $1/(1-u)$ in polyadditions (see Chapter 13). The "co"polymerization of 2-oxazoline and β-propiolactone in N,N-dimethylformamide must therefore be a living chain polymerization since $\overline{X}_n$ continues to increase linearly with u after addition of new monomer (Fig. 12-16). On the other hand, the same joint polymerization in acetonitrile must be an overlap of a living chain polymerization (with polymerization equilibrium) and polyaddition.

12.5.4 Polymerization of CT Complexes

In literature, the term "charge transfer" (CT) refers to three different processes: (1) the partial transfer of electrons from a donor to an acceptor (true charge-transfer with formation of a CT complex), (2) the complete transfer of electrons with formation of ion-radicals, and (3) the complete exchange of charges in the ground state, for example, in $NH_3 + BF_3 \rightarrow [H_3N]^{\oplus}[BF_3]^{\ominus}$.

True **CT complexes** are always in the ground state. They result from the combination of electron donors with low ionization potentials (measuring the lowest binding molecular orbital) and electron acceptors with high electron affinities (measuring the highest binding molecular orbital). The complexes are present as resonance hybrids to which structures $D^{\oplus}A^{\ominus}$ contribute only a little.

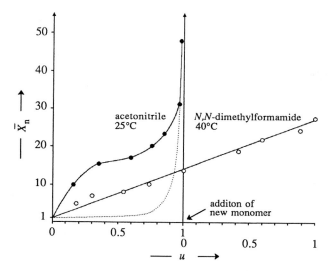

Fig. 12-16 Number-average degree of polymerization, $\overline{X}_n$, as a function of monomer conversion for the 1:1 (mol:mol) copolymerization of 2-oxazoline and β-propiolactone in acetonitrile at 25°C or in N,N-dimethylformamide at 40°C [13]. Degrees of polymerization refer to repeating units, not to monomeric units. ···· Theory for polyaddition.

Exciplexes are CT complexes that are only stable in the excited state (Section 11.2). In these, either the donor or the acceptor is excited.

Ion radicals result from thermal or photochemical excitation. Their formation is promoted by polar solvents since Gibbs energies are lowered by ion-solvent interactions.

The equilibrium constants of the formation of CT complexes can be obtained from either visible or ultraviolet spectra or from chemical shifts in NMR spectra. Observed equilibrium constants are small (Table 12-16),

CT complexes may polymerize either in the ground state or in the excited state. On addition of complexing agents or free-radical initiators, they may also undergo so-called regulated polymerizations (Section 12.5.5).

In the ground state, CT complexes are held together mainly by electrostatic forces. Electrons do not participate in these bonds and CT complexes are thus not very reactive. It is therefore not certain that some of spontaneous joint polymerizations are indeed polymerizations via the ground states of CT complexes.

Table 12-16 Polymerization (PM) of charge-transfer complexes at 25°C. K_{CT} in L/mol. * 60°C.

Monomers	Solvent	K_{CT}	Polymerization
4-Vinyl pyridine + p-chloranil	various	>>1	no polymerization
Vinylidene cyanide + methyl vinyl ether	various	1	spontaneous ionic homo-PM
Maleic anhydride + styrene	heptane *	0.34	radical init. alternating co-PM
Maleic anhydride + dimethoxyethene	CCl_4	0.15	spontaneous alternating co-PM
Vinylidene cyanide + styrene	various	0.1	spontaneous alternating co-PM
Cyclohexene + sulfur dioxide	C_7H_{14}	0.053	spontaneous alternating co-PM
Maleic anhydride + divinyl ether	$CHCl_3$	0.006	
Styrene + methyl α-chloroacrylate + Et_2AlCl	-	0.01	radical init. alternating co-PM

For example, no color formation is observed on mixing vinylidene cyanide and vinyl ethers; the resulting product is a mixture of poly(vinylidene cyanide), poly(vinyl ether), and cycloaddition compounds. The polymerization must be ionic because it is inhibited by both trihydroxy ethylamine and diphosphorus pentoxide. A possible reaction is the primary formation of vinyl ether radical-cations and vinylidene cyanide radical-anions that then dimerize to the respective dications and dianions. The dications start the cationic homopolymerization of vinyl ethers, the dianions the anionic homopolymerization of vinylidene cyanide. It is unclear why these apparently simultaneous anionic and cationic polymerizations do not lead to a neutralization of charges.

On the other hand, the joint polymerization of some electron-accepting and electron-donating monomers do lead to alternating copolymers, mostly as mixtures with the head-to-head cycloaddition products. Examples are the acceptors maleic anhydride, fumaric acid esters, sulfur dioxide, and carbon dioxide combined with the donors butadiene, isobutene, vinyl ether, p-dioxene, and vinyl acetate.

On irradiation with light, CT complexes convert to their excited states and charges are transferred. The excited CT complex is one large molecule with a larger π-electron system than the individual monomer molecules. Such large systems are more easy to polarize which, in turn, requires less activation energy. The excited CT complex is therefore the most reactive species in a mixture of CT complexes and their monomers.

12.5.5 Regulated Copolymerization

Some monomeric electron acceptors A are too weak to form sufficiently stable CT complexes with monomeric electron donors D. In this case, the complex formation may be promoted by the addition of non-monomeric Lewis acids L. Since Lewis acids are themselves electron acceptors, they add to electron donors and form DL complexes. This lowers the electron density at the double bond which in turn allows the electron acceptor A to form a stable ADL complex.

The new CT complex polymerizes to alternating copolymers, sometimes spontaneously, sometimes after addition of free-radical initiators. The copolymerization is inhibited by free-radical inhibitors. It often competes with the cationic homopolymerization of one of the two monomers.

Electron-donating 1-olefins, dienes, unsaturated esters and ethers, and halogen olefins with $e < 1/2$ copolymerize with electron-accepting acrylic monomers and vinyl ketones with high values of e (and also mostly high Q values) if $ZnCl_2$, $VOCl_3$, or $C_2H_5AlCl_2$ are added. Regulated copolymerizations thus allow the polymerization of monomers that do not homopolymerize free-radically, for example, propene and allylic monomers. The proportion of added Lewis acids may be small since their complexations are dynamic equilibria that are only needed for the propagation steps. For example, the free-radical copolymerization of 1,3-butadiene and acrylonitrile requires only 0.01 mol% $(C_2H_5)_3Al$ and 0.0001 mol% $VOCl_3$ for an alternation of units. However, the alternating copolymerization of ethene and vinyl acetate needs equimolar amounts of BF_3.

The complexation by Lewis acids changes both the polarity and the resonance stabilization of the electron-accepting monomer (Table 12-17). The resulting change of Q and e values is not necessarily caused by a change of the bonding state of the double

Table 12-17 Q and e values of monomers M_A after complexation by Lewis acids (L).

| Monomers | | Lewis | Q_A | | e_A | |
A	B	acid	-	+ L	-	+ L
$CH_2=C(CH_3)(COOCH_3)$	$CH_2=CCl_2$	$ZnCl_2$	0.74	26.3	0.4	4.2
$2\ CH_2=C(CH_3)(COOCH_3)$	$CH_2=CHC_6H_5$	$ZnCl_2$	0.74	13.5	0.4	1.74
$CH_2=CHCN$	$CH_2=CCl_2$	$ZnCl_2$	0.6	12.6	1.2	8.2
$CH_2=CHCN$	$CH_2=CHCl$	$SnCl_4$	0.6	2.64	1.2	2.22
$2\ CH_2=CHCN$	$CH_2=CHC_6H_5$	$ZnCl_2$	0.6	24	1.2	2.53

bond. It may also be caused by a change in the direction of the attack. In the copolymerization of butadiene and propene with added VCl_4+Et_3Al, the α-carbon atom of propene is attacked (as usual in coordination polymerizations) and not the β-carbon atom (as in free-radical and conventional ionic polymerizations).

Copolymerizations with addition of a Lewis acid proceed either as bipolymerizations of the electron-donating monomer and the complex electron donor + Lewis acid or as homopolymerizations of the ternary complex electron donor / electron acceptor / Lewis acid.

The addition of free-radical initiators can cause several effects. If added to butadiene + propene + $VCl_4/EtAl_3$, dibenzoyl peroxide (BPO) increases the molecular weight of the resulting alternating copolymers because hydrogen transfer is reduced.

The addition of BPO to vinylidene cyanide + unsaturated ethers generates 1:1 copolymers in addition to the mixture of the two homopolymers + cyclo adduct obtained without BPO. It is not the cycloadduct that undergoes a ring-opening polymerization, however, since the cyclo-adduct has a head-to-head structure while the copolymer has a head-to-tail one.

Free-radical initiated polymerizations of CT complexes are also responsible for joint polymerizations of monomers that do not homopolymerize, for example, maleic anhydride + stilbene or styrene + 1-olefins.

The group of regulated polymerizations also possibly include the copolymerizations of carbon dioxide with epoxides to polyanhydrides, with aziridines to polyurethanes, and with vinyl ethers to polyketo ethers:

(12-67) $CO_2 + 2\ CH_2=\underset{\underset{OR}{|}}{CH} \longrightarrow$ $\sim\!\!\!\sim\underset{\underset{O}{\|}}{C}-CH_2-\underset{\underset{OR}{|}}{CH}-O-\underset{\underset{OR}{|}}{CH}-CH_2\sim\!\!\!\sim$

Copolymer compositions and polymerization rates depend on the solvent since the generally polar solvents shift the position of the equilibrium of the complex. The homopolymerization of the CT complex may thus be converted into a bipolymerization of the CT complex with one of its monomers or into a terpolymerization with both of them. Such effects may be responsible for the variation of the content of acrylonitrile units with the type and concentration of the solvent in terpolymers from acrylonitrile + *p*-dioxane + maleic anhydride. Such variations may also be caused by penultimate effects as well as electrostatic effects that reduce the activation energy of cross-propagations.

Historical Notes to Chapter 12

T.Alfrey, Jr., G.Goldfinger, J.Chem.Phys. **12** (1944) 205
F.R.Mayo, F.M.Lewis, J.Am.Chem.Soc. **66** (1944) 1594
F.T.Wall, J.Am.Chem.Soc. **66** (1944) 2050
I.Sakurada, Kojugo Hanno (1944); *cf.* T.Saegusa, Macromol.Symp. **98** (1995) 1199
    These four groups derived the copolymerization equation independently of each other.

T.Alfrey, Jr., C.C.Price, J.Polym.Sci. **2** (1947) 101
    Derivation of the Q,e scheme.

G.Goldfinger, T.Kane, J.Polym.Sci. **3** (1948) 462
    Derivation of the sequence statistics of copolymers without assuming a steady state.

M.Fineman, S.D.Ross, J.Polym.Sci. **5** (1950) 269
V.Jaacks, Makromol.Chem. **161** (1972) 161
T.Kelen, F.Tüdös, J.Macromol.Sci.-Chem. **A 9** (1975) 1; see also
T.Kelen, F.Tüdös, B.Turcsányi, J.P.Kennedy, J.Polym.Sci.-Polym.Chem.Ed. **15** (1977) 3047
    Development of 3 different (linear) methods for the calculation of copolymerization parameters.

H.J.Harwood, Makromol.Chem., Macromol.Symp. **10/11** (1987) 331
    Bootstrap effect.

M.A.Dube, A.Penlidis, Polymer **36** (1995) 587
    Non-linear optimization.

Literature to Chapter 12

12.1 OVERVIEW
T.Alfrey, Jr., J.J.Bohrer, H.Mark, Copolymerization, Interscience, New York 1952
G.E.Ham, Ed., Copolymerization, Interscience, New York 1964
R.A.Patsiga, Copolymerization of Vinyl Monomers with Ring Compounds, Revs.Macromol.Chem.
    1 (1967) 223
Y.Yamashita, Random and Block Copolymers by Ring-Opening Polymerization, Adv.Polym.Sci.
    28 (1978) 1
W.Bruns, I.Motoc, K.F.O'Driscoll, Monte Carlo Applications in Polymer Science, Springer, Berlin
    1984
G.Allen, J.C.Bevington, Eds., Comprehensive Polymer Science, Vol. 3, G.C.Eastmond, A.Ledwith,
    S.Russo, P.Sigwalt, Eds., Chain Polymerization I, Pergamon, Oxford 1989

12.2 COPOLYMERIZATION THEORY
A.Valvassori, G.Sartori, Present Status of the Multicomponent Copolymerization Theory,
    Adv.Polym.Sci. **5** (1967/68) 28
D.Braun, W.Brendlein, G.Disselhoff, F.Quella, Computer Program for the Calculation of Ternary
    Azeotropes, J.Macromol.Sci.-Chem. **A 9** (1975) 1457
A.Rudin, Calculation of Monomer Activity Ratios from Multicomponent Copolymerization Results,
    Comput.Chem.Ind. **6** (1977) 117
R.Van der Meer, H.N.Linssen, A.L.German, Improved Methods of Estimating Monomer Reactivity
    Ratios in Copolymerization by Considering Experimental Errors in Both Variables, J.Polym.Sci.-
    Chem.Ed. **16** (1978) 2915
C.Hagiopol, Copolymerization: Toward a Systematic Approach, Plenum Press, New York 1999

12.3 FREE-RADICAL COPOLYMERIZATION
P.Wittmer, Kinetics of Copolymerization, Makromol.Chem., Suppl. **3** (1979) 129
K.Plochocka, Effect of the Reaction Medium on Radical Copolymerization, J.Macromol.Sci.-Revs.
    Macromol.Chem. **C 20** (1981) 67

S.M.Samoilov, Propylene Radical Copolymerization, J.Macromol.Sci.-Revs.Macromol.Chem.
  C **20** (1981) 333
A.Sen, The Copolymerization of Carbon Monoxide with Olefins, Adv.Polym.Sci. **73/74** (1986) 125

12.4 IONIC AND COORDINATION COPOLYMERIZATION

M.Szwarc, C.L.Perrin, General Treatment of Equilibrium Copolymerization of Two or More Comon-
  omers Deduced from the Initial State of the System, Macromolecules **18** (1985) 528
J.Luston, F.Vass, Anionic Copolymerization of Cyclic Ethers with Cyclic Anhydrides, Adv.Polym.
  Sci. **56** (1984) 91

12.5 SPONTANEOUS COPOLYMERIZATION

H.Hirai, Mechanism of Alternating Copolymerization of Acrylic Monomers with Donor Monomers
  in the Presence of Lewis Acid, J.Polym.Sci. [Macromol.Revs.] **11** (1976) 47
T.Saegusa, Spontaneous Alternating Copolymerization via Zwitterion Intermediates,
  Angew.Chem.Int.Ed.Engl. **16** (1977) 826
Y.Shirota, H.Mikawa, Thermally and Photochemically Induced Charge-Transfer Polymerizations,
  J.Macromol.Sci.-Revs.Macromol.Chem. C **16** (1977/78) 129
J.Furukawa, E.Kobayashi, Alternating Copolymerization, Rubber Chem.Technol. **51** (1978) 600
I.J.McEwen, Alternating Copolymers from Macrozwitterions, Prog.Polym.Sci. **10** (1984) 317
J.M.G.Cowie, Ed., Alternating Copolymers (= Specialty Polymers **1**), Plenum, New York 1985

References to Chapter 12

[1]   Experimental data of M.Hirooka, Ph.D. thesis, Kyoto University (1971)
[2]   V.Jaacks, Makromol.Chem. **161** (1972) 161, Fig. 2
[3]   P.Wittmer, Makromol.Chem. **103** (1967) 188, Tables 3 and 6
[4]   S.Masuda, T.Tomida, in J.C.Salamone, Ed., Polymeric Materials Encyclopedia **11** (1996)
        8257, Tables 1 and 3
[5]   P.Wittmer, H.Böck, H.Naarmann, B.J.Schmitt, Makromol.Chem. **182** (1981) 2505, Tables 3, 4
[6]   T.Fukuda, Y.-D.Ma, H.Inagaki, Macromolecules **18** (1985) 17, Table V, Figs. 12 and 13;
        Makromol.Chem., Suppl. **12** (1985) 125, Figs. 2 and 3
[7]   T.Fukuda, N.Ide, Macromol.Symp. **111** (1996) 305, Table 1
[8]   H.L.Hsieh, W.H.Glaze, Rubber Chem.Technol. **43** (1970) 22, Fig. 21
[9]   R.Ohlinger, F.Bandermann, Makromol.Chem. **181** (1980) 1935. (a) Data of Tables 1-3, (b) cal-
        culated from the data of Fig. 7
[10] M.Arnold, S.Bornemann, F.Köller, T.J.Menke, J.Kressler, Macromol.Chem.Phys. **199** (1998)
        2647, Tables 1 and 2
[11] W.Kaminsky, Angew.Makromol.Chem. **223** (1994) 101, Fig. 2
[12] M.Dankova, J.M.Petoff, R.M.Waymouth, ACS Polym.Prep. **40/2** (1999) 862
[13] T.Saegusa, Y.Kimura, S.Kobayashi, Macromolecules **7** (1974) 1, Fig. 3; -, Macromolecules **10**
        (1977) 236, Table I

13 Polycondensation and Polyaddition

13.1 Overview

13.1.1 Chain and Step Reactions

The propagation step of a kinetic chain polymerization consists of a reaction between a monomer molecule and an active polymer molecule, either by addition of the monomer molecule to an active center (Chapters 8 and 10) or by insertion of the monomer molecule into the bond between a catalytic center and the polymer chain (Chapter 9). The propagation does not proceed by reaction with molecules other than monomer molecules: at less than 100 % monomer conversion, actual and prospective reaction participants consist only of polymer molecules and unreacted monomer molecules besides initiators. The initiation reaction is always *initiated*; the initiator fragments become end-groups of polymer chains. The various elementary reactions of these chain polymerizations are in succession, i.e., "stepwise" and not in parallel (concurrently, simultaneously) in the classical kinetics of low-molecular weight reactions (p. 156).

Polymer chemistry contrasts "chain polymerization" with "step polymerization" and uses the term "stepwise" exclusively for two other classes of successive reactions, polycondensation and polyaddition (Section 6.1.2). At less than 100 % monomer conversion, reacting species consist here not only of polymer and monomer molecules but also of oligomer molecules. All three types of participants can react with each other. The probability of coupling between reactants is dictated by the concentration of reactive groups; it follows probability laws. There are no initiation reactions; each step may or may be not *catalyzed*. Catalysts, if any, do not become part of polymer chains.

It is irrelevant for the statistics of coupling steps whether they are accompanied by the splitting-off of small molecules (as in polycondensations, Eq.(6-1)) or not (as in polyadditions, Eq.(6-2)). Both polycondensation and polyaddition lead therefore to the same theoretical relations between, e.g., the extent of reaction and the mole fractions of the various participants or the number-average degree of polymerization of the polymer (but not necessarily the weight-average, see below). Many polycondensations and polyadditions are equilibrium reactions. The equilibrium can be shifted by removal of small leaving molecules in polycondensations but obviously not in polyadditions.

Polymer step reactions and polymer chain reactions differ also in other aspects:

- A functionality of at least 2 is required for *monomer molecules* in chain polymerizations but for all *reactants* (monomers, oligomers, polymers) in step polymerizations.
- In non-living chain polymerizations, high molecular weights are obtained at very small monomer conversions. In living chain polymerizations, molecular weights are proportional to monomer conversion. In linear equilibrium step polymerizations, however, very high monomer conversions are required for high molecular weights.
- In step reactions of bifunctional monomer molecules, the formation of linear polymer molecules is often accompanied by cyclization reactions, for example,

$$(13\text{-}1) \quad n\ \text{A—Ar—O—B} \xrightarrow{-\ n\ \text{AB}} \quad +\!\!\left[\text{Ar—O}\right]_{\overline{n}} \quad \text{polycondensation}$$

$$\xrightarrow{-\ n\ \text{AB}} n/2\ \left(\underset{\underline{\phantom{[\text{Ar–O}]_2}}}{[\text{Ar–O}]_2}\right) \longrightarrow +\!\!\left[\text{Ar—O}\right]_{\overline{n}} \quad \begin{matrix}\text{cyclization followed}\\\text{by equilibration}\end{matrix}$$

13.1.2   Types of Polycondensations and Polyadditions

The simplest polycondensations and polyadditions proceed between unlike A and B groups of bifunctional monomers, either as **AB reactions** of AB molecules or as **AA/BB** reactions of AA molecules + BB molecules. An example of an AB polycondensation is the self-condensation of hydroxy acids HO–Z–COOH to polyesters H$\left[\text{O–Z–CO}\right]_n$OH (Section 13.1.4). The polycondensation of hexamethylene diamine, $H_2N(CH_2)_6NH_2$, with adipic acid, $HOOC(CH_2)_4COOH$, to poly(hexamethylene adipamide) (polyamide 6.6, nylon 66), H$\left[\text{NH}(CH_2)_6\text{NHCO}(CH_2)_4\text{CO}\right]_n$OH, is an example of an AA/BB poly-condensation (Section 13.1.5). **AA reactions** of molecules with like functional groups are rare; an example is the formation of poly(ethylene glycol)s, H$\left[\text{OCH}_2\text{CH}_2\right]_n$OH, by **self-condensation** of ethylene glycol, $HOCH_2CH_2OH$ (Eq.(6-1), left).

For historic reasons, AA/BB reactions are *not* called *co*polycondensations; probably, because self-condensations of AA or BB molecules are very rare under polycondensa-tion conditions. There are no homosequences; a-units and b-units always alternate. Tra-ditionally, the term **copolycondensation** refers to reactions of, e.g., A'B + A"B in AB re-actions and, e.g., A'A' + A"A" + BB in AA/BB polycondensations provided they do not lead to periodic sequences of monomeric units. Other examples are AB + AC, AA + BB + CC, and AA + AB + BB polymerizations.

These polycondensations and similar polyadditions proceed intermolecularly and **"linearly"**, i.e., without formation of branched polymer molecules. However, they may be accompanied by intramolecular reactions of endgroups to cyclic molecules (Section 13.6.1). **Cyclopolycondensations** are polycondensations with formations of small rings within otherwise linear polymer chains (Section 13.6.2).

Polymerizations of AB_2, AB_3, etc., monomer molecules lead to so-called **hyper-branched** polymers if A can only react with B and neither A with A nor B with B (Sect-tion 13.8). Hyperbranched polymers are also obtained from AB_2 + AB reactions, etc., at low monomer conversions. At higher conversions, AB_2 + AB reactions, and those of A_3 + B_2, A_3 + B_3, AB_2 + A_3, AB_2 + BA_2, etc., deliver crosslinked polymers (gels) as well as soluble ones (sols) (Section 13.8.2 and 13.9). Polymerizations of monomers with func-tionalities of 3 or more such as AB_2, A_3 + B_2, etc., have been called **nonlinear, polyfunc-tional, multifunctional**, or **three-dimensional**.

Most polycondensations and polyadditions are not "spontaneous" (i.e., self-catalyzed, internally catalyzed) but need external catalysts. The chemistry of the reactions does not differ from those of corresponding monofunctional compounds (nucleophilic, electro-philic, etc. reactions).

Polycondensations and polyadditions may proceed in melts, solutions, and dispersions as well as in crystalline states, in the gas phase, or at interphases, depending on the stabili-ty and reactivity of the reactants. The reactions may occur in a homogeneous or hetero-geneous phase and may be diffusion-controlled or not, resulting in very different de-pendencies of degrees of polymerization on monomer conversion.

Industrially, far fewer standard polymers are produced by polyaddition and polycon-densation than by chain polymerization. Examples are poly(ethylene terephthalate), polyamide 6.6, and phenolic and amino resins. However, these types of polymerization serve for many specialty polymers: aliphatic and aromatic polyamides, polyimides, ep-oxy resins, polyether ketones, alkyd resins, polysiloxanes, and many more (Volume II).

13.1.3 Polyadditions

In polyaddition, functional groups are either added to double bonds or react with rings since rings can be considered as potential double bonds.

Addition to Double Bonds

The most important reaction of this group comprises the formation of polyurethanes from diols and diisocyanates where hydrogen atoms are added to >C=N– double bonds:

(13-2) $HO-U-OH + O=C=N-V-N=C=O \rightarrow \sim O-U-O-CO-NH-V-NH-CO\sim$

The corresponding reaction of diisocyanates with diamines leads to polyureas:

(13-3) $H_2N-U-NH_2 + O=C=N-V-N=C=O \rightarrow \sim NH-U-NH-CO-NH-V-NH-CO\sim$

However, the formation of polyamides, $\sim\sim NH-V-NH-CO-U-CO\sim\sim$ by reaction of diisocyanates, $V(N=C=O)_2$ with dicarboxylic acids $U(COOH)_2$ is a polycondensation since CO_2 is liberated.

Hydrogen atoms are added to carbon-carbon double bonds in the polyaddition of di-thiols to divinyl compounds to form polythioethers,

(13-4) $HS-U-SH + CH_2=CH-V-CH=CH_2 \rightarrow \sim S-U-S-CH_2-CH_2-V-CH_2-CH_2\sim$

in the hydrosilation of vinyl or allyl silanes to yield polycarbosilanes,

(13-5) $HSiR_2-U-SiR_2H + CH_2=CH-SiR_2-V-SiR_2-CH=CH_2$
$$\rightarrow \sim SiR_2-U-SiR_2-CH_2-CH_2-SiR_2-V-SiR_2-CH_2-CH_2\sim$$

or in the self-condensation of acrylic acid to poly(β-propionic acid):

(13-6) $CH_2=CH-COOH \rightarrow \sim CH_2-CH_2-CO-O\sim$

Reaction of bismaleimides with HQ–U–QH results in thermoplastic polyimides

(13-7)

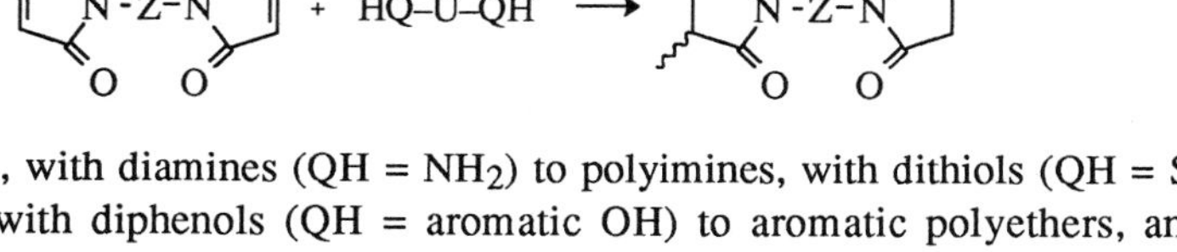

for example, with diamines (QH = NH_2) to polyimines, with dithiols (QH = SH) to poly-thioethers, with diphenols (QH = aromatic OH) to aromatic polyethers, and with ald-oximes (QH = CH=NOH) to polyamides:

Polyadditions also include the so-called **Diels-Alder polymerizations**, for example,

(13-8) 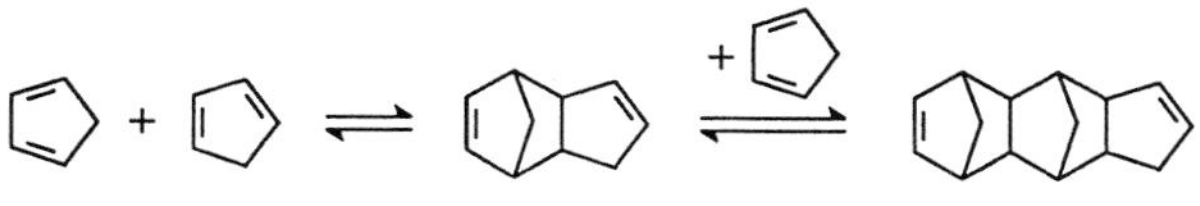   etc.

The first step of **cyclo-polycondensations** is also a polyaddition (see Eq.(13-84)).

Addition to Rings

The second large group of polyadditions comprises the *addition of functional groups to rings*. An important example is the crosslinking ("curing", "hardening") of diepoxides by diamines. The first step involves a ring-opening and the second step a reaction of the newly formed hydroxyl groups with additional epoxy groups.

(13-9) $\triangledown$–Z–$\triangledown$ + H$_2$N–Z'–NH$_2$ $\longrightarrow$ $\sim\!\!\sim$CH$_2$–CH–Z–CH–CH$_2$–NH–Z'–NH$\sim\!\!\sim$
              O   O   OH   OH

(13-10) $\sim\!\!\sim$CH$\sim\!\!\sim$ + $\triangledown$–Z$\sim\!\!\sim$ $\longrightarrow$ $\sim\!\!\sim$CH$\sim\!\!\sim$
              OH   O   O–CH$_2$–CH(OH)–Z$\sim\!\!\sim$

13.1.4   AB Polycondensations

Functional groups of AB molecules must be relatively inert in order to prevent a premature polycondensation on storage. They must have a long shelf life and they must not degrade at elevated reaction temperatures. AB polycondensations are usually only industrially interesting if suitable monomers are available from inexpensive sources.

For example, 11-aminoundecanoic acid is obtained from castor oil. It polycondenses to polyamide 11 (nylon 11, PA 11):

(13-11)   H$_2$N–(CH$_2$)$_{10}$–COOH $\longrightarrow$ ~NH–(CH$_2$)$_{10}$–CO~ + H$_2$O

Agricultural refuse is the source of furfuryl alcohol that polymerizes to soluble furan resins on addition of acids. A major reaction step is

(13-12) $\longrightarrow$ + H$_2$O

Other AB polycondensations comprise polysulfones (see also Table 13-5)

(13-13)   –SO$_2$Cl $\longrightarrow$ $\sim\!\!\sim$–SO$_2$$\sim\!\!\sim$ + HCl

and the polycondensation of diisocyanates to polycarbodiimides:

(13-14)   O=C=N–Z–N=C=O $\longrightarrow$ $\sim\!\!\sim$N–Z–N=C$\sim\!\!\sim$ + CO$_2$

13.1.5   AA/BB Polycondensations

In AA/BB polycondensations of A–V–A and B–W–B monomers, common A groups are COOH, COOR, COCl, SOCl$_2$, and Cl and common B groups are NH$_2$, OH, ONa, and OK (Table 13-1). AA/BB polycondensations can be schematically depicted as

(13-15) n A–V–A + n B–W–B $\rightarrow$ A–(V–W)$_n$–B + (2 n – 1) AB

Table 13-1 Industrially important linear AA/BB polycondensations of the AA/BB type (see generalized Eq.(13-15)). Ar = Bifunctional aromatic group, pPh = para-substituted phenylene group, Z = bifunctional group. * Part of ring system, cyclo-O–CO–CH=CH–CO–, hence 1/2 in Eq.(13-15).

| Polymer | Constitution according to Eq.(13-15) with groups | | | |
	A	V	W	B
Poly(ethylene terephthalate)	H	O(CH$_2$)$_2$O	CO-pPh-CO	OH [1]
	H	O(CH$_2$)$_2$O	CO-pPh-O	OCH$_3$ [1]
Unsaturated polyester	H	O(CH$_2$)$_2$O	CO-CH=CH-CO [2]	1/2 O *
Polycarbonate A	H	O-pPh-C(CH$_3$)$_2$-pPh-O	CO	OC$_6$H$_5$
	Na	O-pPh-C(CH$_3$)$_2$-pPh-O	CO	Cl
Polyarylates	Na	O-Ar-O	CO-Ar-CO	Cl
Poly(p-phenylene sulfide)	Na	S	pPh	Cl
Polysulfides	Na	S	Z	Cl
Polysulfones	K	O-pPh-O [3]	O(pPh-SO$_2$)$_2$ [3]	Cl
Polyamides	H	NH(CH$_2$)$_i$NH	CO(CH$_2$)$_j$CO	OH [4]
Polyetheretherketone	K	O-pPh-CO-pPh-O- [3]	pPh-CO-pPh [3]	F

[1] Two-stage process (see text). [2] From maleic anhydride (on polycondensation, isomerizes mainly to fumaric acid units). [3] Many other similar groups. [4] Industrial polymers include PA 46 ($i = 4$; $j = 4$), 66 ($i = 6$; $j = 4$), 69 ($i = 6$; $j = 7$), 610 ($i = 4$; $j = 8$), 612 ($i = 6$; $j = 10$), and also a copolymer PA 6.66 from ε-caprolactam + hexamethylene diamine + adipic acid.

Many industrial AA/BB polycondensations are conducted in two stages. In the first stage, a monomer is reacted with a great excess of the second monomer, leading to a trimer of the type ABA. An example is the synthesis of poly(ethylene terephthalate):

(13-16)　i HOCH$_2$CH$_2$OH + j CH$_3$OOC(p-C$_6$H$_4$)COOCH$_3$ →

j HOCH$_2$CH$_2$O–OC(p-C$_6$H$_4$)CO–OCH$_2$CH$_2$OH + (i – 2j) CH$_3$OH

This trimer, bis(hydroxyethyl) terephthalate, is the true monomer that self-condenses in an ABA (i.e., AA!) polycondensation with release of ethylene glycol:

(13-17)　j HOCH$_2$CH$_2$O–OC(p-C$_6$H$_4$)CO–OCH$_2$CH$_2$OH →

H[OCH$_2$CH$_2$O–OC(p–C$_6$H$_4$)CO]$_j$–OCH$_2$CH$_2$OH + (j – 1) HOCH$_2$CH$_2$OH

Reactions are more complicated if monomers can not only polycondense but also isomerize or chain polymerize. N-substituted aminobenzoyl lactams with i = 7-15 ring atoms polymerize by ring-opening to $+$NH(p-C$_6$H$_4$)CO–NH(CH$_2$)$_i$CO$+_n$. They may also expel the lactam ring and polymerize the remaining aminobenzoyl entity to poly(p-benzamide), $+$NH(p-C$_6$H$_4$)CO$+_n$. For this series of monomers, the tendency to open the lactam ring increases with increasing ring size. 24 % of the rings are opened at i = 7 but 100 % at i = 11. The higher the percentage of ring-opening, i.e., by chain polymerization, the greater the reduced viscosity and thus the molecular weight.

Carbon-substituted lactams may isomerize during polymerization, resulting in a phantom polymer. This reaction may be (a) a ring-opening polymerization followed by a cyclizing dehydration, (b) a polycondensation followed by a trans-cyclization of the polycondensate, or (c) an isomerization followed by a polycondensation. An example is shown in Eq.(13-18):

(13-18)

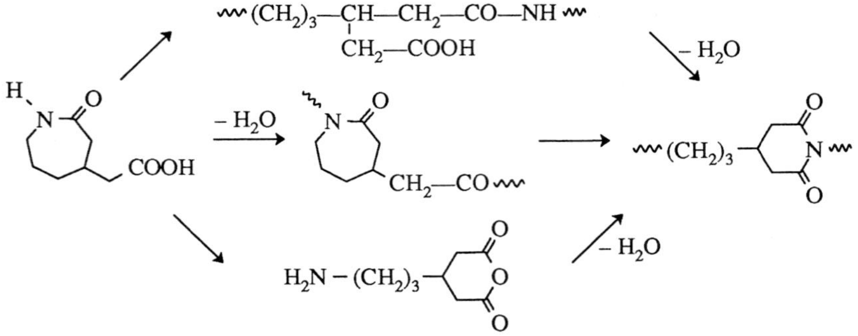

13.2 Stoichiometric Linear Step Reactions

13.2.1 Introduction

Polycondensations were performed as early as in the 19th century but the results were difficult to interpret. Functional group conversions of 90 % were seldom achieved and the resulting molecular weights were much lower than expected. Other polycondensations led to insoluble products at low monomer conversions. Results were difficult to reproduce: small changes in monomer compositions often gave widely varying results whereas in low-molecular weight organic reactions yields differed only slightly.

The situation changed after polycondensations were better understood. Important steps were the recognition of the role of functionality of monomer molecules and the stoichiometry of reactions, both by W. H. Carothers, and the statistical character of polycondensations and the principle of equal chemical reactivity, both by P. J. Flory. These insights allowed one to predict the variation of degrees of polymerization with time and monomer conversion and thus the causes for previously unsuccessful experiments.

13.2.2 Equilibria

Many polycondensations are equilibrium reactions between endgroups and their reaction products. In polycondensations, but not in polyadditions, equilibria are also controlled by leaving molecules L. AB reactions are always stoichiometric because the monomer molecules contain both the leaving groups A and B in equal amounts:

(13-19) n A–U–B $\rightleftarrows$ A–U$_n$–B   (polyaddition)

(13-20) n A–U–B $\rightleftarrows$ A–U$_n$–B + $(n-1)$ L   (polycondensation)

Note that A and B are not functional groups in the meaning of organic chemistry but leaving groups (in ω-amino acids, for example, A = H and not H_2N, B = OH and not COOH, and U = $NH(CH_2)_iCO$ and not $(CH_2)_i$). Monomeric units U retain their identity in polycondensations but change from U to U' in polyadditions (see, for example, Eq.(13-6)).

AA/BB reactions of n_A A–U–A + n_B B–V–B molecules may be either stoichiometric ($n_A = n_B$) or non-stoichiometric ($n_A \neq n_B$). Schematically for *stoichiometric* reactions:

(13-21) $\quad n$ A–U–A + n B–V–B $\rightleftarrows$ A–(U–V)$_n$B $\qquad\qquad$ (polyaddition)

(13-22) $\quad n$ A–U–A + n B–V–B $\rightleftarrows$ A–(U–V)$_n$B + $(2n-1)$ L $\quad$ (polycondensation)

The equilibrium constant $K = [AB][L]/([A][B])$ for the formation of amide groups AB is given by the molar concentrations of the various groups and molecules, respectively: $[A] \equiv [-COOH]$, $[B] \equiv [-NH_2]$, $[AB] \equiv [-CONH-]$, and $[L] \equiv [H_2O]$. Instead of using molar concentrations $[Q_i]$ (where $Q_i = A$, B, AB, or L), one can also work with mole fractions $x_{Q,i} = [Q_i]/\Sigma [Q_i] = [Q_i]/[Q]_{tot}$ where $[Q]_{tot}$ = total molar concentration of all groups: $[Q]_{tot} = [A] + [B] + [AB] + [L] \equiv [R]$.

The expressions for equilibria of stoichiometric reactions ($x_A = x_B$) are given in Eqs.(13-24) and (13-25) and that of non-stoichiometric ones ($x_A \neq x_B$) in Eq.(13-23). The sign before the square root of Eq.(13-25) must be negative since otherwise x_{AB} may be greater than unity (for the derivation of Eq.(13-25), see Appendix A-13).

Polycondensations	*Polyadditions*	*Stoichiometry*
(13-23) $\quad K = \dfrac{[AB][L]}{[A][B]} = \dfrac{x_{AB}x_L}{x_A x_B}$	$K = \dfrac{[AB]}{[A][B]} = \dfrac{x_{AB}}{x_A x_B [R]}$	no
(13-24) $\quad K = \dfrac{[AB][L]}{[A][B]} = \dfrac{x_{AB}x_L}{(1-x_{AB})^2}$	$K = \dfrac{[AB]}{[A][B]} = \dfrac{x_{AB}}{(1-x_{AB})^2 [R]}$	yes
(13-25) $\quad x_{AB} = 1 + \dfrac{x_L}{2K} - \dfrac{x_L}{2K}\left(1 + \dfrac{4K}{x_L}\right)^{1/2}$	$x_{AB} = 1 + \dfrac{1}{2K[R]} - \dfrac{(1+4K[R])^{1/2}}{2K[R]}$	yes

In low-molecular weight chemistry, the mole fraction of the desired entity, e.g., –AB– groups, can be raised by increasing the molar concentrations of either –A groups *or* –B groups. This is not possible in AB step polymerizationss since –A and –B are in the same molecule; the polymerization is always equimolar. In AA/BB step polymerizations, a condition $[A]_0 > [B]_0$ would deprive A groups of B partners after complete reaction of all B groups. This will not only reduce the polymer yield but also the degree of polymerization, X. For example, if 2 moles of $H_2N–Z–NH_2$ + 1 mole of $HOOC–Z'–COOH$ are *completely* reacted, then *formally* a trimer $H_2N–Z–NH–CO–Z'–CO–NH–Z–NH_2$ results with $X = 3$. But X is in reality the number-average degree of polymerization since the "trimer" is a mixture of molecules, $H(NH–Z–NH–CO–Z'–CO)_{n-1}NH–Z–NH_2$, with degrees of polymerization of $X = n = 1, 3, 5, 7$ .... (Section 13.2.5).

High degrees of polymerization can only be obtained by stoichiometric reactions (Section 13.2.6). The equilibrium is then still controlled by the equilibrium constant K and the mole fraction x_L of leaving molecules L in equilibrium (Fig. 13-1). In order to obtain a mole fraction $x_{AB} = 0.99$ of A–B bonds in polymer molecules and thus a number–average degree of polymerization of $\overline{X}_n = 1/(1 - x_{AB}) = 100$, one has to reduce the mole fraction x_L of leaving molecules to $x_L \approx 1.01 \cdot 10^{-2}$ if $K = 100$, to $x_L \approx 1.01 \cdot 10^{-4}$ if $K = 1$, and to $x_L \approx 1.01 \cdot 10^{-6}$ if $K = 0.01$. Table 13-2 shows some equilibrium constants K.

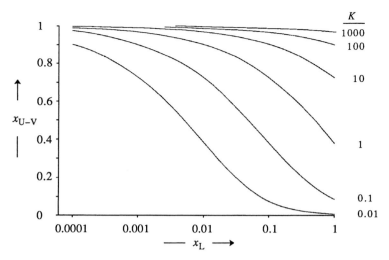

Fig. 13-1 Effect of equilibrium constants K on the mole fraction $x_{A-B} = x_{U-V}$ of U–V bonds (see Eq. (13-22)) as a function of the mole fraction x_L of leaving molecules L. Calculations with Eq.(13-25), left, assuming ideal activity coefficients, $f_{act} = 1$.

Polycondensations and polyadditions are usually exothermic: their equilibrium constants decrease with increasing temperature (examples 8 and 9 in Table 13-2). According to Eqs.(13-23) and (13-24), this will also decrease the concentration of AB groups. In a *closed system*, a temperature increase will thus decrease the degree of polymerization. In *open systems*, higher temperatures remove more volatile leaving molecules L and thus increase the concentration of -AB- groups.

Equilibrium constants of transesterifications are fairly small (examples 1 and 2 in Table 13-2). They are much higher for direct esterifications (examples 3 and 4) and especially for direct amidations (examples 5-7). Very high equilibrium constants were obtained for Schotten-Baumann reactions between acid chlorides and phenols (examples 8 and 9), alcohols, or amines (for interfacial reactions, see Section 13.5.3).

Table 13-2 Equilibrium constants K of polycondensations (examples 1-5), condensations (examples 6-9), and a cyclo-polycondensation (example 10). PMA = pyromellitic dianhydride. * In L/mol.

Monomers	Leaving molecules	$T/°C$	K
1 $HOCH_2CH_2O-OC(p-C_6H_4)CO-OCH_2CH_2OH$	$HOCH_2CH_2OH$	280	0.39
2 $CH_3O-OC(p-C_6H_4)CO-OCH_3 + HOCH_2CH_2OH$	CH_3OH	280	0.42
3 $HOOC(CH_2)_4COOH + HO(CH_2)_5OH$	H_2O	280	6.0
4 $HOOC(p-C_6H_4)COOH + HO(CH_2)_2OH$	H_2O	186	9.6
5 $HOOC(CH_2)_{10}NH_2$	H_2O	280	300
6 $HOOC(o-C_6H_4)COOH + H_2N(CH_2)_8H$	H_2O	200	1360
7 $H(CH_2)_8OC(o-C_6H_4)COOH + H_2N(CH_2)_8H$	H_2O	200	1830
8 $C_6H_5COCl + HOC_6H_5$	HCl	220	220
9 $C_6H_5COCl + HOC_6H_5$	HCl	40	4300
10 PMA + $H_2N(p-C_6H_4)-O-(p-C_6H_4)NH_2$ (to amic acid)	-	250	100000 *
(to polyimide)	H_2O	200	4600

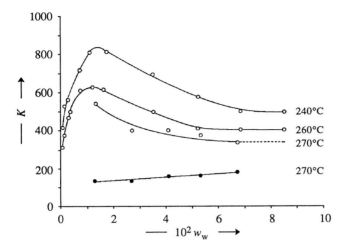

Fig. 13-2 Equilibrium constants K of the polycondensation of ε-aminocaproic acid as a function of the mass fraction w_w of water in equilibrium [1]. O Experimental values, ● corrected for activity.

Very high equilibrium constants are also observed for the two steps of polyimidation (example 10; see Section 13.6.2). Because of their high equilibrium constants, $K > 10^3$, these reactions *appear* as irreversible.

Equilibrium constants from mole concentrations or fractions are *apparent* values if thermodynamic activities do not equal mole concentrations (activity coefficients $f_{act} \neq 1$). Such equilibrium constants vary with concentrations. Values of $f_{act} \neq 1$ are produced by solvation, association, and especially by ionization.

For example, ε-aminocaproic acid is hydrated and ionized by water (from the condensation or deliberately added). The active species is $^{\ominus}[OOC(CH_2)_5NH_3]^{\oplus} \times i\ H_2O$ and not the neutral monomer molecule, $HOOC(CH_2)_5NH_2$. The nature of the amide bond (a partial double bond) is also affected by water. The equilibrium constants of amidation, as calculated by Eqs.(13-23)-(13-25), are therefore only apparent quantities. They pass through a maximum with increasing mass fraction of water before becoming constant (Fig. 13-2). Equilibrium constants calculated from activities instead of concentrations are practically independent of the water concentration.

Lesser activity effects are also observed for polyesterifications. They are probably responsible for the observed dependency of conventional equilibrium constants on the degree of polymerization or molecular weight, respectively, i.e., an apparent violation of the principle of equal chemical reactivity.

Non-ideal activity coefficients are also responsible for the variation of conventionally calculated (apparent) reaction enthalpies with water content (Table 13-3). These apparent enthalpies become constant at low equilibrium concentrations of water.

Since the number of molecules does not change for a polycondenstion in a closed system, one would expect a constant polycondensation entropy. A constancy of the number of molecules also means that the sum of translational and external rotational entropies is practically constant. However, a slight increase of vibrational and internal rotational entropies with temperature can be expected if relatively rigid monomer molecules are transformed into more flexible monomeric units.

Table 13-3 Mass fractions $w_{W,\infty}$ of water in equilibrium and apparent reaction enthalpies, $\Delta H_{p,app}$, as a function of the mass fractions $w_{W,0}$ of water and $w_{AH,0}$ of the nylon salt before the reaction and in the temperature range of the equilibrium of the polycondensation of an aqueous solution of the nylon salt $^{\oplus}[H_3N(CH_2)_6NH_3]^{\oplus\ominus}[OOC(CH_2)_4COO]^{\ominus}$ [2].

Temperature in °C	$10^2 w_{AH,0}$	$10^2 w_{W,0}$	$10^2 w_{W,\infty}$	$\Delta H_{p,app}/(kJ\ mol^{-1})$
210 ± 10	80	20	29.55 ± 0.050	+ 13.0
215 ± 15	90	10	21.1 ± 0.15	+ 4.0
230 ± 20	100	0	12.6 ± 0.20	− 12.5
260 ± 10	precondensate	?	4.8 ± 0.15	− 27.0

The entropic contribution to the Gibbs energy of polycondensation must therefore be relatively small. An equilibrium constant of $K = 260$ was found for the polycondensation of nylon salt at 260°C. With $\Delta H_p - T\Delta S_p = -RT \ln K$ and $\Delta H_p = -27\ 000$ J mol^{-1} (Table 13-3), $T\Delta S_p = -2360$ J mol^{-1} and $\Delta S_p = -4.4$ J K^{-1} mol^{-1} are calculated. The entropic contribution $T\Delta S_p$ is thus only ca. 1/10 of the enthalpic one.

13.2.3   Extent of Reaction, Monomer Conversion, Polymer Yield

For the quantitative description of polycondensations and polyadditions, it is advisable to relate reaction parameters such as the extent of reaction, monomer conversion, and polymer yield to **reactants** (monomers, oligomers, polymers; i.e., $X \geq 1$) and not to polymers ($X \geq 2$). The degree of polymerization of a polymer molecule furthermore always relates to the number N_u of monomeric units in that molecule and not to the number of repeating units (it refers to the process and not to the constitution!).

The degree of polymerization of polymer molecules from the stoichiometric polycondensation of N HOOC(p-C$_6$H$_4$)COOH + N HO(CH$_2$)$_2$OH is therefore $X = 2\ N$ but the degree of polymerization of the same polymer from N HOOC(p-C$_6$H$_4$)COO(CH$_2$)$_2$OH is $X = N$. However, it is $X = 2\ N$ if the latter polycondensation is accompanied by exchange reactions between molecule segments (transesterifications, etc.) since the following statistical considerations refer only to the probability of formation of new bonds.

The course of polycondensations and polyadditions can be described by various parameters. Theoretical considerations usually relate to the **extent of reaction**, p, of functional groups which is a fractional quantity. It is calculated from the amounts $n_{A,0}$ (in mol) of groups A at $t = 0$ and n_A at time $t > 0$ as

(13-26) $p_A \equiv (n_{A,0} - n_A)/n_{A,0}$

and similarly for groups B. p_A and p_B can be obtained directly from the relative contents of A and B endgroups in the reactant at time t.

The **monomer conversion**, u_M, refers similarly to the amount (in mol) of monomer molecules M at time t. It is also a fractional quantity:

(13-27) $u_M \equiv (n_{M,0} - n_M)/n_{M,0}$

Eqs.(13-26) and (13-27) require the knowledge of amounts (in mol). It is much easier to determine the masses m_P (in g) of polymers ($X \geq 2$) or the masses $m_{U,P}$ of monomeric units in polymer molecules. The course of reaction is then presented as relative yields of either polymers P or reactants R ($X \geq 1$) that are related either to the initial mass $m_{M,0}$ of the monomer or to the initial mass, $m_{U,0}$, of monomeric units.

The **relative yield of reactants**, $y_{R(M)} = m_R/m_{M,0}$, **with respect to the monomer** is not a useful quantity. Before the reaction, it is $y_{R(M)} = 1$ since $m_R = m_{R,0} = m_{M,0}$. At complete monomer reaction, it drops to $y_{R,M} = 1 - (m_{L,0}/m_{M,0}) = 1 - (M_L/M_M)$ because $m_R = m_{M,0} - m_{L,0}$ (M_L and M_M: molar masses of leaving molecules and monomer).

The **relative yield of reactants with respect to monomeric units** is always unity since no monomeric units disappear during the polymerization.

Correspondingly, three different **relative yields of polymer** can be defined:

$$(13\text{-}28) \qquad y_{P(M)} = \frac{m_P}{m_{M,0}} = \frac{m_R - m_M}{m_{U,0} + m_{L,0}}$$

$$(13\text{-}29) \qquad y_{P(U)} = \frac{m_P}{m_{U,0}} = \frac{m_R - m_M}{m_{M,0} - m_{L,0}} = \frac{m_{U,P} + m_{L(P)}}{m_{U,0}}$$

$$(13\text{-}30) \qquad y_P = \frac{m_{U,P}}{m_{U,0}} = \frac{m_{U,R} - m_{U,M}}{m_{U,0}} = 1 - \frac{m_{U,M}}{m_{U,0}} = 1 - \frac{n_{U,M}}{n_{U,0}} = 1 - \frac{n_M}{n_{M,0}} \equiv u_M$$

$m_{L,0}$ = mass of L groups that are initially present in the monomer, $m_{L(P)}$ = mass of L groups that are present in the polymer.

These definitions allow one to derive relationships between the extent of reaction and the various relative yields for AB and *stoichiometric* AA/BB polymerizations. Such reactions are characterized by $p_A = p_B = p$; they lead to Schulz-Flory distributions of degrees of polymerization.

The relative yield y_P is calculated as follows. The initial number $N_{M,0}$ of monomer molecules equals the initial number $N_{A,0}$ of A groups or the initial number $N_{B,0}$ of B groups: $N_{M,0} = N_{A,0} = N_{B,0}$ and also $n_{M,0} = n_{A,0} = n_{B,0}$. The mole fraction $x_M = n_M/n_R$ of monomer molecules remaining after some extent of reaction is calculated from the amount n_M of monomer molecules and the amount n_R of all reactant molecules.

According to Eq.(13-37) (see below), one has $n_R = n_{M,0}(1 - p)$. x_M is furthermore given by $x_M = 1 - p$ (see Eq.(13-39) for $i = 1 = M$). The relationship $x_M = n_M/n_R$ is thus transformed into $n_M/n_{M,0} = (1 - p)^2$. Introduction of this equation into Eq.(13-30) and combination with Eq.(13-27) results in

$$(13\text{-}31) \qquad y_P = u_M = 1 - (1 - p)^2 \quad \text{or} \quad p = 1 - (1 - u_M)^{1/2} = 1 - (1 - y_P)^{1/2}$$

The extent of reaction, p, is thus available from either the monomer conversion, u_M, or the relative polymer yield, y_P. The latter quantity is given by the mass of monomeric units and *not* by the masses of reactants, polymers, and/or monomers!

For polyadditions, y_P (Eq.(13-30)) can be replaced by $y_{P(M)}$ (Eq.(13-28)) since $m_{U,P} = m_R - m_M$ and, because $m_{L,0} = 0$, also $m_{M,0} = m_{U,0}$. For polycondensations, one needs to know three molecular weights M_M, M_U, and M_L and also the number-average degree of polymerization of the polymer, $\overline{X}_{P,n}$.

A lengthy derivation (not shown) shows that

$$(13\text{-}32)\qquad y_P = y_{P(M)}\left(\frac{M_M}{M_U+[M_L/\overline{X}_{P,n}]}\right) = y_{P(U)}\left(\frac{M_U\overline{X}_{P,n}}{M_U\overline{X}_{P,n}+M_L}\right)$$

The identity $y_{P(M)}M_M = y_{P(U)}M_U$ in Eq.(13-32) can also be obtained directly from Eqs.(13-28) and (13-29). One can also calculate y_P from $y_{P(U)}$ or $y_{P(M)}$ if the mole fraction $x_M = n_M/n_R$ of monomer molecules in the reactant is known. $\overline{X}_{P,n}$ can then be expressed by $\overline{X}_{R,n}$ (see Eq.(13-48)) and this, in turn, by p, according to Eq.(13-42).

According to Eq.(13-32), relative yields y_P increase non-linearly with p but become $y_P = p$ at $p = 1$ (Fig. 13-3).

An expression for the often used monomer-related relative polymer yields, $y_{P(M)}$, is obtained after a lengthy derivation:

$$(13\text{-}33)\qquad y_{P(M)} = p[2-p-(M_L/M_M)] = p[1-p+(M_U/M_M)]$$

At small extents of reaction, one observes $y_{P(M)} > p$ but at large ones, $y_{P(M)} < y$ (Fig. 13-3). $y_{P(M)}$ thus passes through a maximum which has sometimes been misinterpreted as the onset of polymer degradation. Note that $y_{P(M)}$ cannot exceed a value of $y_{P(M)} = M_U/M_M$ at $p = 1$.

The relative yield $y_{P(U)}$ also passes through a maximum. It may even have values of $y_{P(U)} > 1$ for polycondensations with $M_U > M_L$ because the total mass of polymer molecules ($X \geq 2!$) increases by reactions M + M and P + M but decreases by reactions P + P. The first two reactions dominate at small p, the last reaction at large p. For polyadditions, we have $M_L = 0$ and $M_M/M_U = 1$ and Eqs.(13-32) and (13-33) become

$$(13\text{-}34)\qquad y_P = y_{P(M)} = y_{P(U)} = p(2-p)$$

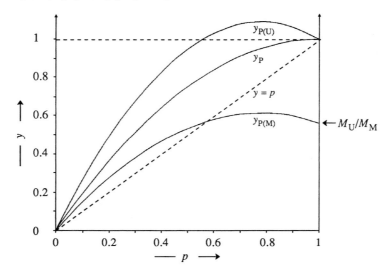

Fig. 13-3 Dependence of the relative yields y_P, $y_{P(U)}$, and $y_{P(M)}$ on the extent of reaction, p, of functional groups in the polycondensation of the AB monomer phenyl p-hydroxybenzoate. These monomer specific relationships reduce to $y = y_P = y_{P(U)} = y_{P(M)}$ for polyadditions.

13.2.4 Degree of Polymerization

The number-average degree of polymerization, $\overline{X}_n$, varies characteristically with the extent of reaction, p_A, of A groups in AB and stoichiometric AA/BB polymerizations. This quantity is defined as the ratio of the amount of monomeric units in all reactants, $n_{U(R)}$, to the amount of all reactant molecules, n_R. Since all monomeric units are retained, $n_{U(R)}$ must equal the amount of originally present monomer molecules, $n_{M,0}$. The number-average degree of polymerization can also be expressed as the ratio of the amount of originally present A groups to the amount of A groups after polymerization, $n_{A,0}/n_A$. Using the definition of the extent of reaction, Eq.(13-26), one arrives at the **Carothers equation** (Carothers function):

$$(13\text{-}35) \qquad \overline{X}_{R,n} \equiv \frac{n_{U(R)}}{n_R} = \frac{n_{M,0}}{n_R} = \frac{n_{A,0}}{n_A} = \frac{1}{1 - p_A}$$

The Carothers equation applies to all reactants ($X_i \geq 1$), including monomer, provided that no intramolecular rings are formed (see Section 13.6.1). It is valid for AB and stoichiometric AA/BB polymerizations (for non-stoichiometric ones, see Section 13.3).

Number-average degrees of polymerization increase only slowly with increasing p_A: at $p_A = 0.9$ (90 %), the value of $\overline{X}_{R,n}$ is only 10, monomer included (see also Fig. 6-1). $\overline{X}_{R,n}$ increases dramatically on approach to $p_A \rightarrow 1$. Since industrial polymers by polycondensation and polyaddition often require $\overline{X}_{R,n} = 200$, extents of reaction of $p_A = 99.5$ % are necessary which is difficult to achieve for most organic reactions.

13.2.5 Distribution of Degrees of Polymerization

The *number* distribution of degrees of polymerization, $X_i = i$, of non-ring-forming polymers from AB and stoichiometric AA/BB step reactions can be calculated by simple statistical means. In such reactions, the probability of connecting one group A with one group B, i.e., 2 molecules, is given by the extent of reaction, $p_A \equiv p$. The probability of the appearance of 2 connections, i.e., 3 molecules, is p^2; of 3 connections, p^3, etc. The number of A–B bonds in a molecule with $X_i \equiv i$ monomeric units is therefore p^{i-1}.

The probability P_i of the appearance of *one* polymer molecule with a degree of polymerization, i, is obtained from the probability p^{i-1} of $i-1$ connections and the probability $1-p$ of non-reacted endgroups:

$$(13\text{-}36) \qquad P_i = p^{i-1}(1 - p)$$

The amount $n_{i,R}$ of *all* molecules of a reactant that have a degree of polymerization of X_i is proportional to both the probability P_i and the amount n_R of all reactant molecules with different degrees of polymerization ($i \equiv X_i \geq 1$):

$$(13\text{-}37) \qquad n_{i,R} = n_R P_i = n_R p^{i-1}(1 - p)$$

The mole fraction $x_{i,R}$ of molecules with a degree of polymerization, $X_i \equiv i$, is thus

$$(13\text{-}38) \qquad x_{i,R} = n_{i,R}/n_R = p^{i-1}(1 - p)$$

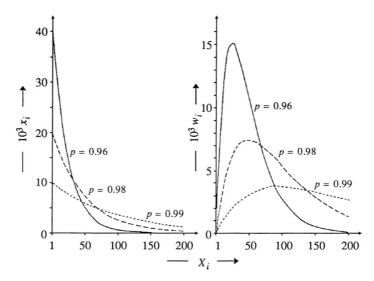

Fig. 13-4 Schulz-Flory distributions of mole fractions (left) and mass fractions (right) of molecules with degrees of polymerization, i, at various extents of reaction, p. Number-distributions are universal for AB and stoichiometric AA/BB polymerizations. Mass distributions are only universal for poly-additions but not for polycondensations. The mass distributions in this graph were calculated for the AB polycondensation of HO-C_6H_4-COOC_6H_5 with $M_U = 120.10$ g/mol and $M_L = 95.11$ g/mol.

At constant p, mole fractions x_i decrease with increasing degree of polymerization, X_i. The distributions become more flat at higher p (Fig. 13-4). Such distributions of the degrees of polymerizations are called **Schulz-Flory distributions** (p. 73).

Figs. 13-3 and 13-4 show that monomers do not disappear *fast* during the first stages of the polymerization. At an extent of reaction of 50 % ($p = 0.5$), the mole fraction of the monomer M ($X_i \equiv i = 1$) is $x_i = x_M = 0.5$ (50 %) and at $p = 0.8$ still 20 %.

The calculation of number distributions of polymers from AB and stoichiometric AA/BB reactions is independent of both the chemical constitution of reactants (monomer, oligomer, polymer) and leaving molecules. For *mass* distributions, one has to consider the molar mass of leaving molecules, M_L. The derivation is given below for AB polycondensations but applies also to stoichiometric AA/BB polycondensations.

The mass fraction (weight fraction) $w_{i,R}$ of reactant molecules with a degree of polymerization i equals the ratio of the mass of i-mers to the mass of all reactant molecules. The masses can be expressed by the products of the amounts, $n_{i,R}$, and the corresponding molar masses, i.e., $M_i = M_L + M_U X_i$ for an i-mer and $\overline{M}_{R,n} = M_L + M_U \overline{X}_{R,n}$ for all reactant molecules where X is the degree of polymerization:

$$(13\text{-}39) \qquad w_{i,R} = \frac{m_{i,R}}{m_R} = \frac{n_{i,R} M_i}{\sum_i n_{i,R} M_i} = \frac{n_{i,R} M_i}{n_R \overline{M}_{R,n}} = \frac{n_{i,R}(M_L + M_U X_i)}{n_R[M_L + M_U \overline{X}_{R,n}]}$$

Introduction of Eqs.(13-35) and (13-38) into Eq.(13-39) leads to

$$(13\text{-}40) \qquad w_{i,R} = \left(\frac{M_L + M_U X_i}{M_U + M_L(1-p)}\right) p^{i-1}(1-p)^2$$

Eq.(13-40) shows for polycondensations that the mass fraction of i-mers in reactants depends on the molecular weights of the monomeric units *and* leaving molecules. For polyadditions ($M_L = 0$), it reduces to

$$(13\text{-}41) \qquad w_{i,R} = X_i(p^{i-1})(1-p)^2 \qquad ; \qquad X_i \equiv i$$

Literature wrongly assigns Eq.(13-41) also to polycondensations, which may lead to great errors. For the polycondensation of phenyl *p*-hydroxybenzoate, for example, the mass fraction w_1 of the monomer ($i = 1$) at an extent of reaction of $p = 0.5$ is $w_{1,R} \approx 0.321$ (Eq.(13-40)) versus 0.25 (Eq.(13-41)); at $p = 0.99$, it is $1.778 \cdot 10^{-4}$ vs. $1.00 \cdot 10^{-4}$.

Mole fractions of polymers are continuously decreasing with increasing degrees of polymerization whereas mass fractions pass through a maximum at approximately the number-average degree of polymerization of the reactant, $\overline{X}_{R,n}$ (Fig. 13-4).

13.2.6 Averages of Degrees of Polymerization

Linear polycondensations and polyadditions are usually monitored by the time dependence of degrees of polymerization or yields of polymer and less often by monomer conversions or extents of reaction of functional groups. One commonly measures *weight*-average degrees of polymerization of *polymers* instead of the much more meaningful *number*-average degrees of polymerization of *reactants*.

The number-average degree of polymerization of reactant molecules is obtained from Table 3-8 and Eq.(3-36) with $X_i = i$ and $\Sigma_i X_i p^{i-1} = 1/(1-p)^2$ as

$$(13\text{-}42) \qquad \overline{X}_{R,n} = \Sigma_i x_i X_i = \Sigma_i (1-p)p^{i-1}X_i = 1/(1-p)$$

The weight-average degree of polymerization of reactant molecules is conventionally defined via the mass of *molecules* as a statistical weight, using the amount of n_i (in mol) of i-mers in the reactant:

$$(13\text{-}43) \qquad \overline{X}_{R,w,conv} = \frac{\Sigma_i (m_{mol})_i X_i}{\Sigma_i (m_{mol})_i} = \frac{\Sigma_i n_i M_i X_i}{\Sigma_i n_i M_i} = \frac{\Sigma_i n_i (M_L + M_U X_i)X_i}{\Sigma_i n_i (M_L + M_U X_i)}$$

For polyadditions, this leads to $\overline{X}_{R,w,conv} = (\Sigma_i n_i X_i^2)/(\Sigma_i n_i X_i) \equiv \overline{X}_{R,w}$ because $M_L = 0$. For polycondensations, the mass-average degree of polymerization of reactants must be defined via the mass $(m_{U,mol})_i$ of all *monomeric units* in a reactant molecule and *not* via the mass of *molecules*; this corresponds to Eq.(13-43) with $M_L/M_U \to 0$:

$$(13\text{-}44) \qquad \overline{X}_{R,w} = \frac{\Sigma_i (m_{U,mol})_i X_i}{\Sigma_i (m_{U,mol})_i} = \frac{\Sigma_i (n_i M_U X_i)X_i}{\Sigma_i n_i M_U X_i} = \frac{\Sigma_i n_i X_i^2}{\Sigma_i n_i X_i} = \Sigma_i w_{i,R} X_i$$

Number-average degrees of polymerization are defined by $\overline{X}_n \equiv \Sigma_i x_i X_i = \overline{X}_{R,n}$ (Eq.(3-50)). Introduction of Eq.(13-38) leads to $\overline{X}_{R,n} = (1-p) \Sigma_i p^{i-1}X_i$. Because $p \leq 1$, the sum can be developed into a series: $\Sigma_i p^{i-1}X_i = 1 + 2p + 3p^2 + \ldots = 1/(1-p)^2$. The number-average degree of polymerization is thus given by the Carothers equation, $\overline{X}_{R,n} = 1/((1-p))$, which applies to AB and AA/BB reactions, provided the latter are stoichiometric. It is independent of leaving molecules (if any).

Mass-average degrees of polymerization are defined by $\overline{X}_w \equiv \Sigma_i w_i X_i = \overline{X}_{R,w}$ (Eq.(3-51)). Introduction of Eq.(13-38) leads to $\overline{X}_{R,w} = (1 - p)^2 \Sigma_i p^{i-1} X_i^2$. Because $p \leq 1$, the sum can be developed into a series, $\Sigma_i p^{i-1} X_i^2 = 1 + 2^2 p + 3^2 p^2 + ... = (1 + p)/(1 - p)^3$, resulting in a mass-average degree of polymerization of reactant molecules of $\overline{X}_{R,w} = (1 + p)/(1 - p) = 2 \overline{X}_{R,n} - 1$ and thus to a polymolecularity index of $(\overline{X}_R)_w / \overline{X}_{R,n} = 2 - (1/\overline{X}_{R,n})$. The latter equation differs from the one given on p. 363, last line, by the term $-(1/\overline{X}_{R,n})$ since the former relates to reactant molecules ($X_i \geq 1$) and the latter to polymer molecules ($X_i \geq 2$) from chain polymerizations with termination by disproportionation. All these equations are universal for step reactions if $M_L/M_U \rightarrow 0$.

The z-average degree of polymerization of *reactant* molecules is obtained similarly as

$$(13\text{-}45) \qquad \overline{X}_{R,z} = \frac{3\overline{X}_{R,w} - 1}{2\overline{X}_{R,w}} = \frac{3\overline{X}_{R,n} - 2}{2\overline{X}_{R,n} - 1}$$

The expressions for $\overline{X}_{R,n,conv}$, $\overline{X}_{R,w}$, and $\overline{X}_{R,z}$ apply to batch reactors (BR) and continuous plug flow reactors (CPFR) (Fig. 10-25) because of their very narrow distributions of residence times (see also Chapter 4 in Volume II). Homogeneous continuous stirred tank reactors (HCSTR) generate broad distributions of residence times and thus broader distributions of degrees of polymerization of *reactants*:

$$(13\text{-}46) \quad \overline{X}_{R,n} = \frac{1}{1-p}; \quad \overline{X}_{R,w} = \frac{1+p}{1-p} = 2\overline{X}_{R,n} - 1; \qquad \frac{\overline{X}_{R,w}}{\overline{X}_{R,n}} = 1+p; \quad \text{BR, CPFR}$$

$$(13\text{-}47) \quad \overline{X}_{R,n} = \frac{1}{1-p}; \quad \overline{X}_{R,w} = \frac{1+p^2}{(1-p)^2} = 2\overline{X}_{R,n}^2 - (2\overline{X}_{R,n} - 1); \quad \frac{\overline{X}_{R,w}}{\overline{X}_{R,n}} = \frac{1+p^2}{1-p}; \quad \text{HCSTR}$$

Average degrees of polymerization of *polymers* can be calculated for equilibria (or BR, CPFR) from average degrees of polymerization of reactants as follows. The number-average degree of polymerization of *reactants*, $\overline{X}_{R,n}$, is obtained from the mole fraction x_M of the monomer with a degree of polymerization of $X_M = 1$ and the mole fraction x_P of the polymer with a number-average degree of polymerization of $\overline{X}_{P,n}$:

$$(13\text{-}48) \qquad \overline{X}_{R,n} = x_M X_M + x_P \overline{X}_{P,n} = x_M + (1 - x_M)\overline{X}_{P,n}$$

Interconversion of these degrees of polymerization thus requires the knowledge of the mole fraction x_M of the monomer in the reactant. For Schulz-Flory distributions, one obtains $x_M = 1 - p$ (Eq.(13-38)) and Eq.(13-48) simplifies for *polymers* to

$$(13\text{-}49) \qquad \overline{X}_{P,n} = \overline{X}_{R,n} + 1 \qquad \text{(also for statistical weights w and z instead of n)}$$

13.2.7   Solid-Stating

Laboratory AB and AA/BB polymerizations are commonly performed in the melt or in solution. Industrial melt polymerizations are often followed by so-called **solid-stating**, i.e., polymerizations that occur after the molecular weight of the polymer becomes so high that the melt solidifies. Solid-stating is thus a polymerization in the solid state at temperatures between the glass temperature and the melting temperature of the polymer.

Solid-stating increases the molecular weight of the polymer. It is commonly used for polyamides and so-called bottle-grade poly(ethylene terephthalate) which crystallizes thermally. Polycarbonates are not subjected to solid stating since their solution polycondensations lead to solvent-induced crystallizations and their melt polycondensations require nucleating agents.

Solid-state reactions can be quite rapid. The solid-state polymerization of the cyclic dimer of bisphenol A carbonate (melting temperature: 330°C) produces at 300°C molecular weights of ca. 2 000 000 within 5 minutes. Continued solid-stating lets the molecular weight drop to less than 100 000.

13.3   Non-Stoichiometric Linear Step Reactions

13.3.1   Degree of Polymerization

In non-stoichiometric AA/BB polymerizations, the amount $n_{U,R}$ of monomeric units in reactants is given by $n_{U,R} = n_{M,0} = (n_{A,0} + n_{B,0})/2$ since every monomer molecule carries either 2 A groups or 2 B groups. The amount of reactant molecules equals the initial amount $n_{M,0}$ of all monomer molecules minus the amount $n_{M,el}$ of molecules that have been eliminated during the coupling reaction. It is further defined that B groups are present in excess. Since each coupling removes an A group, their amount will equal the difference $n_{A,0} - n_A$ of initial and still present A groups. Setting $r_0 \equiv n_{A,0}/n_{B,0} \le 1$, one obtains for the number-average degree of polymerization of AA-BB polymers from non-stoichiometric polycondensations

$$(13\text{-}50) \quad \overline{X}_{R,n} = \frac{n_{U,R}}{n_R} = \frac{n_{M,0}}{n_{M,0} - n_{M,el}} = \frac{(n_{A,0} + n_{B,0})/2}{[(n_{A,0} + n_{B,0})/2] - (n_{A,0} - n_A)} = \frac{1 + r_0}{1 + r_0 - 2r_0 p_A}$$

For stoichiometric polycondensations ($r_0 = 1$), Eq.(13-50) reduces to Eq.(13-35). For *non-stoichiometric reactions* with *complete reaction* of the deficit A groups ($p_A = 1$), Eq.(13-50) becomes

$$(13\text{-}51) \quad \overline{X}_{R,n} = \frac{1 + r_0}{1 - r_0}$$

Absence of stoichiometry lowers the degree of polymerization considerably. For $r_0 = 0.99$, one obtains $\overline{X}_{R,n} = 199$ for $p_A = 1$ (instead of $\overline{X}_{R,n} = \infty$ for $r_0 = 1$) and $\overline{X}_{R,n} \approx 66.8$ for $p_A = 0.99$ (instead of $\overline{X}_{R,n} = 100$ for $r_0 = 1$).

Deviations from stoichiometry appear if amounts of monomers are initially stoichiometric but one monomer is volatile and a part of it leaves with the leaving molecules. In the industrially important polycondensation of dimethyl terephthalate (DMT) and ethylene glycol (E), $CH_3OOC(1,4\text{-}C_6H_4)COOCH_3 + HOCH_2CH_2OH$, a part of the glycol is removed with the leaving compound methanol. In order to obtain high molecular weights, the initial molar ratio $(n_{EG}/n_{DMT})_0$ must be considerably higher than unity, for example, ca. 1.4 according to Fig. 13-5. Still higher values of $(n_{EG}/n_{DMT})_0$ do not increase the degree of polymerization, since the polycondensation first produces the true monomer di(hydroxyethyl) terephthalate, $HOCH_2CH_2OOC(1,4\text{-}C_6H_4)COOCH_2CH_2OH$,

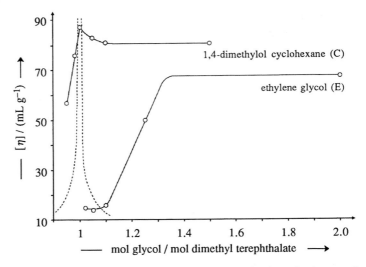

Fig. 13-5 Intrinsic viscosities $[\eta]$ as a measure of the degrees of polymerization as a function of the initial molar ratio of glycol and dimethyl terephthalate for the polycondensation of the latter with either ethylene glycol or 1,4-dimethylol cyclohexane [3]. - - - Theoretical prediction for the polycondensation of two non-volatile monomers (schematic).

which then polycondenses with release of ethylene glycol. If non-volatile monomers are used, the highest molecular weights are obtained for stoichiometric amounts as demanded by theory. An example is the polycondensation of dimethyl terephthalate with the non-volatile cyclohexane 1,4-dimethylol, $HOCH_2(1,4\text{-}C_6H_{10})CH_2OH$ (Fig. 13-5).

13.3.2   Polymer Yield

The following considerations apply to non-stoichiometric linear step reactions with $M_L = 0$ (polyaddition) or $M_L/M_U \rightarrow 0$ (certain polycondensations) with an initial deficit of A groups ($r_0 = n_{A,0}/n_{B,0} < 1$). According to a lengthy derivation, the mass fraction $w_{i,R}$ of i-mers in reactants is given as

$$(13\text{-}52) \qquad w_{i,R} = \frac{X_i r_0^{(i-1)/2}(1-r_0)^2}{1+r_0}$$

for a complete reaction ($p_A = 1$) of deficient A groups since, in this case, reactants consist only of unreacted monomer molecules B–W–B and polymer molecules B-(W-V)$_N$-W-B with odd degrees of polymerization of $X_i = 2N + 1$ with respect to *units* W and V. Expressions are more complicated for incomplete conversions of A groups because the polymer contains molecules with both odd and even numbers of monomeric units.

The complete reaction ($p_A = 1$) of a 1:2 monomer mixture ($r_0 = 1/2$) leads to a reactant with an *average* composition of B–W–V–W–B. However, true trimer molecules are only present with $w_{3,R} = 0.250$ and a mole fraction of $x_{3,R} = w_{3,R} \overline{X}_{R,n}/X_3 = 0.250 \cdot 3/3 = 0.250$. The reactant also contains the pentamer ($w_{5,R} = 0.208$; $x_{5,R} = 0.125$), the heptamer, nonamer, etc., and the monomer B–W–B ($w_{1,R} = 0.167$ and $x_{1,R} = 0.250$).

Very large excesses of B–W–B components can be used to prepare highly uniform oligomers and polymers B(W–V)$_i$-W–B. This **duplication method** utilizes Eq.(13-52) where $w_{i,R}$ is expressed by $w_{i,R} = x_i M_i / \overline{X}_{R,n}$ where $x_i = n_i/n_R$, $M_i = M_U X_i$, and $\overline{X}_{R,n} = \overline{M}_{R,n}/M_U$, i.e., assuming $M_L/M_U \to 0$. Introduction of $n_R = (n_{A,0} + n_{B,0})/2$, $n_{A,0}/2 = n_{AA,0}$, and $n_{B,0}/2 = n_{BB,0}$ leads to

$$(13\text{-}53) \qquad n_i = n_{BB,0}(1-r_0)^2 r_0^{(i-1)/2} \overline{X}_{R,n}$$

The complete reaction of A-groups in a mixture of 1 mol A–V–A and 100 mol B–W–B (100-fold excess of BB groups; $n_{BB,0}/n_{AA,0} = 100$) leads to a reactant with $\overline{X}_{R,n} \approx 1.0202$ (Eq.(13-50)) and a yield of the B–W–V–W–B trimer of $n_{3,R} = 0.9999$ mol (Eq.(13-53)). This trimer can then be reacted with a great excess of the similarly prepared trimer, A–V–W–V–A to the nonamer A(V–W)$_4$V–A. The "duplication method" is thus a **triplication method** that allows one to prepare oligomers in high yields if exchange reactions are absent. The method is less cumbersome than the series of steps B–W–B → B–W–V–A → B–W–V–W–B, etc. However, the separation of steps may be difficult and the use of a solid carrier (**Merrifield method**) is advantageous (Section 14.2.7).

13.3.3 Chain Stabilization

Functional groups A and B never react *completely* in AB and AA/BB *equilibrium* reactions (a p_A =1 in Eq.(13-35) would deliver $\overline{X}_{R,n}$ =1/0!). The remaining endgroups of polymers may react during processing (injection molding, etc.) which increases the melt viscosity and thus the energy consumption; it also slows down processing.

In AA/BB polymerization, the undesired viscosity increase during such post-reactions can be suppressed by a small excess of BB molecules since a complete reaction of A groups would generate polymer molecules with 100 % B endgroups. However, these groups should not undergo self-condensation and exchange reactions.

However, because such reactions cannot be excluded, post-condensations are reduced by addition of **molecular weight stabilizers**, also called **stabilizers, chain terminators,** or **regulators**. These stabilizers are monofunctional compounds B'–Z for AA/BB polymerizations and either monofunctional compounds or bifunctional compounds B'–Z–B' for AB polymerizations. They are especially important for polycondensations with high equilibrium constants. An example is the addition of a diacid in the so-called hydrolytic polymerization of dodecanolactam (laurolactam) to polyamide 12 (PA 12).

For initially present $n_{M,0}$ mol monomer molecules A–U–B and $n_{Z,0}$ mol stabilizer molecules B'–Z–B', the number-average degree of polymerization of reactants, $\overline{X}_{R,n}$, is calculated from the amounts of totally present units, $n_{M,0} + n_{Z,0}$, and totally present molecules, $n_{molecules}$, and fractional monomer conversions, $p = (n_{M,0} - n_M)/n_{M,0}$, as

$$(13\text{-}54) \qquad \overline{X}_{R,n} = \frac{n_{units}}{n_{molecules}} = \frac{n_{M,0} + n_{Z,0}}{n_{M,0} + n_{Z,0} - (n_{M,0} - n_M)} = \frac{1 + (n_{Z,0}/n_{M,0})}{1 - p + (n_{Z,0}/n_{M,0})}$$

Eq.(13-54) is identical to Eq.(13-50) since $r_0 \equiv n_{M,0}/(n_{M,0} + 2\,n_{Z,0})$. Fig. 13-6 shows the effect of the imbalance for various proportions of the stabilizer.

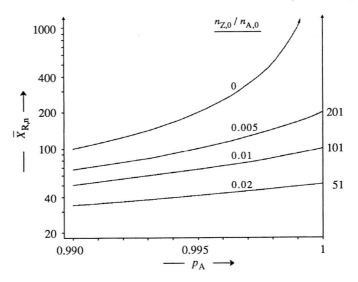

Fig. 13-6 Logarithm of number-average degrees of polymerization of reactants as function of the extent of reaction of A-groups for AB reactions in the presence of various molar ratios of monofunctional chain stabilizers. Numbers to the right indicate attainable maximum number-average degrees of
polymerization of reactants. For the general treatment of AB + B$'_b$ and AA + BB + B$'_b$ reactions with
$b \geq 1$, see Section 16.2.4.

Eq.(13-54) also applies to AA/BB reactions if the initial amount of monomer mole-
cules, $n_{M,0}$, is replaced by the initial amount of A-groups, $n_{A,0}$. In both cases, $\overline{X}_{R,n}$ is
reduced from infinity at $p_A \rightarrow 1$ to much lower values.

13.4 Kinetics of Linear Step Reactions

13.4.1 Experimental Conditions

In polycondensations, leaving molecules are often removed by evaporation at elevated
temperature, in vacuum, or by azeotropic distillation. Their removal is not the rate-determining step, however. For example, lowering the stirring rate slows the release of water
vapor of leaving molecules but does not affect the rate of polyesterification of sebacic
acid and pentaerythritol (Fig. 13-7). Fast stirring always allows the vapor pressure to attain its equilibrium value for that stage of polyesterification. Only a negligible amount
of water remains dissolved in the melt.
For the derivation of kinetic equations in Sections 13.4.2 ff., it is assumed that leaving
molecules do not change the stoichiometric balance between reacting functional groups.
This requirement is not always fulfilled. In the polycondensation of ethylene glycol and
maleic anhydride to unsaturated polyesters, for example, the intermediate trimer,
$HOCH_2CH_2OOCCH=CHCOOCH_2CH_2OH$, may self-condense. The released ethylene
glycol may add to the carbon-carbon double bond of maleic and fumaric acid units.
The resulting monomeric units carry hydroxyl groups that on further esterification lead
to branched polyesters. See also Section 13.8.3 for this polycondensation.

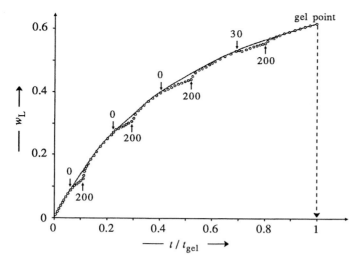

Fig. 13-7 Weight fraction w_L of water in the vapor phase above the melt during the crosslinking polycondensation of sebacic acid, $HOOC(CH_2)_8COOH$, and pentaerythritol, $C(CH_2OH)_4$, as a function of the relative time t/t_{gel} to reach the gel point (= onset of crosslinking) [4].
Solid line: Stirrer speed 200 min^{-1}. Circles: Second experiment in which the stirrer speed was lowered to 0 min^{-1} or 30 min^{-1} and later increased to 200 min^{-1}.

Another example is the polycondensation of diamines, $H_2N-Z-NH_2$, with the activated terephthalic acid ester, $O_2N(p\text{-}C_6H_4)OOC(p\text{-}C_6H_4)COO(p\text{-}C_6H_4)NO_2$, where the split-off p-nitrophenol protonates the amino groups. The protonated species possess a different activity than the non-protonated ones which in turn changes the kinetics.

Rate constants may also change with time if catalysts react with leaving molecules. For example, the esterification catalyst antimony oxide, Sb_2O_3, is deactivated by glycol. The catalyst may also react with reactants. An example is p-toluene sulfonic acid as catalyst for esterifications of dicarboxylic acids. At high monomer conversions, the concentration of remaining COOH groups becomes comparable to that of the catalyst, and the sulfonic acid groups esterify hydroxyl groups. The catalyst thus acts as a monofunctional chain terminator. Its concentration becomes smaller and the rates of reaction and the relative changes of extents of reaction decrease. Washing out the remaining catalyst and replacing it with new catalyst in the orginal concentration restores the reaction rate.

Kinetics is also affected by the change of the polarity of the reacting system with time because the usually highly polar endgroups of monomers are replaced by less polar monomeric units. This change may also affect the activity of reacting groups. These effects can be checked relatively easily by adding polymer to the reacting system, provided the polymer has neither reactive endgroups nor reacting chain units.

13.4.2   Kinetic Equations

Many polycondensations and polyadditions are not controlled thermodynamically but kinetically. They are catalyzed either *externally* by added catalysts or *internally* by their own functional groups (also called self-catalyzed or uncatalyzed reactions).

The term "self-catalysis" seems to contradict the classic definition of a catalyst as an entity that emerges unchanged from the reaction step because self-catalyzing groups such as COOH are ultimately consumed. However, the reaction of $-$COOH $+$ HO$-$ is catalyzed by a *second* COOH group that leaves the intermediary complex of 2 $-$COOH $+$ 1 HO$-$ without change in composition after the reaction. It does not matter that this group may itself be esterified later. "Self" thus refers to the same *type* of group, not the *same* group. Such reactions are certainly not "uncatalyzed" in the usual meaning of the word; they are "uncatalyzed" only in the sense of *externally* uncatalyzed.

The most simple kinetic cases are those of irreversible, stoichiometric, bifunctional reactions between A and B groups without any leaving molecules (polyadditions and polycondensations with $M_L/M_U \rightarrow 0$):

(13-55) j A$-$U$-$B $\rightarrow$ A$-$U$_j$$-$B $+$ $(j-1)$ AB   (AB reaction)

(13-56) j A$-$V$-$A $+$ j B$-$W$-$B $\rightarrow$ A$-$(V$-$W)$_j$$-$B $+$ $(2j-1)$ AB   (AA/BB reaction)

They are described by the differential equations (13-57), the integrated equations (13-58), and, after introducing $p_A \equiv ([A]_0 - [A])/[A]_0$ and $\overline{X}_{R,n} = (1 - p_A)^{-1}$ (Eq.(13-35)), Eqs.(13-59) for the number-average degree of polymerization of reactants:

	(e) *externally catalyzed by* C	(i) *internally catalyzed by* A

(13-57) $-d[A]/dt = k_3[A][B][C]_0 = k_3[A]^2[C]_0;$ $-d[A]/dt = k_{3,i}[A]^3$

(13-58) $[A]^{-1} = [A]_0^{-1} + k_3[C]_0 t$ $[A]^{-2} = [A]_0^{-2} + 2 k_{3,i}t$
                    $= [A]_0^{-1} + k_2 t$ $; k_2 \equiv k_3[C]_0$

(13-59) $\overline{X}_{R,n} = (1 - p_A)^{-1} = 1 + k_2[A]_0 t$ $\overline{X}_{R,n}^2 = (1 - p_A)^{-2} = 1 + 2 k_{3,i}[A]_0^2 t$

Both types are of third order but the externally catalyzed one is a reaction of pseudo-second order because $k_3[C]_0 = k_2 = const.$

Externally catalyzed reactions should show a linear time dependence of $1/(1 - p_A)$ and internally catalyzed ones a linear time dependence of $1/(1 - p_A)^2$. The latter prediction was found to be true for the polycondensation of ethylene glycol + succinic acid (Fig. 13-8). However, the system diethylene glycol + adipic acid followed the predicted 3rd order kinetics only for the range $0.80 < p_A < 1$ but not for $p_A < 0.80$. The latter range is often assumed to be of 2nd order.

A change in reaction order for this system can be checked more easily by considering that an increase of degree of polymerization by 1 corresponds to an increase of the extent of reaction from $p_A = 0$ to $p_A = 50$ % for $\overline{X}_{R,n} = 1 \rightarrow 2$ but only an increase from $p_A = 99$ % to $p_A = 99.0099...$ % for $\overline{X}_{R,n} = 100 \rightarrow 101$. It is thus advantageous to replace the linear scale by a logarithmic one. Eq.(13-59) becomes

(13-60) $\log [(1 - p_A)^{-2} - 1] = \log [2 k_{3,i}[A]_0^2] + \log t$

A plot according to Eq.(13-60) shows that the polycondensation of diethylene glycol + adipic acid is of 3rd order for very small times ($p_A < 0.12$) and very large times ($p_A > 0.9$) but not for the range $0.12 \leq p_A \leq 0.9$ (Fig. 13-9). The third-order rate constants are $k_{3,i}/(g^2 \text{ mol}^{-2} \text{ s}^{-1}) = 21.0$ for small extents of reaction and 7.07 for large ones.

The transition from one kinetic regime to the other may be caused by a dissociation $-$COOH $\rightleftarrows$ $-$COO$^{\ominus}$ $+$ H$^{\oplus}$ that varies with the ratio of ethylene glycol to adipic acid. In-

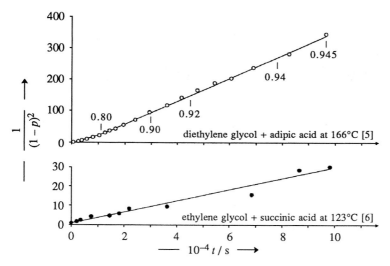

Fig. 13-8 Internally catalyzed polyesterifications in bulk, plotted as 3rd order reactions. Numbers in the upper plot indicate the degree of reaction $p_A = p_{COOH}$. Compare upper part with Fig. 13-10.

deed, the kinetics of polyesterifications with varying glycol concentrations can be described very well by $-d[COOH]/dt = k[COOH][OH]^2$ for the whole conversion range. The experiments of Figs. 13-8 and 13-9 used $-d[COOH]/dt = k[COOH]^2[OH]$ instead.

The initial curvature in 3rd order plots is often interpreted as a start-up effect (see Fig. 13-8, top). They are not real, however, but a consequence of using p_A instead of [A]. This substitution is permissible for polyadditions but not for polycondensations where $M_L/M_U > 0$ because the extent of reaction is defined by amounts, $p_A \equiv (n_{A,0} - n_A)/n_{A,0}$, whereas kinetics uses mole concentrations, $[A] = n_A/V$, with V = total volume.

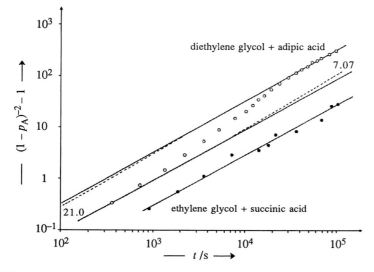

Fig. 13-9 Kinetic data of Fig. 13-8 plotted according to Eq.(13-60) with [A] = [COOH]. Solid lines: slopes of unity (Eq.(13-59)). Broken lines: calculated for $k_{3,i}/(g^2\,mol^{-2}\,s^{-1})$ of 21.0 and 7.07, resp.

The volume V of the system decreases with increasing extent of reaction because leaving molecules are constantly removed.

It is also usually not recognized that the concentration $[A] = n_A/V$ contains the variable n_A in both the numerator and denominator since the volume V is not constant but rather given by $V = V_0 - \Delta V$ in polycondensations where $\Delta V = [(n_{A,0} - n_A)M_L/\rho_L]$ (assuming volume additivity of components). Neither p_A nor $[A]$ or $[B]$ are therefore true variables. Hence, the differential Eqs.(13-57) cannot be integrated directly as is usually done. One rather has to separate the variables first.

The total volume may also change because the volume of monomers is replaced by that of reactants. This problem can be circumvented by using *molal* concentrations, $\{A\}$ = n_A/m, instead of *molar* concentrations, $[A] = n_A/V$, where m = total mass of the system.

An example is the treatment of Eq.(13-57e) for *externally* catalyzed polycondensations. Setting for simplification $\{A\} = \{B\}$, $\{C\}_0 = n_{C,0}/m$, $a = m_0 - n_{A,0}M_L$, $b = M_L$, q_0 = $n_{A,0}$ und $q = n_A$, one arrives at

$$(13\text{-}61) \qquad a^2 q^{-2} dt + 2\, abq^{-1} dq + b^2 dq = -\, k_3 n_{C,0} dt$$

and, after integration and with the definition of p_A:

$$(13\text{-}62) \qquad (a^2/q) - 2\, ab \ln (q/q_0) - b^2 q = (a^2/q_0) - b^2 q_0 + k_3 n_{C,0} t$$

$$(13\text{-}63) \qquad \frac{1}{1-p_A} - \left(\frac{2\,bq_0}{a}\right)\ln(1-p_A) - \left(\frac{bq_0}{a}\right)(1-p_A) = 1 - \left(\frac{bq_0}{a}\right)^2 + \left(\frac{k_3 n_{C,0} q_0}{a}\right)t$$

For *internally* catalyzed polycondensations with $\{A\} = \{B\}$ and $\{A\}_0 = \{B\}_0 = \{C\}_0$, one obtains

$$(13\text{-}64) \qquad a^2 q^{-3} dq + 2\, abq^{-2} dq + b^2 q^{-1} dq = -\, k_{3,i} dt$$

and after integration

$$(13\text{-}65) \qquad \frac{1}{(1-p_A)^2} + \left(\frac{4\,bq_0}{a(1-p_A)}\right) - \left(\frac{2\,b^2 q_0^2}{a^2}\right)\ln(1-p_A) = 1 + \left(\frac{4\,bq_0}{a}\right)^2 + \left(\frac{k_{3,i} q_0^2}{a}\right)t$$

The time dependencies of $(1 - p_A)^{-1}$ (external catalysis, Eq.(13-63)) and $(1 - p_A)^{-2}$ (internal catalysis, Eq.(13-65)), respectively, are thus much more complicated for polycondensations than for polyadditions (Eqs.(13-59)) because the former also have to consider the ratio $bq_0/a = m_{L,0}/(m_0 - m_{L,0})$. For polycondensations at small times, $\overline{X}_{R,n}$ (externally catalyzed) and $\overline{X}_{R,n}^2$ (internally catalyzed) are no longer linear functions of time (Fig. 13-8, top, and Fig. 13-10). The resulting "induction periods" are also not caused by changes in activities because reactivities and thus rate constants were assumed to be constant for the model calculations of Fig. 13-10.

At longer times, the number-average degree of polymerization increases linearly with time for externally catalyzed polycondensations (Fig. 13-10). However, the slope is not the true rate constant (here the assumed rate constant). For example, the slope of the lowest curve in Fig. 13-10 corresponds to $k_3 n_C q_0/a = 1470$ g^2 mol^{-2} s^{-1} whereas the true (assumed) rate constant is $k_3 = 500$ g^2 mol^{-2} s^{-1}.

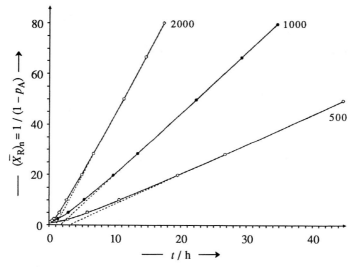

Fig. 13-10 Calculated time dependence of number-average degrees of polymerization of reactants in the polycondensation of $n_{M,0}$ = 0.1 mol phenyl-p-hydroxybenzoate in bulk by $n_{C,0}$ = 0.01 mol of a monofunctional catalyst (M_C = 36 g/mol), assuming rate constants of $k_3/(\text{g}^2\,\text{mol}^{-2}\,\text{s}^{-1})$ = 500, 1000, or 2000 [7]. The slopes at long times are not identical with the true rate constants.

13.5 Reactivities

13.5.1 Principle of Equal Chemical Reactivity

The principle of equal chemical reactivity (see p. 169) is based on the observation that the rate constants k_i of reactions of functional groups Q of molecules RZ_iQ become independent of the number i of groups Z for $i \geq 3$. Some experimental values of k_i for esterifications are shown in Fig. 13-11.

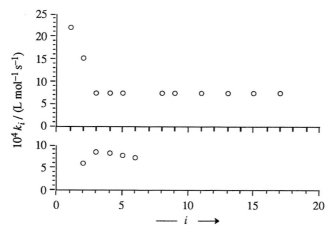

Fig. 13-11 Rate constants of the esterification of monobasic acids, $H(CH_2)_iCOOH$ (top), and dibasic acids, $HOOC(CH_2)_iCOOH$ (bottom), by a large excess of ethanol at 25°C [8].

It has already been pointed out (p. 170) that the principle does not apply if the functional groups are only formally separated from each other such as in conjugated systems (divinyl benzenes, poly(acetylene)s, etc. It also does not apply if the environment of the reacting groups changes as the reaction progresses. An example is the polycondensation of dicarboxylic acids, HOOC–Z–COOH, with glycols, HO–Z'–OH, where the resulting dimers, trimers, ..., polyesters, HOOC–Z–CO$\left[\text{O–Z'–O–OC–Z–CO}\right]_n$O–Z'–OH, associate via the endgroups –OH and –COOH. The concentration of these endgroups, and thus also the association, decreases with increasing degree of polymerization and so do rate constants.

An even stronger effect of the degree of polymerization on the rate constants of irreversible condensations has been observed for the reaction of the two phenolic endgroups of poly(phenolphthalein terephthalate)s with benzoyl chloride (BC) (Fig. 13-12). The effect is probably caused by a change in the local environment of reacting functional groups because it disappears on addition of phenolphthalein dibenzoate.

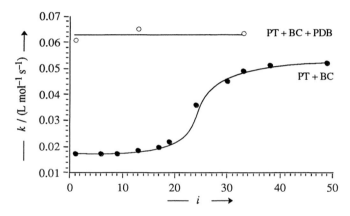

Fig. 13-12 Dependence of rate constants of the triethylamine (TEA) catalyzed reaction of benzoyl chloride (BC) and phenolic endgroups of poly(phenolphthalein terephthalate) (PT) in tetrahydrofuran at 19°C (theta conditions) on the number of repeating groups of PT in the absence and in the presence of various proportions of phenolphthalein dibenzoate (PDB) [9]. The molar ratio of functional groups and triethylamine were $1:1:5 = $ (PT/2) : BC : TEA.

poly(phenolphthalein terephthalate) (PT)   phenolphthalein dibenzoate (PDB)

The effect of microenvironment on reactivities has not been investigated systematically. It should be strong for associating endgroups in theta solvents and relatively weak for reactions in the melt. It may be that the independence of rate constants at $i \geq 3$ in Fig. 13-11 is due to the latter effect (large excess of ethanol). However, microenvironments do change during the course of step reactions, which may be one of the reasons why the order of reaction sometimes varies with increasing monomer conversion (see below).

Table 13-4 Conventionally determined rate constants k_2 of some polycondensations. k_2 includes the concentration of the catalyst (see text). $Ar = p\text{-}C_6H_4$.

Monomer	$\dfrac{T}{°C}$	$\dfrac{10^5 k_2}{L\,mol^{-1}\,s^{-1}}$	Comments
$HO(CH_2)_2OH + HOOC(p\text{-}C_6H_4)COOH$	254	2.3	without catalyst
$HO(CH_2)_2OH + HOOC(p\text{-}C_6H_4)COOH$	251	7.8	with 0.025 wt% Sb_2O_3
$HO(CH_2)_2OH + HOOC(p\text{-}C_6H_4)COOH$	250	110	with 0.001 wt% $Mn(OAc)_2$
$H_2N(CH_2)_{10}COOH$	176	18	in m-cresol
$HO(CH_2)_4OH + OCN(p\text{-}C_6H_4)NCO$	100	90	-
$Cl(p\text{-}C_6H_4)SNa$	250	36	in pyridine
$NaO(Ar)C(CH_3)_2(Ar)ONa + Cl(Ar)SO_2(Ar)Cl$	100	1200	in dimethyl sulfoxide

13.5.2 Structural Effects

Propagation rate constants of polycondensations that are conventionally determined by Eqs.(13-58) and (13-59) are usually in the range $10^{-2} \geq k_2/(\,L\,mol^{-1}\,s^{-1}) \geq 10^{-5}$ (Table 13-4). They also strongly depend on the catalyst. Generalizations are difficult since investigations of systematically varied monomers are lacking. Kinetically effective species are often not identical to the nominal ones (Section 13.5.3). Reactions may also be diffusion controlled (Section 13.5.4).

The principle of equal chemical reactivity may not apply, especially if the two functional groups of a reactant molcules are connected by an aromatic or an ionic group. An example is terephthalic acid, $HOOC(p\text{-}C_6H_4)COOH$, where the reactivity of the second COOH group is changed after the esterification of the first one.

Effect of Solvent

Solvents may affect reactivities in different ways: by changing the polarity of the system, by solvation and/or association of functional groups, and by catalytic action. Solvent effects are therefore inadequately described by macroscopic quantities such as relative permittivities ε_r (Volume IV) or solubility parameters (Volume III). For example, p-aminobenzoic acid, $H_2N(p\text{-}C_6H_4)COOH$, did not polycondense at all with triphenyl-phosphine/hexachloroethane in triethyl amine ($\varepsilon_r = 2.42$) or with hexamethylphosphoric triamide ($\varepsilon_r = 34$) but it did in the mixtures of the latter with tetramethylurea ($\varepsilon_r = 23$). High reduced viscosities (i.e., high molecular weights) were obtained for polymers from polycondensations in mixtures of pyridine and several solvents (Fig. 13-13). Polycondensations in tetramethyl urea or N-methyl-α-pyrrolidone ($\varepsilon_r = 32.2$) gave high polymer yields but the reduced viscosities were much lower (33 mL/g and 15 mL/g, respectively).

Different Rate Constants

Rate constants may depend on the molecular weights of reactants, i.e., the principle of equal chemical reactivity may not apply. For example, the rate constant k_{11} of the reaction of two monomer molecules may differ from the k_P of all other reactants ($M + P_i$, $P_i + P_j$, where $i \geq 2$). For pseudo second-order reactions, the number-average degree of

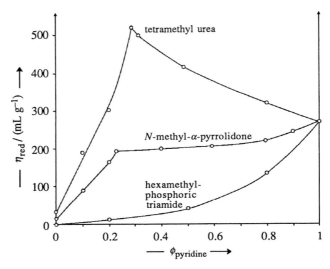

Fig. 13-13 Reduced viscosities of polymers from the polycondensation of 0.4 mol/L p-aminobenzoic acid in mixtures of pyridine and various solvents in the presence of 0.48 mol/L triphenyl phosphine and 0.6 mol/L hexachloroethane after 30 min at room temperature [10].
    With kind permission of the Society of Polymer Science, Japan.

polymerization, $\overline{X}_n$, should increase linearly with time according to Eq.(13-59e) if the principle of equal chemical reactivity applies (Fig. 13-14). If monomer molecules react faster with other monomer molecules than with other reactants, then $k_{11}/k_P > 1$, and $\overline{X}_n$ increases fast at first and then slower with time. At the cross-over point, $\overline{X}_n = 17$ which corresponds to $p = 0.94$. Since monomer molecules react faster with their own kind than with others ($k_{11} > k_P$), polymolecularities $\overline{M}_w/\overline{M}_n$ will be smaller than those for $k_{11} = k_P$.

At least three rate constants are required to describe such reactions: k_{11} for the reaction $M + M$, k_{1i} for $M + P_i$, and k_{jk} for $P_j + P_k$. They may be obtained from polycondensations and model reactions with monofunctional compounds. Table 13-5 shows rate constants for the self-condensation of the potassium salt, $Cl(p\text{-}C_6H_4)SO_2(p\text{-}C_6H_4)OK$, of 4-chloro-4'-hydroxydiphenylsulfone at 165°C.

Reaction 11 has the same rate constant as reaction 21 and reaction 12 the same one as reaction jk (with j, $k \geq 2$). Reactions 12 and 21 are simultaneous. The rate constant k_{ij} of the reaction of one of the monomer molecules and a polymer molecule must therefore be the sum of the rate constants k_{12} and k_{21}.

Table 13-5 Rate constants for the reaction of underlined groups in the formation of a polysulfone [11].

Reacting species		$10^5\, k/(\text{L mol}^{-1}\, \text{s}^{-1})$
Cl–Z–O̲K̲ + C̲l̲–Z–OK		$k_{11} = 0.51$
Cl–Z–O̲K̲ + C̲l̲–Z–O–C$_6$H$_4$–SO$_2$–O–C$_6$H$_5$		$k_{12} = 36.1$
C$_6$H$_5$–SO$_2$–C$_6$H$_4$–O–Z–O̲K̲ + C̲l̲–Z–OK		$k_{21} = 0.51$
C$_6$H$_5$–SO$_2$–C$_6$H$_4$–O–Z–O̲K̲ + C̲l̲–Z–O–C$_6$H$_4$–SO$_2$–O–C$_6$H$_5$		$k_{22} = -$
Cl–C$_6$H$_4$–SO$_2$–C$_6$H$_4$–O–Z–O̲K̲ + C̲l̲–Z–O–C$_6$H$_4$–SO$_2$–O–C$_6$H$_4$–OK		$k_{jk} = 38$

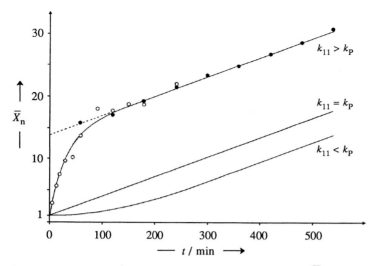

Fig. 13-14 Time-dependence of number-average degrees of polymerization, $\bar{X}_n = 1/(1 - p)$, for second-order reactions with $k_{11} \gtrsim k_P$. Top: externally catalyzed polycondensation of diphenyl terephthalate + (2,2,4+2,4,4)-trimethyl hexamethylenediamine in dimethyl sulfoxide at 90°C [12]; the extent of reaction, p, was determined either via the liberated phenol (O) or via the content of endgroups (●). Data do not indicate whether $k_{22} > k_P$ in addition to $k_{11} > k_P$. Center and bottom: model calculations.

The rate of the decrease of monomer concentration is therefore given by

$$(13\text{-}66) \qquad - d[M]/dt = k_{11}[M]^2 - k_{ij}[M]([R] - [M])$$

where [R] = mole concentration of reactants ($X \geq 1$). The rate of reactant formation is

$$(13\text{-}67) \qquad d[R]/dt = - (1/2)\, k_{11}[M]^2 - k_{1j}[M]([R] - [M]) - (1/2)\, k_{jk}([R] - [M])^2$$

Introduction of parameters $Q \equiv k_{jk}[M]_0 t$, $Y \equiv [R]/[M]_0 = 1/(\bar{X}_R)_n$, $Z \equiv [M]/[M]_0$, $r_1 \equiv k_{11}/k_{jk}$, and $r_2 \equiv k_{1j}/k_{jk}$ leads to differential equations

$$(13\text{-}68) \qquad dZ/dQ = - r_1 Z^2 - r_2 Z(Y - Z)$$

$$(13\text{-}69) \qquad dY/dQ = - (1/2)\, r_1 Z^2 - r_2 Z(Y - Z) - (1/2)(Y - Z)^2$$

that can be solved by a computer if rate constants are known. Eqs.(13-66)-(69) apply to both AB and AA/BB reactions (if $M_L/M_U \to 0$). Since only one type of monomer exists in AB reactions but two types in AA/BB ones and since these equations refer to *effective* concentrations, one has to double the monomer concentrations for AB reactions.

The polycondensation of the potassium salt of 4-chloro-4'-hydroxydiphenylsulfone is accompanied by an exchange reaction with an equilibrium constant k_{1qj}. Eqs.(13-68) and (13-69) thus have to be modified by a term with $r_3 = k_{1qj}/k_{jk}$:

$$(13\text{-}70) \qquad dZ/dQ = - r_1 Z^2 - r_2 Z(Y - Z) - r_3 Z(1 - 2\,Y + Z)$$

$$(13\text{-}71) \qquad dY/dQ = - (1/2)\, r_1 Z^2 - r_2 Z(Y - Z) - (1/2)(Y - Z)^2$$

For the example of Table 13-5, one has $r_1 = (0.51 \cdot 4)/(38 \cdot 4) = 0.0134$, $r_2 = (36.1 + 0.51)/(38 \cdot 4) = 0.241$, and from a plot of $y = (\overline{X}_R)_n = f(Q)$, also $r_3 = 0.11$. The dependence of the number-average degree of polymerization of reactants, $\overline{X}_{R,n}$, on the parameter $k_{jk}[M]_0 t$ is plotted in Fig. 13-15 which allows one to calculate the rate constant $10^5 \, k_{lkj}/(\text{L mol}^{-1} \text{ s}^{-1}) = 16.7$. This rate constant is much larger than the rate constant of dimerization, $10^5 \, k_{11}/(\text{L mol}^{-1} \text{ s}^{-1}) = 0.51$ (Table 13-5); it cannot be neglected for the initial stage of the polycondensation.

The AB reactions of Table 13-5 are second-order reactions in which the chlorine atom is replaced by the nucleophilic phenolate anion. In contrast to classic 2nd order polycondensations, number-average degrees of polymerization of polysulfones do not increase linearly with time, however (Fig. 13-15), because rate constants k_{11} and k_{21} are much lower than rate constants k_{12} and k_{jj} (cf. Fig. 13-14). Here, monomer molecules do not react as fast as polymer molecules. They are therefore present in reactants in greater mole fractions x_M than predicted by classic polycondensation theory. For example, $x_M = 0.715$ was found instead of $x_M = 0.457$ for $\overline{X}_{R,n} = 2.19$. The molecular weight distribution of the reactant is thus broader than in classic polycondensations.

In contrast to the AB reaction, the corresponding AA/BB polycondensation

$$(13\text{-}72) \quad \text{Cl–C}_6\text{H}_4\text{–SO}_2\text{–C}_6\text{H}_4\text{–Cl} + \text{KO–C}_6\text{H}_4\text{–SO}_2\text{–C}_6\text{H}_4\text{–OK} \longrightarrow$$

$$\text{+C}_6\text{H}_4\text{–SO}_2\text{–C}_6\text{H}_4\text{–O–C}_6\text{H}_4\text{–SO}_2\text{–C}_6\text{H}_4\text{–O+}_n$$

is not accompanied by an exchange reaction. Rate constants k_{11} and k_{21} here are much larger than rate constants k_{12}, k_{22}, and k_{ik} (Table 13-6). Comparatively more monomer is consumed than in the classic AB case: the monomer distribution becomes narrower. The number-average degree of polymerization increases first steeply and than less steeply with time (Fig. 13-15; cf. Fig. 13-14).

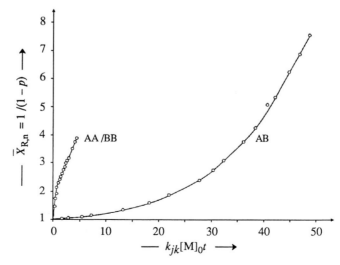

Fig. 13-15 Dependence of the number-average degree of polymerization of the reactant (from endgroup determinations) on the time parameter, $k_{jk}[M]_0 t$, of an AB polycondensation at 165°C and the corresponding AA/BB polycondensation at 120°C, both yielding the same polysulfone ([11], see also Tables 13-5 and 13-6). Circles are experimental, lines are calculated.

Table 13-6 Reactions and rate constants of the formation of the same polysulfone as in Table 13-5 but by an AA/BB reaction instead of AB; $Z = C_6H_4-SO_2-C_6H_4$.

Reactive species				$10^5\ k/(\text{L mol}^{-1}\ \text{s}^{-1})$
Cl–Z–Cl + KO–Z–OK	→ Cl–Z–O–Z–OK	+ KCl		$k_{11} = 16.4$
Cl–Z–Cl + KO–Z–O–Z–H	→ Cl–Z–O–Z–O–Z–H	+ KCl		$k_{12} = 0.90$
C_6H_5–O–Z–Cl + KO–Z–OK	→ C_6H_5–O–Z–O–Z–OK	+ KCl		$k_{21} = 17.6$
H–Z–O–Z–Cl + KO–Z–O–Z–H	→ H–Z–O–Z–O–Z–O–Z–H	+ KCl		$k_{22} = 1.32$
Cl–Z–(O–Z)$_i$–Cl + KO–(Z–O)$_k$–K	→ Cl–Z–(O–Z)$_i$–(O–Z)$_k$–OK	+ KCl		$k_{ik} = 1.36$

Reactions 12 and 21 are consecutive; thus $k_{1j} \approx 0.90 \cdot 10^{-5}$ L mol^{-1} s^{-1} (Table 13-6). Reaction parameters therefore become $r_1 = k_{11}/k_{jk} = 16.4/1.36 = 12.1$ and $r_2 = k_{1j}/k_{jk} = 0.90/1.36 = 0.66$.

Differences in rate constants of monomer, oligomer, and polymer molecules are observed for many polycondensations, for example, esterifications of aromatic dicarboxylic acids (Table 13-7). The simple kinetic equations of Section 13.4.2, based on the principle of equal chemical reactivity, are thus valid only if the following conditions apply:

- High degrees of polymerization, X_i, of reactants (and thus high extents of reaction) because polarities of reactants and the reaction system systematically vary with X_i.
- Absence of reverse reactions, which is true only if the concentration of leaving molecules is smaller than the equilibrium concentration.
- Identity of mole concentrations and activities (activity coefficients of unity), which is usually not true for melts and concentrated solutions.
- Negligible molecular weight M_L of leaving molecules ($M_L/M_U \to 0$).

Table 13-7 Rate constants of the reaction of underlined groups in the formation of poly(ethylene terephthalate) from terephthalic acid and ethylene glycol at 280°C.

Reacting groups			$k/(\text{g mol}^{-1}\ \text{s}^{-1})$
HOOC(p-C_6H_4)COOH	+	HOCH$_2$CH$_2$OH	23.7
HOOC(p-C_6H_4)COOH	+	HOCH$_2$CH$_2$OOC(p-C_6H_4)CO....	22.6
....OCH$_2$CH$_2$OOC(p-C_6H_4)COOH	+	HOCH$_2$CH$_2$OH	13.9
....OCH$_2$CH$_2$OOC(p-C_6H_4)COOH	+	HOCH$_2$CH$_2$OOC(p-C_6H_4)CO....	16.7
....OC(p-C_6H_4)COOCH$_2$CH$_2$OH	+	HOCH$_2$CH$_2$OOC(p-C_6H_4)CO....	2.0

13.5.3 Externally Activated Polycondensations

Rates of step reactions are strongly influenced by the constitution of leaving groups. The amide group –NH–CO– is only slowly formed from –NH$_2$ and HOOC–, faster from –NH$_2$ and ROOC–, and very fast from –NH$_2$ + ClOC–. In esterifications, the reaction rate of –OH + R'OOC– increases with R' in the order H > CH$_3$ > O$_2$N(p-C_6H_4). Molecules with leaving groups that increase reaction rates very strongly are called (internally) **activated**.

Internally activated compounds often have short shelf lives. Since they are too reactive for AB reactions, it is advantageous to activate them *externally*. An example is the reaction of thionyl chloride, $SOCl_2$, with *p*-aminobenzoic acid, $H_2N(p$-$C_6H_4)COOH$. In the first step, the isolable intermediate $OSN(p$-$C_6H_4)COCl$ is formed. The intermediate then converts to *p*-aminobenzoylchloride hydrochloride which subsequently polycondenses. Thionyl chloride is consumed in stoichiometric amounts:

(13-73) $H_2N(p$-$C_6H_4)COOH + 2\ SOCl_2 \longrightarrow O{=}S{=}N(p$-$C_6H_4)COCl + SO_2 + 3\ HCl$
$O{=}S{=}N(p$-$C_6H_4)COCl + 3\ HCl \longrightarrow [H_3N(p$-$C_6H_4)COCl]^{\oplus}\ Cl^{\ominus} + SOCl_2$
$n\ [H_3N(p$-$C_6H_4)COCl]^{\oplus}\ Cl^{\ominus} \longrightarrow +NH(p$-$C_6H_4)CO\frac{1}{n} + 2\ HCl$

Thionyl chloride-activated polycondensations of aromatic hydroxy acids proceed differently. *p*-Hydroxycinnamic acid, $HO(p$-$C_6H_4)CH{=}CHCOOH$, first forms the carboxychlorosulfite with $SOCl_2$. At sufficiently low temperatures, this chloride of a mixed anhydride converts into the carboxylic acid chloride which then self-condenses:

(13-74) $HO(p$-$C_6H_4)CH{=}CHCOOH + SOCl_2 \longrightarrow HO(p$-$C_6H_4)CH{=}CHCOOSOCl + HCl$
$HO(p$-$C_6H_4)CH{=}CHCO{-}O{-}SOCl \longrightarrow HO(p$-$C_6H_4)CH{=}CHCOCl + SO_2$
$n\ HO(p$-$C_6H_4)CH{=}CHCOCl \longrightarrow +O(p$-$C_6H_4)CH{=}CHCO\frac{1}{n} + HCl$

At low monomer conversions, these activated polycondensations of hydroxy acids, HO–Z–COOH, lead to polymers with far higher reduced viscosities (and thus molecular weights) then classic polycondensations (Fig. 13-16). At still higher extents of reactions, increases of molecular weight are even greater.

The catalyst pyridine first forms an intermediate (I) with the activator thionyl chloride which then reacts with the hydroxy acid, HO–Z–COOH, to the reactive mixed ester (II).

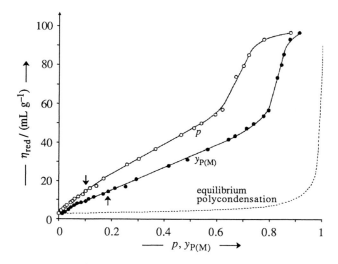

Fig. 13-16 Reduced viscosities (in concentrated sulfuric acid) of polymers from the thionyl chloride activated and pyridine catalyzed polycondensation of $HO(p$-$(2$-$CH_3O)C_6H_4)CH{=}CHCOOH$ (*trans*-ferulic acid) at 21°C as a function of the extent of reaction, *p*, or the relative yield of polymer, $y_{P(M)}$ [13]. - - - Theory for equilibrium polycondensation; arrows indicate onset of heterogeneity.

Such mixed esters are also generated by other activators, for example, $(CH_3CO)_2O$, $SiCl_4$, or poly(phosphoric acid) $HO+P(OH)(O)-O\}_nH$. Similarly to peptide syntheses (Section 14.3.12), other phosphorus containing compounds may also be used, for example, N-phosphonium salts of pyridine (III) or dioxophospholanes (IV). Phosphines R_3P require the presence of hexachloroethane.

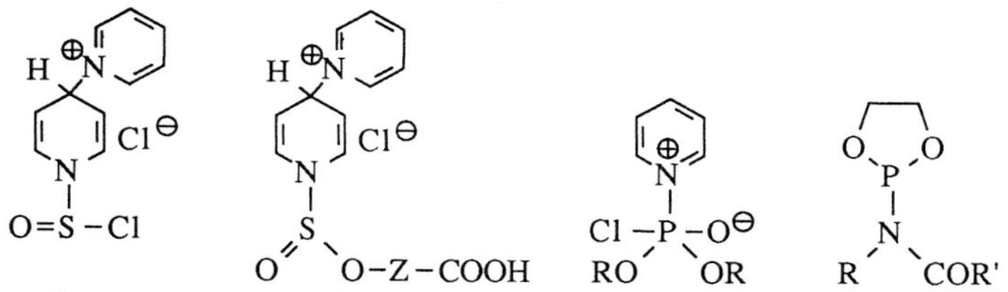

13.5.4 **Heterogeneous Linear Polycondensations**

The synthesis of aliphatic polyamides by direct amidation requires many hours, even at 250°C, because of the strong resonance stabilization of the COOH groups. On the other hand, reactions of diamines and dicarboxylic acid chlorides proceed in minutes.

The acylation of amines, amino acids, alcohols, phenols, etc., in the presence of acid acceptors is known as the **Schotten-Baumann reaction** and, for AA/BB polycondensations in the heterogeneous phase, also as **interfacial polymerization**. Examples are the reactions of dicarboxylic acid dichlorides and diamines to polyamides, of sulfonyl dichlorides and sulfone diphenolates to polysulfones, and of phosgene and the sodium salt of 2,2-*bis*(4-hydroxyphenyl propane) (= bisphenol A) to polycarbonates:

(13-75) $NaO(p\text{-}C_6H_4)\text{-}C(CH_3)_2\text{-}(p\text{-}C_6H_4)ONa + COCl_2 \longrightarrow$

$$+O(p\text{-}C_6H_4)\text{-}C(CH_3)_2\text{-}(p\text{-}C_6H_4)O\text{-}CO\}_x + 2\ NaCl$$

Interfacial polycondensations are also possible for other types of polycondensations, for example, polyaddition of diisocyanates and diols to polyurethanes.

In interfacial polymerizations, a solution of monomer AA in the less dense solvent I is layered over a solution of monomer BB in the more dense solvent II that is immiscible with solvent I; both I and II should be precipitants for the polymer. Examples are solutions of diamines in water and diacid dichlorides in chloroform. The polymerization proceeds at the interface of the two solutions and leads either to a polymer film or to a powder. Mechanically stable films may be withdrawn from the interface. Films remaining at the interface impede the transport of monomer to the interface and the polymerization becomes slower with time.

The highest molecular weights are not obtained from stoichiometric amounts of the two monomers but by a molar excess of diamines (Fig. 13-17). This behavior is controlled by a number of kinetic factors: diffusion rate of the diamine through the film, local concentration of reactants, and hydrolysis of acid chlorides

The polymerization takes place at the organic-solvent side of the film as evidenced by the formation of water droplets. In order to react, the diamine must thus be transported through the film to the organic side. This diffusion through the polymer film can be

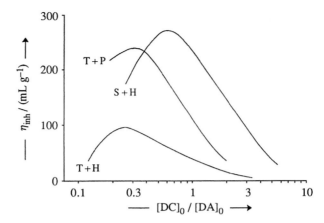

Fig. 13-17 Dependence of inherent viscosities, η_{inh}, as a measure of molecular weights on the initial molar ratio, $[DC]_0/[DA]_0$, in unstirred polycondensations of dicarboxylic acid dichlorides, DC, and diamines, DA, in chloroform [14a]. H = hexamethylene diamine, P = piperazine, S = sebacic acid dichloride, T = terephthaloyl dichloride. $[DA]_0 = 0.4$ mol/L.

greatly enhanced by addition of phase-transfer compounds such as quaternary ammonium and phosphonium salts.

Initially, monomer and oligomer molecules react at the interface as in a classical homogeneous AA/BB polymerization. The initial reaction is of 2nd order as indicated by a plot of lg ($\overline{X}_n - 1$) = f(log t) according to the rearranged and logarithmized Eq.(13-59e) (Fig. 13-18). The initial number-average degrees of polymerization are also directly proportional to $1/(1 - p)$ (Fig. 13-18, insert).

The situation changes after a polymer film is formed. The hydrated monomeric diamine must now permeate from its aqueous solution through the film. On the other side of the film, it will react with the acid chloride endgroups of the already formed polymer, which prevents it from traveling into the interior of the dicarboxylic acid dichloride solution. The reaction continues to be of 2nd order albeit with a smaller rate constant because it is now a reaction between a monomer and a polymer and no longer one between two monomer molecules. The number-average degree of polymerization increases with $1.8/(1 - p)$ instead of $1/(1 - p)$ (Fig. 13-18, insert).

Acid chlorides react faster with diamines than with water because of the greater basicity of the former. Only a small proportion of COCl groups are therefore hydrolyzed in the reaction ~COCl + $H_2O \longrightarrow$ ~COOH + HCl. The rate of this reaction depends on the thickness L of the interphase and a constant k' that is proportional to the rate constant of the hydrolysis. Since this reaction delivers two equivalents of acid that neutralize one equivalent of diamine, hydrochloride from this reaction is often bound by an added non-polymerizing base such as pyridine.

The thickness L of the film increases with time. The growth rate $dL/dt = k(c_A/L) - k'L$ is proportional to the concentration c_A of the diamine and inversely proportional to L minus a correction factor $k'L$ for the hydrolysis that slows the growth of the film. The integration of the rate equation delivers

(13-76) $L = L_\infty[1 - \exp(-2\,k't)]^{1/2}$

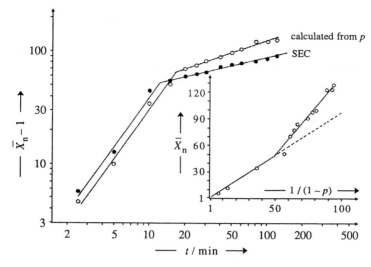

Fig. 13-18 Time-dependence of the number-average degree of polymerization of repeating units in the interfacial polycondensation of the sodium salt of bisphenol A ($x = 0.5$) in aqueous NaOH and the 1:1 mixture of terephthaloyl dichloride ($x = 0.25$) and isophthaloyl dichloride ($x = 0.25$) in dichloromethane at room temperature [15].

Degrees of polymerization were measured by size-exclusion chromatography (SEC; ●) or calculated from the extent of monomer reaction via $\overline{X}_n = 1/(1-p)$ (○).

Insert: $\overline{X}_n$ from SEC (○) as a function of $1/(1-p)$.

with an integration constant of $L_\infty = (k/k')^{1/2} c_A^{1/2}$ that corresponds to the film thickness at infinite time for a growth rate of $dL/dt = 0$.

The diffusion rate of the diamine is smaller than the reaction rate by orders of magnitude. The resulting diffusion control and the loss of acid chloride by hydrolysis insures that the molar ratio of monomers does not need to by exactly equivalent for maximum degrees of polymerization. Indeed, the highest molecular weights are obtained with an excess of diamines (Fig. 13-17). They are regulated by the partition coefficient of the amine between water and organic solvent (Table 13-8).

Similar results are obtained from other heterogeneous systems. In the **emulsion co-polymerization** of oligo(dimethylsiloxane)s with silanol endgroups in dodecylbenzene sulfonic acid micelles, for example, the molecular weight increases strongly with time and then approaches a constant value.

Table 13-8 Effect of partition coefficient K of amines between water and organic solvents on optimum initial ratios of monomers, [diamine]$_0$/[dichloride]$_0$, and inherent viscosities, η_{inh}, as a measure of molecular weights in the interfacial polycondensation of hexamethylene diamine and sebacoyl dichloride in different solvents [14b].

Solvent	K	[Diamine]$_0$/[Dichloride]$_0$	η_{inh}/(mL g^{-1})
Cyclohexane	182	17	86
Xylene	50	8	147
Ethylene chloride	5.6	2.3	176
Chloroform	0.70	1.7	275

The addition of cyclohexanone solutions of isophthaloyl dichloride to aqueous emulsions of 4,4'-diaminodiphenylmethane with the acceptor Na_2CO_3 constitutes a **dispersion copolycondensation**. The degree of polymerization, X, increases here first linearly with time. The polymer then begins to precipitate whereupon X increases first steeply and finally becomes constant.

13.5.5   Reactive Intermediates

Some of the polymerizations with release of leaving molecules are not step reactions in the polymer sense, i.e., polycondensations, but **polyeliminations**, i.e., condensing chain reactions. In these reactions, monomer molecules react with monomer ($X = 1$) or polymer molecules ($X \geq 2$) but polymer molecules do not react with polymer molecules as is the case in true polycondensations. Many of these polymerizations proceed via **reactive intermediates**.

A large group of these reactions comprise nucleophilic aromatic substitutions that lead to aromatic polyethers and polythioethers. 2,6-Dimethylphenol, for example, reacts with oxygen in the presence of copper amines as catalyst in an **oxidative coupling** reaction to poly(oxy-2,6-dimethyl-1,4-phenylene), commonly (and wrongly) called "polyphenylene oxide" (PPO®) or polyphenylene ether (PPE):

(13-77)

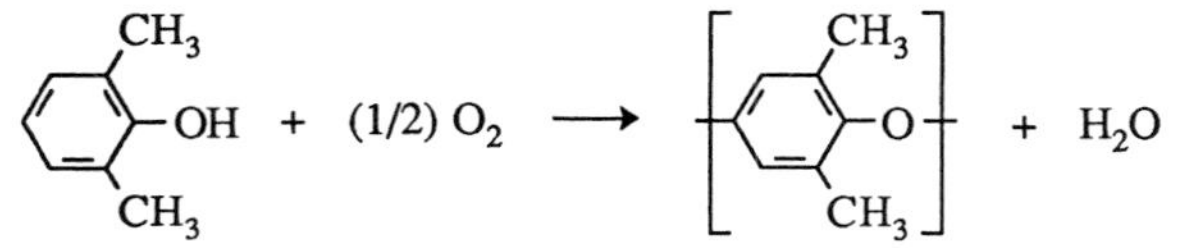

Chromatography of the reaction products showed for a monomer conversion of $u = 0.22$ that the molar ratio of dimers to monomers was $x_2/x_1 = 0.10$, but $x_{i+1}/x_i = 0.61 \pm 0.06$ for i = 2-9. At the higher monomer conversion of 53.4 %, values of $x_2/x_1 = 0.19$ and $x_{i+1}/x_i = 0.71 \pm 0.05$ (for $i \geq 2$) were obtained. Thus, the reaction does not follow the principle of equal chemical reactivity of functional groups of monomer, oligomer, and polymer molecules, which postulates the equality $p = x_{i+1}/x_i$ of extents of reaction, p, and ratios, x_{i+1}/x_i, of mole fractions of reactants that differ by a degree of polymerization of 1 (Eq.(13-38)). It can also not be a reaction where the monomer ($X = 1$) simply has another reactivity than the polymer ($X \geq 2$) because, in this case, the ratio x_{i+1}/x_i should be a constant that is independent of the monomer conversion.

The oxidative coupling to PPO® as well as that of its bromo derivative is mechanistically a 1-electron transfer process. The added base converts the phenol into the phenolate anion and, in turn, by electron transfer into a monomer radical. Addition of a monomer phenolate anion generates a dimer anion:

(13-78)

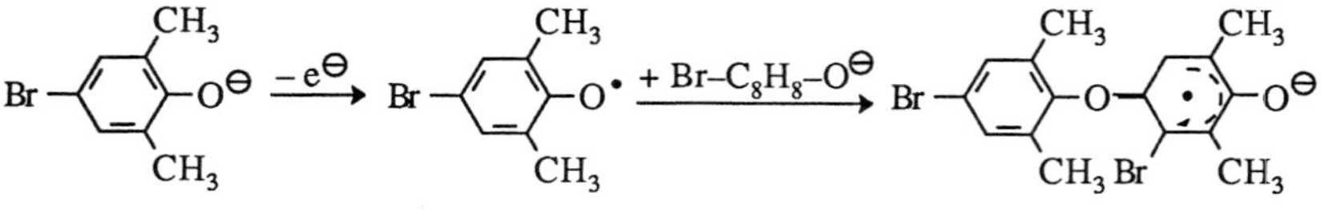

The time dependences of relative yields, $y_{P(U)}$, and degrees of polymerization, $\overline{X}_n$, as well as the dependence of $\overline{X}_n$ on $y_{P(U)}$, are shown in Fig. 13-19.

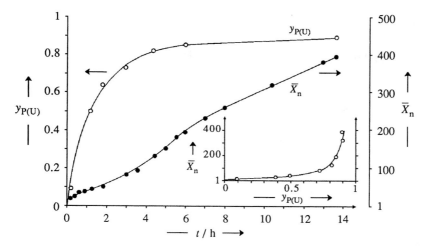

Fig. 13-19 Time dependence of relative yields of polymers, $y_{P(U)}$, and number-average degrees of polymerization, $\bar{X}_n$, in the polymerization of 4-bromo-2,6-dimethylphenol in 6 mol/L solutions of NaOH in water at room temperature in the presence of air [16]. Insert: Dependence of degree of polymerization on relative yields.

Removal of the bromine in the 2-position as bromide anion, $Br^{\ominus}$, leads to the dimer radical, $Br[C_6H_2(CH_3)_2]-O-[C_6H_2(CH_3)_2]-O^{\bullet}$, that adds a new monomer anion, etc. Hence, this reaction is a polyelimination (i.e., a chain reaction), as also evidenced by the variation of the degree of polymerization, X, with time or yield. At small reaction times and small yields, $y_{P(U)} < 0.10$, it increases rapidly to $\bar{X}_n = 20$ whereas it should be only $\bar{X}_n = 1.1$ for polycondensations with $p = 0.10$. The time dependence of $\bar{X}_n$ shows two linear regions that are separated by a steeper increase (Fig. 13-19). This behavior may be caused by an overlap of a living polymerization and a chain coupling (combination).

This class of polymerizations also includes some other polymerizations that are usually classified as polycondensations. Examples are the polymerization of benzene to poly-(phenylene)s with copper dichloride as catalyst

(13-79) $H-C_6H_4-H + 2\ CuCl_2 \longrightarrow +C_6H_4+_n + 2\ CuCl + 2\ HCl$

and the polymerization of 1,4-dichlorobenzene and disodium sulfide to poly(p-phenylene sulfide):

(13-80) $Cl-C_6H_4-Cl + Na_2S \longrightarrow +C_6H_4-S+_n + 2\ NaCl$

13.6   Ring Formation

13.6.1   Cyclic Molecules

Endgroups A and B of linear monomers and polymers can react with other reactants not only intermolecularly to linear polymers but also intramolecularly to form rings (Section 7.5). The ring formation can be thermodynamically or kinetically controlled.

Equilibria

Thermodynamic equilibria are independent of the reaction path. They may be established by chain, step, or exchange reactions and by ionic, free-radical, etc. mechanisms. Equilibria must be defined exactly; it is not permissible to compare equilibria of reactions without leaving molecules with those where molecules L are split off.

In equilibrium, the probability of ring formation is controlled by the distribution of distances between A and B groups. For *infinitely small* AB monomer molecules in *infinitely low* concentrations, the nearest unreacted group to an A-group is always the B-group of its own molecule and the reaction is therefore exclusively intramolecular to rings. The equilibrium concentration, $[u]_{eq}$, of monomeric units in *all* ring compounds increases in direct proportion to the initial monomer concentration, $[M]_0$ (Fig. 7-7). At a certain monomer concentration, the equilibrium concentration of rings is approached and the surplus monomer molecules are converted into linear polymer molecules. The molar concentration of monomeric units in rings becomes constant and independent of the monomer concentration.

Intermolecular and intramolecular reactions compete at *finite* concentrations but the latter is usually assumed to be negligible (Sections 13.2-13.5). Comprehensive experimental data for cyclizations do not exist. However, some indications of the effect of ring formation on degrees of polymerization come from Monte Carlo studies where monomer molecules and monomeric units were placed on the vertices of an isotropic cubic lattice of unit length. Ends of molecules were allowed to form three different bonds: along the lattice vertices (length 1), the diagonals of the face of the cube (length $2^{1/2}$), or the diagonals joining opposite vertices of the cube (length $3^{1/2}$). For this type of simulation, the $\overline{X}_n$ of the total reactant was reduced to 40 from 60 for $1/(1-p) = 60$ because of a considerable fraction of ring molecules with $\overline{X}_{n,ring} \approx 4$ (Fig. 13-20).

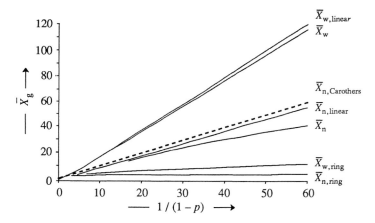

Fig. 13-20 Change of number-average (index g = n) and weight-average (g = w) degrees of polymerization X of reactants (no index), linear chains, and rings with the Carothers parameter, $1/(1-p)$, according to a computer simulation of a polyaddition on a cubic lattice (see text) [17]. Dotted line: Carothers function. With kind permission of the American Chemical Society, Washington (DC).

The probability of formation of stress-free rings containing 1, 2, 3 ... etc. monomeric units depends on the type of monomer and the degree of polymerization. The more chain atoms between the chains ends A and B, the smaller is the probability that an A-group is in the vicinity of a B-group before the reaction. For unperturbed chains, this probability decreases with the $-3/2$th power of the degree of polymerization. Fewer rings are formed at higher X (Fig. 7-7) and the number-average degree of polymerization of rings, $\overline{X}_{n,ring}$, becomes almost constant with increasing $1/(1-p)$ (Fig. 13-20). Freedom of strain is usually observed for organic rings with more than 30-50 chain atoms.

Small rings are not free of strain because of the volume requirements for packing of substituents and the interactions of monomeric units. This leads to a wide variation of yields of rings as a function of ring size for standard reaction conditions (Fig. 13-21).

Interaction of monomeric units with thermodynamically good solvents stiffens segments; chain ends are pushed further apart and the probability of ring formation decreases. For example, the self-cyclization of 0.05 mol/L of $KO(m\text{-}C_6H_4)O(CH_2)_6Br$ resulted in dimer yields of 8.3 % in benzene ($\delta^* = 9.2$), 8.7 % in cyclohexane ($\delta^* = 8.2$), 22.3 % in 1,4-dioxane ($\delta^* = 10$), 52.1 % in 1-butanol ($\delta^* = 11.4$), and 52.9 % in ethanol ($\delta^* = 12.7$) where $\delta^* = \delta/(cal\ cm^{-3})^{1/2}$ is the solubility parameter, a measure of the thermodynamic goodness of the solvent (Volume III).

For entropic reasons, ring formation increases with increasing temperature (Section 7.5). For example, yields of the cyclic tetramer, $[-NH(CH_2)_6NH-OC(CH_2)_4CO-]_2$, from PA 6.6 polycondensations were 1.9 % (275°C), 4.3 % (297°C), and 5.9 % (310°C).

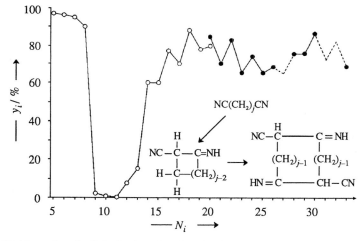

Fig. 13-21 Yields y_i of cycloaliphatic rings (see insert) with N_j ring atoms in the polymerization of 0.067 mol/L of $NC(CH_2)_jCN$ in diethyl ether with 0.67 mol/L sodium alkylaniline and 0.17 mol/L alkylaniline. O Data of [18a], ● data of [18b], - - - hypothetical (yields not reported).

Kinetically Controlled Ring Formation

Most reported polycondensations are not controlled by thermodynamics but by kinetics. According to model calculations for kinetically controlled reactions, relative yields of cyclized monomers, C_1, from linear monomers, M_1, increase with the increasing products of initial monomer concentration, $[M_1]_0$, and ratios $k_1^* = k_{intra,1}/k_{inter}$ of rate constants of intramolecular cyclization, $M_1 \rightarrow C_1$, and linear polymerization $M_j + M_k \rightarrow P$ ($j, k \geq 1$) (Fig. 13-22). The relative yields depend somewhat on the kinetic models I-III discussed below.

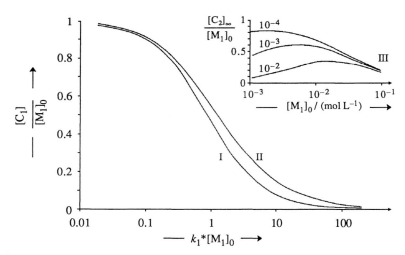

Fig. 13-22 Relative yield, $[C_1]_\infty/[M_1]_0$, of cyclized monomer molecules as a function of the reduced monomer concentration, $k_1{}^*[M_1]_0$, for kinetic models I and II (see text) [19]. Insert: Relative yield, $[C_2]_\infty/[M_1]_0$, of cyclized dimer molecules as a function of initial monomer concentration, $[M_1]_0$, for various reduced rate constants $k_2{}^* = k_\text{inter}/k_{\text{intra},2}$ of 10^{-2}, 10^{-3}, and 10^{-4} (L mol^{-1} s^{-1}) (model III).

All models I-III assume that a linear monomer M_1 cyclizes to C_1 with a rate constant $k_{\text{intra},1}$. They differ with respect to the presence and extent of competing reactions.

Model I allows only for a formation of monomeric rings C_1 but not for larger rings C_2, C_3, etc. This cyclization competes with the step reactions $M_1 + M_1$, $M_1 + M_2$, $M_2 + M_2$, etc., to polymers $M_i = P$ (where $i \geq 2$) that proceeed with a rate constant k_{inter}. According to these calculations, the reduced yield of cyclized monomers at infinite time, $[C_1]_\infty/[M_1]_0$, drops with increasing values of $Y = k_1{}^*[M_1]_0 = k_{\text{inter}}/k_{\text{intra},1}$ to almost zero at high values of Y (Fig. 13-22).

Model II assumes formation of M_2 from $M_1 + M_1$ and the subsequent complete cyclization of M_2 to C_2. No linear polymers M_2, M_3, ... M_N are allowed. The dependence of relative yields of monomeric rings on reduced monomer concentration does not differ very much from that of model I (Fig. 13-22).

The more realistic model III allows not only for the formation of cyclic monomers C_1 with a rate constant $k_{\text{intra},1}$ but also for cyclic polymers C_i with rate constants $k_{\text{intra},i}$ ($i \geq 2$) and linear polymers $P = M_i$ ($i \geq 2$) with a rate constant k_{inter}. According to this model, reduced yields of cyclized monomers at infinite time, $[C_1]_\infty/[M]_0$, depend somewhat on $k_1{}^*[M_1]_0$ but range between the values obtained by the much simpler models I and II for $0.05 > k_1{}^* > 0.001$ (not shown in Fig. 13-22).

No generalized function $[C_i]_\infty/[M_1]_0 = f(k_{\text{inter}}/k_{\text{intra},i})[M_1]_0$ was obtained for $i \geq 2$ (dimers, trimers, etc.). With increasing initial monomer concentration, $[M_1]_0$, reduced concentrations $[C_i]_\infty/[M_1]_0$ pass through maxima, e.g., for dimers with $i = 2$ (Fig. 13-22, insert). These maxima increase with decreasing $k_i{}^*$ because many cyclic compounds are formed if $k_{\text{inter}}/k_{\text{intra}}$ is small. At higher initial monomer concentrations, $[C_2]_\infty/[M_1]_0$ becomes independent of k_2.

Cyclic monomers, oligomers, and polymers are industrially interesting. Cyclic monomers and oligomers may serve as monomers for ring-opening chain polymerizations because, in contrast to polycondensations, they yield higher molecular weights at smaller monomer conversions and do not produce leaving molecules as byproducts that have to be removed and usually recovered at high cost. Cyclic polymers also do not contain endgroups that may either cause additional polymerizations on processing (p. 449) or be attacked by environmental agents.

At low monomer concentrations, cyclic compounds are obtained in high yields. For AA/BB reactions, formation of cyclics is considerably enhanced if dilute solutions of both monomers are simultaneously added dropwise into a large excess of solvent. For example, the polycondensation of a 0.05 mol/L bischlorocarbonate and 0.1 mol/L hexamethylene diamine, both in benzene, resulted in high yields of cyclized compounds: $y =$ 31 % ($i = 1$), 73 % ($i = 2$), 47 % ($i = 3$), 38 % ($i = 4$), 29 % ($i = 5$), 12 % ($i = 6$):

(13-81) $ClOCO(CH_2CH_2O)_iCOCl + 2 H_2N(CH_2)_6NH_2$

$$\longrightarrow Cl^{\ominus}[H_3\overset{\oplus}{N}(CH_2)_6\overset{\oplus}{N}H_3]Cl^{\ominus} + \begin{array}{c} O(CH_2CH_2O)_i \\ O=C \qquad\qquad C=O \\ NH-(CH_2)_6-NH \end{array}$$

Depending on monomers and methods, high yields of very large rings are also sometimes obtained. Proof of the existence of such large rings is usually indirect. For example: far higher molecular weights are found by endgroup determinations than by number-average molecular weights from physical methods if branching is absent.

Thus, the slow addition of a 0.1 mol/L solution of the bisphenol A bischloroformic acid ester in CH_2Cl_2 to an aqueous solution of 0.6 mol/L NaOH + 0.005 mol/L triethylamine at 25°C led to a polycarbonate with number-average molecular weights of 22 900 (osmometry) and 92 000 (endgroups):

(13-82) $(n + m)$ $ClOC-O-(p\text{-}C_6H_4)-C(CH_3)_2-(p\text{-}C_6H_4)-O-COCl$

$$\longrightarrow ClOC\!-\!\!\left[O-(p\text{-}C_6H_4)-C(CH_3)_2-(p\text{-}C_6H_4)-O-CO\right]_n\!\!-\!Cl$$

$$+ cyclo\!-\!\!\left[O-(p\text{-}C_6H_4)-C(CH_3)_2-(p\text{-}C_6H_4)-O-CO\right]_n\!\!+ (m + n - 1)\ COCl_2$$

Some of the polycarbonate molecules must therefore be cyclic. The following endgroups can be expected: chloroformic acid ester groups (determined as Cl) and phenolic groups (determined as OH). Spectroscopy also showed the presence of diethyl carbamate groups (determined as N); these groups resulted from the reaction of OCOCl groups with the impurity diethylamine that was present in triethylamine. The average number $\overline{N}_{end}$ per polymer molecule is calculated from Eq.(13-83) where w_{Cl}, w_{OH}, and w_N are the mass fractions of the endgroups and M_{Cl}, M_{OH}, and M_N their molecular weights:

(13-83) $\overline{N}_{end} = \overline{M}_{r,n,osm}\ [(w_{OH}/M_{r,OH}) + (w_{Cl}/M_{r,Cl}) + (w_N/M_{r,N})]$

Example: A polymer had an osmotically determined number-average moleculer weight of $\overline{M}_{r,n,osm}$ = 22 900 and analytically determined weight fractions of phenolic OH groups ($M_{r,OH}$ = 17.0), chlorine ($M_{r,Cl}$ = 35.5), and nitrogen ($M_{r,N}$ = 14.0) of w_{OH} = 1·10^{-4}, w_{Cl} = 9·10^{-6}, and w_N = 2.2·10^{-4}, respectively [20]. The average number of endgroups per molecule is thus $\overline{N}_{end} = \overline{M}_{r,n,osm}\ [(w_{OH}/M_{r,OH}) + (w_{Cl}/M_{r,Cl}) + (w_N/M_{r,N})]$ = 22 900·10^{-4} [(1/17) + (0.09/35.5) + (2.2/14)] = 0.500. Since linear molecules contain 2 endgroups and cyclic molecules none, one gets from $\overline{N}_{end} = (2\ N_L + 0\ N_R)/(N_L + N_R)$ that N_R/N_L = 3:1: 75 % of the molecules must be cyclic and 25 % linear. The ring molecules were predominantly oligomers with degrees of polymerization of $2 \leq X \leq 20$.

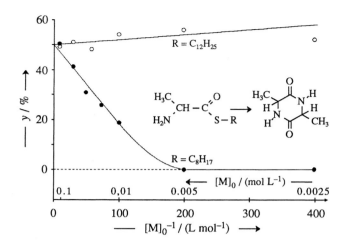

Fig. 13-23 Yield *y* of 3,6-dimethyl-2,5-piperazinedione (DPD) from the pyridine-catalyzed dimeriza-tion of thioalanine-S-alkylesters in water after 7 days at room temperature as a function of the inverse initial molar concentration of DPD [20].

Low monomer concentrations do not always lead to high yields of cyclic molecules. In the polycondensation of activated amino esters, $H_2N-CH(CH_3)-COOR$, the yield of cyclic dimers of alanine, $cyclo$-$(NH-CH(CH_3)-CO)_2$, increases with increasing dilution of the monomer as expected if $R = SC_{12}H_{25}$ but it does not do so if $R = C_8H_{17}$ (Fig. 13-23). Micellization is the suspected cause.

13.6.2   Cyclopolycondensation

Cyclopolycondensations are polycondensations of multifunctional $(f \geq 3)$ monomer molecules that result in practically linear macromolecules with intramolecularly formed rings. Since such monomers may also lead to branched or crosslinked polymers, espe-cially at higher monomer conversions, these reactions have to be carefully controlled. Molded masses need to be shaped before or during crosslinking.

The first step of such cyclopolycondensations is always a polyaddition to linear oligo-mers, for example, to **polyimides**. For example, pyromellitic dianhydride and 4,4-di-aminodiphenylether react in very polar solvents (*N,N*-dimethylformamide, dimethylacet-amide, tetramethylurea, dimethylsulfoxide) to the so-called polyamic acid (I):

(13-84)

In order to avoid crosslinking, the solids content of solutions is restricted to 10-15 % and the monomer conversion to less than 50 %. The molecular weight is substantially influenced by the way the monomers are added. Number-average molecular weights of up to 55 000 and weight-average molecular weights of up to 240 000 are obtainable.

The second step is the intramolecular ring closure to the imide II by splitting off water at 300°C. A competing step is a diamidation in which the free COOH group of the polyamic acid I reacts with another amine group instead of the amide group, which thus leads to branched and later to crosslinked polymers. This undesired diamidation is favored by a small excess of the diamine whereas the desired imidation benefits from a small excess of the dianhydride. The imidation must proceed to very high conversions of COOH groups since the remaining COOH groups would unfavorably affect the use properties of the polyimide.

However, a large extent of COOH reaction increases the probability of intermolecular reactions which, in turn, leads to crosslinking. Imidation and shaping of the polyimide into the desired product must thus occur simultaneously. For laminating resins, solutions of the polyamic acid are therefore applied to the substrate and then heated, whereupon the solvents and water evaporate and the polyamic acid cyclizes and crosslinks. For polyimide films, one first prepares films of polyamic acid. Since the heating of such films would allow already formed imide and amide groups to be hydrolyzed by water from the cyclization reaction, so-called acceptors are added, for example, acetic anhydride or pyridine. Acetic anhydride reacts with water to acetic acid that evaporates while pyridine forms an azeotrope with water; the azeotrope boils off.

Poly(benzimidazole) (PBI) and poly(benzthiazole) (PBTZ) are prepared in an analogous manner (see also Volume II), for example, PBI:

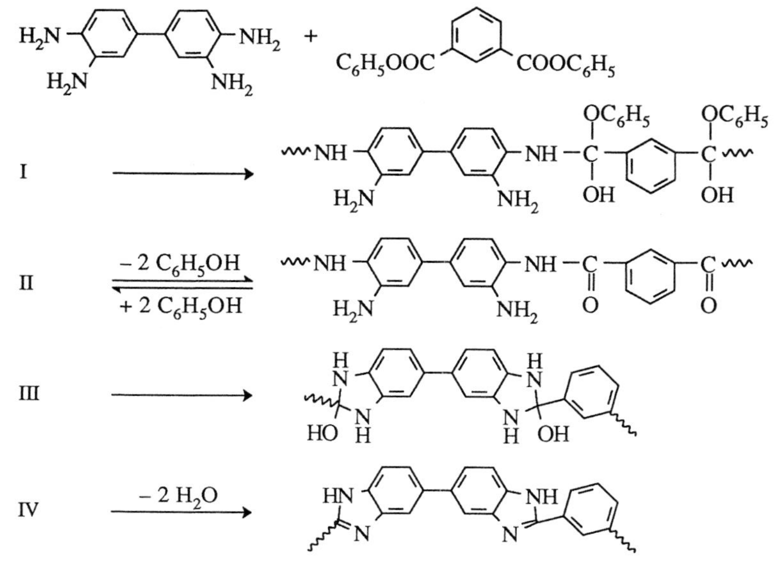

Diphenyl isophthalate, $C_6H_5OOC(i\text{-}C_6H_4)COOC_6H_5$, and the tetrahydrochloride of 3,3'-diaminobenzidine, $(H_2N)_2(C_6H_3)(C_6H_3)(NH_2)_2\cdot4$ HCl, are the industrially preferred monomers. The tetrahydrochloride is used because it is more stable against oxidation than the free diamine. The diphenyl ester is favored because (a) free carboxylic acid groups easily decarboxylate at high reaction temperatures of 250-400°C, (b) acid chlorides react too fast, making ring-closure too difficult, and (c) methyl esters partially methylate amino groups.

The first step I is again a polyaddition to a prepolymer (Eq.(13-85, I)), followed by (II), a reversible splitting-off of phenol. The obtained prepolymer is pulverized, mixed with 2-50 % phenol as plasticizer, and heated to 265-425°C in a nitrogen atmosphere to allow (III) ring-closure, and (IV) elimination of water. Ring closure is never performed to completion because this would lead to intermolecular crosslinking. Industrial poly-(benzimidazole)s thus always contain amino groups.

Polymers with even higher temperature stabilities than poly(benzimidazole)s can be obtained from aromatic tetracarboxylic acids (or their dianhydrides) and tetramines (or their hydrochlorides). These monomers lead to **ladder polymers** (Section 16.3.1) that are stable up to 600°C. Examples are pyrrone and BBB. Some of these polymers possess very stiff chain segments that lead to rodlike behavior and, in strong acids, to lyotropic liquid crystals (Volume III). Nematic phases deliver fibers and films of great stiffness and strength, for example, from poly(*p*-phenylene benzbisoxazole) (PBOX) or poly(*p*-phenylene benzbisthiazole) (PBTZ).

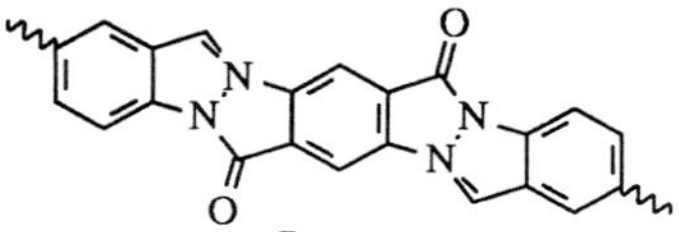

Pyrrone 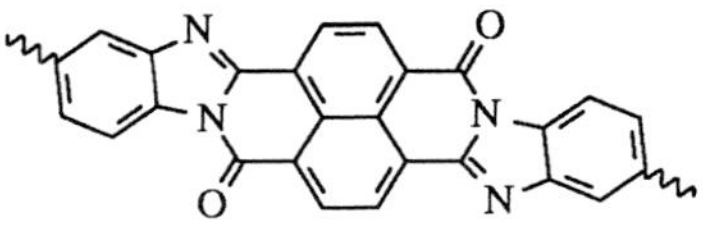   BBB

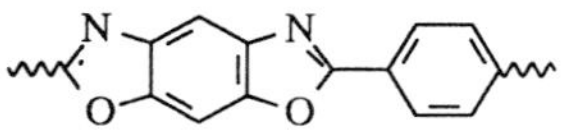

PBOX 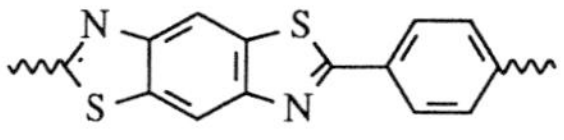   PBTZ

Cyclopolycondensations are also possible by homocyclizations of identical groups. For example, diisocyanates, OCN–Z–NCO, may cyclodimerize to polyuretdiones or cyclotrimerize to polyisocyanurates.

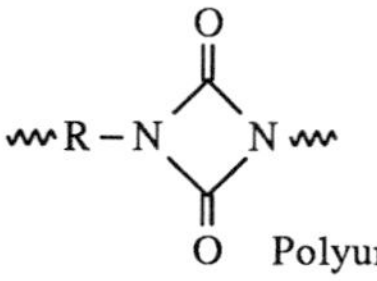

Polyuretdione 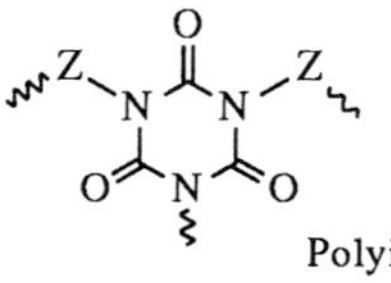   Polyisocyanurate

13.7   Copolycondensation and Copolyaddition

13.7.1   True Copolycondensations and Copolyadditions

True copolycondensations of the AB type proceed between two or more types of monomers such as A'UB + A"VB or A'UB + A"VB + A'WC where A', A", B, and C are leaving groups and U, V, and W monomeric units. By definition, A' and A" react only with B or C but not with themselves; ditto for B and C. Prime ' and double prime " indicate different reactivities of similar groups such as OH in –CH$_2$OH or –CHR–OH.

True copolycondensations of the AA/BB type involve the reaction of two or more types of AA monomers with at least one type of BB monomer. An example of a reaction of the type A'UA' + A"VA" + BWB is the copolyesterification of terephthalic acid, isophthalic acid, and ethylene glycol, $HOOC(p\text{-}C_6H_4)COOH + HOOC(i\text{-}C_6H_4)COOH + HO(CH_2)_2OH$.

Mixed copolycondensations are also used, for example, A'UB' + A"VA" + B"WB". Examples are **polyarylates** (= copolyesters from aromatic dicarboxylic acids + diphenols), for example, from *p*-hydroxybenzoic acid, bisphenol A, and isophthalic acid, $HO(p\text{-}C_6H_4)COOH + HO(p\text{-}C_6H_4)\text{-}C(CH_3)_2\text{-}(p\text{-}C_6H_4)OH + HOOC(i\text{-}C_6H_4)COOH$.

The *extent of reaction* of A'-groups in copolycondensations or copolyadditions of the type A'UA' + A"VA" + BWB is defined as $p_{A'} \equiv [N_{A',0} - N_{A'}]/N_{A',0}$ and that of A"-groups correspondingly as $p_{A''} = [N_{A'',0} - N_{A''}]/N_{A'',0}$. The *mole fraction* of A'-groups is defined as $x' \equiv N_{A'}/(N_{A'} + N_{A''})$ and that of A"-groups as $x'' \equiv N_{A''}/(N_{A'} + N_{A''})$. For stoichiometric reactions, the *extent of reaction* of all A-groups (A' + A") is therefore given by $p_A = x_{A'}p_{A'} + x_{A''}p_{A''} = p_B$.

For example, a polymer molecule by copolycondensation may have the structure

$$A'\text{-}(U\text{-}W)_3\text{-}(V\text{-}W)_2\text{-}(U\text{-}W)\text{-}(V\text{-}W)_4\text{-}(U\text{-}W)_2\text{-}(V\text{-}W)_3\text{-}(U\text{-}W)\text{-}(V\text{-}W)\text{-}B$$

consisting of 2 U-segments U–W, 1 U-segment $(U\text{-}W)_2$, and 1 U-segment $(U\text{-}W)_3$. There is also 1 segment each of V–W, $(V\text{-}W)_2$, $(V\text{-}W)_3$, and $(V\text{-}W)_4$. The number of units per segment is the degree of polymerization X of the *segment* (with respect to the repeating units per segment!).

The mole fraction $x_{U,i}$ of U-units in segments of a certain degree of polymerization, $X_{U,i} = i$, equals their number divided by the total number of U-units in all segments. The mole fraction equals the mass fraction since all U-units have the same mass, resulting in $x_{U,i} = [\sum_i N_{U,i}X_{U,i}]/[\sum_i N_{U,i}] = w_{U,i}$.

The fraction of U-units in *all* fractions of the A-type (i.e., U + V) is defined similarly, i.e., $f_{U,i} = N_{U,i}X_{U,i}/[\sum_i N_{U,i}X_{U,i} + \sum_i N_{V,i}X_{V,i}]$. However, this number fraction equals the mass fraction only if U and V have the same molecular weight.

The mole fraction $f_{U,0}$ of U-units in all units of the AA type is given by $f_{U,0} = N_U/(N_U + N_V)$. It stays constant throughout the polycondensation.

The various types of units are present with the following probabilities:

- Number of possible types that are adjacent to the first chosen unit:
    - segment with symmetry center, for example, A'-U-W-U-A':   *i*
    - segment without symmetry center, since 2 possibilities exist, for example, A'UWUWB and BWUWUA':   2*i*

- probability that a W-unit is followed by some other unit of the AA-type: p

- probability that a U-unit is followed by a W-unit: $p_{A'}$

- probability that a U-unit is *not* followed by a W-unit: $1 - p_{A'}$

- probability that a unit of the AA-type is a unit of the U-type (this probability is therefore $(f_{U,0})^i$ if the segment consists of *i* U-type units): $f_{U,0}$

- probability that a W-unit is followed by a U-unit: $f_{U,0}p_{A'}$

Each sequence of degree of polymerization, X_i, can be realized in six different ways where ~~ indicates any type of segment:

Example for $X_{U,i} = 2$ Probability for $X_{U,i} = i$

A'-U-W-U-A'	i ·	$(p_{A'})^{2i-2}$ ·	$(f_{U,0})^i$ ·	$(1-p_{A'})^2$ ·	
A'-U-W-U-W-B	$2i$ ·	$(p_{A'})^{2i-1}$ ·	$(f_{U,0})^i$ ·	$(1-p_{A'})$ ·	$(1-p)$
A'-U-W-U-W-V~~	$2i$ ·	$(p_{A'})^{2i-1}$ ·	$(f_{U,0})^i$ ·	$(1-p_{A'})$ ·	$(1-x_{U,0}p_{A'})$
B-W-U-W-U-W-B	i ·	$(p_{A'})^{2i}$ ·	$(f_{U,0})^i$ ·	$-$ ·	$(1-p)^2$
B-W-U-W-U-W-V~~	$2i$ ·	$(p_{A'})^{2i}$ ·	$(f_{U,0})^i$ ·	$(1-p)$ ·	$(1-x_{U,0}p_{A'})$
~~V-W-U-W-U-W-V~~	i ·	$(p_{A'})^{2i}$ ·	$(f_{U,0})^i$ ·	$-$	$(1-x_{U,0}p_{A'})$

The summation of these six probabilities leads directly to the distribution of mole fractions of U-units, $f_{U,i}$ in segments of degree of polymerization i with respect to all units of the A-A type:

$$(13\text{-}86) \quad f_{U,i} = f_{U,0} x_{U,i} = f_{U,0} i \, [f_{U,0}(p_{A'})^2]^{i-1} [1 - f_{U,0}(p_{A'})^2]^2$$

The mole fraction $x_{U,i}$ of U-units in U-segments of length i depends on the extent $p_{A'}$ of A'–U–A' molecules but not of that of A"–V–A" molecules.

Many short U-segments are present if the total mole fraction of U-units, $f_{U,0}$, is small (for example, $f_{U,0} = 0.1$ in Fig. 13-24) but their fraction $f_{U,i}$ decreases rapidly with increasing segment length X_i and approaches $f_{U,i} = 0$ at $X_i \approx 5$. The distribution of segment lengths is much broader for high values of $f_{U,0}$; it also runs through a maximum if $f_{U,0} > 0.5$.

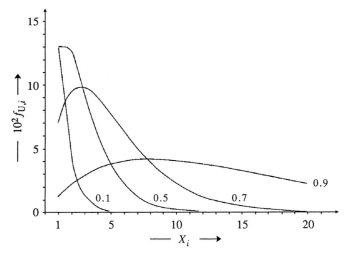

Fig. 13-24 Dependence of mole fractions $f_{U,i}$ of U-units in segments with length i as a function of the degree of polymerization $X_i = i$. Curves were calculated for an extent of reaction of $p_{A'} = 0.99$. Numbers on the curves indicate the fraction $f_{U,0}$ of U-units in all AA-units generated from A'–U–A' and A"-V-A". Calculations according to [21].

At incomplete monomer conversion, segment distributions of U-units and V-units are not only controlled by their relative proportions but also by the rate constants of the reaction of B-groups with A'-groups and A"-groups. The decrease of concentration of free A'-groups with time is given by $-d[\text{A}']/dt = k_{\text{A}'}[\text{A}'][\text{B}]$ and that of A"-groups by $-d[\text{A}"]/dt = k_{\text{A}"}[\text{A}"][\text{B}]$. Division of these rate equations and integration of the result leads to

(13-87) $[\text{A}']/[\text{A}']_0 = \{[\text{A}"]/[\text{A}"]_0\}^r$; $r \equiv k_{\text{A}'}/k_{\text{A}"}$

so that also

(13-88) $p_{\text{A}'} = 1 - (1 - p_{\text{A}"})^r$

The known ratio $k_{\text{A}'}/k_{\text{A}"}$ allows one to calculate values of $p_{\text{A}'}$ from assumed values of $p_{\text{A}"}$ and, with Eq.(13-86), also the total extent of reaction, p. Eq. (13-87) furnishes the mole fractions $f_{\text{U},i}$ of U-units in segments of degree of polymerization i.

Initially, at small degrees of polymerization, polymer molecules contain almost exclusively monomeric units of the faster reacting monomer, i.e., A"–V–A" in Fig. 13-25. With increasing number-average degree of polymerization, mole fractions $f_{\text{V},i}$ of V-diads and V-triads pass through a small maximum and then become constant. No maximum is seen for i-ads of 5 and greater. The corresponding i-ads from the more slowly reacting A'–U–A' monomers approach their final values asymptotically. At high degrees of polymerization (at almost complete monomer conversion), both mole fractions become equal if $f_{\text{U},0}/f_{\text{V},0} = 1$ (as in Fig. 13-25), otherwise, they depend on this ratio. The distribution of units becomes independent of the reactivity of monomers; it follows the Schulz-Flory distribution. This statement applies to initial mixtures of monomers but not to reactions where one monomer is added successively (next section).

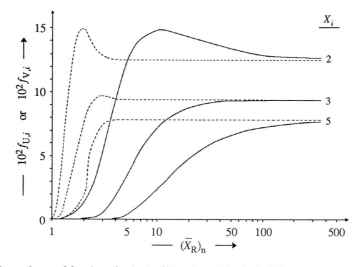

Fig. 13-25 Dependence of fractions $f_{\text{U},i}$ (—) of U-units and $f_{\text{V},i}$ (- -) of V-units, respectively, in *i*-ads of degree of polymerizations X_i on the number-average degree of polymerization of the reactant, $(\bar{X}_{\text{R}})_{\text{n}} = 1/(1 - p)$ [21]. Calculations for $f_{\text{U},0} = f_{\text{V},0} = 1/2$ and $k_{\text{A}'}/k_{\text{A}"} = 0.1$.

13.7.2   Polycondensation of Non-Symmetric Monomers

AA/BB polycondensations of the type A–V–A + B–W–B are not considered copoly-condensations if they lead to alternating polymers A(V–W)$_n$B. However, regioisomerism may arise in polymers if the two A-groups are only formally identical (see below). This case may be treated as copolycondensation.

Such regioisomerisms may even appear in the self-condensation of non-symmetric monomers A–V–V'–A'. The most important case is the copolycondensation of a non-symmetric monomer A–V–V'–A' with a symmetric monomer B–W–W–B, for example:

(13-89) n HOOC–Ar–(CH$_2$)$_3$–COOH + n H$_2$N–CH$_2$–CH$_2$–NH$_2$ $\longrightarrow$

 $+$OC–Ar–(CH$_2$)$_3$–CO–NH–CH$_2$–CH$_2$–NH$\frac{}{n}$ + 2 n – 1 H$_2$O

leading to a polymer with four constitutional repeating units:

—(CH$_2$)$_3$—CO—NH—CH$_2$—CH$_2$—NH—CO—(CH$_2$)$_3$—   –V'–W–W–V'–
—(CH$_2$)$_3$—CO—NH—CH$_2$—CH$_2$—NH—CO—Ar—   –V'–W–W–V–⎫
—Ar—CO—NH—CH$_2$—CH$_2$—NH—CO—(CH$_2$)$_3$—   –V–W–W–V'–⎬ –V–W–W–V*–
—Ar—CO—NH—CH$_2$—CH$_2$—NH—CO—Ar—   –V–W–W–V–⎭

The units V'–W–W–V and V–W–W–V' can usually not be distinguished from each other; they are therefore designated V–W–W–V*.

In stoichiometric equilibrium polycondensations with high monomer conversion, the three randomly distributed constitutional repeating units are present in the ratio 1:2:1. In kinetically controlled polycondensations, the ratio and distribution of constitutionally repeating units depends on the relative reactivities of groups A and A', the monomer conversion, and the manner in which the monomers are added: both monomers completely present, one monomer completely present and the other one added slowly, etc. For example, the very slow addition of 2-methyl-p-phenylene diamine to terephthaloyl dichloride in a 10:1 mixture of N-methylpyrrolidone and tetrahydrofuran led to random copolymers. On the other hand, addition of the diacid dichloride to the diamine gave copolymers with highly ordered sequences.

The chemical structure of such polymers can be described by a symmetry factor s that indicates the mole fraction of non-symmetric constitutional repeating units (CRUs):

(13-90) $$s = \frac{[V-W-W-V*]}{[V-W-W-V]+[V-W-W-V*]+[V'-W-W-V']}$$

This factor becomes $s = 1$ if only head-to-tail structures are present, $s = 0$ if head-to-head structures alternate with tail-to-tail ones, and $s = 1/2$ if CRUs are randomly distributed. Note that s decribes the structure of regular AA/BB polymers from one symmetric and one nonsymmetric monomer. Two additional factors are required if both monomers of AA/BB polymerizations are nonsymmetric, i.e., A-V-V-A' and B-W-W-B'.

The number-average degree of polymerization is given by the ratio of the numbers of monomeric units and molecules or one-half of the number of endgroups (A, A', B):

(13-91) $$\overline{X}_n = \frac{N_{V\text{-}V} + N_{W\text{-}W}}{(1/2)(N_A + N_{A'} + N_B)}$$

The following relationships apply if both monomers are initially present altogether and in stoichiometric amounts: $N_{W-W} = N_{W-W,0} = N_{[V-V],0}$ for monomeric units (where [V–V] indicates the sum of V–V, V–V*, and V'–V') and $N_{A,0} = N_{A',0}$ and $N_{B,0} = 2 N_{A,0}$ for endgroups. Introduction of the amount fractions x with respect to the initially present numbers of each chain end ($N_A = x_A N_{A,0}$, etc.) delivers with $2 x_B = x_A + x_{A'}$ (constancy of material balance) for the number-average degree of polymerization

(13-92) $\overline{X}_n = 4/(x_A + x_{A'} + 2 x_B) = 2/(x_A + x_{A'})$

B-groups react with A-groups with a rate constant k_A and with A'-groups with a rate constant of $k_{A'}$. The larger the proportion of reacted A-groups with respect to the initially present ones, the smaller the proportion of reacted A'-groups and *vice versa*. The mole fractions of groups remaining after a certain time are therefore related via

(13-93) $(x_A)^{k_{A'}} = (x_{A'})^{k_A} \quad \rightarrow \quad x_{A'} = x_A^r \quad ; \quad r \equiv k_{A'}/k_A < 1$

From Eqs.(13-92) and (13-93), one obtains for the degree of polymerization

(13-94) $\overline{X}_n = 2/(x_A + x_A^r)$

The symmetry factor s is calculated from the probability of the reaction of an A-group, $1 - x_A$, and that of an A'-group, $1 - x_{A'}$. The various concentrations in Eq.(13-90) are proportional to the corresponding probabilities so that

[V–W–W–V] ~ $(1 - x_A)^2$,   [V'–W–W–V'] ~ $(1 - x_{A'})^2$,   [V–W–W–V*] ~ $2(1- x_A)(1 - x_{A'})$.

From Eqs.(13-90) and (13-93) one gets therefore for the symmetry factor for sufficiently high degrees of polymerization (ignoring chain ends ~W–B):

(13-95) $s = 2 [1 - x_A][1 - x_A^r]/[2 - x_A - x_A^r]^2$

A plot of $s = f(1/\overline{X}_n)$ shows that the symmetry factor s varies only between 0.48 and 0.50 for number-average degrees of polymerization greater than 10 (Fig. 13-26). The distribution of the orientation of the monomeric units of such polymers is thus always random and independent of the relative magnitude of the rate constants, provided that both monomers are initially completely present.

However, if the symmetric monomer is added "infinitely slowly" to the completely present nonsymmetric monomer, then the mole fraction x_B of the endgroups of the symmetric monomer is vanishingly small. Eq.(13-94) then has to be replaced by

(13-96) $\overline{X}_n = [4/(x_A + x_A^r)] - 1$

The *instantaneous* symmetry factor Δs for infinitely small conversion steps is obtained from the rate equations for the formation of the various constitutional repeating units as

(13-97) $\Delta s = 2 r x_A^{1+r}/[x_A + r x_A^r]^2$

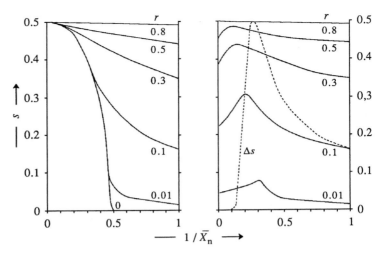

Fig. 13-26 Dependence of the symmetry factor s on the inverse number-average degree of polymerization of polymers from the copolycondensation of a nonsymmetric monomer A–V–V–A' and a symmetric monomer B–W–W–B [22] (reaction from right to left). Numbers indicate ratios $r = k_{A'}/k_A$ of
rate constants of reactions of B + A' and B + A, respectively.
    Left: both monomers completely present at the start of reaction. Right: infinitely slow addition of
B-W-W-B to A-V-V-A' and the reaction mixture. - - - Instantaneous symmetry factor Δs for $r = 0.1$.

The symmetry factor s for all molecules formed between $x_{A,0} = 1$ and x_A leads in this
case to an integral that cannot be solved in a closed form:

$$(13\text{-}98) \qquad s = \frac{2\,r}{2 - x_A - x_A^r} \int_1^{x_A} \frac{dX}{r + X^{1-r}}$$

Symmetry factors can be obtained numerically, however. According to Eq.(13-97),
symmetry factors are $\Delta s = s = 2\,r/(1 - r)^2$ at time zero where $1/\overline{X}_n = 1$ and $x_A = x_{a,0} = 1$.
The more reactive A-groups are consumed faster with increasing monomer conversion
so that the less reactive A'-groups react proportionally more often. The instantaneous
symmetry factor Δs increases strongly (Fig. 13-26, right). The symmetry factor s increases less since it represents an integral value. At a certain monomer conversion (corresponding to $1/\overline{X}_n \approx 0.25$ for $r = 0.1$ in Fig. 13-26), practically all A-groups are consumed. Only A'-groups react and Δs drops sharply. The maximum of the function $s =$
$f(1/\overline{X}_n)$ is at $1/\overline{X}_n \approx 0.1$ (i.e., $\overline{X}_n \approx 10$) for $r = 0.5$.

13.8 Hyperbranching Reactions

Branching polycondensations and polyadditions are step polymerizations in which at
least one of the monomers, the **branching monomer**, has a *total* functionality of $f \geq 3$.
There are two main groups: those with only one type of monomer (Section 13.8.1) and
those with two types of monomers (Section 13.8.2). Branching may also be caused by
side reactions in linear systems (Section 13.8.3).

Several mechanisms may be employed for the synthesis of hyperbranched polymers. Most common are step-growth reactions, i.e., polycondensations and polyadditions. Hyperbranched polymers can also be produced by ring-opening multibranching polymerizations (ROMBP) and self-condensing vinyl polymerizations (SCVP). With respect to the development of hyperbranched structures, the latter two types of syntheses are analogous to those from step-growth reactions although the start and propagation mechanisms are quite different.

Theoretical treatments of the change of monomer conversion with time or degree of polymerization with time or extent of reaction in AB_f, $AA + B_f$, etc., reactions usually assume (1) A groups can only react with B groups but neither A with A nor B with B, (2) groups of the same type have the same reactivity, (3) groups react independently of other groups, and (4) intramolecular reactions are absent, except for the limit $p \rightarrow 1$.

13.8.1 AB_b Type Reactions

The simplest non-linear polycondensations and polyadditions comprise **unitary reactions** of A_aB_b molecules of total functionality $f = a + b$. Unitary branching reactions comprise two subgroups. Self-reactions of A_aB_b molecules with $a = 1$ and $b \geq 2$ lead to **hyperbranching** but never to crosslinking, even at the highest extents of reaction at $p_A \rightarrow 1$ (see below). At $p_A \equiv 1$ (which is thermodynamically impossible), one cyclic hyperbranched molecule would result. Reactions of the fairly rare A_aB_b molecules with $a \geq 2$ and $b \geq 2$ also deliver hyperbranched molecules albeit only up to a certain extent of reaction. Beyond this critical extent of reaction, crosslinking starts (Section 13.9).

This section only discusses AB_b-type reactions of molecules with $b \geq 2$. Such reactions deliver frustrated dendrons (see Fig. 2-11) since the resulting hyperbranched molecules contain dendritic, linear, and terminal units and either a monocore or a dicore unit. An example is the AB_2 selfcondensation of 3,5-diacetoxybenzoic acid ($b = 2$) which proceeds with the release of acetic acid (Fig.13-27). AB_b-type reactions also comprise reactions of the type $AB_b + A'B'$ where $A'B'$ molecules simply act as extenders between branching units. An example is the addition of $CH_3COO(1,4-C_6H_4)COOH$ (acetoxyparaben) to 3,5-acetoxybenzoic acid, $(CH_3COO)_2(C_6H_3)COOH$.

The dependence of the number-average degree of polymerization, $\overline{X}_{R,n}$, of the *reactant* in *equilibrium* reactions (1-pot reactions) is calculated as follows for AB_b polymerizations. All *reactant* molecules (monomer, dimer, trimer, ...) with $f \geq 2$ contain only one A-group but $N_B = 1 + (b-1)X_i$ unreacted B-groups per molecule, regardless of whether the molecule is dendritic, hyperbranched, or linear (cf. Fig. 2-11). For example, a polymer molecule from AB_2 with $X_i = 12$ always contains 13 B-groups.

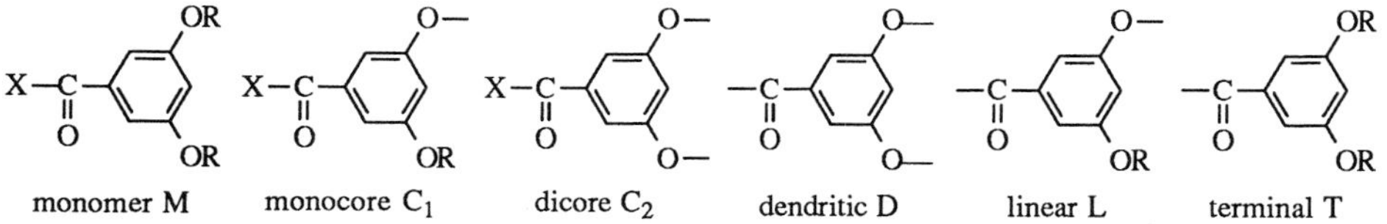

monomer M   monocore C_1 dicore C_2 dendritic D   linear L   terminal T

Fig. 13-27 Monomeric units in hyperbranched polymers from a monomer $XCO(C_6H_3)(OR)_2$, for example, 3,5-diacetoxybenzoic acid with $R = CH_3CO$ and $X = OH$ or 3,5-bis(trimethylsiloxy)benzoyl chloride with $R = (CH_3)_3SiO$ and $X = Cl$.

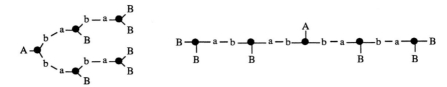

Fig. 13-28 A molecule consisting of 1 dicore unit Abb, 2 linear units abB, and 2 terminal units aBB. The molecule is "linear" in the meaning of polymer chemistry (B groups are side groups in linear units abB and endgroups aBB) but has 4 branch points according to probability theory. Depending on the definition of "branching," different "degrees of branching" can thus be defined (Eqs.(2-6)-(2-8)).

Probability theory defines the reaction of a B-group of AB_b as a branching reaction that leads to a branch point; it thus treats core units AbB and Abb, linear units abB, and dendritic units abb as branching units but not terminal units aBB (see Fig. 13-28). The extent of reaction of B-groups, $p_B \equiv (n_{B,0} - n_B)/n_{B,0}$, thus equals the probability α of a branching point ($p_B = \alpha$).

There are b times more B groups than A groups in AB_b molecules. The extent of reaction of B-groups is thus $1/b$ times the extent of reaction of A groups, $p_B = (1/b)p_A$. The number-average degree of polymerization of *reactant molecules* is controlled by the extent of reaction of the lesser functional group, i.e., the A-group in AB_b. Thus

$$(13\text{-}99) \qquad \overline{X}_{R,n} = \frac{1}{1 - p_A} = \frac{1}{1 - bp_B}$$

Eq.(13-99) applies to all functionalities b. For $p_A \to 1$, it predicts that $p_B \to 1/b = \alpha_{crit}$ and $(\overline{X}_R)_n \to \infty$. The number-average degree of polymerization thus approaches infinity at $p_B \to 1$ for AB, at $p_B \to 1/2$ for AB_2, at $p_B \to 1/3$ for AB_3, etc. (Fig. 13-29). However, p_A can never attain unity in a thermodynamic equilibrium. If it could, the polymer would be *one* giant *cyclic* hyperbranched molecule because no A-group is left. But it is still not a crosslinked one in the common meaning of the term.

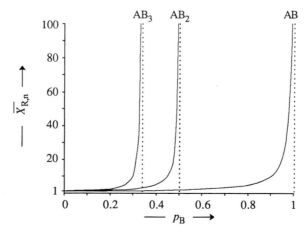

Fig. 13-29 Dependence of the number-average degree of polymerization of reactants on the extent of reaction of B-groups in the hyperbranching reaction of AB_b. $\overline{X}_{R,n}$ approaches infinity for $bp_B \to 1$.

The calculation of the weight-average degree of polymerization is much more complicated. Flory (see reference in "Literature 13.8") obtained it by a lengthy calculation for reactions without leaving molecules (i.e., $M_L/M_U \rightarrow 0$) as

$$(13\text{-}100) \qquad \overline{X}_{R,w} = \frac{1-(p_A^2/b)}{(1-p_A)^2} = \frac{1-p_B^2 b}{(1-p_B b)^2}$$

Division of Eq.(13-100) by Eq.(13-99) delivers the polymolecularity index:

$$(13\text{-}101) \qquad \frac{\overline{X}_{R,w}}{\overline{X}_{R,n}} = \frac{1-(p_A^2/b)}{1-p_A} = \frac{1-p_B^2 b}{1-p_B b}$$

For example, for an extent of reaction of $p_A = 0.99$ of monomer AB_2, one obtains $\overline{X}_{R,n} = 100$, $\overline{X}_{R,w} = 5099.5$, and $\overline{X}_w/\overline{X}_n \approx 51$. On approaching complete conversion of A-groups, not only do $\overline{X}_w$ and $\overline{X}_n$ approach infinity individually but so also does their ratio. The differential distribution becomes infinitely broad and infinitely low.

Comprehensive experimental data for the variation of composition and molecular weight averages as a function of time and extent of reaction do not seem to exist for polycondensations. However, there are some scattered data for polycondensations as well as some other hyperbranching polymerizations which collectively attest to the validity of some of the theoretical predications as well as the experimental problems.

Polycondensation

For example, weight-average and number-average degrees of polymerization increase practically linearly with time in the polycondensation of the AB_2 monomer 3,5-diacetoxybenzoic acid after a sluggish beginning (Fig. 13-30). Such kinetic behavior may indicate a large difference between the rate constants for the monomer + monomer reaction on one hand and that of polymer + monomer and/or polymer + polymer on the other (see also Fig. 13-14).

Contrary to theoretical predictions, the polymolecularity index is small ($\overline{M}_w/\overline{M}_n \approx$ 2.7 at $\overline{M}_w \approx 18\ 000$). It also increases only slightly with increasing molecular size (Fig. 13-30, insert) but this may have to do with cyclization and/or the vagueness of molecular weight determinations. For both compositional and architectural reasons, size-exclusion chromatography with universal calibration is not appropriate for these systems.

The same discrepancy between theory and experiment is seen for polymers from the polycondensation of $NaO(C_6H_3)(Z-C_6R_4)_2F$ (Fig. 13-31) where either $R = H$ and $Z = CO$ or $R = F$ and no Z. These experiments showed that the resulting molecular weights and polymolecularity indices increase with the initial monomer concentration, strongly for the $R = F$, $Z = $ nil series and less strongly for the $R = H$, $Z = CO$ series.

Again, polymolecularity indices are much lower than demanded by theory. For the highest number-average molecular weight of 35 500 ($\overline{X}_n \approx 92$), Eq.(13-99) delivers an extent of reaction of $p_B = 0.4946$ (theory for infinite molecular weights: $p_B \rightarrow 1/2$) that should lead to a polymolecularity index of $\overline{X}_w/\overline{X}_n = 47.3$. Experimental data indicate a polymolecularity index of only $\overline{M}_w/\overline{M}_n = 3.77$, however. Again, polymolecularities by SEC are much lower than demanded by theory.

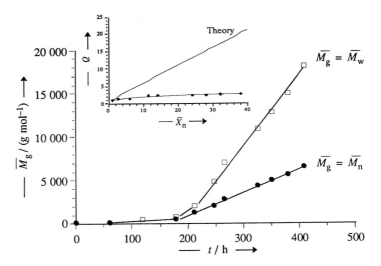

Fig. 13-30 Time dependence of mass-average and number-average molar mass during the polycondensation of 3,5-diacetoxybenzoic acid at 195°C [23]. $\overline{M}_w$ and $\overline{M}_n$ were determined by size-exclusion chromatography coupled with low-angle laser light scattering and dilute solution viscometry.
These measurements were restricted to the low-molecular weight region although polymers with molecular weights of up to one million were also reported. It is not clear whether the molecular weights refer to those of reactants (polymer molecules + monomer molecules, $X \geq 1$) or only to polymer molecules ($X \geq 2$) or fractions thereof.
Insert: Polymolecularity index $Q = \overline{X}_w / \overline{X}_n$ as a function of the number-average degree of polymerization $\overline{X}_n$ (calculated average of the extremal values of linear and almost dendritic compositions).

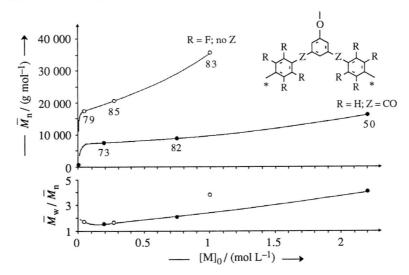

Fig. 13-31 Top: Number-average molecular weight and polymolecularity index of hyperbranched polymers with the indicated dendritic unit (O: R = F, no Z; ●: R = H, Z = CO) as a function of the initial monomer concentration [24]. One * is replaced by F in linear molecules and both * by F in terminal units. Numbers indicate the polymer yield in percent. Molecular weights were obtained from size-exclusion chromatography using poly(styrene) standards. Bottom: Polymolecularity index of both types of polymers as a function of the initial monomer concentration.

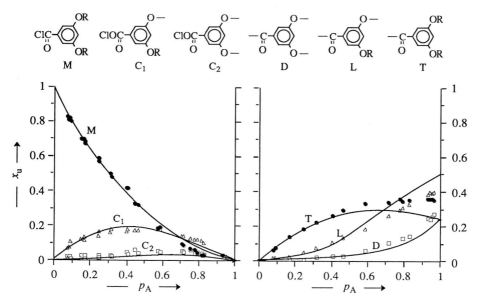

Fig. 13-32 Mole fractions x_u of monomer molecules (u = M) and monomeric units (C$_1$ = monocore; C$_2$ = dicore; L = linear; D = dendritic; T = terminal) as a function of the extent of reaction p_A of the COCl group during the polycondensation of M (where R = Si(CH$_3$)$_3$) [25]. Data points are experimental (^{13}C NMR); lines were calculated from simulation, using 1000 iterations.

The proportions of the various types of monomeric units follow theoretical models if the reactive sites are well separated as in, for example, (HOC$_6$H$_4$)$_2$CH(CH$_2$)$_3$COOH. Steric and electronic nearest neighbor effects come into play in irreversible reactions such as the chloride ion catalyzed polycondensation of ClOCC$_6$H$_3$(OR)$_2$ where R = Si(CH$_3$)$_3$ (Fig. 13-32). This reaction proceeds in three steps: two nucleophilic substitution reactions are followed by the rate-determining elimination of Cl$^\ominus$:

(13-102)

$$\text{\textcircled{·}}-O-SiR_3 \underset{-\,ClSiR_3}{\overset{+\,Cl^\ominus}{\rightleftharpoons}} \text{\textcircled{·}}-O^\ominus \underset{}{\overset{+\,ClOC-\square}{\rightleftharpoons}} \text{\textcircled{·}}-O-\overset{\overset{Cl}{|}}{\underset{\underset{O^\ominus}{|}}{C}}-\square \overset{-\,Cl^\ominus}{\longrightarrow} \text{\textcircled{·}}-O-\overset{}{\underset{\underset{O}{\|}}{C}}-\square$$

The results show that at medium extents of reaction of ca. 0.4–0.5 of the lesser functional group (ClOC–) relatively more dicore units than monocore units are formed than in an ideal situation (Fig. 13-32, left). In turn, this leads at higher extents of reaction of ca. 0.7–0.9 to a relative preponderance of terminal and dendritic units compared to linear ones (Fig. 13-32, right).

Lack of suitable experimental methods so far has prevented the detection and determination of intramolecular cyclization during hyperbranching reactions. Cyclization reduces the number of functional groups per molecule that are available for further hyperbranching reactions as shown for a dimer in Eq.(13-103):

(13-103)

$$B-Y\overset{A}{\underset{B}{\diagdown}}\quad+\quad\overset{B}{\underset{A}{\diagup}}Y-B\quad\longrightarrow\quad B-Y\overset{a-b}{\underset{B\ \ A}{\diagdown}}Y-B\quad\longrightarrow\quad B-Y\overset{b-a}{\underset{a-b}{\diagdown}}Y-B$$

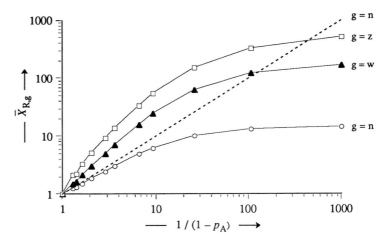

Fig. 13-33 Logarithm of degrees of polymerization of reactants, $\overline{X}_{R,g}$, as a function of the logarithm of the Carothers parameter, $(1 - p_A)^{-1}$, for AB_2 polycondensations [26]. The dotted line indicates the Carothers function for linear polycondensations and polyadditions, $\overline{X}_{R,n} = 1/(1 - p_A)$.

The same computer model that indicated noticeable ring formation in linear polycondensations (Fig. 13-20) also showed that intramolecular ring formation prevents the development of higher-molecular weight molecules in AB_2 step reactions (Fig. 13-33). The various g-average degrees of polymerization (g = n, w, z) become constant at infinitely high degrees of polymerization. At $p_A = 2p_B = 0.98$, the polymolecularity index was lowered to $\overline{X}_{R,w}/\overline{X}_{R,n} \approx 8.9$ (Fig. 13-33) from $\overline{X}_{R,w}/\overline{X}_{R,n} \approx 50$ for AB_2 reactions without intramolecular cyclization (p. 483).

Functions $\overline{X}_{R,g} = f[(1 - p_A)^{-1}]$ for step reactions with intramolecular cyclization are not universal because they depend on the probability density of reaction (p. 215) and thus on molecular parameters. However, a simple relationship exists between the number $N_{R,0}$ of monomer molecules at time $t = 0$, the number N_R of reactant molecules at time $t > 0$ (monomer molecules + linear polymer molecules + branched polymer molecules + cyclic molecules), the number N_{bd} of bonds formed after time t, and the number N_c of resulting cyclic molecules at time t.

In reactions without cyclizations, formation of one bond reduces the number of reactant molecules by one, thus $N_{R,0} = N_R + N_{bd}$. Cyclization increases the number of bonds by one but does not alter the number of molecules, thus

$$(13\text{-}104) \quad N_{R,0} = N_R + N_{bd} - N_c$$

which, after rearranging and introducing the extent of reaction, $p_A = N_{bd}/N_{R,0}$, and the number-average degree of polymerization, $\overline{X}_{R,n} = N_{R,0}/N_R$, becomes

$$(13\text{-}105) \quad 1/\overline{X}_{R,n} = N_R/N_{R,0} = 1 - p_A + (N_c/N_{R,0})$$

In the limit of complete reaction of the lesser type of functional groups ($p_A \to 1$), the number-average degree of polymerization becomes $\overline{X}_{R,n} = N_{R,0}/N_c$. Cyclization thus controls the molecular weight (Fig. 13-33).

Self-Condensing Vinyl Polymerization

The group of self-condensing vinyl polymerizations (**SCVP**) comprises the polymerization of AB* monomers such as p-(chloromethyl)styrene, $CH_2=CH(1,4-C_6H_4)CH_2Cl$, where the A group, $CH_2=CH(1,4-C_6H_4)–$, contains a polymerizable double bond and the B* group, $–CH_2Cl$, is a latent initiator moiety. The polymerization is initiated by an activated species that may be a "living" free radical, a carbanion, or an electrophilic cationic species, for example, $SnCl_4$, in molar ratios of $[SnCl_4]/[monomer] \approx 1$. The initially formed carbocation is converted into a radical that adds a new monomer molecule:

(13-106)

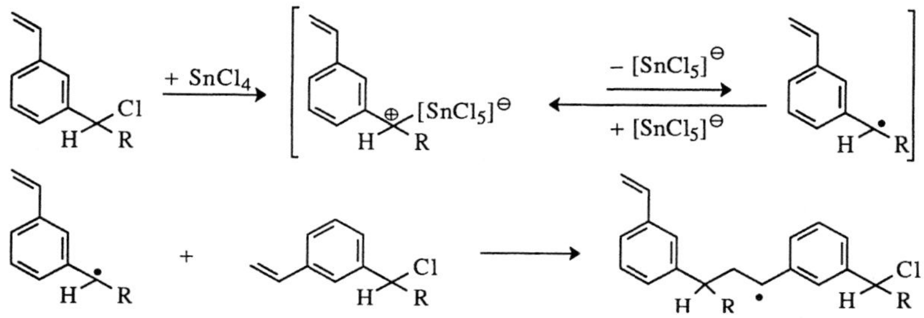

The resulting dimer, symbolized by A–b–a*–B*, now contains two active sites and a vinyl double bond. Reaction of the dimer with another monomer molecule A–B* can thus deliver two different trimers: a linear one from the addition to an A or terminal B* group and a branched one from the addition to a side group B* or a center group "a":

(13-107)

The extent of branching increases with increasing molecular weight and so does the polymolecularity index, $\overline{M}_w/\overline{M}_n$ (Table 13-9). Assuming equal reactivities of A and B* sites, kinetic analysis predicts that the number-average and weight-average degrees of polymerization of *reactants*, $\overline{X}_{R,n}$, are both exponential functions of the time parameter, $\tau \equiv k[M]_0 t$, where k = rate constant, $[M]_0$ = initial monomer concentration, and t = time (Table 13-10). The predicted dependences of $\lg M = f(t)$ were indeed found (Fig. 13-34) albeit with too high intercepts that may be caused by the work-up ($\overline{M}_{R,w} \neq \overline{M}_w$).

Table 13-9 Results of the self-condensing vinyl polymerization of m-(chloromethyl)styrene (CMS) in CH_2Cl_2 initiated by $[(C_4H_9)_4N]^{\ominus} Br^{\ominus} + SnCl_4$ at $-15°C$ to $-20°C$ [27]. Molecular weights by SEC with either poly(styrene) standards (PS) or universal calibration (UC). U = unfractionated, F = fraction.

No.	$\dfrac{[CMS]}{mol\ L^{-1}}$	$\dfrac{[SnCl_4]}{[CMS]}$	$\dfrac{t}{h}$	$\dfrac{yield}{\%}$	$\dfrac{\overline{M}_n}{PS}$	$\dfrac{\overline{M}_n}{UC}$	$\dfrac{\overline{M}_w}{UC}$	$\dfrac{\overline{M}_w/\overline{M}_n}{UC}$
6 U	0.25	3.1	8	62		3 600	17 000	4.7
5 U	0.33	2.2	8	88		3 400	17 000	4.9
1 U	0.35	0.3	24	89		4 900	33 000	6.6
2 F	0.53	1.1	8	80	33 000	46 000	130 000	2.9
3 F	0.53	1.1	10	89	32 000	50 000	250 000	5.1
4 F	0.53	1.1	12	80	47 000	68 000	660 000	9.8

Table 13-10 Theoretical predictions for number-average molecular weights ($\overline{X}_{R,n}, \overline{X}_n$) and weight-average molecular weights ($\overline{X}_{R,w}, \overline{X}_w$) of reactants ($\overline{X}_{R,n}, \overline{X}_{R,w}$) and polymers ($\overline{X}_n, \overline{X}_w$) and the corresponding polymolecularity indices ($\overline{X}_{R,w}/\overline{X}_{R,n}, \overline{X}_w/\overline{X}_n$) of reactants ($X_i \geq 1$; index R) and polymers ($X_i \geq 2$) as a function of the conversion $u_A = ([M]_0 - [A])/[M]_0$ of double bonds in SCVP [28a]. $[M]_0$ = initial monomer concentration; $[A]$ = concentration of A-groups at time t; $\tau = k[M]_0 t$.

Equations for reactants	Equations for polymers

$$\overline{X}_{R,n} = \frac{1}{1-u_A} = \exp(\tau)$$

$$\overline{X}_{R,w} = \frac{1}{(1-u_A)^2} = \exp(2\tau)$$

$$\frac{\overline{X}_{R,w}}{\overline{X}_{R,n}} = \frac{1}{1-u_A} = \exp(\tau) = \overline{X}_{R,n}$$

$$\overline{X}_n = \frac{1-(1-u_A)\exp(-u_A)}{(1-u_A)[1-\exp(-u_A)]}$$

$$\overline{X}_w = \frac{1-(1-u_A)^3\exp(-u_A)}{(1-u_A)^2[1-(1-u_A)\exp(-u_A)]}$$

$$\frac{\overline{X}_w}{\overline{X}_n} = \frac{(1-\exp(-u_A)[1-(1-u_A)^3\exp(-u_A)]}{(1-u_A)[1-(1-u_A)\exp(-u_A)]^2}$$

Kinetic analysis showed theoretically for batch processes that molecular weight distributions from SCVPs become extremely broad with increasing group conversion. Like conventional step reactions, a reactant molecule can react with any other reactant molecule. Unlike bifunctional step reactions, functionalities of reactants are not constant, however, but rather proportional to the degree of polymerization. The larger molecules grow faster than the smaller ones: the polymolecularity index equals the number-average degree of polymerization of the reactants.

Kinetic analysis also indicates that molecular weight distributions become even more broad for SCVPs with unequal reactivities of A and B* ($\overline{X}_{R,w}/\overline{X}_{R,n} > \overline{X}_{R,n}$). The expression $\overline{X}_{R,n} = 1/(1-u_A)$ for the number-average degree of polymerization of reactants is maintained but no analytical solution for the weight-average and the distribution of degrees of polymerization was found.

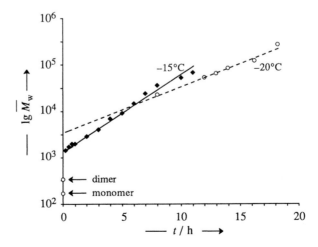

Fig. 13-34 Dependence of the logarithmus of the experimental weight-average molecular weight on time at two polymerization temperatures. At $t = 0$, experimental lines do not approach the molecular weight of the monomer, probably because of a loss of monomer and oligomer molecules during work-up. Reprinted with permission from [28b]. Copyright (1997) American Chemical Society.

Ring-Opening Multibranching Polymerization

Hyperbranched polymers can also be obtained by ring-opening multibranching poly-merizations (ROMBP) of monomers with a reactive cyclic group (i.e., a latent double bond) and an additional functional group. An example is the anionic polymerization of glycidol (Table 13-11), which leads to four different units:

| glycidol | linear 1,3 | linear 1,4 | dendritic | terminal |

The highest number-average degree of polymerization corresponds to a number-average degree of polymerization of ca. 90. Polymolecularity indices were fairly low be-cause the polymerizations were performed by slow addition of glycidol to a partially de-protonated initiator, 1,1,1-tris(hydroxymethyl)propane, and not as a batch operation.

Table 13-11 Number-average molecular weights, polymolecularity indices, and mole fractions of the four possible monomeric units from the anionic polymerization of glycidol by a 10 % deprotonated 1,1,1-tris(hydroxymethyl)propane as monofunctional core unit [29]. x_{br} = degree of branching.

Property	Method	PGly-1	PGly-2	PGly-3	PGly-4
$\overline{M}_n$	MALDI-TOF	1298	1571	3583	
	^{13}C NMR	1245	1779	3423	6314
	vapor-phase osmometry	1078	1532	3491	6494
$\overline{M}_w/\overline{M}_n$	mass spectroscopy	1.15	1.19	1.15	
	size-exclusion chromatography	1.18	1.23	1.13	1.47
$x_{terminal}/\%$	NMR	41	36	33	32
$x_{linear\ 1,3}/\%$	NMR	10	11	11	12
$x_{linear\ 1,4}/\%$	NMR	28	29	29	28
$x_{dendritic}/\%$	NMR	21	24	26	28
x_{br} [29]	$= 2\,x_{dendritic}/(2\,x_{dendritic} + x_{linear})$	0.525	0.545	0.565	0.583
x_{br} (Eq.(2-5)	$x_{dendritic}/(x_{dendritic} + x_{linear})$	0.356	0.375	0.394	0.412

13.8.2 $A_a + B_b$ Type Reactions

This group of reactions comprises **binary reactions** of A_a with B_b molecules with functionalities $a \geq 2$ and $b \geq 2$ where at least one of the functionalities is greater than 2. An example is the reaction of a dicarboxylic acid with a tetrol such as the polycondensa-tion of adipic acid, $HOOC(CH_2)_4COOH$, with pentaerythritol, $C(CH_2OH)_4$.

Such $A_a + B_b$ step reactions are performed with initial stoichiometric amounts of *groups* A and B; they deliver insoluble polymers in addition to soluble ones if the reac-tion is driven beyond the so-called critical extent of reaction, a quantity whose minimum value can be calculated theoretically by probability theory (Section 13.9). Only hyper-branched molecules result if the reaction is stopped just before the gel point.

As shown in Section 13.9, gelation does not occur in a system $A_a + B_b$ as long as the condition $r_0 p^2 \leq [(a-1)(b-1)]^{-1}$ is fulfilled. Expressing the initial molar ratio of re-acting *groups*, r_0, by $r_0 = an_{A,0}/(bn_{B,0}) \leq 1$ where a and b are the functionalities and

$n_{A,0}$ and $n_{B,0}$ are the initial amounts (in moles) of *molecules* A_a and B_b, respectively, one obtains for the extent of reaction $p = p_A$ of the minority component A_a:

$$(13\text{-}108) \quad p_A \leq \left(\frac{b}{a(a-1)(b-1)} \frac{n_{B,0}}{n_{A,0}} \right)^{1/2}$$

Table 13-12 lists initial amounts required for a complete reaction of all groups ($p = 1$) and the theoretical critical values of p for initial stoichiometries of groups and molecules, respectively. Experimental critical values of p_A are usually higher because of cyclization (Section 13.9); they depend on molar ratios and on molar concentrations.

Weight-average degrees of polymerization approach infinity at the gel point but number-average degrees are quite small (Table 13-12 and Section 13.9). They were calculated with the help of Eq.(13-109) from the initial amounts, $n_{A,0}$ and $n_{B,0}$, of A_a and B_b *molecules* with functionalities a and b, respectively, and the extent of reaction, p_A, of the minority component with respect to *groups*, assuming absence of cyclization:

$$(13\text{-}109) \quad \overline{X}_n = \frac{n_{A,0} + n_{B,0}}{n_{A,0} + n_{B,0} - p_A a n_{A,0}}$$

Table 13-12 Theoretical critical values, p, for the onset of crosslinking of systems $n_{A,0}A_a + n_{B,0}B_b$ with varying initial molar amounts $n_{A,0}$ and $n_{B,0}$ (in moles) of *molecules* and varying functionalities a and b of molecules A_a and B_b. A_a is the minority component except for amounts indicated by *.

| Functionalities | | Initial amounts | | Stoichiometry with respect to | | p_A | $\overline{X}_n$ |
a	b	$n_{A,0}$	$n_{B,0}$	groups	molecules		
2	2	1	1	+	+	1.000	∞
2	3	3	1 *	–	–	1.000	4.00
		1.5	1	+	–	≈ 0.707	≈ 6.60
		1	1	–	+	≈ 0.577	≈ 2.36
2	4	6	1 *	–	–	1.000	≈ 2.33
		2	1	+	–	≈ 0.577	≈ 4.34
		1	1	–	+	≈ 0.408	≈ 1.69
3	3	4	1 *	–	–	1.000	2.50
		1	1	+	+	0.500	4.00
3	4	8	1 *	–	–	1.000	2.25
		4	3	+	–	≈ 0.408	≈ 3.33
		1	1	–	+	≈ 0.354	≈ 2.13
4	4	9	1 *	–	–	1.000	≈ 1.67
		1	1	+	+	≈ 0.333	≈ 2.99

13.8.3 Branching by Side Reactions

Branches may also be introduced into polymer molecules by side reactions which may subsequently either increase or decrease the functionality of polymer molecules.

An example is the industrial synthesis (I) of *unsaturated polyester* molecules from maleic anhydride (MA) and ethylene glycol, Eq.(13-110), in which the cis double bonds of maleic anhydride monomer molecules isomerize predominantly to trans. The resulting fumaric acid units of the polymer may add hydroxy endgroups, either intramolecularly (II) with formation of lactone rings (causing a decrease of functionality of the polyester molecules), or, if [OH] > [COOH], intermolecularly (III) to branched molecules (increase of functionality of polyester molecules):

(13-110) $(n + 1)$ ⟨MA⟩   + $(n + 1)$ $HOCH_2CH_2OH$

$$\downarrow I$$

$$HO-\underset{\underset{O}{\|}}{C}-CH=CH-\underset{\underset{O}{\|}}{C}\!\!\left[O-CH_2-CH_2-O-\underset{\underset{O}{\|}}{C}-CH=CH-\underset{\underset{O}{\|}}{C}\right]_n\!\!O-CH_2-CH_2-OH$$

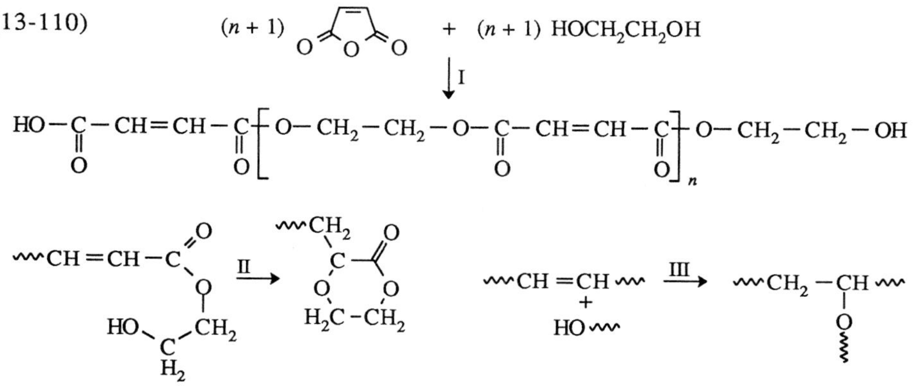

The resulting unsaturated polyester molecules possess many double bonds per chain. They serve therefore as crosslinking agents in the free-radical polymerization of styrene to unsaturated polyester resins (Volumes II and IV; in technical lingo, the bifunctional styrene molecules are the crosslinkers!).

13.9   Crosslinking Step Polymerization

13.9.1   Phenomena

Crosslinked polymers are produced by chain polymerizations (Section 10.3.12) or step reactions (see below) of monomer molecules with functionalities of 3 or higher or by after-reactions of linear or branched macromolecules (Section 2.5.10).

An example is the polycondensation of 3 moles of adipic acid, $HOOC(CH_2)_4COOH$, with 2 moles of glycerol, $HOCH_2–CHOH–CH_2OH$, that proceeds with the release of water. The viscosity of the system increases first slowly and then very rapidly (Fig. 13-35). At a certain time, the system becomes so highly viscous that bubbles of water vapor no longer raise to the surface and the stirrer gets stuck. The transition from a very viscous solution to a **gel** is so sharp that one can speak of a **gel "point."** The gel contains a fraction that is insoluble in all solvents (**gel fraction**) and a soluble fraction (**sol fraction**) whose proportion decreases with increasing extent of reaction.

The reason for these phenomena is the formation of branched molecules that contain more and more unreacted functional groups with increasing extent of reaction and degree of polymerization. The probability that a functional group of a branched molecule reacts with a functional group of another branched molecule increases strongly with in-creasing degree of polymerization. At the gel point, an infinitesimally small additional reaction suffices to unite a small fraction of polymer molecules to an "infinitely large" (and insoluble) polymer molecule that extends from wall to wall of the reaction vessel. The resulting gel consists of this "infinitely" large molecule plus soluble, branched poly-mer molecules and unreacted monomer molecules (for a quantitative description, see be-low). Theoretical treatments all *assume* that the infinitely large molecule(s?) is not intra-molecularly crosslinked since otherwise the mathematics would be intractable.

Shortly before gelation, the crosslinking system consists of many monomer and oli-gomer molecules and a few very high-molecular weight polymer molecules. Hence, it has a low number-average degree of polymerization and a steeply increasing viscosity which is heavily influenced by the weight-average degree of polymerization (Fig. 13-35). Shortly after gelation, the diffusion of the uncrosslinked molecules and especially that of the segments of the network to reactive sites is strongly hindered. New cross-linking sites are thus formed at sites that do not correspond to those that would be formed by equilibrium systems. The polymer networks formed by crosslinking reactions are thus not equilibrium systems except immediately after their first formation.

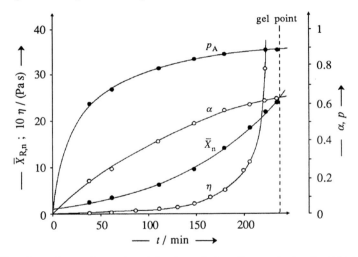

Fig. 13-35 Time dependence of experimental values of viscosity η, extents of reaction p_A (by titra-tion), number-average degrees of polymerization $\overline{X}_n$ (calculated from experimental extents of reaction, p_A), and branching coefficients α (Eq.(13-117)) in the p-toluene sulfonic acid catalyzed polycondensa-tion of diethylene glycol with a mixture of succinic acid and tricarballylic acid ($q_0 = 1.002$; $x_A{}^\Delta = 0.404$) at 109°C [30]. With kind permission by the American Chemical Society, Washington (DC).

For the same reasons, the resulting polymer networks are also not homogeneous. Their inhomogeneities are caused by unreacted functional groups, loose chain ends, entanglements of chain sections, formation of catenane structures, etc. They are caused by differing reactivities of functional groups of monomer molecules, pre-ordering of monomer molecules in the uncrosslinked state, dilution effects, demixing by diffusion-controlled reactions, and phase separation of polymer molecules.

Crosslinking during step reactions of low-molecular weight non-linear monomer sys-tems is usually observed at fairly high group conversions. Diffusion effects are usually small and the resulting networks are fairly homogeneous. However, crosslinking during chain polymerization of low-molecular weight polyfunctional monomers and crosslinking of preformed macromolecules takes place at fairly low conversions. The crosslinking sites are fixed in space and it is from these centers that the polymerization proceeds. The result is more tightly crosslinked regions in a less tightly crosslinked matrix.

This effect, in combination with phase separation, is utilized for the synthesis of **macroporous** (or **macroreticular**) **networks** by chain polymerization of monomers in solvents that are non-solvents for the polymers. An example is the free-radical chain polymerization of styrene with divinylbenzenes (DVB) as crosslinking agents in hexane, a solvent for the monomers but a non-solvent for the polymers.

Phase separation and crosslinking occur simultaneously at small monomer conversions. Further polymerization takes place near the precipitated growth centers. Depending on the reaction conditions, either pores in a continuous matrix or interconnected globular structures with porous interstitial volumes are formed. At the same degree of crosslinking, macroreticular networks are much more permeable for solvents and solutes than homogeneous ones.

13.9.2   Thermosets

The knowledge of the gel point (Section 13.9.4) is industrially very important since crosslinked polymers can only be processed with difficulty. The shaping of parts has thus to be performed before gelation sets in. Crosslinked polymers are highly valued in industry because of their high thermal and solvent stabilities.

Highly crosslinked polymers with high glass temperatures are called **thermosets** in industry. They result from the chemical, irreversible, *narrow mesh* crosslinking of liquid monomers or oligomers called **thermoset(ting) resins**. Monomers with higher molecular weights or intermediates between monomers and thermosets are also known as **prepolymers** or **reaction polymers**. The crosslinking reaction is usually called **hardening** or **curing**. All these terms are sometimes used interchangeably.

Thermosets are obtained by polycondensation, polyaddition, or chain polymerization. Their chemistry is discussed in Volume II, their processing and application in Volume IV. Section 13.9 of this volume is only concerned with the principles of crosslinking.

Phenolic resins are the oldest thermosetting resins. The acid or base catalyzed reaction of phenol and formaldehyde (and occasionally other aldehydes) delivers ortho and para substituted methylols that react with release of water to form molecules with methylene, ether, or formal bridges. Further reactions lead to thermosets.

(13-111)

$$i = 0 : \text{methylene bridge}$$
$$i = 1 : \text{ether bridge}$$
$$i = 2 : \text{formal bridge}$$

(structures: a benzene ring with CH$_2$OH and OH; a benzene ring with CH$_2$(OCH$_2$)$_i$ and OH; a benzene ring with OH)

Alkyd resins result from the pregelation polycondensation of multifunctional *alco*hols (glycerol, trimethylolpropane, etc.) and di*acids* (phthalic acid, succinic acid, etc.) or acid anhydrides. They are usually also modified with fatty acids as comonomers. Alkyd resins from glycerol and phthalic acids are known as **glyptal resins**. The products are used as paint resins that are subsequently hardened (crosslinked).

Amino resins result from the reaction of NH_2 or NH groups containing compounds such as urea or melamine with carbonyl compounds (usually formaldehyde). Crosslinking proceeds via >NH– containing intermediates:

(13-112)

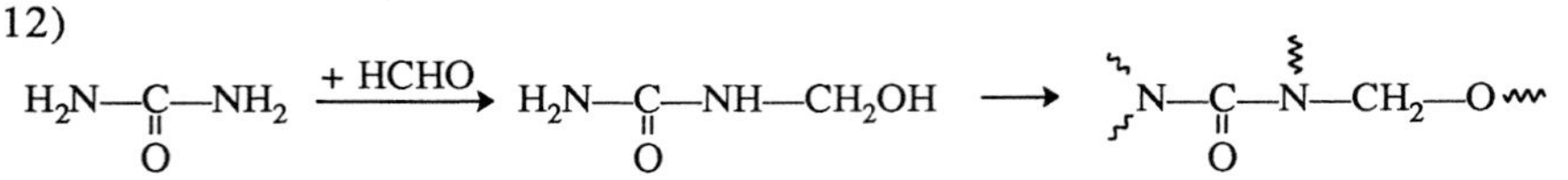

Epoxy resins contain both oxirane groups and hydroxy groups. They are crosslinked by reaction with diacid anhydrides, diamines or triamines. The standard epoxy resin (Eq.(13-111)) is obtained from bisphenol A, $HO(p\text{-}C_6H_4)\text{-}C(CH_3)_2\text{-}(p\text{-}C_6H_4)OH$, and epichlorohydrin. The average number of repeating groups per molecule, N, is usually small, for example $0.1 \leq N \leq 0.6$ in liquid resins.

(13-113)

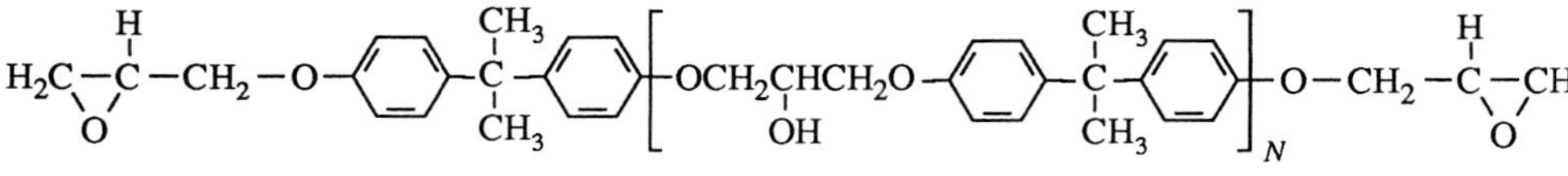

Polyurethanes with the characteristic urethane group –NH–CO–O– are synthesized from isocyanates (toluene diisocyanate, diphenylmethane diisocyanate, etc.) and f-functional hydroxyl compounds ($f \geq 2$), often called **polyols**, by polyadditions, ~NCO + HO~ → ~NH–CO–O~. Monomers react so fast that crosslinking polymerization and shaping of articles proceed simultaneously with high speed in the **reaction injection molding (RIM) process**.

13.9.3 Functionalities

In multifunctional step reactions, the number of endgroups per polymer molecule, i.e., its functionality, increases with increasing degree of polymerization, X. If two or more different types of monomers participate (for example, A_a and $A_{a'}$ with functionalities $a \neq a'$), then their contributions must be weighted according to their mole fractions. Examples of *number-average functionalities* of reactants with degrees of polymerization, X_i, are shown in Table 13-13. For example, the number-average functionality of a dimer >—< ($X_i = 2$) from an A_3 molecule >– ($f = 3$) is $2 + (3 - 2)2 = 4$, etc.

Table 13-13 Number-average functionalities of i-mers. a, a', b, f = Functionality of monomer molecules; x_{Aa} = fraction of A-groups in all molecules A_a with respect to all A-groups in the system.

Example	Monomers	Functionality $\bar{f}_j$ of i-Mers	Comments
I	A_a	$2 + (a - 2)X_i$	
II	$A_a + A_{a'}$	$2 + [a' + (a - a')x_{Aa} - 2]X_i$	$x_{Aa} = N_{Aa}/(N_{Aa} + N_{Aa'})$ if stoichiometric
III	$A_a + A_{a'} + B_b$	$2 + \{(1/2)[a + b + (a' - b)x_{Aa}] - 2\}X_i$	if stoichiometric
IV	$A_f + B_f$	$2 + (f - 2)X_i$	

Sometimes, so-called *weight-average functionalities* $\bar{a}_k$ are used if functional groups of the same type reside in monomers A_{a_i} with different functionalities a_i:

(13-114) $\bar{a}_k = \sum_i n_{A_{a_i}} a_i^2 / (n_{A_{a_i}} a_i)$

However, $\bar{a}_k$ is not a weight average (see Section 3.5.6) since the weighing parameter is not the mass m (or weight fraction w) but the amount n, a number-related quantity. $\bar{f}_k$ is in reality the ratio of the second and the first moment of the distribution of functionalities a_i (see also Volume III). The true weight-average functionality of molecules with various numbers of A groups per molecule is rather $\bar{a}_w = \sum_i m_{A_{a_i}} a_i / m_{A_{a_i}}$

13.9.4   Gel Points

The onset of gelation (gel point) is a critical point that can be calculated from molecular data using combinatorial theory, stochastic branching, recursive approach, percolation, and kinetics. The general A + B process involves the reaction of n_A molecules A_a with different functionalities a and n_B molecules B_b with different functionalities b such as $n_{A,2}A_2 + n_{A,3}A_3 + ... + n_{B,2}B_2 + n_{B,3}B_3 + ...$; it is mathematically quite complicated.

The classic probability theory computes the probabilities of occurrence of certain bonds and repeating units, respectively. The famous Flory example is the polycondensation of succinic acid, $HOOC(CH_2)_2COOH$, with diethylene glycol, $(HOCH_2CH_2)_2O$, and tricarballylic acid, $(HOOCCH_2)_2C(H)COOH$ (Fig. 13-35), i.e., that of a bifunctional monomer UB_2 with a mixture of monomers VA_2 and WA_3. The monomeric groups $-W<$ are branching units that terminate linear repeating units of the general structure

$$\begin{array}{c}\mathrm{\sim\sim A} \\ \quad \diagdown \\ \mathrm{\sim\sim A} \diagup \end{array} \mathrm{W{-}(A{-}B){-}[U{-}(B{-}A){-}V{-}(A{-}B)]_i{-}U{-}(B{-}A){-}W} \begin{array}{c}\diagup \mathrm{A \sim\sim} \\ \diagdown \mathrm{A \sim\sim} \end{array} \qquad ; \ 0 \le i \le \infty$$

$$\qquad\qquad\quad \mathrm{I} \qquad\quad \mathrm{II} \qquad\quad \mathrm{III} \qquad \mathrm{IV}$$

The probability of occurrence, p_I, of the bond I equals the extent of reaction of A-groups, p_A. The probability of occurrence, p_{II}, of bond II is determined by the extent of reaction, p_B, of B-groups and by the mole fraction $x_{A,lin}$ of A-groups in bifunctional AVA molecules, or, according to $x_{A,lin} = 1 - x_A^\Delta$, by the mole fraction x_A^Δ of A-groups in trifunctional WA3 molecules. The repeating unit contains $(B-A)_i$ bonds of type II with a total probability of $p_{II} = [p_B(1 - x_A^\Delta)]^i$. The probability of occurrence of the i bonds of type III is $p_{III} = p_A^i$ whereas that of bond IV is $p_{IV} = p_B x_A^\Delta$. The probability of occurrence of the repeating unit is then the product of all these probabilities:

(13-115) $(p_A)_{br,i} = p_I p_{II} p_{III} p_{IV} = p_A [p_B(1 - x_A^\Delta)]^i p_A^i p_B x_A^\Delta$

The **branching coefficient** is defined as the probability that a functional group at a branching unit is bound to another branching unit via a repeating unit. The branching coefficient of the whole system therefore equals the sum of all probabilities, $(p_A)_{br,i}$:

(13-116) $\alpha = \sum\limits_{i=0}^{i=\infty} p_{A,br,i} = \sum\limits_{i=0}^{i=\infty} p_A \left[p_B(1 - x_A^\Delta) \right]^i p_A^i p_B x_A^\Delta = Q \sum_i Z^i = Q(1 + Z + Z^2 + ...)$

where $Q \equiv p_A p_B x_A{}^\Delta$ and $Z \equiv p_A p_B (1 - x_A{}^\Delta)$. The quantity Z is always smaller than unity. The series $1 + Z + Z^2 + \ldots$ can therefore be converted into a closed expression, i.e., $\Sigma_i Z_i = 1/(1 - Z)$. With $r_0 = n_{B,0}/n_{A,0} \leq 1$, Eq.(13-116) converts to

$$(13\text{-}117) \qquad \alpha = \frac{p_A p_B x_A{}^\Delta}{1 - p_A p_B (1 - x_A{}^\Delta)} = \frac{r_0 p_A{}^2 x_A{}^\Delta}{1 - r_0 p_A{}^2 (1 - x_A{}^\Delta)} = \frac{p_B{}^2 x_A{}^\Delta}{r_0 - p_B{}^2 (1 - x_A{}^\Delta)}$$

Crosslinking occurs at a critical value of the branching coefficient, α_{crit}, that is given by the functionality f_{br} of the *branching points* ($f_{br} = 3$ in WA$_3$ + VA$_2$ + UB$_2$). The number a of arms at a branching point increases by $a - 1$ if a new a-functional molecule is added. The probability that an arm carries another branching point is $\alpha(a - 1)$. Crosslinking occurs if $\alpha(a - 1) \geq 1$. The branching coefficient α thus has a critical value of

$$(13\text{-}118) \qquad \alpha_{crit,theor} = \frac{1}{f_{br} - 1} = \frac{r_0 (p_{A,crit,theor})^2 x_A{}^\Delta}{1 - r_0 (p_{A,crit,theor})^2 (1 - x_A{}^\Delta)}$$

The theoretical extent of reaction of A-groups at the gel point, $p_{A,crit,theor}$, depends therefore only on the functionality f_{br} of branching molecules, the initial molar ratio of B-groups and A-groups, $r_0 = n_{B,0}/n_{A,0}$, and the mole fraction $x_A{}^\Delta$ of A-groups in branching molecules with respect to all A-groups. The critical extent of reaction should be independent of reaction temperature and reaction time, which was confirmed experimentally. For example, the same experimental value of $p_{A,crit} = 0.796 \pm 0{,}001$ was independent of reaction temperature (160-215°C) and time (50-860 min) for the polycondensation of 2 moles of glycerol with 3 moles of phthalic anhydride. However, the experimental gel point of $p_A = 0.796$ was considerably higher than $p_{A,crit,theor} = 0.707$ as calculated from Eq.(13-118) with $f_{br} = 3$, $r_0 = 1$, and $x_A{}^\Delta = 1$.

Higher than predicted gel points are also found for many other systems (Table 13-14). This deviation from theory is caused by intramolecular reactions that waste functional groups which could have reacted intermolecularly. As a consequence, the onset of gelation is shifted to higher extents of reaction.

Table 13-14 Comparison of predicted, $p_{A,crit,theor}$, and experimentally found gel points, $p_{A,gel}$, in stoichiometric (with respect to functional groups) nonlinear polyesterifications [31].

Type of reaction	A$_a$	B$_b$	$p_{A,crit,theor}$	$p_{A,gel}$	$\dfrac{p_{A,gel}}{p_{a,crit,theor}}$
3 A$_2$ + 2 B$_3$	phthalic acid	glycerol	0.707	0.765	1.082
	phthalic anhydride	glycerol	0.707	0.795	1.124
	succinic acid	glycerol	0.707	0.765	1.082
	adipic acid	glycerol	0.707	0.770	1.089
	adipic acid	trimethylol propane	0.707	0.765	1.082
	adipic acid	trimethylol ethane	0.707	0.721	1.020
	1,10-decanediol	1,3,5-brenztriacetic acid	0.707	0.720	1.018
2 A$_2$ + B$_4$	adipic acid	pentaerythritol	0.577	0.620	1.075
A$_3$ + B$_3$	tricarballylic acid	glycerol	0.500	0.580	1.160
	tricarballylic acid	trimethylol propane	0.500	0.580	1.160
4 A$_3$ + 3 B$_4$	tricarballylic acid	pentaerythritol	0.400	0.495	1.238

Intramolecular reactions increase with increasing dilution (Section 6.1.5). Extrapolation of gel points observed for systems with finite initial monomer concentrations to infinite dilution indeed delivers the theoretical gel point (Fig. 13-36).

The deviation of $p_{a,gel}$ from $p_{A,crit,theor}$ is a measure of cyclization. It increases with the decrease in the distance between the functional groups of the same monomer molecule. The cyclization is thus larger for polycondensations with phthalic acid than for those with isophthalic acid (1,2 versus 1,3 position) and for those with pentaerythritol instead of trimethylol propane (tetrasubstituted versus trisubstituted alcohol).

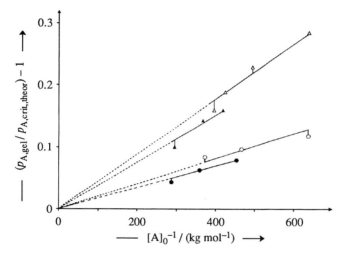

Fig. 13-36 Ratio of experimental ($p_{A,gel}$) and theoretical ($p_{A,crit,theor}$ from Eq.(13-118)) extent of reaction at the gel point for the crosslinking polycondensation of isophthalic acid with trimethylol propane (●) or pentaerythritol (O) and phthalic anhydride with trimethylol propane (▲) or pentaerythritol (△) in toluene as a function of dilution of alcohol (in kg mixture/mol alcohol) [32].

13.9.5 Degrees of Polymerization

Gel points indicate the formation of a very small **gel fraction** which is usually assumed to consist of *one* "infinitely" large macromolecule. At the gel point, most of the monomeric units of the system reside in oligomer and unreacted monomer molecules that can be extracted from the gel as **sol fraction**.

Since the number-average degree of polymerization, $\overline{X}_{R,n}$, is controlled by the (great) *number* of molecules; it is small at the gel point. The weight-average degree of polymerization, $\overline{X}_{R,w}$, on the other hand, depends very strongly on the (great) *mass* of molecules: the weight-average degree of polymerization is high at the gel point and so is the melt viscosity (Fig. 13-35), a quantity that depends on $\overline{X}_{R,w}$ (volume III).

This apparent paradox is easily illustrated by a numerical example. Consider a crosslinking polycondensation just beyond the gel point where the mass fraction of the gel fraction is $w_{gel} = 0.0001$ and that of the sol fraction is $w_{sol} = 0.9999$; the corresponding molar masses are $M_{sol} = 10^3$ g/mol and $M_{gel} = 10^{26}$ g/mol, respectively. The molar mass averages of the reactants are therefore

$$\overline{M}_{R,n} = \frac{1}{(w_{sol}/M_{sol})+(w_{gel}/M_{gel})} = \frac{g\,mol^{-1}}{0.9999\cdot10^{-3}+10^{-4}\cdot10^{-26}} \approx 10^3\,g\,mol^{-1}$$

$$\overline{M}_{R,w} = w_{sol}M_{sol} + w_{gel}M_{gel} = (0.9999\cdot10^3 + 10^{-4}\cdot10^{26})\,g\,mol^{-1} \approx 10^{22}\,g\,mol^{-1}$$

The molar masses and degrees of polymerization, respectively, can be calculated as follows, for example, for the system $A_2 + A_3 + B_2$ which corresponds to the system in Fig. 13-35 (succinic acid (A_2) + tricarballylic acid (A_3) + diethylene glycol (B_2)). Initially, there are $N_{B,0}$ B-groups and $N_{A,0}$ A-groups from which the mole fraction x_A^{Δ} is in branching molecules with the functionality f_{br}. This system contains

$$(13\text{-}119) \quad N_{mer} = \frac{N_{B,0}}{2} + \frac{N_{A,0}x_A^{\Delta}}{f_{br}} + \frac{N_{A,0}(1-x_A^{\Delta})}{2} = N_{A,0}\left(\frac{1}{2q_0}+\frac{x_A^{\Delta}}{f_{br}}+\frac{1-x_A^{\Delta}}{2}\right)$$

monomeric units where $q_0 = N_{A,0}/N_{B,0} \geq 1$. The number N_{mol} of molecules in the system at an extent of reaction, p_A, is given by the difference of the numbers N_{mer} of monomeric units and the number $N_{bd} = N_{A,0}p_A$ of the bonds that were formed:

$$(13\text{-}120) \quad N_{mol} = N_{mer} - N_{bd} = (1/2)\,N_{A,0}(1 - x_A^{\Delta} - 2\,p_A + 2\,x_A^{\Delta}f_{br}^{-1} + q_0^{-1})$$

The *number-average* degree of polymerization of reactants, $\overline{X}_n$, is defined as the ratio of the number N_{mer} of monomeric units and the number N_{mol} of all reactant molecules:

$$(13\text{-}121) \quad \overline{X}_{R,n} = \frac{N_{mer}}{N_{mol}} = \frac{2\,x_A^{\Delta} + f_{br}[1 - x_A^{\Delta} + (1/q_0)]}{2\,x_A^{\Delta} + f_{br}(1 - x_A^{\Delta} + (1/q_0) - 2\,p_A)}$$

The number-average degree of polymerization as a function of time t (or extent of reaction, p_A) can thus be calculated from the functionality of branching molecules, f_{br}, the proportions of the ingredients, q_0 and x_A^{Δ}, and the experimentally determined extent of reaction, p_A. For the data shown in Fig. 13-35, a value of $\overline{X}_{R,n} = 24.7$ is calculated from Eq.(13-121) for the experimental value of $p_{A,crit} = 0.894$ ($\overline{X}_{R,n}$ was not measured independently!) whereas the one calculated with the theoretical value of $p_{A,crit,theor} = 0.843$ (from Eq.(13-118) with $\alpha_{crit,theor} = 1/2$) was only $\overline{X}_{R,n} = 10.5$. Of course, both values of $\overline{X}_{R,n}$ are fictitious since intramolecular reactions were neglected.

Weight-average molecular weights are more difficult to calculate. The theoretical weight-average degree of polymerization, $\overline{X}_{R,w}$, for the binary system $n_{A,0}\,A_a + n_{B,0}\,B_b$, consisting of initial amounts $n_{A,0}$ and $n_{B,0}$ of *molecules* A_a and B_b with functionalities a and b has been calculated by probability theory (derivation not shown) for extents of reaction of groups A and B, p_A and p_B, below the gel point as

$$(13\text{-}122) \quad \overline{X}_{R,w} = 1 + \frac{p_A p_B ab[2 + p_A(a-1) + p_B(b-1)]}{[ap_A + bp_B][1 - p_A p_B(a-1)(b-1)]}$$

if cyclization is absent and no leaving molecules are formed (polycondensations with $M_L/M_U \rightarrow 0$, polyadditions). At the gel point, weight-average degrees of polymerization of reactants approach infinity and Eq.(13-122) reduces to $p_A p_B(a-1)(b-1) = 1$.

For A_a reactions and group-stoichiometric systems $A_a + B_b$, weight average degrees of polymerization approach infinity near the gel points (Fig. 13-37). Gels are formed at lower extents of reaction, the higher the functionalities f and g of monomer molecules (series $a = 4 \rightarrow 3 \rightarrow 2$ or $a/b = 4/4 \rightarrow 3/3 \rightarrow 2/2$). Bifunctional monomers act as diluents for multifunctional ones (cf. $a/b = 3/3$ with $a/b = 2/4$ and $a/b = 4/4$ with $a/b = 2/6$). Such systems start to crosslink at higher extents of reaction than those derived solely from multifunctional monomers ($a, b \geq 3$) with the same overall functionality.

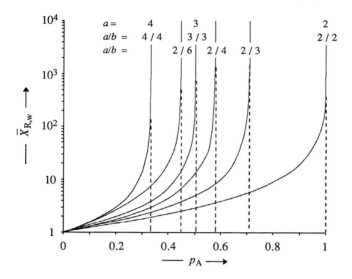

Fig. 13-37 Weight-average degrees of polymerization of reactants as a function of extent of reaction of A-groups for the polyaddition and polycondensation (with $M_L/M_U \rightarrow 0$) of A_a (numbers of first line)) and stoichiometric systems (with respect to groups) $A_a + B_b$, respectively (numbers of the second and third line). a, b = functionalities of monomer molecules.

At time t, amounts $an_{A,0}p_A$ of A_a-molecules and amounts $bn_{B,0}p_B$ of B-molecules have reacted. In a stochiometric system with respect to *groups*, one has equal amounts of groups and therefore $an_{A,0}p_A = bn_{B,0}p_B$ and $p_B = [an_{A,0}/(bn_{B,0})]p_A = r_0p_A$ where $r_0 \equiv 1$. In a non-stoichiometric system, the imbalance parameter is defined as $r_0 \leq 1$. Substitution of p_B in $p_Ap_B(a - 1)(b - 1) = 1$ for a non-stoichiometric systems thus leads to $r_0p_A^2 = (a - 1)^{-1}(b - 1)^{-1}$ and therefore also to Eq.(13-108).

13.9.6   Post-Gel Relationships

The equations discussed in previous sections apply only to reactions up to the gel point. After surpassing the gel point, higher-molecular weight oligomers are more likely to be incorporated into the crosslinked gel than low-molecular weight ones because of the higher proportion of functional groups in the former. The "sol" will thus contain less and less high-molecular weight oligomers until only a few monomer molecules are present besides the "infinitely large" gel shortly before the reaction is nearly complete. As a result, theories expect molecular weights of the sol to drop sharply in the region between

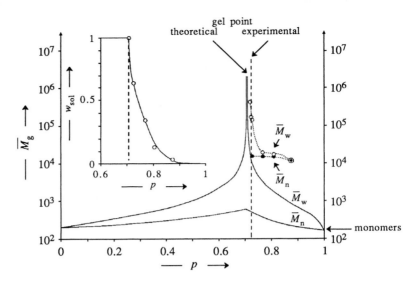

Fig. 13-38 Dependence of the logarithm of weight-average and number-average molecular weights of sol fractions as a function of the extent of reaction, p, in the stoichiometric polycondensation of 1,3,5-benzenetriacetic acid and 1,10-dihydroxydecane, 2 $C_6H_3(CH_2COOH)_3$ + 3 $HO(CH_2)_{10}OH$ [33].
Lines: Predictions of theories: Flory probability theory for $p < p_{crit}$, Macosko-Miller recursion theory and Gordon percolation theory for $p > p_{crit}$ (both give practically the same result). O,● Experimental data [33].
Insert: Weight fraction w_{sol} of sol as a function of the extent of reaction after surpassing the gel point; O experiment [34]. Line calculated with $w_{sol} = (1 - p^2)^3/p^6$ (Flory's probability theory; this approach implicates intramolecular reactions according to Stockmayer (see Historical Notes).

the gel point at p_{crit} and complete reaction at $p = 1$ (Fig. 13-38). Indeed, experimental weight-average molecular weights of sol fractions do drop sharply near the experimental gel point but then remain considerably higher than predicted by theories. Number-average molecular weights are also higher and vary little with p.

The weight fraction w_{sol} of the sol at $p > p_{crit}$ also decreases sharply with the extent of reaction once the gel point is passed (Fig. 13-38, insert). The experimental points follow probability theory very well. The theory predicts w_{sol} to depend on the branching probability $\alpha = p^2$ according to $w_{sol}^* = (1 - p^2)^3/p^6$ for trifunctional branching molecules albeit for the case of $M_L = 0$, i.e., polyaddition.

The weight fraction $w_{1,sol}$ of monomer molecules in the sol fraction at $p_c > p_{crit}$ decreases with increasing extent of reaction, passes through a minimum at the gel point and then increases to $w_{1,sol} \to 1$ at $p \to 1$ where $w_{sol} \to 0$ (Fig. 13-39). The weight fraction $w_{2,sol}$ of dimer molecules exhibits the same behavior except that it falls to $w_{1,sol} \to 0$ at $p \to 1$. Both functions qualitatively agree with theory. However, the experimental values of both $w_{1,sol}$ and $w_{2,sol}$ are larger than the theoretical ones because they were calculated for a polyadditon ($M_L = 0$) and not for a polycondensation ($M_L > 0$). The calculations also assume equilibria but equilibria do not exist beyond the gel point.

Agreement between theory and experiment is thus excellent for the weight fraction of sol (Fig. 13-38), good for the dimer fraction (Fig. 13-39), reasonable for the monomer fraction (Fig. 13-39), and poor for the weight-average and number-average molecular weights of the sol fractions (Fig. 13-38).

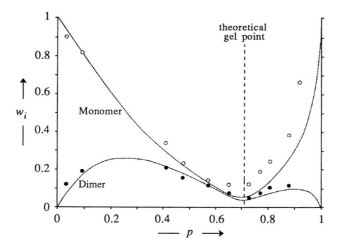

Fig. 13-39 Weight fractions w_i of monomer molecules (O [34]; $i = 1$) and dimer molecules (● [35]; i = 2) in sol fractions as function of the extent of reaction, p, of the polycondensation of 1,3,5-benzenetriacetic acid and 1,10-dihydroxydecane. Lines calculated for the absence of leaving molecules.

There are several reasons for the discrepancies:

1. The ratio of intramolecular to intermolecular reactions increases with increasing p > p_{gel}. Since the extent of reaction, p, is calculated from the total amount of liberated water, measured p values are higher than those from intermolecular reactions.

2. Chromatography detected the presence of microgel particles in the sol both before and after the onset of gelation. Microgels are very small and highly crosslinked entities with correspondingly superhigh molecular weights. They cannot be removed completely from the sol by conventional means; hence the much too high weight-average molecular weights of the sol fractions. The presence of microgel particles shows that the insoluble material at the gel point is not *one* supersized molecule but actually an intertwined mixture of several molecular entities.

3. Sol molecules are entrapped in the network of the gel. Their absence in the postgel sol increases the weight-average molecular weight of the sol molecules.

Gel and sol formation in multifunctional step-growth reactions have so far been discussed and interpreted mainly by using **probability theory**. This theory is a **mean-field theory** that assumes that the same average force field acts on all segments: the probability of a reaction of an A-group with a B-group is assumed to be independent of the local environment (fringes or interior of the hyperbranched sol molecules or the crosslinked gel). This may be correct for the innate chemical reactivity but is very questionable for the accessibility of the groups. The model relies on a strong overlap of segments, but segments are not infinitely thin (excluded volume effect) and cannot pass through each other. Loops (cyclics) are also not taken into account.

An alternative theory is based on **percolation**. Monomer molecules are placed on a lattice; bonds between the neighboring pairs are assumed to be formed at random. The resulting network strands should not overlap; rather, they should be just at the critical overlap concentration (see Volume III) according to an often used theorem. As in the probability model, there is a critical branching point called *percolation threshold*.

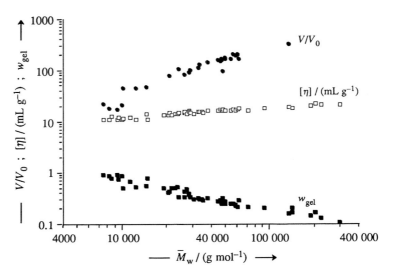

Fig. 13-40 Dependence of the logarithm of swelling ratios V/V_0 of gels after the removal of sol, in-trinsic viscosities $[\eta]$ of sols, and weight fractions w_{gel} of gels in post-gel reactants on the logarithm of the weight-average molecular weight $\overline{M}_w$ of sols [36]. Sols and gels were obtained from polyesters that were prepared by post-reaction of a precursor polymer.

Probability and percolation theories differ in their predictions for **scaling**. Sol mole-cules and gel particles of different sizes are **self-similar objects** that scale with a certain dimensionality d (Volume III). For example, masses m of regular objects scale with lengths L according to $m \sim L^d$ ($d = 3$ for spheres; $d = 1$ for rods, etc.). Irregular objects such as molecules scale with fractions of physical properties such as length. Such objects are thus called **fractals**.

Masses or molecular weights also scale with other properties. In crosslinking polymer-izations, for example, they scale with the gel fraction according to $w_{gel} \sim \overline{M}_w^{-\beta}$, with the swelling ratio of the gel according to $V/V_0 \sim \overline{M}_w^{-\beta z} = w_{gel}^{-z}$, and with the intrinsic viscosity of the sol according to $[\eta] \sim \overline{M}_w^{\alpha}$ (Fig. 13-40).

Table 13-15 Critical exponents for the post gelation of polyesters of Fig. 13-40 according to the de-gree of swelling,V/V_0, intrinsic viscosity, $[\eta]$, and weight fractions of gels, w_{gel} [36]. d = Spatial di-mension, d_f = fractional dimension, τ = Fisher exponent. * Including pregelation.

		$V/V_0 \sim w_{gel}^{-z}$	$[\eta] \sim \overline{M}_w^{\alpha}$	$w_{gel} \sim \overline{M}_w^{-\beta}$
Predictions,	3D percolation	$z = \dfrac{(d/d_f)-1}{\tau-2}$	$a = \dfrac{(d/d_f)-\tau+1}{3-\tau}$	$\beta = \dfrac{\tau-2}{3-\tau}$
	3D percolation	$z = 2.5$	$a = 0.38$	$\beta = 0.25$
	mean field	$z = 1$	$a = 0$	$\beta = 1$
Experiment	(using τ from w_{gel})	$z = 1.70 \pm 0.07$	$a \approx 0.21*$	$\beta = 0.56 \pm 0.04$
Exponents τ,	mean field	$\tau = 2.5$	$\tau = 2.5$	$\tau = 2.5$
	percolation	$\tau = 2.2$	$\tau = 2.2$	$\tau = 2.2$
	experiment	$\tau = 2.29 \pm 0.02$	$\tau = 2.34 \pm 0.03$	$\tau = 2.36 \pm 0.02$

The observed exponents were compared with those predicted by the mean-field and 3-dimensional percolation theories, using a spatial dimension of d = 3 and a fractal dimension of $d_f = 2$ (Table 13-15).

Data on polyesters show that the experimental Fisher exponents of $2.29 \leq \tau \leq 2.36$ are intermediate between the one predicted by the mean field theory ($\tau = 2.5$) and the percolation theory ($\tau = 2.2$). This may be caused by experimental restrictions (cut-off at high molecular weight in SEC experiments). It may also be due to the structure of the investigated polymers since the average number of monomeric units between branching points was $N_u \approx 30$ which is intermediate between the requirements of percolation theory ($N_u = 1$) and probability theory ($N_u \gg 1$).

13.9.7 Crosslinking Post-Reactions

Crosslinked polymers are also obtained by post-reactions of uncrosslinked primary macromolecules with crosslinking agents which may be multifunctional molecules, free radicals, etc. These reactions are usually not step reactions but are treated here because they clearly show that the onset of crosslinking is indeed controlled by the *weight*-average molecular weight as postulated by the probability theory of multifunctional step reactions and not by the number-average molecular weight.

Macromolecules can only be joined to a single network if each primary molecule is connected with at least two other primary chains. Depending on the functionality of crosslinking sites, different numbers of junctions per primary molecule are required: trifunctional crosslinking junctions need at least two junctions per primary molecule but tetrafunctional ones need only one (Fig. 13-41).

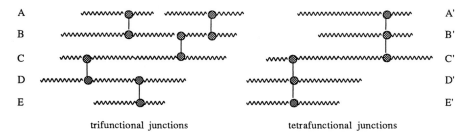

trifunctional junctions   tetrafunctional junctions

Fig. 13-41 Trifunctional and tetrafunctional junctions of chains.

In the simplest case, all primary molecules have the same degree of polymerization, X. The expectation ε to find a crosslinking site in a primary molecule must be proportional to the probability p_u to find such a site in any given monomeric unit. The expectation is also higher, the greater the degree of polymerization of the primary molecule. However, no crosslinking is possible for a monomer with X = 1. The expectation is thus

$$(13\text{-}123) \quad \varepsilon = p_u(X - 1)$$

An expectation $\varepsilon < 1$ does not correspond to a crosslinking of *all* primary molecules of a polymeric substance since some primary macromolecules will not become part of a network but will rather remain branched. An expectation $\varepsilon > 1$, on the other hand, corresponds to crosslinked structures. The critical condition for a crosslinking of all primary molecules is therefore $\varepsilon_{crit} = 1$. Eq.(13-123) becomes

(13-124) $p_{u,crit} = 1/(X - 1) \approx 1/X$

The critical probability for the onset of crosslinking in polymers with a molecular weight distribution of primary molecules is obtained from the following considerations. To become a part of the network, each primary molecule must have at least one tetrafunctional crosslinking site (or two trifunctional ones) per primary molecule.

If a system contains equal numbers of primary molecules and tetrafunctional crosslinking sites and if the crosslinking sites are distributed at random among the primary molecules, then the larger primary molecules will have a chance to have more than one crosslinking site per molecule. The smaller primary molecules, on the other hand, may not possess a crosslinking site at all. Hence, it is not the number fraction x (= mole fraction) of crosslinked primary molecules that is important but the mass fraction w (= weight fraction).

Example: 2 crosslinkable monomeric units are distributed at random among a polymer consisting of 1 primary molecule of degree of polymerization $X = 20$ and 1 primary molecule with $X = 10$. The expectation to find one of the crosslinkable sites in the molecule with $X = 20$ is twice as large as the expecation for the molecule with $X = 10$. The total number of monomeric units in the system is $10 + 20 = 30$. The expectations are therefore 20/30 for the large primary molecule and 10/30 for the small one. These expectations correspond to the mass fractions of *molecules* and not to their number fractions since for the large molecule

$$x_{20} = N_{20}/(N_{20} + N_{10}) = 1/(1 + 1) = 1/2$$
but $$w_{20} = m_{20}/(m_{20} + m_{10}) = 20/(20 + 10) = 2/3$$

The expectation ε_i to find a crosslinking site in a primary molecule with a degree of polymerization X_i must therefore also be proportional to the weight fraction of that molecule. Eq.(13-124) becomes

(13-125) $\varepsilon_i = p_u w_i(X_i - 1)$

and the mean expectation for the polymeric substance is now

(13-126) $\bar{\varepsilon} = \sum_i \varepsilon_i = \sum_i p_u w_i (X_i - 1) = p_u \sum_i (w_i X_i - w_i)$

Introduction of $\sum_i w_i \equiv 1$ and $\sum_i w_i X_i \equiv \bar{X}_w$ delivers for the critical probability $\bar{\varepsilon} = \bar{\varepsilon}_{crit} = 1$ of formation of crosslinkable monomeric units

(13-127) $p_{u,crit} = 1/(\bar{X}_w - 1) \approx 1/\bar{X}_w$

This critical probability equals the proportion of crosslinkable monomeric units in all units since the mass fraction of *units* equals their mole fraction. The critical fraction is therefore determined by the weight-average degree of polymerization and not the number average.

A-13   Appendix: Derivation of Eq.(13-25)

Eq.(13-24, left) is transformed to $-x_{AB} + (K/x_L)(1 - x_{AB})^2 = 0$. After adding "1" to each side of this equation, both sides are multiplied by $(4\ K/x_L)$. Again, "1" is added to both sides. After setting $2\ K(1 - x_{AB})/x_L = b$, the left side of the equation is transformed to $1 + 2\ b + b^2 = (1 + b)^2$. Taking the square root of both sides and solving for x_{AB} delivers Eq.(13-25).

Historical Notes to Chapter 13

Synthesis of Polymers by Polycondensation and Polyaddition

L.H.Baekeland, US-Pat. 942 699 (1907); -, Ind.Eng.Chem. **1** (1909) 149
Phenolic resins: First industrially important synthetic crosslinked polymer (thermoset) from phenol and formaldehyde by the heat-and-pressure process.

W.H.Carothers, G.A.Arvin, J.Am.Chem.Soc. **51** (1929) 2560
First controlled synthesis of linear condensation polymers: saturated aliphatic polyesters.

C.Ellis, US-Pat. 2 225 313 (6 August 1937; to Ellis-Foster)
Synthesis of unsaturated polyesters for thermosetting materials.

W.H.Carothers, US-Pat. 2 067 172 (16 February 1937; to DuPont)
Synthesis of linear "superpolymers" for fibers.

O.Bayer, H.Rinke, W.Siefken, L.Orthner, H.Schild, DP 728 981 (1937; to IG Farben)
First polyaddition (synthesis of polyurethanes).

W.H.Carothers, US-Pat. 2 130 523, 2 130 947, 2 130 948 (all dated 20 September 1938; to DuPont)
Synthesis of the first commercially successful fiber (polyamide 6.6).

J.R.Whinfield, J.T.Dickson, Brit.Pat. 578 079 (1941; to Calico Printers)
J.R.Whinfield, Nature **158** (1946) 930
Synthesis of poly(ethylene terephthalate). The use of terephthalic acid and its ester was not covered by the Carothers-DuPont umbrella patents which list very many monomers for linear polycondensations. Terephthalic acid was not an industrial chemical compound at that time.

Theory of Linear Polycondensation and Polyaddition

W.H.Carothers, J.Am.Chem.Soc. **51** (1929) 2548
Survey of literature and new classification of polymers as A-polymers (now: polymers by chain polymerization) and C-polymers (now: polymers by polycondensation)

W.Kuhn, Ber.Dtsch.Chem.Ges. **63** (1930) 1503
Theoretical derivation of the mass distribution of degrees of polymerization by random chain scission of linear polymer chains. The distribution generated by this equilibrium process is identical with the one derived later by Flory for polycondensation equilibria (see below).

W.H.Carothers, Chem.Revs. **8** (1931) 353; Trans.Faraday Soc. **32** (1936) 39
Conflicting results in the literature (crosslinking, low molar masses) are traced to the functionality of monomers. Linear polymers are possible only if the monomer functionality is 2 and high molecular weights only at high monomer conversions in stoichiometric reactions of AA and BB monomers.

P.J.Flory, J.Am.Chem.Soc. **58** (1936) 1877; Chem.Revs. **39** (1946) 137
Chemical reactivities of functional groups are independent of chain lengths; polycondensations have statistical character.

H.Jacobson, W.H.Stockmayer, J.Chem.Phys. **18** (1950) 1600
Theory of cyclization in linear polycondensations. Rings are small (usually $X_{ring} \leq 4$). Their proportion increases with increasing dilution.

For the treatment of both linear and non-linear polycondensations, see below.

Interfacial Polycondensation

A.Einhorn, Liebigs Ann.Chem. **300** (1898) 135
First interfacial polycondensation to a polycarbonate by the Schotten-Baumann reaction of an alcoholic solution of hydroquinone with a solution of phosgene in toluene.

H.Schnell, Angew.Chem. **68** (1956) 633; Ind.Eng.Chem. **51** (1959) 57
Industrial Schotten-Baumann reaction of phosgene and the sodium salt of bisphenol A to polycarbonate A.

A.J.Conix, Ind.Chim.Belge **22** (1957) 1457; Ind.Eng.Chem. **51** (1959) 147
Schotten-Baumann reaction to thermoplastic polyesters.

W.L.Wittbecker, P.W.Morgan, J.Polym.Sci. **40** (1959) 289;
P.W.Morgan, S.L.Kwolek, J.Polym.Sci. **40** (1959) 299
Polyamides, polyurethanes, and polycarbonates by interfacial polycondensation.

Theory of Branching and Crosslinking Step Reactions

P.J.Flory, J.Am.Chem.Soc. **63** (1941) 3083, 3091, 3096; **69** (1947) 30; J.Phys.Chem. **46** (1942) 132
Probability theory for relationships between functionality, degree of reaction, and monomer conversion in multifunctional polycondensations. Prediction of gel points.

W.H.Stockmayer, J.Chem.Phys. **11** (1943) 45; **12** (1944) 125; J.Polym.Sci. **9** (1952) 69; **11** (1953) 424
General statistical theory for condensation polymers from monomers of any functionality. The first paper shows in particular that Flory's expressions for the weight fractions of gel and sol fractions include the inherent assumption of intramolecular reactions in the gel although no provision for intramolecular reactions was made by the theory.

M.Gordon, Proc.R.Soc.London, Ser. *A*, **268** (1962) 240; D.S.Butler, G.N.Malcolm, M.Gordon, Proc.R.Soc.London, Ser. *A*, **295** (1966) 29; M.Gordon, T.G.Parker, Proc.R.Soc.Edinburgh, Sect. *A*, **69** (1970/71) 13; M.Gordon, T.C.Ward, R.S.Whitney, in A.J.Chompf, S.Newman, Eds., Polymer Networks, Plenum Press, New York 1971
Percolation theory of multifunctional polymerizations.

C.W.Macosko, D.R.Miller, Macromolecules **9** (1976) 199 (a); D.R.Miller, C.W.Macosko, Macromolecules **9** (1976) 206 (b); Macromolecules **11** (1978) 656 (c)
Recursive method for calculating properties of nonlinear polymers. (a) Molecular weights and gel points; (b) post-gel properties of networks; (c) properties of polymers from monomers with unequal reactivities.

D.Durand, C.-M.Bruneau, J.Polym.Sci.-Phys.Ed. **17** (1979) 273; Macromolecules **12** (1979) 1216; Polymer **23** (1982) 69; Makromol.Chem. **183** (1982) 1007, 1021; Polymer **24** (1983) 587, 592; M.Adam, M.Delsanti, D.Durand, Macromolecules **18** (1985) 2285
Application of graph theory to the calculation of pre-gel and post-gel properties of polymers from multifunctional step polymerizations.

Literature to Chapter 13

13.1 OVERVIEW
G.J.Howard, The Molecular Weight Distribution of Condensation Polymers, in J.C.Robb,
    F.W.Peaker, Eds., Progress in High Polymers 1 (1961) (early competing theories)
G.F.Ham, Ed., Kinetics and Mechanism of Polymerizations, Vol. 3, Condensation Polymerization,
    Dekker, New York 1967
I.B.Sokolov, Synthesis of Polymers by Polycondensation (in Russian), Nauka, Moskau 1966; Israel
    Program for Scientific Translations, Jerusalem 1968
J.H.Peebles, Jr., Molecular Weight Distributions in Polymers, Wiley, New York 1971
J.K.Stille, T.W.Campbell, Eds., Condensation Monomers, Wiley-Interscience, New York 1972
D.H.Solomon, Ed., Step-Growth Polymerizations, Dekker, New York 1972
G.Allen, J.C.Bevington, Eds., Comprehensive Polymer Science, Vol. 5, G.C.Eastmond, A.Ledwith,
    S.Russo, P.Sigwalt, Eds., Step Polymerization, Pergamon Press, Oxford 1989
M.E.Rogers, T.E.Long, Eds., Synthetic Methods in Step-Growth Polymers, Wiley-Interscience,
    Hoboken (NJ) 2003
C.E.Carraher, Jr., G.G.Swift, Functional Condensation Polymers, Kluwer Academic-Plenum, New
    York 2003

13.2-13.3 LINEAR STEP REACTIONS
A.M.Kotliar, Interchange Reactions Involving Condensation Polymers, J.Polym.Sci.-Macromol.
    Revs. 16 (1981) 367
S.Fakirov, Ed., Transreactions in Condensation Polymers, Wiley-VCH, Weinheim 1999

13.4 KINETICS OF LINEAR STEP REACTIONS
V.V.Korshak, S.V.Vinogradova, Irreversible Polycondensation (in Russian), Nauka, Moskau 1972
J.H.Saunders, F.Dobinson, The Kinetics of Polycondensation Reactions, in C.H.Bamford,
    C.F.H.Tipper, Eds., Comprehensive Chemical Kinetics, Vol. 15, Non-Radical Polymerisation,
    Elsevier, Amsterdam 1976
A.Fradet, E.Maréchal, Kinetics and Mechanisms of Polyesterifications, Adv.Polym.Sci. 43 (1982) 51
S.K.Gupta, A.Kumar, Reaction Engineering of Step Growth Polymerization, Plenum Press, New
    York 1987

13.5 REACTIVITIES
P.W.Morgan, Condensation Polymers: By Interfacial and Solution Methods, Interscience,
    New York 1965
A.S.Hay, Polymerization by Oxidative Coupling - A Historical Review, Polym.Eng.Sci. 16
    (1976) 1
N.Yoda, M.Kurihara, New Polymers of Aromatic Heterocycles by Polyphosphoric Acid Solution
    Methods, J.Polym.Sci. D [Macromol.Revs.] 5 (1971) 109
F.Millich, C.E.Carraher, Eds., Interfacial Synthesis, Dekker, New York 1977 (2 vols.)
N.Yamazaki, F.Higashi, New Condensation Polymerizations by Means of Phosphorus Compounds,
    Adv.Polym.Sci. 38 (1981) 1
C.E.Carraher, Jr., J.Preston, Interfacial Synthesis, Vol. III, Dekker, New York 1982 (reprint of papers
    in J.Macromol.Sci.-Chem. **A 15/5**)
V.Percec, J.H.Wang, R.S.Clough, Mechanisms of the Aromatic Polyetherification Reactions,
    Makromol.Chem., Macromol.Symp. 54/55 (1992) 275

13.6 RING FORMATION
N.Yoda, M.Kurihara, N.Dokoshi, New Synthetic Routes to High Temperature Polymers by Cyclo-
    condensation Reactions, Progr.Polym.Sci. 4 (1972) 1
V.V.Korshak, The Principal Characteristics of Polycyclotrimerization, Vysokomol.Soyed. A 16
    (1974) 926; Polym.Sci.USSR 16 (1974) 1066
J.A.Semlyen, Ed., Cyclic Polymers, Elsevier, New York 1986

13.7 COPOLYCONDENSATIONS AND COPOLYADDITIONS
S.I.Kuchanov, Distribution of Monomer Units in Products of Homogeneous Irreversible Copoly-
    condensation, Vysokomol.Soyed. A 15 (1973) 2140; Polym.Sci.USSR 15 (1973) 2434

V.V.Korshak, S.V.Vinogradova, S.I.Kuchanov, V.A.Vasney, Non-Equilibrium Copolycondensation in Homogeneous Systems, J.Macromol.Sci. [Revs.] C 14 (1976) 27

J.-C.Bollinger, Characterization of Block Structures in Copolycondensates: A Review, J.Macromol. Sci. [Revs.] C 16 (1977/78) 23

J.-C.Bollinger, Synthesis and Properties of Block Copolycondensates: A Review of Recent Advances, Progr.Polym.Sci. 9 (1983) 59

H.A.Nguyen, E.Maréchal, Synthesis of Reactive Oligomers and Their Use in Block Polycondensation, J.Macromol.Sci.-Revs.Macromol.Chem.Phys. C 28 (1988) 187

13.8 HYPERBRANCHING REACTIONS

J.R.Schaefgen, P.J.Flory, Synthesis of Multichain Polymers and Investigation of Their Viscosities, J.Am.Chem.Soc. 70 (1948) 2709 (system AB + A_f)

P.J.Flory, Principles of Polymer Chemistry, Cornell University Press, Ithaca (NY), 1953, p. 347 ff.

C.W.Macosko, D.R.Miller, A New Derivation of Average Molecular Weights of Nonlinear Polymers, Macromolecules 9 (1976) 199

H.Tobita, Random Sampling Technique to Predict the Molecular Weight Distribution in Nonlinear Polymerization, Macromol.Theory Simul. 5 (1996) 1167

D.Hölter, H.Frey, Degree of Branching in Hyperbranched Polymers. 2. Enhancement of the DB: Scope and Limitations, Acta Polymerica 48 (1997) 298

U.Beginn, C.Drohmann, M.Möller, Conversion Dependence of the Branching Density for the Polycondensation of AB_n Monomers, Macromolecules 30 (1997) 4112

W.Radke, G.Litvinenko, A.H.E.Müller, Effect of Core-Forming Molecules on Molecular Weight Distribution and Degree of Branching in the Synthesis of Hyperbranched Polymers, Macromolecules 31 (1998) 239 (this paper treats self-condensing vinyl polymerizations)

13.9 CROSSLINKING STEP POLYMERIZATIONS

P.J.Flory, Introductory Lecture. Gels and Gelling Processes, Faraday Disc.Chem.Soc. 57 (1974) 7

M.Gordon, S.Ross-Murphy, The Structure and Properties of Molecular Trees and Networks, Pure Appl.Chem. 43 (1975) 1

D.R.Miller, C.W.Macosko, A New Derivation of Post Gel Properties of Network Polymers, Macromolecules 9 (1976) 206

D.Stauffer, Introduction to Percolation Theory, Taylor and Francis, London 1985

C.J.Brinker, G.W.Scherer, Sol-Gel Science, Academic Press, San Diego 1990

13.9.7 CROSSLINKING POST-REACTIONS

W.W.Graessley, Entangled Linear, Branched and Network Polymer Systems - Molecular Theories, Adv.Polym.Sci. 30 (1979) 89

L.H.Sperling, Interpenetrating Polymer Networks and Related Materials, Plenum, New York 1981

J.E.Mark et al., Polymer Networks, Adv.Polym.Sci. 44 (1982) 1

K.Dusek, Formation and Structure of End-Linked Elastomer Networks, Rubber Chem.Technol. 55 (1982) 1

D.Stouffer, A.Coniglio, M.Adam, Gelation and Critical Phenomena, Adv.Polym.Sci. 44 (1982) 103

S.C.Temin, Recent Advances in Crosslinking, J.Macromol.Sci.-Revs.Macromol.Chem.Phys. C 22 (1982/83) 131

S.P.Pappas, Ed., UV Curing: Science and Technology, Technology Marketing Corp., Norwalk (CT), 2 vols., 1978 and 1985

D.J.P.Harrison, W.R.Yates, J.F.Johnson, Techniques for the Analysis of Crosslinked Polymers, J.Macromol.Sci.-Revs.Macromol.Chem.Phys. C 25 (1985) 481

O.Kramer, Ed., Biological and Synthetic Polymer Networks, Elsevier Appl.Sci., New York 1988

A.Baumgärtner, C.Picot, Eds., Molecular Basis of Polymer Networks, Springer, Berlin 1989

O.Guven, Crosslinking and Scission in Polymers, Kluwer, Dordrecht 1990

W.Burchard, S.B.Ross-Murphy, Eds., Physical Networks, Elsevier Appl.Sci., Amsterdam 1990

C.J.Brinker, G.W.Scherer, Sol-Gel Science, Academic Press, San Diego 1990

D.Stauffer, A.Aharony, Introduction to Percolation Theory, Taylor and Francis, London 1992

M.Sahimi, Appliction of Percolation Theory, Taylor and Francis, London 1994

S.C.Kim, L.H.Sperling, Eds., IPNs Around the World, Wiley, New York 1997

R.F.T.Stepto, Ed., Polymer Networks-Principles of Their Formation, Structure, and Properties, Blackie Academic, Glasgow, UK, 1997

References to Chapter 13

[1]   C.Giori, B.T.Hayes, J.Polym.Sci. [A-1] **8** (1970) 335 (Fig. 1), 351 (Fig. 8, Table I)
[2]   P.Matthies, Polyamide, in Ullmanns Enzyklopädie der Technischen Chemie, Verlag Chemie, Weinheim, 4th ed., **19** (1980) 41
[3]   H.-G.Elias, unpublished
[4]   M.Gordon, W.B.Temple, T.G.Parker, J.A.Love, J.Prakt.Chem. **313** (1971) 411, Fig. 2
[5]   P.J.Flory, J.Am.Chem.Soc. **61** (1939) 3334, Table I
[6]   H.Dostal, R.Raff, Monatsh.Chem. **68** (1936) 188, Table 7
[7]   H.-G.Elias, Makromol.Chem. **186** (1985) 847, Fig. 1
[8]   B.V.Bhide, J.J.Sudborough, J.Indian Inst.Sci. **8A** (1925) 89; as reported by P.J.Flory, Principles of Polymer Chemistry, Cornell Univ. Press, Ithaca (NY) 1953, p. 71
[9]   S.I.Kuchanov, M.L.Keshtov, P.G.Halatur, V.A.Vasnev, S.V.Vinogradova, V.V.Korshak, Makromol.Chem. **184** (1983) 105, Fig. 1
[10] G.-C.Wu, H.Tanaka, K.Sanui, N.Ogata, Polym.J. **14** (1982) 571, Fig. 4
[11] T.-y.Yu, S.-k.Fu, S.-j.Li, C.-g.Ji, W.-z.Cheng, Polymer **25** (1984) 1363, Figs. 1 and 2
[12] K.Weisskopf, G.Meyerhoff, Eur.Polym.J. **10** (1985) 859, selective data of Fig. 2
[13] H.-G.Elias, J.-H.(L) Tsao, J.Palacios, Makromol.Chem. **186** (1985) 893, Fig. 6
[14] P.W.Morgan, S.L.Kwolek, J.Polym.Sci. **40** (1959) 299, (a) Fig. 10, (b) Table III
[15] H.-B.Tsai, Y.-D.Lee, J.Polym.Sci. A-Polym.Chem. **25** (1987) 1505, 2195
[16] V.Percec, T.D.Shaffer, J.Polym.Sci. C (Polymer Letters) **24** (1986) 439, Fig. 1-3
[17] A.H.Fawcett, R.A.W.Mee, F.V.McBride, Macromolecules **28** (1995) 1481, Fig. 3
[18] (a) K.Ziegler, R.Aurnhammer, Liebigs Ann.Chem. **513** (1934) 113; (b) K.Ziegler, W.Hechelhammer, Liebigs Ann. Chem. **528** (1937) 114
[19] G.Ercolani, L.Mandolini, P.Mencarelli, Macromolecules **21** (1988) 1241, Fig. 2 and additional calculated data
[20] Y.Kawabata, M.Kinoshita, Makromol.Chem. **176** (1975) 2797, data of Table 1
[21] L.F.Beste, J.Polym.Sci. **36** (1959) 313
[22] U.W.Suter, P.Pino, Macromolecules **17** (1984) 2248, Figs. 1 and 2
[23] S.R.Turner, B.I.Voit, T.H.Mourey, Macromolecules **26** (1993) 4617, Fig. 1
[24] T.M.Miller, T.X.Neenan, E.W.Kwock, S.M.Stein, Macromol.Symp. **77** (1994) 35, data of Table 1
[25] D.Schmaljohann, H.Komber, J.G.Barratt, D.Appelhans, B.I.Voit, Macromolecules **36** (2003) 97, data of Fig. 4
[26] C.Cameron, A.H.Fawcett, C.R.Hethrington, R.A.W.Mee, F.C.McBride, ACS Polymer Preprints **38/1** (1997) 56, Fig. 1
[27] J.M.J.Fréchet, M.Henmi, I.Gitsov, S.Aoshima, M.R.Leduc, R.B.Grubbs, Science **269** (25 August 1995) 1080, data of Table 1
[28] A.H.E.Müller, D.Yan, M.Wulkow, Macromolecules **30** (1997) 7015; (a) Eqs.(21-(23); (b) data of Fig. 1 of [27]
[29] A.Sunder, R.Hanselmann, H.Frey, R.Mülhaupt, Macromolecules **32** (1999) 4240, Tables 1 and 2
[30] P.J.Flory, J.Am.Chem.Soc. **63** (1941) 3083, Fig. 2
[31] J.W.Stafford, J.Polym.Sci.-Chem.Ed. **19** (1901) 3219, Table IX
[32] J.J.Bernardo, P.F.Bruins, J.Paint Technol. **40** (1968) 558, Table 3-6
[33] D.S.Argyropoulos, R.M.Berry, H.I.Bolker, J.Polym.Sci., Part B, Polym.Phys. **25** (1987) 1191; Makromol.Chem. **188** (1987) 1985
[34] D.S.Argyropoulos, R.M.Berry, H.I.Bolker, Macromolecules **20** (1987) 357 $(p > p_{crit})$
[35] N.S.Clarke, C.J.Devoy, M.Gordon, Brit.Polym.J. **3** (1971) 194 $(p < p_{crit})$
[36] R.H.Colby, M.Rubinstein, J.R.Gillmor, T.H.Mourey, Macromolecules **25** (1992) 7180, Table II

14 Biological Polymerization

14.1 Cell Biology

14.1.1 Overview

Biopolymers are naturally occurring macromolecular compounds. The same name is also used for synthetic compounds if they are composed of the same or similar monomeric units as those of natural biopolymers. The most important classes of biopolymers are nucleic acids, proteins, polysaccharides, polydienes, lignins, and aliphatic poly(α-hydroxy alkanoate)s. Practically all types of molecular architectures are represented: homopolymers, periodic and random copolymers, block and graft copolymers and linear, branched, and crosslinked polymers, not only within a class but also as combinations of members of two classes. Macroconformations comprise random coils, helical structures, spheres, sheets, and various complex structures.

Most biological polymers are synthesized *intracellularly* in plant or animal cells; only a few are produced *extracellularly*. The synthesis of biopolymers does not markedly disturb the osmotic equilibrium between the interior and exterior of cells because the high molecular weights of biopolymers produce only small osmotic pressures (see Volume III). Some biosyntheses proceed "stepwise" at molecular matrices, for example, those of nucleic acids and proteins. However, they are not "step reactions" in the sense of the polycondensations of Chapter 13 but rather matrix-controlled polymerizations. Other biosyntheses are chain reactions in which small or large leaving molecules are produced; they are polyeliminations (Chapter 6).

All biosyntheses are highly specific in the selection of monomers that are present in cells. These monomers are called **substrates** like all starting materials in biological reactions. Biosyntheses are also highly regioselective and stereoselective. They proceed with high speed at the "normal" temperatures of 0-40°C for living beings. Biosyntheses are catalyzed by four types of **biocatalysts**:

Vitamins (L: *vita* = life; "amine" because thiamine (= vitamin B_1) was the first substance that was recognized as vitally necessary) are a group of chemically very different chemical compounds that are mainly used for the synthesis of so-called co-enzymes. They are usually produced by plants. Animals acquire them with food. Microorganisms produce vitamins only in rare cases, for example, bacteria in the colon.

Hormones also have very different chemical structures. They do not catalyze directly but interact with biological receptors so that the latter can act enzymatically (G: present participle of *horman* = to urge on, to stimulate). Hormones are transported by blood from their loci of synthesis to their places of action in living cells and tissues.

Enzymes (G: *en* = in; *zyme* = to leaven, from their action in fermented dough (sour dough)) are a special class of proteins = copolymers of α-amino acids and imino acids. Extracellular enzymes reside ouside of biological cells. Intracellular enzymes are not distributed homogeneously in cells but are enriched in certain cell compartments or cell organelles, where they are present either free (cytosol enzymes) or bound to membranes.

Nucleic acid enzymes are catalytically active nucleic acids. s (**ribozymes**) catalyze the cleavage and reassembly of aminoacyl bonds in nucleic acids. **DNA enzymes** catalyze the scission of bonds between RNA and phosphoric acid groups.

14.1.2   Cells

Most biological polymerizations proceed in cells but studies of these polymerizations are usually done in vitro and not in vivo because chemical reactions in cells are very difficult to monitor. However, bioreactions are very much influenced by the structure of cells so that a short and very simplified review of cell structures may be helpful. Special emphasis is placed on macromolecular aspects.

All living organisms consist of one or more **cells** (L: *cella* = chamber) which are the smallest structural units that, in principle, can function independently. A cell consists of a **cytosol** (= **cytoplasm** (G: *kytos* = hollow vessel; *plassein* = to mold)) that is surrounded by a **cell membrane** (= **cytoplasmic membrane, plasma membrane** (L: *membrana* = skin covering a body, from *membrum* = member)) (Fig. 14-1). Higher cells incorporate **organelles** of which the **nucleus** (= **karyon**; L: *nucleus*, G: *karyon* = nut, kernel) is the most important. The nucleus contains chromatin (Section 14.2.4).

Following English tradition to Latinize Greek names and words (e.g., Heraclitus instead of Heraklitos) and Anglicize Roman names (e.g., Pliny instead of Plinius), one finds several spellings for these terms. For example, the etymologically correct term "proka̱ryo̱ntic" seems to be spelled "proka̱ryotic" in more recent literature but "procaryotic" in most older books (there is no "c" in Greek!).

Depending on the number of cells per organism and the presence or absence of a nucleus, all living organisms can be grouped in five or six kingdoms or three empires (Table 14-1). There is a tendency to subdivide the kingdom of *Monera* into the empires of *Bacteria* and *Archaea* because the genes of the latter resemble those of *Eukaryotae*.

Animalia (animals), *Plantae* (plants), *Fungi* (fungi), and *Protista* (algae, molds, protozoa) have a nucleus that is surrounded by a cell membrane; they are thus called **eukaryotic** (G: *eus* = true, good, beautiful). *Bacteria* and *Archaea* (formerly: *Archaebacteria*) have no nucleus but only a nucleus-like region called **nucleoid**; they are named **prokaryotic** because they appeared geologically earlier (L: *pro* = before). *Eukaryotae* contain various **cell organelles** (mitochondria, lysosomes, etc.) that are bound to a **cytoskeleton**. The cytoplasms of *Prokaryotae* do not contain cell organelles but so-called **inclusion bodies** which may be gas vacuoles or embedded particles composed of fat, proteins, polysaccharides, poly(hydroxyalkanoic acid)s, etc.

Prokaryotic cells are primitive cells that are 1-10 μm long. Eukaryotic cells are much more complex. They are usually about 10 times as large as prokaryotic cells but may be even several meters long in spurges (*Euphorbiaceae*).

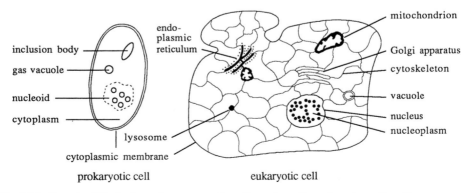

Fig. 14-1 Oversimplified schematic representation of a prokaryotic and a eukaryotic cell.

Table 14-1 Classification of living organisms.

	Five kingdoms Whittaker (1959)		Six kingdoms	Three empires or domains Woese (1990)	
uni-cellular	*Monera*	{	*Bacteria* *Archaea*	*Bacteria* *Archaea*	no nucleus or other organelles
	Protista		*Protista*		
multi-celluar	*Fungi* *Plantae* *Animalia*		*Fungi* *Plantae* *Animalia*	*Eukaryota*	nucleus and organelles

Cell Membranes

Cell membranes enclose cells and, in eukaryotic cells, also nuclei and cell organelles. They consist of 7-10 nm thick, highly mobile (η = 0.1-1 Pa s), liquid-crystalline double layers of lipids and, in part, glycolipids. The double layers embed 20-80 % glycoproteins. Many of these glycoproteins function as channels for the passage of ions and smaller molecules by osmosis, diffusion, or carriers, thus allowing the transport of nutrients into the cell and that of wastes out of it.

Lipids are mainly phopholipids such as phosphatidyl choline. The polar (ionic) headgroups face toward the exterior of the membrane and the two long hydrophobic tails the interior. Outer cell membranes are stabilized by layers of polymers, usually only on one side, for example, crosslinked spectrin on the inner surface of the membrane of erythrocytes. The cell membranes of plants, fungi, and bacteria are fortified by more or less solid cell walls. Animal cells are stabilized by an interior cytoskeleton (no cell wall).

$$CH_3(CH_2)_iCOOCH_2 \diagdown$$
$$CH_3(CH_2)_jCOO \diagup CH-CH_2-O-\overset{\overset{O}{\|}}{\underset{O^\ominus}{P}}-CH_2-CH_2-\overset{\oplus}{N}(CH_3)_3 \quad \text{phosphatidyl choline}$$

Cell Walls

The cell walls of cells of *plants* consist of cellulose. The interstitial space between plant cells is filled with pectates and sometimes also with hemicelluloses. These cell walls can lignify by incorporation of lignin or cork by deposition of suberic acid.

The cell wall of cells of *fungi* is composed of chitin.

Cell walls of *prokaryotic cells* have much more complex structures. In cyanobacteria (formerly called blue-green algae), they consist of three layers: cellulose in the innermost one, a gelatinous membrane (= middle lamella) consisting of Mg and Ca pectates in the center, and mucopolysaccharides on the outside.

Cell walls of *bacteria* have an inner layer of proteins and an outer layer of murein, a bag-like giant molecule of the size of the bacterium. Murein consists of polysaccharide chains that are crosslinked by short peptide chains. All bacteria have the same polysaccharide chains but different peptide components. Two types of bacterial cell walls can be distinguished by the differential staining method of H.C.J.Gram: Gram-positive bacteria are colored dark-blue to blackish blue but Gram-negative bacteria appear red.

In Gram-positive bacteria (*pneumococcus, streptococcus, staphylococcus*), teichoic acids (Section 14.2.1) are covalently bound to murein via phosphodiester bridges. Gram-negative bacteria (*Escherichia coli, salmonella, klebsiella*) have additional exterior membranes composed of lipids, proteins, and lipid-containing proteins. The cell walls of these bacteria contain hydrophobically bound lipopolysaccharides that are diblock polymers composed of a lipid A–block and a polysaccharide B–block. The lipid A–blocks are oligomers that are structurally similar for all bacteria. The polysaccharide blocks differ strongly; they often contain uncommon types of sugars.

Bacteria are also often encapsulated by polysaccharides, usually acidic ones. In many cases, these polysaccharides are segmented block copolymers of four different blocks composed of seven different types of sugar residues. These cell layers make Gram-negative bacteria very resistant against external influences, for example, pharmaceuticals.

Archaea differ from both bacteria and eukaryotae; they live under harsh conditions (geysers, black smokers in the deep seas) and also in marshes and sewage. Their membranes are lipidic; they consist of branched hydrocarbons that are linked to glycerol by ether groups. The cell walls of *archaea* lack peptidoglycans.

Protoplasma

The jellylike material inside a cell (**protoplasma**) consists of the **nucleoplasma** inside the nucleus and the **cytoplasm** between the nucleus and the cell wall (G: *kytos* = cave). The cytoplasm contains the **cytosol** in which various cell organelles, the cytoskeleton, and different granular materials are embedded (Fig. 14-1).

Prokaryotae have few, if any, cytoplasmic structures. In *Eukaryotae*, **cytoplasmic structures** comprise intercellular membranes similar to the cytoplasmic membrane and the **cytoskeleton** (**endoplasmic reticulum**; G: *endo* = within; L: *reticulum* = diminutive of *rete* = net) that supports higher cells. The cytoskeleton connects the plasma membrane with the cell organelles. It consists of filaments, microtubules, etc., of proteins such as actin (in muscle cells) or spectrin (in erythrocytes).

Cell organelles are responsible for various biological functions:

Ribosomes are ca. 25 nm in diameter. They are nucleoproteins (Section 14.2.4) that are responsible for the protein synthesis (*ribo* → ribose; G: *soma* = body).

Lysosomes are membrane bags that contain lysozymes, DNA, RNA, and carbohydrates. They digest proteins and dead cells (G: *luein* = to loosen; *zyme* = leaven).

The **Golgi apparatus** consists of a system of membrane bags that contain carbohydrates and digesting enzymes. The bags then move to the outer cell membrane where the content is emptied.

Mitochondria are microscopic bodies in eukaryotes that are assumed to be degenerated prokaryotes. A mitochondrion is about 500-1000 nm wide and 5000-10 000 nm long (G: *mitos* = thread; *chondrion* = small grain). Mitochondria contain various enzymes, for example, for the oxidation of sugars to pyruvic acid.

Composition of Cells

Cells contain very many molecules (Table 14-2); their molecules must therefore be tightly packed. For example, the rod-like cell of the bacterium *Escherichia coli* has a length of 4 μm, a diameter of 1.4 μm, and a volume of $V_{cell} \approx 6.2 \cdot 10^9$ nm^3. Its nucleic acid molecules are cyclic molecules that form double helices. As linear, randomly coiled molecules, they would require a volume of $38 \cdot 10^9$ nm$^3 > V_{cell}$. As linear rigid rods with a diameter of 2 nm and a volume of 0.018 nm$^3 < V_{cell}$, their lengths would be 1.3 mm $> V_{cell}$. The rings of DNA must therefore be substantially coiled.

Table 14-2 Composition of a cell of *Escherichia coli*. DNA = deoxyribonucleic acid, RNA = ribonucleic acid (m = messenger, r = ribosomal, t = transfer); M_r = molecular weight (* average per class); w = mass fraction; 23 S, 16 S = sedimentation coefficient as a measure of molecular weight.

Class	Number of types	M_r		molecules per cell	w in %
DNA	1	2 500 000 000		4	1.11
23 S r-RNA	1	1 000 000		30 000	3.32
16 S r-RNA	1	500 000		330 000	1.66
t-RNA	60	25 000		400 000	1.11
m-RNA	1000	1 000 000		1 000	0.11
Nucleotides	200	300	*	12 000 000	0.40
Proteins	2500	40 000	*	1 000 000	4.42
Amino acids	20	100	*	30 000 000	0.33
Lipids	50	750	*	25 000 000	2.07
Carbohydrates	150	200	*	200 000 000	4.43
Other organic molecules	250	150	*	17 000 000	0.28
Inorganic ions	20	40	*	250 000 000	1.11
Water	1	18		40 000 000 000	79.65
Total	4354			40 500 000 000	100.00

14.1.3 Biological Reactions

A human consists of 10^{14} cells, each of which is about 1000 times larger than a bacterial cell. Each human cell contains about 100 000 enzyme molecules that catalyze a total of 1000-2000 different chemical reactions. Each type of reaction thus involves only 50-100 enzyme molecules per cell. The reactions take place in a confined space that is crowded with molecules which do not participate directly in the particular reaction but may exert nearest-neighbor effects on the substrates (educts) and the products.

In biochemistry, a **substrate** is an essential carrier of certain biological, chemical, and/or physical properties (L: *sub* = under, *sternere* = to spread out flat), for example, a culture medium. In the more narrow sense of biological *reactions*, it is a starting substance (educt) that is converted by a specific enzyme, for example, a monomer.

All educts must be brought into the cell, all products out of it. However, lipid membranes such as cell membranes only let pass through oil-soluble hydrocarbon molecules with less than 5 carbon atoms by simple permeation. More complicated low-molecular weight compounds such as sugars, amino acids, or nucleotides are transported through the cell walls of primitive membranes by coupling to special carriers whereas the **activated transport** through cell membranes of higher cells employs special proteins.

Polymer molecules are brought into cells by neither diffusion nor by activated transport but by **endocytosis**, either by **pinocytosis** of liquids (G: *pinein* = to drink) or by **phagocytosis**, the capture of solids by specialized cells (macrophages and monocytes) (G: *phagein* = to eat) which involves envelopment of the solid by the membrane.

All endocytosed molecules and particles move into lysosomes which contain about 50 types of lysing enzymes (G: *lyein* = to dissolve) that hydrolyze nucleic acids, proteins, and polysaccharides to their monomers. Substrate-specific carriers then transport the monomers through the cell membrane of lysosomes into the cytoplasm.

14.2   Nucleic Acids

14.2.1   Chemical Structure

Nature produces two types of organic esters of phosphoric acid: teichoic acids and nucleic acids. **Teichoic acids** are primitive precursors of nucleic acids that reside in the cell walls of bacteria (G: *teichos* = wall). They are polyesters of phosphoric acid and gly-cerol or ribitol (Section 14.4.2) which are, in part, substituted by D-alanine, *N*-acetylgly-cosamine, or sugars. Teichoic acids are often crosslinkend by murein (p. 513).

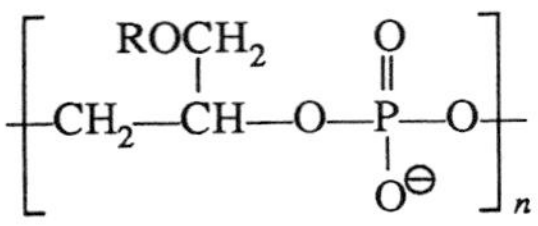

R = H, D-alanyl, sugars 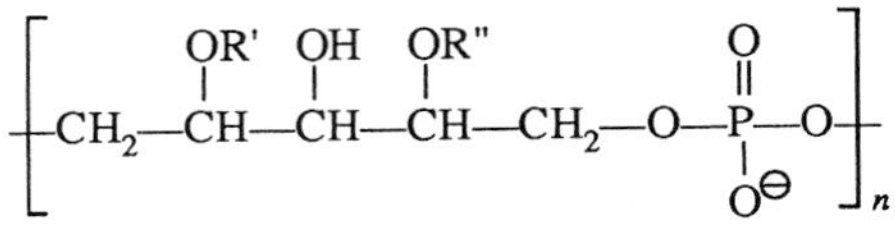   R' = H, D-alanyl; R" = H, *N*-acetylglucosamine

Nucleic acids reside in the nuclei and nucleoids of cells. They are copolyesters of phosphoric acid with the sugars ribose or 2'-deoxyribose. These two pentoses are present as furanoses that are substituted by purine or pyrimidine bases. The chemical com-pounds of these bases with sugars are called **nucleosides. Nucleotides** are **nucleoside phosphates**, i.e., phosphoric acid esters of nucleosides.

Ribonucleic acids (RNAs) are phosphoric acid esters of ribose that are substituted by *four* types of bases: the purine bases adenine and guanine and the pyridine bases cyto-sine and uracil. In prehistoric times, these bases probably arose from hydrocyanic acid (hydrogen cyanide, HCN); for example, adenine is a pentamer of HCN. The four bases are bound β-glycosidically to ribose, resulting in the nucleosides adenosine, guanosine, cytidine,and uridine (Fig. 14-2, Table 14-3).

Deoxyribonucleic acids (DNAs) have 2'-deoxyribose units instead of ribose units. Like RNAs, they contain the three bases adenine, guanine, and cytosine. However, the uracil of RNAs is replaced in DNAs by its 3-methyl-substituted derivative, thymine. DNAs of several mono-celled algae contain 5 different types of bases, i.e., additional 5-hydroxymethyl uracil or even 6 different ones, additional N^6-methyladenine.

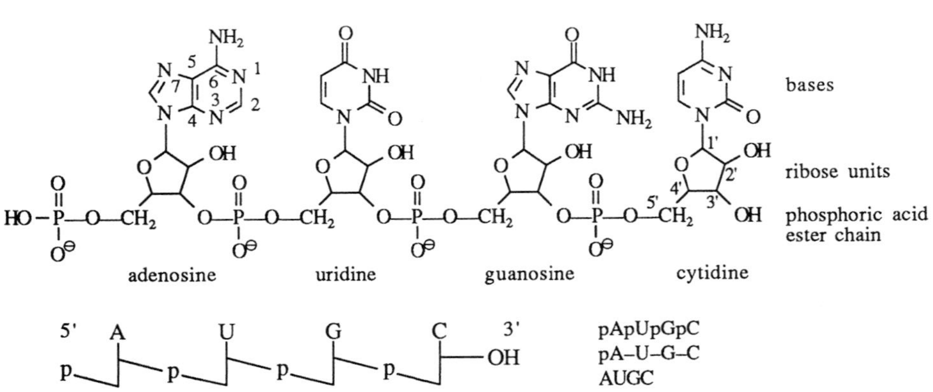

Fig. 14-2 Top: Tetrameric RNA with two purine nucleosides (adenosine, guanosine) and two pyr-imidine nucleosides (uridine, cytidine). Bottom: Various short-hand notations for this tetramer.

Table 14-3 Names of units of nucleic acids and their presence in RNAs and DNAs.

Presence	Bases		Nucleosides		Nucleotides	
RNA, DNA	adenine	(Ade)	adenosine	(A)	adenylic acid	(Ado)
RNA, DNA	guanine	(Gua)	guanosine	(G)	guanidylic acid	(Guo)
RNA, DNA	cytosine	(Cyt)	cytidine	(C)	cytidylic acid	(Cyd)
RNA	uracil	(Ura)	uridine	(U)	uridylic acid	(Urd)
DNA	thymine	(Thy)	thymidine	(T)	thymidylic acid	(Thd)

Cells of higher organisms may also contain small proportions of base derivatives in addition to the five basic types adenine, guanine, cytosine, uracil, and thymine (= 5-methyluracil). 5-Methylcytosine is present in plant RNAs (up to 6 %) and in certain mammals (up to 1.5 %). RNAs may also have 1-methylguanine or dihydrouracil. The two latter derivatives are produced by modifying the bases *after* in vivo synthesis.

Nucleotides are connected in 5',3'-positions, leading to one hydroxy endgroup at both the 5' and 3' positions. In representations of the structure, the 5' end of the molecule is always to the left. A phosphoric acid group in the 5' position is indicated by a p to the left of the symbol for the nucleoside and one in the 3' position by a p to the right of the symbol. Nucleosides of deoxyribonucleic acids carry a d in front of their names.

The internucleotide bond is sensitive to hydrolysis. The glycosidic bond between furanose groups and base groups can be attacked by acids, especially in purine nucleotides. RNAs are futhermore alkali labile because of the OH groups in 2' position.

According to the **Chargaff rule**, each nucleic acid contains the same number of adenine and thymine groups (or adenine and uracil) and the same number of guanine and cytosine groups (incl. 5-methylcytosine) (Table 14-4), resulting in **base pairs A/T, A/U, G/C, and G/5MC**. DNAs generally obey this rule very well but RNAs sometimes depart from it significantly. This behavior is caused by the pronounced structural differences between DNA and RNA molecules (Sections 14.2.2 and 14.2.3).

The Chargaff rule arises because purine bases are complementary to pyrimidine bases because of their ability to form hydrogen bridges, either intermolecularly (as in most DNAs) or intramolecularly (as in most RNAs) (see below).

Table 14-4 Composition of various DNAs and RNAs. * Hydroxymethylcytosine.

Source	Mole fraction of bases					Molar ratio	
DNA source	A	T	G	C	5MC	A/T	G/(C+5MC)
Human thymus	0.309	0.294	0.199	0.198	-	1.05	1.00
Calf thymus	0.282	0.278	0.215	0.212	0.013	1.01	0.96
Wheat germ	0.265	0.270	0.235	0.172	0.058	0.98	1.02
T2-Bacteriophages	0.325	0.325	0.182	-	0.168*	1.00	1.08
RNA source	A	U	G	C	-	A/U	G/C
Calf liver	0.195	0.164	0.350	0.291	-	1.19	1.20
Chicken liver	0.195	0.207	0.333	0.265	-	0.94	1.25
Baker's yeast	0.251	0.246	0.302	0.201	-	1.02	1.50
Tobacco mosaic virus	0.299	0.263	0.254	0.185	-	1.14	1.37

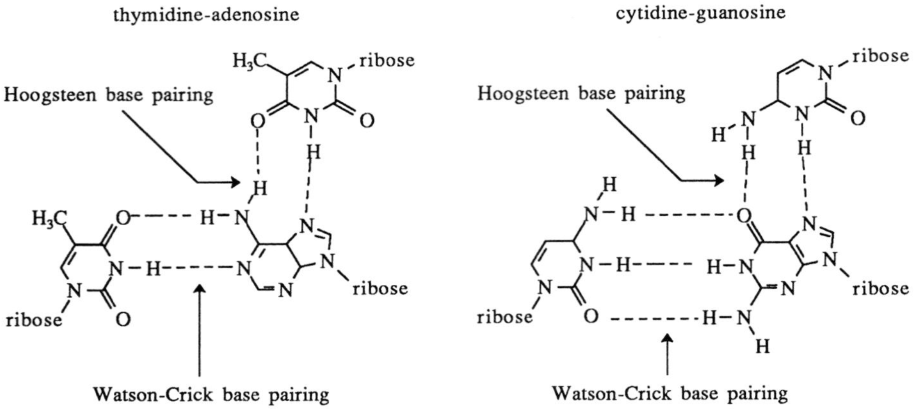

thymidine-adenosine

cytidine-guanosine

Fig. 14-3 Watson-Crick and Hoogsteen base pairings. Watson-Crick: lengths of hydrogen bonds are 0.280 nm (T/A) and 0.290 nm (C/G) for =O···H–N< and 0.300 nm (T/A and C/G) for >N···H–N<.

The **Watson-Crick model** proposes two hydrogen bonds for the base pair A-T (or A-U) and three hydrogen bonds for the base pair G-C (or G-5MC) whereas the **Hoogsteen model** envisages two hydrogen bonds for each base pair (Fig. 14-3). The higher stability of the pair G-C is also reflected in the stability constants $K = [XY]/([X][Y])$ of nucleosides in benzene at 25°C: $K_{G/C} = 30\,000$ L/mol versus $K_{G/A} = K_{G/U} = K_{G/G} = 1200$ L/mol and $K_{A/U} = 150$ L/mol versus $K_{A/C} = 28$ L/mol or $K_{A/A} = 8$ L/mol.

14.2.2 Deoxyribonucleic Acids

Depending on the species, deoxyribonucleic acids (**DNA**) contain approximately between $4 \cdot 10^3$ and 10^{11} base pairs. Their molecular weights accordingly range between millions and many trillions (Table 14-5). Some DNA molecules are open chains, others are rings. Their contour lengths vary between ca. 1 μm and 35 m.

Table 14-5 Number N_{base} of base pairs, molecular weights M, and contour lengths L_{cont} of DNA molecules of selected species.

Kingdom	Species	N_{base}	M	L_{cont}/mm	Shape
Viruses	*Polyoma* SV 40	4 500	$3 \cdot 10^6$	0.0015	cyclic
	Herpes simplex	155 000	$105 \cdot 10^6$	0.0530	linear
Bacteria	*Bacillus subtilis*	3 000 000	$2\,000 \cdot 10^6$	1.0	linear
	Escherichia coli	3 700 000	$2\,500 \cdot 10^6$	1.3	cyclic
Eukaryota	Yeast	13 500 000	$9\,000 \cdot 10^6$	4.6	linear
	Human	2 900 000 000	$1\,933\,000 \cdot 10^6$	990	linear
	Lung fish	102 000 000 000	$68\,000\,000 \cdot 10^6$	34700	

In general, the number of base pairs per nucleic acid increases with decreasing geological age of the species (exception: lung fish). For example, one of the simplest life forms, the bacteriophage φX 174, consists of only 5375 nucleotides. Assuming an expo-

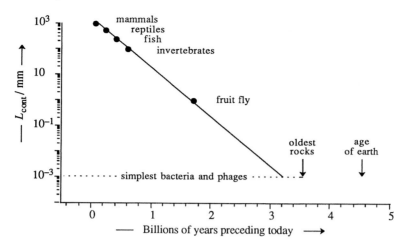

Fig. 14-4 Contour lengths L_{cont} of DNA molecules as a function of the first appearance of species.

nential decrease of the contour length L_{cont} with geological time t, $L_{cont} = A \exp(-Bt)$, one arrives at an age of ca. $3.2 \cdot 10^9$ years for the most primitive bacteria and phages with contour lengths of 10^{-3} mm = 1 μm, corresponding to 3000 base pairs (Fig. 14-4).

The shape of DNA molecules varies with their geological age. For example, the DNA of the bacteriophage φX 174 consists of just one single-stranded, cyclic molecule. DNAs of bacteria and mitochondria of higher species are also cyclic but double-stranded; these DNAs are called **plasmids**. DNAs of all higher species are linear and double-stranded. Some synthetic DNAs are known to exist in triple-strands and even quadruple ones.

The native macroconformation (the B-form, see below) of DNAs of higher organisms was revealed by models for the size, shape, and hydrogen bonding of the nitrogen bases (James D. Watson and Francis H.C. Crick (Nobel prizes 1962) and X-ray data (Maurice Wilkins (Nobel prize 1962); Rosalind Franklin with Raymond Gosling) of high-quality DNA samples (Rudolf Signer): there must be two complementary strands (Watson and Crick) since purine bases are always complementary to pyrimidine bases (Chargaff rule); the sugar-phosphate units must be at the exterior (Franklin and Gosling) and the bases must be near the fiber axis (Watson and Crick). The 3'5'-internucleotide bonds of the two strands have opposite directions. The strands are thus antiparallel and form helices (G: *hélix* = screw, spiral) that are right-handed in B-DNA (P-helix; the screw turns away clockwise from the viewer if viewed along the fiber axis). The screwlike structure creates alternating larger and smaller "dimples" that are called **grooves** (Fig. 14-5).

DNAs are polymorphous (A, B, C, Z, etc., types) (Table 14-6). The A-DNA ("dry DNA") contains ca. 20 wt% water; it is best prepared at 75 % relative humidity. The B-DNA ("wet DNA") with ca. 40 wt% water is best obtained at 90 % relative humidity. It converts to A-DNA at lower relative humidities or by addition of alcohol and certain ions, probably via the C-DNA. B-DNA and A-DNA are both right-handed.

The extra hydration of the B-type keeps the strands further apart than in the A-type. Thus, the "swollen" B-type is less ordered than the A-type: the distance between the free oxygens of the phosphate groups increases to 0.66 nm (B) from 0.53 nm (A); it can no longer be bridged by a single water molecule. Stretching converts the paracrystalline B-type into the A-type where the bases are now perpendicular to the fiber axis.

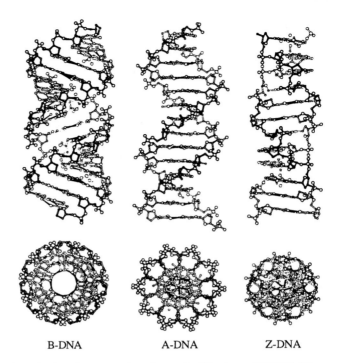

B-DNA   A-DNA   Z-DNA

Fig. 14-5 Side views (top) and cross-sections (bottom) of B-, A-, and Z-DNA [1].

DNAs may also exist as D and Z helices as well as some other types (B', C', C'', E) with slightly varying structures. D-Types are formed by DNAs with long, alternating nucleotide sequences, for example, poly[d(A-*alt*-T)]. The Z-DNA is the only known left-handed DNA double helix (M-helix). It was observed for poly[d(C-*alt*-G)] and also for native DNAs with methylated cytosine and/or guanosine units.

Some DNA data are shown in Table 14-6. The pitch gives the length of helical axis for one complete turn. The helical rise is the distance between two subsequent base pairs along the helical axis. The slope of base pairs is negative if it is clockwise.

Table 14-6 Structure of some DNA double helices. The repeating unit of Z-DNA is the dimer C-G.

	A	B	C	D	Z
Handedness of helix	right	right	right	right	left
Slope of base pairs relative to helix axis	19°	− 6°	− 8°	− 16°	− 7°
Number of nucleotides per turn	11	10	9	8	12
Pitch in nm	2.8	3.4			3.1
Helical rise in nm	0.256	0.338	0.307	0.24	0.74
Turn angle per nucleotide in degrees	+ 32.7	+ 36			+ 30
Width of grooves in nm: major	0.27	1.17	broad	broad	broad
minor	1.10	0.57	narrow	narrow	very narrow
Depth of grooves in nm: major	1.35	0.85			
minor	0.28	0.75			

In general, the nitrogen bases in the interior of the double helices are protected against enzymatic attacks by their hydrogen bridges, the π-π interactions between bases (staple effect), and the ionized sugar phosphates at the exterior of the helices. However, some enzymes are able to recognize, bind, and cut specific sequences of 4-8 base pairs (Section 14.2.5). This attack seems to proceed at conformational disorders (Fig. 14-6) since these are more open than regular DNA double-strands and/or are under stress.

Conformational errors differ for the various nucleotide sequences. For example, T/A pairs have hydrogen bonds of different lengths (Fig. 14-3); long T/A sequences thus cause the double helix to bend. Long sequences of –A–G–/–C–T– pairs lead to A-DNA conformations. They may also create triple strands in the large grooves of B-DNA conformations if a homopurine-homopyrimidine double-strand attaches another homopyrimidine strand via Hoogsteen pairing (Fig. 14-3). Double-strands of alternating purine-pyrimidine copolymers, $+G–C+_n/+C–G+_n$, may exist as Z-DNA and thus in left-handed macroconformations. Crosses are generated by **palindrome sequences**, i.e., the same sequences if the structure is read 5'→3' instead of 3'→5'.

Stresses are generated by **superhelices** (helices composed of helices, Fig. 14-6) or by Z-conformations in circular plasmids. Local Z-forms are essential for some biological functions since DNA sections that regulate genes for immunological defenses must first be converted into the Z-form before they can act. This necessity may have to do with evolution since there are some viruses that contain Z-DNAs.

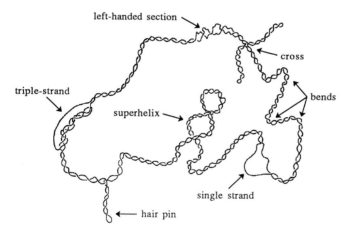

Fig. 14-6 Some conformational disorders in plasmids (small circular deoxyribonucleic acids).

14.2.3 Ribonucleic Acids

The linear and single-stranded molecules of ribonucleic acids (**RNAs**) transfer the genetic information of DNAs in the biosynthesis of proteins (Section 14.3.8). RNAs are subdivided according to their biological functions into

m-RNA = messenger RNA, informational RNA (carriers of genetic information)
r-RNA = ribosomal RNA (catalysts for the synthesis of peptide bonds)
t-RNA = transfer RNA (carry activated amino acids to the site of protein synthesis)
- viral RNA

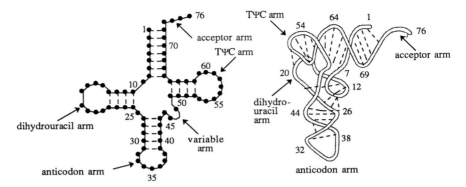

Fig. 14-7 Structure of the phenylalanine-specific t-RNA from yeast with a degree of polymerization of 76. Left: Sequence of nucleotides and projection of the macroconformation. Right: Spatial structure. At the pseudouridyl 55, the base is connected via its C^5 atom with the sugar. - - - - Hydrogen bridges between nucleotides.

All known **t-RNA**s have very similar structures (Fig. 14-7). Their two-dimensional projections are clover-like but their three-dimensional structures are elliposoidal. In contrast to DNAs, not all but only certain sections of t-RNAs show base-pairings. Not every base therefore finds a partner, which explains why the compositions of RNAs deviate from the Chargaff rule (Table 14-4). However, some viruses do have A-DNAs that are completely double-stranded.

The clover-structures of the t-RNAs have always five arms:

- The *TΨC-arm* has always of the same number of bases in all t-RNAs. It also always starts with the same base sequence.
- The *anticodon arm* is always of the same size. It contains the anticodon which is a base triplet that is bound to the m-RNA in protein synthesis.
- The *dihydrouracil arm* has either 3 or 4 bases in its helical section. Its loop always contains dihydrouracil units.
- The number of nucleotides in *variable arms* of RNAs range only between 3 and 14. All t-RNAs thus have similar degrees of polymerization (65-110).
- The *amino acid arm* consists of the two chain ends. The 3'-end of one arm always ends with the nucleotide triplet ~C–C–A that carries an α-amino acid residue at A (*acceptor arm*).

m-RNAs have molecular weights of ca. 500 000. Their base sequences correspond to the sequences of the DNAs of nuclei, i.e., they also have coding sections (**exons**) that are separated from each other by non-coding sections (**introns**).

r-RNAs comprise about 50 % of the dry weight of ribosomes where they are bound to proteins. The r-RNAs of these nucleoproteins have molecular weights ranging between 600 000 and 2 000 000.

Viroids are ribonucleic acids that are infectious for plants. They can be considered "naked viruses" because they lack the hull proteins of true viruses. The single-strand, cyclic RNA molecules of viroids have degrees of polymerization of ca. 3600. The molecules are assembled in rods of ca. 37 nm length and axial ratios of ca. 20. Such viroids are therefore 10^4-10^5 times smaller than conventional viruses.

14.2.4   Nucleoproteins

Nucleoproteins are "complexes" of nucleic acids and proteins with very different compositions (Table 14-7). The nucleic acids are DNAs in the nuclei of higher cells and in animal viruses but RNAs in the cytoplasm of cells and in plant viruses. Nucleoproteins influence chemical reactions in nuclei and cytoplasms; these reactions do not proceed between free molecules but between structurally bound ones.

Table 14-7   Composition and molecular weight M of some nucleoproteins. N_{NA}= number of nucleic acid molecules, N_{Pr} = number of protein molecules, T = number of protein types; all per nucleoprotein. S = Svedberg unit, a measure of the sedimentation rate and the molecular weight (Vol. III).

Nucleoprotein	wt% protein	N_{NA}	+	N_{Pr}	T	M
70 S-Ribosome from *E.coli*	33	3 RNA	+	57	53	2 600 000
Bacteriophage ϕX 174 (*E. coli*)	74	1 DNA	+	?	?	6 200 000
Tomato bushy stunt virus (TBS virus)	85	1 RNA	+	?	?	10 600 000
Tobacco mosaic virus (TMV)	94	1 RNA	+	2130	1	40 600 000

Centrifugation of cell contents delivers first nuclei, then mitochondria, and finally microsomes. **Microsomes** consist of polysomes that are bound to a matrix of lipoproteins. **Polysomes**, in turn, are composed of ribosomes that are held together *in vivo* by m-RNA. **Ribosomes** are nucleoproteins in which fairly low-molecular weight protein molecules are bound to helices of high molecular weight RNA molecules by magnesium ions.

Viruses (L: *virus* = poison, slime) have structures similar to ribosomes. *Plant viruses* are true nucleoproteins but *animal viruses* contain lipids in addition to nucleoproteins. All viruses consist of a core of nucleic acids (called **virion**) and a hull of proteins (called **capsid** (L: *capsa* = box)). The outer shapes of viruses differ widely. The tomato bushy stunt virus (TBS) is spherical but the tobacco mosaic virus (TMV) is rod-like. In TMV, protein molecules stud the helical RNA molecules.

Chromatin of the nuclei of eukaryotic cells (Section 14.1.2) is much more complex. Its nucleic acid does not form a simple spiral as in TMV but is connected with the chromatin proteins in a far more sophisticated way. Chromatin proteins are subdivided into well researched histones and far less explored non-histone proteins (see below).

Histones (G: *histos* = beam, mast, web) have relatively low molecular weights between 11 236 and 22 500 (Table 14-8). Some of of their amino acid units form a strongly hydrophobic, globular structure, the others very flexible, strongly basic "fingers."

Based on electrophoresis, histones can be (somewhat simplifying) subdivided into five types (Table 14-8). Histones H3 and H4 have practically the same composition in most eukaryotes. For example, only two of the α-amino acid units of histones H4 of calf thymus and pea germs are different. In H1, H2A, and H2B histones, on the other hand, basically only the globular parts were preserved during evolution.

In vivo, histones form dimers H3·H4 and H2A·H2B. Two dimers H3·H4 and two dimers H2A·H2B combine to form a protein complex of 8 histone molecules with a hydrophobic interior and a "sticky" surface caused by their basic "fingers". The histone octamer is wrapped by a 146 base pair long section of an acidic DNA molecule (core DNA), resulting in a histone-DNA complex called a **nucleosome** (Fig. 14-8).

Table 14-8 Histones of mammals. Birds, reptiles, and fish possess an additional histone H5.
N_H = Number of histone molecules per protein complex; X = degree of polymerization of histone
molecules; M = molecular weight of histone molecules.

Type	N_H	X	M	Composition
H1	1	224	22 500	rich in lysine
H2A	2	129	13 960	less lysine than H1
H2B	2	125	13 774	less lysine than H1
H3	2	135	15 273	rich in arginine and cysteine
H4	2	102	11 236	rich in arginine and glycine

Nucleosomes are connected by "free" (non-histone bound) DNA sections (**linkers**)
and spaced at regular distances along the DNA chain. The length of the linkers depends
on the species; it varies between 20 and 80 base pairs.

The resulting pearl-like structure is wound up in a regular spiral with 5 nucleosomes
per turn. The spiral is held together by H1 histones (1 per histone octamer; Table 14-8)
that reside in the interior of the spiral. The resulting **solenoids** (G: *solenoeides* = tubular,
pipe-shaped) form various loops (domains) that are held together by a complex of scaf-
folding proteins. Each loop contains between 20 000 and 80 000 base pairs, i.e., 100-
500 nucleosomes. The resulting overall structure is the chromosome.

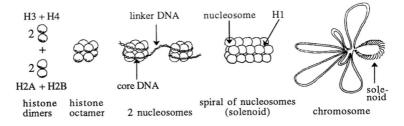

Fig. 14-8 Schematic representation of the subunits of a chromosome.

14.2.5   Function of Nucleic Acids

Living organisms differ in the number, size, and shape of **chromosomes** (G: *chroma*
= color (because they can be dyed); G: *soma* = body). Bacteria possess only 1 chromosome (haploid). All animals have pairs (diploids) of chromosomes; their diploid chromosome numbers range from 2 in horse ascarids, 8 in fruit flies, 38 in cats, 46 in humans,
48 in apes, and 78 in chickens to 100 in some birds. Higher plants are haploid-diploid.
Chromosomes are always generated by replication and division, never *de novo* (p. 513).

The single double-strand DNA of a chromosome contains the whole hereditary information. However, the transfer of this information during the creation of a new protein or
RNA molecule does not involve the whole chromosome but only certain DNA segments
that are called **genes** (G: *genos* = race). The bacteriophage φX174 has 9 genes, the bacterium *Escherichia coli* ca. 4300 (with a total of 4 638 858 base pairs), and a human ca.
20 000-25 000 (formerly estimated as 80 000). Only ca. 2 % of the nucleotides of hu-

man DNAs thus reside in genes. These protein-encoding parts are called **exons**. Exons are separated by non-encoding **introns** (so-called "junk DNA") whose role is unclear.

Sizes of genes for the various biological tasks vary widely. On average, they consist of ca. 600 base pairs. The totality of genes in chromosomes is called the **genome**. Ca. 20 % of the genome consists of highly condensed and repetitive non-coding DNA.

Replication

DNAs are involved in several biological tasks: replication, transcription, mutation, and function. Replication is the doubling of a DNA strand. It requires opening of the double helix into two single strands that become the matrices for the new strands (Fig. 14-9). After replication, each new double strand contains one old and one new single strand. The mechanism is thus semi-conservative.

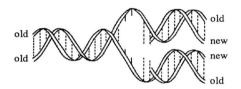

old
new
new
old

Fig. 14-9
Semiconservative replication
(doubling) of DNA

The opening of the double strand has been proven for the double-stranded cyclic bacterial DNA where the enzyme DNA gyrase attaches itself to the DNA and cuts a piece of the DNA chain that is not part of the DNA/enzyme complex. Both strands are then cut stepwise at a distance of four base pairs. The enzyme pulls the piggy-backed part of the double helix through the gap which it then closes.

This process reduces the number of base pairs per turn from 8 to 4. The resulting deviation from the original B-macroconformation is compensated by a twist of the cyclic DNA molecule. This "supertwist" exerts a torsional force on the helix that unwinds at several positions into two single strands (see Fig. 14-6). The replicating enzymes dock at the single strands which stabilizes the macroconformation. The intact sections of the double strand then engage in replication (DNA doubling), transcription (reading of genetic information), and recombination (exchange of strands). The fraction of twisted rings is small but only these are active. This process is also used in gene technology to multiply animal and plant genes in bacteria.

The ends of the linear DNA of chromosomes are called **telomeres** (compare the different meaning of "telomer" in Section 1.2). Telomeres protect chromosomes from cell enzymes that devour DNA. The telomeres thus guarantee the correct replication of chain ends. However, they become a little bit shorter after each replication process, except for stem cells and some others which are restored by the enzyme telomerase.

Transcription

According to the **central dogma** of molecular biology, genetic information is transferred from the DNA to RNA (transcription) and from the RNA to proteins (translation):

(14-1)   DNA $\rightleftharpoons$ RNA $\longrightarrow$ Protein

However, information is transferred not only from DNA to RNA and RNA to protein but also from DNA to DNA ($\curvearrowright$), RNA to RNA ($\curvearrowright$), and RNA to DNA ($\leftarrow$).

The genetic information is contained in the **genetic code** which is given by sequences of triplets of bases (**triplet code**). In the transcription, the 4-letter alphabet of the DNA (A, G, C, T) is converted into the different but complementary 4-letter alphabet of the RNA (U, C, G, A). The transcription is chemically a polyelimination with "stepwise" addition of one of the four different d-p-nucleoside triphosphates (ATP, GTP, CTP, TTP) to m-RNA with the help of the enzyme DNA-dependent RNA-polymerase. About 1000 monomeric units are added per minute.

Nucleoside triphosphates are more "energy rich" than the corresponding monophosphates. "Energy rich" in biochemistry does not refer to chemical compounds with high thermal dissociation energies but to compounds with an easily activated hydrolysis. According to quantum-chemical calculations, adenosine triphosphate (ATP) and the other three nucleoside triphosphates are such energy-rich compounds because the triphosphate segments contain five successive positively charged chain atoms that are easily attacked by the negative ions of phosphatases.

adenosine triphosphate

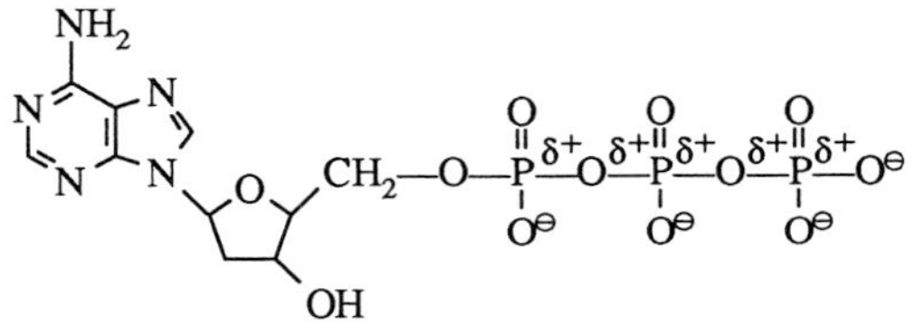

Mutation

Mutations are sudden changes of the chemical composition of DNAs (L: *mutare* = to change) and thus of the genetic code, the m-RNA, and the corresponding proteins.

Mutations sometimes occur by endogenic actions (oxidation, radiation, etc.) but are mostly spontaneously without any recognizable action at a rate of. ca. 10 000 per day and cell. Most of these mutations are immediately repaired by special enzymes. Mutations in the junk sections of DNA have no effect on the genetic code.

Mutations may occur in a gene, a larger section of a chromosome, or even in the number of chromosomes. *Point mutations* happen by erroneous base pairings or by chemical reaction at a nucleotide, for example, by UV light or mutagenic agents. For example, adenine is deaminated by HNO_2 to hypoxanthine that behaves like guanine during transcription. As a consequence, the newly formed DNA chain will not incorporate thymine at this point but cytosine:

(14-2)

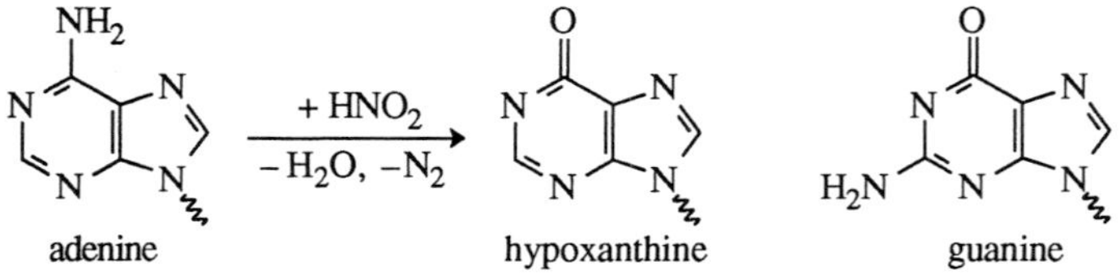

adenine   hypoxanthine   guanine

$+ HNO_2$
$- H_2O, -N_2$

In *chromosomal mutations*, whole DNA sections are illegitimately exchanged during meiotic cell divisions (divisions with formation of sexual cells). The chromosomes are then either longer or shorter. The increase or decrease of the number of chromosomes relative to those of normal cells is also a chromosomal mutation. Mistakes can also occur during the transfer of a chromosome from one cell to another.

Mutations are responsible for evolution, i.e., for the change of composition and sequences of protein chains. For example, one end of the serum albumin molecules of higher animals is always asparagine but the other end differs from animal to animal:

asp..val	turkey
asp..ala	chicken, duck
asp-thr....................................leu-ala	sheep, cow
asp-ala....................................leu-ala	horse, donkey, mule
asp-ala..................gly-val-ala-leu	ape
asp-ala.................. lys-val-ala-leu	human

The magnitude and rate of the changes depends strongly on the structure of proteins. For example, the protein sequences of histones are buried deep in the nucleosomes and are therefore fairly well protected against attacks (Fig. 14-8): only 1 α-amino acid was changed in 600 million years. The active site of the enzyme cytochrome C sits in a cleft and is therefore also relatively well protected: one amino acid was exchanged every 29 million years. The structure of hemoglobin is much more open: on average, exchanges took place every 5.8 million years. In fibrinopeptides, on the other hand, such exchanges occured every 1.1 million years because they involved amino acids on the surface of the protein molecule.

14.2.6 Cell-Free Polynucleotide Syntheses

In cell-free polynucleotide syntheses, nucleoside-5'-triphosphates are coupled linearly with the help of enzymes from microorganisms or cells of higher organisms. The 3'-hydroxy group of the last sugar unit of the growing chain is connected here with the 5'-hydroxy group of the new nucleotide by a phosphorus diester bridge. Most of the enzymes used are insensitive toward the chemical nature of the nitrogen bases.

Cell-free syntheses are either de novo, primer-dependent, or matrix-dependent:

In **de novo** syntheses (L: *de* = from, *novus* = new), nucleoside triphosphates are polymerized by certain enzymes; neither matrices, primer, nor Mg^{2+} ions are present. The method is used for the synthesis of single-strand homopolymers such as poly(dA). The resulting polynucleotides have a Schulz-Flory distribution of molecular weights.

De novo syntheses are also used for the synthesis of copolymers. Whether or not true copolymers are obtained depends on the structure of nucleoside-5'-triphosphates as well as the conditions during synthesis. For example, a mixture of d-GTP and d-CTP delivers a mixture of two homopolymers, poly(dG) and poly(dC), that form double helices poly(dG)·poly(dC). Such complexes of two homopolymers are assigned a **complexity** 1. Complexities are defined as the number of monomeric units in the smallest relevant sequence, i.e., the repeating unit. Natural DNAs have complexities of 10^4-10^5.

Matrix-dependent syntheses are enzymatically catalyzed polymerizations of nucleoside triphosphates in the presence of natural or synthetic nucleic acids as matrices. In these reactions, the nucleotide sequence of the matrix is copied according to the principle of base pairing. However, the matrix is not built into the new polynucleotide.

For example, the mixture of monomers ATP and UTP delivers with the enzyme RNA polymerase (from bacteria) and synthetic poly[d(A-*alt*-T)] as the matrix the alternating copolymer poly[d(U-*alt*-A)]. However, the same monomers and the same matrix lead to random copolymers poly[d(U-*ran*-A)] if the enzyme polynucleotide-phosphorylase is used instead. This reaction is therefore not a matrix-dependent synthesis.

Natural DNAs may also serve as matrices. The semiconservative replication of the four nucleoside-5'-triphosphates d-ATP, d-TTP, d-GTP, and d-CTP by the enzyme DNA-polymerase (**Kornberg enzyme**) leads to complementary strands. However, the doubling does not involve the whole length of the matrix but only relatively short segments of ca. 1000 units, called **Okazaki fragments**. These fragments are then united to larger molecules by the enzyme polynucleotide ligase.

The molecular weight of the matrix usually determines the molecular weight of the resulting polynucleotides. However, the matrix may also slip along the growing poly-nucleotide chain:

$$(14\text{-}3) \qquad \text{TATATA} \xrightarrow{\text{replication}} \begin{array}{c} \text{TATATA} \\ \text{5'-p-ATATAT} \end{array} \xrightarrow{\text{slipping}} \begin{array}{c} \text{TATATA} \\ \text{5'-p-ATATAT} \end{array}$$

For example, the heptamer d(pA)$_7$ may act as a starter for high-molecular weight poly(dA) if (pT)$_4$ is present as the matrix. Sometimes, oligonucleotides with as little as three triplets are sufficient.

Completely intact DNA double strands are slow-acting matrices because they require an opening of the double strand before replication. For the circular double strands of plasmids, this is brought about by the enzyme DNA gyrase. For linear double strands, the same effect is obtained if the double strands are slightly digested by the enzyme pan-creas DNA polymerase and then denatured by a short heating to 77°C. These oberva-tions let one suspect that DNA polymerase is in fact a repairing enzyme. One can indeed repair gaps in DNA molecules with the help of DNA polymerase, the corresponding nucleoside triphosphates, and a DNA matrix. At elevated temperatures, this method generates branched polynucleotides, however.

These phenomena are utilized in the **polymerase chain reaction** (**PCR**; Kary B. Mul-lis, Nobel prize 1993). This automated process produces microgram quantities of DNA from nanogram quantities of the same DNA without requiring the insertion of genetic material into phages, viruses, or plasmids. A double-strand sequence of the DNA to be copied serves as the matrix which is mixed with two primers (see below), the four de-oxyribonucleoside triphosphates as monomers, and a heat-stable DNA polymerase. The primers are single-strand oligonucleotides with degrees of polymerization of 20-40 that correspond to the start sequences of codons and nonsense-codons of DNAs (see Section 14.3.8). The mixture is heated briefly to 94°C in order to denature ("melt") the double helix. After cooling to 37°C, primer molecules attach to the single DNA strands. On re-newed heating to 72°C, the polymerase causes the nucleoside triphosphate molecules to react with the primer strands. After 25 cycles of 7 minutes each, the amount of the origi-nal DNA can be multiplied by a factor of 100 000. This process can be called a matrix-controlled stepwise polycondensation in the lingo of polymer science.

Primer-dependent syntheses are cell-free nucleotide syntheses in which the matrix is incorporated into the polynucleotide. The term is also used for syntheses with unknown reaction mechanisms. In these syntheses, the action of the enzymes PN-pase and/or ad-dase causes nucleotide units to add stepwise to an existing polynucleotide. Addase leads to polynucleotides of the type ddd-r(ddd)$_n$ whereas the mixture of addase and PN-ase causes the synthesis of rrr-d(ddd)$_n$. The polymerization corresponds to a living one. It leads to narrow molecular weight distributions.

14.2.7 Chemical Polynucleotide Syntheses

In principle, polynucleotides can be synthesized by successive additions of nucleotide units to polynucleotide chains. The method involves the reaction of a nucleoside with an activated one, the subsequent reaction of the dimer with another activated nucleoside, etc. The activating substituents are usually phosphorus compounds such as phosphorus diesters or triesters, phosphites, or phosphonates. The method is difficult because the participating bonds are very labile. Several functional groups must be protected for certain reaction steps or even for the whole synthesis.

Chemical polynucleotide syntheses are now practically always conducted on solid car-riers (**Merrifield method**, see Section 14.3.12). The carriers were originally crosslinked polystyrenes or polyamides for the triester method and silica for the phosphite method. Today, however, all methods use glass beads with controlled pore widths. The beads are functionalized; the functionalized groups are then reacted with the 5'-groups of the nu-cleosides and nucleotides, respectively. As a rule, spacers are needed between the glass and the nucleoside units, for example, long-chained amines. The first bound nucleotide unit is then reacted with the second nucleoside. Because of the large excess of the latter, the yield of the dinucleotide unit is high. Surplus reagents and undesired byproducts are then removed by simple washing of the solid carrier.

14.3 Proteins

Proteins are naturally occurring copolymers with α-amino acid units –NH–CHR–CO– and imino acid units –NR'–CH(R")–CO– where R' and R" may be part of a common ring (Table 14-9). Imino acids are therefore also called heterocyclic amino acids.

14.3.1 Amino Acids

Under prehistoric conditions, α-amino acids can be synthesized from NH_3, CH_4, H_2, and H_2O by repeated electric discharges. However, such a process delivers racemic α-amino acids whereas naturally occurring proteins of higher animals contain exclusively L-α-amino acid units (with few exceptions in the [S]-configuration). Even the proteins of lower animals are predominantly composed of L-units whereas those of bacteria may contain up to 15 % D-α-units. See p. 115 for the predominance of one antipode.

Today, nature generates α-amino acids by amination of α-keto acids to imino acids (R–CO–COOH + NH_3 → R–C(=NH)–COOH + H_2O) that are subsequently hydrogena-ted to R–CH(NH_2)–COOH. For the synthesis of proteins, see Section 14.3.12.

Nature produces ca. 260 amino acids but only 22 of these are genetically encoded (Table 14-9). Mammals encode just 19 α-amino acids and 1 imino acid. Selenocysteine is encoded by some other organisms and pyrrolysine apparently only by the microbe *Methanosarcina barkeri*. Encoded amino acids are given single-letter symbols.

All of the other of the ca. 240 naturally occurring amino acids are produced by post-translational reactions such as cysteinylation (cysteine $\rightarrow$ cystine), hydrogenation (glutamine $\rightarrow$ ornithine), oxidation (lysine $\rightarrow$ hydroxylysine), hydroxylation (proline $\rightarrow$ 4-hydroxyproline), methylation, etc. Some of these reactions are specific for the α-amino acids of a certain class of living organisms, for example, ornithine to birds (class *Aves*).

All common amino acids are also produced industrially, either by extraction of protein hydrolysates, by fermentation (uncontrolled enzymatic processes), enzymatically by addition of enzymes), or by total synthesis (Table 14-9). Glutamic acid, methionine, and lysine are used as nutritional additives while all other amino acids serve only for special products, including some polymers (see Section 10.2.2 in Volume II).

14.3.2 Classification and Occurrence

Peptides are homopolymers or copolymers of α-amino acids with various degrees of polymerization, X. They are subdivided into **oligopeptides** ($X \approx 2$-10) and **polypeptides** ($X > 10$) (G: *pepsis* = digestion, because enzymes, a subgroup of proteins, "digest" chemical compounds (see Section 14.3.10)). Copolymers whose sequence of amino acid units is the reverse of that of normal proteins are called **retropeptides**. The amide group –NH–CO– of peptides and proteins is called a **peptide group**.

The peptide bond of **isopeptides** is not between the α-NH$_2$ and the COOH of the α-amino acid units as in peptides, $-\text{NH–CHR–CO}-_n$, but between one of these groups and a corresponding group in the substituent R of multifunctional α-amino acids. Examples are the groups –NH–CH(COOH)–CH$_2$–CH$_2$–CO– (γ-unit derived from glutamic acid) and –NH–(CH$_2$)$_4$–CH(NH$_2$)–CO– (ε-unit derived from lysine). **Polydepsipeptides** are copolymers of α-amino acids and α-hydroxy acids.

Poly(α-amino acid)s are synthetic or natural homopolymers $-\text{NH–CHR–CO}-_n$ with L and/or D units of α-amino acids. They are usually obtained synthetically by polyelimination of *N*-carboxy anhydrides (Section 14.3.12). In Japan, a mixture of poly(γ-D-glutamic acid) and poly(γ-L-glutamic acid) is produced by the action of *Bacillus subtilis natto* on soy beans or gluten (proteins) of wheat. "Natto" from soy is a traditional Japanese food. These polymers are also present in the webs of the tussah spider (tussah silk) and are probably also produced by other spiders. Poly(γ-D-glutamic acid) is found in anthrax bacteria.

Proteins are naturally occurring copolymers with α-amino acid and imino acid units. In nature, they are often not present as such but as **conjugated proteins**, i.e., chemical compounds of proteins and non-protein components (**prosthetic groups**). This group comprises chromoproteins (compounds with dye molecules), flavoproteins (with flavine derivatives), hemoproteins (with iron-porphyrin compounds), lipoproteins (with lipids), nucleoproteins (with nucleic acids), metalloproteins (with metals), glycoproteins (having less than ca. 4 % polysaccharides), and mucoproteins (containing more than 4 % polysaccharides). About 50 % of the proteins of human cells are glycosylated.

Table 14-9 Structures, trivial names, and symbols of selected amino acids, their conformations in proteins (PR) and in poly(α-amino acid)s (PAS), and the 1977 industrial production of amino acids (methods and tons per year) ([2]; newer data do not seem to be available).

α = α-Helix; β = folded sheet structure; 10_3, 3_1 = other helix types. Values in [ ] indicate conformations after stretching. h = helix forming, o = indifferent, b = helix breaking. glx (or Z) is used if one cannot distinguish between glu (E) and gln (Q), ditto for asx (B) = asp (D) vs. asn (N). E = enzymatic, X = extraction, F = fermentation, S = synthetic. * D or L configurations (all others only L).

Substituent R	Trivial name	Symbols		Conformation PAS	PR	Production Method	t/a [2]

I. Genetically encoded amino acids in mammals
A. α-Amino acids NH_2-CHR-COOH with substituents R

Substituent R	Trivial name	Symbols		PAS	PR	Method	t/a [2]
$-H$	glycine	gly	G	β	b	S	3 000
$-CH_3$	alanine	ala	A	α [β]	h	E	100
$-CH_2C_6H_5$	phenylalanine	phe	F	α	h	S	100
$-CH_2(p-C_6H_4)COOH$	tyrosine	tyr	Y	α	b	X	50
$-CH(CH_3)_2$	valine	val	V	β	h	S,F	100
$-CH_2CH(CH_3)_2$	leucine	leu	L	α	h	X	100
$-CH(CH_3)CH_2CH_3$	isoleucine	ile	I	β	(h)	F	50
$-CHOH-CH_3$	threonine	thr	T	β	o	F,S	100
$-CH_2OH$	serine	ser	S	β	o	E,S	50
$-CH_2SH$	cysteine	cys	C	$-$	o	S	100
$-CH_2CH_2SCH_3$	methionine	met	M	α	h	S	100 000
$-CH_2COOH$	aspartic acid	asp	D (B) *	α	o	E	1 000
$-CH_2CH_2COOH$	glutamic acid	glu	E (Z) *	α	h	F	250 000
$-CH_2CONH_2$	asparagine	asn	N (B) *	$-$	b	S	50
$-CH_2CH_2CONH_2$	glutamine	gln	Q (Z) *	β	h	F	300
$-CH_2CH_2CH_2CH_2NH_2$	lysine	lys	K	α	h	F	25 000
$-CH_2CH_2N=C(NH_2)_2$	arginine	arg	R	$-$	o	F,X	300
$-CH_2-C_3H_3N_2$, see below	histidine	his	H	α	h	F,X	100
$-CH_2-C_7H_7N$, see below	tryptophan	trp	W	α	h	S	100

B. Imino acid

Substituent R	Trivial name	Symbols		PAS	PR	Method	t/a [2]
see below	proline	pro	P	10_3,3_1	b	F	100

II. α-Amino acids that are genetically encoded in other living organisms

Substituent R	Trivial name	Symbols		PAS	PR	Method	t/a [2]
$-CH_2SeH$	selenocysteine			$-$	o	S	100
see below	pyrrolysine						

histidine   tryptophan   pyrrolysine   proline

III. Important α-amino acids by post-translational modifications
A. α-Amino acids NH_2-CHR-COOH with substituents R

Substituent R	Trivial name	Symbols		PAS	PR	Method	t/a [2]
$-CH_2S-SCH_2-$	cystine	cyS-Scy	$-$	$-$	$-$	X	50
$-CH_2CH_2CHOHCH_2NH_2$	hydroxylysine	hyl	$-$	$-$	$-$	$-$	$-$
$-CH_2CH_2CH_2NH_2$	ornithine	$-$		$-$	$-$	$-$	$-$
CH_3NHCH_2COOH	sarcosine	sar	$-$	$-$	$-$	$-$	$-$

B. Imino acid

Substituent R	Trivial name	Symbols		PAS	PR	Method	t/a [2]
see above	4-hydroxyproline	hyp	$-$	3_1	$-$	X	50

Many proteins have very large complexities; in enzymes, amino acid sequences are often not repeated at all. Many proteins are also not chemical molecules but physical molecules consisting of 2, 4, etc., true chemical molecules, called **subunits**.

The number of amino acid units per subunit (the degree of polymerization) is on average just 60. A random placement of 20 different amino acid units in a chain with a degree of polymerization of 60 would thus generate $N = 20^{60} \approx 1.2 \cdot 10^{78}$ different protein molecules. However, the maximum number of naturally occurring protein types is just 10^9-10^{10} because there cannot be more protein types than DNA types. In nature, a non-statistical selection of protein types must thus have occurred.

In organisms, proteins serve as biocatalysts (enzymes), transport agents, structural elements, or immunological defense molecules. These functions are determined by the constitution, configuration, and micro and macroconformation of the protein molecules. Transport proteins and many enzymes are compact and spheroidal, which allows rapid transport to the loci of their action in the organism. Scleroproteins, on the other hand, should provide the organ with high structural stability. Hence, they usually form fibrilles.

The lifetime of proteins *in vivo* varies considerably. For example, the half-life of the enzyme RNA-polymerase from rat liver is only 78 minutes while the scleroprotein myosin from rabbit muscle has a half-life of 30 days and the transport protein hemoglobin from erythrocytes lives practically forever.

Proteins are used by man mainly as food, to a smaller extent for clothing (wool, silk), and in even smaller amounts for other purposes, for example, as biocatalysts. Food proteins are found in animal products such as eggs (11 %), milk (4 %), meat (15-23 %), cheese (14-28 %), and in plants such as wheat grains (14 %), rice (8 %), nuts (5-21 %), lentils (26 %), and soy beans (45 %). Proteins for animal fodder are also produced synthetically (see below).

Wool and **silk** are protein fibers that are basically used as such. **Leather** consists of chemically or physically crosslinked collagen fibers of animal skins. Degraded collagen from bones is used as **bone glue** and for photographic films and human consumption **(gelatin)**. Crosslinked **casein**, a milk protein, serves for haberdashery (see Volume II).

14.3.3 Primary Structure

In principle, chemical and physical structures of proteins can be described with the same structural parameters as other macromolecular compounds: constitution, configuration, microconformation, macroconformation, association, and other supramolecular structures. However, in part for historical reasons and in part for usefulness, another classification system is used: primary, secondary, tertiary, and quaternary structure. The latter system does not conform to the usual structural parameters since covalent bonds are responsible for primary and tertiary structures, conformations for secondary and tertiary ones, etc. For example, –S–S– bonds are not considered part of the primary structure.

The **primary structure** is dictated by the number, type, sequence, and configuration of amino acid units, i.e., the chemical structure. The N-terminal amino acid is always written to the left and the C-terminal one always to the right. gly-ala-ser is therefore a short-hand notation for $NH_2CH_2CO–NHCH(CH_3)CO–NHCH(CH_2OH)COOH$.

The chemical composition of proteins is determined by total acidic hydrolysis. The resulting amino acids are chromatographically separated on ion exchange columns. They are identified by their retention volumes and quantitatively determined after dyeing with ninhydrin. Such an analysis requires less than 24 hours.

Targeted enzymatic cleavages allow the determination of amino acid sequences. The enzyme trypsin splits peptide groups at the –CO– group of arginine and lysine whereas pepsin and chymotrypsin hydrolyze unspecifically. Carboxypeptidase A chops amino acid units from the COOH end of peptide chains, etc. The chemical composition of the fragments is determined by total hydrolysis, the C and N-terminal amino acid units by endgroup analysis, and the sequence of peptide units by targeted degradation. The combination of "overlapping" peptides from various enzymatic cleavages then allows one to determine the sequence of amino acid units in the protein.

14.3.4 Secondary and Tertiary Structure

Peptide bonds are 0.132 nm long and thus shorter than the 0.146-0.150 nm long amide bonds of aliphatic amides. However, the length of the C=O double bond is enlarged to 0.124 nm (peptides) from 0.1215 nm (aliphatic amides). The C–N bond must therefore have about 60 % double bond character so that peptide bonds are present in mesomeric forms, $\sim$NH–CO$\sim$ $\longleftrightarrow$ $\sim^{\oplus}$NH=CO$^{\ominus}\sim$.

The partial double bond character of the peptide bond leads to planar arrangements of peptide groups. Conformational changes of peptide chains therefore cannot occur around the –NH–CO– bond but only around other chain bonds. The peptide group can furthermore form hydrogen bonds to the carbonyl oxygens of peptide units, hydroxyl groups, and amino groups.

These hydrogen bonds lead to pleated-sheet structures and helical sequences (Fig. 14-10). Such spatially ordered conformational sequences are called **secondary structures**. Conformationally disordered structures are not counted as secondary.

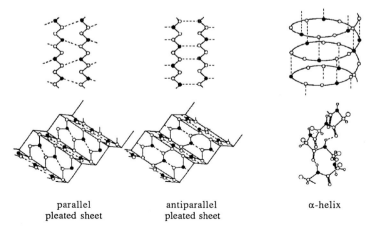

parallel   antiparallel   α-helix
pleated sheet   pleated sheet

Fig. 14-10 Secondary structures of poly(α-amino acid)s, polypeptides, and proteins. Left, center, and top right: -O- CHR group, >O- - C=O group, -●- NH group, $---$ hydrogen bond >NH- - C=C<. Bottom right: O- substituent R in CHR, -O- CH group, o ··· >C=O··· group, >●·· >NH··· group.

Pleated-sheet structures extend in proteins over only relatively short distances because they require the absence of steric hindrances by substituents R of peptide units –NH–CHR–CO–. In peptide chains, they are usually antiparallel.

Helices are present in peptides and proteins in many different types (Table 14-10). Most common is the so-called α-helix with a repeating period of 18 peptide units per 5 turns. It is now assumed that hydrogen bonds contribute relatively little to the stability of helices and that α-helices are mainly held together by other secondary bonds between non-bonded atoms (atoms at different chain atoms, see Section 3.2.1 in Volume III). The tendency to engage in helix formation is often (but not always) the same for an amino acid unit in both poly(α-amino acid)s and proteins (Table 14-9).

The fairly rigid structure of many proteins was originally assumed to be caused by many intersegmental hydrogen bonds. This hypothesis was based on the observation that urea can form hydrogen bonds between peptide units and thus lessen the hydrogen bonds between peptide units themselves. However, urea also increases the solubility of alkanes in water, i.e., it increases the tendency to form hydrophobic bonds.

The combined actions of the partial double-bond character of the peptide bond, the intramolecular hydrogen bonds, the hydrophobic bonds, and the L-configuration of the peptide units force most amino acid sequences into right-handed helices. There are exceptions to this rule, however. For example, poly(L-β-benzylaspartate) forms a left-handed helix despite the L-configurations of its peptide units.

In some proteins, helical axes of peptide sequences are furthermore somewhat twisted. Some helices are also combined with a second or third helix to form a kind of **supersecondary structure** that is called a **coiled coil** in the biochemical literature (Fig. 14-11). In **superhelices** consisting of double helices, 2 regular helices are wound around each other; they are interconnected by bridges between side chains, for example, in the scleroproteins keratin, myosin, paramyosin, and tropomyosin. In another supersecondary structure, three pleated-sheet structures are combined with two α-helices, for example, in the enzymes phosphorylase, phosphoglycerate-kinase, and some dehydrogenases.

A "coil" in common parlance is an individual spiral or ring or "a series of connected spirals or concentric rings formed by gathering or winding" (American Heritage Dictionary). In molecular biology, the term "coil" is used for a regular succession of spirals (i.e., a helix) whereas in polymer science it refers to an irregular succession of individual spirals (i.e., a *random* coil).

Table 14-10 Secondary structures of proteins and poly(α-amino acid)s. The number of peptide units per complete turn of a helix is positive for right-handed helices and negative for left-handed ones.

Structure		Type of helix	Number of units per turn	Rise per residue in nm	Radius of helix in nm
α-Helix		18_5	3.6	0.150	0.23
β-Structure (folded sheet), parallel		2_1	2.0	0.325	
	antiparallel	2_1	2.0	0.35	
γ-Helix			5.1	0.098	
π-Helix			4.4	0.115	0.28
ω-Helix		4_1	4.0	0.1325	
2_1-Helix		2_1	2.0	0.280	
Poly(glycine),	II-helix	3_1	± 3.0	0.310	
Poly(L-proline),	II-helix	3_1	-3.0	0.312	
	I-helix	10_3	3.3	0.20	0.19

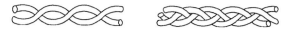

Fig. 14-11 Coiled coils with two and three strands of α-helices.

In **tertiary structures**, macroconformations are stabilized by interactions between side groups of peptide units. These interactions may be covalent bonds such as disulfide bridges of cystine or ionic and hydrophobic bonds. The latter bonds may be strengthened or loosened by changes in the temperature or environment (electrolyte addition, change of pH, etc.).

Nature certainly did not find the native macroconformations of proteins by random trials. A polypeptide with 150 peptide units in two rotatory states each can adopt $2^{150} \approx 1.4 \cdot 10^{45}$ different macroconformations. If the change from one macroconformation into the other needs 10^{-12} s (see page 142), the random trial of all possible macroconformations would require $1.4 \cdot 10^{33}$ seconds $\approx 4.5 \cdot 10^{25}$ years. The universe is only $13 \cdot 10^9$ years old, however. The energy is rather minimized by first generating relatively small conformational sequences (helix, pleated sheet, random coil) by short-range interactions which then arrange themselves in supersecondary, tertiary, and quaternary structures.

14.3.5 Quaternary Structure

Quaternary protein structures are defined associates of two or more, equal or different proteins, called **subunits**. For example, hemoglobin consists of 2 so-called A-chains and 2 B-chains. Tobacco mosaic virus, on the other hand, is composed of 1 nucleic acid chain and 2100 peptide chains. Quaternary structures are often so stable that they appear to be one molecule; they can be considered **supramolecules**. Proteins with quaternary structures often have molecular weights of several millions (Table 14-10) whereas those of true protein *molecules* (those without quaternary structures) rarely exceed 200 000.

Proteins with two or more peptide chains that are connected via cystine units can be split into the primary chains by reagents that sever the S–S bonds; an example is insulin. The quaternary structures of most other proteins are held together by non-covalent bonds. These bonds can dissociate if the pH, ion strength, solvent composition, etc., is changed, leading to smaller protein structures.

Table 14-11 Molecular weights of some proteins with quaternary structures. * Plus nucleic acid.

Protein	Molecular weight of protein	Number of subunits	Molecular weight of subunit
Insulin	11 466	2	5 733
β-Lactoglobulin	35 000	2	17 500
Hemoglobin	64 500	4	16 000
Glyceraldehyde-3-phosphate dehydrogenase	140 000	4	37 000
Catalase	250 000	4	60 000
Urease	483 000	6	83 000
Myosin	620 000	3	200 000
Glutamic acid dehydrogenase	2 000 000	8	250 000
	250 000	5	50 000
Turnip yellow mosaic virus	5 000 000	150 *	21 000
Tobacco mosaic virus	40 000 000	2 130 *	17 500

14.3.6 Denaturation

At physiological conditions, protein molecules with chemical functions such as enzymes adopt their native macroconformations. Because of the many intra and intermolecular bonds, they are fairly rigid and compact with almost spherical or ellipsoidal shapes (Section 14.3.9). Their outer structure is revealed by hydrodynamic measurements and/or electron microscopy, their inner structure by nuclear magnetic resonance spectroscopy or by X-ray analysis of protein crystals that are doped with heavy metals.

Most living organisms are **mesophilic**: their physiology requires temperatures of ca. 20°C, atmospheric pressures of 1 bar (0.1 MPa), and osmotic pressures of ca. 5 bar. But there are also **extremophiles**: thermophiles living in deserts, hot springs, and compost bins may endure 100°C, psychrophiles of the arctic and antarctic endure –40°C, and barophiles of the deep seas survive pressures up to 1100 bar.

The stability of all these organisms is essentially determined by the stability of their proteins and conjugated proteins. In general, primary, secondary, and tertiary structures of mesophilic and extremophilic organisms do not differ very much. Thermophilic organisms contain higher proportions of multivalent metal cations, though.

Above a certain temperature, proteins change their macroconformations because of the dissociation of secondary bonds. The highly ordered native structures convert to disordered states. In this **denaturation**, all intramolecular physical bonds are severed so that the protein molecule finally adopts the macroconformation of a random coil. In principle, denaturations are reversible since they are only determined by the constitution and configuration of the molecule. However, they are usually accompanied by a fairly fast irreversible aggregation of the protein molecules that leads to coagulation. This process is also called denaturation in industry and in the older scientific literature.

14.3.7 Prebiotic Syntheses

The first natural peptides and proteins may have been synthesized by either a condensation of prebiotically generated α-amino acids (Section 14.3.1) or by polymerization of hydrocyanic acid (exists in the universe). A possible reaction sequence may be

α-Amino acids can condense thermally, for example, in the presence of hot lava. Their derivatives such as D,L-alanine adenylate (the mixed anhydride of alanine and adenylic acid) even polycondense to poly(D,L-alanine) in the water-swollen layers of the clay mineral montmorillonite. This mineral has been shown to exist on Jupiter, a planet that is now in the same state as Earth was in prebiotic times.

14.3.8 Biological Syntheses

Biosyntheses of proteins require the presence of nucleoside triphosphates (p. 526). The first nucleosides were probably condensation products of HCN (p. 516) whereas the phosphate units originated from the mineral apatite, $Ca_5[(F,Cl)|(PO_4)_3]$. Under volcanic conditions, apatite produces P_4O_{10} which then hydrolyzes to soluble phosphates.

The biosynthesis of proteins is a highly complex process that requires encoded matrices. It starts in the cell nucleus with the transcription of the DNA code into the RNA code (Section 14.2.5). The so synthesized m-RNA molecules move to the cytoplasm where the RNA code is translated into the signals for the incorporation of the various α-amino acid units into the protein. In this way, the 4-letter alphabet of nucleic aids of higher mammals is translated into the 20-letter alphabet of proteins.

This synthesis of polypeptides or proteins by transcription and translation is called **(gene) expression**. The smallest enzymes are encoded for gene sequences of ca. 300 nucleotides which translates into 10^{180} statistical possibilities!

The DNA code is a triplet code: three nucleotide units work together to give the signal for the incorporation of one α-amino acid unit. A triplet code is the simplest way to recognize 20 different α-amino acids by just 4 different base types of DNA or RNA. A doublet code would provide only $4^2 = 16$ possibilities, thus 4 less than required. A triplet code, on the other hand, generates $4^3 = 64$ different ways, thus $64 - 20 = 44$ more than necessary. A quadruplet code would not only provide $256 - 20 = 236$ "surplus" ways to incorporate an amino acid but also too strong bonds via 4 bases instead of 3. There must be therefore various triplets (**codons**) that code for the same amino acid (Table 14-12).

Table 14-12 First (1), second (2 = U, C, A, G), and third (3) bases of codons of m-RNA for polypeptide syntheses and bases for the 5' end, the central position, and the 3' end of anticodons of t-RNA. Codons: [ ] = start codons (see text); amber, ochre, opal = end codons (nonsense codons). Opal sometimes codes for selenocysteine. Anticodons: the 5' end sometimes deviates; this wobble position is often occupied by inosine = 6-hydroxypurine-9-D-ribofuranoside (ino or I).
Selenocysteine is incorporated by UGA, pyrolysine by UAG.

| Codons of m-RNA | | | | | | Anticodons of t-RNA | | | | |
1	U	C	A	G	3	5'	U	C	A	G	3'
U	phe	ser	tyr	cys	U	U	lys	arg	met	thr	U
U	phe	ser	tyr	cys	C	C	lys	arg	met	thr	U
U	leu	ser	ochre	opal	A	A	asn	ser	ile	thr	U
U	leu	ser	amber	trp	G	G	asn	ser	ile	thr	U
C	leu	pro	his	arg	U	U	glu	gly	val	ala	C
C	leu	pro	his	arg	C	C	glu	gly	val	ala	C
C	leu	pro	gln	arg	A	A	asp	gly	val	ala	C
C	leu	pro	gln	arg	G	G	asp	gly	val	ala	C
A	ile	thr	asn	ser	U	U	end	end	leu	ser	A
A	ile	thr	asn	ser	C	C	end	trp	leu	ser	A
A	ile	thr	lys	arg	A	A	tyr	cys	phe	ser	A
A	[met]	thr	lys	arg	G	G	tyr	cys	phe	ser	A
G	val	ala	asp	gly	U	U	gln	arg	leu	pro	G
G	val	ala	asp	gly	C	C	gln	arg	leu	pro	G
G	val	ala	glu	gly	A	A	his	arg	leu	pro	G
G	[val]	ala	glu	gly	G	G	his	arg	leu	pro	G

In this strongly degenerated code, the first two bases usually determine which amino acid is incorporated. For example, alanine is coded by the triplets GCU, GCC, GCA, and GCG. The resulting **wobble effect** decreases the probability of mutation; it thus stabilizes the species.

Synthetic homopolymeric polynucleotides were used to identify which amino acids are coded by the homotriplets UUU, CCC, AAA, and GGG. For heterotriplets, synthetic random nucleotide copolymers are employed. Examples are matrices of random co-polymers of U, G, and C with average compositions of U_5G_1 and $U_6C_1G_1$, respectively. Both poly(U_5G_1) and poly($U_6C_1G_1$) contain many UUU triplets that code for phenyl-alanine (Table 14-12). Both matrices also incorporate glycine and tryptophan more often than in blind trials. The matrix poly($U_6C_1G_1$) leads to more arginine units and the matrix poly(U_5G_1) to fewer ones than blind trials. The codon for arginine must there-fore not only contain U and G but also C.

For protein syntheses, COOH groups of α-amino acids are first esterified with the 3' end (acceptor arm) of the t-RNA with the help of the enzyme aminoacetyl t-RNA syn-thetase. Special t-RNAs and special synthetases are responsible for each type of amino acid. These **synthetases** are enzymes with 1-4 subunits and molecular weights of 46 000 to 290 000. They are present in cells in concentrations of ca. 10^{-6} mol/L.

The resulting aminoacetyl-t-RNA possesses in its anticodon arm the necessary base sequence which allows this anticodon to bind specifically to the bases of the complemen-tary codon of the m-RNA via hydrogen bridges (see Fig. 14-7 for the structure of a t-RNA). All known anticodons are shown in Table 14-12. Since the base sequence of m-RNAs is read from the 5' end to the 3' end and every third base of a condon is less speci-fic than the other two, the first base of the anticodon of the t-RNA must be the wobble base (for example, base no. 34 in Fig. 14-7). Adenosine is never the first anticodon base, but inosine may be. The m-RNAs then unite with the ribosomes (consisting of RNA molecules and specialized protein molecules). The resulting polysomes are the loci in which the protein synthesis takes place.

Protein chains are built from N-terminal ends since amino acids are bound to t-RNAs via their COOH groups. An example is the synthesis of met-ala-ser-tyr (Eq.(14-5)):

(14-5)

The genetic code is universal for all Eukaryotae and, with the exception of the initi-ation step, also for Prokaryotae: The initiation is by methionine for the former but by *N*-formylthionine for the latter. According to experiments with higher cells, methionine is split off after 15-20 amino acids have been assembled. The end of the gene is signaled by the nonsense codons. The protein chain is terminated with the help of the enzyme peptidyl transferase, for example, by the codons UGA, UAG, and UAA of m-RNAs in animal cells.

The genetic code does not apply to mitochondria whether they are from yeasts or from human placenta. This indicates that mitochondria are probably the oldest elements of life. The start codons of human placenta are AUU and AUC instead of AUG and GUG. AUA leads to "ile" in prokaryotes and eukaryotes but to "met" in animal mitochondria. AGA normally leads to "arg" but signals "ser" in mitochondria of the fly *Drosophila* and "stop" in human mitochondria.

14.3.9 Spheroidal Proteins

The locally ordered amino acid sequences of enzymes and many blood and milk proteins produce water-soluble molecules with spheroidal shapes. Examples are proteins that transport oxygen: myoglobin (Fig. 14-12) in muscles and hemoglobin in blood.

The protein chains produced in the cytoplasm are initially not in their native macroconformation although the latter is determined by the primary structure (Christian B. Anfinsen, Nobel prize 1972). The conversion into the native macroconformation is "spontaneous" *in vitro* and in intracellular enzyme molecules but needs assistance for extracellular enzymes *in vivo*.

The conversion into the native macroconformation is called **protein folding**. It is a cooperative effect that proceeds in several steps. First, secondary structures are formed locally in less than 0.01 seconds. In ca. 1 second, these structures then convert local tertiary folds into "molten globules" which are water-containing fluid particles that are held together by hydrophobic bonds. Intracellular enzyme molecules than fold into the native tertiary structures within 1-1000 seconds.

Extracellular enzymes cannot pass through the membrane in the folded state, however. Their molten globules are therefore stabilized by "chaperones" that guide the enzymes through the membrane. Such chaperones are proteins that bind non-stoichiometrically in cells to newly synthesized protein molecules and prevent the formation of nonnative macroconformations. Most of them do not increase the speed of folding.

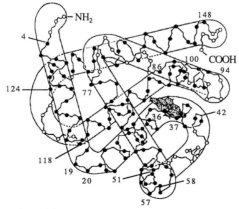

Fig. 14-12 Macroconformation of the protein myoglobin, consisting of 153 amino acids in 8 helical sequences of peptide units (●) 4-19, 20-35, 36-42, 51-57, 58-77, 86-94, 100-118, and 124-148. The other peptide units (○) reside in non-helical sequences. The active (oxygen binding) heme group (dotted) sits in a cleft.

14.3.10   Enzymes

All biochemical polymer syntheses are catalyzed by **enzymes** (G: *en zyme* = in yeast, in sour dough) Most of the enzymes are proteins although some are nucleic acids (p. 511). Enzymes were formerly called **ferments** (L: *fermentum* = sour dough). The term "fermentation" is still used for enzymatic processes that use enzyme-containing natural products such as yeasts.

Enzymes are present in all living organisms: animals, plants, and microorganisms. Extracellular enzymes are excreted by cells; they can be relatively easily isolated and investigated. The much more common intracellular enzymes are bound to cell membranes or organelles which makes them much more difficult to study.

Enzymes comprise ca. 2/3 of the total protein of liver cells and 1/3 of the non-structural proteins of muscle cells. The most common enzyme is ribulose-1,5-bisphosphate carboxylase/oxidase, the most important plant enzyme for the photochemical conversion of carbon dioxide.

Isoenzymes (isozymes) are multiple enzymes that catalyze the same chemical reaction but are produced by different genes. They are subdivided into isozymes that originate in different locations (e.g., malate dehydrogenases from mitochondria and from the cytosol of heart muscle cells), polymers with various peptide subunits (e.g., 5 electrophoretically different isozymes of the lactate dehydrogenase), or genetically caused **enzyme variants (alleloenzymes)** such as the ca. 50 genetic variants of human glucose-6-phosphate dehydrogenase. **Pseudoenzymes (metazymes)** are secondary enzyme modifications that are usually caused by changes in macroconformations.

About 2000 enzymes are known, including isozymes. Ca. 150 of them are produced commercially in milligram to kilogram quantities and used in medicine, analysis, and biochemical research. Only 17 enzymes are employed industrially for the production of foodstuffs and other products.

All enzymes, except nucleic acid enzymes, are globular proteins. Their fairly compact structure arises from their unique combination of helical segments, pleated-sheet structures, intramolecular associations, etc. (cf. Fig. 14-12). A large fraction of all amino acid residues is usually in helical structures with an average of 12 peptide units. Each end of the helix contains 4 peptide units that are not in the helical conformation. These end sequences are usually more hydrophilic than the peptide units of the helix themselves. Consequently, most helical sections of globular proteins are at the surface of molecules since this guaranties the best interaction with surrounding water molecules.

Peptide units of proteins are thus in different surroundings than dissolved free amino acids. This is reflected in the pK values of amino acids in the α-position, which are usually 7.6-8.4 in proteins but 9.0-10.6 in free amino acids.

Only a part of the enzyme, the **active center**, is catalytically active; the remaining peptide units are probably evolutionary junk. It is speculated that the protein accepted in geological times was the first "randomly" formed structure that was enzymatically effective for the function required at that time. However, this structure need not be the most rational and optimal one for those times nor for the present time.

The active center is always forming in clefts of the spheroidal structure, never in extruded parts. Clefts are formed either by the macroconformation of the primary chain as in most respiratory enzymes or by association of subunits in most regulating ones. The metal atom of metallo enzymes always sits in the interior of the cleft that is formed by two globular parts. Near the metal atom is a hydrophobic depression that accommodates the substrate; this design is thus responsible for the specificity of the enzyme.

For example, the active center of the enzyme chymotrypsin is composed of one serine and two histidine units. The remaining peptide units of the enzyme keep these three groups in defined positions relative to each other. Only certain substrates can thus react with the active center. Other substrates either do not fit into the cleft or cannot bring their reactive groups into sterically and/or electronically favorable positions for a reaction. An optimal efficiency is obtained if substrate and active center behave like **key and lock**. In this case, the substrate is "adsorbed" by the enzyme until saturation.

The binding of the substrate by the active center dramatically enhances the effective substrate concentration which in turn strongly increases reaction rates over that for free substrates in solution. Equilibrium constants for the binding of a substrate to an active center are usually ca. 10^4 L/mol whereas those for the complexation of the same groups in solution are only 10^{-8} L/mol. In enzymes, the effective substrate concentration is thus increased by a factor of $10^4/10^{-8} = 10^{12}$. The reaction rate rises accordingly.

Enzymes are subdivided according to their action as oxidoreductases (class 1), transferases (class 2; transfer various chemical groups), hydrolases (class 3), lyases (class 4; remove groups from the substrate, except by hydrolysis), isomerases (class 5), and ligases (= synthetases (class 6); combine two molecules by cleavage of a pyrophosphate bond). Each enzyme is characterized by up to 4 numbers that are separated by dots. The enzyme 1.1.3 is therefore an oxidoreductase (1...), that oxidizes CHOH groups (1.1...) with the help of molecular oxygen as acceptor (1.1.3).

The efficiency of an enzyme is often characterized by a **turnover number**

$$(14\text{-}6) \qquad TN = \frac{\text{number of reacted substrate molecules}}{\text{minutes} \times \text{number of enzyme molecules}}$$

that is sometimes related to seconds instead of minutes. Turnover numbers range from ca. 10^2 s^{-1} (carboxy peptidase) to 10^6 s^{-1} (carbonic acid anhydrase).

The concentration of enzyme molecules is unknown for many enzyme preparations. The enzyme activity is then characterized by the so-called **enzyme unit** or **international unit (IU)**. The IU is that amount of enzyme that converts 1 μmol of substrate per minute. Also used is the **katal** (1 kat = 1 mol of converted substrate per second; i.e., 1 IU = 16.67 nmol/s = 16.67 nkat). These values are usually given for standard conditions (37°C; pH = 7.8-8.0; substrate concentration of 10 %). For immobilized enzymes, one refers to the gram of biomass (dry or wet!).

14.3.11 Scleroproteins

Scleroproteins (G: *skleros* = hard, dry) are the proteins of muscles and tissues. They form fibrils and fibers and are thus also called **fiber proteins** or even **linear proteins** (as opposed to globular proteins) although some of them are chemically crosslinked. This group of proteins comprises the collagens and elastins of the tissue of skin, bones, and cartilage, the keratins of hair, feathers, and nails, and the elastins of connective tissues.

Industrially used are the fibers **silk** (secretion of silkworms) and **wool** (hair of sheep, goats, rabbits, etc.), hides and skins of animals (after chemical and/or physical cross-linking to **leathers**), and **gelatin** (from collagen) (see Volume IV).

Scleroproteins are insoluble in body fluids. Their peptide chains are therefore synthe-sized separately, transported to the loci of their actions, and then assembled. An example is the biosynthesis of **collagen** where the α-amino acid units from aminoacetyl t-RNA monomers are assembled at the ribosomes to pro-α_1 chains and pro-α_2 chains in the ratio 2:1 (Fig. 14-13). Each pro-α chain contains ca. 1300 peptide units. Each chain consists of a large "core segment" with mainly proline, leucine, phenylalanine, glutamic acid, and arginine units and two shorter end segments.

Some of the proline and lysine units are subsequently hydroxylated. Hydroxyproline units are therefore generated by a chemical after-reaction and not by a coded peptide synthesis (collagen is the only protein with hydroxyproline units). The hydroxyl groups of the hydroxylysine units are then substituted by glucose and galactose. Two α_1 chains and one α_2 chain then combine to form trimeric **procollagen** that forms a triple helix in the center and is crosslinked via disulfide bridges at the ends.

The procollagen molecules are secreted and move into the extracellular regions. Most of peptide units at the ends of the chains are here split off. The remaining 41 peptide units per end (**telopeptides**) form short coiled sections without triplet structures.

The remaining **tropocollagen** is the true collagen molecule. Its α_1 and α_2 chains dif-fer slightly from species to species but possess basically the same chemical structure. For example, the α_1 chains of calf hides consist of 1052 peptide units, 1011 of which are in triplets of the type gly-Z-Q. Z is either prolin, leucine, phenylalanine, or glutamic acid and Q mainly hydroxyprolin or arginine.

α_1 and α_2 chains form left-handed helices. They combine in the ratio 2:1 to form the right-handed superhelix of tropocollagen. The tropocollagen forms a **protofibril** with the outer shape of a rod of 300 nm length and 1.2 nm diameter.

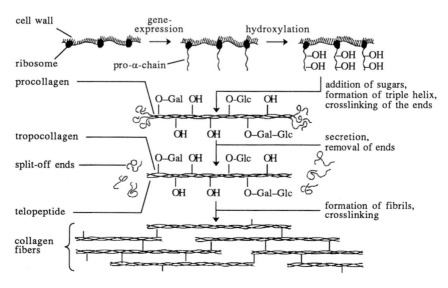

Fig. 14-13 Biosynthesis of collagen (see text). Gal = Galactose, Glc = glucose.

Some protofibrils combine spontaneously to form **subfibrils** of ca. 3.5 nm diameter. Several subfibrils then assemble to **collagen fibrils** of 10-200 nm diameter. In these collagen fibrils, disordered polar segments with predominantly positively charged groups face disordered polar segments with predominantly negatively charged groups. This ordered arrangement of the subfibrils with alternating polar and apolar regions creates the characteristic vertical stripes that are seen after dyeing collagen fibrils with uranyl salts. Dark stripes indicate amorphous polar regions, light stripes crystalline nonpolar ones.

Some of the lysine and hydroxylysine units of collagen fibrils are deaminated to aldehydes. Some of these aldehyde groups convert to aldols whereas others react with lysine or histidine units to Schiff bases. Aldols and Schiff bases crosslink the collagen fibrils to **collagen fibers**.

Collagen does not participate in metabolism. Instead, tropocollagen molecules remain intact during the whole lifetime of the organism; they are not constantly broken down and reassembled like other proteins. With increasing age of the organism, they become more and more crosslinked, though, which is thought to be one of the reasons for aging.

Heating of water-swollen collagen to temperatures of 40-60°C causes tropocollagen to dissociate into the α_1 and α_2 chains that are subsequently partially hydrolyzed to **gelatin** (L: *gelare* = to freeze, congeal (because dilute solutions of gelatin congeal on lowering the temperature)). Further heating produces **bone glue** from which collagen got its name (G: *kolla* = glue; *genes* = to cause).

14.3.12 Peptide and Protein Syntheses

Poly(α-amino acid)s

Homopolypeptides $+NH-CHR-CO+_n$ are usually obtained by ring-opening polymerization (polyelimination) of N-carboxyanhydrides (Leuchs anhydrides) of α-amino acids (Section 8.3.5). An industrial synthesis of *rac*-poly[(α-aspartic acid)-*stat*-(β-aspartic acid)] with a ratio of $\alpha{:}\beta = 25{:}75$ proceeds from D,L-aspartic acid via intermediary formed poly(N,α-succinimide):

(14-7)

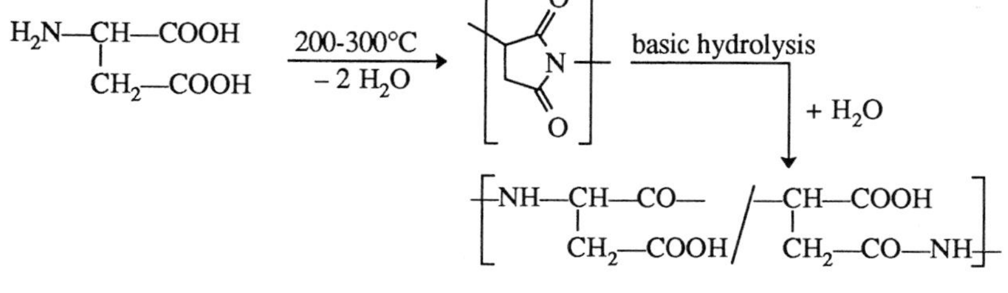

$$\underset{\underset{CH_2-COOH}{|}}{H_2N-CH-COOH} \xrightarrow[-2\ H_2O]{200\text{-}300°C} \left[\vcenter{\hbox{structure}} \right]_{\ N\ } \xrightarrow[+\ H_2O]{\text{basic hydrolysis}}$$

$$\left[\underset{\underset{CH_2-COOH}{|}}{-NH-CH-CO-} \Big/ \underset{\underset{CH_2-CO-NH}{|}}{-CH-COOH} \right]$$

Recombinant gene technology allows the synthesis of periodic α-amino acid copolymers such as $+(\text{ala}-\text{gly})_i-\text{glu}-\text{gly}+_n$ with $i = 3, 4, 5,$ or 6 and $n \leq 36$ by gene expression in *Escherichia coli*.

Synthetic Peptides

Peptide bonds are formed *in vitro* in three steps: Introduction of one protecting group Q or Q' per amino acid or peptide, respectively, reaction of the remaining unprotected groups (e.g., QNHCHRCOOH + H₂NCHR'COOQ' → QNHCHRCO–NHCHR'COOQ'), removal of one of the protecting groups of the new molecule (for example, QNHCHRCO–NHCHR'COOQ' → QNHCHRCO–NHCHR'COOH), reaction of the reactive group of the resulting molecule with another, single-protected molecule, etc. Each step requires the removal of the unreacted species; otherwise, mixtures of polypeptides would result.

This problem was resolved by the **Merrifield method** (Robert Bruce Merrifield, Nobel prize 1984) of providing protection by coupling one endgroup of the peptide molecule to a solid matrix. The matrix was originally chloromethylated and crosslinked poly(styrene). Today, it is usually functionalized porous glass.

The functional groups of this matrix are reacted with the amino group of that α-amino acid that will become the N-terminal end of the polypeptide, for example:

(14-8)   matrix–CH₂Cl + H₂N–CHR–COOZ → matrix–CH₂–NH–CHR–COOZ + HCl

Suitable protecting groups OZ are ester groups (methyl, ethyl, benzyl, *p*-nitrobenzene, *t*-butyl) or substituted hydrazides (OZ = NHNHQ). The subsequent reaction sequence is the same as in free syntheses except that the unreacted materials can be easily removed by washing the solid matrix. Each reaction must be carefully controlled for completion, though. Merrifield syntheses today are performed in computer-controlled machines. An example is the synthesis of the enzyme ribonuclease whose 124 peptide units were assembled in 11 911 steps involving 369 chemical ractions.

Industrial Proteins

Proteins are produced industrially by various microbiological syntheses from ammonia and several substrates. These reactions lead to **single-cell proteins (SCP)** that are mixtures of proteins, fats, carbohydrates, salts, and water. Substrates used include for

yeasts:	straight-chain paraffins, gas oils, ethanol, cellulose;
bacteria:	paraffins, gas oils, methanol, ethanol, methane, cellulose;
fungi:	sugars, starch, or other carbohydrates;
algae:	carbon dioxide.

An example is the production of the SCP with the composition $CH_{1.7}O_{0.5}N_{0.2}$ from paraffins, ammonia, and atmospheric oxygen in the presence of mineral salts. The reaction produces ca. 32 000 kJ per kilogram of dry mass. The resulting thick mass is centrifuged for the removal of paraffins and very carefully washed. The yellowish endproduct is used as animal fodder.

14.4   Polysaccharides

14.4.1   Sources and Importance

Polysaccharides are homopolymers or copolymers composed of sugar units (G: *sakcharon* = sugar). Their chains may be linear, branched, or crosslinked. Many native polysaccharides are not "pure" but contain a few percent of covalently bound peptides that are rich in hydroxy group containing α-amino acids.

Polysaccharides are present in animals, plants, algae, bacteria, and fungi. In algae and higher plants, they are constituents of cell walls and cell interiors, in bacteria and fungi also metabolic products. They serve in plants and microorganisms as structural and storage materials and in animals mainly for the production of energy.

Polysaccharides are divided according to their structure, function, or use:

Cell walls contain **structural polysaccharides** which are fiber-like or plane-like arrangements of linear chains. Structural polysaccharides include the celluloses and xylans of plants and the chitin of the shells of arthropods, e.g., insects and crustaceans. Lobster shells are very strong composites of chitin and calcium carbonate.

Reserve polysaccharides are food and/or energy reserves. They are lightly to strongly branched and form compact macromolecules that are easily transported in organisms because of the low viscosity of their solutions (Volume III). This class comprises the amyloses and amylopectins of the starch of plants and the glycogen of animals. Bacteria, on the other hand, use linear polyesters as food reserves (Section 14.5.2).

Gel-forming polysaccharides take up large amounts of water. They consist of linear chains. This group comprises the mucopolysaccharides of connective tissues and some plant gums, such as agar and pectin. Some plant gums are used by plants to seal wounds.

Very large numbers of polysaccharides are used commercially, either directly or after chemical conversion (Volumes II and IV). Examples are lumber (a composite of cellulose, lignin, and water); paper and cardboard (mainly cellulose); cellulose fibers such as cotton ramie, hemp, or sisal; cellulose derivatives (for fibers, plastics, thickeners); starch (for food, paper additives, fiber sizes); plant gums (thickeners, emulsifiers, stabilizers); dextran as a blood plasma expander; etc. The largest applications (by weight) are in the construction business and in the food industry.

14.4.2　Monosaccharides

The monomeric units of all polysaccharides are based on monooxo-polyhydroxy compounds (**sugars**) with the general formula $C_nH_{2n}O_n$. The names of the classes with various n are formed by the numeral and the ending -ose ($n = 4 \rightarrow$ tetrose; $n = 5 \rightarrow$ pentose, $n = 6 \rightarrow$ hexose, etc.). Sugars result formally from a dehydrogenation of alcohols, for example, the hexoses:

(14-9)

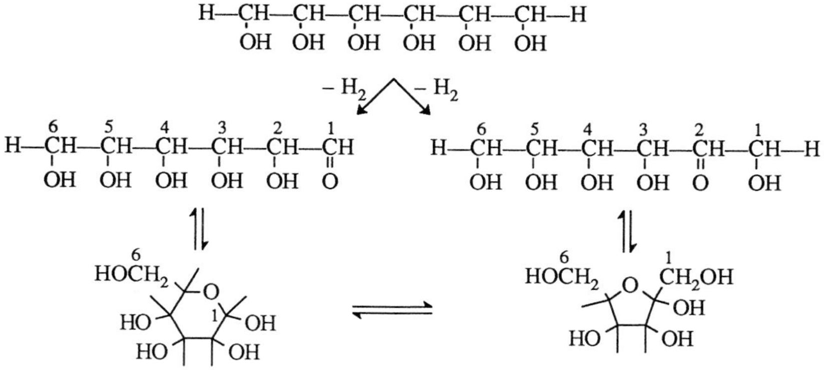

pyranose (aldehyde sugar)　　　　　　　　furanose (keto sugar)

Nature produces only those sugars that carry the oxo group at carbon atom 1 (**alde-hyde sugars = aldoses**) or at carbon atom 2 (**keto sugars = ketoses**) and not at other carbon atoms. Sugars must furthermore have an oxo-cyclo tautomerism such as the one between pyranoses and furanoses of hexoses (Eq.(14-9)).

The C^1 of aldoses belongs to an aldehyde group, the C^2 of ketoses to a keto group. Since an oxo-cyclo tautomerism requires rings with at least 5 ring atoms, aldehyde sugars must have at least 4 carbon atoms and keto sugars at least 5. Simple sugars are thus cyclic half-acetals of monooxo-polyhydroxy compounds with at least 4 carbon atoms in an unbranched chain.

Aldehyde sugars possess $j = n - 2$ and keto sugars $j = n - 3$ asymmetric carbon atoms. There are therefore $2^2 = 4$ aldotetroses, $2^3 = 8$ aldopentoses, $2^4 = 16$ aldohexoses, etc., and $2^2 = 4$ ketopentoses, $2^3 = 8$ ketohexoses, etc. The prefix of the trivial names of these sugars characterizes the configuration and the syllable "ose" the sugar (Fig. 14-14). The systematic names maintain the trivial names but insert the syllables "furan" or "pyran." The cyclic form of D-glucose is thus called D-glucopyranose.

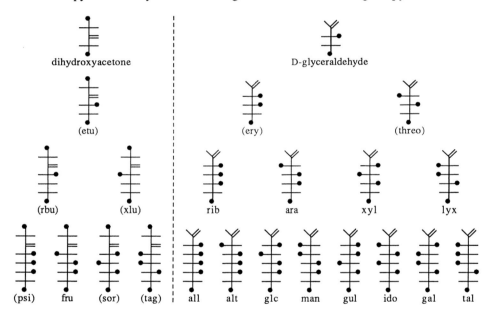

Fig. 14-14 D-Keto sugars (left) are derived from dihydroxyacetone and D-aldehyde sugars (right) from D-glyceraldehyde. 2nd line: tetroses; 3rd line: pentoses; 4th line: hexoses. OH-groups are symbolized by ● and C=O groups by =. Sugars are shown in the Fischer projection. The official acronyms are

all	allose	gal	galactose	man	mannose	(tag)	tagatose
alt	altrose	glc	glucose (dextrose)	(psi)	psicose (allulose)	tal	talose
ara	arabinose	gul	gulose	(rbu)	ribulose (araboketose)	(threo)	threose
(ery)	erythrose	ido	idose	rib	ribose	(xlu)	xylulose
(etu)	erythrulose	lyx	lyxose	(sor)	sorbose	xyl	xylose
fru	fructose (levulose)						

The ending "itol" characterizes sugars in which the CHO group is replaced by a CH_2OH group. There are three pentitols: ribitol = adonitol (from ribose), lyxitol = arabi(ni)tol (from lyxose or arabinose), and xylitol (from xylose). Hexitols are allitol, altritol (= talitol), galactitol (= dulcitol (dulcose, dulcin, dulcite, euonymit, melampyrit)), glucitol (= sorbitol (sorbite)), iditol, and mannitol (mannite).

One-half each of the tetroses, pentoses, hexoses, etc., are in the D-configuration, the other half in the L-configuration. The configurations refer to those carbon atoms that are next to the CH$_2$OH group that is farthest away from the CO group. D-Sugars are defined as those in which the carbon atom adjacent to the CH$_2$OH group has the same configuration as D-glyceraldehyde.

Sugars form different types of **epimers**, i.e., chemical compounds that differ only in the configuration around a single carbon atom. D-Glucose and D-mannose are such epimers with respect to the C^2 carbon, D-glucose and D-galactose with respect to C^4, etc.

Ring formation at C^1 leads to a special kind of epimerism that is called **anomerism** in sugars. In α-anomers, the anomeric OH group is oriented axially and the anomeric hydrogen equatorially. In β-anomers, it is just the opposite: the anomeric OH is equatorial (eq) and the anomeric H axial (ax). Equatorial positions are thermodynamically favored. Glucose is therefore the most common sugar in nature since all of its OH groups are equatorial.

The β-anomer of D-glucopyranose, but not the α-anomer, can form a hydrogen bridge between the CH$_2$OH group and equatorial anomeric OH group. The β-anomer is thus more stable than the α-anomer: crystalline D-glucopyranose is exclusively in the β-anomeric form. On dissolution, some of the hydrogen bonds are severed which leads to an equilibrium between β and α anomeric forms via the open-chain structure (Eq.(14-10)). The resulting **mutarotation** changes the optical activity of the glucose (see below) which, in turn, can be used to determine the proportions of the various isomers of sugars, for example, D-aldohexoses (Table 14-12). Mutarotations are also observed for many pentoses.

(14-10)

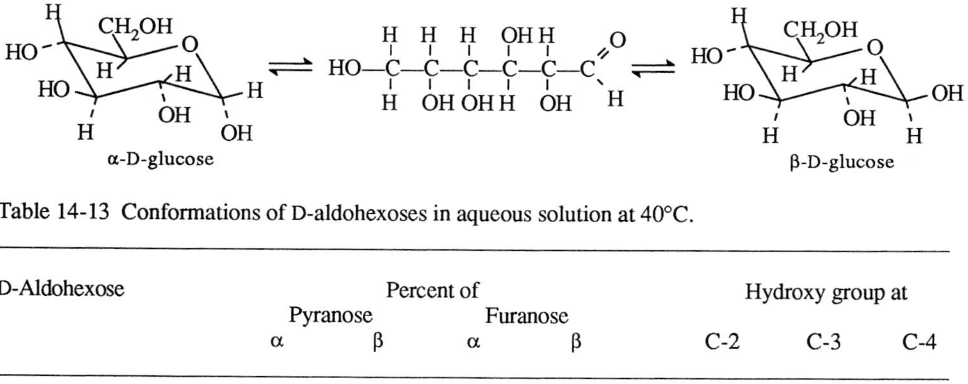

α-D-glucose ⇌ ⇌ β-D-glucose

Table 14-13 Conformations of D-aldohexoses in aqueous solution at 40°C.

| D-Aldohexose | Percent of | | | | Hydroxy group at | | |
| | Pyranose | | Furanose | | | | |
	α	β	α	β	C-2	C-3	C-4
D-Configurated C^2							
Glucose	32	68	0	0	eq	eq	eq
Galactose	27	73	0	0	eq	eq	ax
Gulose	21	79	0	0	eq	ax	ax
Allose	18	70	5	7	eq	ax	ax
L-Configurated C^2							
Mannose	67	33	0	0	ax	eq	eq
Talose	58	42	0	0	ax	eq	eq
Altrose	28	39	20	13	ax	ax	eq
Idose	31	37	16	16	ax	ax	ax

The hydrogen bond between the CH_2OH group and the β-OH group causes simple sugars to exist as rings, exclusively in the crystalline state and predominantly in aqueous solutions. For steric reasons, the voluminous CH_2OH must always be equatorial. All D-sugars therefore have the same chair conformation.

Intramerar hydrogen bonds lower the asymmetry of the molecule and therefore also the optical activity. The optical activity of β-anomers is thus always lower than that of its α-anomers, for example, 18.7° (β) versus 122.2° (α) for the D-glucopyranose. Mutarotation leads to 52.3°. In aqueous solution, the β-anomer is therefore present in 68 %.

Process	Derivative	Example	Name of example
Acetalization of anomeric hydroxy groups	glycoside	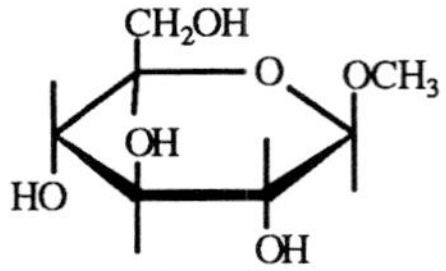	1-*O*-methyl-β-D-glucopyranose
Etherification of non-anomeric hydroxy groups	alkyl sugar	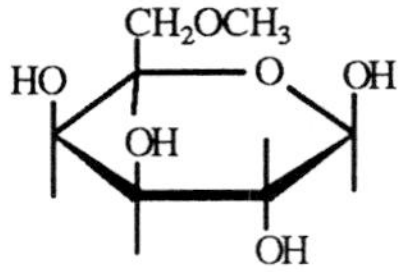	6-*O*-methyl-β-D-galactopyranose
Esterification of non-anomeric hydroxy groups	acyl sugar	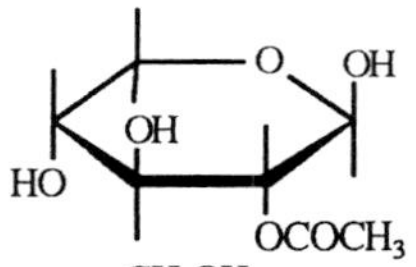	2-*O*-acetyl-β-D-xylopyranose
Esterification of OH groups to the half-ester of sulfuric acid	sulfate sugar	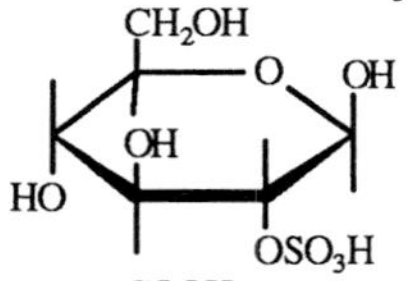	β-D-glucopyranose-2-sulfate
Replacement of CH_2OH by COOH	uronic acid	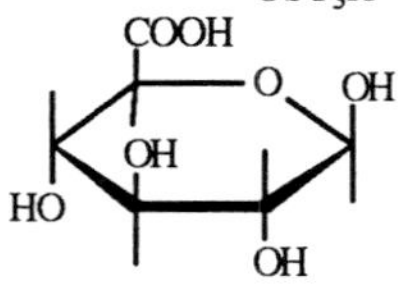	β-D-glucopyranosylic acid (glucuronic acid)
Replacement of CH_2OH by CH_3	6-deoxy sugar	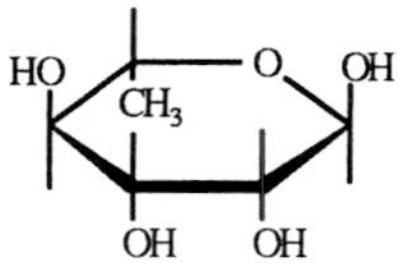	6-deoxy-α-L-mannopyranose (L-rhamnose)
Replacement of OH by NH_2	amino sugar	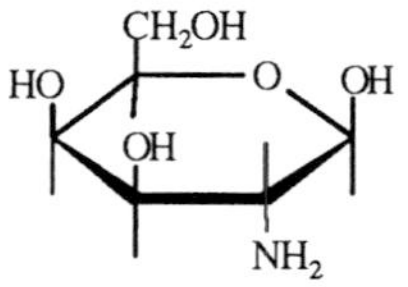	2-amino-2-deoxy-β-D-galactopyranose (galactosamine)
Intramerar etherification	anhydrosugar	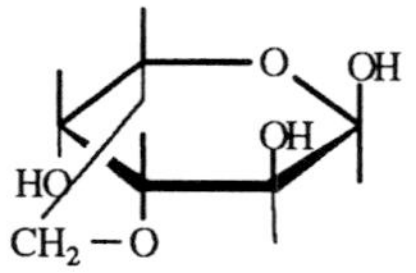	3,6-anhydro-α-L-galactopyranose

Fig. 14-15 Some sugar derivatives.

For the same reason, β-anomers dominate in aldoses with D-configurated C^2 atoms (allose, glucose, gulose, galactose; Fig. 14-15). In aldoses with L-configurated C^2 atoms, either the α-anomer dominates (mannose, talose) or α and β anomers are present in approximately equal proportions (Table 14-13).

The C^1 atoms of aldoses are in aldehyde groups in open structures and in hemiacetal groups in rings (Eq.(14-9)). Acetalization of the OH group at C^1 removes the reducing end. The resulting alkyl glucopyranosides (= **glycosides**) can no longer mutarotate. The acetal bond is correspondingly called a **glycosidic bond**. Some other substituted or replaced sugar groups are shown in Fig. 14-15.

14.4.3   Oligosaccharides

Oligosaccharides are composed of a few (2 or more) glycosidically connected monosaccharides. Some disaccharides and trisaccharides carry trivial names, for example, the disaccharides saccharose and maltose from the enzymatic hydrolysis of starch or glycogen (Fig. 14-16). Oligosaccharides may be linear (only glycosidic bonds between sugar units), branched, or cyclic. Examples of the latter are the cycloamyloses with rings composed of 6-12 glucose units (Section 16.4.2 and Volume II). They are obtained from starch by *Bacillus macerans*.

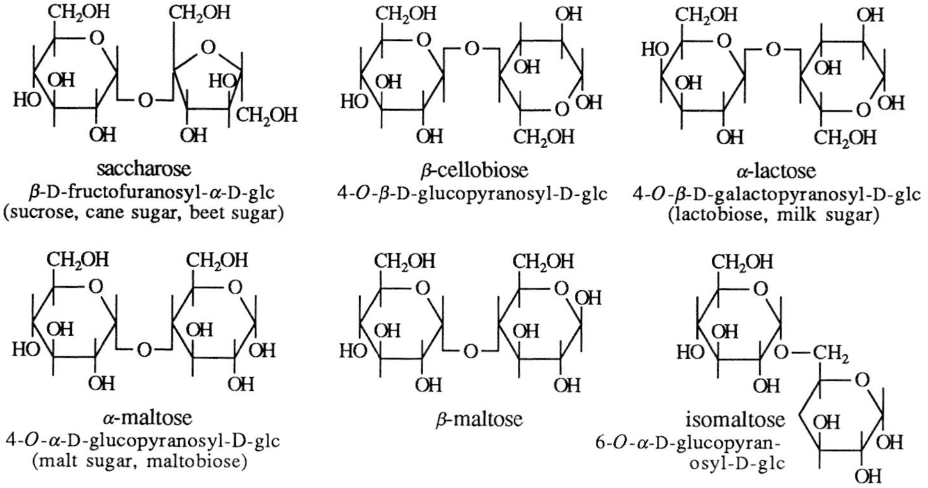

saccharose
β-D-fructofuranosyl-α-D-glc
(sucrose, cane sugar, beet sugar)

β-cellobiose
4-*O*-β-D-glucopyranosyl-D-glc

α-lactose
4-*O*-β-D-galactopyranosyl-D-glc
(lactobiose, milk sugar)

α-maltose
4-*O*-α-D-glucopyranosyl-D-glc
(malt sugar, maltobiose)

β-maltose

isomaltose
6-*O*-α-D-glucopyran-
osyl-D-glc

Fig. 14-16 Some importat disaccharides derived from glucose (glc).

14.4.4   Polysaccharides

Polymers composed of sugar units are commonly called **polysaccharides** or, more systematically, **glycans** (because of the many glycosidic bonds). Homopolymers with a single type of sugar units are **homoglycans**, those with two or more different types, **heteroglycans** or **heteropolysaccharides** (the terms **copolyglycan** and **copolysaccharide** are not used). In the food industry, they are known as **complex polysaccharides**.

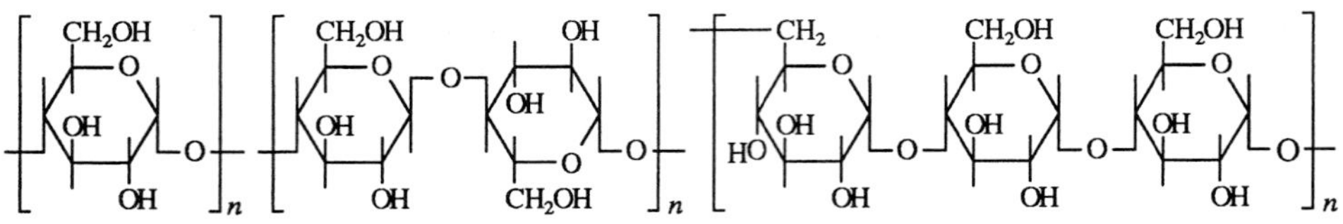

Amylose	**Cellulose**	**Pullulan**
poly(α-D-glucose)	poly(β-D-cellobiose]	poly(α-D-maltotriose]
α-(1→4) bonds	β-(1→4) bonds	α-(1→6') bonds

Polysaccharide	Mer	Repeating unit	Bond
Amylose	α-D-glucose	α-D-glucose	α-1,4'
Cellulose	β-D-glucose	β-D-cellobiose	β-1,4'
Pullulan	D-glucose	α-D-maltotriose	α-1,6'

Fig. 14-17 Poly(anhydro-D-glucopyranose]s. Bond designations are, for example, either 1→4 or 1-4'.

Some heteroglycans are alternating or periodic copolymers in the lingo of polymer science. Examples are cellulose and pullulan in the series of poly(glucose)s (Fig. 14-17).

The systematic nomenclature takes into account the glycosidic α and β acetalic bonds between the anomeric hydroxy group of one sugar residue and the non-anomeric hydroxy group of the next sugar unit that lead to anhydro structures. Since these bonds are between different sugar units and not between OH groups of the same glucose unit, position numbers of the carbon atoms of the glycosidic bonds are connected by an arrow. It must further be noted whether the sugar unit is in the pyranose or furanose form. Cellulose thus gets the systematic name poly[β-(1→4)-anhydro-D-glucopyranose].

The systematic names are often shortened, especially, if the type of glycosidic bond is not known with certainty. Cellulose is then called a poly[β-(1-4')-anhydro-D-glucose], a poly(anhydroglucose) or poly(anhydroglucan), or even a poly(glucose) or a poly(glucan). The neglect of the terms "pyranose" or "furanose" is often somewhat justified since naturally occurring sugars are usually aldehyde sugars that are almost always in the pyranose form.

The designations α and β for the anomeric bonds have been more recently replaced by the much more meaningful terms **axial** (ax) and **equatorial** (eq). Cellulose is correspondingly a poly[(1eq-4eq)-anhydro-D-glucopyranose] or a poly[(1eq-4eq)glucan].

Natural polysaccharides occur in many more types than nucleic acids or proteins. Homopolysaccharides may be linear, regioperiodic, or branched. Copolymers are most often linear and alternating but there are also linear periodic quaterpolysaccharides (tetrapolysaccharides) and random copolymers with two or more types of units (Table 14-14). **Conjugated polysaccharides** are compounds of polysaccharide units and nucleic acids, lipids, or proteins; in the latter case, they may also be non-covalent.

Many polysaccharides are branched. **Starch** is a mixture of two polyglucoses, amylose and amylopectin (Volume II). **Amylose** is an almost linear poly[α-(1-4')-glucopyranose] according to permethylations. **Amylopectin** has 1 glucose side unit per 18-27 glucose chain units with α-(1-4') bonds. **Glycogens** (in the liver) contain 1 branching unit per 8-16 glucose units. The highly branched **dextrans** have α-(1-6') bonds in the main chain and branches via α-(1-4') bonds.

Table 14-14 Some polysaccharides. For abbreviations and acronyms of sugar residues, see Fig. 14-14. A = Uronic acid, NAc2 = *N*-acetyl-2-amino, Su6 = 6-sulfate, Su4 = 4-sulfate, p-Est = partially esterified; ax = axial, eq = equatorial. * Between cells.

Polymer	Main unit	Chain bonds	Occurrence
Linear homopolysaccharides			
Amylose	D-glc	1ax-4ax	starches
Pustulan	D-glc	1eq-6eq	lichens
Cellulose	D-glc	1eq-4eq	cell walls of plants
Callose	D-glc	1eq-3eq	higher plants
-	D-glc	1eq-2eq	agrobacteria
Chitin	D-glc(NAc2)	1eq-4eq	shells of arthropods
Pectic acid	D-gal(A)(p-Est)	1ax-4ax	higher plants *
Laminarin	D-xyl	1eq-3eq	green algae
Inulin	D-fru	2eq-1eq	plants
Guluronic acid	L-gul(A)	1ax-4ax	brown algae *
Linear regio-periodic homopolysaccharides			
Pullulan	D-glc	(1ax-4ax)-alt-(1ax-6ax)	fungus *A. pullulans*
Lichenin	D-glc	(1eq-3eq)-alt-(1eq-4eq)	lichens
Branched homopolysaccharides			
Amylopectin	D-glc	1ax-4ax/1ax-6ax	starch
Glycogen	D-glc	1ax-4ax/1ax-6ax	animals, microorganisms
Dextran	D-glc	1ax-6ax/1ax-3ax	bacteria
-	D-ara	1ax-3ax/1ax-5ax	brown algae
-	D-man	1ax-2ax/1ax-6ax	yeasts
-	D-gal	1eq-3eq/1eq-6eq	cow lung
Levan	D-fru	2eq-6eq/2eq-1eq	bacteria
Alternating polydisaccharides			
Hyaluronic acid	D-glc(NAc2)	(1eq-4eq)-D-glcA-(1eq-3eq)	tears, joints
Chondroitin-6-sulfate	D-gal(NAc2,Su6)	(1eq-4eq)-D-glcA-(1eq-3eq)	matrix of cartilage
Chondroitin-4-sulfate	D-gal(NAc2,Su4)	(1eq-4eq)-D-glcA-(1eq-3eq)	matrix of cartilage
Dermatan sulfate	D-gal(NAc2,Su4)	(1eq-4eq)-L-idoA-(1eq-3eq)	matrix of skin
Pneumococcus III	D-glcA	(1eq-4eq)-D-gluA-1eq-3eq)	pneumococcae
Periodic polytetrasaccharide			
Heparin	D-glc(NAc,Su)-(1ax-4eq)-D-glcA-(1eq-4eq)- D-glc(NAc,Su)-(1ax-4ax)-L-idoA-(1ax-4eq)		liver tissue
Multiblock polydisaccharide			
Alginic acid	blocks of (1eq-4eq)-D-manA with blocks of (1ax-4-ax)-L-gulA and alternating copolymers of D-manA and L-gulA		brown algae
Linear homopolysaccharide with sugar sidegroups			
Guaran	[D-man-(1eq-4eq)]-*alt*- [D-man(D-gal-(1ax-6ax))-(1ax-4eq)]		seeds of guar plants
Random copolymer			
Xanthan	main chain: D-gluA-*stat*-D-man-*stat*-D-glc side groups: D-man and 4,6-*O*-(1-carboxyethylidene)-D-glu		from dextrose by *Bacillus Xanthomonas* *campestris*

Polysaccharides from microbes are either homopolymers or periodic copolymers of up to seven different types of sugars; they are usually regularly branched. Their structure is described by a special type of shorthand notation, for example, for the Klebsiella types K-4 and K-17:

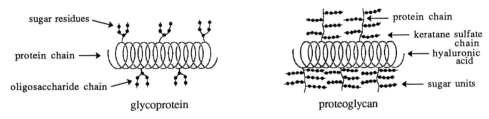

$$\underset{\alpha}{\overset{3}{-}}\text{glc}\,\underset{\alpha}{\overset{1\quad 2}{}}\text{glcA}\,\underset{\alpha}{\overset{1\quad 3}{}}\text{man}\,\underset{\alpha}{\overset{1\quad 3}{}}\text{glc}\,\underset{\beta}{\overset{1}{}}\qquad\qquad\underset{}{\overset{4}{-}}\text{glcA}\,\underset{\beta}{\overset{1\quad 3}{}}\text{rha}\,\underset{\alpha}{\overset{1\quad 4}{}}\text{glc}\,\underset{\alpha}{\overset{1\quad 2}{}}\text{rha}\,\underset{\beta}{\overset{1}{}}$$

K–4   K–17

These polysaccharide form the hull of bacteria cells. Because they are frequently secreted into cultures, they are also called **exopolysaccharides** or **slime polysaccharides**. These polysaccharides play important roles in immune reactions.

The conjugates of polysaccharides and proteins comprise two main types. **Glycoproteins** are graft copolymers of sugars on a protein helix that is part of a globular protein molecule (Fig. 14-18). The grafted sugars act as a kind of antenna for biological recognition. In **proteoglycans**, polysaccharide chains such as keratane sulfates (sulfates of mucopolysaccharides = glycosaminoglycans) are grafted on protein chains that in turn are grafted on a helix of hyaluronic acid.

sugar residues

protein chain

oligosaccharide chain

glycoprotein

protein chain

keratane sulfate chain

hyaluronic acid

sugar units

proteoglycan

Fig. 14-18 Schematic structure of glycoproteins and proteoglycans.

Like those of synthetic polymers, glycans with molecular weight distributions of their molecules are called **polymolecular**. **Polydisperse** glycans are mixtures of constitutionally similar molecules that differ only slightly in other constitutional features such as the number and length of side-chains, the degree of acetylation, etc. **Polydiversity** referss to mixtures of polysaccharide molecules that differ in the constitutions of monomeric or repeating units.

14.4.5   Biosyntheses

Natural polysaccharides are synthesized intracellularly: starches in the chloroplasts of green plants and algae, galactomannans in the endoplasmic reticulum, pectins within the cisterns of the Golgi apparatus, and cellulose in the plasma membranes. The loci of biosyntheses differ from the loci of storage or use (see below).

The biosynthesis of polysaccharides is not a simple reversal of hydrolysis because the direct polycondensation of monosugars has a positive Gibbs energy in aqueous solutions. Instead, low-molecular weight sugars are first converted in several steps to compounds with higher energy contents before they are added to the non-reducing ends of growing chains with simultaneous release of low-molecular weight leaving molecules.

The total biosynthesis can be subdivided into three groups of reactions: synthesis of true monomers, polymerization, and production of energy.

Monomer synthesis consists of a reaction of the sugar S (for example, glucose) with adenosine triphosphate (ATP) to sugar-6-phosphate (S-6-P) and adenosine diphosphate. The intermediary S-6-P is converted into the sugar-1-phosphate (S-1-P) with the help of an enzyme (EG). The glucose-1-phosphate is known as **Cori ester**.

The subsequent reaction of sugar-1-phosphate with another nucleoside triphosphate (NTP) to a nucleoside diphosphate-sugar (NDP-S) and pyrophosphoric acid is facilitated by another enzyme (EN); NTPs and ENs differ for the various sugars (Table 14-15). The NDP-S is then converted into a phosphorylated carrier lipid (NDP-S-lipid) which is the true monomer:

$$(14\text{-}11) \quad S \underset{-\text{ ATP}}{\overset{+\text{ ATP}}{\rightleftharpoons}} S\text{-}6\text{-}P \underset{\text{EG}}{\overset{\text{EG}}{\rightleftharpoons}} S\text{-}1\text{-}P \underset{\text{EN; } -\text{ H}_4\text{P}_2\text{O}_7}{\overset{\text{EN; } +\text{ NTP}}{\rightleftharpoons}} NDP\text{-}S \longrightarrow NDP\text{-}S\text{-}lipid$$

The *polymerization* consists of the reaction of NDP-S-lipids (monomers) with the non-reducing end of the growing polysaccharide chain, $\sim S_n\text{-}OR$, or, more general, by a primer, under the action of an enzyme EQ. The sugar residue S of S-NDP is inserted into the bond $\sim S_n\text{-}OR$ and the nucleoside diphosphate NDP is released (Eq.(14-12)). It is not clear whether R is a sugar unit or another group.

$$(14\text{-}12) \quad \sim S_n\text{-}OR + S\text{-}NDP \overset{\text{EQ}}{\longrightarrow} \sim S_{n+1}\text{-}OR + NDP$$

This reaction has been called a chain polymerization (because of the reaction between a polymer molecule and a monomer molecule), an insertion polymerization (because the monomer is inserted and not added), or a polycondensation (because low-molecular weight molecules are released). It should be called an **inserting polyelimination** (see Table 6-1) because monomer units are *inserted successively* between the chain and its end while releasing leaving molecules. The name "inserting polyelimination" is analogous to the name "inserting chain polymerization" for polymerizations that are commonly called "coordination polymerizations" or "insertion polymerizations".

Table 14-15 Sugars S and NTP nucleosides for the biochemical syntheses of polysaccharides.

Polysaccharide	Sugar S	NTP of	Occurence
Cellulose	D-glucose	guanosine	plants
Cellulose	D-glucose	uridine	*acetobacter xylinum*
Callose	D-glucose	uridine	beans
Slime polysaccharides	D-glucose + glucuronic acid	uridine	type III pneumococcae
Chitin	N-acetylglucosamine (as dextrin)	uridine	crustacea, insects
Hyaluronic acid	N-acetylglucosamine + glucuronic acid	uridine	fluids of joints
Pectic acid	D-galacturonic acid	uridine	higher plants
Xylan	D-xylose	uridine	grasses
Glycogen	D-glucose	uridine	liver
Glycogen	D-glucose	adenosine	bacteria
Amylose	D-glucose	adenosine	plants

In order to proceed, these inserting polyeliminations need *energy* which is provided by ATP. The true monomers are not the sugars (S) themselves but their nucleoside diphosphates (NDP-S). The insertion steps release nucleoside diphosphates (NDP) that are then converted back into nucleoside triphosphates (NTP) by ATP. The reaction NDP + ATP → NTP + ADP delivers adenosine diphosphate (ADP) that is subsequently converted back into ATP by oxidative phosphorylation. ATP is therefore always present in catalytic amounts; it is the energy source for all biosyntheses.

The resulting polysaccharides are either stored in plastids (starches), in vacuoles (fructosans), at cell walls (mannans, xyloglucans), or are excreted from the cell by exocytosis (celluloses).

The biosynthesis of **amylose** is fairly well known. Here, the enzyme EG is phosphoglucomutase, the enzyme EN is ADPG-pyrophosphorylase (ADPG = adenosine diphosphate glucose), and the enzyme EQ is adenosine diphosphate transglucosylase. The polymerization proceeds via the non-reducing end of the polymer chain. The branching is probably caused by another special enzyme that is as not yet known.

The true monomer for **plant celluloses** is guanosine diphosphate-D-glucose. Glucose, glucose-1-phosphate, UDP-glc, TDP-glc, ADP-glc, and CTP-glc do not deliver celluloses. Cellulose chains of cotton probably grow only via the non-reducing ends; cellulose chains therefore run parallel in cotton (and also in ramie (China grass)). **Bacteria celluloses**, on the other hand, require UDP-glc-isoprenylphosphoryl as monomer. The monomer for **chitin** is not the uridine diphosphate derivative of the simple *N*-acetylglucosamine but that of its dextrin (degradation product of starch).

14.4.6 Enzymatic Polymerizations

The same polysaccharide can be synthesized enzymatically *in vivo* and *in vitro* from different monomers. **Cellulose** is produced in plants from guanosine diphosphate D-glucose but *in vitro* also from adenosine diphosphate D-glucose or from the Cori ester, glucose-1-phosphate.

Amylose is produced in plants from ADP-glucose by the enzyme transglucosylase. In vitro, one can also use UDP-glucose and the enzyme phosphorylase. The achievable degree of polymerization depends very much on the quality of the enzyme preparation. Phosphorylases in juices from cold-pressed potatoes lead only to amyloses with degrees of polymerization, X, between 30 and 250. After heating, the same juices produced amyloses with $X \approx 10\,000$. The enzyme preparation must therefore have contained a thermolabile hydrolyzing enzyme.

These enzymatically catalyzed polymerizations proceed like all enzyme reactions via an intermediary enzyme-substrate complex, ES, that is in equilibrium with the monomer (the substrate S) and the enzyme E. The complex reacts in a second, rate-determining step to release the enzyme and the product S*:

(14-13) $\mathrm{E} + \mathrm{S} \underset{k_{-1}}{\overset{k_1}{\rightleftharpoons}} \mathrm{ES} \overset{k_2}{\longrightarrow} \mathrm{E} + \mathrm{S}^*$

The reaction follows **Michaelis-Menten kinetics**. In simple Michaelis-Menten kinetics (conversion of a low-molecular weight substrate S to another low-molecular weight product S*), the enzyme-substrate complex ES is formed with a rate of $R_f = d[ES]/dt = k_1[E][S] = k_1([E]_0 - [ES])[S]$ where $[E]_0 - [ES]$ is the difference of concentrations of the initially present and the complexed enzyme. The complex disappears with the rate $R_d = k_{-1}[ES] + k_2[ES]$. In the steady state, one has $R_f = R_d$ and therefore

$$(14\text{-}14) \qquad \frac{([E]_0 - [ES])[S]}{[ES]} = \frac{k_{-1} + k_2}{k_1} = K_m$$

The Michaelis-Menten constant K_m equals the equilibrium constant of the dissociation of the enzyme-substrate complex if $k_2 \ll k_{-1}$. In this case, no product is obtained.

For very small enzyme concentrations, one has $[E] \ll [S]$. Practically all enzyme molecules are then bound in the enzyme-substrate complex and the reaction rate adopts its maximum value of $R_{max} = k_2[E]_0$. The initial reaction rate R_0 of the rate-determining reaction, on the other hand, is given by $R_0 = k_2[ES]$ regardless of the enzyme concentration. Introduction of the concentration $[ES] = [E]_0[S]/(K_m + [S])$ from Eq.(14-14) leads to $R_0/R_{max} = [S]/(K_m + [S])$.

Contrary to this simple Michaelis-Menten kinetics, the second step of **enzymatic polymerizations** is not simply a cleavage of the enzyme-monomer complex EM into the free enzyme E and a converted substrate molecule M*. Instead of M*, a polymer molecule P is obtained that has been extended by one monomeric unit.

The enzyme-monomer complex EM must rather react with the polymer molecule P_n and extend the polymer chain by one monomeric unit. This reaction proceeds by release of low-molecular weight molecules L, for example, nucleoside diphosphates from NDP-sugars as monomers or, in the enzymatic synthesis of dextran, fructose from saccharose. Instead of Eq.(14-13), one rather has

$$(14\text{-}15) \qquad E + M \underset{k_{-1}}{\overset{k_1}{\rightleftharpoons}} EM \xrightarrow{+ P_n;\ k_2} E + P_{n+1} + L$$

The second reaction is a propagation in the terminology of macromolecular chemistry. In biochemistry, it is called a *transfer reaction* since a monomer molecule is "transferred" to a polymer molecule, resulting in an increase of the degree of polymerization of the polymer by 1.

The polymerization rate of the second, rate-determining step is

$$(14\text{-}16) \qquad R_{gross} = k_2[EM][P_n]$$

The released enzyme molecule E (Eq.(14-15)) forms an EM complex with another monomer molecule. The enzyme molecule thus moves from one chain to another one **(multiple chain mechanism)**.

The second step, $EM + P_n \rightarrow E + P_{n+1} + L$, requires polymer molecules which are however not present at the start of the reaction. In this step, P_n cannot be replaced by the monomer M because this would lead at small monomer conversions to the formation of oligomers, similarly to polycondensations and polyadditions. However, oligomers are not found. Instead, very high molecular weights are observed even at low monomer conversions; these molecular weights do not change appreciably during the further course of the reaction.

One could expect high initial molecular weights if the dimer is considerably more re-active than the monomer. In that case, the rate should initially be small, show an "in-duction period", and increase with the progress of reaction. However, high polymeriza-tion rates are observed even at low monomer conversions. The rates then stay constant or decrease with time.

Substances must therefore be present at the beginning of polymerization that act sim-ilarly to polymers P_n. Examples of such substances (**primers**; L: *primus* = first) are oligosaccharides and polysaccharides such as maltotriose in the polymerization of UDP-glucose by phosphorylase. However, there are "phosphorus-free" polysaccharide synthe-ses where neither was a primer added nor were nucleoside triphosphates, nucleoside di-phosphates, or sugar-1-phosphates present. In such cases, one often assumes that the en-zyme preparation was contaminated with very small, undetectable concentrations of a primer P* that acts as an initiator. The number-average degree of polymerization in-creases with the monomer conversion u according to $\overline{X}_n = u[M]/[P]$.

The action of polymers and oligomers as primers points toward the possibility that the enzyme is complexed by the polymer and not by the monomer, leading to the reaction sequence Eq.(14-17) instead of Eq.(14-15):

$$(14\text{-}17) \qquad EP_n + M \underset{k_{-1}}{\overset{k_1}{\rightleftharpoons}} EP_nM \xrightarrow{k_2} E + P_{n+1} + L$$

Here, the monomer must be inserted into the enzyme-polymer bond. The degree of polymerization, $\overline{X}_n = u[M]/[E]$, is then given by the enzyme concentration [E] and not by the primer concentration. The polymerization rate is independent of the polymer concentration if the second step is rate-determining:

$$(14\text{-}18) \qquad R_p = k_2[EP_nM] = (k_2k_1/k_{-1})[EP_n][M] = k_2K[EP_n][M]$$

Examples are the polymerizations of saccharose (= β-D-fructofuranosyl-α-D-gluco-pyranoside) by the enzyme dextran saccharase to the poly(glucose) **dextran** and the leaving compound fructose and by the enzyme levan saccharase to the poly(fructose) levan and the leaving compound glucose. It was found for the dextran synthesis that the intermediary complex EP_nM not only reacts according to Eq.(14-17) but can also add another monomer molecule to give MEP_nM. This complex then dissociates into a mono-mer-enzyme complex ME and a dextran molecule MP_n that carries a saccharose mole-cule M as endgroup:

$$(14\text{-}19) \qquad EP_nM + M \underset{k_{-3}}{\overset{k_3}{\rightleftharpoons}} MEP_nM \xrightarrow{k_4} ME + P_{n+1}$$

The ME complex starts a new chain. The growth of the chain is stopped at a certain degree of polymerization by a transfer reaction (in the meaning of macromolecular chemistry) to saccharose. Such transfer agents (in the terminology of macromolecular chemistry) are called **acceptors** in biochemistry. Other sugars act similarly: the higher the concentration of the acceptor, the lower the degree of polymerization.

The enzyme in the acceptor complex MEP_nM is no longer available for the propaga-tion reaction. The polymerization rate is thus determined by two factors (Eq.(14-20)).

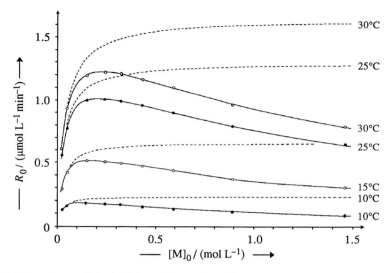

Fig. 14-19 Dependence of the initial rate R_0 on the initial saccharose concentration, $[M]_0$, of the polymerization of saccharose to dextran by the enzyme dextran saccharase at various temperatures. Circles: experimental [3]; solid lines: calculated by Eq.(14-20) using adjustable constants K_a; broken lines: calculated with simple Michaelis-Menten kinetics ($K_a = 1$).

The first multiplicand of this equation is the Michaelis-Menten term whereas the second one describes the inhibition by the acceptor:

$$(14\text{-}20) \qquad R_p = \frac{k_2[E]_0[M]}{K_m + [M]} \cdot \frac{K_a}{K_a + [M]}; \qquad K_m \equiv \frac{k_{-1} + k_2}{k_1}; \qquad K_a \equiv \frac{k_3 + k_4}{k_3}$$

With increasing initial monomer concentration, experimental initial rates of polymerization, R_0, first increase with increasing monomer concentration as predicted by simple Michaelis-Menten kinetics (Fig. 14-19)). The experimental rates then do not become approximately constant as predicted by simple Michaelis-Menten kinetics (first term of the right side of Eq.(14-20)); dotted lines in Fig. 14-19) but rather decrease because of the action of the acceptor. A correction for the acceptor effect according to Eq.(14-20) does indeed adequately describe the experimental findings (solid lines in Fig. 14-19).

14.4.7 Stepwise Syntheses

The stepwise synthesis of polysaccharides is an acetal synthesis, formally between the acetal function at the C^1 of a sugar molecule and any other hydroxy group of another sugar molecule. It may be either a polycondensation or a ring-opening chain polymerization. Because monomers must be coupled stereospecifically at the C^1, both types of polymerization are considerably less robust than those of nucleic acids and proteins. The C^1 carbon atom is adjacent to the ring oxygen which destabilizes equatorial electronegative leaving groups and stabilizes neighboring carbonium ions. Stereospecific syntheses are also made more difficult by ring flexibilities, neighboring group effects, and steric hindrances.

The *stepwise synthesis* requires appropriately substituted monosugars that have a reactive leaving group Q at C^1, an easy to remove protecting group B at the prospective hydroxy group (for example, at C^4), and stable substituents R at the remaining OH groups. Examples are Q = Br, B = CO–(p-C_6H_4)NO_2, and R = CH_2–C_6H_5. The B-protected monomer is then coupled with the unprotected chain, the B group is removed before the next coupling step, etc.:

(14-21)

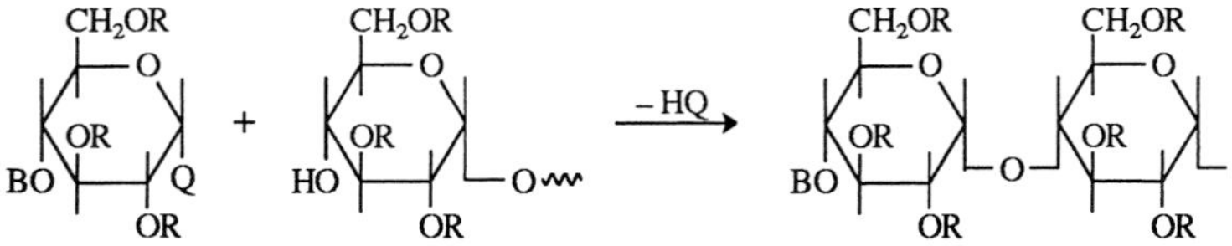

Because of dipole-dipole interactions, electronegative leaving groups are only stable in the axial position (α position). However, the methanolysis of completely etherified α-D-glucopyranosylchloride (cis-1,2-configuration) leads to an almost complete stereospecific inversion at C^1, resulting in the etherified methyl-β-D-glucopyranoside (trans-1,2-configuration). At the corresponding bromide, the same reaction results in racemization. This method is therefore only suitable for the preparation of β-D-glucosides.

Because of neighboring group effects, solvolyses of sugar derivatives with non-reacting groups at C^2 lead to high trans-1,2-stereospecificities, for example, with benzoylated glucosyl or mannosyl bromides. This method is therefore usually not suitable for the synthesis of 1,2-glycosides with exclusive cis-configurations.

Glycoside syntheses of these monomers depend strongly on the type of substituents. The methanolysis of benzoylated α-D-glucopyranosylbromides with a *p*-nitrobenzoate group at the C^6 delivers the corresponding methyl-α-glucosides in 90 % yield. Replacement of the *p*-nitrobenzoate group by the *p*-methylbenzoate group results in the formation of β-glucosides, however. In general, stepwise syntheses by successive condensations of polymer and monomer molecules do not lead to completely stereoregular polysaccharides. Furthermore, polymer yields are often low.

14.4.8 Ring-Opening Polymerizations

In general, two types of ring-opening polymerizations are used for the syntheses of oligosaccharides and polysaccharides.

In *orthoester syntheses*, a cyclic orthoester with protected OH groups is polymerized by $HgBr_2$ to polysaccharides in yields of up to 50 %. The degree of polymerization is determined by the molar ratio of monomer to initiator. The polymerization must therefore proceed via activated monomers:

(14-22)

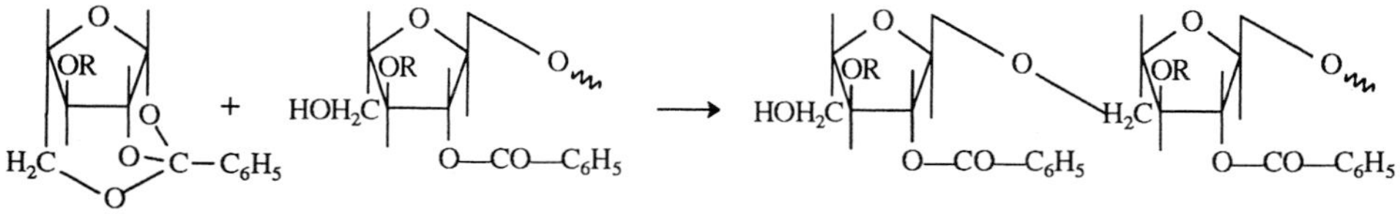

Anhydrosugar polymerizations proceed presumably cationically via activated chains. This happens probably by an attack of the bridge oxygen of a monomer molecule on C^1 of the trialkyloxonium ion at the active end of the growing chain and simultaneous ring-opening of the oxonium-ion ring:

(14-23)

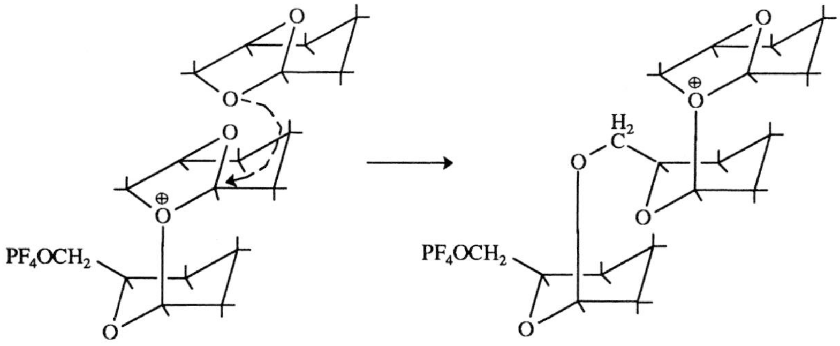

The stereospecific polymerization at low temperatures (e.g., $-70°C$) with PF_5 as catalyst results in polymer yields of up to 95 % and degrees of polymerization of up to 2000. Molecular weights do not change much with monomer conversion which points to a chain growth with chain transfer, probably to the catalyst.

A typical homopolymerization is that of 1,6-anhydromaltose benzylether to poly[α-(1-6)-mannopyranan]. Polymerization rates and degrees of polymerization decrease in the order manno > gluco > galacto. Copolymerizations seem to follow the classical copolymerization theory.

Partial degradation of cellulose with completely substituted OH groups allows one to prepare cellulose chains with C^1-Cl endgroups. The subsequent activation of endgroups by $AgSbF_6$ delivers cellulose monocations that can initiate the polymerization of tetrahydrofuran, thus leading to block copolymers of cellulose and poly(tetrahydrofuran).

14.5 Other Biopolymers

14.5.1 Poly(isoprene)s

Polyprenes are produced by more than 2000 plants, mostly as mixtures of polymers with terpenes and waxes (see Volume II). The most important polyprene is **natural rubber** which is today mostly obtained as plantation rubber from the latex of the tree *Hevea brasiliensis*. Natural rubber now provides ca. 40 % of all consumed rubbers (2000), most of it for tires (Volume IV). Natural rubber is an isoprene polymer with the idealized structure $+CH_2-C(CH_3)=CH-CH_2+_n$ that contains ca. 95 % *cis*-1,4 and 3 % 3,4-isoprene groups plus units with aldehyde and epoxy groups.

Trans-1,4-poly(isoprene)s are obtained from the latices of *Palaquium gutta* and *Mimusops balata*. They are mainly used for cable sheatings (**guttapercha, gutta**) and conveyor belts (**balata**). **Chicle** of the plant *Achras sapota* is a mixture of guttapercha and triterpenes. It is the base of some chewing gums.

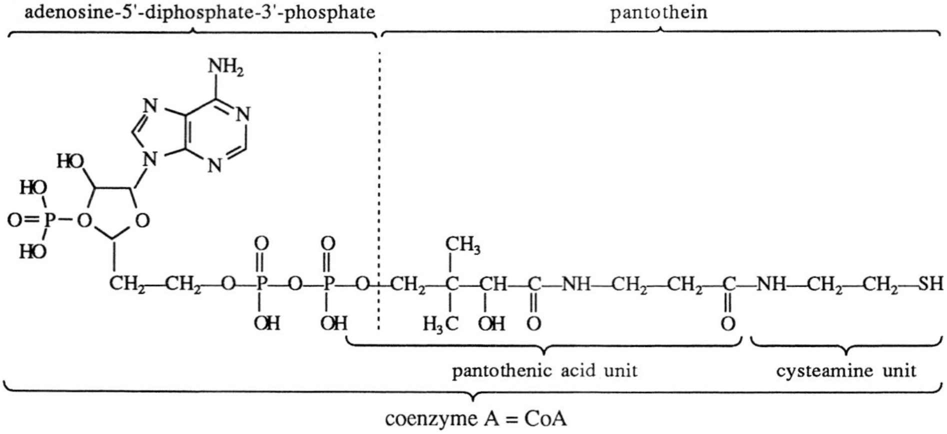

adenosine-5'-diphosphate-3'-phosphate   pantothein

$$NH_2$$

HO
HO
$$O=P-O$$
HO

$$CH_2\text{—}CH_2\text{—}O\text{—}P\text{—}O\text{—}P\text{—}O\text{:}CH_2\text{—}C\text{—}CH\text{—}C\text{—}NH\text{—}CH_2\text{—}CH_2\text{—}C\text{—}NH\text{—}CH_2\text{—}CH_2\text{—}SH$$

$$\underset{OH}{|}\quad\underset{OH}{|}\qquad \underset{H_3C}{|}\ \underset{OH}{|}\ \underset{O}{\|}\qquad\qquad\qquad \underset{O}{\|}$$

                  pantothenic acid unit   cysteamine unit

coenzyme A = CoA

Plants produce polyprenes from derivatives of phosphoric acid, the so-called **activated acetic acid (acetyl-CoA)**, the S-acetic acid ester of **coenzyme A** (CoA, CoA-SH) (see above). Two acetyl-CoA molecules **I**, $CH_3CO\text{–}S\text{–}(CoA)$, react by splitting off HS-(CoA) to give acetoacetyl-CoA (**II**). Further addition of acetic acid delivers β-hydroxy-β-methylglutaryl-S-(CoA) (**III**) which is then reacted with enzymatically reduced pyridine-nucleosidetriphosphate, NTPH, to mevalonic acid (**IV**):

(14-24) $2\,CH_3CO\underset{\textbf{I}}{\text{—}S\text{—}}(CoA)\ \xrightarrow[-\ HS\text{-}CoA]{}\ CH_3CO\text{—}CH_2\text{—}CO\underset{\textbf{II}}{\text{—}S\text{—}}(CoA)$

$$+\ CH_3COOH\ \downarrow$$

$$HOOC\text{—}CH_2\text{—}\underset{\underset{\textbf{III}}{OH}}{\overset{\overset{CH_3}{|}}{C}}\text{—}CH_2\text{—}CO\text{—}S\text{—}(CoA)\ \longrightarrow\ HOOC\text{—}CH_2\text{—}\underset{\underset{\textbf{IV}}{OH}}{\overset{\overset{CH_3}{|}}{C}}\text{—}CH_2\text{—}CH_2OH$$

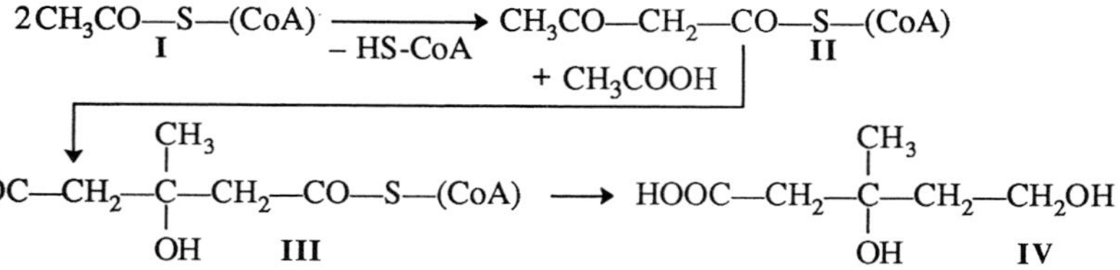

Mevalonic acid is the intermediate for all polyprenoids, including terpenes, sterols, and carotenoids. It is phosphorylated by ATP, then decarboxylated, and finally dehydrated to 3-(3-methylbutenyl)-2-pyrophosphate:

(14-25) $HOOC\text{—}CH_2\text{—}\underset{OH}{\overset{\overset{CH_3}{|}}{C}}\text{—}CH_2\text{—}CH_2OH\ \longrightarrow\ CH_3\text{—}\overset{\overset{CH_3}{|}}{C}\text{—}CH=CH_2$

$$O\text{—}P(OH)\text{—}O\text{—}P(OH)\text{—}OH$$
$$\quad\|\qquad\qquad\ \|$$
$$\quad O\qquad\qquad\ O$$

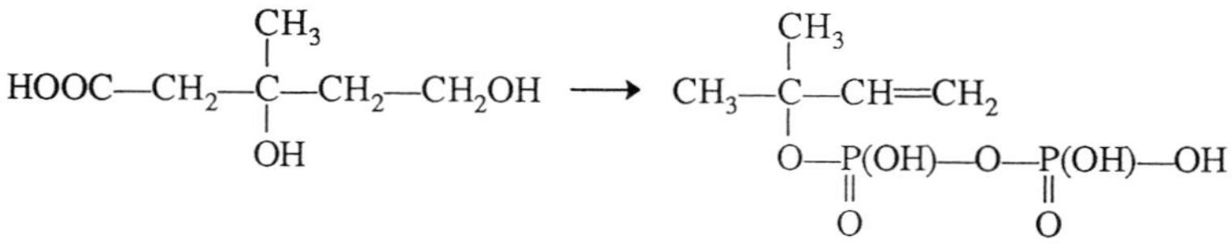

An isomerase then converts 3-(3-methylbutenyl)-2-pyrophosphate into isopentenyl-diphosphate:

(14-26) $CH_3\text{—}\overset{\overset{CH_3}{|}}{C}\text{—}CH=CH_2\qquad\qquad \longrightarrow\ CH_2=\overset{\overset{CH_3}{|}}{C}\text{—}CH_2\text{—}CH_2\text{—}O\text{—}P_2O_6^{3\ominus}$

$$O\text{—}P(OH)\text{—}O\text{—}P(OH)\text{—}OH$$
$$\|\qquad\qquad\ \|$$
$$O\qquad\qquad\ O$$

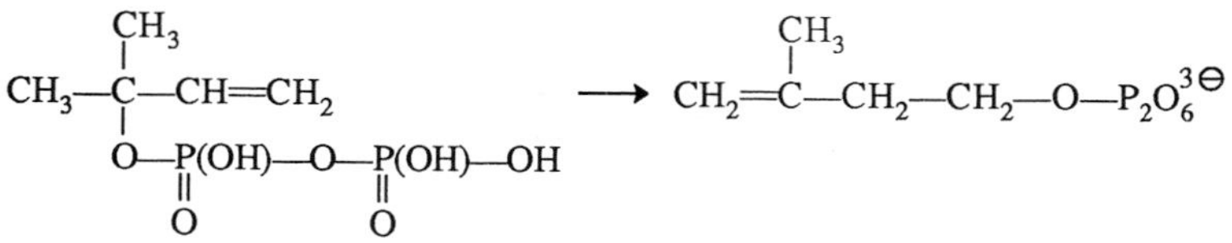

a part of which is further converted into dimethylallylpyrophosphate:

$$(14\text{-}27) \quad CH_2{=}\overset{\underset{\displaystyle |}{CH_3}}{C}{-}CH_2{-}CH_2{-}O{-}P_2O_6^{3\ominus} \; \rightleftharpoons \; CH_3{-}\overset{\underset{\displaystyle |}{CH_3}}{\underset{\displaystyle |}{\underset{\displaystyle S-enzyme}{C}}}{-}CH_2{-}CH_2{-}O{-}P_2O_6^{3\ominus}$$

$$+ \text{ HS-enzyme} \bigg\Vert - \text{HS-enzyme}$$

$$CH_3{-}\overset{\underset{\displaystyle |}{CH_3}}{C}{=}CH{-}CH_2{-}O{-}P_2O_6^{3\ominus}$$

Poly(isoprene) chains are then generated from isopentenylpyrophosphate as monomer and dimethylallylpyrophosphate as primer:

$$(14\text{-}28) \quad CH_3{-}\overset{\underset{\displaystyle |}{CH_3}}{C}{=}CH{-}CH_2{-}O{-}P_2O_6^{3\ominus} \; + \; CH_2{=}\overset{\underset{\displaystyle |}{CH_3}}{C}{-}CH_2{-}CH_2{-}O{-}P_2O_6^{3\ominus} \longrightarrow$$

$$CH_3{-}\overset{\underset{\displaystyle |}{CH_3}}{C}{=}CH{-}CH_2{-}CH_2{-}\overset{\underset{\displaystyle |}{CH_3}}{C}{=}CH{-}CH_2{-}O{-}P_2O_6^{3\ominus} + H^{\oplus} + P_2O_6^{4\ominus}$$

The formation of cis-units is controlled by the enzyme farnesylpyrophosphate-synthase which is combined with the so-called lengthening factor. Natural and technical syntheses of poly(isoprene)s are therefore quite different (see Volume II).

Synthases catalyze the joining of two molecules (reversal of the reaction of lyases). Synthetases (now called ligases) catalyze the formation of C–C, C–O, C–S, C–N, etc., bonds.

14.5.2 Polyesters

Plant and Animal Polyesters

Nature produces not only polyesters of phosphoric acid (= nucleic and teichoic acids) but also of various ω-hydroxycarboxylic acids and dicarboxylic acids.

Ground-dwelling bees line their nests with copolyesters of 18-hydroxyoctadecanoic acid, $HO(CH_2)_{17}COOH$, and 20-hydroxyeicosanoic acid (= 20-hydroxyeicosic acid)= 20-hydroxyarachic acid (= 20-hydroxyarachidic acid), $HO(CH_2)_{19}COOH$.

Cork (L: *cortex* = bark), the bark of cork oaks (*Quercus suber*), consists mainly of **suberin** (L: *suber* = cork), a high-molecular weight polyester with ester and lactone structures from various hydroxycarboxylic acids and dicarboxylic acids, especially

$HOOC–(CH_2)_{20}–COOH$	phellogenic acid	$C_{22}H_{42}O_4$
$HOOC–(CH_2)_{21}–OH$	phellonic acid	$C_{22}H_{42}O_3$
$HOOC–(CH_2)_7–CHOH–CHOH–(CH_2)_7–COOH$	phloionic acid	$C_{18}H_{34}O_6$
$HOOC–(CH_2)_7–CHOH–CHOH–(CH_2)_7–CH_2OH$	phloionolic acid	$C_{18}H_{36}O_5$

Cutin is a waxlike polyester from dihydroxy and trihydroxy C_{18} fatty acids that is crosslinked by ester or peroxide bridges. It is present in the walls of some plant cells and is also the essential component of the cuticle (L: *cuticula* = diminutive of *cutis* = skin), the protective layer of the epidermis of plants (G: *derma* = skin; *epi* = upon, over, at).

Several bacteria produce **poly(β-D-hydroxybutyrate)**, $+O–CH(CH_3)–CH_2–CO+_n$, from D(–)-β-hydroxybutyryl-coenzyme A. This polyester is stored in bacteria as hydrophobic granules of 500 nm diameter (see also Volume II).

The polymers have degrees of polymerization of ca. 20 000 and very narrow molecular weight distributions. They serve bacteria as carbon and energy reserves. The biologically degradable, thermoplastic polymer is now produced biotechnologically, not only as a homopolymer but also as a copolymer with 80 % β-hydroxybutyrate and 20 % β-hydroxyvalerate units, $-O-CH(C_2H_5)-CH_2-CO-$.

Bacterial and Fungal Polyesters

Over 100 bacteria and fungi (Ae, etc., see below) produce linear **poly(hydroxyalkanoic acid)s (PHA)** from three straight-chain ω-hydroxyalkanoic acids (HA)

$HO-CH_2-CH_2-COOH$	3-hydroxypropionic acid	3HP	(Ae)
$HO-CH_2-CH_2-CH_2-COOH$	4-hydroxybutyric acid	4HB	(Ae)
$HO-CH_2-CH_2-CH_2-CH_2-COOH$	5-hydroxyvaleric acid	5HV	(Ae)

and more than 40 PHAs from substituted hydroxyalkanoic acids, for example,

$HO-CH_2-\underset{\underset{CH_3}{|}}{CH}-CH_2-COOH$ 4-hydroxyvaleric acid   4HV   (Ae)

$HO-\underset{\underset{CH_3}{|}}{C}=CH-COOH$ 3-hydroxy-2-butenoic acid
                                                            3-hydroxycrotonic acid   3HB:2en   (*Nocarda*)

$HO-\underset{\underset{CH_3}{|}}{CH}-CH_2-COOH$ 3-hydroxybutyric acid   3HB   (Bm)

$HO-\underset{\underset{CH_2-CH_3}{|}}{CH}-CH_2-COOH$ 3-hydroxyvaleric acid   3HV   (Ae)

$HO-\underset{\underset{CH=CH_2}{|}}{CH}-CH_2-COOH$ 3-hydroxy-4-pentenoic acid   3HV:4en   (Rr)

$HO-\underset{\underset{CH_2-CH(CH_3)_2}{|}}{CH}-CH_2-COOH$ 3-hydroxy-5-methylhexanoic acid   3HHx5Me   (Po)

$HO-\underset{\underset{(CH_2)_7-CH_2Br}{|}}{CH}-CH_2-COOH$ 3-hydroxy-11-bromoundecanoic acid   3HVUD11Br   (Po)

$HO-\underset{\underset{CH_2-CH=CH-CH_2-CH=CH-(CH_2)_4-CH_3}{|}}{CH}-CH_2-COOH$ 3-hydroxy-5,8-tetradecenoic acid   3HTD:5,8dien (Po)

The names of these acids are often abbreviated, using their chemical names (e.g., HV for "hydroxyvaleric") and position numbers for the substituents: number and abbreviation H for the OH group before the abbreviation of the acid and after that abbreviation for all other substituents. An example is 3HHx5Me with a 3-hydroxy group (3H), a hexanoic group (Hx), and a methyl group in the 5-position (5Me).

The only known HA with 4 C-atoms in the aliphatic segment is 5HV. Also, only two HAs are known with 3 C-atoms in that segment, 4HB and 4HV. There are many HAs that derive from 3HP, for example, 3HB, 3HV, 3HB:2en, etc.

Aliphatic PHAs are synthesized by the bacteria *Alcaligenes eutrophus* (Ae), *Bacillus megaterium* (Bm), *Nocarda*, *Pseudomonas oleovorans* (Po), and *Rhodosprillum rubrum* (Rr). *Pseudomonas oleovorans* (Po) also produces an aromatic PHA, poly(3-hydroxy-5-phenylpentanoic acid), $+CO-CH_2-CH(CH_2CH_2C_6H_5)-O\frac{}{}_n$, from 5-phenylvaleric acid, $C_6H_5(CH_2)_4COOH$, as the sole carbon source. Fungi synthesize the poly(malic acid) of $+OCH(COOH)CH_2CO\frac{}{}_n$. PHAs are also produced by a few plants.

PHAs are only synthesized by these organisms if the carbon source (sugars, alkanes, etc.) is present in excess and if an important nutrient (N, S, P, Mg, Fe, K, etc.) for the production of new cells is missing. The biological synthesis requires several enzymes. In the synthesis by *Alcaligenes eutrophus*, the carbon source is first converted into the corresponding activated acetic acid (see formula on p. 560).

Two activated acid molecules are then dimerized to the acetoacetyl-CoA by the enzyme β-ketothiolase in a kind of Claisen condensation. The CH₃CO group of this molecule is subsequently reduced to a HOCH(CH₃) group by a NADPH-dependent acetoacetyl-CoA reductase. The resulting D-(–)-3-hydroxybutyryl-CoA is the true monomer (see CoA on p. 560 with –S–CH₂–CH(OH)–CH₃ instead of –SH).

The monomer is polymerized to poly(3-hydroxybutyric acid) by P(3HB) synthase, a PHA-synthase:

$$(14\text{-}29) \quad 2\ H_3C-\underset{O}{\overset{\|}{C}}-SCoA \quad \xrightarrow[-\ HSCoA]{\beta\text{-ketothiolase}} \quad H_3C-\underset{O}{\overset{\|}{C}}-CH_2-\underset{O}{\overset{\|}{C}}-SCoA$$

$$\xrightarrow[-\ NADP^{\oplus}]{\substack{\text{acetoacetyl-CoA-reductase} \\ +\ NADPH + H^{\oplus}}} \quad HO-\underset{CH_3}{\overset{|}{C}H}-CH_2-\underset{O}{\overset{\|}{C}}-SCoA$$

$$\xrightarrow[-\ HSCoA]{P(3HB)\ \text{synthase}} \quad \left[\!\!-O-\underset{CH_3}{\overset{|}{C}H}-CH_2-\underset{O}{\overset{\|}{C}}-\!\!\right]$$

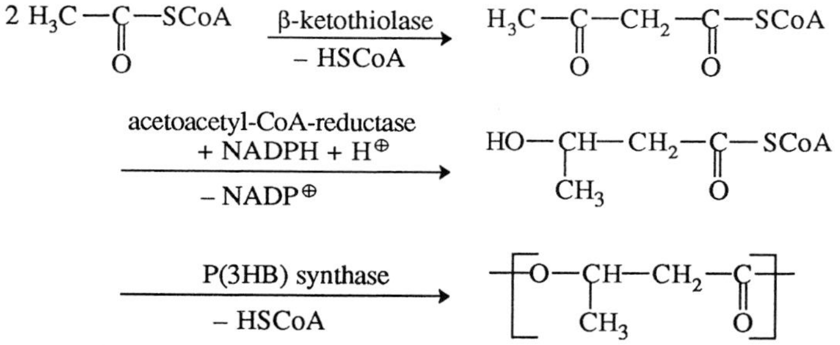

NADP$^{\oplus}$ = nicotinamide-adenine-dinucleotidephosphate; NADP$^{\oplus}$ + 2 H $\rightleftarrows$ NADPH + H$^{\oplus}$.

PHAs are stored in bacteria cells as hydrophobic granules of ca. 500 nm diameter. Their composition depends on both the type of bacterium and the carbon source. Molecular weights may reach 3 million. Polymers of 3HP, 4HB, and 5HV do not have a chirality center. All other natural PHAs consist of D-(–) enantiomers; they are 100 % isotactic. For industrial (biotechnological) syntheses, see Volume II.

14.5.3 Lignins

Wood is a naturally occurring composite that is plasticized by water and "foamed" by air. The solids vary with the type of tree; they contain 42-60 % celluloses, 16-34 % lignins, 14-32 % wood polyoses (see Volume II), and 0.3-3 % minerals. Wood technologists call "lignin" all those parts of wood that are insoluble in dilute acids and organic solvents. Chemically, lignins are a group of crosslinked polymers of coniferyl alcohol.

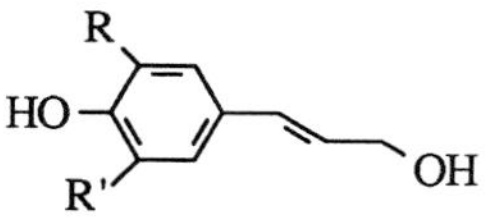

R = H	R' = H	*p*-cumaryl alcohol
R = H	R' = OCH₃	coniferyl alcohol
R = OCH₃	R' = OCH₃	sinapyl alcohol

According to spectroscopy and controlled degradation, lignins contain many different monomeric units (Fig. 14-20). Ca. 28 % of these units are branching units.

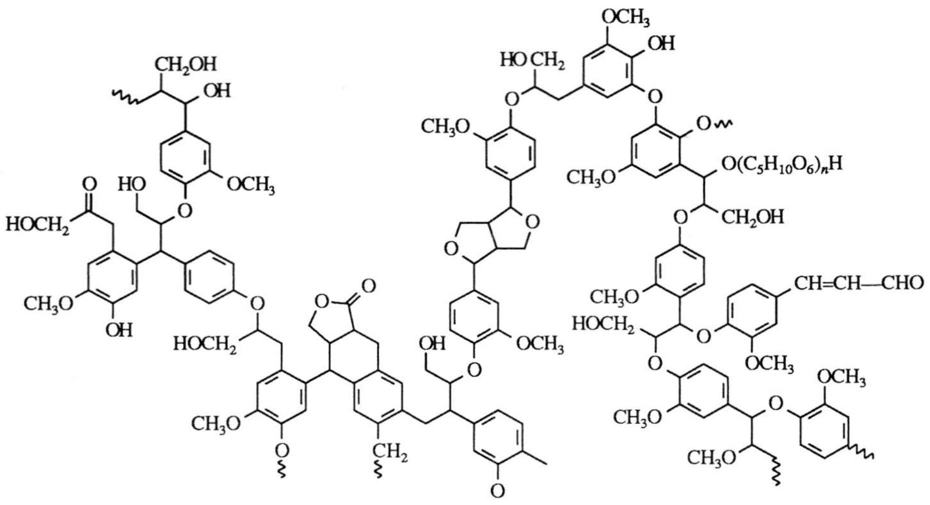

Fig. 14-20 Schematic representation of a lignin from needle trees.

Heating wood with an aqueous solution of NaOH, Na_2CO_3, and Na_2S for the isolation of celluloses (called **Kraft process** because it produces especially strong papers (German: Kraft = strength)) degrades lignins and converts them into lignin sulfonates with molecular weights up to ca. 100 000 (world production: $40 \cdot 10^6$ t/a). 99.9 % of these lignins are burned for energy. The remaining 0.1 % is either used directly (for example, as dispersion agents for cements) or serves as raw material for the synthesis of vanillin and other organic chemicals.

In vivo, coniferyl alcohol is synthesized from glucose via trioses and tetroses. The first intermediate is shikimic acid I (first observed in the fruits of the tree *Illicium religiosus* (Japanese: shikimi)). Addition of a triose to I results in prephenic acid II. The acid II is then converted into *p*-hydroxyphenyl pyruvic acid III and further, in various steps, to coniferyl alcohol IV.

(14-30)

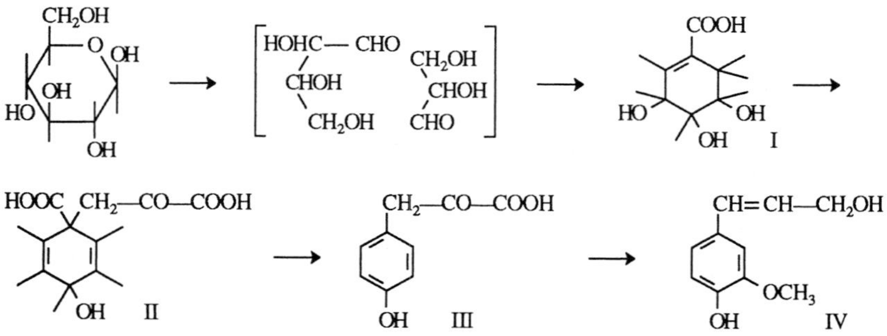

The polymerization of coniferyl alcohol starts with dehydrogenation by the enzyme laccase. The resulting radical of dehydroconiferyl alcohol reacts further via various mesomorphic structures to form branched and crosslinked structures. Renewed dehydrogenation and addition of radicals then produces the very complex structure of lignins:

(14-31)

Literature to Chapter 14

There is an enormous number of books and review articles on biological polymers. The Internet is no help since information on books is buried in the entries for keywords, book titles do not match the contents, publishers do not list older books or "forget" to give the publication date or the names of coauthors, etc. The following is a very selective list of publications which I have found useful, including older ones (for historical aspects, etc.).

14.1a BIOLOGICAL POLYMERS (general biology and biochemistry)
C.R.Cantor, P.R.Schimmel, Biophysical Chemistry, Freeman, San Francisco 1980 (3 vols.)
M.J.Ostro, Liposomes, Dekker, New York 1987
J.Stenesh, Dictionary of Biochemistry and Molecular Biology, Wiley, Somerset (NJ), 2nd ed. 1989
D.D.Lasic, Liposomes from Physics to Applications, Elsevier, Amsterdam 1993
B.Alberts, D.Bray, J.Lewis, M.Raff, K.Roberts, J.D.Watson, Molecular Biology of the Cell, Garland
    Publ., New York, 3rd ed. 1994
D.Voet, J.G.Voet, Biochemistry, Wiley, New York, 2nd ed. 1995
N.A.Campbell, J.B.Reece, L.G.Mitchell, Biology, Addison-Wesley, Menlo Park (CA), 5th ed. 1999
A.L.Lehninger, D.L.Nelson, M.M.Cox, Principles of Biochemistry, Worth Publishing, New York,
    3rd ed. 2000
H.Lodish, A.Berk, L.S.Zipurski, P.Matsudaira, D.Baltimore, J.Darnell, Molecular Cell Biology,
    Freeman, New York, 4th ed. 2000
B.Alberts, A.Johnson, J.Lewis, M.Raff, K.Roberts, P.Walter, Molecular Biology of the Cell,
    Garland Science, New York, 4th ed. 2002

14.1b BIOLOGICAL POLYMERS (general)
C.Helene, Structure, Dynamics, Interactions, and Evolution of Biological Macromolecules, Reidel,
    Dordrecht 1983
O.Kramer, Ed., Biological and Synthetic Polymer Networks, Elsevier, New York 1988
A.Steinbüchel, Ed., Biopolymers, Wiley, New York 2001-2003 (12 volumes)
    Vol. 1: A.Steinbüchel, M.Hofrichter, Eds., Lignin, Humic Substances and Coal (2001)
    Vol. 2: T.Koyama, A.Steinbüchel, Eds., Polyisoprenoids (2001)
    Vol. 3a: Y.Doi, A.Steinbüchel, Eds., Polyesters I - Biological Syntheses and Biotechnological
        Products (2001)
    Vol. 3b: Y.Doi, A.Steinbüchel, Eds., Polyesters II - Properties and Chemical Structures (2001)
    Vol. 4: A.Steinbüchel, Y.Doi, Eds, Polyesters III - Applications and Commercial Products (2002)

Vol. 5: A.Steinbüchel, E.J.Vandamme, S.De Baets, Eds., Polysaccharides I - Polysaccharides from
Prokaryotes (2002)
Vol. 6: A.Steinbüchel, S.De Baets, E.J.C.Vandamme, Eds., Polysaccharides II - Polysaccharides
from Eukaryotes (2002)
Vol. 7: A.Steinbüchel, S.R.Fahnestock, Eds., Polyamides and Complex Proteinaceous
Materials I, (2002)
Vol. 8: S.R.Fahnestock, A.Steinbüchel, Eds., Polyamides and Complex Proteinaceous
Materials II, (2003)
Vol. 9: A.Steinbüchel, C.Matsumura, Eds., Miscellaneous Biopolymers and Biodegradation
of Synthetic Polymers (2002)
Vol. 10: A.Steibüchel, Ed., General Aspects and Special Applications (2003)
-, A.Steinbüchel, Ed., Cumulative Index (2003)
A.Steinbüchel, Y.Doi, Eds., Biotechnology of Biopolymers - From Synthesis to Patents, Wiley-
VCH, Weinheim 2004

14.2 NUCLEIC ACIDS
www.chem.qmul.ac.uk/iupac/misc/naabb.html, Abbreviations and Symbols for Nucleic Acids,
Polynucleotides and Their Constitutents (IUPAC-IUB Joint Commission of Biochemical
Nomenclature (CBN)
J.D.Watson, The Double Helix. A Personal Account of the Discovery of the Structure of DNA,
Atheneum, New York 1968
(see also G.S.Stent, What they are saying about honest Jim, Quart.Rev.Biol. **43** (1968) 179
(a review of "The Double Helix"))
J.N.Davidson, The Biochemistry of Nucleic Acids, Academic Press, New York, 7th ed. 1972
J.S.Cohen, F.H.Portugal, The Search for the Chemical Structure of DNA, Connecticut Medicine
38/10 (1974) 551
P.R.Stewart, D.S.Latham, Eds., The Ribonucleic Acids, Springer, New York, 2nd ed. 1978
S.Altman, Ed., Transfer RNA, MIT Press, Cambridge (MA) 1979
P.Klosinski, S.Penczek, Teichoic Acids and Their Models, Adv. Polym.Sci. **79** (1986) 139
W.Saenger, Principles of Nucleic Acid Structure, Springer, Berlin 1988
A.Vologodskii, Topology and Physics of Circular DNA, CRC Press, Boca Raton (FL) 1992
V.N.Soyfer, V.N.Potaman, Triple-Helical Nucleic Acids, Springer, Berlin 1996
M.Meili, Signer's Gift–Rudolf Digner and DNA, Chimia (Aarau) **57** (2003) 735

14.2.4 NUCLEOPROTEINS
F.A.Jurnak, A.McPherson, Eds., Biological Macromolecules and Their Assemblies, Vol. 1: Virus
Structures, Wiley-Interscience, New York 1984
K.E. van Holde, Chromatin, Springer, Berlin 1989
L.S.Hnilica, G.S.Stein, J.L.Stein, Histones and Other Basic Nuclear Proteins, CRC Press, Boca
Raton (FL) 1989

14.2.5 FUNCTION OF NUCLEIC ACIDS
A.Kornberg, T.Baker, DNA Replication, Freeman, San Francisco, 2nd ed. 1992
R.F.Gesteland, J.F.Atkins, Eds., The RNA World, Cold Spring Harbor Laboratory Press,
Plainnen (NY) 1993
C.Calladine, H.Drew, Understanding DNA. The Molecule and How It Works, Academic Press, San
Diego, 2nd ed. 1997
F.Eckstein, D.M.J.Lilley, Catalytic RNA, Springer, Berlin 1997

14.2.6 CELL-FREE POLYNUCLEOTIDE SYNTHESES
K.B.Mullis, F.Ferré, R.A.Gibbs, Eds., The Polymerase Chain Reaction, Springer, Berlin 1994
J.Gee, Ed., In Situ Polymerase Chain Reaction and Related Technology, Springer, Berlin 1995

14.2.7 CHEMICAL POLYNUCLEOTIDE SYNTHESES
A.Kornberg, DNA Synthesis, Freeman, San Francisco 1974
L.B.Townsend, Ed., Nucleic Acid Chemistry: Improved and New Synthetic Procedures, Methods
and Techniques, Wiley-Interscience, New York 1978 (2 vols.)
K.Takemoto, Y.Inaki, Synthetic Nucleic Acid Analogs, Adv.Polym.Sci. **41** (1981) 1
G.Bannwarth, Gene Technology: A Challenge for a Chemist, Chimia **41** (1987) 302

14.3 PROTEINS: OVERVIEW

H.Neurath, R.L.Hill, Ed., The Proteins, Academic Press, New York, 3rd ed. 1975-82 (5 vols.)

G.D.Fasman, Ed., CRC Handbook of Biochemistry and Molecular Biology, Section A: Proteins (3 vols.), CRC Press, Baton Rouge (FL), 3rd ed. (1976)

R.E.Dickerson, I.Geis, The Structure and Action of Proteins, W.A.Benjamin, New York 1981

R.Scopes, Protein Purification - Principles and Practice, Springer, Berlin 1982

Th.E.Creighton, Proteins: Structures and Molecular Properties, Freeman, New York, 2nd ed. 1993

J.Drenth, Principles of Protein X-Ray Crystallography, Springer, Berlin 1994

C.L.Brooks III, M.Karplus, B.Montgomery Pettitt, Proteins. A Theoretical Perspective of Dynamics, Structure, and Thermodynamics, Wiley, New York 1998

14.3.3-14.3.6 STRUCTURE OF PROTEINS

G.Walton, Polypeptide and Protein Structure, Elsevier, Amsterdam 1981

C.Frieden, L.W.Nichol, Protein-Protein Interactions, Wiley-Interscience, New York 1981

C.C.Ghélis, J.Yon, Protein Folding, Academic Press, New York 1982

A.McPherson, Preparation and Analysis of Protein Crystals, Wiley, New York 1982

P.M.Harrison, Ed., Metalloproteins, VCH, Weinheim 1985 (2 parts)

J.B.C.Findlay, M.J.Geisow, Eds., Protein Sequencing, IRL Press, Oxford 1988

G.D.Fasman, Ed., Prediction of Protein Structure and the Principles of Protein Conformation, Plenum, New York 1989

K.M.Merz, Jr., S.M.Legrand, Eds., The Protein Folding and Tertiary Structure Prediction, Birkhäuser, Basel 1994

J.B.Smith, Ed., Protein Sequencing Protocols, Humana Press, Totowa (NJ) 1997

Y.Lvov, H.Möhwald, Eds., Protein Architecture, Dekker, New York 1999

M.Kinter, N.E.Sherman, Protein Sequencing and Identification Using Tandem Mass Spectrometry, Wiley-Interscience, New York 2000

14.3.7-14.3.8 PREBIOTIC AND BIOLOGICAL SYNTHESES

H.Weissbach, S.Peska, Eds., Molecular Mechanisms of Protein Biosynthesis, Academic Press, New York 1977

J.L.Cleland, C.S.Craik, Protein Engineering. Principles and Practice, Wiley, New York 1996

A.L.Fink, Y.Goto, Eds., Molecular Chaperones in the Life Cycle of Proteins, Dekker, New York 1997

14.3.10 ENZYMES

-, Enzyme Nomenclature: Recommendations (1984) of the Nomenclature Committee of the International Union of Biochemistry, Academic Press, Orlando (FL) 1985

P.D.Boyer, Ed., The Enzymes, Academic Press, New York, 3rd ed. 1970-1982 (15 vols.)

M.Dixon, E.C.Webb, C.J.R.Thorne, K.F.Tipton, Enzymes, Academic Press, New York, 3rd ed. 1980

N.C.Price, L.Stevens, Fundamental Enzymology, Oxford University Press, New York, 2nd ed. 1989

H.Uhlig, Ed., Industrial Enzymes and Their Applications, Wiley, New York 1998

W.Aehle, Ed., Enzymes in Industry, Wiley-VCH, Weinheim 2003

A.S.Bommarius, B.R.Riebel, Biocatalysis. Fundamentals and Applications, Wiley-VCH, Weinheim 2004

K.Buchholz, V.Kasche, U.T.Bornscheuer, Biocatalysts and Enzyme Technology, Wiley-VCH, Weinheim 2005

14.3.11 SCLEROPROTEINS

D.A.D.Pery, L.K.Creamer, Fibrous Proteins, Academic Press, New York 1979 (2 vols.)

J.Woodhead-Galloway, Collagen, The Anatomy of a Protein, Arnold, London 1980

J.F.V.Vincent, Structural Biomaterials, Wiley, New York 1982

M.Nimni, Ed., Collagen: Biochemistry, Biotechnology and Molecular Biology, CRC Press, Boca Raton (FL), 1988 (3 vols.)

14.3.12 PEPTIDE AND PROTEIN SYNTHESES

H.Gounelle de Pontanel, Ed., Protein from Hydrocarbons, Academic Press, New York 1972

E.C.Blossey, D.C.Neckers, Eds., Solid-Phase Syntheses, Dowden, Hutchinson & Ross, Stroudsburg (PA) 1975 (reprints of benchmark papers in the field)

M.Bohdanszky, Y.S.Klausner, M.A.Ondetti, Peptide Synthesis, Wiley, New York 1976
R.E.Offord, Semisynthetic Proteins, Wiley, New York 1980
P.J.Kocienski, Protecting Groups, Thieme, Stuttgart 1994
B.Gutte, Ed., Peptides. Synthesis, Structures, and Applications, Academic Press, Orlando 1995

14.4 POLYSACCHARIDES: OVERVIEW

G.M.W.Cook, R.W.Stoddard, Surface Carbohydrates of the Eucaryotic Cell, Academic Press, New York 1973
I.W.Sutherland, Ed., Surface Carbohydrates of the Procaryotic Cell, Academic Press, New York 1977
J.Preiss, Ed., The Biochemistry of Plants, vol. 3, Carbohydrates, Structure and Function, Academic Press, New York 1980
P.M.Collins, Ed., Carbohydrates, Chapman and Hall, London 1987

14.4.2-14.4.3 MONOSACCHARIDES AND OLIGOSACCHARIDES

IUPAC, Abbreviated Terminology of Oligosaccharide Chains (Recommendations 1980), Pure Appl. Chem. **54** (1982) 1517
W.W.Pigman, D.Horton, Eds., The Carbohydrates, Academic Press, New York, 2nd ed. 1970
H.S. El Khadem, Carbohydrate Chemistry, Academic Press, San Diego, CA 1988
J.F.Kennedy, Carbohydrate Chemistry, Oxford Science Publ., New York 1988
A.Lipták, P.Fügedi, Z.Szurmai, J.Harangi, CRC Handbook of Oligosaccharides, CRC, Boca Raton, 3 vols.: I. Disaccharides (1990); II. Trisaccharides (1990); III. Higher Oligosaccharides (1991)

14.4.4 POLYSACCHARIDES

IUPAC, Polysaccharide Nomenclature (Recommendations 1980), Pure Appl.Chem. **54** (1982) 1523
www.chem.qmul.ac.uk/iupac/misc/psac.html, Symbols for Specifying the Conformation of Polysaccharide Chains (IUPAC-IUB Joint Commission of Biochemical Nomenclature (CBN)
N.Sharon, Complex Carbohydrates: Their Chemistry, Biosynthesis, and Function, Addison-Wesley, Reading (MA) 1975
D.A.Rees, Polysaccharide Shapes, Chapman and Hall, London 1977
G.O.Aspinall, Ed., The Polysaccharides, Academic Press, New York, vol. 1 (1982), vol. 2 (1983)
V.Creszenzi, I.C.M.Dea, S.S.Stivala, New Developments in Industrial Polysaccharides, Gordon and Breach, New York 1985
M.Yalpani, Ed., Industrial Polysaccharides, Elsevier Sci.Publ., New York 1987
B.A.Stone, A.E.Clarke, Chemistry and Biochemistry of (1→3)-β-Glucans, Academic Press, New York 1988
J.F.Kennedy, G.O.Phillips, P.A.Williams, Ed., Wood and Cellulosics, Wiley, New York 1988
R.L.Whistler, J.N.BeMiller, Eds., Industrial Gums - Polysaccharides and Their Derivatives, Academic Press, San Diego (CA), 3rd ed. 1993
K.Hill, W. von Rybinski, G.Stoll, Eds., Alkyl Polyglycosides, Wiley-VCH, Weinheim 1996
S.Dumitriu, Ed., Polysaccharides: Structural Diversity and Functional Versatility, Dekker, New York 1998
D.-N.S.Hon, N.Shiraishi, Eds., Wood and Cellulosic Chemistry, Dekker, New York, 2nd ed.. 2001

14.4.5-14.4.8 SYNTHESES OF POLYSACCHARIDES

K.H.Ebert, G.Schenk, Mechanism of Biopolymer Growth: The Formation of Dextran and Levan, Adv.Enzymol. **30** (1968) 179
C.Schuerch, Systematic Approaches to the Chemical Synthesis of Polysaccharides, Acc. Chem.Res. **6** (1973) 184
M.R.Brown, Jr., Ed., Cellulose and Other Natural Polymer Systems: Biogenesis, Structure, and Degradation, Plenum, New York 1982
R.W.Stoddart, The Biochemistry of Polysaccharides, Croom Helm, Beckenham (UK) 1984
C.H.Haigler, P.J.Weimer, Biosynthesis and Biodegradation of Cellulose, Dekker, New York 1991
I.A.Tarchevsky, G.N.Marchenko, Cellulose: Biosynthesis and Structure, Springer, Berlin 1991

14.4.x TRANSFORMATIONS

H.A.Krässig, Cellulose - Structure, Accessibility, and Reactivity, Gordon and Breach, Amsterdam 1993
D.-N.S.Hon, Ed., Chemical Modification of Lignocellulosic Materials, Dekker, New York 1996

D.Klemm, B.Philipp, T.Heinze, U.Heinze, W.Wagenknecht, Comprehensive Cellulose Chemistry, Wiley-VCH, Weinheim 1998 (2 vol.) (regenerated celluloses and their derivates)

14.5.1 POLY(ISOPRENE)S (see literature in Volume II)

14.5.2 POLYESTERS
D.P.Mobley, Ed., Plastics from Microbes, Hanser, München 1995
W.Babel, A.Steinbüchel, Biopolyesters, Springer, Berlin 2001

14.5.3 LIGNINS
K.Freudenberg, A.C.Neish, Constitution and Biosynthesis of Lignin, Springer, Berlin 1968
K.V.Sarkanen, C.H.Ludwig, Eds., Lignins, Occurrence, Formation, Structure and Reactions, Wiley, New York 1971
J.F.Kennedy, G.O.Phillips, P.A.Williams, Eds., Wood and Cellulosics, Wiley, New York 1988
D.Fengel, G.Wegener, Wood Chemistry, Ultrastructure, Reactions, De Gruyter, Berlin 1989
M.Lewin, I.S.Goldstein, Eds., Wood Structure and Composition, Dekker, New York 1991
J.F.Kennedy, G.O.Phillips, P.A.Williams, Lignocellulosics: Science, Technology, Development and Use, Prentice-Hall, Englewood Cliffs (NJ) 1992
E.Sjöström, Wood Chemistry: Fundamentals and Applications, Academic Press, San Diego (CA), 2nd ed. 1993
D.-N.S.Hon, Ed., Chemical Modification of Lignocellulosic Materials, Dekker, New York 1996
D.-N.S.Hon, N.Shiraishi, Eds., Wood and Cellulosic Chemistry, Dekker, New York, 2nd ed. 2001

References to Chapter 14

[1]   W.Saenger, Nachr.Chem.Techn. **30**/1 (1982) 8, Fig. 5 (A and B are here mistakenly transposed, see Nachr.Chem.Techn.Lab. **30**/3 (1982) 186)
[2]   Y.Izumi, I.Chibata, T.Itoh, Angew.Chem. **90** (1978) 187, Table 1
[3]   K.H.Ebert, G.Schenk, Adv.Enzymol. **30** (1968) 179, Fig. 1

15   Reactions of Macromolecules

15.1   Overview

Macromolecules are reacted with low or high molecular weight chemical compounds in order to determine the chemical structure, modify the composition, or remove unwanted molecules. These reactions always change the molecular weight but the degree of polymerization may remain constant, increase, or decrease. Macromolecules may also serve as catalysts or as carriers for catalysts.

Undesired reactions of polymeric materials are called **chemical aging**. These reactions may be caused by environmental compounds (oxygen, nitroxides, water, etc.) or by catalyst residues from the polymerization process.

Reactions and properties of polymers are controlled by the constitution, configuration, conformation, and molecular weight of their macromolecules. It is expedient to use these structural quantities and not reaction mechanisms to classify the reactions. The following sections thus discuss catalyses (Section 15.2), isomerizations (Section 15.3), polymer-analog reactions (Section 15.4), and degradation reactions (Section 15.5). Cross-linking reactions have already been discussed in previous chapters.

15.2   Polymer Catalysts

By definition, **catalysts** are materials that accelerate chemical reactions by lowering the activation energy (G: *kata* = down, *lyein* = to loosen, release. See also p. 163). They do not influence the reaction equilibrium and emerge unchanged from the reaction.

Polymer catalysts are therefore polymers with catalytically active groups. They differ from **polymer reagents** which are chemical compounds that change their chemical composition on reaction (Section 15.4.6).

Polymer catalysts are expected to have three advantages over their low-molecular weight counterparts: greater selectivity of substrates, increased specificity with respect to the type of reaction, and better separability of reaction products. All three properties are also exhibited by solid non-polymeric catalysts such as metals on carriers. However, polymer catalysts make it easier to identify and control catalytically active centers than other heterogeneous catalysts. Compared to enzymes and enzymatic preparations, polymer catalysts offer lower costs and the possibility to catalyze reactions that cannot be catalyzed by enzymes.

Enzymes are, of course, also polymer catalysts, albeit mostly soluble ones. They may catalyze not only biochemical reactions but also some that are in the realm of synthetic chemistry. Examples are the formation of siloxane bonds by the enzyme hydrolase trypsin ($2 \sim SiR_2-OQ \rightarrow 2 \sim SiR_2-OH \rightarrow \sim SiR_2-O-SiR_2\sim$) and the ring-opening polymerization of ε-caprolactam by novozym 435.

Polymer catalysts may serve for many types of reactions. Ion exchange resins have been used for hydrolyses, hydrations and dehydrations, alkylations of phenols, esterifications and transesterifications, aldol condensations, Cannizzaro reactions, cyanoethylations, and the Prins reaction of aldehydes with olefins to form 1,3-dioxanes. Chelated poly(amino acid)s, poly(vinylpyridine)s, and poly(phthalocyanine)s as well as conjugated polymers such as poly(acetylene)s act as catalysts for mild oxidations, dehydrogenations, and dehydrations.

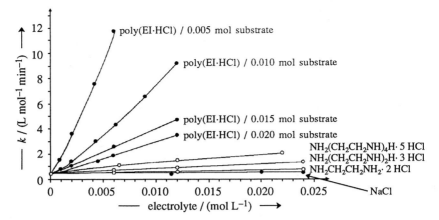

Fig. 15-1 Rate constants k of the reaction between bromoacetic acid and thiosulfate (Eq.(15-1)) at 25°C as a function of the concentration of the "catalyzing" electrolytes poly(ethyleneimine hydrochloride) (poly(EI·HCl)), tetraethylene pentamine hydrochloride, diethylene triamine hydrochloride, ethylene diamine hydrochloride, and sodium chloride. The substrate concentrations were always 0.01 mol/L for the added low-molecular weight compounds but differed for the polyelectrolyte (see graph).

With kind permission of the American Chemical Society, Washington (DC) [1].

In many cases, reaction rates in the presence of polymers are greater than in the presence of their low-molecular weight analogs. However, such rate increases alone do not prove a catalytic action. Rather, one has to make sure that equilibria are not shifted.

An example is the reaction between bromoacetic acid and thiosulfate

$$(15-1) \qquad Br{-}CH_2{-}COO^{\ominus} + S_2O_3{}^{2\ominus} \; \rightleftarrows \; {}^{\ominus}S_2O_3{-}CH_2{-}COO^{\ominus} + Br^{\ominus}$$

that is more strongly accelerated by polycations than by low-molecular weight analogs (Fig. 15-1). However, these added compounds also shift the equilibrium. The strong acceleration is probably caused by the high local concentration of reactants near the "catalytically effective" groups of polymer molecules. These high local concentrations increase the probability of cooperative effects (compare titration of polyacids, Section 15.4.2).

Polyions accelerate such reactions because polyelectrolytes in aqueous solution form clusters with lattice-like structures (Volume III). The resulting high local ion concentrations can cause rate increases by factors of up to 10^9, i.e., enzyme-like effects (p. 541). Cation-cation reactions are accelerated by anions (and vice versa) but cation-anion reactions are decelerated by both cations and anions.

In cooperative catalyses, two different catalyzing groups act simultaneously on the substrate, e.g., an electrophilic and a nucleophilic group. Such bifunctional actions are responsible for the high activity of many enzymes. The **cooperativity** of such groups is determined by their relative spatial positions (Fig. 15-2). The groups may be in the backbone (A, E, F) or the substituents (B, C, D), randomly distributed along the chain (A, B) or in neighboring positions (C, D, F), present as isolated pairs (A, B, C, D) or in double pairs (E, F), etc. Catalyzing groups may be introduced into the polymer catalyst by conventional copolymerization or by graft copolymerization. They may be directly bound to the chain or separated from it by long segments, called **spacers**.

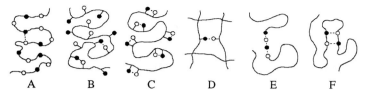

Fig. 15-2 Schematic representation of cooperatively active groups O and ● (see text).

Polymer catalysts from conventionally prepared graft and random copolymers are usually not very effective since their catalytically active groups are not present in optimal positions (A and B in Fig. 15-2). The pairwise arrangement C is more favorable because of the side-by-side position. The highest catalytic action is obtained if the two different groups oppose each other (D).

Position D can be generated by using crosslinkable precursors of functional groups. For example, the 2,3-*O*-*p*-vinyl phenylboric ester of D-glyceric acid *p*-vinylanilide can be copolymerized with divinylbenzene in acetonitrile. Hydrolysis of the crosslinked polymer removes D-glyceric acid and the free amino and boric acid groups are now held in "correct" positions by the crosslinked carrier (**molecular imprinting**):

(15-2)

$$CH_2=CH-\bigotimes-NH-CO \diagdown O \diagup B-\bigotimes-CH=CH_2$$

polymerization
hydrolysis $\longrightarrow$

$$CH_2-CH-\bigotimes-NH_2 \qquad \overset{OH}{\underset{OH}{B}}-\bigotimes-CH-CH_2$$

The polymer can separate D,L-glyceric acid and D,L-glyceraldehyde into their antipodes. Similar effects are obtained with catalyzing groups. In each case, the close proximity of the two acting groups produces entropically more favorable conditions for the reaction with the substrate. Termolecular reactions appear to be bimolecular, etc.

The specificity of a catalyst cannot be caused exclusively by a binding of the substrate to two points, however, since two points are not sufficient to produce an asymmetry. F in Fig. 15-2 is thus a more effective arrangement than D. The "cage" for the active groups must also be neither too small nor too large for the substrate.

Catalytic activities can even be reduced if the active groups are in close proximity, for example, the endgroups E on the "surface" of a dendrimer with the following units:

$$-(CH_2)_3-\underset{\underset{(CH_2)_3-}{\overset{(CH_2)_3-}{|}}}{Si}-(CH_2)_3-$$

core

$$-Si\diagup\diagdown\begin{matrix}(CH_2)_3Si(CH_3)_2O(CH_2)_4OCO-\\(CH_2)_3Si(CH_3)_2O(CH_2)_4OCO-\\(CH_2)_3Si(CH_3)_2O(CH_2)_4OCO-\end{matrix}$$

branching unit with repeating unit

$$\bigotimes\begin{matrix}N(CH_3)_2\\\downarrow\\Ni-Br\\\uparrow\\N(CH_3)_2\end{matrix}$$

endgroup E

The selectivity of a polymer catalyst depends on the size of the substrate as well as hydrophobic effects. Larger substrates find it more difficult to contact the catalyzing groups. For this reason, cyclododecene is hydrogenated five times more slowly than cyclohexene by the polymeric rhodium catalyst I. The low-molecular weight analog II does not show this effect:

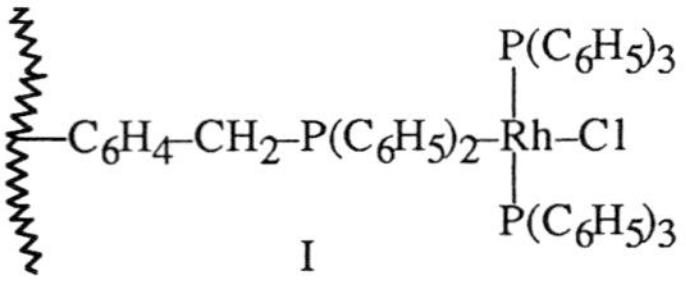

$$\text{--C}_6\text{H}_4\text{--CH}_2\text{--P(C}_6\text{H}_5)_2\text{--Rh--Cl} \qquad (\text{C}_6\text{H}_5)_3\text{P--Rh--Cl}$$

with P(C₆H₅)₃ groups, labeled I and II.

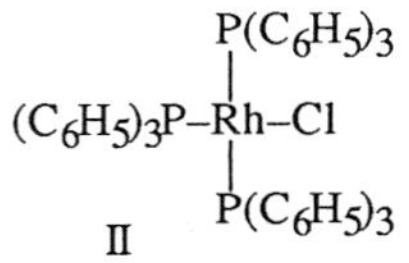

On the other hand, poly(vinyl imidazole) IV hydrolyzes esters III 1000fold faster than the imidazole V itself. IV also acts auto-catalytically. The larger the number of methylene groups in the ester group of III, the faster the rate of hydrolysis (factor 25 from $n = 1$ to $n = 17$). The effect is possibly caused by increasing intramolecular association of the polymer which forms intermediate acyl structures on hydrolysis of III.

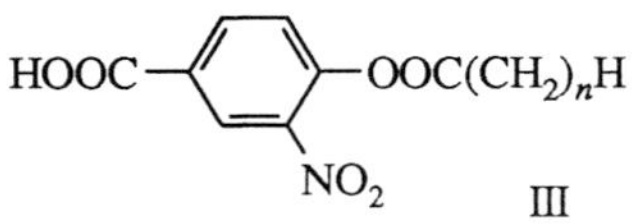

$$\text{HOOC} - \langle\text{ring}\rangle - \text{OOC(CH}_2)_n\text{H}$$

with NO₂, labeled III; structures labeled IV and V containing N and NH imidazole rings.

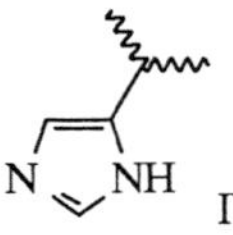

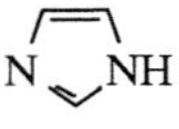

Semiconducting polymers are good catalysts for dehydrogenations, oxidations, and fragmentations but their mode of action is not very clear. Dehydrogenations of aromatic polymers seem only to proceed catalytically if an intermediate quinoidal structure can be formed. The oxidation of cyclohexanol to cyclohexanone by semiconducting polymers proceeds at 250°C without side reactions. At 350°C, cyclohexene is dehydrogenated to benzene without the disproportionation of cyclohexene into benzene and cyclohexane that is common with palladium catalysts.

15.3 Isomerization

Isomerizations are defined as changes of constitution and/or configuration without changes in composition or number-average degrees of polymerization. This group of reactions also comprises exchange equilibria between molecular segments.

15.3.1 Exchange Equilibria

Exchange reactions are equilibrium reactions in which a macromolecular segment is exchanged for another one, for example, **transamidation** between two amide groups (**scrambling reaction**, see also next page):

$$(15\text{-}3) \qquad
\begin{array}{cc}
\text{Z--NH} & \text{CO--Z''} \\
| & | \\
\text{Z'--CO} & \text{NH--Z'''}
\end{array}
+
\quad \rightleftharpoons \quad
\begin{array}{c}
\text{Z--NH--CO--Z''} \\
+ \\
\text{Z'--CO--NH--Z'''}
\end{array}$$

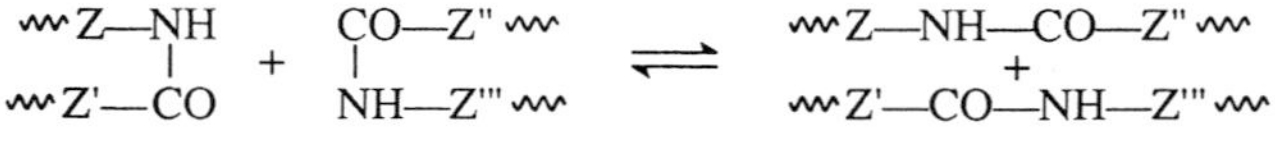

Exchange reactions leave unchanged the number of molecules in the system and therefore also the number-average degrees of polymerization and molecular weight. All other averages change, however, for example, the weight-average degrees of polymerization and molecular weight, because the reaction may proceed between segments of different lengths. A reaction between two macromolecules with the same degree of polymerization, X, leads to one molecule with a higher X and one with a lower X. The weight-average degree of polymerization of the mixture increases because more statistical weight is given to the mass of molecules than their number. In polymolecular polymers, the weight-average degree of polymerization increases until an equilibrium between all segments is established. Since the position of an equilibrium is independent of the paths leading to it, all exchange reactions of polymers will finally lead to the polymolecularities that are obtained in equilibrium polyadditions and polycondensations: $\overline{X}_{R,w} = 2\,\overline{X}_{R,n} - 1$ for reactants (Eq.(13-46)) and $\overline{X}_{P,w} = 2\,\overline{X}_{P,n}$ for polymers.

Exchange reactions are not "spontaneous". They are always catalyzed, either externally by added catalysts or catalyst residues from the preceding polymerization, or internally by catalytically acting groups in the macromolecules themselves. For example, the scrambling reaction of Eq.(15-3) plays little or no role in transamidations where an acid-catalyzed aminolysis by amine endgroups is the main (or sole?) reaction:

$$(15\text{-}4) \qquad \begin{array}{c} \text{\scriptsize www}Z\text{---}CO\text{---}NH\text{---}Z''\text{\scriptsize www} \\ + \\ \text{\scriptsize www}Z'\text{---}NH_2 \end{array} \quad \rightleftharpoons \quad \begin{array}{c} \text{\scriptsize www}Z\text{---}CO \\ | \\ \text{\scriptsize www}Z'\text{---}NH \end{array} \; + \; H_2N\text{---}Z''\text{\scriptsize www}$$

No transamidation is observed if amino endgroups are removed by reaction with acetic acid, acetanhydride, or ketene. However, scrambling reactions between two chain segments are present in the equilibration of poly(dimethylsiloxane)s (Eq.(7-3)).

Heterochains (p. 34) are especially prone to equilibrations because of the low activation energies for attacks on heteroatoms. Polyesters, polyacetals, and polyurethanes are examples in addition to polyamides and polysiloxanes. These exchange reactions are also called **trans reactions** (transesterifications, etc.; in the special case of polysiloxanes also **equilibrations**). Carbon chains do not undergo equilibrations because their main-chain carbon atoms have neither free electron pairs nor electron gaps or unsaturated electron shells. Attacks on carbon-carbon single bonds thus need high activation energies. Carbon-carbon double bonds, on the other hand, are easily activated, for example, in metathesis reactions (Section 9.4).

Exchange reactions proceed between similar bonds. The enthalpy of reaction is therefore zero. The reaction is rather propelled by the reaction entropy, i.e., an increase in the number of combinations of segments.

Exchange equilibria are shifted if one component is removed from the system, for example, by crystallization of the newly formed sequences. An example is the polyester PCT from terephthalic acid (TA) and 1,4-cyclohexanedimethylol (CHD). Commercial PCTs contain random distributions of the two possible diol units in the ratio of cis:trans 2:1. The melting temperature depends only on the content of the trans units. Addition of an ester exchange catalyst and heating to temperatures just below the melting temperature lowers the solubility of the polymer. The melting temperature increases because the ester exchange results in the formation of longer trans sequences, i.e., a segmented copolymer.

Exchange reactions lead to a time-dependence of physical properties. For example, the anionic polymerization of ε-caprolactam results in polyamides with fairly narrow molecular weight distributions. The subsequent melt processing (extrusion, etc.) is accompanied by transamidation which causes an increase of the weight-average molecular weight and thus the melt viscosity (Volume III). The speed of processing must therefore be constantly readjusted in order to maintain a good quality product.

15.3.2 Constitutional Isomerizations

Isomerizations of skeleton chains are fairly rare. An example is the photo-Fries reaction of aromatic polycarbonates (Section 11.2.4) where the carbonate group of the chain is transformed to a keto group. Much more frequent are transformations of side groups (Section 15.4).

15.3.3 Configurational Isomerizations

Certain chemical reactions may change the tacticities of polymers, especially that of geometric isomers (torsional isomers) and less that of configurational isomers. The configurational equilibrium is controlled by the difference in the energy content of the two isomers A and A':

$$(15\text{-}5) \qquad \Delta G_{\text{iso},0} = G_{\text{A}'} - G_{\text{A}} = - RT \ln ([\text{A}']/[\text{A}])$$

Equal energy contents lead to **racemizations** where both enantiomers are present in equal amounts (Section 4.1.4).

Cis-Trans Isomerizations

Addition of a radical R$^\bullet$ to a double bond in the backbone of polydienes generates single bonds and, in turn, a more or less free rotation of chain segments around the new single bond. The removal of the radical reconstitutes a double bond albeit not necessarily one with the same geometric isomerism:

$$(15\text{-}6)$$

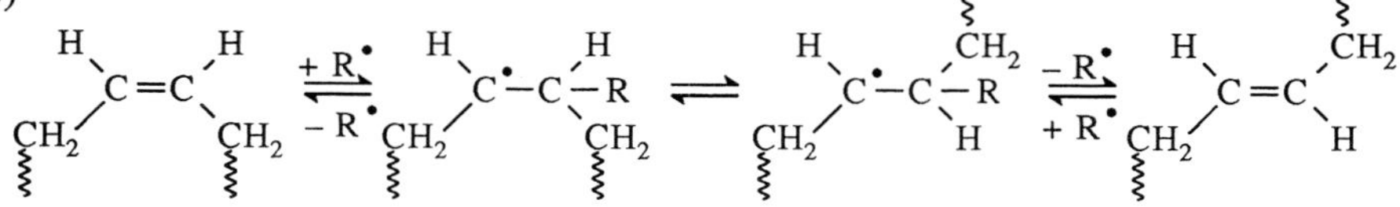

Radicals R$^\bullet$ may be generated by irradiation of bromine or organic bromides, sulfides, or mercaptans with ultraviolet light. Isomerizations may also occur on formation of charge-transfer complexes of polydienes with sulfur or selenium, for example, in the vulcanization of natural rubber (Volume II). At 25°C, cis-1,4-poly(butadiene) is isomerized in this way to a configurational equilibrium with 77 % of the double bonds in the trans configuration, i.e., $K_{\text{iso},0} = 77/23 \approx 3.35$. According to Eq.(15-5), the free enthalpy of the reaction is thus $\Delta G_{\text{iso},0} = -3.0$ kJ/mol.

Isomerizations of Tactic Polymers

Isotactic and syndiotactic carbon-carbon chains isomerize by breaking the bond between a stereogenic center (see p. 115) and a substituent or an adjacent chain atom. The subsequent reconstitution of these bonds leads to polymer chains with a ratio of isotactic and syndiotactic diads that corresponds to the *conformational* equilibrium.

Tactic polymers with C-C chains are difficult to isomerize because of the very high activation energy. Their isomerization is also often accompanied by a chain degradation.

Isomerizations without side reactions are rare; an example is the isomerization of it-poly(isopropyl acrylate) by catalytic amounts of sodium isopropylate in dry isopropanol. The base temporarily produces a carbanion at the α-carbon atom of the monomeric units that is mesomeric to its enolate. The return to the original constitution is accompanied by configurational changes at the stereogenic carbon atom which leads to "atactic" chains (i.e., chains that are not predominantly tactic, see p. 122):

(15-7)

$$\text{⋀CH}_2\overset{\text{it}}{\underset{\text{O=C−OR}}{\text{CH}}}\text{⋀} \quad \xrightarrow[-\text{BH}]{+\,\text{B}^{\ominus}} \quad \left[\text{⋀CH}_2\overset{\ominus}{\underset{\text{O=C−OR}}{\text{C}}}\text{⋀} \quad \longleftrightarrow \quad \text{⋀CH}_2\overset{}{\underset{\ominus\text{O−C−OR}}{\text{CH}}}\text{⋀}\right] \quad \xrightarrow[+\text{BH}]{-\,\text{B}^{\ominus}} \quad \text{⋀CH}_2\overset{\text{at}}{\underset{\text{O=C−OR}}{\text{CH}}}\text{⋀}$$

The configurational equilibrium is shifted if one of the isomers is removed from the equilibrium, for example, by crystallization. An example is an isomerization process in which a small amount of a 100 % *trans*-1,4-poly(butadiene) is added to a 1,4-poly(butadiene) with a high (but less than 100 %) content of trans-configurations. In this case, the trans content decreases but then increases again with time. This behavior is probably caused by a trans isomerization at the crystalline-amorphous border which leads to an incorporation of longer trans sequences into the crystal lattice. These blocks of trans sequences are removed from the isomerization equilibrium which then tries to reestablish itself by generation of new trans configurations.

15.4 Polymer-Analog Reactions

15.4.1 Overview

Polymer-analog reactions are reactions in which substituents A of a starting polymer are replaced by substituents C through the action of a compound B with or without the formation of leaving molecules D:

(15-8)
$$\overset{}{\underset{\text{A A A A A A A A}}{\text{⋀⋀⋀⋀⋀⋀⋀⋀⋀⋀}}} \quad \xrightarrow[-\text{D}]{+\,\text{B}} \quad \overset{}{\underset{\text{C C C A C N C C}}{\text{⋀⋀⋀⋀⋀⋀⋀⋀⋀⋀}}}$$

In the ideal case, conversions A → C are complete without remaining A groups or formation of unwanted N groups. By definition, polymer-analog reactions change the composition and the molecular weight but not the degree of polymerization. They comprise acid-base reactions (Section 15.4.2), ion-exchange reactions (Section 15.4.3), polymer transformation reactions (Section 15.4.4), intramolecular ring-closure reactions (Section 15.4.5), and reactions of polymer reagents (Section 15.4.6).

Polymer-analog reactions are called **functionalization** if they provide the polymer molecules with "functional groups". The term "functionalization" is sometimes used exclusively for the modifiction of chain ends, i.e., for **chain analog reactions**, especially those of living polymers (Section 8.3.7).

Polymer analog reactions are subdivided according to the nature of the groups, A, C, and N, molecules B and D, and the type of the desired endproduct:

- **Polymer analog reactions** are classic **polymer transformations** where poly(C) is the desired endproduct of the reaction of poly(A) with B.
- **Reactive resins** are crosslinked polymers with reactive groups that are used to synthesize D. Poly(C) is here an intermediate product that is recycled to Poly(A).
- **Polymer reagents** are electrically uncharged reactive resins.
- **Ion exchangers** (**ion exchange resins**) are charged polymer reagents that can exchange ions with low-molecular weight compounds.
- **Polymer complexes** result from the addition of B to poly(A) without release of D and without formation of covalent bonds between A and B; B may also be a polymer group. In most cases, B–A bonds are hydrogen bonds, dipole-dipole interactions, or hydrophobic bonds. However, complexes may also be formed via coordinative bonds or electron-deficient bonds, i.e., chemical bonds.
- **Stereocomplexes** are polymer complexes between differently configured poly(A) and poly(A') molecules where A and A' are monomeric units with the same constitution but different configuration. An example is the stereocomplex of isotactic and syndiotactic poly(methyl methacrylate). Such complexes neither need to be formed in the ratio 1:1 of constitutional groups nor do they necessarily crystallize. In the general case, they are not "racemic crystallizations".

15.4.2   Acid-Base Reactions

Acid-base reactions between polymer molecules and low-molecular weight compounds differ in several aspects from those between two low-molecular weight compounds. For the latter, a simple relationship exists between the pH, the degree of dissociation, α, and the degree of neutralization, β. The logarithm of the equilibrium constant, $K = [H^{\oplus}][A^{\ominus}]/[HA]$, of an acid HA is

$$(15\text{-}9) \quad \lg K = \lg [H^{\oplus}] + \lg \{[A^{\ominus}]/[HA]\}$$

Introduction of the definitions pH $\equiv - \lg [H^{\oplus}]$ and $pK_a \equiv - \lg K$ transforms Eq.(15-9) into the **Henderson-Hasselbalch equation**:

$$(15\text{-}10) \quad pH = pK_a + \lg \{[A^{\ominus}]/[HA]\} = pK_a + \lg \{(1 - \beta)/\beta\} = pK_a + \lg \{\alpha/(1 - \alpha)\}$$

where $\beta = [HA]/([HA] + [A^{\ominus}])$ = **degree of neutralization** of acid groups and $\alpha = 1 - \beta$ the **degree of dissociation**. β is usually calculated from the amount of added base.

The Henderson-Hasselbalch equation must be modified for polymers because of additional electrostatic and statistical effects. An example is the different behavior of poly-(acrylic acid), $+CH_2-CH(COOH)+_n$, and its low-molecular weight counterpart, propi-

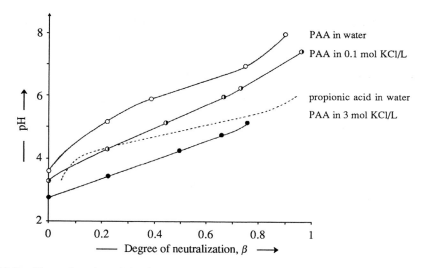

Fig. 15-3 pH as a function of the degree of neutralization, β, for the titration of 0.01 mol mer/L poly(acrylic acid), PAA, in water with or without addition of KCl. - - - pH = f(β) for 0.01 mol/L propionic acid. With kind permission by the American Chemical Society, Washington (DC) [2].

onic acid, H–CH$_2$–CH(COOH)–H, on neutralization (Fig. 15-3). The pH of aqueous solutions of poly(acrylic acid) (PAA) is higher at all degrees of neutralization than that of propionic acid (ppA): poly(acrylic acid) is the weaker acid. At high concentrations of neutral salts such as KCl, it is just the opposite.

This increase of acid strength is caused by an electrostatic effect. In dilute aqueous solutions, many of the –COOH groups of PAA are dissociated to –COO$^{\ominus}$ and H$^{\oplus}$. The intramolecular repulsion between –COO$^{\ominus}$ groups leads to a stiffening of PAA chains. Addition of a neutral salt pushes the protons more to the –COO$^{\ominus}$ groups. The repulsion between neighboring –COO$^{\ominus}$ groups is reduced and the poly(acrylic acid) molecules become less stiff (smaller radii of gyration). However, the closer proximity of –COO$^{\ominus}$ and H$^{\oplus}$ causes the latter to have a smaller effect on the dissociation of neighboring –COO$^{\ominus}$ groups. Because of the resistance of the strong electrostatic forces, additional electrical work, $\Delta G_{el,0}$, is now required to bring a proton from the chain backbone to an infinite distance from it. An additional Gibbs energy is required which can be written as $\Delta G_{el,0} = -2.303\ RT \lg K'$. With $\lg K' = -pK_a'$, one obtains

(15-11) $pH = pK_a + \lg \{(1 - \beta)/\beta\} + pK_a' = pK_a + \lg \{(1 - \beta)/\beta\} + 0.434\ \Delta G_{el,0}/(RT)$

In addition, there is also a statistical effect because the protons of the acid groups compete with the cations of the added base for the available positions near –COO$^{\ominus}$ groups. Such an effect is known for low molecular weight dicarboxylic acids. Below a degree of neutralization of $\beta = 0.5$, more possibilities exist for dissociation than for association of protons; at $\beta > 0.5$, it is just the opposite. In the absence of the electrostatic effect, polyvalent acids should be stronger acids than the corresponding monovalent acids.

The presence of the statistical effect can be seen if the curves - - – and o––o are shifted parallel to the β-axis. However, the electrostatic effect dominates. At all degrees of neutralization, poly(acrylic acid) is therefore a weaker acid than propionic acid.

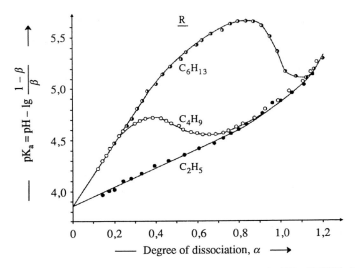

Fig. 15-4 Apparent values of pK_a of $-[CH(COOH)-CH(COOH)-alt-CH_2-CH(OR)-]_n$ in water at 30°C. Data points for C_2H_5 are symbolized by black dots, for C_4H_9 by open circles, and for C_6H_{13} by half-filled circles. With kind permission by Kluwer Publications, New York [3].

Acid strengths of polyacids increase with decreasing size of nonhydrated counterions. As a consequence, polyacids bind more strongly to smaller lithium cations (higher charge density) than to larger rubidium ions (smaller charge density). The titration of a polyacid with LiOH solutions thus leads to a greater apparent acid strength.

Experiments have shown that the titration of polyacids with bases containing large cations requires ca. 20 % more base than one calculates for stoichiometric amounts. In osmotic, diffusion, and electrophoretic measurements, one also observes that polyions behave as if they bind great proportions of counter ions (sometimes up to 70 %!), i.e., as if a great proportion of counterions does not dissociate at all.

Additional effects are caused by hydrophobic bonds. For example, the polyacid from the saponified alternating copolymer of maleic anhydride and ethyl vinyl ether behaves completely normally in titrations (Fig. 15-4). However, maxima in the plots of apparent pK_a values as a function of the degree of dissociation, α, were observed in the corresponding copolymers with butyl or cyclohexyl groups.

15.4.3 Ion Exchange

Ion-exchangers are inorganic or organic materials, usually in particle form, that are capable of reversibly exchanging their ions. Important inorganic ion exchangers include zeolites, montmorillonite, bentonites, etc. (Volume II). The bulk of industrially produced organic ion-exchange resins consists of crosslinked copolymers (PSX) from styrene S and divinyl benzenes DVB that have been functionalized.

Treatment of the PSX with SO_3 delivers very acidic cation exchange resins of the type $(poly)^{\ominus}$ (I). Reaction of PSX with chlorodimethyl ether and subsequent quaternation leads to II and that of PSX with N-chloromethyl phthalimide followed by saponification to III. Both II and III are strongly basic anion exchange resins of the type $(poly)^{\oplus}$.

Weakly acidic cation exchange resins are obtained from copolymers of divinyl benzenes and acrylic esters, followed by saponification with alkali. Many other ion exchange resins are known, for example, based on cellulose, phenolic resins, etc.

$$-CH_2-CH- \qquad -CH_2-CH- \qquad -CH_2-CH-$$

I   SO$_3$H   II   CH$_2$N(CH$_3$)$_2$ Cl$^{\ominus}$ III   CH$_2$NH$_2$

All ion exchange resins swell strongly in water whereupon dissociable groups become accessible. The dissociating low-molecular weight counterions are exchanged in equilibrium reactions. Salty water thus becomes desalted after passing through a cation exchange resin, (poly)$^{\ominus}$ H$^{\oplus}$, and an anion exchange resin, (poly)$^{\oplus}$ OH$^{\ominus}$:

(15-12) $\quad$ (poly)$^{\ominus}$ H$^{\oplus}$ $\quad$ + Na$^{\oplus}$ $\qquad\qquad$ = (poly)$^{\ominus}$ Na$^{\oplus}$ + H$^{\oplus}$

$\qquad\qquad$ (poly)$^{\oplus}$ OH$^{\ominus}$ + Cl$^{\ominus}$ $\qquad\qquad$ = (poly)$^{\oplus}$ Cl$^{\ominus}$ + OH$^{\ominus}$

$\overline{\qquad\qquad\qquad\qquad\qquad\qquad\qquad\qquad\qquad\qquad\qquad\qquad\qquad\qquad\qquad}$

$\quad$ (poly)$^{\ominus}$ H$^{\oplus}$+ (poly)$^{\oplus}$ OH$^{\ominus}$ + NaCl $\;$ = (poly)$^{\ominus}$ Na$^{\oplus}$ + (poly)$^{\oplus}$ Cl$^{\ominus}$ + H$_2$O

Regeneration of (poly)$^{\ominus}$ Na$^{\oplus}$ by acids and (poly)$^{\oplus}$ Cl$^{\ominus}$ by bases restores the original ion exchange resins.

The ion exchange capacity is determined by the crosslinking density and the number of dissociable groups per monomeric unit. The higher the crosslinking density, the less accessible are the ionic groups and the higher the pK$_a$. Especially high exchange capacities are furnished by macroreticular ion exchange resins (Section 16.5.1) because of their relatively open but rigid and less swellable structures.

The acid strength of crosslinked polyacids decreases with increasing size of the *hydrated* counterions. For example, hydrated lithium ions are larger than hydrated rubidium ions. The apparent acid strength of crosslinked polyacids is therefore greater in the presence of hydrated rubidium ions than in the presence of hydrated lithium ions. This effect is the opposite of that of *un*crosslinked polyacids where the size of the *nonhydrated* ions is important (nonhydrated Rb$^{\oplus}$ are larger than nonhydrated Li$^{\oplus}$).

The effect is probably caused by the different physical structures of water in crosslinked and uncrosslinked polymers. According to proton magnetic resonance measurements, water is less ordered in gels of poly(styrene sulfonic acid) than in uncrosslinked poly(styrene sulfonic acid). The degree of order varies with the degree of crosslinking. This is possibly the reason why the selectivity of charged gels toward ions passes through a maximum with increasing crosslink density.

15.4.4   Polymer Transformations

Polymer transformations are **polymer modifications** in which groups of the parent polymer are transformed into other groups in a manner similar to the corresponding reactions in lowmolecular weight compounds. They are performed in industry for two reasons: (1) conversion of inexpensive natural polymers into more useful ones and (2) modification of synthetic polymers because the desired chemical structures cannot be

Table 15-1 Industrial polymer transformations (without cyclizations, see Section 15.4.5). cell = Cellulose, E = ethylene unit, PVC = poly(vinyl chloride), VAc = vinyl acetate unit; * = Chlorine content in %, ** % SO_2Cl groups per ethylene group.

No.	Monomeric unit of parent polymer	Reagent	Conversion in %	Transformed monomeric unit	Application
1	CH_2–$CH(OOCCH_3)$	ROH	98	CH_2–CHOH	thickener
2	ditto, E/VAc units	CH_3OH	99	CH_2–$CHOH$/CH_2–CH_2	engineering polymer
3	cell–$NHCOCH_3$	H_2O		cell–NH_2	paper additive
4	cell–OH	CH_3COOH	83-100	cell–$OOCCH_3$	fibers, films
5	cell–OH	HNO_3	67-97	cell–ONO_2	plastics, etc. (see text)
6	cell–ONa	Oxirane	53-87	cell–OCH_2CHRCH_3	thickener
7	cell–OH	Oxirane	100-400	cell–$OCH_2CH(OR')CH_3$	thickener
8	CH_2–CH_2	Cl_2	25-40*	CH_2–CHCl	elastomers
			> 40*	CH_2–CHCl + CHCl–CHCl	impact improver for PVC
9	CH_2–CHCl	Cl_2	64*	ditto	adhesive, lacquer
10	CH_2–CH_2	Cl_2/SO_2	< 42*	CH_2–CHCl	coatings
			< 2**	+ CH_2–$CHSO_2Cl$	
11	CH_2CH=$CHCH_2$ + CH_2–CHCN	H_2		$(CH_2)_4$ + CH_2–CHCN	elastomers
12	N=PCl_2	RONa		N=$P(OR)_2$	elastomers

obtained by direct polymerization or only with high cost of monomers and processes. Table 15-1 lists 12 important examples:

1. Poly(vinyl alcohol) is not produced by polymerization of vinyl alcohol (or acetaldehyde, Section 7.4.2) but by transesterification of poly(vinyl acetate) with methanol or butanol. Methyl acetate and butyl acetate are sought after industrial solvents.

2. The partial ester exchange of acetate groups of poly(vinyl acetate-*co*-ethene)s delivers packaging films that are very impermeable to oxygen.

3. Saponification of the mucopolysaccharide chitin from the shells of *Crustacae* produces chitosan (see Volume II) which is used as a paper additive.

4. Cellulose is esterified to cellulose (2 1/2)-acetate and cellulose triacetate by acetic acid. Both serve as fibers, the former also for cigarette filters.

5. Treatment of cellulose by nitrating acid (various mixtures of nitric acid and sulfuric acid) delivers cellulose nitrates with different degrees of nitration (formerly called "nitrocelluloses") that are used for nitro lacquers, guncotton, celluloid (in combination with camphor), etc.

6-7. Etherification of cellulose by ethylene oxide and/or propylene oxide leads to cellulose ethers. The etherification generates new OH groups. The degree of substitution may thus exceed 100 %.

8-10. Poly(ethylene), ethene-propene rubbers, and poly(vinyl chloride) are chlorinated and sulfochlorinated.

11. Butadiene-acrylonitrile rubbers (= nitrile rubbers) are hydrogenated which improves their resistance against aging.

12. Poly(dichlorophosphazene) (= phosphonitrile chloride) hydrolyzes easily. Reaction of the polymer with fluorinated alcohols delivers excellent specialty elastomers.

Polymer transformations deviate from the corresponding transformations of low-molecular weight compounds in two respects. "By-products" become part of the macromolecular structure; they cannot be separated from the "main product" as in low-molecular weight chemistry. The resulting "wrong" chain units almost always lead to worse polymer properties. Industry therefore uses only those polymer transformations that proceed without side reactions or only with those that do not deliver unwanted ones. This is one of the main reasons why industry uses only four types of polymer transformations besides cyclization (Section 15.4.5): transesterification/ saponification (1-5), chlorination/sulfochlorination (8-10), etherification (6, 7, 12), and hydrogenation (11) (see Table 15-1). Oxidation, for example, always leads to degradation (Volume IV).

The second peculiarity is the often strong nearest neighbor effect. For this reason, reactivities of polymeric units should not be compared with those of their monomers or their hydrogenated monomers but with more appropriate compounds. An example is the hydrolysis of poly(vinyl acetate) and some model compounds by sodium hydroxide in acetone/water (75/25) at 30°C. The rate constants k indicate that isopropyl acetate is the correct model compound for vinyl acetate units but not ethyl acetate (or vinyl acetate).

$\text{—CH}_2\text{—CH—CH}_2\text{—}$ $\quad\quad$	$\quad$	$H\text{—CH}_2\text{—CH—CH}_2\text{—H}$	$H\text{—CH}_2\text{—CH—H}$
$\quad\quad$ O—CO—CH$_3$		O—CO—CH$_3$	O—CO—CH$_3$

| k / (L mol^{-1} min^{-1}) = | 0.37 | 0.57 | 3.5 |

Rate constants of polymer groups are furthermore only similar to those of appropriate low-molecular weight compounds if the polymer solution is very dilute (absence of intermolecular nearest neighbor effects) and the reactive polymeric units form only a very small proportion of the total ones (absence of intramolecular nearest neighbor effects). The magnitude of the first factor is often underestimated as can be seen from the following example.

A 1 % solution of ethyl acetate ($M \approx 88$ g/mol; degree of polymerization $X_M = 1$) and a 1 % solution of poly(vinyl acetate) of $M = 10^6$ g/mol ($X_P \approx M/M_u = 10^6/86.5 \approx 11\ 560$) are both solutions with a concentration of $[Ac]_{soln} = 0.115$ acetate groups per liter. Poly(vinyl acetate) molecules form random coils in solution (Volume III). Approximation of coils as homogeneous spheres of volume $V_{coil} = 4\ \pi\ r_3 M_u$ delivers a group concentration of $C_{coil} = X/V_p = (3\ M)/(4\ \pi\ r^3 M_u)$. The radius of such a coil is $r \approx 20$ nm. The coil thus contains $C_{coil} = 3.45 \cdot 10^{20}$ groups/mL. The molar concentration of groups in the coil is therefore $[Ac]_{coil} = C_{coil}/N_A \approx 0.57$ mol/L and thus 5 times as high as the overall molar concentration $[A]_{soln} = 0.115$ mol/L of groups in solution.

In homopolymers with A-groups, reacting A-groups may be between two other A-groups in $–A–A'–A–$, between an A-group and a C-group (= already reacted A-group) in $–A–A''–C–$, or between two C-groups in $–C–A'''–C–$. The reaction rate of a pseudo-monomolecular reaction is thus

$$(15\text{-}13) \quad\quad - \text{d}[A]/\text{d}t = k'[A'] + k''[A''] + k'''[A'''] = k_{app}[A]$$

Depending on the relative reactivity of the central A-group in the three different triads AAA, (AAC + CAA), and CAC, polymers with less than 100 % conversion of A-groups may have A-groups alternating with C-groups, randomly distributed A- and C-groups, or A-groups in long blocks. Of course, the two limiting cases, poly(A–*alt*–C) and poly(A-*block*-C) as well as the limiting case of a mixture of poly(A) and poly(C) are relatively rare. But they are not impossible as the following two examples show.

The saponification of poly(alkyl methacrylate)s delivers first polymers with randomly distributed $-COO^{\ominus}$ groups. These groups then induce the saponification of their nearest neighbors which leads to pseudocopolymers with blocks of methacrylic acid and alkyl methacrylate units.

The hydrogenation of poly(styrene) with Raney nickel as a catalyst leads initially to a conversion of only those phenyl groups that are in the vicinity of catalyst particles, i.e, to blocks of vinyl cyclohexane units and poly(styrene) units. If the vinyl cyclohexane blocks exceed a certain length, they will become incompatible with the styrene blocks (Volume IV) and phase separation results. If the chains with vinyl cyclohexane blocks are still near the catalyst, all styrene units in that chain will be hydrogenated. The result is a mixture of poly(styrene) chains and poly(vinyl cyclohexane) chains, i.e., a polymer blend of these two polymers and not partially hydrogenated poly(styrene) *molecules*.

More or less alternating pseudocopolymers are obtained only in those rare cases where suitable intermediates are formed. For example, the esterification of syndiotactic poly(methacrylic acid) by alcohols R'OH with the help of carbodiimides does not lead directly to the ester. Rather, an anhydride is formed first which then reacts with the alcohol to form a pseudocopolymer with alternating acid and ester groups:

(15-14)

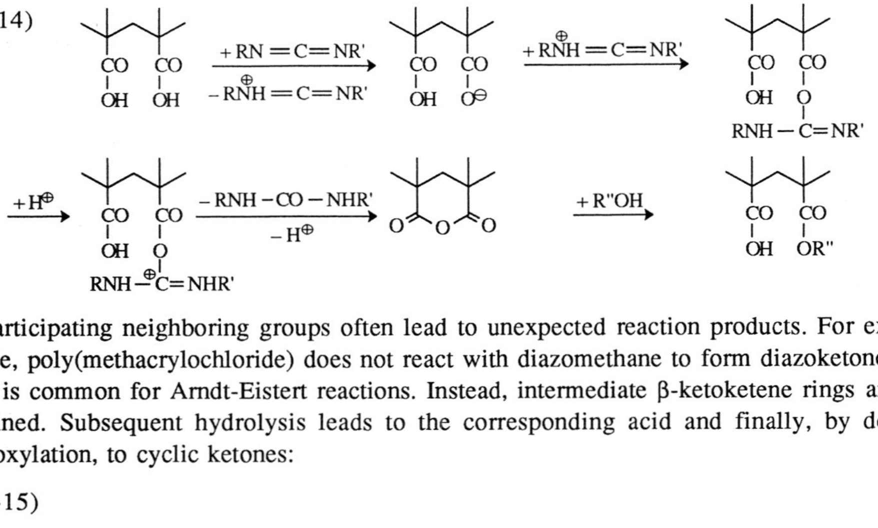

Participating neighboring groups often lead to unexpected reaction products. For example, poly(methacrylochloride) does not react with diazomethane to form diazoketones as it is common for Arndt-Eistert reactions. Instead, intermediate β-ketoketene rings are obtained. Subsequent hydrolysis leads to the corresponding acid and finally, by decarboxylation, to cyclic ketones:

(15-15)

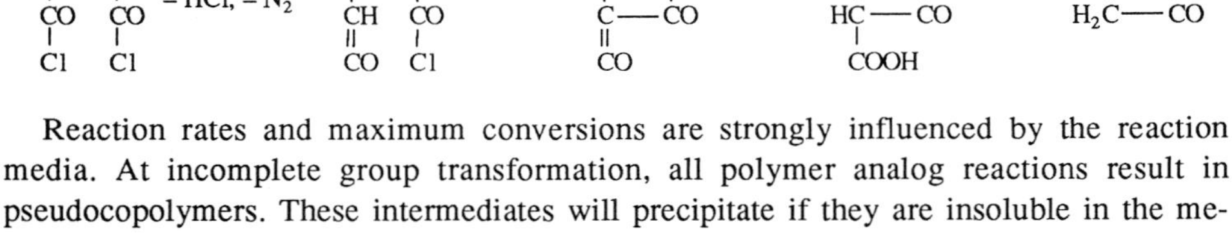

Reaction rates and maximum conversions are strongly influenced by the reaction media. At incomplete group transformation, all polymer analog reactions result in pseudocopolymers. These intermediates will precipitate if they are insoluble in the medium. The remaining unreacted groups are buried and cannot react. The resulting product is a pseudopolymer with a non-random distribution of substituents. In order to obtain complete (or nearly so) polymer analog reactions, solvents must be used that dissolve at least the completely converted reaction product.

Another complication can be the formation of partially crystalline polymers where only the amorphous regions are accessible for further reactions. The maximum conversion of groups here depends very much on the reaction conditions.

15.4.5 Intramolecular Ring-Closing Reactions

Intramolecular ring-closing reactions are special cases of polymer analog reactions. Depending on the constitutions of starting polymers and reagents as well as the type of reaction, either condensed or isolated rings may result.

Ladder polymers with (ideally) totally condensed ring systems are obtained directly by polycondensation or polyaddition in 1-step processes (Chapter 16.3.1). 2-step procedures, on the other hand, produce first long chains with suitable substituents that are then polymerized intramolecularly. Intermolecular reactions must be excluded, i.e., the reactions must be conducted at low polymer concentrations.

An example is the polymerization of 1,3-butadiene to 1,2-poly(butadiene) and subsequent cyclization of the pendant vinyl groups:

(15-16)

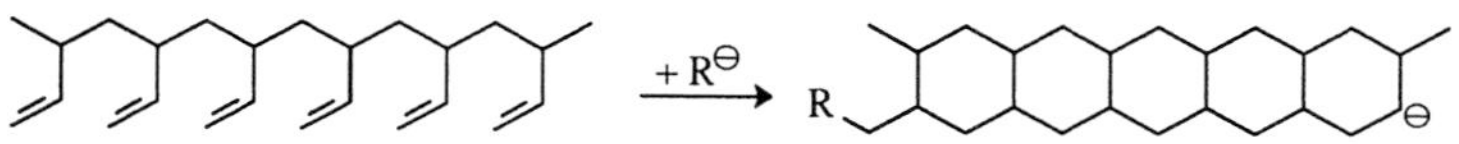

Another example is the polymerization of acrylonitrile, $CH_2=CH(CN)$, to poly-(acrylonitrile), $+CH_2–CH(CN)\frac{}{}_n$, with subsequent intramolecular polymerization (cyclization) of adjacent nitrile groups, $–C\equiv N$, to $–C=N–$ groups. Ladder polymers also result from the intramolecular polymerization of the $–N=C–$ double bonds of poly(vinyl isocyanate), $+CH_2–CH(N=C=O)\frac{}{}_n$.

In all these cases, free-radical and anionic polymerizations of pendant groups are preferred over cationic ones since the latter tend to transfer reactions that interrupt the annellation (L: *anellus* = small ring). In principle, infinite ladder structures should be possible by living polymerizations.

Such 100 % annellated rings can only be obtained if the ring-closure reactions begin exclusively at one end of the chain and neither at both ends nor in the middle. An example of the latter is the dehydratation of poly(vinyl methylketone) that starts randomly at various keto groups. The remaining keto groups then face other keto groups (instead of methyl groups) and the remaining methyl groups other methyl groups instead of keto groups (Eq.(15-17)). The intramolecular dehydration stops at a fraction of $f = 1/(2\ e)$ of unreacted keto groups if the reaction is totally random (Table 15-2):

(15-17)

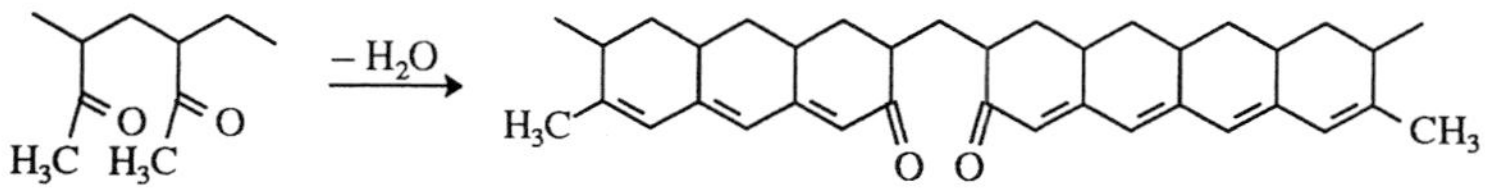

The situation is different for poly(methyl vinyl ketone)s where head-to-head and tail-to-tail structures alternate. In this case, only 50 % each of all CH_3 and CO groups react in *intra*molecular reactions between nearest neighbors and one obtains a phantom polymer with 5-membered rings as part of the backbone:

(15-18)

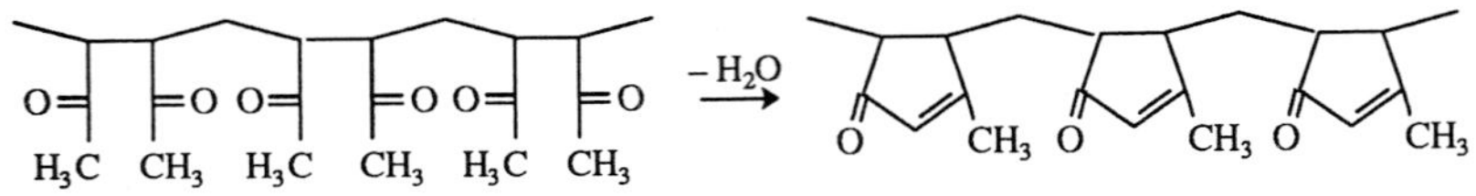

Similar statistical restrictions apply to *irreversible inter*molecular reactions of side groups of polymer molecules with bifunctional low molecular weight compounds. If all groups are exclusively in a 1,3-position (head-to-tail), a fraction of $1/e^2 = 0.135$ cannot react according to statistical calculations (Appendix A-15; Table 15-2). In *equilibrium reactions*, however, the maximum conversion may reach almost 100 %.

An example is the acetalization of poly(vinyl alcohol) by aldehydes RCHO (Eq.(15-19)). For irreversible reactions of head-to-tail polymers, theory predicts a maximum group conversion of $1 - (1/e^2) = 1 - (1/2.718^2) = 1 - 0.135 \rightarrow 86.5$ %. Experimental values were 85.8 % for chloroacetaldehyde, 85.0 % for palmitic aldehyde, and 83 % for benzaldehyde. The maximum conversion was considerably lower for aldehydes with ionizable groups, for example, 44 % for *o*-benzaldehyde sulfonic acid and 26 % for benzaldehyde disulfonic acid. Additional intermolecular reactions boost the conversion to more than $1/e^2$, leading to branched and crosslinked polymers.

(15-19)

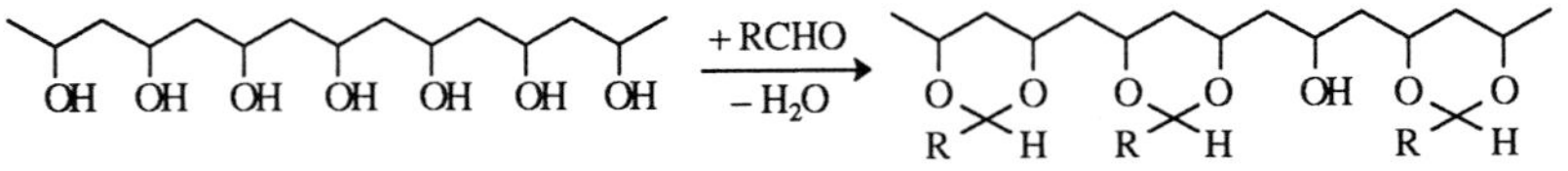

Similar effects were also observed for the ketalization of poly(vinyl alcohol) by ketones. After 100 hours in dimethylsulfoxide at 25°C, maximum degrees of ketalization were 84.2 % for butanone, 80 % for cyclohexanone, 70.8 % for acetone, 59.5 % for dibutylketone, and only 34.7 % for di-*t*-butylketone.

Only a few cyclization reactions are used industrially. Poly(vinyl butyral) with a content of 80 % butyral groups, 18% hydroxy groups, and 2 % acetate groups is plasticized with 30 % dibutyl sebacate and used as an interlayer film for safety glass and as a primer for lacquer and varnishes. Cellulose acetate-*co*-butyrate contains 17-48 % butyral groups and 6-29 % acetyl groups. It serves for corrosion resistant packaging materials.

At elevated temperatures, natural rubber cyclizes by cationic polymerization of the double bonds; it simultaneously degrades. The resulting mono, di, and tricyclic structures are separated from each other by methylene groups of non-cyclic isoprene units. The resulting cyclorubber serves as a binder for printing inks, lacquers, and adhesives.

Table 15-2 Maximum theoretical fractions f of unreacted groups from the irreversible bifunctional reaction of substituents R with a probability p of constitutional diads of reactive groups. $0 \le p \le 1$ for azeotropic copolymers but $p \equiv 1$ for homopolymers. H = head, T = tail.

Arrangement of monomeric units	Groups per mer	Fraction f of unreacted groups if		
		$0 \le p \le 1$	$p = 0$	$p = 1$
H-T	1	$\exp(-2p)$	1	$\exp(-2) = 1/e^2 = 0.135$
H-T	2	$1 - p + (2/9)p^3 - ...$	1	$1/(2\,e) = 0.184$
H-T, H-H, and T-T at random	2	$1 - (3/4)p + (5/72)p^3 - ...$	1	0.312
H-H and T-T alternating	2	$1 - (1/2)p$	1	0.500

On heating with ammonia, copolymers of methacrylic acid and methacrylonitrile form methacrylimide units. The evaporating water expands the polymer to a hard foam:

(15-20)

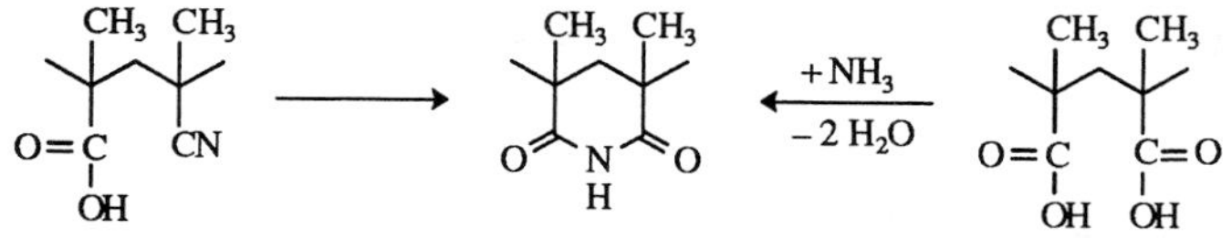

15.4.6 Polymer Reagents

Polymer reagents are reagents composed of soluble or insoluble polymers. Examples are ion exchange resins (Section 15.4.3) and the resins for the Merrifield syntheses of proteins and peptides (Section 14.3.12) and nucleic acids (Section 14.2.7). Polymer reagents also include **redox polymers** (reduction-oxidation polymers) for electron transfers. Examples of polymer reagents and their reactions are

~(p-C$_6$H$_4$)–ICl$_2$	cis-chlorination of olefins;
~(p-C$_6$H$_4$N)–BH$_3$	hydrogenation and reduction of aldehydes and ketones;
~(p-C$_6$H$_4$)–P=CRR'	Wittig reaction of aldehydes;
~(p-C$_6$H$_4$)–I(OOCCH$_3$)$_2$	oxidation of aniline to azobenzene;
~N(Cl)–CO~~	oxidation of alcohols.

Most polymer reagents are crosslinked polymers that carry the desired functional groups. They can be synthesized by various strategies: direct crosslinking polymerization of the functional group-carrying monomer or, much more commonly, crosslinking polymerization to a "neutral" polymer with subsequent functionalization. A common example of a crosslinking polymerization is the free-radical suspension polymerization of styrene and divinylbenzenes (Volume II). An example of functionalization is the reaction of the resulting polymer beads with chlorodimethylether which delivers ~CH$_2$Cl groups that are subsequently reacted further to the desired functional group, for example, according to ~CH$_2$Cl + LiP(C$_6$H$_5$)$_2$ → ~CH$_2$P(C$_6$H$_5$)$_2$ + LiCl.

The use of crosslinked polymer reagents offers an easy separability of products and educts. The large concentration gradients in columns of such polymer reagents and the ability to apply a large excess of the educt increases the yield of the desired product. Efficient polymer reagents are easy to regenerate, stable against degradation, and non-adsorbent for the product. The latter can be achieved by using macroreticular networks.

Neighbor group effects do not always let reactions at polymer reagents proceed like those of low-molecular weight reagents. General rules obviously do not exist. An example is the reaction of cyclohexene with N-bromosuccinimide and its polymeric analog. The former causes a substitution reaction to bromocyclohexene whereas the latter leads to an addition of bromine to 1,2-dibromocyclohexane:

(15-21)

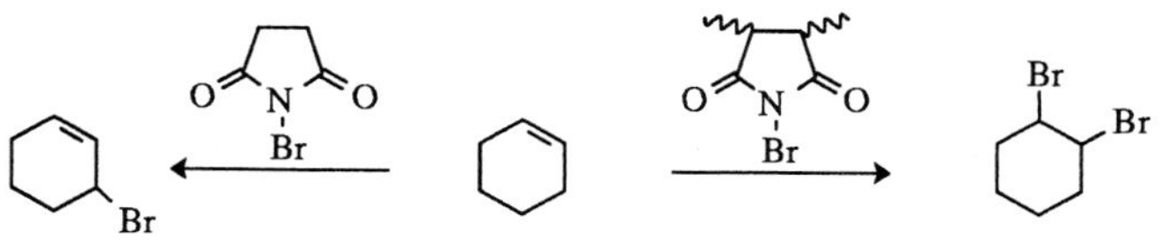

Polymeric **phase transfer catalysts** (PPTCs) are low-molecular weight PTCs that are bound to a crosslinked polymer such as crosslinked poly(styrene) X-PS. Examples are quaternary ammonium and phosphonium ions, crown ethers, and even poly(oxyethylene)s that complex a reactant (mostly an anion) and transport it through the interface to the second reactant in the other phase. An example is the alkylation of phenylacetonitrile by 1-bromobutane. Since two immiscible solvents and a solid body are involved, one also talks about *three-phase catalysis*.

15.5   Polymer Degradation

15.5.1   Survey

"Degradation" is an ill-defined term that is used in the polymer literature for reactions which decrease degrees of polymerization *or* an uncontrolled change of polymer constitution *or* both. IUPAC recommends restricting "degradation" to those chemical modifications that lead to an undesired change of the use properties of a polymeric material.

Reactions that decrease the degree of polymerization, but do not change the constitution of the chain units, are the reverse of the corresponding polymerizations, i.e., they are **retro-polymerizations** (L: *retro* = backwards). In analogy to the general classification of polymerization reactions (Section 6.1.3), one can thus distinguish four different types of such reactions:

chain scissions at any bond between two chain atoms:
– **retro-polyadditions** without participation of low-molecular weight molecules,
– **retro-polycondensations** with participation of low-molecular weight molecules;

depolymerizations with release of monomer molecules from a chain end:
– **retro-chain polymerizations** without participation of low-molecular weight molecules,
– **retro-polyeliminations** with participation of low-molecular weight molecules.

In addition, there may be unwanted polymer-analog reactions that change the constitution of chain units, i.e., **partial decompositions** of the chains. Retro-polymerizations and partial decompositions can be induced chemically, thermally, and/or mechanically and also by light or ultrasound. The reactions may proceed via radicals, cations, anions, or charge-transfer complexes. Their presence may be desired as in the removal of plastics refuse or undesired as in the chemical aging of plastics or the causation of flammability of fibers. This chapter describes the basics of degradations (in the common sense); applications are discussed in Volumes II and IV.

15.5.2   Chain Scissions

Degree of Scission
The random scission of polymer chains with formation of larger or smaller chain fragments is the reversal of their (sometimes hypothetical) polycondensations or polyadditions. Such scissions occur especially at polymer chains with easily activated chain units

such as heteroatoms. For example, the thermal chain scission of Diels-Alder polymers is a retro-polyaddition (see Eq.(13-8)) that can be schematically described as

$$(15\text{-}22) \qquad M_{i+j} \rightarrow M_i + M_j \rightarrow M_{i-k} + M_k + M_{j-m} + M_m, \qquad \text{etc.}$$

Retro-polycondensations require the participation of small molecules. Examples are the saponification of aliphatic polyamides by water (with the end-products diamine and dicarboxylic acid) and the transesterification of poly(ethylene terephthalate) by methanol (with the end-products ethylene glycol and dimethyl terephthalate).

Chain scissions sever N_b bonds per chain. The number-average degree of polymerization of the polymer, $\overline{X}_n$, after the scissions is therefore related to the number-average degree of polymerization, $\overline{X}_{n,0}$, by $\overline{X}_{n,0} = (1 + N_b)\,\overline{X}_n$. The **degree of chain scission** is defined as the fraction f_b of severed chain bonds:

$$(15\text{-}23) \qquad f_b = \frac{N_b}{\overline{X}_{n,0} - 1} = \frac{\overline{X}_{n,0} - \overline{X}_n}{\overline{X}_n(\overline{X}_{n,0} - 1)} \approx \frac{1}{\overline{X}_n} - \frac{1}{\overline{X}_{n,0}}$$

where the approximation relates to large degrees of polymerization. Degrees of polymerization count *all reactants* including released monomer molecules. Of course, retro-polycondensations are true reversals of polycondensations if both types of reaction proceed under the same conditions, for example, if both reactions are equilibrium reactions.

The probability of a chain scission equals the extent of reaction (= fractions of scissions) $q = 1 - p$ where p = fraction of bonds. The mole fraction $x_{i,R}$ of reactants that is obtained after the scission is given by Eq.(13-38) as $x_{i,R} = q(1 - q)^{i-1}$ for both retro-polyadditions and retro-polycondensations. Taking the natural logarithm delivers

$$(15\text{-}24) \qquad \ln x_{i,R} = \ln q + [\ln (1 - q)]\,(i - 1)$$

An example is shown in Fig. 15-5 (see description on page 590).

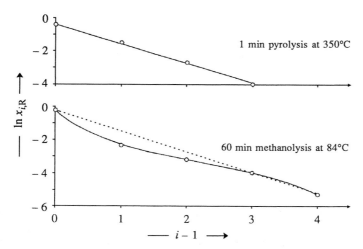

Fig. 15-5 Chain scission of poly(β-hydroxybutyrate) [4]. The dotted line corresponds to the solid line for pyrolysis. Both intercept $\ln q$ and slope $\ln (1 - q)$ of the pyrolysis plot deliver $q = 0.71 \pm 0.01$.

For random chain scissions, a plot of $\ln x_{i,R} = f(i-1)$ should thus lead to the same q from both intercept $\ln q$ and slope $\ln(1-q)$. This is indeed found for the rapid pyrolysis of poly(β-hydroxybutyrate) but not for the methanolysis of the same polymer (Fig. 15-5), probably because the principle of equal chemical reactivity does not apply to the latter at low degrees of polymerization ($i \leq 5$).

The mass fraction of reactants from *retro-polyadditions* is obtained from Eq.(13-42), $w_{i,R} = X_i(p^{i-1})(1-p)^2$ by setting $p = 1 - q$. The resulting equation applies also to retro-polycondensations in the limit of $M_L/M_U \rightarrow 0$.

Retro-polycondensations include not only reversals of linear polycondensation polymers but also those of crosslinked ones, provided diffusion effects are absent. Retro-polycondensations of crosslinked polymers thus require highly swollen networks. In this case, there is a great likelihood that all chain bonds are split with the same probability, for example, ester bonds in crosslinked polyesters.

At the beginning of the scission, most split chain sections still remain bound to the network. Only a few chain sections will be totally severed and dissolved. At a later stage, a single scission may remove a part of the crosslinked molecule that will also go into solution. Both the soluble fraction as well as its molecular weight therefore strongly increase with increasing time (Fig. 15-6). Equilibrium theory is fairly well obeyed.

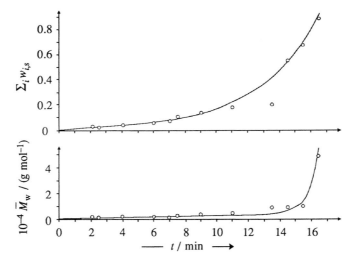

Fig. 15-6 Time dependence of cumulative mass fractions and weight-average degrees of polymerization of soluble reactants (polymers, oligomers, and monomers) from the hydrolytic chain scission of a swollen (in tetrahydrofuran) crosslinked polymer from 1,3,5-benzene triacetic acid and 1,10-decamethylene glycol. Solid lines: Theory (as reverse polycondensation); O experiment [5].

Rate Constants

At low molecular weights, rate constants are not necessarily independent of the degree of polymerization, i.e., the principle of equal chemical reactivity does not apply. At this range of molecular weights, experimentally found (initial!) rate constants are to be interpreted as arithmetic averages of the rate constants for the scission of chain bonds of oligomers with different degrees of polymerization:

(15-25) $\bar{k} = \sum_i N_i k_i / \sum_i N_i$

In the simplest case, two identical rate constants k_t have to be considered for the two terminal bonds and $(N_i - 2)$ rate constants k_m for the remaining "middle" bonds of the molecule. Eq. (15-25) thus becomes

(15-26) $\bar{k} = \dfrac{2 k_t + (N_i - 2)k_m}{N_i} = k_m + \dfrac{2(k_t - k_m)}{N_i} = k_m + \dfrac{k_s}{N_i}$

where $k_s \equiv 2 (k_t - k_m)$.

The proposed dependence of average initial rate constants on the inverse of the initial number of cleavable bonds is indeed found for hydrolyses (Fig. 15-7). The slope is negative for poly(glycine)s ($k_t < k_m$) but positive for poly(glucose)s ($k_t > k_m$).

However, Eq.(15-26) is not the only possible explanation for the empirically found relationship $\bar{k} = k_m + (k_s / N_i)$. For example, the two terminal bonds may be split with different rate constant so that $k_t = (k_t' + k_t'')/2$.

An identical function to Eq.(15-26) is also observed for equal rate constants k_t of the two terminal bonds, equal rate constants k_{tt} for the two penultimate bonds, and equal rate constants k_m for the remaining $(N_i - 4)$ "middle" bonds:

(15-27) $\bar{k} = k_m + \dfrac{2(k_t + k_{tt}) - 4 k_m}{N_i} = k_m + \dfrac{k_s'}{N_i}$

Eqs.(15-26) and (15-27) thus say only that there are *at least two different* types of bonds per molecule. Only at high degrees of polymerization do gross rate constants $\bar{k}$ become practically independent of the degree of polymerization. For $k_s = 3 \, k_m$, this is approximately fulfilled within 1 % for $N_{i,0} > 300$.

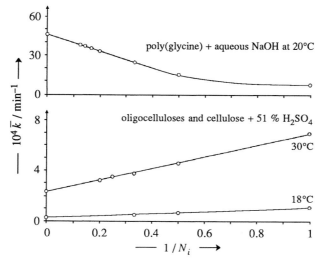

Fig. 15-7 Initial rate constant of hydrolysis as a function of the inverse number of initial hydrolyzable bonds per molecule [6].

Kinetics

Simple relationships for the time dependence of degrees of polymerization can be derived if the rate constants are independent of molecular weights. The rate of chain scission, df_r/dt, is here proportional to the concentration [C] of the catalyst and the fractional proportion of the remaining bonds, $f_r = 1 - f_b$:

(15-28) $- df_r/dt = k[C]f_r$

Integration delivers Eq.(15-29) with $f_{r,0} = 0$ since no bonds are broken at time $t = 0$:

(15-29) $f_r = f_{r,0} \exp(- k[C]t) = \exp(- k[C]t)$

At *high degrees of polymerization*, very few bonds are broken at small times so that $f_r/f_{r,0} \approx 1$ and thus $\exp(- k[C]t) \approx 1$ and also $k[C]t \to 0$. $\exp(- k[C]t)$ can thus be approximated by $1 - k[C]t$ and Eq.(15-29) becomes

(15-30) $f_r = f_{r,0}(1 - k[C]t)$

Introduction of $f_r + f_b \equiv 1$ and Eq.(15-23) results in

(15-31) $1/\overline{X}_n = 1/\overline{X}_{n,0} + k[C]t$

The inverse number-average degree of polymerization should thus increase linearly with time (Fig. 15-8). Higher catalyst concentrations lead to more chain scissions.

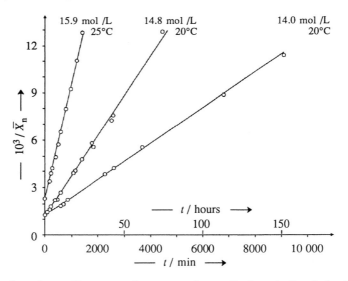

Fig. 15-8 Time dependence of inverse number-average degree of polymerization during the hydrolysis of cellulose by different concentrations of phosphoric acid [7].

In many cases, the weight-average degree of polymerization or even the viscosity-average is measured instead of the number-average, which leads to a much more complex general equation for the time-dependence of average degrees of polymerization.

However, distributions of degrees of polymerization of polymers resulting from degradations with more than 5 chain scissions per molecule can already be approximated by a Schulz-Flory distribution of degrees of polymerization. In this case, $\overline{X}_w \approx 2\,\overline{X}_n$ and Eq.(15-31) becomes

$$(15\text{-}32) \qquad 1/\overline{X}_w = (1/\overline{X}_{w,0}) + (1/2)\,k[C]t$$

Random chain scissions convert originally molecularly homogeneous polymers ($\overline{X}_w/\overline{X}_n \equiv 1$) into molecularly inhomogeneous ones. At infinite times, all chain bonds are severed and the resulting monomer is again molecularly homogeneous. The polymolecularity index, $\overline{X}_w/\overline{X}_n$, must thus pass through a maximum.

Eqs.(15-31) and (15-32) are approximations for small and medium degrees of scission since they assume that the concentration of the originally present cleavable bonds remains practically constant. This is no longer true at higher degrees of scission and Eq.(15-31) then has to be replaced by

$$(15\text{-}33) \qquad \ln\left(1 - \frac{1}{\overline{X}_n}\right) = \ln\left(1 - \frac{1}{\overline{X}_{n,0}}\right) - k[C]t$$

Eq.(15-31) must also be modified if a mole fraction x_M of monomer (or oligomer) is lost in the process:

$$(15\text{-}34) \qquad (1 - x_M)/\overline{X}_n = (1/\overline{X}_{n,0}) + k[C]t$$

The preceding equations apply to all statistically proceeding chemical and thermal scission processes of polymers with identical and regularly arranged monomeric units. Deviations can be expected for polymers with "weak links" (p. 45).

For example, poly(vinyl alcohol) $-[CH_2-CH(OH)]_n-$ contains 1-2 % head-to-head or tail-to-tail structures. CrO_3 oxidizes hydroxy groups in head-to-tail positions $-H-T-$ to $-CH_2-CO-$ but not those in $-H-H-$ or $-T-T-$. The $-CH_2-CO-$ units are further oxidized to acetic acid, CH_3COOH, while oligomers with $-H-H-$ and $T-T$ structures remain. Periodic acid HIO_4, on the other hand, selectively oxidizes head–to–head structures of poly(vinyl alcohol) to oxalic acid, ($-CHOH-CHOH- \rightarrow HOOC-COOH$), and $-tail-to-tail-$ structures to succinic acid, ($-CHOH-CH_2-CH_2-CHOH- \rightarrow HOOC-CH_2-CH_2-COOH$).

Chain scission proceeds only to a degree of polymerization of $\overline{X}_{n,\infty}$ and not to $X = 1$ if one type of chain bond is cleaved much faster than the other one. Eq.(15-31) has to be modified to

$$(15\text{-}35) \qquad (1/\overline{X}_n) - (1/\overline{X}_{n,\infty}) = (1/\overline{X}_{n,0}) + k[C]t$$

Non-Random Chain Scissions

Treatment of polymers with **energy-rich radiation** may lead to either chain scission or crosslinking, depending on the G-values (Section 11.3.2). Chain scission is observed for poly(α-methylstyrene), poly(methyl methacrylate), poly(vinylidene chloride), and poly(tetrafluoroethylene) but crosslinking is obtained for poly(ethylene), poly(vinyl chloride), poly(styrene), and polyamides.

Macromolecules may also undergo chain scission by mechanical means such as grinding, sawing, drilling, freeze-drying, ultrasonics, and flowing. Chain-scissions by these means are not random like the ones during chemical and thermal degradation.

Elongational flow (**extensional flow**) causes rigid, rodlike molecules to orient in the flow direction and random coil molecules to deform to ellipsoidal shapes above a critical flow rate (Volume III). The whole molecule moves with the flowing liquid but both magnitude and sign of the relative movement of the liquid changes along the molecule, except for the center. Schematically:

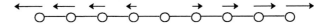

Elongational flow causes the molecule to split in the center so that the resulting molecules have degrees of polymerization that corrspond to 1/2, 1/4, 1/8, etc., of the original ones (Volume III).

Shear flow causes chain scission only if random coil molecules are entangled, i.e., above critical molecular weights and concentrations (Volume III). In thermodynamically good solvents, chain scission occurs only in turbulent flow, i.e., above the critical Reynolds number. In thermodynamically bad solvents, where the coils are much more contracted, chain scission is also observed for laminar flow at temperatures near the phase separation. The higher the degree of coiling, the less a molecule can avoid shear stress and the greater the probability of chain scission. High-molecular weight poly(isobutylene) undergoes chain scission even by pushing its solutions through capillaries. Very stiff molecules cannot convert the energy supplied by shearing into rotations of their chain elements around the main chain. They are degraded very easily, for example, high-molecular weight deoxyribonucleic acids during pipetting!

Chain scission by **ultrasonics** resembles that by shear flow. Ultrasonic waves in solution generate periodically changing tensions and pressures which tear the solution at interphases with gaseous or solid nuclei. The process generates cavities with diameters of the size of several solvent molecules. The cavities collapse fast and generate during this process considerable pressures and shear forces with energies that exceed the bond energy of covalent bonds. Polymer molecules cannot evade these forces fast enough, so their chains break.

15.5.3 Depolymerizations

Depolymerization is the reverse of chain polymerization. It proceeds without side reactions only if the bonds within substituent groups and from substituent groups to the main chain are much more stable than the chain bonds themselves.

In living polymers, depolymerization starts spontaneously from the chain ends. It then proceeds anionically or cationically. In all other polymers, depolymerizations can start only after a chain bond has been split homolytically. The chain then depolymerizes free-radically. Heterolytic chain scission with subsequent ionic depolymerization is not very probable for carbon chains but very prominent for polysiloxanes.

Depolymerizations are zip reactions in which one monomer molecule after another is removed from the end(s) of the main chain. Before depolymerization, polymers are present with a number-average degree of polymerization of $\overline{X}_{n,0} = N_{u,0}/N_{P,0}$ where $N_{u,0}$ = total number of monomeric units and $N_{P,0}$ = number of initially present polymer chains. The degree of polymerization reduces to $\overline{X}_n = N_u/N_P$ after a time t. Since the degree of polymerization in chain polymerizations does not include the monomer (i.e., $X \geq 2$ and not $X \geq 1$ as in polycondensations and polyadditions), it is expedient to use the same convention for depolymerizations. The number N_M of liberated monomer molecules thus has to be subtracted from the numbers N_u and N_P.

Depolymerizations can furthermore start only after chain bonds are broken. The number of polymer chains therefore has to be corrected for the number N_b of broken bonds per polymer molecule. The number-average degree of polymerization of a depolymerized polymer is therefore

$$(15\text{-}36) \qquad \overline{X}_n = \frac{N_u}{N_P} = \frac{N_{u,0} - N_M}{N_{P,0}(1 + N_b) - N_M}$$

Only one chain bond is broken for each liberated monomer molecule. The number N_b of broken bonds is therefore $N_b = N_M/N_{P,0}$ if the primary homolysis is neglected because it involves only a small fraction of all broken bonds. Eq.(15-36) thus converts to

$$(15\text{-}37) \qquad \overline{X}_n = \overline{X}_{n,0} - [\,\overline{X}_{n,0}/N_{u,0}]N_M = \overline{X}_{n,0}(1 - x_M) \quad ; \quad x_M \equiv N_M/N_{u,0}$$

The degree of polymerization should thus decrease linearly with the mole fraction x_M of liberated monomer molecules which is indeed observed for high-molecular weight polymers (Fig. 15-9). However, no linearity is found for lower initial molecular weights. At very low initial molecular weights, $\overline{X}_n/\overline{X}_{n,0}$ is even independent of x_M. These deviations arise because neither termination nor transfer reactions were considered while deriving Eq.(15-37). The reactions create dead polymer chains that cannot depolymerize unless they undergo another homolysis. They therefore change the degree of polymerization and also lower the probability of formation of new monomer molecules.

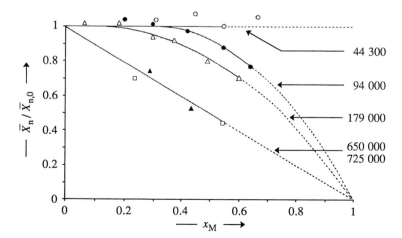

Fig. 15-9 Normalized number-average degrees of polymerization of poly(methyl methacrylate)s as a function of the mole fraction of the liberated monomer at various initial number-average degrees of polymerization. With kind permission of the Royal Society of Chemistry, London [8].

It is therefore expedient to introduce a **zip length** Ξ which is defined as the number of liberated monomer molecules per kinetic chain (see Table 15-4). Zip lengths equal the ratio of the probability of retrogrowth to those of transfer and termination. Small zip lengths indicate small proportions of liberated monomer but not necessarily little polymer degradation. For example, the zip length is only 3.1 if poly(styrene) is thermally decomposed at 330°C and normal pressure but this degradation produces 63 wt% styrene, 19 wt% dimers (2,4-diphenyl-1-butene and 1,3-diphenyl propane) as well as 4 wt% trimers (2,4,6-triphenyl-1-hexene and 1,3,5-triphenylpentane).

The behavior of a depolymerizing polymer depends strongly on the ratio of molecular weight and zip length and the initial molecular weight distribution (Fig. 15-9). At high molecular weights of *molecularly homogeneous* polymers, $\overline{X}_{n,0}$ is greater than Ξ and only some polymer molecules depolymerize. The resulting polymer thus consists of original and decomposed polymer molecules. Its number-average degree of polymerization decreases linearly with the mole fraction of liberated monomer (Fig. 15-9).

If the zip length is larger than the initial degree of polymerization of a molecularly homogeneous polymer, then some chains are totally decomposed but others maintain their original degree of polymerization. The number-average degree of polymerization thus does not vary with the mole fraction of the liberated monomer (Fig. 15-9).

Table 15-3 Zip lengths Ξ, mass fractions w_M and mole fractions x_M of monomer yields from the thermal decomposition of polymers in vacuum at ca. 300°C.

Polymer	Ξ	w_M	x_M
Poly(methyl methacrylate)	> 200	1.00	1.00
Poly(α-methyl styrene)	> 200	1.00	1.00
Poly(deuterostyrene)	11.8		
Poly(styrene)	3.1	0.42	0.65
Poly(isobutylene)	3.1	0.32	0.78
Poly(ethylene)	0.01	0.03	0.21

In *molecularly inhomogeneous* polymers, the degree of polymerization is higher than the zip length for some chains but lower for others. The probability of scission is thus higher for the high-molecular weight chains than for the low-molecular weight ones. A completely depolymerized low-molecular weight chain ($\Xi > \overline{X}_{n,0}$) disappears from the polymer and the degree of polymerization of the remaining polymer increases. A high-molecular weight chain, on the other hand, will not be completely depolymerized ($\Xi < \overline{X}_{n,0}$) and $\overline{X}_n$ of the remaining polymer decreases. The two effects cancel each other at certain combinations of Ξ, $\overline{X}_{n,0}$, and $\overline{X}_{w,0}/\overline{X}_{n,0}$ and $\overline{X}_n$ stays constant at small mole fraction x_M ($\overline{M}_{n,0}$ = 179 000 in Fig. 15-9). However, the molecular weight distribution is shifted with increasing progress of depolymerization toward the low-molecular weight species. The degree of polymerization must then start to decrease with increasing x_M.

In contrast to chain scission, kinetics of depolymerization cannot be generalized since it is very specific for the type of initiation reaction, molecular weight distribution, etc. It is dependent on the type of endgroups. In order to obtain a monomer yield of 50 % within 45 minutes, poly(methyl methacrylate) requires a temperature of 283°C if it was polymerized by dibenzoyl peroxide but 325°C if it was polymerized thermally.

Not all depolymerizations are unwanted. They may serve for the preparation of ultra-pure monomers and they are also essential in the "dry" preparation of resists (see below and Volume IV). They are not used, however, for the disposal of plastics since unmixed scrap from processors can be easily recycled whereas refuse from households is often a mixture of very different polymers.

Acrylic acid is one of the monomers that is prepared by depolymerization. The addition of ketene $CH_2=C=O$ to formaldehyde delivers β-propiolactone that polymerizes immediately to poly(β-propiolactone). This polyester depolymerizes at 150°C to a very pure acrylic acid:

$$(15\text{-}38) \qquad (n+1)\,\underset{O}{\overset{O}{\bigsqcup}} \longrightarrow CH_2=CHCO(OCH_2CH_2CO)_nOH \longrightarrow (n+1)\,CH_2=CHCOOH$$

Depolymerization is also used for the dry preparation of electronic wafers. Silicon is coated with a thin layer of silicon dioxide and then a layer of poly(1-butene-*alt*-sulfur dioxide), which has a very low ceiling temperature. The desired pattern is generated by depolymerizing the polymer with X-ray or electron beams. The SiO_2 at the unprotected areas is then etched away and the now accessible Si is doped (see Volume IV). The protecting polymers must resist the etching, hence the term "resist."

15.5.4 Pyrolyses

Chain scissions and depolymerizations without side reactions can only be expected at relatively low temperatures. Additional reactions occur at elevated temperatures, especially reactions of substitutents, leading to degradation in the common sense. Thermal degradation is often called **pyrolysis** (G: *pyr* = fire; *lyein* = to loosen, to dissolve). It is unwanted in the case of disastrous fires but useful for the disposal of plastics waste and the manufacture of, for example, heat-resistant carbon fibers and certain ceramics.

The chemistry of pyrolyses should be similar to that of low-molecular weight chemical compounds, but one often finds degradation to other products and/or start of the reaction at much lower temperatures. For example, poly(ethylene) decomposes at a temperature, which is ca. 200 K lower than that for hexadecane. Low-molecular weight primary esters pyrolyze to olefins and acids but poly(methyl methacrylate) depolymerizes at 300-450°C practically completely to its monomer (Table 15-3). Pyrolyses in vacuum and under a nitrogen blanket deliver different products, etc.

Several factors are responsible for these differences. First, macromolecules often do not have the ideal structures their common constitutional formulas imply. They always have endgroups (unless cyclic) and frequently branching units and weak links. Second, polymers from their preparation often contain extraneous materials such as initiator residues, solvent, emulsifiers, etc., which are difficult to remove. It is at these materials where the decomposition can start.

Decompositions will proceed at lower temperatures and to higher monomer yields if the Gibbs polymerization energy is low (Chapter 7). It is for this reason that poly(α-methyl styrene) and poly(methyl methacrylate) depolymerize practically quantitatively to their monomers. Polymers with higher polymerization energies depolymerize only at

Table 15-4 Activation energies of homolysis of chain bonds (b), depolymerization of chains (dp), and crosslinking (x) of polymers that decompose with crosslinking (X) or depolymerization (D) [9].

Polymer		$\Delta E^{\ddagger} / (\text{kJ mol}^{-1})$			Decomposition
	b	dp	x	x – b	
Poly(ethylene)	118	118	162	44	X
Poly(vinyl acetate), at	117	118	163	46	X
Poly(styrene), at	100	102	163	61	X
Poly(propylene), it-		102	163		X
Poly(vinyl chloride), at	87	87	163	78	X
Poly(methyl methacrylate), at	85	98	180	95	D
Poly(vinylidene chloride)		82	186	104	D
Poly(α-methyl styrene), at	95	52	238	143	D
Poly(tetrafluoroethylene)	147	132	322	175	D

higher temperatures where the bonds between substituents and chain atoms are often weaker than those between the chain atoms themselves. Poly(vinyl acetate) decomposes thermally into polyene structures and acetic acid for this reason.

Some polymers decompose at elevated temperatures completely to low-molecular weight products. Other polymers deliver crosslinked materials and even carbon (Table 15-4). Whether a polymer decomposes to low-molecular weight products or with formation of crosslinked structures depends essentially on the difference between the activation energy $\Delta E_b^{\ddagger}$ for the homolysis of bonds between chain atoms, for example,

$$(15\text{-}39) \qquad \text{\wave}CH_2\text{—}CHR\text{—}CH_2\text{—}CHR\text{\wave} \longrightarrow \text{\wave}CH_2\text{—}\overset{\bullet}{C}HR + \overset{\bullet}{C}H_2\text{—}CHR\text{\wave}$$

and the activation energy $\Delta E_x^{\ddagger}$ for crosslinking. An example is the crosslinking of poly-(vinyl chloride), PVC, that is initiated by polymer radicals which are formed by the abstraction of HCl from PVC (Eq.(15-40)):

$$(15\text{-}40) \qquad \text{\wave}CH=\overset{\bullet}{C}-CH_2\text{\wave} + \text{\wave}CH_2\text{—}CHCl\text{—}CH_2\text{\wave} \longrightarrow \begin{array}{c} \text{\wave}\overset{\bullet}{C}H \quad\quad CH_2\text{\wave} \\ \diagdown \quad\quad\quad \diagup \\ CH\text{—}\overset{|}{C}Cl \\ \diagup \quad\quad\quad \diagdown \\ \text{\wave}CH_2 \quad\quad CH_2\text{\wave} \end{array}$$

Decompositions with formation of volatile products are especially easy to monitor because of the weight loss of the specimen (**thermogravimetry**, Fig. 15-10). Measurements may be performed under nitrogen or other inert gases for thermal stability or under oxygen for combined thermal and oxidative degradations.

Dynamic thermogravimetry (**TGA**) determines the change of weight as a function of temperature at constant heating rates, usually 2-30 K/min. TGA measures the short-time temperature stability of the polymer because the temperature changes fast.

Isothermic thermogravimetry (**IGA**) measures the time dependence of weight loss as a function of time. It is therefore a measure of long-time temperature stability.

TGA curves differ characteristically for the various types of thermal decomposition. With increasing temperature, polymers with large zip length ($\Xi \gg 1$) maintain their mass until the temperature is high enough to provide the energy that is necessary for the homolysis of a chain bond. Monomer molecules are removed one after the other from

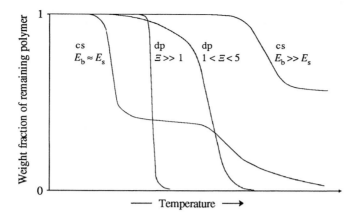

Fig. 15-10 Characteristic TGA curves at constant heating rate (see text) [10].
cs = Chain scission; dp = depolymerization. E_b = energy of bonds between chain atoms; E_s = energy of bonds between substituents and chain atoms. Ξ = zip length.

macroradicals and the mass of the specimen drops to zero within a small temperature interval. Examples are poly(α-methyl styrene) and poly(oxymethylene).

Depolymerizations with small zip lengths ($1 < \Xi < 5$) lead at lower temperatures to volatile but stable oligomer molecules but at higher temperatures to practically only monomer molecules. The weight fraction, w_P, of the polymer decreases first slowly and then fast. The function $w_P = f(T)$ shows a distinct inflection point. Examples are poly(styrene) and poly(isobutylene).

In degradations with chain scissions, the shape of the function $w_P = f(T)$ depends on the relative strengths of main chain bonds (c-c) and bonds between substituents and chain atoms (c-s). If the energy for the scission of c–s is smaller than that of c–c ($E_b \gg E_s$), a few substituents are split off and the mass of the polymer is slightly reduced. The main chain is not severed, however, but reacts intermolecularly and intramolecularly by cyclization and crosslinking to thermally stable products: the weight fraction of the polymer becomes independent of temperature (right curve of Fig. 15-10). Examples are the formation of carbon and graphite fibers from cellulose and acrylic fibers and the dehydrogenation of 1,2-poly(butadiene) to double-strand (ladder) polymers.

If the bond between substituents and chain atoms is weak and that between chain atoms not much stronger ($E_c \approx E_s$), substituents will be split off first and chain bonds later. The weight fraction of the remaining polymer drops first strongly. It reaches a plateau after most of the substituents have been stripped off. Finally, chain bonds are severed and all polymer is decomposed (left curve of Fig. 15-10). An example is poly(vinyl chloride).

Susceptible to pyrolysis are branching points, electron-accepting groups, long methylene sequences, and all groups that can easily form five and six-membered rings. Thermostability is enhanced by aromatic rings, ladder structures, fluorine atoms as substituents, and the absence of hydrogen (see Volume IV). Examples of decomposition temperatures by thermogravimetry under a nitrogen blanket are 415°C for poly(ethylene), 450°C for $+(p\text{-}C_6H_4)\text{–}CH_2\text{–}CH_2 \, \frac{}{}_n$ vs. 535°C for $+(p\text{-}C_6H_4)\text{–}CF_2\text{–}CF_2 \, \frac{}{}_n$, and 660°C for $+p\text{-}C_6H_4 \, \frac{}{}_n$ vs. 720°C for $+p\text{-}C_6F_4 \, \frac{}{}_n$. The decomposition temperatures are lower under oxygen. Note that these temperatures also depend on the speed of the experiment.

A-15 Appendix: Intramolecular Cyclization

The maximum conversion of substituents in kinetically controlled (irreversible) intra-molecular ring formations between neighboring groups of head-to-tail polymers with degrees of polymerization of X_i is calculated as follows.

N_i is the number of unreacted groups. It equals $N_1 = 1$ since the single group of a molecule with $X_1 = 1$ cannot form a ring. Both groups of a dimer will react, hence $N_2 = 0$ for $X_2 = 2$. Only two of the three substituents of a trimer can form a ring. One group is left and therefore $N_3 = 1$.

Tetramers may have three possible reactions: 1-2, 2-3, and 3-4. If units 1+2 react first, units 3+4 will also form a ring and the situation is the same as for the dimer: all groups of the tetramer have reacted to rings. The same is true for the reaction of 4+3 and a sub-sequent reaction of units 2+1. If units 2+3 react first, units 1 and 4 can no longer react and the situation is the same as for the monomer.

After the reaction of the first pair of substituents of a tetramer (1+2, 2+3, or 3+4), two possibilities exist for the reaction of a dimer unit (3+4 if 1+2 reacted first; 1+2 if 3+4 was first) and two probabilities for left-over units (1 and 4 if 2+3 was formed). The total number of combinations is $X - 1 = 3$. The number N_4 of isolated groups in a tetrameric unit is therefore

$$(\text{A-15-1}) \quad N_4 = (2\,N_1 + 2\,N_2)/3 = 2\,(N_1 + N_2)/(X - 1) = 2\,(1 + 0)/(4 - 1) = 2/3$$

In pentamer molecules, the primary formation of 1+2 may be followed by a reaction of 3+4 (with unit 5 remaining) or of 4+5 (with unit 3 remaining). It is similar for the pri-mary reaction of 5+4 and the secondary reactions of units 3+2 (unit 1 remains) or units 2+1 (unit 3 remains). The situatation for the formation of a second pair is thus the same as for a trimer molecule. However, if pairs 2+3 or 3+4 react first, both a monomeric unit and a dimeric unit remain. Totally,

$$(\text{A-15-2}) \quad N_5 = (2\,N_3 + 2\,N_2 + 2\,N_1)/(X - 1) = (2{\cdot}1 + 2{\cdot}0 + 2{\cdot}1)/(5 - 1) = 1$$

and by analogy for arbitrary degrees of polymerization:

$$(\text{A-15-3}) \quad N_i = 2\,(N_1 + N_2 + N_3 + \ldots N_{i-3} + N_{i-2})/(X - 1)$$

$$(\text{A-15-4}) \quad N_{i-1}(X - 2) = 2\,(N_1 + N_2 + \ldots N_{i-3})$$

Subtraction of Eq.(A-15-4) from Eq.(A-15-3) delivers

$$(\text{A-15-5}) \quad N_i(X - 1) - N_{i-1}(X - 2) = 2\,N_{i-2}.$$

Introduction of Eq.(A-15-5) into the definitions $\Delta_i \equiv N_i - N_{i-1}$, $\Delta_{i-1} \equiv N_{i-1} - N_{i-2}$, etc., leads to $(X - 1)\Delta_i + \Delta_{i-1} = N_{i-2}$. The analogous expression $(X - 2)\Delta_{i-1} + \Delta_{i-2} = N_{i-3}$ is subtracted and the resulting equation transformed to

$$(\text{A-15-6}) \quad \Delta_i - \Delta_{i-1} = [(-2)/(X - 1)][\Delta_{i-1} - \Delta_{i-2}]$$

Repeated substitution delivers

$$(\text{A-15-7}) \qquad \Delta_i - \Delta_{i-1} = \frac{(-2)^{X-1}}{(X-1)!}(\Delta_1 - \Delta_0)$$

For $X = i = 2$ one has $\Delta_2 - \Delta_1 = -2\,[\Delta_1 - \Delta_0]$. Since $\Delta_2 = N_2 - N_1 = 0 - 1 = -1$ and $\Delta_1 = N_1 - N_0 = 1 - 0 = 1$, one obtains $-1 - 1 = -2\,[\Delta_1 - \Delta_0]$ or $[\Delta_1 - \Delta_0] = 1$.

Eq.(A-15-7) can be developed in a series

$$(\text{A-15-8}) \qquad \Delta_i = 1 - \frac{2}{1!} + \frac{4}{2!} - \frac{8}{3!} + \ldots \frac{(-2)^{X-1}}{(X-1)!}$$

which corresponds to $1/e^2$ for $X \to \infty$, i.e., $\Delta_i = 1/e^2$. A molecule with a *high* degree of polymerization, X, contains therefore

$$(\text{A-5-9}) \qquad N_i \approx X/e^2$$

unreacted groups. Analogous considerations deliver for shorter chains

$$(\text{A-15-10}) \qquad N_i = X - 2(X-1) + \frac{4(X-2)}{2!} - \ldots + (-2)^{X-1}\left(\frac{X-(X-1)}{(x-1)!}\right)$$

Historical Notes to Chapter 15

Polymer-Analog Reactions (Polymer Transformations)

S.S.Pickles, J.Chem.Soc. [London] **97** (1910) 1085
Used bromination as proof of the constitution of natural rubber.

H.Staudinger, H.Fritschi, Helv.Chim.Acta **5** (1922) 785
Polymer analog reactions (hydrogenation of natural rubber).

Chain Scissions

W.Kuhn, Ber.Dtsch.Chem.Ges. **63** (1930) 1503
First statistical treatment of polymers (chain scission of cellulose).

Literature to Chapter 15

15.1 OVERVIEWS
E.M.Fetters, Ed., Chemical Reactions of Polymers, Interscience, New York 1964
R.W.Lenz, Organic Chemistry of Synthetic High Polymers, Interscience, New York 1967
J.A.Moore, Ed., Reactions on Polymers, Reidel, Dordrecht 1973
G.Allen, J.C.Bevington, Eds., Comprehensive Polymer Science, Vol. 6, G.C.Eastmond,
    A.Ledwith, S.Russo, P.Sigwalt, Eds., Polymer Reactions, Pergamon Press, Oxford 1989

15.2 POLYMER CATALYSTS

N.K.Mathur, C.K.Narang, R.E.Williams, Polymers as Aids in Organic Chemistry, Academic Press, New York 1980

P.Hodge, D.C.Sherrington, Polymer-Supported Reactions in Organic Synthesis, Wiley, New York 1980

M.Kaneko, E.Tsuchida, Formation, Characterization and Catalytic Activities of Polymer-Metal Complexes, J.Polym.Sci.-Macromol.Revs. **16** (1981) 397

W.T.Ford, M.Tomoi, Polymer-Supported Phase Transfer Catalysts: Reaction Mechanisms, Adv.Polym.Sci. **55** (1984) 49

G.Wulff, Molecular Recognition in Polymers Prepared by Imprinting with Templates, in W.T.Ford, Ed., Polymeric Reagents and Catalysts (ACS Symp. Ser. **308**), ACS, Washington (DC) 1986,

E.A.Bekturov, S.E.Kudaibergenov, Polymer Catalysis, Hüthig and Wepf, Zug 1994

B.Sellergren, Imprinted Polymers with Memory for Small Molecules, Proteins, or Crystals, Angew.Chem. **112** (2000) 1069; -, Angew.Chem.Int.Ed.Engl. **39** (2000) 1029

M.Komiyama, T.Takeuchi, T.Mukawa, H.Asanuma, Molecular Imprinting: From Fundamentals to Applications, Wiley-VCH, Weinheim 2003

15.3.1 EXCHANGE EQUILIBRIA

B.A.Rosenberg, W.I.Irzak, N.S.Enikolopian, Interchain Exchange Reactions in Polymers, Chimia, Moscow 1975 (in Russian)

15.4.2 ACID-BASE REACTIONS

S.A.Rice, M.Nagasawa, Polyelectrolyte Solutions, Academic Press, New York 1961

15.4.3 ION EXCHANGE RESINS

R.Kunin, Ion Exchange Resins, Wiley, New York, 2nd ed. 1971

K.Dorfner, Ion Exchangers, W. de Gruyter, Berlin 1991

D.Muraviev, V.Gorshov, A.Warshawsky, Ion Exchange, Dekker, New York 1999

H.Strathmann, Ion Exchange Membrane Separation Processes, Elsevier, Amsterdam 2004

15.4.4 POLYMER TRANSFORMATIONS

N.A.Platé, O.V.Noah, A Theoretical Consideration of the Kinetics and Statistics of Reactions of Functional Groups of Macromolecules, Adv.Polym.Sci. **31** (1979) 133

N.Ise, I.Tabushi, An Introduction to Specialty Polymers, Cambridge University Press, Cambridge 1983

D.C.Sherrington, P.Hodge, Eds., Syntheses and Separations Using Functional Polymers, Wiley, Chichester 1988

A.Akelah, A.Moet, Functionalized Polymers, Routledge (Chapman and Hall), New York 1991

E.Tsuchida, Macromolecular Complexes, VCH, Weinheim 1991

K.Takemoto, R.M.Ottenbrite, M.Kamachi, Eds., Functional Monomers and Polymers, Dekker, New York, 2nd ed. 1997

R.Arshady, Ed., Desk Reference of Functional Polymers - Synthesis and Applications, American Chemical Society, Washington (DC) 1997

J.J.Meister, Ed., Polymer Modification, Dekker, New York 2000

T.C.M.Chung, Functionalization of Polyolefins, Academic Press, London 2002

R.Arshady, Ed., Desk Reference of Functional Polymers - Synthesis and Applications, American Chemical Society, Washington, DC, 1997

15.4.5 INTRAMOLECULAR RING-CLOSING REACTIONS

W.de Winter, Double Strand Polymers, Revs.Macromol.Sci. **1** (1966) 329

V.V.Korshak, Heat-Resistant Polymers, Israel Progr.Sci.Transl., Jerusalem 1971

Y.Imanishi, Intramolecular Reactions on Polymer Chains, Macromol.Revs. **14** (1979) 1

15.4.6 POLYMER REAGENTS

E.C.Blossey, P.C.Neckers, Solid-Phase Synthesis, Halsted Press, New York 1975

N.K.Mathur, C.K.Narang, R.E.Williams, Polymers as Aids in Organic Chemistry, Academic Press, New York 1980

M.A.Kraus, A.Patchornik, Polymeric Reagents, J.Polym.Sci.-Macromol.Revs. **15** (1980) 55

P.Hodge, D.C.Sherrington, Eds., Polymer-Supported Reactions in Organic Synthesis, Wiley, New York 1980

N.Ise, I.Tabushi, An Introduction to Specialty Polymers, Cambridge University Press, Cambridge 1983

D.C.Sherrington, P.Hodge, Eds., Syntheses and Separations Using Functional Polymers, Wiley, Chichester 1988

M.R.Buchmeiser, Ed., Polymeric Materials in Organic Synthesis and Catalysis, Wiley-VCH, Weinheim 2003

15.5.1 POLYMER DEGRADATION

IUPAC, Macromolecular Division, Commission on Macromolecular Nomenclature, Definitions of Terms Relating to Degradation, Aging, and Related Chemical Transformations of Polymers, Pure Appl.Chem. **68** (1996) 2313

H.H.G.Jellinek, Degradation of Vinylpolymers, Academic Press, New York 1955

S.L.Madorsky, Thermal Degradation of Organic Polymers, Interscience, New York 1964

N.Grassie, Chemistry of High Polymer Degradation Processes, Butterworths, London, 2nd ed. 1966

L.Reich, S.Stivala, Elements of Polymer Degradation, McGraw-Hill, New York 1971

C.H.Bamford, C.F.H.Tipper, Eds., Comprehensive Chemical Kinetics **14** (Degradation of Polymers), Elsevier, Amsterdam 1975

A.M.Basedow, K.H.Ebert, Ultrasonic Degradation of Polymers in Solution, Adv.Polym.Sci. **22** (1977) 84

H.H.G.Jellinek, Ed., Aspects of Degradation and Stabilization of Polymers, Elsevier, Amsterdam 1978

T.Kelen, Polymer Degradation, Hanser, Munich 1981

S.S.Stivala, J.Kimura, L.Reich, The Kinetics of Degradation Reactions, Degrad.Stab.Polym. **1** (1983) 1

W.L.Hawkins, Polymer Degradation, Springer, Berlin 1984

N.M.Emanuel, A.L.Buchachenko, Chemical Physics of Polymer Degradation and Stabilization, VSP, Utrecht 1988

N.Grassie, G.Scott, Polymer Degradation and Stabilisation, Cambridge Univ.Press, Cambridge, UK 1988

S.J.Huang, Ed., Biodegradable Polymers, Hanser, Munich 1989

M.Vert, S.Feijen, A.Albertsson, G.Scott, E.Chiellini, Biodegradable Polymers and Plastics, CRC Press, Boca Raton (FL) 1992

S.Halim Hamid, Ed., Handbook of Polymer Degradation, Dekker, New York 1992

O.Guven, Crosslinking and Scission in Polymers, Kluwer, Dordrecht 1990

G.Griffin, Chemistry and Technology of Biodegradable Polymers, Blackie (Chapman and Hall), New York 1993

J.F.Rabek, Polymer Photodegradation: Mechanisms and Experimental Methods, Chapman and Hall, New York 1994

G.Scott, D.Gilead, Ed., Degradable Polymers: Principles and Applications, Chapman and Hall, New York 1995

J.F.Rabek, Photodegradation of Polymers. Physical Characterization and Applications, Springer, Berlin 1996

A.J.Domb, J.Kost, D.M.Wiseman, Handbook of Degradable Polymers, Harwood Academic Publ., Newark (NJ) 1998

S. Halim Hamid, Ed., Handbook of Polymer Degradation, Dekker, New York, 2nd ed. 2000

G.Scott, Ed., Degradable Polymers, Kluwer, Dordrecht, 2nd ed. 2003

15.5.4 PYROLYSIS

A.H.Frazer, High Temperature Resistant Polymers, Wiley, London 1968

R.T.Conley, Thermal Stability of Polymers, Dekker, New York 1969

V.V.Korshak, Heat-Resistant Polymers, International Scholarly Book Services, Portland (OR) 1971

C.J.Hilado, Pyrolysis of Polymers, Technomic Publ., Westport (CT) 1978

E.Stahl, V.Brüderle, Polymer Analysis by Thermofractography, Adv.Polym.Sci. **30** (1979) 1

C.Arnold, Jr., Stability of High-Temperature Polymers, Macromol.Revs. **14** (1979) 265

P.E.Cassidy, Thermally Stable Polymers, Dekker, New York 1980

J.P.Critchley, G.J.Knight, W.W.Wright, Heat-Resistant Polymers, Plenum, New York 1983

A.Davies, D.Sims, Weathering of Polymers, Elsevier Appl.Sci.Publ., New York 1983

K.J.Wynne, R.W.Rice, Ceramics via Polymer Pyrolysis, Ann.Rev.Mater.Sci. **14** (1984) 297
R.M.Aseeva, G.E.Zaikov, Combustion of Polymer Materials, Hanser, Munich 1986

References to Chapter 15

[1]   N.Ise, F.Matsui, J.Am.Chem.Soc. **90** (1968) 4242, Fig. 1
[2]   H.P.Gregor, L.B.Luttinger, E.M.Loebl, J.Phys.Chem. **59** (1955) 34, Fig. 1
[3]   P.I.Dubin, U.P.Strauss, in A.Rembaum, E.Selegny, Eds., Polyelectrolytes and Their Applications, Reidel (Kluwer Academic Publishers), Dordrecht 1975, Fig. 3
[4]   R.Lehrle, R.Williams, C.French, T.Hammond, Macromolecules **28** (1995) 4408, Figs. 4, 11
[5]   D.S.Argyropoulos, H.I.Bolker, Macromolecules **20** (1987) 2915, Table II
[6]   H.-G.Elias, Makromol.Chem. **187** (1986) 2209, Figs. 1 and 2
[7]   H.F.Mark, A.V.Tobolsky, Physical Chemistry of High Polymer Systems, Interscience, New York 1950, Fig. XIII-3
[8]   N.Grassie, H.W.Melville, Proc.R.Soc. [London] **A 199** (1949) 14, Fig. 2
[9]   V.T.Kagiya, K.Takemoto, M.Hagiwara, J.Appl.Polym.Sci.-Appl.Polym.Symp. **35** (1979) 95, Tables III and IV
[10]   D.O.Hummel, H.J.Düssel, H.Rosen, in Z.Jedlinski, Ed., Advances in the Chemistry of Thermally Stable Polymers, Panst.Wydawn Nauk, Warsaw 1977, p. 99, Fig. 1

16   Molecular Engineering

16.1   Polymer Architecture

Molecular polymer engineering comprises the syntheses of macromolecules of defined constitution and configuration, especially the controlled sequences of monomeric units and their "three-dimensional" connections. In addition, the term "molecular engineering" is sometimes used for the preparation of supramolecular structures and also for the controlled preparation of polymer blends and polymer composites although these morphologies do not stem from molecular entities *per se* (see Volume IV).

This chapter discusses true *molecular* engineering, i.e., the **strategies** for the synthesis of polymeric building blocks and their combination to larger structures. Syntheses of building blocks employ all basic types of polymerization: chain polymerization, polycondensation, polyaddition, and polyelimination, including their subgroups with respect to active centers such as free-radical, anionic, cationic, coordinative, etc., polymerizations. They have been discussed in the previous Chapters 6 to 15.

Molecular polymer engineering leads to different **architectures** of macromolecules, i.e., orderly arrangements of monomeric units in basic structures and combination of these structures to orderly arrangements in one, two, or three directions (architecture: any design or orderly arrangement perceived by man (American Heritage Dictionary)). One can view these arrangements in two ways: (A) "chemically" with respect to monomeric units or (B) "physically" with respect to basic structures composed of monomeric units. Scheme A arranges architectures with respect to their ring character or branching and their chemical dimensionality (Fig. 16-1). Scheme B (which is used in this book) looks at the synthesis of basic structures such as blocks, stars, combs, and dendrons (Section 16.2) and their chemical or physical assemblage in "straight arrangements" such as ladders, layers, and plates (Section 16.3) or in "curved arrangements" such as rings, tubes, and spheres (Section 16.4), i.e., as ordered geometric structures.

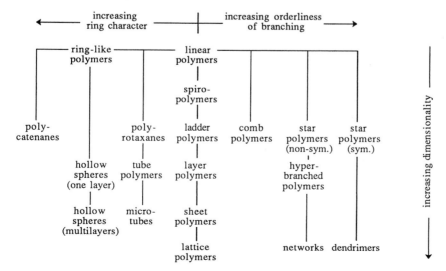

Fig. 16-1 Schematic depiction of macromolecular architectures with respect to monomeric units.

16.2   Basic Structures

The smallest architectural unit of a polymer molecule is the **segment** consisting of like monomeric units. Like or unlike segments can be combined in different ways to entities that may serve as such or as building blocks for architectures of various complexities. This group of polymers comprises suprapolymers (Section 16.2.1), block polymers (Section 16.2.2), comb polymers (Section 16.2.3), star polymers (Section 16.2.4), and dendritic polymers (Section 16.2.5).

16.2.1   Suprapolymers

Supramolecules are physical molecules that result from the ordered self-association of chemical molecules via several adjacent non-covalent bonds such as hydrogen bonds. The sum of the bond strengths of these clusters of bonds equals or surpasses that of single covalent bonds so that the resulting entities behave in melts and suitable solvents like true chemical molecules, hence the name (L: *supra* = above, beyond, earlier).

The formation of supramolecules is a special kind of a self-association process (Fig. 16-2). Most often employed is hydrogen bonding between like groups such as two OH groups or two unlike groups such as >C=O and HN<. Supramolecular bonding is not restricted to hydrogen bonds, however. Another example is the π-π interaction between extended aromatic ring systems.

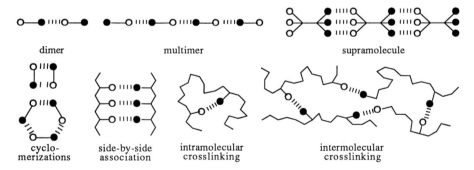

Fig. 16-2 Self-associations ⅠⅠⅠ between like (if O = ●) or unlike (if O ≠ ●) groups such as hydrogen bond-forming ones (OH, COOH, etc.), ions, extended aromatic ring systems, etc. Examples:

- intermolecular association of monofunctional compounds such as the dimerization of monocar-boxylic acids or butyl lithium (Section 8.3.4 and Eq.(8-4)). The association may lead to open structures (top left) or ring structures (dimers, trimers, etc., bottom left);
- intermolecular end-to-end associations of bifunctional unimers to linear multimers, for example, the self-association of poly(ethylene glycol)s (p. 25), see also Volume III;
- intermolecular end-to-end association of unimers with multifunctional ends to linear suprapoly-mers (example: p.25);
- intermolecular side-by-side association of linear chains. Examples: (a) Base pairing in double helices of deoxyribonucleic acids (Section 14.2.2); (b) ion association of oppositely charged polyelectrolyte molecules (Volume III);
- intramolecular physical crosslinking. Example: Hydrogen bonds or ionic bonds within a protein molecule including α-helix structures (Section 14.3.4);
- intermolecular physical crosslinking. Example: Physical gels, p. 60 and Volume III.

The nomenclature of associations is not uniform (Fig. 16-2). Association products of molecules with monofunctional ends such as single OH endgroups in poly(ethylene glycol)s, $HO[CH_2CH_2O]_{n-1}CH_2OH$, are called **multimers** (Volume III) but "supramolecules" if several adjacent groups are involved (p. 25). Molecules with *end-to-end* self-association via monofunctional groups show only some properties of true supramolecules such as a high melt viscosity or liquid-crystalline behavior, for example,

Intramolecular and intermolecular *side-by-side* self-associations have been known for a long time but these are neither called self-associates nor supramolecules. Examples of ordered intramolecular selfassociates are the α-helices of poly(α-amino acid)s and proteins. The double helices of deoxyribonucleic acids are intermolecular self-associates.

Quaternary structures of certain enzymes are also associates (Section 14.3.5) which may be called supersuprapolymers. An example is hemoglobin which consists of two identical A-subunits and two identical B-subunits. The subunits are held together by various hydrogen bonds, ionic bonds, and hydrophobic bonds in such a way that hemoglobin appears as a true chemical molecule down to its smallest measurable concentrations. Hemoglobin does dissociate into its four subunits in aqueous solutions of urea.

16.2.2 Block Polymers

Block polymers can be obtained by various procedures: block-forming copolymerizations, block copolymerizations, block polymerizations, or block couplings. The applicable strategy depends on the structure of monomers and desired block polymers as well as suitable polymerization mechanisms.

Block Copolymerizations

Block *co*polymerizations are polymerizations where the different monomers are already present in the reaction mixture at the start of the polymerization. Joint free-radical bipolymerizations of two monomers A and B with copolymerization parameters of $0 < r_a \neq 1/r_b < 1$ deliver polymers with long blocks of b-units in addition to randomly sized and distributed a-segments (p. 397). Both long a-blocks and long b-blocks result if $1 < r_a < \infty$ and $1 < r_b < \infty$. In contrast to other types of block formation, block-forming copolymerizations result in multiblock copolymers with various proportions and statistical distributions of the different types of blocks per polymer molecule.

Much larger blocks are obtained by the joint anionic bipolymerization of a very rapidly polymerizing monomer I (e.g., isoprene) and very slowly polymerizing monomer S (e.g., styrene). In these block *co*polymerizations, both monomers are initially present. An added initiator, for example, a small dianion $^\ominus Z^\ominus$, causes the polymerization of first the monomer I and then, after I is practically exhausted, the second monomer S:

$$(16\text{-}1) \qquad ^\ominus Z^\ominus \xrightarrow{+\,2\,n\,\text{I}\,+\,2\,m\,\text{S}}\ ^\ominus \text{I}_n Z\text{I}_n^\ominus \longrightarrow\ ^\ominus \text{S}_m\text{I}_n Z\text{I}_n\text{S}_m^\ominus$$

The living polymers are then terminated by capping their ends. This type of bipolymerization reduces the probability of chain termination. Ideally, it delivers triblock copolymers with two S_m blocks and one I_nZI_n block. In practice, gradient copolymers result because reactivities do not differ sufficiently (Figs. 12-10 and 12-12).

Block copolymerization is used industrially to produce thermoplastic styrene-isoprene-styrene triblock copolymers. Initiators are thermally stable aromatic lithium compounds that are soluble in hydrocarbons, for example, from the reaction

(16-2)

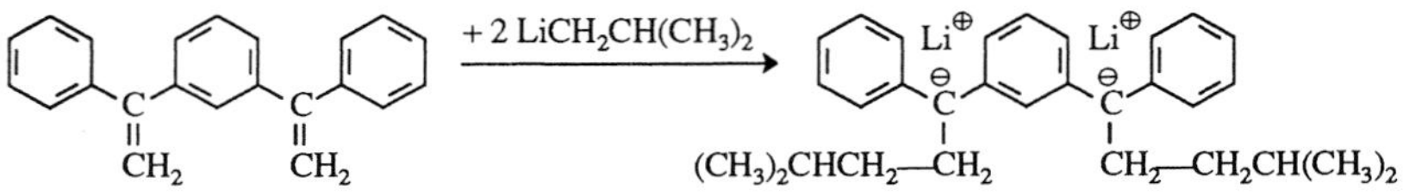

Ionic Block Polymerizations

Diblock polymerizations are chain polymerizations in which the second monomer is only added and polymerized after the first monomer is completely polymerized. Triblock polymerizations proceed similarly. Block polymerizations are not joined polymerizations and hence not block *co*polymerizations. Various strategies can be used.

Block polymerizations work most often with *monofunctional initiators* and successive living anionic polymerizations. For example, an A-block is generated first, then a B-block, and finally a C-block (or another A-block) in triblock polymerizations:

(16-3) $\qquad R^{\ominus} \xrightarrow{+\,m\,A} RA_m^{\ominus} \xrightarrow{+\,n\,B} RA_mB_n^{\ominus} \xrightarrow{+\,p\,C} RA_mB_nC_p^{\ominus}$

Each step requires complete monomer conversion before the next monomer is added. This strategy involves three separate growth processes and thus an increased probability of undesired termination processes. Also, the formation of homopolymers and diblock polymers cannot be excluded.

Chain ends $\sim A_m^{\ominus}$, $\sim B_n^{\ominus}$, and $\sim C_p^{\ominus}$ have different polarities and it is therefore important in which succession the various blocks are created. For example, poly(methyl methacrylate) anions initiate the polymerization of acrylonitrile but not that of styrene. However, the desired diblock polymers of methyl methacrylate and styrene can be obtained from the initiation of methyl methacrylate by poly(styrene) anions.

The growing chain ends $\sim A^{\ominus}$, $\sim B^{\ominus}$, $\sim C^{\ominus}$, etc., always possess the same countercation, e.g., $Li^{\oplus}$. The propagation reactions thus require the chain ends $\sim$ALi, $\sim$BLi, $\sim$CLi to be sufficiently reactive to initiate the polymerization of other types of monomers. This condition is often fulfilled in anionic polymerizations but not in cationic ones. For example, diblock polymers of styrene and isobutene are not obtainable by an anionically initiated two-step block polymerization since isobutene does not polymerize anionically.

However, such diblock polymers can be synthesized by an *initiator change* without a mechanism change after formation of the first block. For the synthesis of poly(isobutene-*block*-styrene), isobutene is polymerized cationically by the monofunctional initiator BCl_3-H_2O. After completion of isobutene polymerization, excess initiator is distilled off under vacuum. The second block is then obtained by addition of styrene and a subsequent cationic polymerization with $(C_2H_5)_2AlCl$ as initiator. An alternative strategy is to maintain the initiator but to *change the solvent* for the second step.

The second block can also be prepared by a *change of mechanism*. Examples are the conversion of $\sim A_m{}^{\ominus}$ to $\sim A_m{}^{\oplus}$ (or vice versa) or that of $\sim A_m{}^{\ominus}$ to $\sim A_m{}^{\bullet}$. These conversions often require functionalized chain ends as intermediates. Examples are the conversions of macroanions into macrocations or macroradicals:

$$(16\text{-}4) \qquad \text{\tiny{MM}} A^{\ominus} Na^{\oplus} \xrightarrow[-\,NaBr]{+\,Br_2} \text{\tiny{MM}} ABr \xrightarrow[-\,AgBr]{+\,AgPF_6} \text{\tiny{MM}} A^{\oplus} PF_6^{\ominus}$$

$$(16\text{-}5) \qquad \text{\tiny{MM}} A^{\ominus} Na^{\oplus} \xrightarrow[-\,Na^{\oplus}]{+\,O_2\,;\,+\,H^{\oplus}} \text{\tiny{MM}} AOOH \longrightarrow \text{\tiny{MM}} AO^{\bullet} + {}^{\bullet}OH$$

Functionalization of macroanions may be by oxidation, carbonylation (with CO_2 to $\sim COO^{\ominus}$), hydroxylation (with oxirane or ketones), amination (with $Cl(CH_2)_3NR_2$ or $C_6H_5CH{=}NSi(CH_3)_3$), or sulfonation (with sultones). Functionalizations are sometimes incomplete. For example, the reaction of macroanions $RM_n{}^{\ominus}$ with CO_2 (+ $H_3O^{\oplus}$) delivers not only RM_nCOOH but also $(RM_n)_2CO$ and $(RM_n)_3COH$.

Triblock polymers are also obtained from *bifunctional initiators* in two steps:

$$(16\text{-}6) \qquad {}^{\ominus}Z^{\ominus} \xrightarrow{+\,2\,m\,B} {}^{\ominus}B_mZB_m{}^{\ominus} \xrightarrow{+\,2\,n\,S} {}^{\ominus}S_nB_mZB_mS_n{}^{\ominus}$$

This strategy is used for the synthesis of the thermoplastic elastomer poly(styrene-*block*-butadiene-*block*-styrene). The bulk polymerization of butadiene by highly soluble aromatic dilithium compounds in the presence of small proportions of aromatic ethers is preferred because it leads to poly(butadiene) blocks with relatively high contents of cis-1,4 units. Sodium naphthalene, NaNp, which also forms dianions (p. 223, p. 238), cannot serve as the initiator because it requires tetrahydrofuran (THF) as a solvent. The system NaNp + THF, however, produces butadiene blocks with low contents of cis-1,4 structures and thus triblock polymers with far higher glass temperatures of the elastomeric butadiene blocks, making them unsuitable as thermoplastic elastomers.

After formation of the central butadiene blocks, styrene in dimethoxyethane is added and polymerized. The polymerization is terminated by addition of methanol. An unwanted termination reaction is the one to butadiene-styrene diblock polymers since these are detrimental to the desired properties of thermoplastic elastomers. Termination to homopolymeric poly(butadiene) delivers only an additional elastomeric content.

Free-Radical Block Polymerizations

Living block polymerizations by ionic or coordination mechanisms are limited to relatively few monomers. They also require high purities (Section 8.3) and suffer from the relatively slow build-up of molecular weights. For these reasons, *quasi-living free-radical block polymerizations* are attractive, for example, stable free-radical polymerizations using nitroxide radicals such as TEMPO ($T^{\bullet}$) (Section 10.5.2). However, they are restricted to styrene monomers since, for example, propagating tertiary radicals such as those in methyl methacrylate polymerizations disproportionate to a macromonomer and the hydride of the nitroxide radical, TH. The latter species then terminates additional macroradicals by chain transfer. Such polymerizations are also prohibitively expensive for industrial purposes since they require equimolar amounts of expensive nitroxides.

More promising are strategies that use *chain transfer agents*. An example is the block polymerization of butyl methacrylate, $CH_2=C(CH_3)COOC_4H_9$, and the macromonomer of phenyl methacrylate, $H[CH_2-C(CH_3)(COOC_6H_5)]_iCH_2-C(=CH_2)COOC_6H_5$, as addition-fragmentation chain transfer agent (Fig. 16-3). The strategy even succeeds with methyl methacrylate as monomer and the macromonomer of methacrylic acid.

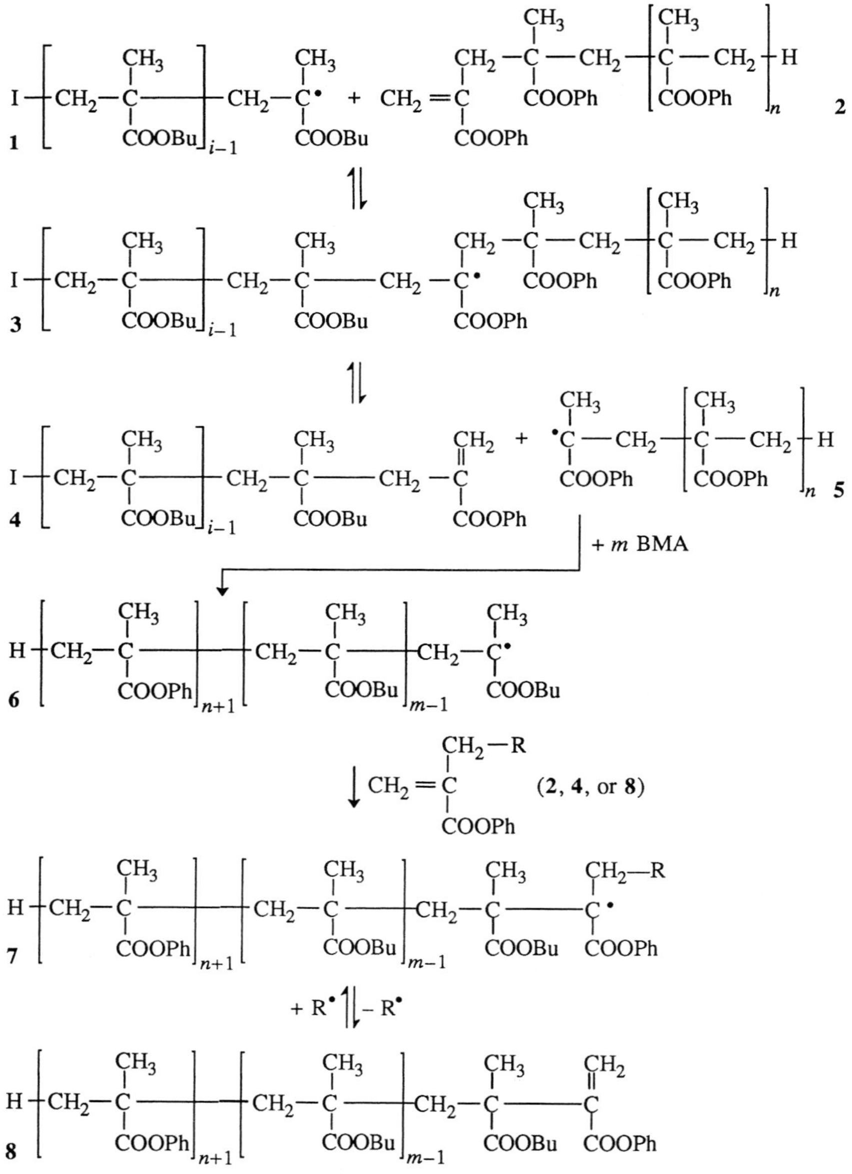

Fig. 16-3 Example of a block polymerization with a macromonomer as chain transfer agent [1].
I = initiator fragment, for example, $[SO_4]^{\ominus}$ from $K_2S_2O_8 \rightleftarrows K^{\oplus} + 2\ ^{\bullet}[SO_4]^{\ominus} + K^{\oplus}$. Bu = C_4H_9, Ph = C_6H_5. The radical $R^{\bullet}$ is a fragment of macromonomers **2**, **4**, or **8**.

In this strategy (Fig. 16-3), a monomer (butyl methacrylate) and a free-radical initiator ($K_2S_2O_8$) were added to the heated aqueous emulsion of the macromonomer **2**. The resulting poly(butyl methacrylate) radicals **1** form a reversible adduct **3** with the macromonomer **2**. The adduct dissociates reversibly to the new macromonomer **4** and the radical **5** of the macromonomer **2**. The macroradical **5** initiates the polymerization of butyl methacrylate to a diblock macroradical **6** which then reacts with one of the macromonomers **2**, **4**, or **8**,

2 $CH_2=C(COOPh)-CH_2-[C(CH_3)(COOPh)-CH_2]_i-H$,
4 $CH_2=C(COOPh)-CH_2-[C(CH_3)(COOC_4H_9)-CH_2]_j-H$,
8 $CH_2=C(COOPh)-CH_2-[C(CH_3)(COOC_4H_9)-CH_2]_k-[C(CH_3)(COOPh)-CH_2]_i-H$

to form the diblock macroradical **7** and finally the diblock macromonomer **8**.

The strategy is successful if chain transfer to macromonomer dominates. Termination reactions leading to dead polymers must be subdued, for example, the mutual deactivation between any two radicals **1**, **3**, **5**, **6**, or **7**. All these termination reactions stem ultimately from initiator-generated radicals. The rate of initiation must therefore be small but must still lead to resonable polymerization rates.

The molar fraction of double bonds stayed constant during the polymerization according to endgroup analysis by NMR, indicating the absence of grafting which would have reduced the content of double bonds. Fragmentation of type **3** macroradicals is therefore always dominant.

Number-average molecular weights increased linearly with increasing proportions of monomer added to the macromonomer, indicating a living polymerization. These molecular weights were considerably lower than the ones obtained in experiments without added macromonomer, however, which again shows the control by chain transfer.

Much more simple is the use of low-molecular weight chain transfer agents, especially that of highly efficient sulfur compounds (Table 10-9). An example is the use of mixed esters of thiocarbonic acid such as $RS-C(S)-OEt$ for the successive polymerization of monomers M' and M" that is initiated by a free-radical initiator IJ:

(16-7)

$$RS-\underset{\underset{S}{\|}}{C}-OEt \xrightarrow[+\ IJ]{+\ n\ M'} R-(M')_n-S-\underset{\underset{S}{\|}}{C}-OEt \xrightarrow{+m\ M''} R-(M')_n-(M'')_m-S-\underset{\underset{S}{\|}}{C}-OEt$$

Block Couplings

Block couplings are couplings of separately prepared blocks, for example,

(16-8) $\quad R^* + n\,A \longrightarrow RA_n{}^* \xrightarrow{+\ ^*B_mR''} RA_nB_mR'' \xleftarrow{+\ ^*A_nR} R''B_m{}^* \longleftarrow R''^* + m\,B$

The "active" site * may be a functional endgroup of a condensation polymer. In ionic polymerization, $RA_n{}^*$ may be a macrocation $RA_n{}^\oplus$ and $R''B_m{}^*$ a macroanion $R''B_m{}^\ominus$. Alternatively, $RA_n{}^*$ and $^*B_mR''$ may be coupled via an additional molecule. In all cases, coupling must be complete because incomplete couplings lead to homopolymers that are usually undesired.

Multiblock polymers are produced if the starting blocks are bifunctional. An example is the coupling of poly(styrene) dianions, $^\ominus S_n{}^\ominus$, and poly(tetrahydrofuran) dications,

$^{\oplus}T_m{}^{\oplus}$. Polyperoxides and polyazo compounds deliver biradicals that generate block couplings with yields of up to 80 %.

Blocks may also be coupled by polycondensation. An example is the thermoplastic elastomer with "hard" blocks (high T_G) of poly(ethylene terephthalate) and "soft" blocks (low T_G) of poly(tetrahydrofuran), i.e., a polyether ester:

$$\mathrm{+OCH_2CH_2CH_2CH_2OOC(\textit{p}\text{-}C_6H_4)CO\text{+}_n\text{+}OCH_2CH_2CH_2CH_2)_nOOC(\textit{p}\text{-}C_6H_4)CO\text{+}_n}$$

All these reactions usually deliver polymers with distributions of block numbers (see next section), including couplings of macrodications and macrodianions which belong to the class of polyadditions.

Block couplings can also be combined with block polymerizations. For example, one can first generate diblock polymers by a two-step polymerization with monofunctional initiators. The diblock polymers are then coupled to a triblock polymer with the help of a coupling agent. An example is the anionic diblock polymerization of monomers A and B by $C_4H_9{}^{\ominus}Li{}^{\oplus}$ as initiator and the subsequent coupling of the resulting diblock polymers by $COCl_2$ ($^{\oplus}Z^{\oplus} = {}^{\oplus}CO^{\oplus}$):

$$(16\text{-}9) \qquad 2\,R^{\ominus} + 2\,m\,A \longrightarrow 2\,RA_m^{\ominus} \xrightarrow{+\,n\,B} 2\,RA_mB_{n/2}^{\ominus} \xrightarrow{+\,{}^{\oplus}Z^{\oplus}} RA_mB_{n/2}\text{--}Z\text{--}B_{n/2}A_mR$$

Combinations of block polymerizations and block couplings for the synthesis of triblock polymers usually deliver more diblock polymers as byproducts than diblock syntheses via block polymerizations with bifunctional initiators.

Degrees of Polymerization

The *number-average degree of polymerization*, $\overline{X}_n$, of a diblock polymer RA_nB_mR' from two molecularly inhomogeneous blocks $I \equiv RA_n$ and $II \equiv B_mR'$ is simply the sum of the two number averages $\overline{X}_{I,n}$ and $\overline{X}_{II,n}$ of degrees of polymerization of the two blocks I and II, i.e., $\overline{X}_n \equiv \overline{X}_{I-II,n} = \overline{X}_{I,n} + \overline{X}_{II,n}$.

If N_I blocks I and N_{II} blocks II with the same number-average degree of polymerization of both blocks I and II are coupled via *their ends*, the number-average degree of the resulting multiblock polymer is therefore $\overline{X}_n = N_I\overline{X}_{I,n} + N_{II}\overline{X}_{II,n}$. If the *ends* of N_i monofunctional blocks with number-average degrees of polymerization, $\overline{X}_{I,n}$, are coupled to $f = \Sigma_i\,N_i$ sites of one f-functional nucleus, the resulting star polymer has a number-average degree of polymerization of

$$(16\text{-}10) \qquad \overline{X}_n = \Sigma_i\,N_i\overline{X}_{I,n,i}$$

which is independent of the type of distribution of the degrees of polymerization (logarithmic normal, Schulz-Zimm, Wesslau, etc.).

Eq.(16-10) does *not* apply to *side-by-side* couplings (see also the discussion of self-association in Volume III. The equations given in Volume III apply to both physical and chemical couplings!). The reason for the difference between end-to-end and side-by-side couplings is the following. In couplings of ends, the number of coupling sites per primary block (or primary molecule in associations) is constant and independent of the chain length. In side-by-side couplings, however, the number of coupling sites is proportional to the degree of polymerization and the coupling is weighted according to the length of the molecule. It therefore depends on the type of distribution of X or M.

For side-by-side couplings of N unimolecular blocks of type I, the *number-average degree* of polymerization of these block in the polymer is given by

(16-11) $\overline{X}_n = \overline{X}_{I,n} + (N-1)\overline{X}_{I,w}$

The *weight-average degree* of polymerization of coupling products of molecularly inhomogeneous blocks is more difficult to calculate. Suppose three different types of blocks I, II, and III with number-average degrees of polymerization of $\overline{X}_{I,n}$, $\overline{X}_{II,n}$, and $\overline{X}_{III,n}$, weight-average degrees of polymerizations of $\overline{X}_{I,w}$, $\overline{X}_{II,w}$, and $\overline{X}_{III,w}$, and mole fractions of $x_I \equiv \overline{X}_{I,n}/\overline{X}_n$, $x_{II} \equiv \overline{X}_{II,n}/\overline{X}_n$, and $x_{III} \equiv \overline{X}_{III,n}/\overline{X}_n$. The blocks are coupled via their *ends* to either linear triblock polymers or star molecules with three arms. The weight-average degree of polymerization of the triblock polymer is

(16-12)

$$\overline{X}_w = x_I\overline{X}_{I,w} + x_{II}\overline{X}_{II,w} + x_{III}\overline{X}_{III,w} + 2(x_Ix_{II} + x_Ix_{III} + x_{II}x_{III})(\overline{X}_{I,n} + \overline{X}_{II,n} + \overline{X}_{III,n})$$

which is independent of the type of distribution. For N blocks with the same distributions of X, Eq.(16-12) reduces to $\overline{X}_w = \overline{X}_{I,w} + (N-1)\overline{X}_{I,n}$ since $\overline{X}_n = \sum_i N_i\overline{X}_{n,i}$.

For *side-by-side* couplings of blocks with identical types of distributions, one obtains $\overline{X}_w = N\overline{X}_{I,w}$, independently of the type of distribution. Note that the corresponding expression for the number-average does depend on the type of distribution type!

z-averages differ for end-to-end (E-E) and side-by-side (S-S) couplings, assuming the same distributions of blocks ($\overline{X}_{I,n} = \overline{X}_{II,n} = \overline{X}_{III,n} = ...$, $\overline{X}_{I,w} = \overline{X}_{II,w} = \overline{X}_{III,w} = ...$, etc.):

(16-13) $\overline{X}_z = \overline{X}_{I,z} + (N-1)\overline{X}_{I,n}\{1 + [2\overline{X}_{I,w} - \overline{X}_{I,n} - \overline{X}_{I,z}][\overline{X}_{I,w} + (N-1)\overline{X}_{I,n}]^{-1}\}$ (E-E)

(16-14) $\overline{X}_z = \overline{X}_{I,z} + (N-1)\overline{X}_{I,w}$ (S-S)

16.2.3   Comb Polymers

Polymerization of macromonomers (p. 3) leads to *regular* comb macromolecules with side chains of the same length at each monomeric unit.

Irregular comb macromolecules are obtained by graft polymerization, usually with relatively short side chains. Grafting of one type of monomer leads to **graft polymers** and grafting of two or more monomers to **graft copolymers**. Grafting may proceed *from* an active site at the backbone or *to* the backbone from active molecules that will become side chains (Fig. 16-4). Main chains and side chains usually have different constitutions.

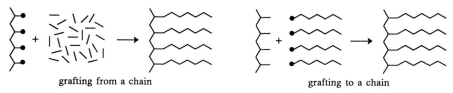

grafting from a chain   grafting to a chain

Fig. 16-4   Schematic representation of grafting. ● Active site, e.g., a radical or a functional group.

Copolymerization of conventional monomers and macromonomers generates **pseudo-graft copolymers**. If the endgroup R of a sidechain ~~~R of a pseudo-graft polymer is a reactive group, it may be used for further grafting to true graft polymers.

The primary polymer chain, poly(A), is called the **main chain, backbone**, or **graft-(ing) substrate**. The newly introduced segments with monomeric units B are referred to as **side chains** or **grafts**.

Grafting of monomer B on substrates poly(a) usually leads to mixtures of ungrafted primary polymers poly(a), graft polymers (poly(a)-*graft*-poly(b)), and sometimes also homopolymers poly(b). Grafting is therefore characterized by four parameters: **success, yield, degree**, and **intensity** of grafting.

Suppose the grafting of B on the substrate poly(a) delivered a product that contains a weight fraction $w_a = 1 - w_b = 0.40$ of a-units. Chromatography or preparative fractionation indicated the following weight fractions: $w_a = 0.10$ of ungrafted poly(a), $w_{agb} = 0.70$ of poly(a)-*graft*-poly(b), and $w_b = 0.20$ of homopolymer B. Therefore, the graft polymer contained a weight fraction $w_{bg} = w_B - w_b = 0.60 - 0.20 = 0.40$ of all b-groups present in the product and a weight fraction $w_{ag} = w_A - w_a = 0.40 - 0.10 = 0.30$ of all a-groups. The grafting is thus characterized by the following parameters:

Success of grafting w_{ag}/w_a ;   Example: w_{ag}/w_a = 0.30/0.40 = 0.75
Degree of grafting w_{bg}/w_a ;   Example: w_{bg}/w_a = 0.40/0.40 = 1.00
Yield of grafting w_{bg}/w_b ;   Example: w_{bg}/w_B = 0.40/0.60 = 0.667
Intensity of grafting w_{bg}/w_{ag} ;   Example: w_{bg}/w_{ag} = 0.40/0.30 = 1.33

Grafting from Main Chains

Grafting from the main chain (backbone) is the most common method since backbone chains contain grafting sites either per monomeric units *per se* or per post-reactions including irradiation or radical transfer.

For example, hydroxy groups of cellulose start the polymerization of ethylene imine (aziridine) which results in graft polymers, cell–O–$(CH_2CH_2NH)_nH$.

Active sites may be introduced in backbones by *post-reactions* of polymer chains. For example, the reaction of poly(vinyl alcohol), $+CH_2-CH(OH)+_n$, with Ce^{4+} generates monomeric units with the radical site $-CH_2-^\bullet C(OH)-$ that can induce free-radical graft polymerizations of other monomers. The reaction is fairly mild and does not lead to severe chain degradation.

Grafting sites can also be obtained by *copolymerization* of monomers with small proportions of a comonomer that contains reactive groups or groups that can be activated. Such comonomers may be macromonomers, i.e., low-molecular weight polymer chains with one or two functional groups, for example, $CH_2=CH$~~~R.

Reactive sites can also be produced by *irradiation*. However, only a few polymers can be activated directly by UV light. Others are degraded, for example, poly(vinyl methyl ketone). Grafting by irradiation is also accompanied by homopolymerization.

Formation of polyradicals by γ-rays is also not specific. Radicals here are produced in both amorphous and crystalline (if any) regions. At temperatures $T \leq T_G$, resulting radical sites may combine in amorphous regions, leading to crosslinking. Irradiation in presence of monomers delivers both monomer radicals and backbone radicals and thus both grafting to and from the main chain as well as homopolymerization.

Unwanted homopolymerization can be reduced if certain polymer-monomer pairs are used. Halogen compounds have high $G(R^\bullet)$ values and aromatic compounds small ones (Section 11.3.1). Thus, irradiation of poly(vinyl chloride) in the presence of styrene delivers graft polymers of styrene on poly(vinylchloride) in high yields but only a little poly(styrene).

Irradiation of backbone chains in the presence of grafting monomer is required if macroradicals combine easily. If the probability of combination of backbone radicals is small, one can also irradiate the backbone polymer alone and then add the monomer. In this case, the temperature has to be chosen in such a way that the rate of deactivation of radicals is lower than that of initiation of monomer polymerization. For this reason, grafting yields are especially high near the glass temperature where polymer segments are not very mobile and thus radicals are unlikely to combine.

A universal method is grafting by free-radical *chain transfer*. Here, polymer radicals $P^\bullet$ and/or initiator radicals $I^\bullet$ abstract atoms from polymer molecules, for example, Cl from poly(vinyl chloride), $+CH_2-CHCl\rightarrow_n$. The resulting macroradicals $+CH_2-{}^\bullet CH\rightarrow_n$ initiate the polymerization of the added monomer.

However, chain transfer constants of macroradicals are relatively small (p. 349) which leads to small grafting yields. Polyradicals of the main chain are therefore more effectively generated by transfer of initiator radicals, i.e., by high initiator concentrations. Not all initiator molecules are suitable for all polymers, however. For example, AIBN produces radicals on poly(vinyl acetate) but not on poly(styrene) (see also p. 349).

Whether or not the resulting polyradicals initiate the polymerization of added monomers depends on the resonance stabilization and the polarity of polyradicals and monomers (p. 410). This method is used industrially in the graft copolymerization of styrene and acrylonitrile on elastomeric poly(butadiene-*co*-acrylonitrile) (ABS polymers) (see Volumes II and IV).

Grafting to Main Chains

Grafting to chains is used if grafting from chains is impractical or impossible. It usually involves the exposure of the backbone chain to simultaneously introduced monomers and reaction carriers such as radicals. Grafting to chains always applies drastic reaction conditions. It can therefore be used for many different polymers but it is also very unspecific. Chain degradations can often not be avoided and the grafting itself may be questionable.

16.2.4   Star Polymers

Star polymers can be considered the simplest type of branched polymers. Their molecules $Q(A_nR)_f$ consist of a core Q from which $f \geq 3$ *linear* arms $-A_nR$ with endgroups R radiate (Fig. 2-5). The core Q may be a single atom such as $-N<$ ($f = 3$), derived from simple molecules such as $C(CH_2OH)_4$ ($f = 4$) or C_6H_6 ($3 \leq f \leq 6$), or a large entity such as a latex particle with very many connection points ($f \gg 3$). Arms $-A_nR$ may consist of like or unlike monomeric units A and they may be of equal or unequal length n per arm (Section 2.5.4).

Attaching of Arms to a Core

Star polymers can be synthesized by two strategies: attaching pre-existing arms to a core or growth of arms from a core. The attachment of f arms of type A_nR with one reactive A-site per molecule to a core Q

(16-15) $f\,RA_n\ +\ Q_f\ \rightarrow\ (RA_n)_fQ_f$

is a special case of block coupling. There are four possibilities:

Case I   All core sites react with arms of equal length.
Case II   All core sites react but with arms of unequal length.
Case III   Arms of equal length react with some but not all of the core sites.
Case IV   Arms of unequal length react with some but not all of the core sites.

Case I leads to "monodisperse" star molecules, i.e., molecules with the same number and lengths of arms.

Case II. If all arms have the same chance to react with the f active sites of the core, then the resulting star polymers will have a molecular weight distribution because of the unequal lengths of the arms. However, the molecular weight distribution of the star molecules is *more narrow* than the molecular weight distribution of the arms before the coupling process since a distribution of a few arms is always narrower than a distribution of many arms. According to the equations in Section 16.2.2 (p. 612), one obtains from $\overline{X}_n= f\overline{X}_{I,n}$ and $\overline{X}_w = \overline{X}_{I,w} + (f-1)\overline{X}_{I,n}$ that $\overline{X}_w/\overline{X}_n = 1 + (1/f)[(\overline{X}_{I,w}/\overline{X}_{I,n}) - 1]$. A polymolecularity $\overline{X}_{I,z}:\overline{X}_{I,w}:\overline{X}_{I,n} = 3:2:1$ of arms is thus reduced to $\overline{X}_z:\overline{X}_w:\overline{X}_n = 6:5:4 = 1.5:1.25:1$ for star molecules with $f = 4$.

Case III. The star polymer will have a *broader* molecular weight distribution than the arm polymer before the attachment if the f core sites are not completely reacted with f "monodisperse" arms.

Case IV. In reality, arm polymers $Q(A_nR)_f$ are never really molecularly homogeneous and core sites almost never react completely. The two effects II and III then cancel each other to some extent and the resulting star polymer may have the same breadth of molecular weight distribution as the arm polymer (if II and III cancel each other completely), a narrower one (if Case II dominates), or a broader one (if Case III dominates).

The situation is much more complicated for star polymers with miktoarms (p. 52) as one can see from Eq.(16-13).

Growth of Arms from a Core

f-functional core molecules, Q_f, with $f \geq 3$ active sites can seed the growth of arms by polycondensation with bifunctional monomer molecules A–U–B, where A and B are leaving groups and A can react with B but neither A nor B can react with themselves. The same scheme applies to polyadditions where A and B react to a group that is incorporated into the polymer. Essentially only f-star molecules $Q(U_iA)_f$ will be obtained if the initial amount of A–U–B is much greater than that of $Q(A)_f$ and *practically* all B-groups have reacted. Note that an equilibrium reaction can never yield *exactly* 100 % f-star molecules because there can neither be $x_{AB} = 1$ nor $x_L = 0$ in Eq.(13-24).

The following discussion assumes that QA_f and i A–U–B are completely present at the start of the reaction (portion-wise additions are more complicated). At less than complete reaction of B-groups, the resulting polymer consists of a mixture of linear molecules with one arm, $A_{f-1}QU_iA$, and two arms, $A_{f-2}Q(U_iA)_2$, and star molecules with $f \geq 3$ arms. The probability that a certain *arm* contains i units U is $P_i = p^i(1-p)$ where p is the extent of reaction of B-groups. The arms may be of different lengths so that the polymer has a total of $j = 1 + i_1 + i_2 + i_3 + \ldots$ units, counting the central unit Q as 1. The probability that the f arms have lengths of $i_1, i_2 \ldots$ is thus

(16-16) $\qquad P = p^{i_1} p^{i_2} \ldots (1-p)^f = p^{j-1}(1-p)^f$

The probability that a certain star *molecule* contains U-units in all of its arms j equals the mole fraction x_j of j-mers. $x_j = P/N_{star}$ is given by the ratio of the probability P and the number N_{star} of combinations that satisfy Eq.(16-17). Since there are $(j-1) + (f-1) = j + f - 2$ possibilities for the $j-1$ monomeric units that are distributed among $f-1$ core sites, the number N_{star} becomes $N_{comb} = (j+f-2)!/(f-1)!(j-1)!$. The mole and weight fractions of the j units are therefore given by

(16-17) $\qquad x_j = \dfrac{(j+f-2)!}{(f-1)!(j-1)!} p^{j-1}(1-p)^f \ ; \qquad w_j = \left[\dfrac{j(j+f-2)!}{(f-1)!(j-1)!} \right] p^{j-1} \left[\dfrac{(1-p)^{f+1}}{1-p(f-1)} \right]$

The number-average and weight-average degrees of polymerizations are calculated as

(16-18) $\qquad \overline{X}_{R,n} = \dfrac{1+(fN_{port}-1)p}{1-p} \ ; \qquad \overline{X}_{R,w} = \dfrac{1+(fN_{port}-1)^2 p^2 + (3fN_{port}-2)p}{(1-p)[1+(fN_{port}-1)p]}$

where N_{port} is the number of portion-wise additions of A–U–B molecules. These equations apply to batch reactors, continuous flow reactors, and homogeneous stirred tank reactors. For one-pot polymerizations, they reduce to Eqs.(13-46). Calculated number-average and weight-average degrees of polymerization of reactants, $\overline{X}_{R,n}$ and $\overline{X}_{R,w}$, as well as polymolecularity indices, $\overline{X}_{R,w}/\overline{X}_{R,n}$, are shown in Table 16-1.

Table 16-1 Theoretical number-average and weight-average degrees of polymerization of reactants R and polymolecularity indices, $\overline{X}_{R,w}/\overline{X}_{R,n}$, of polymers from the reaction of monomers A–U–B without (–) or with addition of a small proportion of a f-functional core molecule $Q(ZA)_f$ after an extent of reaction of B-groups of $p = 0.99$ (data of [2]-[4]). Data are given for reactions in batch reactors BR with single addition ($N_{ab} = 1$) or successive ($N_{ab} > 1$) additions of A–U–B to $Q(ZA)_f$. HCSTR: reaction in a homogeneous continuously stirred tank reactor.

Monomers	f	N_{port}	Reactor	$\overline{X}_{R,n}$	$\overline{X}_{R,w}$	$\overline{X}_{R,w}/\overline{X}_{R,n}$	Equation
A–U–B	-	1	BR	100	199	1.99	(13-46)
A–U–B + QA	1	1	BR	100	199	1.99	(16-18)-(16-19)
A–U–B + QA$_2$	2	1	BR	199	≈ 299	≈ 1.50	(16-18)-(16-19)
A–U–B + QA$_4$	4	1	BR	397	≈ 497	≈ 1.25	(16-18)-(16-19)
A–U–B + QA	1	10	BR	991	$\approx 1\ 091$	≈ 1.10	(16-18)-(16-19)
A–U–B + QA$_2$	2	10	BR	1981	$\approx 2\ 081$	≈ 1.05	(16-18)-(16-19)
A–U–B + QA$_4$	4	10	BR	3961	$\approx 4\ 061$	≈ 1.025	(16-18)-(16-19)
A–U–B	-	continuous	HCSTR	100	$\approx 19\ 800$	≈ 198	(13-47)

Table 16-1 shows that for one-pot polymerizations to high monomer conversions in batch reactors the number-average degree of polymerization is ca. $(f+1)/2$ times greater than that of homopolymers (A–U–B), regardless of the functionality f of the core molecule QA_f. For a large number of portion-wise additions, it becomes slightly less than $N_{port}f$ times as large so that considerably higher degrees of polymerizations can be obtained at the same extent of reaction of A–U–B molecules if a small proportion of a seed molecule is used. This is even true for linear polymers prepared with QA or QA_2.

At the same conditions, polymolecularities calculated from

$$(16\text{-}19) \qquad \frac{\overline{X}_{R,w}}{\overline{X}_{R,n}} = 1 + \frac{fN_{port}\,p}{[1+(fN_{port}-1)p]^2} \approx 1 + \frac{1}{fN_{port}p}$$

are predicted to become much more narrow since the statistical distribution of chain lengths is now intramolecular instead of intermolecular as in linear polymerizations. The same phenomenon is observed for the closed self-association of polymolecular polymers P to micelles P_N according to $N\ P \rightleftarrows P_N$ (Volume III). Note that the polymolecularity index becomes $\overline{X}_{R,w}/\overline{X}_{R,n} \approx 1 + (1/f)$ for $N_{port} = 1$ and $p \to 1$.

Experimentally, polymolecularity indices of linear molecules QU_iA assume the predicted value of $\overline{M}_{R,w}/\overline{M}_{R,n} = 2$ for the whole molecular weight range (Fig. 16-5). Linear molecules $Q(U_iA)_2$ follow the predictions for low molecular weights but not for higher ones. Deviations from theory increase with increasing number of arms of star molecules; they are more pronounced at higher weight-average degrees of polymerization. It is not clear whether these deviations result from the approximations made in the calculation of molecular weights $\overline{M}_{R,w}$ and $\overline{M}_{R,n}$ or are due to an increased crowding of arms at the core that prevents a truly random propagation at the various arms of the same molecule.

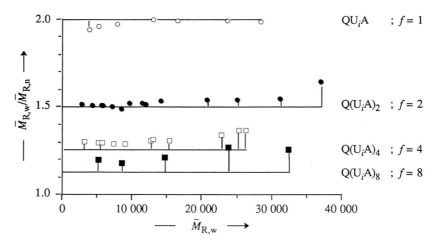

Fig. 16-5 Polymolecularity index as a function of the weight-average molecular weight for polymers from the hydrolytic polymerization of ε-caprolactam (A–U–B monomer) with various proportions of the A_f monomers stearic acid ($f = 1$), sebacic acid ($f = 2$), cyclohexanone tetrapropionic acid ($f = 4$), and dicyclohexanone octapropionic acid ($f = 8$) [5]. Molecular weights $\overline{M}_n$ and $\overline{M}_w$ were not determined independently by physical methods but rather calculated from amino endgroups and the molar ratios of components. Vertical lines indicate theoretical polymolecularity indices for $p_B \to 1$.

16.2.5 Dendrons and Dendrimers

Dendrimers can be viewed as "completely ordered hyperbranched polymers" (Section 13.8.2) or "star molecules with orderly branch-upon-branch" topology (Section 2.5.5). They are usually prepared by repeated controlled step reactions, either from the core outwards (divergent synthesis) or from surface groups inwards (convergent synthesis). Both methods employ molecules with protected functional groups that have to be removed before the synthesis of the next generation. All syntheses are in solution, probably because wide-pored carriers for Merrifield-type syntheses are unavailable.

Divergent Syntheses

In these syntheses, core molecules QB_g are reacted with an excess of monomer molecules $A–V–(BZ)_{f-1}$. B and A are leaving groups (not functional groups in the common meaning of the term). B-groups of monomers are protected by groups Z so that A-groups of monomers can react only with B-groups of core molecules and not with B-groups of other monomer molecules. The functionalities f and g are usually 3 or 4.

The reaction of QB_g with $A–V–(BZ)_{f-1}$ delivers $Q[V(BZ)_{f-1}]_g$ and BA. Protecting groups Z are then removed from the first generation products. The resulting unprotected molecules $Q[VB_{f-1}]_g$ are subsequently reacted with a excess of molecules $A–V(BZ)_{f-1}$. The protecting groups are removed, etc. In this way, dendrimer molecules with more and more generations can be built (Fig. 16-6). Divergent syntheses are **exponential-growth syntheses** since the addition of a new generation of branching molecules increases the power i of the multiplicity F of the branch units by one.

Three different "degrees of branching" can be defined, depending on whether the core (⊛ in Fig. 16-6) and/or the surface units ($–V(BZ)_3$ in Fig. 16-6) are counted as a branching unit, i.e., whether "branching" relates to the process or to the constitution:

(16-20) $N_V = F_{core}(F^i - 1)/(F - 1)$ Degree of polymerization (number of V units)

(16-21) $N_{br,0} = 1 + N_V$ number of branch junctures (incl. core and surface units)

(16-22) $N_{br} = N_{br,0} - (N_{end}/F)$ degree of branching (excl. surface units)

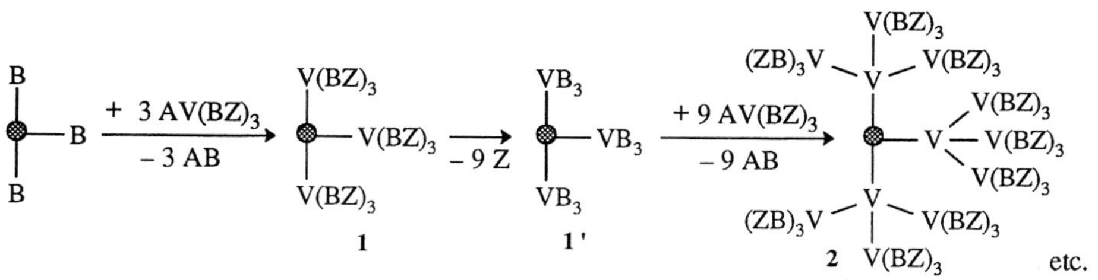

Fig. 16-6 Schematic representation of a divergent synthesis of dendrimers with branch units V of multiplicity $F = 3$ and a core of multiplicity $F = 3$. Multiplicities indicate the number of units that can be added to form a new generation (p. 53).
Endgroups (B or BZ) number $N_{end} = 3 \cdot 3 = 9$ in dendron **1**; $3 \cdot 3^2 = 27$ in dendron **2**, $3 \cdot 3^3 = 81$ in dendron **3**, etc., thus $N_{end} = F_{core}F^i$. Numbers of surface units (VB_3 or $V(BZ)_3$) are $N_{end}/F = 3$ (dendron **1**), 9 (dendron **2**), 27 (dendron **3**), etc. Numbers of V-units are calculated as $N_V = 3$ (dendron **1**), $3 + 3^2 = 12$ (dendron **2**), $3 + 3^2 + 3^3 = 39$ (dendron **3**), etc.

Depending on the constitution of the core and monomer molecules, many variations of divergent syntheses are possible. Protecting groups are sometimes not necessary. For example, the synthesis of commercial PAMAM dendrimers (Section 2.5.5) starts with the reaction of 1,2-diaminoethane (= ethylene diamine) $H_2NCH_2CH_2NH_2$ (instead of NH_3 as in Fig. 2-8) with methyl acrylate $CH_2=CH-COOCH_3$ to form the tetrafunctional molecule $(H_3COOC-CH_2CH_2)_2NCH_2CH_2N(CH_2CH_2-COOCH_3)_2$. The $-COOCH_3$ groups are then converted into amidoamine units $-CONHCH_2CH_2NH_2$ by reaction with a large excess of 1,2-ethylene diamine. The resulting tetramine is again reacted with methyl methacrylate, etc., leading to higher and higher generations of amidoamine dendrimers with an increasing number of amine end-groups and $-NHCH_2CH_2NH_2$ surface groups.

Another commercial product adds acrylonitrile, $CH_2=CHCN$, to a primary amine, RNH_2. The resulting dinitrile $RN[CH_2CH_2CN]_2$ is hydrogenated to the core molecule $RN[CH_2CH_2CH_2NH_2]_2$ with 4 possible branching sites. The core molecule is reacted with acrylonitrile to build the first generation dendrimer, etc.

However, it is not possible to prepare dendrimers with an "infinite" number of generations because the surface groups of higher generation dendrimers become so crowded that some proportion cannot react with new monomer molecules. It has been calculated that the maximum number of generations of dendrimers of Section 2.5.5 is given by $N_{max} = 1 + 2.88 (1.5 + \ln N_u)$ where N_u is the number of monomeric units between two branch points. For $N_u = 2$, one obtains $N_{max} \approx 7.3$, i.e. a maximum of ca. 7 generations. No perfect dendrimers with generations higher than 10 seem to have been reported.

Convergent Syntheses

Convergent methods work from surface cells S toward the core, building successively dendrons with increasing multiplicity. They are less stringent than divergent methods but are usually only applied for the synthesis of dendrimers with a few generations.

For example, two molecules SA are reacted with a trifunctional branching molecule B_2VB' to the dendron S_2VB' (1) (Fig. 16-7). The functional group B' of S_2VB' is then converted into a functional group A. Two of the resulting dendrons S_2VA (molecules 1') are then reacted with a branching unit B_2V_2B' to the second generation dendron 2 with four surface cells S (for numbers N_{end} and N_V, see Table 16-2).

The synthesis of poly(benzyl ether) dendrimers is an example. First, two molecules $R(p-C_6H_4)CH_2Br$ are condensed with 1 molecule $(HO)_2(i-C_6H_3)CH_2OH$ to the branched $[R(p-C_6H_4)CH_2O_2]_2(i-C_6H_3)CH_2OH$ (1) that is converted to the bromide (1'). 2 molecules of 1' react with 1 molecule of $(HO)_2(i-C_6H_3)CH_2OH$ to form the dendron 2, etc.:

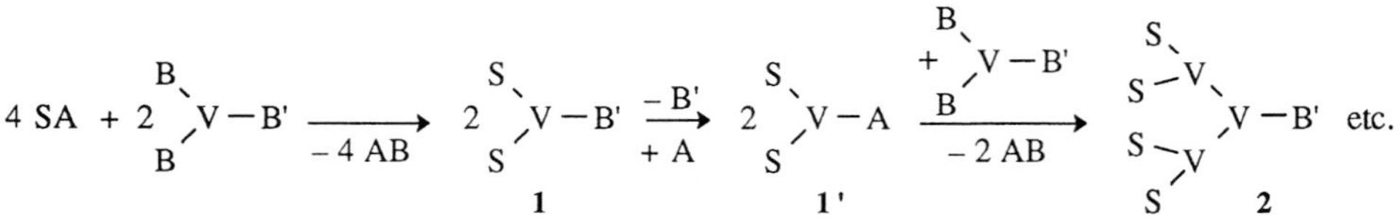

Fig. 16-7 Schematic representation of a convergent dendrimer synthesis. Branching molecules B_2VB' are first provided with surface groups S to form **1** which is converted to **1'**. Reaction of two branching molecules **1'** with one branching molecule B_2VB' delivers the dendron **2** with 4 surface groups S and 3 branch units V. Subsequent reactions result in dendron **3** (not shown) with the structure $[(S_2V)_2V]_2VB'$ with 8 surface groups S and 7 branch units V, etc. (see Table 16-2).

Table 16-2 Comparison of the numbers N_{end} of end-groups and N_V of branching units in ideal dendrons with different numbers i of generations and various multiplicities F_{core} of core molecules and F of branching molecules by divergent and convergent strategies. na = not applicable.

Example	Strategy	F_{core}	F	N_{end}			N_V		
				$i=1$	$i=2$	$i=3$	$i=1$	$i=2$	$i=3$
Fig. 16-5	divergent	3	3	9	27	81	3	12	39
Fig. 2-7	divergent	3	2	6	12	24	3	9	21
Fig. 16-6	convergent	2	2	4	8	16	3	7	15
Fig. 16-7	convergent, double-exponential	na	2	4	16	64	3	17	49

Double-exponential convergent syntheses start with molecules $ZAVB_2$ (**1**) with a functional group A protected by a group Z and molecules $AV(BZ')_2$ (**1'**) with functional groups B protected by Z'. Reaction of **1** with 2 molecules of **1'** generates the dendron **2** with 3 branch units V (Fig. 16-8). Removal of Z from **2** leads to **2'** and removal of Z' to **2"**. The partially unprotected compounds **2'** and **2"** are then reacted to the fully protected dendron **3**. The procedure is repeated for the synthesis of dendron **4** (not shown), etc. The dendrons are finally coupled to a core molecule with a functionality of $f^* \geq 2$ to form the desired dendrimer.

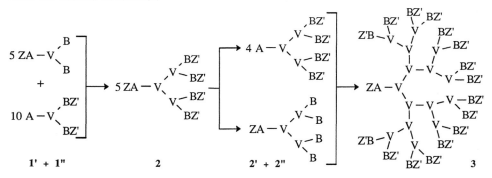

Fig. 16-8 Double-exponential convergent synthesis. A, B = reactive groups; V = branch units; Z, Z' = protecting groups. The numbers of endgroups and branch units increase rapidly (Table 16-2).

16.3 Straight Hierarchical Structures

16.3.1 Ladder Polymers

Ladder polymers (**double-strand polymers**) consist of linearly fused rings such as pyrone and BBB (p. 474) or two parallel chains that are interconnected at regular distances (Fig. 16-9). The simplest structures of the latter are composed of six-membered rings that are joined in 1,2:4,5 positions. They can be synthesized by different strategies, for example, **intermerar cyclization**. Examples are the chain polymerization of vinyl groups of 1,2-poly(butadiene) (I) or nitrile groups of poly(acrylonitrile) (II) and the polycondensation of methylketo groups of poly(vinyl methyl ketone) (III).

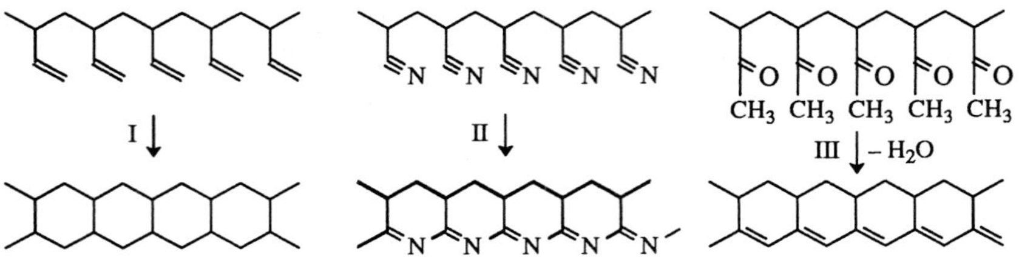

Fig. 16-9 Synthesis of ladder polymers by intermerar-intramolecular polymerization.

This strategy delivers complete cyclization of, for example, poly(vinyl methyl ketone) if (a) monomer units are exclusively present in head-to-tail position, (b) intermolecular reactions are absent, and either (c) equilibrium reactions are driven to 100 % or (d) kinetically controlled reactions start exclusively at one chain end and continue in a zip reaction to the other end, a condition that is never fulfilled. In kinetically controlled reactions III of head-to-tail polymers a fraction $f = 1/(2\ e) \approx 0.184$ of substituents never react (Eq.(15-17)). Polymers with head-to-head and tail-to-tail structures yield only maximum conversions of 50 % (Eq.(15-18)) and no ladder structures at all.

A second strategy employs multifunctional monomers that are converted into ladder polymers by polyaddition or polycondensation. This strategy succeeds if the primary bond formation is followed by an intramolecular ring-closure reaction. Such intramolecular reactions are favored if (a) ring formation is favored and/or (b) resonance energy is gained. Examples are

(16-23)

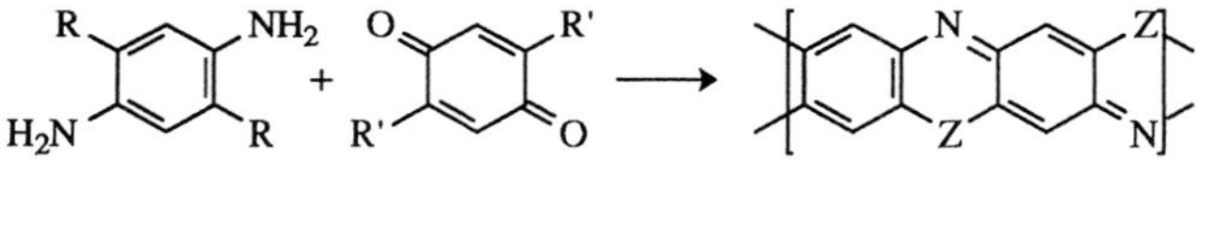

$R = NH_2$, OH, SH, Cl
$R' = NH$, Cl, $OCOCH_3$
$Z = NH$, O, S

These step reactions may proceed as either AB reactions or AA/BB reactions. An example of an AB reaction is the Diels-Alder self-addition of cyclopentadiene (Eq.(13-8)). However, the Diels-Alder "(co)polymerization" of 2-vinylbutadiene with *p*-quinone, a repetitive cycloaddition, is the AA/BB polyaddition of a trisdiene and a bisdienophile:

(16-24)

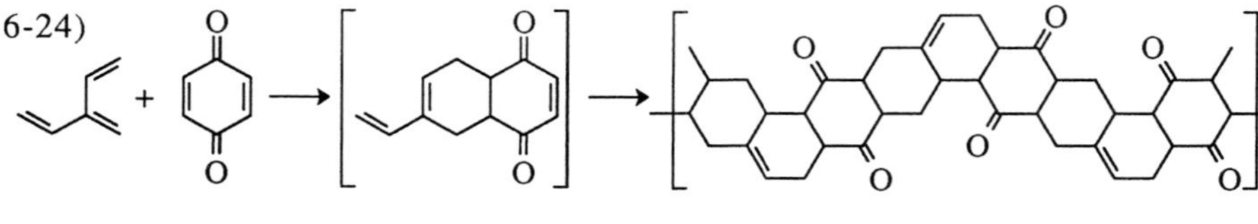

The reaction may produce branched and crosslinked polymers, especially at higher monomer conversions (higher degrees of polymerization, similar to cyclopolycondensations to polyimides and polybenzimidazoles (Section 13.6.2). Another problem is the decrease of solubility with increasing degree of polymerization which can sometimes be prevented by working with substituted monomers.

Fully aromatic ladder polymers are of interest because their conjugated ring systems are expected to lead to outstanding optical and electronic properties such as photo-luminescence and electroluminescence (Volume IV). However, all attempts to aromatize polymers such as the one shown in Eq.(16-24) have failed.

Fully aromatic polymers must therefore be synthesized by coupling aromatic monomers to polymers with subsequent ring-closure to ladder structures. The problem here is to select from the many possible organic reactions those that proceed to high yields with a minimum of side reactions.

An example is the AA-type coupling of aromatic dibromo diketones to form a soluble linear polyketone followed by a carbonyl olefination with the help of B_2S_3 (Fig. 16-10). The condensation products precipitate from solution despite the long side groups that usually promote solubility.

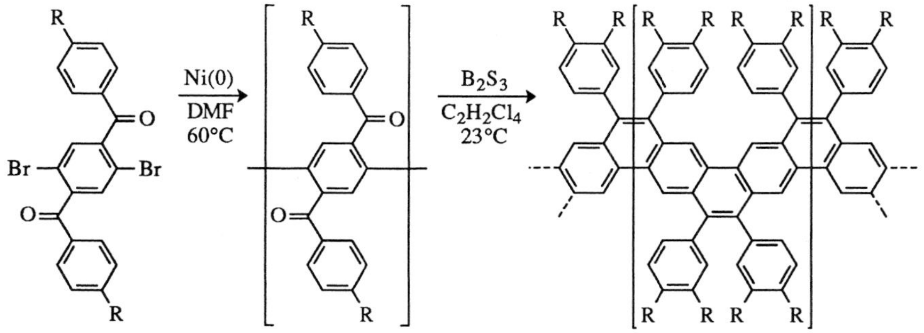

Fig. 16-10 Preparation of fully aromatized ladder polymer with $R = C_{10}H_{21}$ and SEC-based number-average molecular weights of 3700-4100 (see p. 80!) [6]. The coupling agent B_2S_3 is formed *in situ* from BCl_3 and $[(C_6H_{11})_3Sn]_2S$. Molecular weights correspond to degrees of polymerization of 3-4.

Stable **pseudo-ladder** structures are formed by supramolecular polymers provided that they have a regular array of four hydrogen bonds as shown, for example, in Fig. 16-11, left. They show the typical high melt viscosity of true polymers but dissociate back to the monomer at elevated temperatures. Supramolecular structures with a regular array of three hydrogen bonds show liquid crystalline behavior (p. 25). Crosslinked polymers result if the hydrogen bonds are irregular (Fig. 16-11, right). Such polymers may be useful for hot melt adhesives and coatings since they have low viscosities for processing but good mechanical properties as solids.

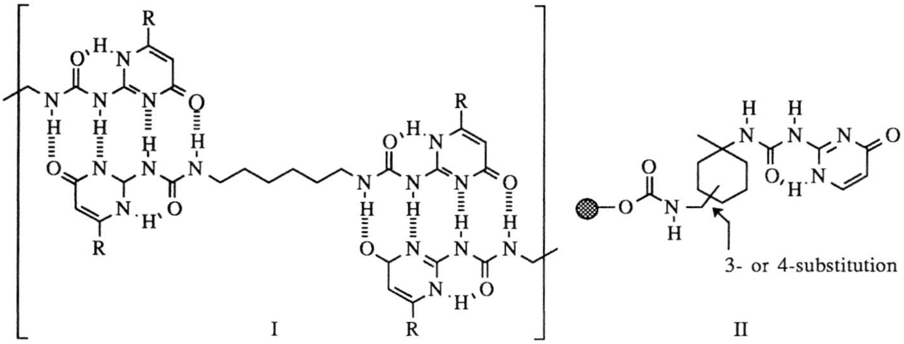

Fig. 16-11 Stable linear suprapolymer (I) and a network-forming suprapolymer (II) [7].

16.3.2 Layer Polymers

The repetition of ladder structures in two dimensions leads to **layer polymers (parquet polymers)**. Well-known examples are layers of graphite, $[C]_n$, and silicon monochloride, $[SiCl]_n$. These polymers are usually depicted as single-layer structures (Fig. 16-12) although they are conventionally obtained only as multilayers.

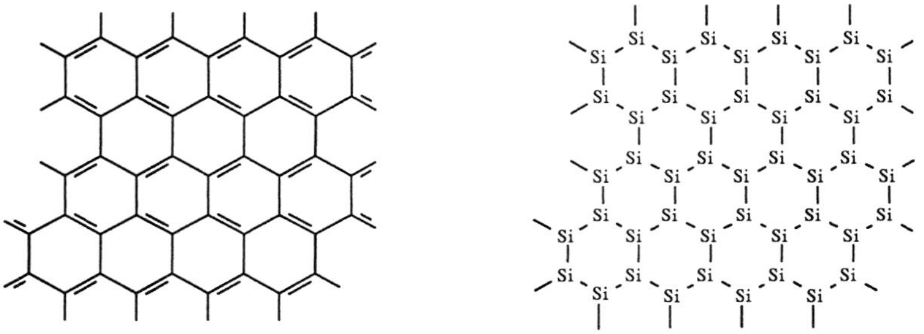

graphene

silicon monochloride
(each Si atom is substituted with one Cl atom)

Fig. 16-12 Structures of the sheet polymers graphene, $[C]_n$, and silicon monochloride, $[SiCl]_n$.

The *material* **graphite** consists of stacked layers of carbon molecules called **graphene**. Planar graphene sheets with thicknesses between one and several carbon atoms have been made by peeling off layers from highly oriented pyrolytic graphite. At ambient temperatures, these sheets transport electrons at ultrafast speeds. The diameters and thicknesses of these graphene sheets cannot be readily controlled and there have therefore been several attempts to produce layer polymers, usually called **sheet polymers**.

The terms layer, sheet, films, etc., are either not well or differently defined in science and technology. Science often uses "film" for layers that are a few atoms thick (example: monomolecular film). ASTM defines "film" as a two-dimensional entity with thicknesses between 40 μm and 400 μm and as "sheetings" or "sheets" those with thicknesses between ca. 250 μm and 1500 μm.

One strategy involves the synthesis of dendrimers with a high density of aromatic units as precursors for large graphene-like structures. An example is the 1st generation dendrimer **3** of Fig. 16-13 with a biphenyl core that could be successfully converted into a 2nd generation dendrimer **4**.

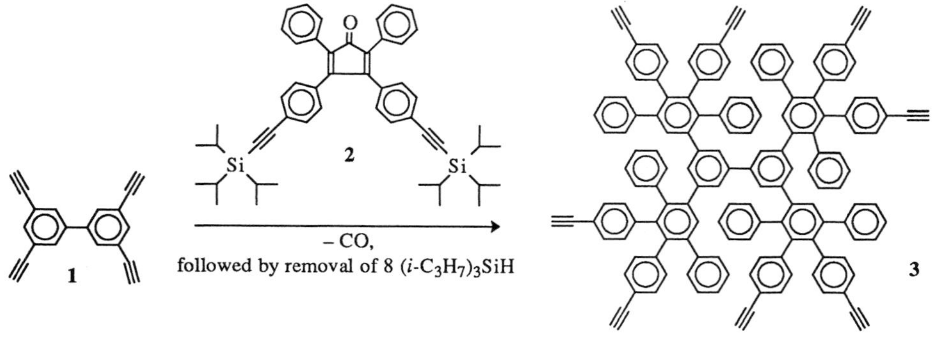

$-$ CO,
followed by removal of 8 $(i\text{-}C_3H_7)_3SiH$

Fig. 16-13 Preparation of the first generation of a graphene-like aromatic dendrimer [8].

The first step of the synthesis of **3** is a [2+4]-cycloaddition reaction of the free acetylene groups of the core molecule **2** and the diene function of a 3,4-bis(4-triisopropylsilylethynylphen-1-yl)-2,5-diphenylcyclopenta-2,4-dienone **1** (Fig. 16-12). The elimination of carbon monoxide is followed by the removal of the protecting triisopropylsilyl groups. The resulting 1st generation (G-1) dendrimer **3** with free acetylene groups can then be reacted with 8 molecules of **2** to give the 2nd generation (G-2) dendrimer with a total of 62 benzene rings (not shown). By the same procedure, a 3rd-generation dendrimer (G-3) with 142 benzene rings was obtained from the G-2 dendrimer. The G-3 dendrimer showed considerable overlap of the benzene rings whereas the G-1 (**3**) and G-2 dendrimers were planar.

3 was converted into **3'** and planarized by oxidative dehydrogenation (loss of all 56 H-atoms) to a fully aromatic graphene **5** (Fig. 16-14). The aromatization of G-2 failed.

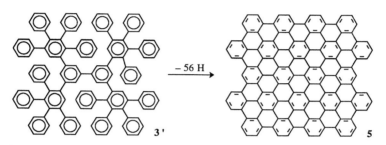

Fig. 16-14 Oxidative dehydrogenation of the 1st generation dendrimer G-1 (**3'**) to the graphene **4** [8].

The phenylene dendrimers **3**, **3'**, G-2, G-3 and the graphene **4** are true monolayers with the thickness of one benzene ring. Thicker layers ("sheets") with thicknesses of one molecule (Fig. 11-8, II) or two and more (Fig. 11-8, IV) can be produced from amphiphilic monomers by the Langmuir-Blodgett method (p. 386). The monomers can then be polymerized to layers (Fig. 16-15). The layers are not very stable, however, since each monomeric unit is connected with other monomers only linearly and not by ladder or network structures and all layers are only held together by hydrophobic bonds.

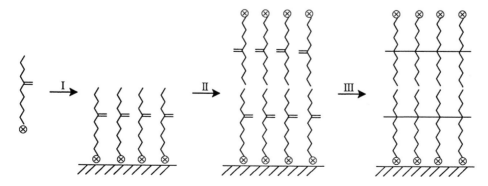

Fig. 16-15 Double-layer polymers via the Langmuir-Blodgett technique. In a Pockels-Langmuir trough, amphiphilic monomers with a polar head group ⊗ and a vinyl or acrylic group ⊁ form (I) ordered monolayers on polar surfaces. A second layer is deposited on the hydrophobic ends (II). The central carbon-carbon double bonds are finally polymerized (III).

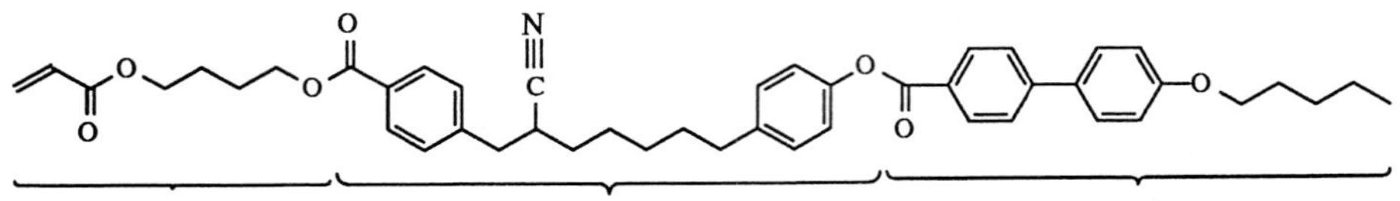

1st polymerizable segment	2nd polymerizable segment	smectogenic group
contains polymerizable group for the layer's surface	promotes molecular recognition and provides reactivity for one of the stitching reactions	promotes layering of the oligomers

Fig. 16-16 A macromonomer **7** and its segments from precursor monomers [9].

Much more stable **sheet polymers** consist of long macromonomers that are stitched together at two or more interior positions. An example is the macromonomer **7** of Fig. 16-16 which contains a smectogenic group, a polymerizable acrylic ester endgroup, and a polymerizable nitrile group in its center. In the melt, these macromonomers dimerize spontaneously, form double layers, and finally smectic liquid crystals (Fig. 16-17). The acrylic groups in the center of the double layers can be polymerized free-radically to monomer conversions of 90 % but the subsequent polymerization of nitrile groups is only successful to 30-50 %. The many resulting constitutional mistakes lead to soluble polymers with molecular weights between 10^7 and 10^9. The triple "stitching" of the double layers produces sheet polymers with thicknesses of ca. 10 nm.

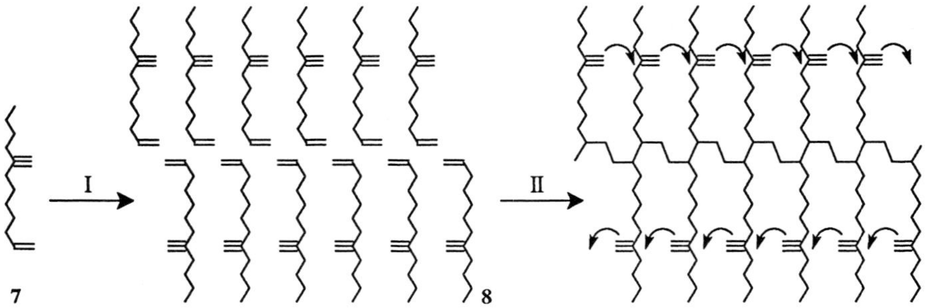

Fig. 16-17 The macromonomer **7** forms smectic liquid crystals consisting of double layers **8** that can be polymerized via their acrylic groups (II) and finally ($\frown$) to two-dimensional sheet polymers **9**.

16.3.3   Plate-Like Polymers

Self-organization of macromonomers can be used not only for the synthesis of polymers consisting of one or two polymerized layers (Section 16.3.2) but also for that of lattice polymers consisting of many layers. The self-organization can be brought about by smectogenic segments (as in macromonomer **7**) or by ordered arrangements of hydrogen-bond forming groups (as in Fig. 16-11). Fixation of the structure is usually by polymerization of suitable groups.

Layer-upon-layer structures are usually assumed to require two types of groups or segments that fit together like a lock and key such as the hydrophobic and hydrophilic ends of molecules in bilayer cell membranes. It is therefore remarkable that molecules **10** with a flexible ("coil-forming") hydrophobic segment and a liquid-crystal-generating ("rod-forming") somewhat hydrophilic segment such as the one shown in Fig. 16-18 are said to assemble themselves on substrates from solution upon evaporation of the solvent whereby "coils" face "rods" and "hydrophilics" face "hydrophobics."

Films with 5-150 layers i.e., with thicknesses of 50-1500 nm, have been prepared from **10**. Upon free-radical polymerization of the isoprene units, they become plate-like polymers with irreversible crosslinks between the layers.

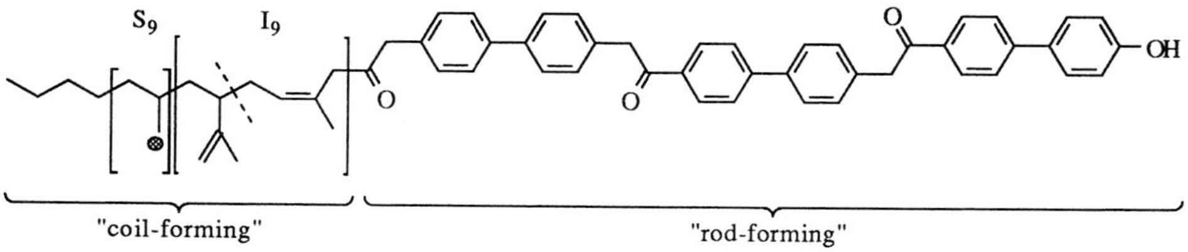

Fig. 16-18 Macromonomer **10** with nine styrene units (S₉) and nine isoprene units (I₉) in its coil-forming part and a long smectogenic ("rod-forming") tail leads to plate-like polymers [10].

Plate-like structures with reversible connections between the layers are obtained from muconate dianions, $^\ominus OOC–CH=CH–CH=CH–COO^\ominus$, and alkylammonium cations that control the distances between the muconate layers via the length of their alkyl residues R (Fig. 16-19). The muconic acid residues polymerize upon exposure to ultraviolet light and form a molecule-thick polymer layer (not shown). The alkylammonium cations are neutralized by acid which results in free uncharged polymer layers. They can again be "glued" together by addition of alkylammonium cations, thus forming a polymer sand-wich that can be filled with different cations similarly to clay (see Volume II).

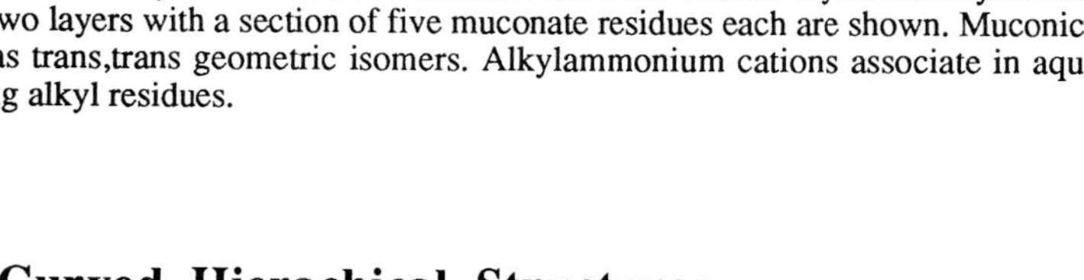

Fig. 16-19 Vertical layers of muconate dianions between vertical layers of alkylammonium cations [11]. Only two layers with a section of five muconate residues each are shown. Muconic acid residues are shown as trans,trans geometric isomers. Alkylammonium cations associate in aqueous solution via their long alkyl residues.

16.4 Curved Hierachical Structures

16.4.1 Polymer Rings

Nature produces cyclic deoxyribonucleic acids with millions of chain atoms by matrix-controlled step reactions (Section 14.2.2). Synthetic cyclic macromolecules result from ring-extension reactions of small cyclic molecules or from intramolecular ring-closure reactions between the chain ends of linear molecules.

Ring-extension reactions are very often equilibrium reactions that lead to homologous series of cyclic oligomer and polymer molecules (Section 7.5). Large rings are only obtained in small proportions since the concentration of larger rings decreases with the 5/2th power of the ring size. In kinetically controlled ring-opening polymerizations, ring-opening often competes with ring-extension (Section 7.2.1).

High yields of high-molecular weight rings are obtained by kinetically controlled polycondensations and polyadditions. **Ring closure** is promoted if local monomer concentrations are kept low so that intramolecular reactions are preferred over intermolecular ones (Section 13.6.1).

Industry prefers high-molecular weight cyclic macromolecules over linear ones of the same constitution. Endgroups of linear polymers often cause trouble since they may either react on processing (Section 13.3.4) or are prone to attack by atmospheric molecules such as oxygen, ozone, nitroxides, etc. (Volume IV).

16.4.2 Linked Rings

Cyclic molecules can be interconnected topologically (but not chemically) with other cyclic molecules to **catenanes** (L: *catena* = chain, shackle) (Fig. 16-20). The catenane in Fig. 16-20 contains five rings; it is thus called a [5]-catenane. Catenanes may be natural or synthetic products. For example, leukocytes of leukemia contain up to 26 % [2]-catenanes of deoxyribonucleic acids.

polycatenane   polyrotaxane   polymer tube

Fig. 16-20 Some combinations of rings with themselves and of rings and chains (schematic).

Polycatenanes contain many interlocked rings. They are synthesized from bifunctional linear compounds with two different functional groups such as A[M]$_i$Z[M]$_i$B. The central atoms Z are complexed by metal ions Mt (Eq.(16-25)). The selective reaction of B groups with bifunctional compounds D~~~~D delivers chains. Ring-closure is obtained by reaction of A-groups with bifunctional compounds E~~~~E.

(16-25)

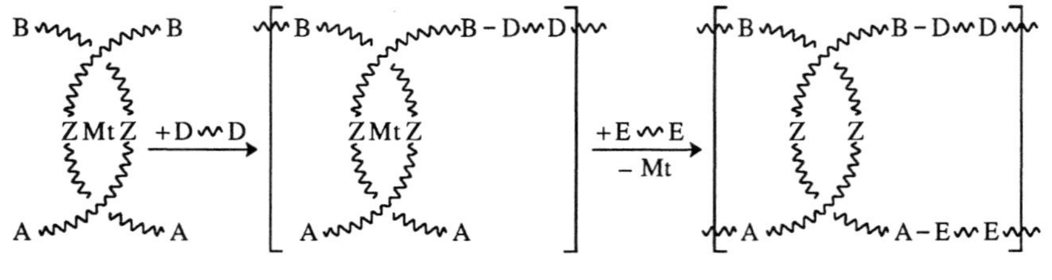

Threading of small ring molecules on linear chains without chemical bond formation creates **polyrotaxanes** that resemble wheels on axes (L: *rota* = wheel; rotax*ane* in analogy to caten*ane*). The ends of the axes are secured by cotters.

The simplest polyrotaxanes consist of cyclodextrin molecules (Schardinger dextrins, starch gums) that are threaded on water-soluble poly(oxyalkane) chains. Cyclodextrins are water-soluble cyclic oligoglucoses with 6 (α), 7 (β) or 8 (γ) glucose units in α-1,4-connections (Fig. 16-21; see also Volume II). The holes in these molecules have diameters of 0.45 nm (α-cyclodextrin = cyclohexaamylose), 0.70 nm (β-cyclodextrin = cycloheptaamylose), and 0.85 nm (γ-cyclodextrin = cyclooctaamylose).

α-Cyclodextrin threads on poly(oxyethylene)s, $RCH_2CH_2\text{+}OCH_2CH_2\text{+}_nR$ but not on the more bulky poly(oxypropylene)s, $\text{+}OCH(CH_3)CH\text{+}_n$, which can accommodate β-cyclodextrin, however. The driving force is the formation of hydrogen bonds between interior OH groups of cyclodextrin molecules and ether groups of poly(oxyalkane)s. These hydrogen bonds are especially stable because the interior OH groups of cyclodextrins are arranged in head-to-head and tail-to-tail positions, leading to hydrogen bonds of the type $-OH\text{-}\text{-}O\text{-}\text{-}HO-$.

The ends of the chain molecules are subsequently sealed by bulky groups. Sealed poly(oxyethylene)s with $\bar{X}_n = 78$ and an α-cyclodextrin ring at each ether group do not dissolve in water but do so in N,N-dimethylformamide.

Cyclodextrins can also be threaded on unbranched polyimines, $\text{+}NH(CH_2)_i\text{+}_n$, with $i = 11$; equilibrium is reached after ca. 2 hours. They do not form polyrotaxanes with polyimines if $i = 7$.

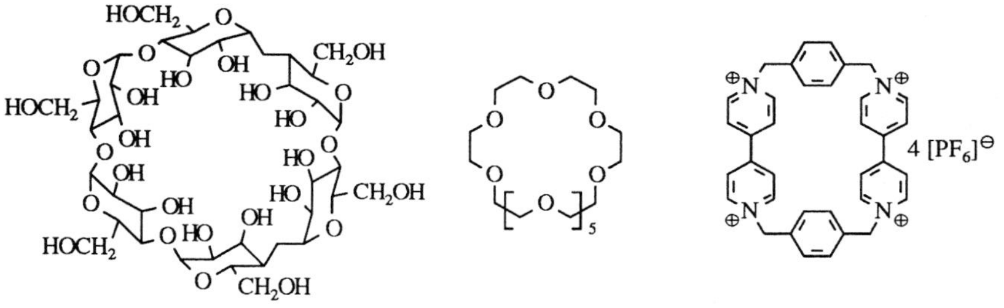

α-cyclodextrin
(cyclohexaamylose)

30(O$_6$-coronand-6)

cyclobis(*paraquat-p*-phenylene)

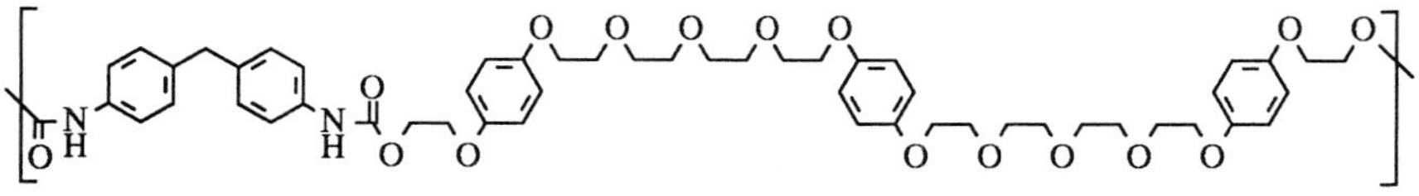

polyether chain from the polyaddition of $OCN(p\text{-}C_6H_4)CH_2(p\text{-}C_6H_4)NCO$ and $HOCH_2CH_2O(p\text{-}C_6H_4)O(CH_2CH_2O)_4(p\text{-}C_6H_4)O(CH_2CH_2O)_4(p\text{-}C_6H_4)OCH_2CH_2OH$

Fig. 16-21 Some compounds that have been used for the synthesis of polyrotaxanaes (see [12]).

Polycondensation of sebacoyl chloride, $ClOC(CH_2)_8COCl$, with a slight excess of 1,10-decanediol in the presence of the 30-crown ether (see Fig. 16-21) delivers polyrotaxanes with 1 crown ether per 6 repeating units. The driving force is probably the interaction of the ether groups of the crown ether with the ester groups of the polyester. The chains are sealed at both ends by the bulky 3,3,3-triphenylpropionic acid.

Another strategy threads electron-poor cyclic compounds such as the paraquat derivative shown in Fig. 16-21 top, right) on electron-rich polyether chains (Fig. 16-21, bottom). The ring compound is a derivative of paraquat (= 1,1'-dimethyl-4,4'-bipyridinium dichloride, $^\ominus Cl^\oplus [H_3C(NC_5H_4\text{-}C_5H_4N)CH_3]^\oplus Cl^\ominus$.

16.4.3   Polymer Tubes

Various strategies are employed to produce polymer tubes (**nanotubes, microtubules, tubules, molecular tubes,** etc.). The best known examples of tube polymers are seamless carbon nanotubes with graphite-like hulls (Fig. 16-22) that are produced by vaporization of carbon in an inert atmosphere (Volume II). The tubes can then be transformed (modified, functionalized, etc.) by polymer-analog reactions.

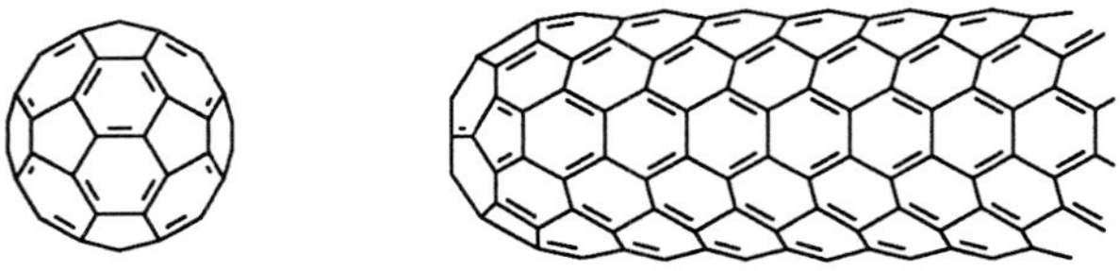

Fig. 16-22 Left: C_{60} fullerene. Right: Left section of a single-walled carbon nanotube in armchair arrangement of carbon atoms, capped at the end by a half-spherical C_{60} fullerene structure.

Several strategies are known for the synthesis of tube polymers that combine known chemical reactions with physicochemical mechanisms. Most common is the use of self-associating cyclic oligomers, macromonomers, triblock polymers, or dendrons to tube-like entities that may or may not be subsequently stabilized by polymerization. Another strategy employs polyrotaxanes as precursors. A third strategy employs molecular matrices as scaffolds for the synthesis of tube polymers. Several strategies deliver isolable tube polymers whereas others lead to assemblies of tube polymers. Some types of nanotubes are inherently stable whereas others need to be crosslinked for stability.

Nanotubes via Preformed Structures

Cyclic octapeptides with alternating D and L peptide units, *cycl*-[D-(NH–CHR–CO)-*alt*-L-(NH–CHR–CO)]$_4$, assemble themselves in long tubes upon acidification of their alkaline aqueous solutions. The tubes have lengths of 100-1000 nm and interior diameters of ca. 0.8 nm. The octapeptide molecules **I** selfassociate via eight hydrogen bonds per octapeptide. Such *supramolecular* entities **II** behave like true molecules because the sum of the strengths of the hydrogen bonds per octapeptide molecule exceeds that of a true covalent bond. Ions pass through these nanotubes faster than through the pentadeka-peptide gramicidin A, a constituent of cell membranes.

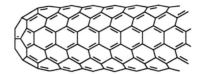

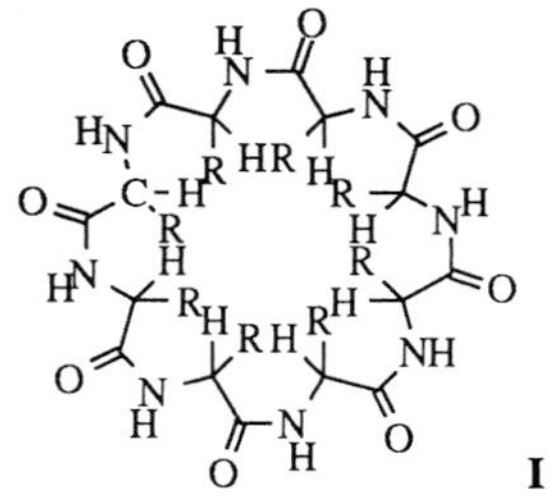

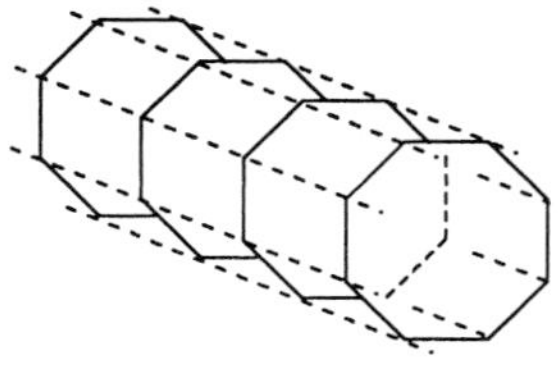

Fig. 16-23 **I**: cyclic octapeptide with peptide units –NH–CHR–CO–. **II**: schematic representation of a section of a tube composed of supramolecules **I**. - - - symbolizes hydrogen bonds between rings.

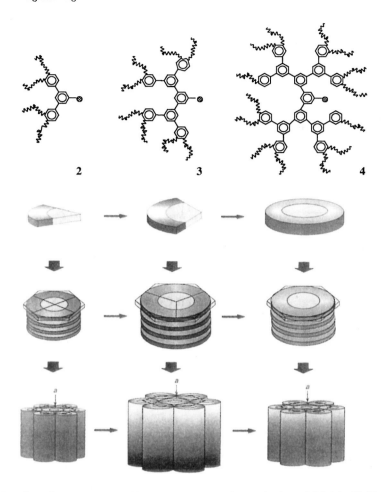

Fig. 16-24 Bundles of nanotubes by self-association and self-organization of G2 (**2**), G3 (**3**), and G4 (**4**) dendrons [13]. Row 1: schematic representation of G2, G3, and G4 monodendrons with core units >C_6H_3–R as ⬡–⊗ ; connecting units –CH_2–O– as —; 1,3,5-C_6H_3– in dendritic units and 1,3,4-C_6H_3- in end units as ⬡ ; and end units 3,4-($C_{12}H_{25}O$)$_2$– as ∿∿. Rows 2-4: Schematic representation of the shape of monodendrons as a section of discs (row 2), assembly of monodendrons as stacked discs (row 3), and the resulting bundles of stacked discs as nanotubes (row 4).

Certain monodendrons self-assemble in the solid state to disk-like entities (Fig. 16-24) that self-organize into lattices as shown by X-ray diffraction experiments at 49°C (**2**), 36°C (**3**), and 51°C (**4**). Examples are the 2nd, 3rd, and 4th generation monodendrons with core units >(CH_2O)$_2C_6H_3$–R (R = CH_2OH (**2**) or $COOCH_3$ (**3**)), dendritic units >(CH_2O)$_2C_6H_3$– and terminal units ($C_{12}H_{25}O$)$_2C_6H_3$– (Fig. 16-24, top). Four monodendrons **2** assemble to a disk with a radius of 2.16 nm whereas only three of the larger monodendrons **3** were required for a disk with a radius of 2.66 nm and only one monodendron **4** for a disk with a radius of 2.28 nm; the disks were 0.47 nm thick. The disks are assembled in supramolecular structures. It is not clear whether there are indeed channels as Fig. 16-24 suggests, and if so, how large they are (see also Section 16.4.4).

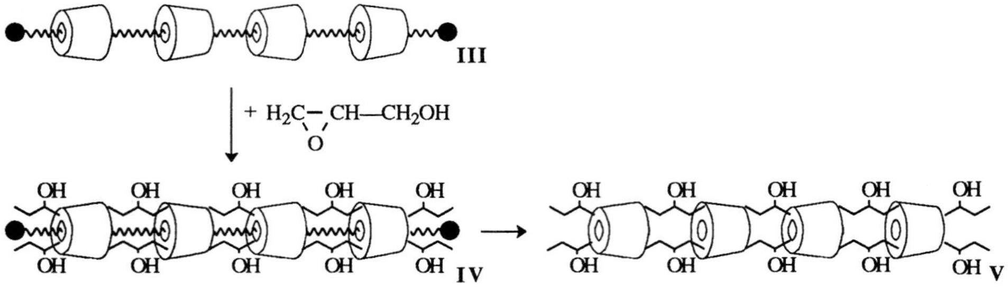

III

$$+ \; H_2C\!\!-\!\!CH\!\!-\!\!CH_2OH$$
$$O$$

IV → V

Fig. 16-25 Synthesis of a molecular tube from a polyrotaxane [14]. ⌇⌇⌇ Poly(oxyethylene) chain, ● bulky endgroup.

Another strategy is to synthesize polymer tubes by connecting the preexisting holes of α-cyclodextrins in a polyrotaxane (Fig. 16-25). The first step is the threading of cyclodextrins on a poly(oxyethylene) chain and sealing the chain with bulky groups. The OH groups of cyclodextrin molecules of the resulting polyrotaxane **III** are reacted with epichlorohydrin + NaOH which leads to)–O–CH₂–CH(OH)–CH₂–O–(bridges between adjacent cyclodextrins (**IV**). Finally, endcaps are broken and the poly(oxyethylene) threads are removed to give water-soluble α-cyclodextrin nanotubes **V** with diameters of ca. 1.5 nm (external) and ca. 0.5 nm (internal. The highest molecular weights of nanotubes were ca. 17 000 which corresponds to ca. 15 interconnected cyclodextrins.

Higher cyclodextrins have larger internal diameters of 0.65 nm (β) and 0.8 nm (γ), respectively. Nanotubes from such interconnected cyclodextrins may act as conduits for the passage of ions or molecules, or, with an encapsulated electrically conducting polymer chain, electrons.

The interior of cyclodextrin molecules is relatively hydrophobic which causes rodlike *all-trans*-1,6-diphenyl-1,3,5-hexatriene molecules (DPH) to insert themselves spontaneously from aqueous solution into the cavities of β and γ-cyclodextrin (CD) molecules (Fig. 16-26). The resulting aggregates (nanotubules) of 20-35 cyclodextrin molecules are fairly rigid since a cavity contains two or three sections of a DPH molecule and the relatively large DPH molecules span several CD units.

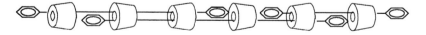

Fig. 16-26 Schematic representation of nanotubules consisting of cyclodextrin molecules threaded on *all-trans*-1,6-diphenyl-1,3,5-hexatriene molecules, C_6H_5–CH=CH–CH=CH–CH=CH–C_6H_5 [15].

Nanotubes via Association of Block Polymers

Block polymer molecules self-assemble in the amorphous state mainly to spherical, cylindrical, and lamellar structures, depending on the relative length of the various blocks (see Section 8.5 in Volume III). In solution, block polymers usually form spherical micelles. More recently, clever design of chemical structures and block lengths of diblock and triblock polymers has led to the discovery of a variety of other morphologies, especially polymer tubules (polymer tubes).

Fig. 16-27 Schematic representation of spherical and cylindrical micelles of diblock polymers.

Block polymer molecules such as A_n-block-B_m self-assemble in liquids that are solvents for A_n blocks but non-solvents for B_m blocks. The shape of the resulting "micelles" depends on both the space requirements of the blocks and the surface area of the insoluble "phase". Spherical micelles with an "insoluble" core of B_m blocks and a "soluble" corona of A_n blocks will result if the A_n blocks are much larger than the B_m blocks (Fig. 16-27, left). Layer structures, on the other hand, are obtained from neutral diblock polymer molecules if the two blocks have similar space requirements and from charged ones if attractive and repulsive forces balance each other (see Figs. 16-15, 16-18, 16-19).

Cylindrical micelles can be viewed as either "stretched spheres" or "shrunk lamellae". Like spheres, their formation requires some (but not too strong) imbalances of space and force requirements. Like lamellae, surface energy is minimized along the sides of the cylinders. The difference between the small energy of curvature along the cylinder and the larger energies of curvature of the hemispherical ends is a driving force for the formation of long cylinders instead of spheres. It is not easy to determine whether the resulting cylinders are simply rods with a more or less "homogeneously filled interior" or tubules with a longitudinal "empty space". The incorporation of metal atoms or organic molecules does not prove the existence of hollowness (see Section 16.4.4). Holes seen in electron micrographs may be artifacts caused by experimental techniques, etc.

For example, poly(ferrocenyldimethylsilane)-*block*-poly(dimethylsiloxane) (PFS-*b*-PDMS) formed tubules in alkanes (Table 16-3) consisting of an inner shell of PFS and an outer corona of PDMS. Tube lengths were considerably and tube diameters slightly higher in *n*-decane than in *n*-hexane whereas the wall thickness remained constant.

Table 16-3 Lengths L_{tube}, diameters d_{tube}, wall thicknesses t_{wall}, and diameters d_{cavity} of tubules from poly(ferrocenyldimethylsilane)-*block*-poly(dimethylsiloxane) [16]. * No tubules formed.

	(PFS)$_{40}$-*block*-(PDMS)$_{480}$				(PFS)$_{80}$-*block*-(PDMS)$_{960}$			
	23°C	61°C	61°C	151°C	23°C	61°C	61°C	151°C
	n-hexane	*n*-hexane	*n*-decane	*n*-decane	*n*-hexane	*n*-hexane	*n*-decane	*n*-decane
L_{tube}/nm	600	4000	100 000	100 000	4000	4000	100 000	*
d_{tube}/nm	23	23	21	21	25-26	25-26	25-26	*
t_{wall}/nm	7	7	7	7	7	7	7	*
d_{cavity}/nm	9	9	7	7	11-12	11-12	10-21	*

The tube structures of PFS-*b*-PDMS diblock polymers existed for about one month. More stable tubules were be obtained by fixing the structures through polymerization. Examples are triblock polymers PI-*block*-PCEMA-*block*-P*t*BA with poly(isoprene) (PI) and poly(*t*-butyl acrylate) (P*t*BA) endblocks and poly(2-cinnamoylethyl methacrylate) (PCEMA) center blocks with block lengths of n-p-q = 133-129-800 ($\overline{M}_w$ = 9000 g/mol) or n-p-q = 370-420-550 ($\overline{M}_w$ = 25 000 g/mol).

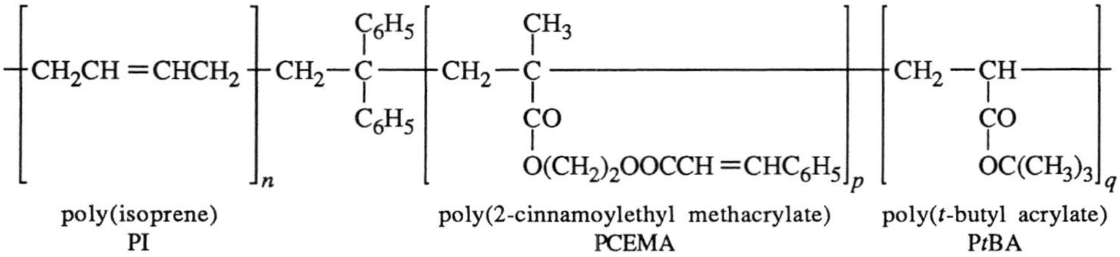

poly(isoprene)   poly(2-cinnamoylethyl methacrylate)   poly(*t*-butyl acrylate)
PI   PCEMA   P*t*BA

These triblock polymers micellize to cylinders of micrometer length with a core of PI blocks that is surrounded by a shell of PCEMA blocks and a corona of large P*t*B A blocks (Fig. 16-28). Photolysis crosslinks the PCEMA blocks and subsequent ozonolysis destroys the PI blocks, delivering nanofibers of ca. 22 nm diameter. Transmission electron micrographs showed light white stripes in the centers of cylinders which indicate the formation of nanotubes.

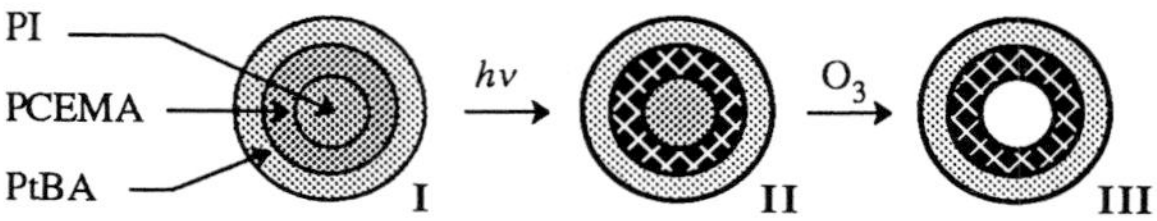

M_w/(g mol^{-1})		9 000	25 000
PI	n =	133	370
PCEMA	p =	129	420
P*t*BA	q =	800	550

Fig. 16-28 Structure of cylindrical triblock micelles (**I**) and the formation of nanotubes (**III**) by crosslinking of PCEMA blocks to **II** and subsequent ozonolysis of the PI blocks to **III** [17].

Many variations of this strategy are obviously possible, for example, use of a hydrophobic outer corona and a crosslinked hydrophilic shell for the incorporation of water-soluble molecules, hydrophilic coronas and cores and a crosslinked shell, etc.

Microtubules

Tube polymers should not be confused with **microtubules** which have much thicker walls. Microtubular *channels* exist in **ionotropic gels**. Such gels are formed by coagulation of alginates if bivalent cations diffuse into aqueous solutions of alginic acid.

Pyrrole can be polymerized in preformed channels to yield *isolable* microtubules. These microtubules are obtained from an aqueous solution of pyrrole that is separated from an aqueous FeCl$_3$ solution by a membrane such as an 8 μm thick Nuclepore® polycarbonate filter with 0.5 μm wide pores. The interdiffusion leads to the formation of an FeCl$_3$ salt of pyrrole which converts its aromatic character into an olefinic one. The monomer then polymerizes cationically on the anionic interior walls of the pores. At a certain time, the polymerization is suppressed and the microtubules are isolated by destroying the membrane. Microfibers result if the polymerization is allowed to proceed without interruption.

Such microtubules can also be obtained by an electrochemical polymerization in a Nuclepore membrane that is mounted onto the surface of a platinum disk electrode which is immersed in a CH_3CN solution that contains 0.5 mol/L N-methylpyrrole and 0.2 mol/L $[(C_2H_5)_4N]^{\oplus} [BF_4]^{\ominus}$. The polymerization proceeds cationically (Fig. 16-29).

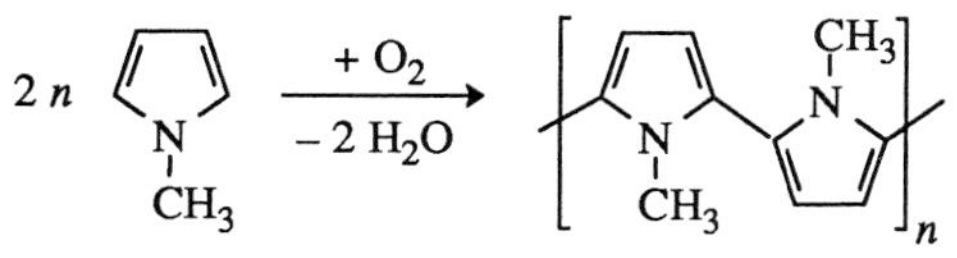

$$2\,n \quad \underset{\underset{CH_3}{|}}{\overset{\displaystyle\bigcap}{N}} \quad \xrightarrow[-\,2\,H_2O]{+\,O_2} \quad \left[\cdots \right]_n$$

Fig. 16-29 Electrochemical polymerization of N-methyl pyrrole [18a] to microtubules [18b].

16.4.4 Polymer Spheres

Nature produces hollow spheres and spheroids, using yet unknown mechanisms. The hollow sphere of the protein molecule **apoferritin** consists of 24 subunits (i.e., chemical protein molecules) with molecular weights of ca. 19 500. The sphere has an outer diameter of 12.2 nm and an inner diameter of 7.3 nm. It can take up micelles of iron(III) hydroxide phosphate with up to 4500 iron atoms. The resulting enzyme **ferritin** has the same diameter as apoferritin but a 20 % larger molecular weight. Even larger are the bag-like protein hulls of certain bacteria (p. 513).

Synthetic hollow spheres result from the polymerization of vesicles composed of monomer molecules (p. 384). The diameters of these nanospheres are usually ca. 3 nm.

Certain higher generation dendrons of the type of Fig. 16-24 assemble to form globular shapes (Fig. 16-30) whereas the lower generation ones deliver stacks of polymer tubes (Fig. 16-24). The globulization begins in solution with the G4 monodendron but in the bulk state with the G5 monodendron.

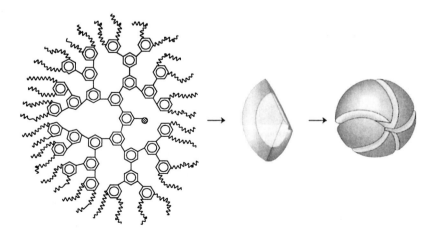

Fig. 16-30 Schematic representation of a G5 monodendron (left), its sectional shape in spheres (center), and the resulting spheres ($d = 5.42$ nm) [13]. The chemical structure (left) uses the following symbols: $>C_6H_3$–R as ⬡—⊗ ; connecting units $-CH_2-O-$ as —; $1,3,5$-C_6H_3– in dendritic units and $1,3,4$-C_6H_3- in end units as ⬡ ; and end units $3,4$-$(C_{12}H_{25}O)_2$– as 〰.

Spheres obtained from the self-assembly of monodendrons are not hollow and neither are dendrimers although the latter may incorporate molecules and metal atoms. In a *first approximation*, the overall shape of such dendrimers can be assumed as spherical, especially for hydrodynamic measurements, similar to the "spherical" shapes of globular protein molecules. The interior of these spheroids is not filled homogeneously with mass as shown by the comparison of their radii in solution, R_{soln}, and their radii in the dry state, $R_{dry} = (3\ \tilde{v}M/(4\ \pi\ N_A)^{1/3}$ ($\tilde{v}$ = specific volume).

A-16 Appendix: Joining Primary Molecules

Primary chain molecules can be joined via their ends or side-by-side to larger entities that may be called "übermolecules" following the latest linguistic craze (the term "supramolecule" is already used for a special class of "übermolecules" and "supermolecule" has a special connotation).

Ends of linear molecules may be joined chemically via covalent bonds to form linear block polymers (Section 16.2.2) or physically via secondary bonds (e.g., hydrogen bonds) to open associations (Volume III). The chemical joining of three or more linear chain molecules via one end each results in star molecules (Section 16.2.4) or via physical bonds to closed associations, for example, micelles (Volume III). Side-by-side joinings may also be chemical or physical. Whether the joining is chemical or physical, the mathematical formalism for the calculation of size distributions of "übermolecules" from substances with molecular weight distributions of primary molecules is always the same.

The effect of joining molecules of polymolecular polymers is calculated as follows. In the simplest case, the various primary polymers have the same number-average degree of polymerization, $\overline{X}_{I,n}$, and the same polymolecularity index, $\overline{X}_{I,w}/\overline{X}_{I,n}$. If the molecules of N primary polymers are joined via their ends, then the number-average degree of polymerization, $\overline{X}_n$, of the resulting "übermolecules" (block polymers, star polymers, end-to-end associates, etc.) is just given by N times the number-average molecular weight of the primary polymers, i.e., $\overline{X}_n = N\ \overline{X}_{I,n}$. This expression has to be replaced by $\overline{X}_n = \sum_i N_i \overline{X}_{I,n,i}$ (Eq.(16-10)) if the i primary polymers differ in their number-average degrees of polymerization, $\overline{X}_{I,n}$.

The calculation of the weight-average degree of polymerization, $\overline{X}_w$, of "übermolecules" resulting from the joining of N primary polymers with the same number-average degree of polymerization, $\overline{X}_{I,n}$, and the same polymolecularity index, $\overline{X}_{I,w}/\overline{X}_{I,n}$, requires a considerable mathematical effort. It is even more complicated if the primary molecules have different number-average degrees of polymerization and different polymolecularity indices.

The principle can be seen for the special case of joining the molecules of two primary polymers I and II, i.e., $N = 2$. The primary polymer I consists of primary molecules with $N_i = 1, 2, 3, \dots i$ monomeric units and the primary polymer II of primary molecules with $N_j = 1, 2, 3, \dots$ units. Since each primary molecule can couple with another primary molecule of the same or another size, the following weight fractions w_{ij} ($i \neq j$) and w_{ij} are present:

$$w_{11} \quad w_{12} \quad w_{13} \quad .... \quad\quad w_{1N}$$
$$w_{22} \quad w_{23} \quad .... \quad\quad w_{2N}$$
$$w_{33} \quad .... \quad\quad w_{3N}$$
$$\vdots$$
$$w_{NN}$$

The weight-average degree of polymerization, $\overline{X}_w$, of "übermolecules" is therefore given by

(A-16-1) $\displaystyle \overline{X}_w = \sum_{i=1}^{N} w_{ii}X_{ii} + \sum_{1\le i \le j \le N} w_{ij}X_{ij} = \frac{1}{\overline{X}_{I,n}}\left[2\sum_{i=1}^{N} x_i^2 X_i^2 + \sum_{1\le i \le j \le N} x_i x_j (X_i + X_j)^2 \right]$

The quantities in the first equation are evaluated as follows. The weight fractions w_{ii} are given by $w_{ii} = x_{ii}X_{ii}/\overline{X}_n$ according to Section 3.3.3 (p. 71). This expression can be replaced by $w_{ii} = x_i^2 X_i / \overline{X}_{I,n}$ since $x_{ii} = x_i^2$ is the mole fraction of "übermolecules" of type i with a degree of polymerization of $X_{ii} = 2\,X_i$ and the number-average degree of polymerization of "übermolecules" is given by $\overline{X}_n = 2\,\overline{X}_{I,n}$ according to Eq.(16-10).

Weight fractions are obtained from $w_{ij} = x_{ij}X_{ij}/\overline{X}_n = x_i x_j(X_i + X_j)/\overline{X}_{I,n}$. Since i and j are exchangeble, the mole fraction x_{ij} can be replaced by $x_{ij} = x_i x_j + x_j x_i = 2\,x_i x_j$. Also:

(A-16-2) $\displaystyle (\overline{X}_{I,n})^2 = \left(\sum_{i=1}^{N} x_i X_i\right)^2 = \sum_{i=1}^{N} x_i^2 X_i^2 + 2\sum_{1\le i \le j \le N} x_i x_j X_i X_j$

Introduction of Eq.(A-16-2) in Eq.(A-16-1) delivers

(A-16-3) $\displaystyle \overline{X}_w = \frac{1}{\overline{X}_{I,n}}\left[(\overline{X}_{I,n})^2 + \sum_{i=1}^{N} x_i^2 X_i^2 + \sum_{1\le i \le j \le N} x_i x_j X_i^2 + \sum_{1\le i \le j \le N} x_i x_j X_j^2 \right]$

The three sums of this equation are developed into series and the various terms added:

(A-16-4) $\displaystyle \sum_{i=1}^{N} x_i^2 X_i^2 + \sum_{1\le i \le j \le N} x_i x_j X_i^2 + \sum_{1\le i \le j \le N} x_i x_j X_j^2 =$

$$= \quad x_1 x_1 X_1^2 + x_1 x_2 X_1^2 + x_1 x_3 X_1^2 + ... + x_1 x_N X_N^2$$
$$+ \quad x_2 x_1 X_2^2 + x_2 x_2 X_2^2 + x_2 x_3 X_2^2 + ... + x_2 x_N X_N^2$$
$$+ \quad x_3 x_1 X_3^2 + x_3 x_2 X_3^2 + x_3 x_3 X_3^2 + ... + x_3 x_N X_N^2$$
$$+ \quad ....$$
$$+ \quad x_N x_1 X_N^2 + x_N x_2 X_N^2 + x_N x_3 X_N^2 + ... + x_N x_N X_N^2$$
$$= \sum_{i=1}^{N} x_i X_i^2 = \overline{X}_{I,n}\overline{X}_{I,w}$$

Introduction of Eq.(A-16-4) into Eq.(A-16-3) leads to $\overline{X}_w = \overline{X}_{I,w} + \overline{X}_{I,n}$, i.e., the special case of $\overline{X}_w = \overline{X}_{I,w} + (N-1)\overline{X}_{I,n}$ for $N = 2$.

Literature to Chapter 16

16.1 POLYMER ARCHITECTURE
B.Voit, Synthesis of Defined Polymer Architectures, Wiley-VCH, Weinheim 2002
K.E.Geckeler, Ed., Advanced Macromolecular and Supramolecular Materials and Processes, Kluwer Academic/Plenum, New York 2003

16.2.1 SUPRAPOLYMERS
J.-M.Lehn, Supramolecular Chemistry. Concepts and Perspectives, VCH, Weinheim 1995
A.Ciferri, Ed., Supramolecular Polymers, Dekker, New York 2000

16.2.2 BLOCK POLYMERS
R.J.Ceresa, Block and Graft Copolymers, Wiley, New York, 2 Vols. (1973 and 1976)
M.J.Folkes, Ed., Processing, Structure, and Properties of Block Copolymers, Elsevier, New York 1985
I.W.Hamley, Block Copolymers, Oxford Univ. Press, Oxford 1999
F.J.Baltá Calleja, Z.Roslaniec, Block Copolymers, Dekker, New York 2000
P.Alexandridis, B.Lindman, Eds., Amphiphilic Block Copolymers, Elsevier Science, Amsterdam 2001
N.Hadjichristidis, S.Papas, G.A.Floudas, Block Copolymers, Wiley-Interscience, New York 2003
volumes II-Iv?

16.2.3 COMB POLYMERS
H.A.J.Battaerd, G.W.Tregear, Graft Copolymers, Interscience, New York 1967
R.J.Ceresa, Block and Graft Copolymers, Wiley, New York, 2 Vols. (1973, 1976)
J.P.Kennedy, Cationic Graft Copolymerization, Wiley, New York 1978
Y.Ikeda, Characterization of Graft Copolymers, Adv.Polym.Sci. **29** (1978) 47
T.Corner, Free Radical Polymerization. The Synthesis of Graft Copolymers, Adv.Polym.Sci. **62** (1984) 95
N.A.Platé, V.P.Shibaev, Eds., Comb-Shaped Polymers and Liquid Crystals, Plenum, New York 1987
Y.Yamashita, Ed., Chemistry and Industry of Macromonomers, Hüthig and Wepf, Basel 1993

16.2.4 STAR POLYMERS
S.Bywaters, Preparation and Properties of Star-Branched Polymers, Adv.Polym.Sci. **30** (1979) 89
W.H.Dickstein, Rigid Rod Star Block Copolymers, Technomic, Lancaster (PA) 1990
M.K.Mishra, S.Kobayashi, Eds., Star and Hyperbranched Polymers, Dekker, New York 1999

16.2.5 DENDRONS AND DENDRIMERS
D.A.Tomalia, A.M.Naylor, W.M.Goddard III, Starburst-Dendrimere, Angew.Chem. **102** (1990) 119; -, Starburst Dendrimers, Angew.Chem.Int.Ed.Engl. **29** (1990) 138
G.R.Newkomme, Ed., Advances in Dendritic Macromolecules, JAI Press, Greenwich (CT), Vol. 1 (1994), Vol. 2 (1995)
G.R.Newkomme, C.N.Moorefield, F.Vögtle, Dendritic Molecules. Concepts, Syntheses, Perspectives, VCH, Weinheim 1996
P.R.Dvornic, D.A.Tomalia, Recent Advances in Dendritic Polymers, Current Opinions in Colloid and Interface Sci. **1/2** (1996) 221
W.Devonport, C.J.Hawker, Architectural Control in Dendritic and Hyperbranched Macromolecules, Polymer News **21** (1996) 370
O.A.Matthews, A.N.Shipway, J.F.Stoddart, Dendrimers–Branching Out from Curiosities into New Technologies, Progr.Polym.Sci. **23** (1998) 1
A.D.Schlüter, J.P.Rabe, Dendronized Polymers, Angew.Chem. **112** (2000) 860; -, Angew.Chem. Int.Ed.Engl. **39** (2000) 864
F.Vögtle, S.Gestermann, R.Hesse, H.Schwierz, B.Windisch, Functional Dendrimers, Progr.Polym.Sci. **25** (2000) 987
G.R.Newkomme, C.N.Moorefield, F.Vögtle, Dendrimers and Dendrons. Concepts, Syntheses, Applications, Wiley-VCH, Weinheim 2001
J.M.J.Fréchet, D.A.Tomalia, Eds., Dendrimers and Other Dendritic Polymers, Wiley, New York 2002

16.3.1 LADDER POLYMERS
W.De Winter, Double Strand Polymers, Revs.Macromol.Sci. **1** (1966) 329

C.G.Overberger, J.A.Moore, Ladder Polymers, Adv.Polym.Sci. **7** (1970) 113
U.Scherf, K.Müllen, The Synthesis of Ladder Polymers, Adv.Polym.Sci. **123** (1995) 1

16.3.2 LAYER POLYMERS
S.I.Stupp, S.Son, H.C.Lin, I.S.Li, Synthesis of Two-Dimensional Polymers, Science **259** (1993) 59
T.Kunitake, Synthesis of Ultrathin Polymer Films by Self Assembly, Macromol.Symp. **98** (1995) 45

16.4.1 POLYMER RINGS
J.A.Semlyen, Ed., Cyclic Polymers, Elsevier, New York 1986

16.4.2 LINKED RINGS
G.Schill, Catenanes, Rotaxanes, and Knots, Academic Press, New York 1971
G.Wenz, B.Keller, Polyrotaxanes, Angew.Chem.Int.Ed.Engl. **31** (1992) 325
H.W.Gibson, M.C.Bheda, P.T.Engen, Rotaxanes, Catenanes, Polyrotaxanes, Polycatenanes, and
    Related Materials, Prog.Polym.Sci. **19** (1994) 843
D.B.Amabilino, I.W.Parsons, J.F.Stoddart, Polyrotaxanes, Trends Polym.Sci. **2** (1994) 146
A.Harada, Design and Construction of Supramolecular Architectures Consisting of Cyclodextrins and
    Polymers, Adv.Polym.Sci. **133** (1997) 141
J.-P.Sauvage, C.Dietrich-Buchecker, Eds., Molecular Catenanes, Rotaxanes, and Knots, Wiley-VCH,
    Weinheim 1999
A.Takata, N.Kihara, Y.Furusho, Polyrotaxanes and Polycatenanes: Recent Advances in Syntheses and
    Applications of Polymer Comprising of Interlocked Structures, Adv.Polym.Sci. **171** (2004) 1

16.4.3 POLYMER TUBES
T.W.Ebbesen, Ed., Carbon Nanotubes: Preparation and Properties, CRC Press, Boca Raton (FL)
    1996
S.E.Webber, P.Munk, Z.Tuzar, Eds., Solvents and Self-Organization of Polymers, Kluwer,
    Dordrecht 1996 (NATO ASI Series, Series E)
K.E.Geckeler, S.Samal, Syntheses and Properties of Macromolecular Fullerenes, A Review,
    Polym.Int. **48** (1999) 743
K.E.Geckeler, S.Samal, Macrofullerenes and Polyfullerenes: New Promising Polymer Materials,
    J.Macromol.Sci.-Rev.Macromol.Chem.Phys. **C 40** (2000) 193

References to Chapter 16

[1]   J.Krstina, G.Moad, E.Rizzardo, C.L.Winzor, C.T.Berge, M.Fryd, Macromolecules **28** (1995)
      5381
[2]   P.J.Flory, Principles of Polymer Chemistry, Cornell University Press, Ithaca (NY) 1953,
      p. 333, Eqs.23-24
[3]   Y.Yang, H.Zang, J.He, Macromol.Theory Simul. **4** (1995) 1995, Eqs.46-47
[4]   A.Echte, Handbuch der technischen Polymerchemie, VCH, Weinheim 1993, Eqs.755-757
[5]   J.R.Schaefgen, P.J.Flory, J.Am.Chem.Soc. **70** (1948) 2709, Tables I-IV
[6]   K.Chmil, U.Scherf, Macromol.Chem., Rapid Commun. **14** (1995) 217
[7]   R.P.Sijkesma, F.H.Beijer, L.Brunsveld, B.J.B.Folmer, J.H.K.K.Hirschberg, R.F.M.Lange,
      J.H.K.Lowe, E.W.Meijer, Science **278** (1997) 1601
[8]   A.J.Berresheim, F.Morgenroth, U.-M.Wiesler, K.Müllen, ACS Polym.Prep. **39** (1998) 721
[9]   S.I.Stupp, S.Son, H.C.Lin, L.S.Li, Science **259** (1993) 59; see also E.Thomas, Science **259**
      (1993) 49
[10] S.I.Stupp, V.LeBonheur, K.Walker, ACS Polym.Prepr. **36**/1 (1995) 562
[11] A.Matsumoto, T.Odani, K.Sada, M.Miyata, K.Tashiro, Nature **405** (2000) 328; A.Matsumot,
      S.Oshita, D.Fujioka, J.Am.Chem.Soc. **124** (2002) 13749
[12] J.F.Stoddart, J.A.Preece, F.M.Raymo, in J.C.Salamone, Polymeric Materials Encyclopedia,
      CRC Press, Boca Raton (FL) 1996, p. 6695
[13] V.Percec, W.-D.Cho, G.Ungar, D.J.P.Yeardley, Angew.Chem.Int.Ed. **39**/9 (2000) 1598,
      Scheme 3
[14] A.Harada, J.Li, M.Kamachi, Nature **364** (1993) 516

[15] G.Li, L.B.McGown, Science **264** (1994) 249

[16] J.Raez, I.Manners, M.A.Winnik, J.Am.Chem.Soc. **124** (2002) 10381, data of Table 2

[17] G.Liu, S.Stewart, Polym.Mater.Sci.Eng. **81** (1999) 10; S.Stewart, G.Liu, Angew.Chem. **112** (2000) 348; -, Angew.Chem.Int.Ed. **39** (2000) 340

[18] (a) Z.Cai, C.R.Martin, J.Am.Chem.Soc. **111** (1989) 4138; (b) C.R.Martin, L.S. Van Dyke, Z.Cai, W.Liang, J.Am.Chem.Soc. **112** (1990) 8976

17 Appendix

17.1 Physical Quantities and Units

According to **Maxwell**, many physical properties can be described in quantitative terms by *quantity calculus*. The value of a physical quantity (symbol in *italics*) equals the product of a numerical value (Roman (upright) type) and a physical unit (symbol in upright letter):

physical quantity = numerical value × physical unit

This equation can be manipulated by the ordinary rules of algebra. If, for example, a certain item has a length L of 0.002 meters (m) = 2 millimeters (mm), then this may be written as

$$L = 0.002 \text{ m} \quad \text{or} \quad L = 2 \cdot 10^{-3} \text{ m} \quad \text{or} \quad 10^3 \, L/\text{m} = 2 \quad \text{or} \quad L = 2 \text{ mm} \quad \text{or} \quad L/\text{mm} = 2$$

but *not* as $10^{-3} \, L/\text{m} = 2$. A column head $10^2 \, F/(\text{N m}^{-2})$ for a column entry of 7.35 thus indicates a force $F = 7.35 \cdot 10^{-2} \text{ N/m}^2$. Literature data often do not follow these SI (Systéme International) rules by the International Standardization Organization (ISO), which are adopted by IUPAP (International Union of Pure and Applied Physics), IUPAC (International Union of Pure and Applied Chemistry), etc. Instead one finds various nonrational notations such as F, N m^{-2} or F [N m^{-2}] for a column entry of $7.35 \cdot 10^{-2}$, often with wrong algebraic statements such as $10^{-2} \, F$, N m^{-2}.

The International System of Units (systéme international; **SI system**) uses seven **base physical quantities** and seven **SI base units** (Table 17-1). It replaces the formerly used CGS and MKS systems; it is *not* "the metric system". The SI system also defines several **derived units** (Table 17-2) which now include the radian and steradian that were formerly considered "supplementary units." Several older units may still be used for the time being (Table 17-3).

Names of units derived from persons' names are not capitalized (exception: degree Celsius). In order to avoid cumbersome numbers, symbols of units may be prefixed with SI prefixes (Table 17-4). The decimal sign between digits in a number is either a point (US) or a comma (e.g., Europe). No commas or periods should be used to separate groups of three digits for better reading. Instead, spaces should be applied (exception: groups of four digits may be written without space). Example: 1 234.567 8 or 1234.5678 or 1234,5678 or 1 234,567 8 but *not* 1,234.5678. Numbers and units are separated by a gap, groups of quantities or units by spaces or multiplication signs (· or ×).

All countries except the United States of America, Liberia, and Myanmar have adopted the SI system. In many countries, the SI system is the only system of weights and measures that can be lawfully used in trade and commerce. American technical literature and sometimes also American scientific literature still uses American and old Imperial British units although the metric system was introduced in 1896 by an act of Congress and by law all U.S. government units were supposed to convert to the SI system by the end of 1992. The following tables list names and symbols of base and derived SI units, temporarily allowed non-SI units, and SI prefixes. Conversions of older units into SI units can be found in Volumes II-IV. This book follows IUPAC/IUPAP recommendations; exceptions are noted in the list of symbols for physical quantities (front matter).

Table 17-1 Names and symbols for physical quantities and their SI base units.

Physical quantity SI symbol	SI name	SI base unit English name	American name	SI symbol
l	length	metre	meter	m
m	mass	kilogramme	kilogram	kg
t	time	second	second	s
I	electric current	ampere	ampere	A
T	thermodynamic temperature	kelvin	kelvin	K
n	amount of substance	mole	mole	mol
I_v	luminous intensity	candela	candela	cd

Table 17-2 Derived SI units and their symbols recommended by IUPAC-IUPAP.

| Physical quantity | | SI unit | | | |
Symbol	Name	SI name	SI symbol and unit(s)		
α, β, γ	plane angle	radian	rad $= $ m/m	$= 1$	
ω, Ω	solid angle	steradian	sr $=$ m^2/m^2	$= 1$	
ω	angular velocity	-	rad/s		
	angular acceleration	-	rad/s^2		
ν	frequency	hertz [1]	Hz $= $ s^{-1}		
v, u, w	speed [2], velocity [3]	-	m s^{-1}		
$a, (g)$	acceleration	-	m s^{-2}		
P	power, radiant flux	watt	W $= $ V A	$= $ J s^{-1}	$= $ m^2 kg s^{-3}
E	energy, work, heat	joule	J $= $ N m		$= $ m^2 kg s^{-2}
F [3]	force	newton	N $= $ J m^{-1}		$= $ m kg s^{-2}
	impact strength (US)	newton	J m^{-1}		$= $ m kg s^{-2}
G	weight	newton	N $= $ J m^{-1}		$= $ m kg s^{-2}
	impact strength (Europe)	-	J m^{-2}		$= $ kg s^{-2}
γ	interfacial tension	-	J m^{-2}	$= $ N m^{-1}	$= $ kg s^{-2}
p, σ	pressure, stress	pascal	Pa $= $ N m^{-2}	$= $ J m^{-3}	$= $ m^{-1} kg s^{-2}
	impulse, momentum	-	N s		$= $ m kg s^{-1}
Q	electric charge	coulomb	C $= $ A s		
U	electric potential, electromotive force	volt	V $= $ W A^{-1}	$= $ J C^{-1}	$= $ m^2 kg s^{-3} A^{-1}
R	electric resistance	ohm	Ω $= $ V A^{-1}		$= $ m^2 kg s^{-3} A^{-2}
G	electric conductance	siemens	S $= \Omega^{-1}$		$= $ m^{-2} kg^{-1} s^3 A^2
C	electric capacitance	farad	F $= $ C V^{-1}		$= $ m^{-2} kg^{-1} s^4 A^2
ε	relative permittivity [4]	-	1		
Φ	magnetic flux	weber	Wb $= $ V s		$= $ m^2 kg s^{-2} A^{-1}
L	magnetic inductance	henry	H $= $ V s A^{-1}		$= $ m^2 kg s^{-2} A^{-2}
B	magnetic flux density	tesla	T $= $ Wb m^{-2}		$= $ kg s^{-2} A^{-1}
	magnetic field strength	-	A m^{-1}		
Φ_v	luminous flux	lumen	lm $= $ cd sr		
E_v	illuminance	lux	lx $= $ cd sr m^{-2}		
A	radioactivity	becquerel	Bq $= $ s^{-1}		
D	absorbed dose (radiation)	gray	Gy $= $ J kg^{-1}		$= $ m^2 s^{-2}
$\dot{D}$	(absorbed dose rate)	-	Gy s^{-1}	$= $ W kg^{-1}	$= $ m^2 s^{-3}
x	(exposure)	-	C kg^{-1}		
$\dot{x}$	(exposure rate)	-	A kg^{-1}		
-	dose equivalent	sievert	Sv $= $ J kg^{-1}		$= $ m^2 s^{-2}
t, θ [5]	Celsius temperature	degree Celsius [6]	°C		

[1] The physical unit "hertz" should *only* be used for "frequency" in the sense of "cycles per second". Radial (circular) and angular velocities have the unit rad/s, which can be written as s^{-1} but *not* as Hz.

[2] Nonvectorial; the velocity of light usually has the symbol c.

[3] Vectorial; the symbols are then in bold letters (u, v, w).

[4] Formerly: dielectric constant.

[5] The recommended symbol t can easily be confused with the same, much more common symbol t for time and the recommended symbol θ with the common polymer science symbol Θ for the theta temperature. This book thus also uses the symbol T for the Celsius temperature. A confusion with T for kelvin is unlikely since physical equations use only kelvins.

[6] The SI unit of the Celsius temperature *interval* is the degree Celsius, which equals the kelvin (*not*: degree kelvin or degree Kelvin). Both temperatures are related by $\theta/°C = (T/K) - 273.15$.

Table 17-3 Older units. * Units that may be used with SI prefixes or SI units.

Physical quantity	Physical unit Name	Physical unit Symbol	Physical unit Value in SI units	Notes	
Time	minute	min	$= 60$ s	1)	
Time	hour	h	$= 3600$ s	1)	
Time	day	d	$= 86\ 400$ s	1)	
Length	ångstrøm	Å	$= 10^{-10}$ m $= 0.1$ nm	2)	
Area	barn	b	$= 10^{-28}$ m^2		
Volume	liter	l, L	$= 10^{-3}$ m$^3 \equiv 1$ L		*
Mass	ton(ne)	t	$= 10^3$ kg	3)	*
Mass	unified atomic mass unit 4)	u $= m_a(^{12}C)/12$	$\approx 1.660\ 54 \cdot 10^{-27}$ kg	4,5)	
Energy	electronvolt	eV	$\approx 1.602\ 18 \cdot 10^{-19}$ J	6)	*
Pressure	bar	bar	$= 10^5$ Pa	2)	*
Plane angle	degree	°	$= (\pi/180)$ rad		
Plane angle	minute	'	$= (\pi/10\ 800)$ rad		
Plane angle	second	''	$= (\pi/648\ 000)$ rad		
Temperature	Celsius temperature	°C	$= \theta/°C = (T/K) - 273.15$	7)	

1) IUPAC allows the use of the non-SI units "minute", "hour", and "day "in appropriate contexts although these three physical units are not part of the SI system. These units should not be used with SI prefixes.

"Month" and "year" are not scientific units. In commercial data, the symbol for "month" is often "mo" and the symbol for "year" either "yr" or, preferably, "a" (L: *annus* = year).

2) This unit is approved for "temporary use with SI units" in fields where it is presently used.

3) IUPAC allows the use of the physical unit "ton(ne)" = 1000 kg (especially for technical and commercial data) which is, however, not an SI unit. "Ton(ne)" (symbol: ton) is not to be confused with "long ton" ($\approx$ 1016.047 kg) and "short ton" ($\approx$ 907.185 kg); both "ton(ne)s" are often used in commerce without the adjectives "long" and "short".

4) The value of this unit depends on the experimentally determined value of the Avogadro constant N_A; the value of the corresponding SI unit is therefore not exact.

5) The unified atomic mass (physical unit: kg) is sometimes called the dalton (symbol Da). In the biosciences, and recently also in polymer science, "dalton" has erroneously come to mean the relative molecular mass (physical unit: 1) or the molar mass (physical unit: g/mol)!

6) The value of this unit depends on the experimentally determined value of the elementary charge e; the value of the corresponding SI unit is therefore not exact.

7) The SI unit of the Celsius temperature *interval* is the degree Celsius (symbol of the unit: °C), which is equal to the kelvin (*not* "degree kelvin"). The *symbol* of the unit kelvin (small k!) is K, *not* °K. Celsius is always written with a capital C.

When quoting temperatures in kelvin, a space should be written between the numerical value and the symbol K as it is customary for all physical properties. However, if temperatures are given in degrees Celsius, no space should exist between the numerical value and the symbol. Thus: $T = 298$ K but $\theta = 25°C$.

IUPAC now recommends the symbol θ for the physical property "Celsius temperature". In polymer science, a capital theta (Θ) is the traditional symbol for the property "theta temperature". Experience of this author has shown that θ and Θ are easily mixed up, even by experienced polymer scientists. This book thus uses the symbol T for both thermodynamic temperatures and Celsius temperatures; the possibility of a mix-up is remote since physical units are always given.

Table 17-4 SI prefixes for SI units.
    Origin: D = Danish, G = Greek, I = Italian, L = Latin, N = Norwegian.
a) US finance and gas business use the following prefixes for dollar, cubic feet of gas, etc.: $10^3 \equiv$ M;
   $10^6 \equiv$ MM $\equiv \overline{\text{M}}$; $10^9 \equiv$ B; $10^{12} \equiv$ T.
b) ISO adds the letter "y" because a prefix "o" would be misleading.
c) ISO replaced "s" by "z" in order to avoid the double use of "s" ("s" is also the symbol for "second").

Factor	Prefix	Symbol a)	Common name			Origin of prefix
			American	European 1)		
10^{24}	yotta b)	Y	septillion	quadrillion 2)	L:	*octo* = eight [$10^{24} = (10^3)^8$]
10^{21}	zetta c)	Z	sextillion	1000 trillion 3)	L:	*septem* = seven [$10^{21} = (10^3)^7$]
10^{18}	exa	E	quintillion	trillion 4)	G:	*hexa* = six [$10^{18} = (10^3)^6$]
10^{15}	peta	P	quadrillion	1000 billion 5)	G:	*penta* = five [$10^{15} = (10^3)^5$]
10^{12}	tera	T	trillion	billion 6)	G:	*teras* = monster
10^9	giga	G	billion	1000 million 7)	G:	*gigas* = giant
10^6	mega	M	million	million	G:	*megas* = big
10^3	kilo	k	thousand	thousand 8)	G:	*khilioi* = thousand
10^2	hekto 9)	h	hundred	hundred	G:	*hekaton* = hundred
10^1	deka 10)	da	ten	ten	G:	*deka* = ten
10^{-1}	deci	d	one tenth	one tenth	L:	*decima pars* = one tenth
10^{-2}	centi	c	one hundredth	one hundredth	L:	*pars centesima* = one hundredth
10^{-3}	milli	m	one thousandth	one thousandth	L:	*pars millesima* = one thousandth
10^{-6}	micro 11)	μ	one millionth	one millionth	G:	*mikros* = small
10^{-9}	nano	n	one billionth	one milliardth	G:	*nan(n)os* = dwarf
10^{-12}	pico	p	one trillionth	one billionth	I:	*piccolo* = small
10^{-15}	femto	f	one quadrillionth	one billiardth	D, N:	*femten* = fifteen
10^{-18}	atto	a	one quintillionth	one trillionth	D, N:	*atten* = eighteen
10^{-21}	zepto c)	z	one sextillionth	one trilliardth	L:	*septem* = seven [$10^{-21} = (10^{-3})^7$]
10^{-24}	yocto b)	y	one septillionth	one quadrillionth	L:	*octo* = eight [$10^{-24} = (10^{-3})^8$]

1) Most European countries (Germany: see 3), 5), 7)). England reverted to the US system in 1974.
2) France: called "septillion" before 1948.
3) France: called "sextillion" before 1948. Germany: Trilliarde.
4) France: called "quintillion" before 1948.
5) France: called "quadrillion" or "quatrillion" before 1948. Germany: Billiarde.
6) France: called "trillion" before 1948.
7) France: "milliard"; was called "billion" besides "milliard" in France before 1948. Germany: Milliarde.
8) France: "mille".
9) NIST recommends "hekto" (etymologically correct, see right column) but ISO uses "hecto".
10) NIST recommends "deka" (etymologically correct, see right column) but ISO uses "deca".
11) USA: μ as the symbol for "micro" is neither known to the general public nor to newspapers and magazines; it is also not on typewriter keyboards. The prefix "μ" is therefore sometimes replaced by the non-SI prefix "mc" (from "micro"; 1 mcg = 1 microgram = 1 μg). The non-SI prefix "ml" is then substituted for the SI prefix "m" = "milli" (1 mlg ≡ 1 milligram = 1 mg).

Table 17-5 Fundamental constants used in this book.

Physical quantity	Symbol	= number × physical unit	IUPAC symbol
Boltzmann constant	k_B	$= 1.380\ 658 \cdot 10^{-23}$ J K^{-1}	k
Avogadro constant	N_A	$= 6.022\ 136\ 7 \cdot 10^{23}$ mol^{-1}	N_A or L
Loschmidt constant	N_A/V_m	$= 2.686\ 763 \cdot 10^{25}$ m^{-3}	n_o
Molar gas constant	R	$= 8.314\ 510$ J K^{-1} mol^{-1}	R
Unified atomic mass constant	m_u	$= 1.660\ 540\ 2 \cdot 10^{-27}$ kg (= $m(^{12}C)/12$)	m_u

17.2 Numerals

Table 17-6 Classical Greek (G) and Latin (L) numerals and Greek (G), Latin (L), and chemical (C) multiplicative prefixes. The latter ones are in part artificial modifications of Greek and sometimes Latin words, the former ones are often latinized (for example, written with "c" instead of "k" (no "c" in Greek!)). Multipl. = multiplicative.

Num-eral	Cardinal numbers G	L	Multipl. prefixes G [1]	C [1]	Number adverbs G [2] x times	C [2] complex	L [7] x times	C [3] x-times
1	heis,mia,hen	unus,una,unum	mono	mono [4]	haplo		uni	
2	dyo	duo, ae, -o	di	di [5]	dis	bis	bis	bi
3	treis, tria	tres, tres, tria	tri	tres	tris	tris	ter	ter
4	tettares, tettara	quattuor	tetra	tetra	tetrakis	tetrakis	quater	quater
5	pente	quinque	penta	penta	pentakis	pentakis	quinter	quinque
6	hex	sex	hexa	hexa	hexakis	hexakis	sexi	sexi
7	hepta	septem	hepta	hepta	heptakis	heptakis	septi	septi
8	okto	octo	okta	octa	oktakis	oktakis	octa	octi
9	ennea	novem	en(ne)a	nona	enakis	nonakis	nona	novi
10	deka	decem	deka	deca	etc.	etc.	deca	deci
11		undecim		undeca				
12		duodecim		dodeca [6]				
20	eikosi	viginti		icosa [6]				
30		triginta		triaconta				
40		quadraginta		tetraconta				
50		quinquaginta		pentaconta				
60		sexaginta		hexaconta				
100	hekaton	centum	hecta	hecta				
200		ducenti, -ae, -a		dicta				
300		trecenti		tricta				
400		quadringenti		tetracta				
500		quingenti		pentacta				
1000	khilioi	mille, milia	khilia	kilia				
2000		duo milia		dilia				
3000				trilia				
4000				tetralia				
5000				pentalia				
many	polys	multi	poly	poly			multi	

[1] Multiplicative prefixes for multiples of the same kind.
[2] Multiplicative prefixes for multiples of the same kind in complex chemical compounds, for example, the enumeration of substituted substituents.
[3] Multiplicative prefixes for unbranched chemical compounds consisting of two or more identical repeating units, for example, in copolymers.
[4] Only for "one" in the meaning of "single" (for example, "monosubstituted"). "hen" instead of "mono" is used in other numerals (example: "henicosan" (= 21)). An exception is "undeca" (= 11).
[5] Only if not combined with other numerals; examples: 2 ("di"), 200 ("dicta"), 2000 ("dilia"), except for 20 ("icosa"). "do" is used in combination with other numerals, for example "dodeca" (= 12) and "docosan" (= 22).
[6] Multiplicative prefixes for whole numbers of 13 and higher are composed of the multiplicative prefix of the simple whole number 1-9 and a root syllable for the powers of ten. The root syllables are "conta" for the tens (except "icosa" for "20" (Beilstein and CAS use eicosa)); "icta" for the hundreds (except "hecta" instead of "hencta" for "100"); "lia" for the thousands.
[7] x of each: *bini, terni, quaterni, quini, seni, septeni, octoni, noveni, deni*, etc.

Note that digits are given in reverse order for figures composed of many numbers. The prefix for a compound of 537 identical units is therefore "heptatriacontapentacta".

Table 17-7 Greek cardinal numbers (one, two, three, ...), ordinal numbers (first, second, third, ...), number adverbs (once, twice, thrice, ...), and multiplying numbers (single, double, triple, ...).

Cardinal numbers		Ordinal numbers		Number adverbs		Multiplying numbers
1	heis, mia, hen	1st	protos	1×	haplo-	
2	dyo	2nd	deuteros	2×	dis	diplo-
3	treis, tria	3rd	tritos	3×	tris	triplo-
4	tettares, tettara	4th	tetartos	4×	tetra(kis)	
5	pente	5th	pentos	5×	pentakis	
6	hex					
7	hepta					
8	okto					
9	ennea					
10	deka	10th	dekatos			
20	eikosi	20th	eikostos			
		100th	hekatostos			
		1000th	khiliostos			

Table 17-8 Roman and Arabic (i.e., Indian!) numerals.

Roman	Arabic	Roman	Arabic	Roman	Arabic	Roman	Arabic
I	1	X	10	C	100	M	1000
V	5	L	50	D	500		

Numerals are added if read from left to right (example: XXVI = 26). However, a smaller numeral before a larger one is subtracted (example: MCMXLIV = 1944).

The Roman numerals V, X, L, C, D, and M are probably derived from Greek letters. They are *not* abbreviations of names of cardinal numbers.

Table 17-9 Origin of Roman Numerals.

Origin		Early version		Later version	Arabic
v	upper half of X	V	= V		= 5
		X	= X		= 10
Ψ	(ps)	L	= L		= 50
Θ	(th)	C	= C (*not* the first letter of *centum*!)		= 100
D	as right half of φ	I)	= D		= 500
φ	(ph) as φ for 1000	(I)	= M (*not* the first letter of *mille*!)		= 1 000
		((I))	= X̄		= 10 000
		(((I)))	= C̄		= 100 000
		((((I))))	= M̄		= 1 000 000

17.3 Concentrations

Concentrations measure the abundance of substance 1 in all substances i present; $i = 1, 2, 3, ...$

Mass fraction $= w_1 = m_1/\Sigma_i\, m_i = m_1/m = c_1/c$. Mass m_1 of substance 1 divided by the sum of masses m_i of all substances i. Since all masses reside in the same gravity field, a mass fraction can also be called a **weight fraction** ("weight" is not an accepted ISO term; it was formerly the name of a mass in a gravity field).

The value of 100 w_1 is called weight per cent (wt-%) and the value of 1000 w_i is called weight pro-mille (wt-‰). The English language literature also uses part per million (1 ppm $= 10^{-4}$ %), part per (American) billion (1 ppb $= 10^{-7}$ %), and part per (American) trillion (1 ppt $= 10^{-10}$ %).

Volume fraction $= \phi_1 = V_1/(\Sigma_i V_i = V_i/V$. Volume V_1 of substance 1 divided by the total volume of all substances i *before* the mixing process.

Mole fraction, amount fraction, number fraction $= x_1 = n_1/(\Sigma_i n_i) = n_1/n$. Amount n_1 of substance divided by the sum of amounts n_i of all substances i. "Amount-of-substance" (short: "amount") is measured in moles, never in kilograms; it is not a mass. The amount-of-substance was (and still is) erroneously called "mole *number*" in the literature.

Mass concentration = mass density. In polymer science, generally as $c_1 = m_1/V$, i.e., as mass m_1 of substance 1 per volume V of mixture *after* mixing. IUPAC recommends the symbols γ_1 or ρ_1 instead of c_1 but ρ_1 may be confused with the same symbol for the mass density of a *neat* substance. Mass concentrations are usually called "concentrations" in the literature.

Number concentration = number density of entities. $C_1 = N_1/V$. Number N_1 of entities of type 1 (molecules, atoms, ions, etc.) per volume V of mixture *after* mixing. IUPAC recommends the name "concentration" for this physical quantity but this may be confused with the much more common "concentration" = mass concentration.

Amount-of-substance concentration = amount concentration. In polymer science, it is usually defined as $[1] = n_1/V$, i.e., amount-of-substance 1 per volume V of mixture *after* mixing. IUPAC recommends $c_1 = n_1/V$, which may be confused with the symbol c_1 for the mass concentration. The amount concentration is often called "**mole concentration**" or **molarity** and given the symbol M; the latter symbol is not recommended by IUPAC and should not be used with SI prefixes (i.e., not mM for an amount concentration of "millimole" per Liter).

Molality of a solute. $a_1 = n_1/m_2$, i.e., amount n_1 of substance 1 per mass m_2 of solvent 2. Molalities are often denoted by m, which should not be used as the symbol for the unit mol kg^{-1}.

17.4 Ratios of Physical Quantities

The terms "normalized", "relative", "specific", and "reduced" are sometimes used with different meanings in the literature although they are clearly defined.

Normalized requires that the quantities in the numerator and denominator are of the same kind. A normalized quantity is always a fraction (quantity of the subgroup divided by the quantity of the group); the sum of all normalized quantities equals unity.

Relative also refers to quantities that are of the same kind in the numerator and denominator but the quantity in the denominator may be in any defined state. Example 1: relative viscosity $\eta_r = \eta/\eta_2$ as the ratio of the viscosity η of a solution to the viscosity η_2 of the solvent 2. Example 2: relative humidity = ratio of moisture content of air to moisture content of air saturated with water (both at the same temperature and pressure).

Specific refers to a physical quantity divided by the mass. The symbol of a specific quantity is the lower case form of the symbol of the quantity itself. Example: specific heat capacity $c_p = C_P/m =$ = heat capacity (in heat per temperature) divided by mass.

The so-called specific viscosity $\eta_{sp} = (\eta - \eta_1)/\eta_1$ is *not* a *specific* quantity. Indices are italicized only if they refer to a constant physical quantity.

Reduced refers to a quantity that is divided by a specified other quantity. Example: reduced osmotic pressure Π/c = osmotic pressure Π divided by the mass concentration c.

Dimensionless quantity: a product or ratio of two or more different physical quantities that are combined in such a way that the resulting physical quantity has the physical unit of unity (i.e., is "dimensionless").

18 Subject Index

Entries are listed in strict alphabetical order; they may consist of a single word, abbreviations, acronyms, or combinations thereof. For alphabetization, technical terms consisting of two nouns were considered to be one word, whether written as two words (example: acetal polymer), with a hyphen (for example, tension-thinning), or in parentheses, brackets, or braces (example: "Catalyst, def.", comes before "Catalyst efficiency"). Qualifying numbers and letters as well as hyphens, parentheses, brackets, and braces in names of chemical compounds such as 1-, 1,4-, α-, β–, o-, m-, p-, L-, D-, etc., also have been disregarded for alphabetization. Terms consisting of an adverb and a noun, are arranged according to the noun (example: Molar mass $\rightarrow$ Mass, molar).

The following abbreviations are used: abbr. = abbreviation, def. = definition; eqn. = equation; ff. = and following; PM = polymerization, ZN = Ziegler-Natta.